# CONVECTIVE HEAT TRANSFER

# CONVECTIVE HEAT TRANSFER

Louis C. Burmeister

Mechanical Engineering Department
University of Kansas, Lawrence

**A Wiley-Interscience Publication**

**JOHN WILEY & SONS**

**New York ● Chichester ● Brisbane ● Toronto ● Singapore**

Copyright © 1983 by John Wiley & Sons, Inc.

All rights reserved. Published simultaneously in Canada.

Reproduction or translation of any part of this work
beyond that permitted by Section 107 or 108 of the
1976 United States Copyright Act without the permission
of the copyright owner is unlawful. Requests for
permission or further information should be addressed to
the Permissions Department, John Wiley & Sons, Inc.

*Library of Congress Cataloging in Publication Data:*

Burmeister, Louis C.
  Convective heat transfer.

  "A Wiley-Interscience publication."
  Includes bibliographical references and index.
  1. Heat—Transmission. 2. Heat—Convection. I. Title.

QC320.B87 1982     536'.25     82-8338
ISBN 0-471-09141-3            AACR2

Printed in the United States of America

10    9    8    7    6    5    4    3    2    1

*To Rosalyn, Elise, Amanda, and my parents*

# Preface

This book evolved from notes prepared over a period of years and was used in a graduate course in convective heat and mass transfer. Heat transfer by convection is heavily emphasized, with mass transfer by convection included primarily by analogy. Expositions of events at the molecular level first give insight into the physical origin of the transport properties. Proceeding from this sound base, the equations describing convective transport on the continuum level are derived next. Examples of physical situations described by one-dimensional formulations and their solutions follow, giving insight into important conclusions and solution techniques. Only after these basics have been presented are more complex applications, such as laminar and turbulent duct and boundary-layer flows, treated. Problems, many with answers, are given at the end of each chapter. Learning is best accomplished by practice, as every teacher knows.

Every author has an intended audience. This book was written primarily for the beginning engineering graduate student. Having had prior application courses at the undergraduate level, such a student is ready for a more sharply defined course. Applications, although always the ultimate motivation, need not be emphasized as a result. The resurgence of interest in energy and its transport in thermal form, most effectively done by convection, maintains the importance of detailed knowledge of convective heat transfer and the courses offered in that subject in most schools of engineering.

When presenting a subject as rich in physical phenomena as convective transport, a class schedule must be judiciously balanced between breadth and depth. Almost all transport properties can be viewed as tabulated quantities. Little classroom time need be devoted to Chapters 1–3 if that view is adopted, although the presence of the first three chapters gives the reader a choice. Knowledge in depth of the equations describing convective transport in Chapter 4 is important since their use without understanding is dangerous. Exposition of one-dimensional problems in Chapter 5 is thorough, so little classroom time is needed. Duct and boundary-layer flows are usually given scant attention in undergraduate courses, so Chapters 6–11 should be given full consideration. Integral methods covered in Chapter 8 need only brief class time since the student can read them unaided. Numerical methods for boundary layers at the ends of Chapters 7 and 10 can be given special attention; however, detailed discussion of numerical methods is usually best reserved for a following course (e.g., in computational fluid mechanics). Driving forces for convection not externally imposed are the common denominators of Chapters 12–14. A substantial coverage of them is recommended since most undergraduate courses do not treat natural convection, boiling, and condensation in detail despite their technological importance. Realizing that only a guide is possible, I suggest the following schedule for a 14-week semester.

| Chapter | 1–3 | 4 | 5 | 6 | 7–8 | 9 | 10 | 11 | 12–14 | Examinations and Special Topics |
|---|---|---|---|---|---|---|---|---|---|---|
| Class periods | 5 | 4 | 2 | 3 | 6 | 3 | 4 | 3 | 8 | 4 |

Key topics are covered and, especially if examinations are of the take-home variety, there is time for special topics.

Numerous journals and reference books exist. End-of-chapter references suggest many of them as sources of additional information. Seldom is it possible to satisfy every interest in a single book. Students and practitioners seeking grounding in the basics will find the present book helpful and, it is hoped, they will appreciate the benefits of a clear separation between textbooks such as this one and the archival and reference literature. Huge amounts of relevant information exist. A coherent assemblage of this information with sufficient, but not overwhelming, detail is a formidable undertaking. Very often many sources of differing levels of complexity and detail must be consulted. Eliciting the coherent base of the subject is the first and most difficult step, and it is that base that the present book is intended to provide.

Systems of units used are the Systéme International primarily and the English system secondarily. Even though the SI system has been officially adopted almost everywhere, much information is in the English system and facility in conversion of units should be maintained. Eventually the SI system will be in common use, but that time has not yet come. Numerical conversion factors are provided to expedite conversions.

Acknowledgment of influential preceding works is appropriate here. Gifted writers serve us better than is commonly realized. Realistically, few writings do more than report and organize information developed by others. Exceptional, though, are the earlier textbooks entitled *Transport Phenomena* by Bird, Stewart, and Lightfoot and *Heat and Mass Transfer* by Eckert and Drake. A reading of the present book will show their subtle influence.

There are many who assisted in the preparation of this book. Little exaggeration is possible in saying that the students in the graduate classes, who were taught from the notes out of which this book evolved, were immensely helpful. It is with pleasure that the contribution of Mrs. Georgia Porter in typing the manuscript is recognized. Great support was provided by my family. Help of indispensable nature was provided by the publisher through careful editing and preparation of final drawings. Thanks are also due to anonymous reviewers who made valuable suggestions for improvements.

Inevitably, a few errors will have escaped repeated proofreading. Should they be found, please communicate them to me. No amount of care is sufficient, it seems, to ensure perfect copy. It does help to eliminate those that are found, though. No bounty is offered.

Effort is usually accompanied by gain. That is what is hoped for the reader. When this book is finished, a working understanding of convective heat transfer should have been acquired. Only the reading and studying remain for the reader to supply.

Louis C. Burmeister

*Lawrence, Kansas*
*July*, 1982

# Contents

CONTENTS

# 1

# INTRODUCTION

The area of principal interest in this book is the convective transport of heat, mass, and momentum. It is recognized that heat can be transferred by the three modes of conduction, convection, and radiation. Usually all three modes act simultaneously, but there are many important cases in which conduction and convection are predominant, and these are considered to the near exclusion of radiation.

The restricted scope of an interest regarding convective transport offers the intended audience of beginning graduate students and advanced undergraduate students the advantage of a depth of coverage that is not possible in a book with broader interests. The textbooks typically used in beginning courses in heat transfer, fluid mechanics, and thermodynamics provide ample examples and studies of the breadth of the fields of heat transfer and fluid mechanics and may be referred to for background information. The aim of this book is to provide the reader with an understanding of convective transport to a depth that includes phenomena at the molecular level.

In any discussion of convective heat transfer it is appropriate to recall the definition of heat. Heat is defined as energy flowing across a boundary as a result of a temperature difference across that boundary. In the strictest sense, therefore, convection is not necessarily a means of heat transfer. For example, a container of hot coffee might be moved from place to place by hand; its internal energy moves with it, of course. But because the "convection" of the coffee's internal energy was not directly caused by a temperature difference, this convection cannot in the strictest sense be termed a heat-transfer mechanism.

Since convective transport requires fluid motion, it is necessary to consider the relations that describe the flow of fluids. Because a flowing fluid transports any properties it has (e.g., internal energy, a species of mass, or momentum), it will also be necessary to consider the relations that describe the manner in which the velocity distributions affect temperature and concentration distributions. It can be seen from these describing relations that there are many similarities between heat, mass, and momentum transfer, although some differences may also be apparent.

## 1.1 CONSERVATION PRINCIPLES

The basic relationships on which an analysis of convective transport phenomena can be based are rather few in view of the complexity of the ultimate problems to be solved. The basic relationships known may be divided into the two categories of conservation principles and rate equations.

Conservation principles are fundamental to any study of physical phenomena. Physical quantities that are conserved are mass, energy, species of mass (in the absence of chemical reactions), and electrical charge. In most instances it is not necessary to apply the conservation of electrical charge principle. However, in problems that involve the interaction of electric and magnetic fields with fluids, relations involving electromagnetic quantities (e.g., the conservation of electrical charge) are needed.

Energy can be converted into different forms without violating the energy conservation principle, and it is also possible for one kind (species) of mass to be converted into another kind (species); the only requirement of the mass conservation principle is that the total amount of mass remain constant. In any such conversion the mass of each chemical element remains constant, of course, disregarding relativistic effects. For mass to be converted from one species to a different species, a chemical reaction is required; if there is no chemical reaction, it is not possible for conversion into different mass species to occur, thus validating the mass species conservation principle. For example, a container filled with a stoichiometric mixture of gaseous hydrogen and oxygen contains what may be said to be two species of mass. If no chemical reaction occurs, the mass of each of the two original species, the hydrogen and the oxygen, remains constant. A chemical reaction such as might be initiated by a spark would cause the hydrogen and oxygen to combine into a third species of mass, namely, water, destroying the original two species of hydrogen and oxygen. The masses of hydrogen and oxygen, the two chemical elements involved in this process, are unchanged, of course.

Newton's laws of motion are needed for a description of the manner in which a fluid can move. Although they are not conservation laws in the strictest sense of the term, they can still be usefully considered as such. For example, Newton's second law of motion for a particle of constant mass $m$ is

$$\mathbf{F} = \frac{m}{g_c} \frac{d\mathbf{v}}{dt} \tag{1-1}$$

where $\mathbf{F}$ is the net vector force acting on the particle, $\mathbf{v}$ is the vector velocity of the center of mass, and $g_c = 32.2 \text{ lbm ft/lb}_f \text{ sec}^2 = 1 \text{ kg m/N s}^2$ is a constant of proportionality. This may be rewritten, since $m$ is constant for this special circumstance, as

$$F = \frac{d(m\mathbf{v}/g_c)}{dt} \tag{1-2}$$

In the form given by Eq. (1-2), Newton's second law can be interpreted as a conservation law if $\mathbf{F}$ is considered as representing the rate at which the momentum of the solid particle is generated rather than as a net external force. Then, since $m\mathbf{v}/g_c$ is the momentum of the particle, in the absence of a net external force the momentum of the particle is conserved.

## 1.2 RATE EQUATIONS

It is necessary to be able to predict the rate at which a quantity can diffuse relative to the medium through which it is passing. The predictive equations are appropriately termed rate equations; they are sometimes also called *phenomenological relationships* because one pertains to the phenomenon of heat conduction, one pertains to the phenomenon of electrical conduction, and so forth. A basic recounting of these relations follows; Appendix B gives more details.

### Fourier's Law For Heat Conduction

Consider a large slab of homogeneous material as depicted in Fig. 1-1. The slab is not necessarily a solid (e.g., a layer of liquid between two plates), but there should be no motion of one part of the slab relative to another part. A small temperature difference is steadily imposed across the slab, and it is found by experiment that the steady-state heat flow rate by conduction in the $x$ direction $Q_x$ is predictable by

$$Q_x = kA \frac{T(x) - T(x + \Delta x)}{\Delta x} \tag{1-3}$$

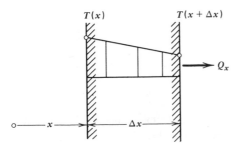

**Figure 1-1**  One-dimensional steady heat conduction.

where the thermal conductivity $k$ is a property of the material, $A$ is the area across which heat flows, and $T$ is temperature. The heat flows perpendicularly to the faces of the slab (in the $x$ direction) from the high to the low temperature. In addition, it is experimentally determined that the temperature distribution in the slab varies linearly, provided the temperatures of the two slab faces do not differ greatly. Equation (1-3) yields accurate predictions for all slab thicknesses encountered in practical situations. (Of course, the slab thickness cannot be allowed to be so thin as to approach the distance between molecules.) In the limit as $\Delta x \to 0$ (it may be helpful to envision a small slice being taken from the original slab), Eq. (1-3) becomes

$$Q_x = -kA \frac{dT}{dx} \tag{1-4}$$

which is the differential form of Eq. (1-3). The heat flux is given from Eq. (1-4) as

$$q_x = \frac{Q_x}{A} = -k \frac{dT}{dx} \tag{1-5}$$

which is commonly called *Fourier's law*.

### Fick's Law For Binary Mass Diffusion

Consider the large slab of homogeneous material, material 2, in Fig. 1-2. Through this slab a different material, material 1, diffuses steadily as a consequence of steadily maintaining the amount of material 1 at a level that is slightly higher on one side of the slab than on the other. The slab could consist of a stagnant layer of oxygen through which hydrogen gas passes by diffusion, for example, but it could be solid; it is required here only that the parts of the slab not move relative to one another. It is found by experiment that the steady diffusion rate of species 1 in the $x$-direction $\dot{M}_{1x}$, is predictable by

$$\dot{M}_{1x} = \rho D_{12} A \frac{\omega_1(x) - \omega_1(x + \Delta x)}{\Delta x} \tag{1-6}$$

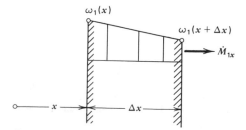

**Figure 1-2**   One-dimensional steady binary mass diffusion.

where $\rho = \rho_1 + \rho_2$ is the total density of the slab and $D_{12}$ is the mass diffusivity of species 1 through species 2. The cross-sectional area across which flow occurs is $A$, and $\omega_1 = \rho_1/\rho$ is the mass fraction of species 1. The diffusion of species 1 is perpendicular to the slab faces and from the slab face that has the amount of species 1 maintained at a high level to the slab face that has a low level of species 1. Because Eq. (1-6) applies with equal accuracy to slabs of all thicknesses (so long as the slab thickness does not decrease to such an extent as to approach the dimensions of the spacing between molecules), the limit of Eq. (1-6) as $\Delta x \to 0$ is

$$\dot{m}_{1x} = \frac{\dot{M}_{1x}}{A} = -\rho D_{12} \frac{d\omega_1}{dx} \qquad (1\text{-}7)$$

which is commonly called *Fick's law for binary diffusion*. Fick's law in the form of Eq. (1-6) is restricted to cases in which the total density $\rho$ of the slab does not vary much in the distance $\Delta x$. Also, it should not be applied across an interface between two materials or two phases. When the density $\rho$ is nearly constant, Eq. (1-6) can be rewritten as

$$\frac{\dot{M}_{1x}}{A} = D_{12} \frac{\rho_1(x) - \rho_1(x + \Delta x)}{\Delta x}$$

To illustrate the difficulty encountered in applying Eq. (1-7) across an interface, consider a pool of liquid water in equilibrium with a mixture of its vapor and air. No net mass transfer occurs; yet there is a difference in mass fraction across the liquid–vapor interface in the amount of

$$\frac{\rho_{liq}}{\rho_{liq}} - \frac{\rho_{vap}}{\rho_{vap} + \rho_{air}} \neq 0$$

Substitution of this difference into Eq. (1-7) would erroneously predict a net mass flow across the interface.

## Newton's Viscosity Law

Consider a slab of fluid contained between two plates as shown in Fig. 1-3*a*. The top plate steadily moves to the right, relative to the bottom plate and parallel to it in response to the steady net force $F_{ext}$ imposed on the top plate. The relative motion of the two plates demonstrates the fact that a fluid is unable to support shear forces without undergoing continuous displacement; in other words, a fluid flows when a shear force is imposed on it. The resultant flow can be smooth, with a particle of fluid moving steadily in a smooth line parallel to the plates—since a thin layer of fluid then moves as a lamination, such a flow is called *laminar flow*. A second major possible flow is one that is erratic and chaotic with a particle of fluid moving unsteadily in an unpredictable zigzag path—such a flow is descriptively called *turbulent flow*. Turbulent

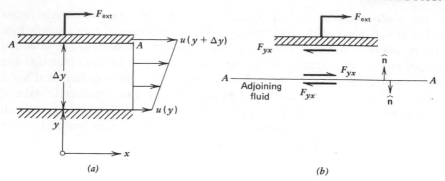

**Figure 1-3**   One-dimensional steady laminar flow.

flow is generally expected to occur when velocities are high; laminar flow is expected when velocities are low. As velocities increase from low values, the fluid flow undergoes a gradual transition from a state of laminar flow to one of turbulent flow, the velocity at which transition occurs depending strongly on such imposed conditions as the specific geometry and the pressure gradient.

   Because it is desired to avoid convective transport of any quantity in the direction perpendicular to the slab faces, attention is restricted to the laminar flow case for the moment. For laminar flow it is experimentally found that the net external force that must be steadily applied to the top plate is predictable (provided the slab thickness substantially exceeds the molecular spacing) by

$$F_{ext} = \mu A \frac{u(y + \Delta y) - u(y)}{\Delta y} \tag{1-8}$$

where the dynamic viscosity $\mu$ is a property of the fluid and $A$ is the surface area of the plate. The free-body diagram of the top plate shown in Fig. 1-3$b$ illustrates that the fluid must exert an opposing leftward force on the plate of magnitude $F_{yx} = F_{ext}$ in order for the plate to move at constant velocity; by Newton's third law of equal action and reaction, the plate exerts a rightward force of magnitude $F_{yx} = F_{ext}$ on the adjoining fluid surface.

   At this point it is necessary to stipulate the side of a fluid surface on which the acting shear force is to be calculated. As shown in Fig. 1-3$b$ for surface $A-A$, Newton's third law of equal action and reaction requires forces of equal magnitude but opposite direction on each side of a fluid surface. This choice did not arise in the heat conduction or binary mass diffusion cases, but rather is arbitrary, dictated by both the convenience of the results obtained from it and the desirability of adhering to previously established conventions.

   The fluid mechanics convention is to identify the force acting on a surface by two subscripts, the first one designating the face on which the force acts and the second one designating the direction in which the force acts, since complete description of a force requires specification of both its point of application and direction of action. If the force acts on a positive face (a face whose surface normal $\hat{n}$ points in the positive direction denoted by the first subscript), the magnitude of the force is positive if the force is directed in the positive direction denoted by the second subscript. If the force acts on a negative face (a face whose surface normal $\hat{n}$ points in the negative direction denoted by the first subscript), the magnitude of the force is positive if the force is directed in the negative direction denoted by the second subscript.

Figure 3-1$b$ shows that the force $F_{yx}$ on the positive $y$ face of surface $A$–$A$ acts in the positive $x$ direction. The magnitude of this force, in the limit as $\Delta y \to 0$, is

$$F_{yx} = F_{\text{ext}}$$

$$= \mu A \frac{du}{dy}$$

The shear stress on the upper face of surface $A$–$A$, $\tau_{yx} = F_{yx}/A$, is then

$$\tau_{yx} = \mu \frac{du}{dy} \tag{1-9}$$

which is commonly called *Newton's law of viscosity*.

## Ohm's Law

Consider a slab of electrically conductive material across whose two faces is steadily imposed a small voltage difference. Figure 1-4 shows this situation. It is experimentally observed that a steady electrical current $I_x$ flows perpendicular to the two constant voltage faces of a magnitude predictable by

$$I_x = k_e A \frac{V(x) - V(x + \Delta x)}{\Delta x} \tag{1-10}$$

and in the direction leading from the high to the low voltage. Here the electrical conductivity $k_e$ is a property of the material, $A$ is the area across which current flows, and $V$ is voltage. In the limit $\Delta x \to 0$, Eq. (1-10) gives the

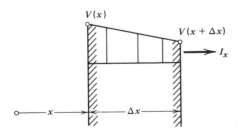

**Figure 1-4**   One-dimensional steady electrical conduction.

flux as

$$i_x = \frac{I_x}{A} = -k_e \frac{dV}{dx} \qquad (1\text{-}11)$$

which is commonly called *Ohm's law.*

## Newton's Law of Cooling

Consider a hot solid wall that is exposed to a cool flowing fluid as shown in Fig. 1-5. It is experimentally observed that heat flows from the hot solid wall into the cooler fluid and that the rate of heat transfer is proportional to both the surface area $A$ and to the temperature difference $T_w - T_f$. These observations are incorporated in the predictive equation for the convective heat-transfer rate

$$Q = hA(T_w - T_f)$$

or, in terms of a heat flux,

$$q = \frac{Q}{A} = h(T_w - T_f) \qquad (1\text{-}12)$$

which is referred to as Newton's law of cooling. The proportionality factor $h$ is interchangeably called a *film coefficient of heat transfer*, or a *heat-transfer coefficient.* The film coefficient term is descriptive of the observed thin fluid film near the solid wall in which velocity and temperature vary between the values of the ambient fluid and the wall. Experiments under normal conditions show that at the wall there is no relative velocity between the fluid and the wall and that the fluid temperature equals the wall temperature. In other words, at a solid wall a fluid normally experiences no slip and no temperature jump.

The form of Eq. (1-12) is convenient and simple, with heat flux proportional to the imposed temperature difference rather than to a temperature gradient as in the case for diffusion as exemplified by Fourier's law in Eq. (1-5). It is experimentally observed that the heat-transfer coefficient $h$ in Eq. (1-12) is not a constant; it is sensitive to the flow conditions in the ambient fluid flowing by. On physical grounds it can be understood how this might be so. The

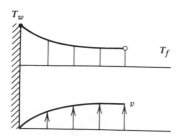

**Figure 1-5** Convective heat transfer from a solid surface to a flowing fluid.

thickness of the fluid film near the wall, or boundary layer (since it is also a layer at a boundary), is observed to decrease as the fluid flows by more rapidly. Since the constant imposed temperature difference occurs over the largely stagnant film that is of diminished thickness, the heat flux should increase, thus requiring an increased heat transfer coefficient $h$. It is apparent that convection is not itself a heat-transfer mechanism; rather, it is a combination of several other more fundamental transport mechanisms.

It is evident that the magnitude of the heat-transfer coefficient is dependent on both the rate at which heat enters the fluid from the wall and the rate at which the fluid can convect a particle away from the wall after it has been heated. Since the fluid experiences no slip at the wall, heat can flow from the wall into the fluid only by conduction. (Thermal radiation is often negligible at the moderate temperatures envisioned here.) As the heat diffuses further into the fluid where velocity is high, convection of heated particles becomes significant. Finally, near the outer edge of the boundary layer, almost all the energy transport occurs by convection. The simple form of Eq. (1-12) tends to obscure the basic phenomena that are active, but its convenience is responsible for its continued use.

## Thermal Radiation

The physical situation in which thermal radiation is an important mechanism of heat transfer is briefly discussed here. This is done for completeness and to ensure an awareness of the major role of thermal radiation in many significant processes. Examples of processes in which radiative transport plays a major, and even predominant, role are heat transfer from spacecraft and in the fireboxes of boilers of central electrical generating plants and warming of objects (e.g., solar collectors) by the sun.

The rate at which a perfect emitter, called a *blackbody* since it also absorbs all incident thermal radiation, emits energy by the mechanism of thermal radiation is given by

$$Q = A\sigma T^4 \tag{1-13}$$

where $\sigma = 0.1714 \times 10^{-8}$ Btu/hr ft$^2$ R$^4$ = $5.66961 \times 10^{-8}$ W/m$^2$ K$^4$ is the Stefan–Boltzmann constant, which is a constant of nature; $T$ is the emitter absolute temperature; $Q$ is the heat flow rate; and $A$ is the surface area. This law was deduced independently from experimental measurements and by analytical means. A nonblackbody emits at a lesser rate as expressed by insertion of a multiplicative constant in Eq. (1-13) to give

$$Q = \epsilon A\sigma T^4 \tag{1-14}$$

where $\epsilon$ is the emittance $0 \leqslant \epsilon \leqslant 1$. The emitted energy can be regarded as propagating either as an electromagnetic wave at the speed of light or as packets of photons.

The fraction of thermal radiation incident on a body that is absorbed is denoted by the absorptance $\alpha$ where $0 \leqslant \alpha \leqslant 1$. The symbols $\tau$ and $\rho$ denote the fraction of incident radiation that is transmitted and reflected, respectively, with $0 \leqslant \tau \leqslant 1$ and $0 \leqslant \rho \leqslant 1$. Conservation of energy requires that the relationship $\tau + \rho + \alpha = 1$ always be satisfied. It can be additionally shown that at thermal equilibrium $\epsilon = \alpha$, a relationship referred to as *Kirchhoff's law*.

When the space between two solid surfaces is transparent (because of either the absence of intervening matter as in a vacuum or the inability of the intervening matter to be substantially affected by the frequencies present in the electromagnetic wave), the net rate of heat transfer is given by

$$Q_{1-2} = \mathscr{F}_{1-2}\sigma\left(T_1^4 - T_2^4\right)A_1 \qquad (1\text{-}15)$$

in which the factor $\mathscr{F}_{1-2}$ contains the combined effects of the radiative properties and geometric orientation of the surfaces. The geometric orientation is accounted for by a geometric shape factor $F_{1-2}$ that can be interpreted as being the ratio of the diffuse radiation leaving surface 1 that strikes surface 2.

Equations (1-14) and (1-15) show the rate of heat transfer to be proportional to a difference of the fourth power of absolute temperatures, a relationship that is unique to thermal radiation and is largely accurate when the space between the absorber and the emitter is transparent. When matter that is capable of absorbing and emitting radiation intervenes between the two surfaces, the rate of heat transfer is greatly influenced by the properties of that matter and is characterized more by diffusion as a transport process than by thermal radiation. This fact is used in the Rosseland diffusion approximation [1, 2] of radiative energy transport.

## 1.3 ANALOGIES FOR ONE-DIMENSIONAL DIFFUSION

The foregoing discussions of the rate equations show them to have a remarkably similar form. Each is the result of exhaustive experimentation, of course, and constitutes an independent relationship between the flux of a fundamental, but often difficult to measure, quantity such as energy in thermal form and a more easily measured quantity such as temperature. For convenience and ease of perception, the four rate equations (or, phenomenological relations) are displayed in Table 1-1. There it is apparent in the third column that all the rate equations are of the same form—flux = constant $\times$ potential gradient. The proportionality constant is a function of the material involved in the transport process and is called a *transport property*. The transport properties in Table 1-1 are $k$ (thermal conductivity), $D_{12}$ (mass diffusivity), $\mu$ (viscosity), and $k_e$ (electrical conductivity). The presence of a positive sign in Newton's law of viscosity is due solely to the choice of sides of a fluid surface on which to evaluate the acting stress; if the other side had been chosen, a negative sign would appear in the rate equation.

Table 1-1  One-Dimensional Rate Equations

| Diffusing Quantity | Process | Rate Equation | Name |
|---|---|---|---|
| Thermal energy | Heat conduction | $q_x = -k\, dT/dx$ | Fourier's law |
| Species 1 Mass | Binary mass diffusion | $\dot{m}_{1_x} = -\rho D_{12}\, d\omega_1/dx$ | Fick's law |
| Momentum | Viscous fluid shear | $\tau_{xy} = \mu\, du/dx$ | Newton's law of viscosity |
| Electrical charge | Electrical conduction | $i_x = -k_e\, dV/dx$ | Ohm's law |

The left-hand side of each rate equation is a flux, having the units of the diffusing substance/area per time. This point is less clear for the case of viscous fluid shear for which the shear stress $\tau_{xy}$ has units of $N/m^2$ in SI units ($1b_f/ft^2$ in English units). Conversion of units by multiplication by $g_c$ then gives the units of shear stress as $kg\ m\ s^{-1}/s\ m^2$ in SI units ($1b_m\ ft\ sec^{-1}/sec\ ft^2$ in English units) more clearly signifying that shear stress has units of momentum/area per time and showing that the diffusing quantity is momentum (horizontal momentum for the situation shown in Fig. 1-3).

To strengthen the suggestion of analogies for the cases of one-dimensional diffusion to which Table 1-1 pertains, the four rate equations can be recast as shown in Table 1-2 under the assumption of constant properties—$c'''$ is electrical capacitance per unit volume.

Consideration of the second column of Table 1-2 shows the common form

$$\frac{\text{Diffusing quantity}}{\text{Time area}} = \pm (\text{diffusivity})\, \frac{d(\text{diffusing quantity/volume})}{d(\text{distance})}$$

suggesting that the flux of a diffusing quantity is proportional to the concentration gradient of that quantity. The property of diffusivity that serves as a coefficient has the same units, $(\text{distance})^2/\text{time}$, regardless of the physical phenomenon considered. The ratio of diffusivities is a common dimensionless number that indicates the relative rate of diffusion of the corresponding physical quantities. The Prandtl number is defined as $Pr = \nu/\alpha$, the Schmidt number is defined as $Sc = \nu/D_{12}$, and the Lewis number is defined as $Le = D_{12}/\alpha$.

The common functional form of the rate equations demonstrated in Table 1-2 and the similar nature of the several conservation principles suggest that analogies between heat, mass, momentum, and electrical transport can be found. Of course, final proof that analogies exist must be deferred until the full mathematical descriptions for a specific physical situation are shown to be of

**Table 1-2  Recast One-Dimensional Rate Equations**

| Process | Recast Rate Equation | Diffusivity | Rate Equation in SI Units |
|---|---|---|---|
| Heat conduction | $q_x = -\alpha\, d(\rho C_p T)/dx$ | $\alpha = k/\rho C_p$ | $J/s\ m^2 = -(m^2/s)d(J/m^3)/d(m)$ |
| Binary mass diffusion | $\dot{m}_{1_x} = -D_{12}\, d\rho_1/dx$ | $D_{12}$ | $kg_1/s\ m^2 = -(m^2/s)d(kg_1/m^3)d(m)$ |
| Viscous fluid shear | $\tau_{xy} = \nu\, d(\rho u)/dy$ | $\nu = \mu/\rho$ | $kg\ m\ s^{-1}/s\ m^2 = (m^2/s)d(kg\ m\ s^{-1}/m^3)d(m)$ |
| Electrical conduction | $i_x = -\alpha_e\, d(c'''V)/dx$ | $\alpha_e = k_e/c'''$ | $C/s\ m^2 = -(m^2/s)d(C/m^3)d(m)$ |

the same form. When an analogy does occur, and it does not always, it is most useful because it allows information gained with an experimentally convenient physical phenomenon to be applied whole to the analogous physical phenomena whose experimental treatment is more difficult.

## 1.4 OVERVIEW OF FOLLOWING CHAPTERS

The discussion of convective transport begins in Chapter 2 at the molecular level for the simplest of all fluids, a gas. An abbreviated exposition of the kinetic theory of gases gives insight into such basic quantities as molecular velocity distributions, average distance traveled between molecular collisions that is related to diffusive migration by a *Drunkard's Walk*, and the rate at which molecules cross an area.

Building on these insights in Chapter 3, it is shown that the transport properties of mass diffusivity, thermal conductivity, and viscosity naturally arise. The nature of their dependence on temperature and molecular weight is discussed for both the liquid and the gaseous states, imparting a fundamental appreciation for transport phenomena at the molecular level.

In Chapter 4 fluids are considered to be continuous. Their molecular makeup is tacitly acknowledged by the use of transport properties, of course. The partial differential equations that describe their convective behavior are derived, showing how the effects of nonuniformities in temperature, velocity, and mass fraction can be taken into account.

In Chapter 5 the essential features of convective transport by forced convection are ascertained by relatively easy solutions to a few illustrative one-dimensional problems. In Chapters 6–8 these insights are extended by detailed consideration of laminar flow in ducts and boundary layers. Technically important conclusions and analytical methods are presented. In Chapters 9–11 the insights acquired from study of transport properties and laminar flow, together with additional empirical information, are applied to forced turbulent flow in ducts and boundary layers.

In Chapters 12–14 situations are considered in which flow is not forced. Natural convection, boiling, and condensation are treated in a manner that merges previously developed insights and techniques with pertinent new information.

## REFERENCES

1  R. Siegel and J. R. Howell, *Thermal Radiation Heat Transfer*, McGraw-Hill, New York, 1972, pp. 469–474.

2  S. Rosseland, *Theoretical Astrophysics; Atomic Theory and the Analysis of Stellar Atmospheres and Envelopes*, Clarendon Press, Oxford, 1936.

# 2

# KINETIC THEORY
# OF GASES

The transport properties of viscosity, thermal conductivity, and mass diffusivity are of particular interest. Their values are needed for use in the rate equations discussed in Chapter 1 and needed for both the liquid and gaseous states.

Since the gaseous phase is the simplest and best understood state of matter and since its analysis yields information that is valuable for extension to other areas, the important elements of the kinetic theory of gases are presented next. Modern understanding of the gaseous state certainly extends beyond the somewhat limited kinetic theory, but kinetic theory does provide an informative picture of the basic transport mechanisms and does yield reasonably accurate and qualitatively correct results. The kinetic theory presentation that follows is patterned after the exposition given by Lee et al. [1]—see Truesdell and Muncaster [11] for a complete treatment.

## 2.1 WORKING HYPOTHESES

In the kinetic theory of gases it is assumed that the laws of mechanics as deduced from the behavior of large objects are also applicable to such small objects as molecules. Also, it is assumed that any finite volume contains a very large number of molecules, an assumption that can be tested by estimating the volume occupied by a molecule under room conditions. At the standard temperature of 0°C and 1 atm of pressure, 1 kgmol of gas is known to occupy 22.4 m³. Avogadro's number $N_A$ is additionally known to be

$$N_A = 6.022 \times 10^{26} \text{ molecules/kgmol}$$

As a result, a molecule can be thought of as occupying a cube having dimension

$$\delta = \left( \frac{22.4 \text{ m}^3/\text{kgmol}}{6.022 \times 10^{26} \text{ molecules}/\text{kgmol}} \right)^{1/3}$$

$$\approx 3 \times 10^{-9} \text{ m/molecule}$$

A space of any appreciable size does indeed have many molecules—a cube 1 mm on a side contains about $10^{17}$ molecules—and it is reasonable to take the gas to be a continuous medium. Furthermore, if a gas is a continuous medium, the denser liquids and solids (with much closer molecular spacings) certainly will also be.

The mass $m$ of a gas molecule can be simply computed from the molecular weight $M$ and Avogadro's number $N_A$ as

$$m = \frac{M}{N_A}$$

Hydrogen with $M = 2$ kg/kgmol is accordingly found to have a molecular mass of $m = 0.33 \times 10^{-23}$ g.

Molecules are assumed to exert no forces on one another until they actually collide. Actually, they do attract one another when slightly removed and repel one another on a slight interpenetration. When later taking a molecule to be a rigid sphere (which would be more appropriate for a monatomic than a polyatomic gas), the molecular diameter should not be regarded as a rigorously defined quantity. Figure 3-3 may be consulted for an illustration of intermolecular forces. All collisions are assumed to be elastic, whether with one another or with a wall. This is intuitively reasonable since inelastic collisions would eventually result in all molecules having zero velocity.

It is presumed that molecules are uniformly distributed throughout their containing volume. If the total number of molecules is $N$ and their containing volume is $V$, the molecular density

$$n = \frac{N}{V}$$

is a constant that is independent of position in the volume. Accordingly, a volume $dV$ contains the number of molecules $dN$ given by

$$dN = n \, dV \tag{2-1}$$

in which the first derivative notation denotes a quantity of the first order of smallness.

Finally, all directions of motion are assumed to be equally probable. This is quite reasonable on physical grounds since molecules would otherwise tend to

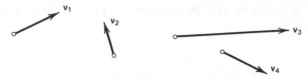

**Figure 2-1**   Velocities of molecules.

preferentially drift to one side of the container. If bulk flow of the gas takes place, this is considered to be just a local shifting of the zero-velocity reference and has no effect on the results.


## 2.2   MOLECULAR FLUX

An important quantity for later use in estimating transport properties is the rate at which molecules cross a surface. To begin the determination of this quantity, imagine that to each molecule is attached a vector that represents the magnitude and direction of its velocity as illustrated in Fig. 2-1. Now each vector is transferred to a common origin about which a sphere of radius $r$ is constructed as shown in Fig. 2-2. One velocity vector, or its extension if its length is less than $r$, pierces the sphere surface for each molecule; and, because of the assumed uniform directional distribution of velocities, these piercing points are uniformly distributed over the sphere surface. The average density of these piercing points is simply $N/4\pi r^2$ if there are $N$ molecules. Thus the number of piercing points in an elemental area $dA$ on the sphere surface is just $N\,dA/4\pi r^2$.

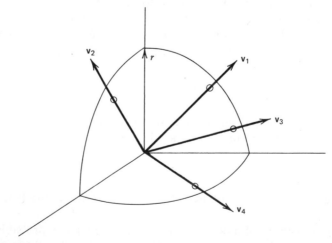

**Figure 2-2**   Velocities of molecules transferred to a common origin.

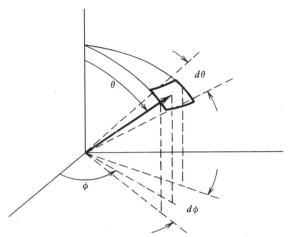

**Figure 2-3**  Area element on a sphere.

Now an area element on the sphere surface, expressed in spherical coordinates as shown in Fig. 2-3, is

$$dA = r^2 \sin \theta \, d\theta \, d\phi$$

The number of molecules with velocity vectors pointing in the $\theta$, $\phi$ direction (lying within $d\theta/2$ of $\theta$ and $d\phi/2$ of $\phi$) is

$$d^2 N_{\theta\phi} = N \sin \theta \, d\theta \frac{d\phi}{4\pi} \qquad (2\text{-}2)$$

which is termed the number of $\theta$, $\phi$ molecules, where the second derivative notation on the left-hand side denotes a quantity of the second order of smallness. The number of $\theta$, $\phi$ molecules per unit volume is found to be

$$d^2 n_{\theta\phi} = \frac{d^2 N_{\theta\phi}}{V} = n \sin \theta \, d\theta \frac{d\phi}{4\pi} \qquad (2\text{-}3)$$

Next to be determined is the number of molecules moving near a specified speed $v \pm dv/2$ and directions of travel $\theta \pm d\theta/2$ and $\phi \pm d\phi/2$ that strike a surface element $dA$ from one side in the time interval $dt$. This is called a $\theta\phi v$ *collision*. For this purpose, a prismlike element holding all the molecules of interest is constructed about the surface element as sketched in Fig. 2-4. The prism side is of length $v \, dt$, which is the distance traveled by the molecules of speed $v$ in the time interval $dt$ and has its axis in the $\theta\phi$ direction. It contains all the $\theta\phi v$ molecules that will strike the surface element in the time interval $dt$, which is sufficiently small that intermolecular collisions are unlikely to deflect molecules out of the prism.

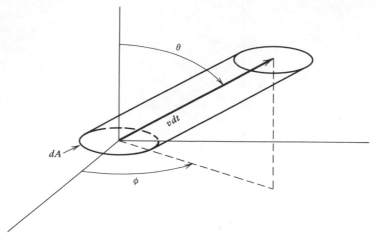

**Figure 2-4** Prismlike element holding all molecules that experience a $\theta\phi v$ collision with the prism base.

There are molecules in the prism with speeds other than $v$ and directions of travel other than $\theta\phi$. Some of them will not collide with $dA$ because they either are not moving in the $\theta\phi$ direction or are not moving with sufficient speed. Those colliding with $dA$ but not having speed $v$ are not counted among the $\theta\phi v$ collisions. Similarly, molecules initially outside the prism may collide with $dA$; but, because they are not moving in the $\theta\phi$ direction, they are not counted among the $\theta\phi v$ collisions, either. For these reasons, only molecules initially in the prism can experience a $\theta\phi v$ collision with $dA$.

As long as there are a large number of molecules in the prism, it can be assumed that the molecules have a velocity distribution the same as those in a larger containing volume, allowing use of previously derived relationships. The number of prism molecules per unit volume with speeds of $v \pm dv/2$ is to be denoted by $dn_v$, and $d^3n_{\theta\phi v}$ denotes the number of prism molecules per unit volume with speeds of $v \pm dv/2$ and travel directions of $\theta \pm d\theta/2$ and $\phi \pm d\phi/2$. It is seen that

$$d^3n_{\theta\phi v} = d\left(d^2n_{\theta\phi}\right)$$

where the third derivative notation signifies a quantity of the third order of smallness. Substitution of Eq. (2-3) into the preceding relationship gives

$$d^3n_{\theta\phi v} = dn_v \sin\theta \, d\theta \, \frac{d\phi}{4\pi} \qquad (2\text{-}4)$$

The number of molecules experiencing a $\theta\phi v$ collision—$d^3N_{\theta\phi v}$—is obtained by multiplying Eq. (2-4) by the prism volume $\cos\theta \, v \, dt \, dA$ to yield

$$d^3N_{\theta\phi v} = \sin\theta\cos\theta \, v \, dn_v \, d\theta \, d\phi \, dA \, \frac{dt}{4\pi} \qquad (2\text{-}5)$$

The rate at which molecules strike $dA$ from one side in the time interval $dt$ is found by integrating Eq. (2-5) over a hemisphere, following division by $dA\ dt$. Letting this molecular flux be represented by the symbol $\dot{N}$, one finds

$$\dot{N} = \int_{v=0}^{\infty} \int_{\theta=0}^{\pi/2} \int_{\phi=0}^{2\pi} v \sin\theta \cos\theta \, d\theta \, d\phi \, \frac{dn_v}{4\pi}$$

$$\dot{N} = \frac{n\bar{v}}{4} \qquad\qquad\qquad (2\text{-}6)$$

where $\bar{v}$ is the average speed of a molecule as given by

$$\bar{v} = \int_{v=0}^{\infty} v \, \frac{dn_v}{n} \qquad\qquad\qquad (2\text{-}7)$$

The average speed cannot be explicitly evaluated until the manner in which molecular speeds are distributed among the molecules has been determined.

## 2.3   GAS PRESSURE

The pressure exerted by a gas on the walls of its container is determined next. All collisions are considered to be perfectly elastic, or specular, so that the angle of incidence always equals the angle of reflection—the plane containing both the reflected trajectory and the surface normal is considered to be coincident with the plane containing the incident trajectory and surface normal, thus implying that a molecule has no spin of importance. The possibility of diffuse reflections has been considered by Schrage [3], who shows that the Maxwell velocity distribution still results. Figure 2-5 illustrates a collision.

On colliding with a wall, a molecule undergoes no change in tangential momentum, but its normal momentum does change. Before collision a molecule has normal momentum $-mv\cos\theta$, where $m$ is the mass of a molecule. After reflection, a molecule has normal momentum $mv\cos\theta$. The net result is that

$$\Delta\frac{\text{normal momentum}}{\text{molecule}} = 2mv\cos\theta \qquad\qquad (2\text{-}8)$$

Since many molecules collide with an elemental area $dA$ in the time interval $dt$ from the $\theta\phi$ direction and with speed $v$, the total change in $\theta\phi v$ normal momentum is

$$\Delta(\theta\phi v \text{ normal momentum}) = \Delta\frac{\text{normal momentum}}{\text{molecule}}$$

$$\times (\text{number of } \theta\phi v \text{ colliding molecules})$$

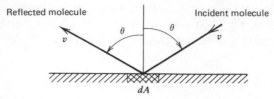

**Figure 2-5**   Side view of molecular path on collision with a wall.

Introduction of Eqs. (2-5) and (2-8) into this relationship gives

$$\Delta(\theta\phi v \text{ normal momentum}) = mv^2 \, dn_v \sin\theta \cos^2\theta \, d\theta \, d\phi \, dA \frac{dt}{2\pi} \quad (2\text{-}9)$$

Since molecules come from all directions and speeds to impact on the elemental area, the normal momentum change expressed by Eq. (2-9) must be integrated over a hemisphere and over all speeds. This procedure gives

$$\Delta(\text{normal momentum}) = \frac{m \, dA \, dt}{2\pi} \int_{v=0}^{\infty} \int_{\theta=0}^{\pi/2} \int_{\phi=0}^{2\pi} \sin\theta \cos^2\theta \, v^2 \, dn_v \, d\theta \, d\phi$$

$$= \frac{mn\,\overline{v^2}}{3} dA \, dt \quad (2\text{-}10)$$

where $\overline{v^2} = n^{-1}\int_{v=0}^{\infty} v^2 \, dn_v$ is the mean-square speed of a molecule.

The change in normal momentum was produced by a force $dF$ exerted by the wall on the molecules during the time interval $dt$. In other words, the wall exerted an impulse on the molecules that, by the laws of mechanics, must equal the change in normal momentum. Therefore,

$$dF \, dt = \frac{mn\,\overline{v^2}}{3} dA \, dt$$

or

$$p = \frac{dF}{dA} = \frac{mn\,\overline{v^2}}{3}$$

Rewriting this expression for pressure gives, since $n = N/V$,

$$pV = \frac{mN\,\overline{v^2}}{3} \quad (2\text{-}11)$$

It has been determined by independent experiments that a perfect gas such as has been assumed to be present has as its equation of state

$$pV = \frac{NRT}{N_A} \quad (2\text{-}12)$$

where $T$ is absolute temperature, $R = 1545.32$ ft $1b_f/1b_m$ mole $= 8314.34$ J/kg mole K is the universal gas constant, and $N_A$ is Avogadro's number as before. For the kinetic theory relation given by Eq. (2-11) to be correct, Eq. (2-12) reveals that it is necessary to have

$$\frac{m\overline{v^2}}{3} = \frac{R}{N_A}T$$

or

$$\frac{m\overline{v^2}}{2} = \frac{3KT}{2} \qquad\qquad (2\text{-}13)$$

where $K = R/N_A = 1.38062 \times 10^{-23}$ J/molecule K is Boltzmann's constant, which may be interpreted as the universal gas constant per molecule. The root-mean-square speed is given by Eq. (2-13) as $v_{rms} = (\overline{v^2})^{1/2} = (3KT/m)^{1/2}$.

Equation (2-13) relates the average translational kinetic energy of a molecule to thermodynamic temperature. Note particularly that the simple billiard ball molecule that has been considered can move in any of the three directions $x$, $y$, and $z$. These three directions will each have a velocity component so that $\overline{v^2} = \overline{v_x^2} + \overline{v_y^2} + \overline{v_z^2}$. Because of the equal probability of each direction of motion, $\overline{v_x^2} = \overline{v_y^2} = \overline{v_z^2}$. This fact, when inserted into Eq. (2-13), leads to the equipartition law—the energy in each energy storage mode equals $KT/2$—which is here expressed as

$$\frac{m\overline{v_x^2}}{2} = \frac{m\overline{v_y^2}}{2} = \frac{m\overline{v_z^2}}{2} = \frac{KT}{2}$$

If the energy in each energy storage mode corresponds to $KT/2$, one wonders why the spin of a billiard ball molecule does not store as much energy as do the translational modes. Intuitively, it seems plausible that a tiny molecule would require such a fantastically high angular velocity to store appreciable energy that energy storage in spin is unlikely. The absence of spin-stored kinetic energy in the present rigid sphere model is due primarily to the fact that its possible existence was excluded from the beginning; the detailed description provided by quantum mechanics as readably outlined by Feynman et al. [4] resolves this dilemma and shows spin-stored energy to indeed be negligible at normally encountered temperatures. The equipartition of energy law is true only if the energy associated with a degree of freedom is a quadratic function of the variable specifying the degree of freedom.

## 2.4  MOLECULAR VELOCITY DISTRIBUTION

In the foregoing developments information was derived without knowledge of the molecular velocity distribution. For example, the pressure exerted on a surface was found to depend on $m n \overline{v^2}/3$ according to Eq. (2-11). Also, the rate at which molecules collide with a surface or cross it from one side equals $n \overline{v}/4$ as given by Eq. (2-6). A great deal more could be deduced if the velocity distribution were known in addition.

Recall once again that with each molecule is associated a velocity vector as illustrated in Fig. 2-1. Again, let all the velocity vectors be moved to a common origin, thereby constructing a velocity space. As Fig. 2-6 shows, $v$ is the magnitude, or speed, of the velocity vector so that

$$v^2 = v_x^2 + v_y^2 + v_z^2$$

where $v_x$ is the $x$ component of the velocity vector $\mathbf{v}$, for example. Each velocity vector is defined by the coordinates of its endpoint and each endpoint corresponds to one molecule. The number of points within an elemental volume $dv_x \, dv_y \, dv_z$ in velocity space is then merely the number of molecules with velocity components in the range of $v_{x,\,y,\,z} \pm \frac{1}{2} dv_{x,\,y,\,z}$.

Let $dN_{v_x}$ be the number of molecules out of a total of $N$ that have velocities with $x$ components in the range $v_x \pm \frac{1}{2} dv_x$, allowing $v_y$ and $v_z$ to be any value. The fraction $dN_{v_x}/N$ of the total lying in this slice of velocity space as illustrated in Fig. 2-7 should depend on the slice thickness $dv_x$ and may also depend on the slice location $v_x$. Thus the simple guess is made that

$$\frac{dN_{v_x}}{N} = f(v_x) \, dv_x \tag{2-14a}$$

where $f(v_x)$ is a velocity distribution function yet to be determined. Because all

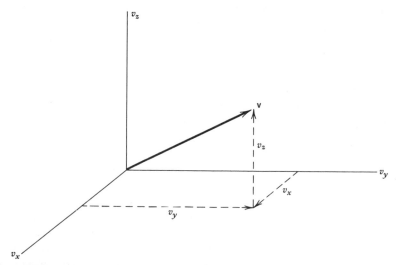

**Figure 2-6**   Decomposition of molecular velocity into components.

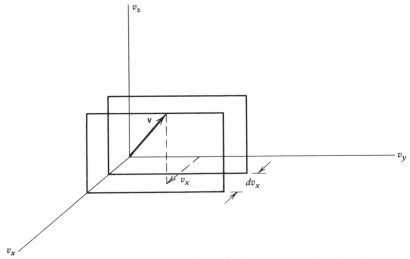

**Figure 2-7**  Molecular velocity with one component specified.

directions of motion are equally probable (bulk motion merely has the effect of shifting the zero-velocity reference point), it is similarly guessed that

$$\frac{dN_{v_y}}{N} = f(v_y) \, dv_y \qquad\qquad (2\text{-}14\text{b})$$

and

$$\frac{dN_{v_z}}{N} = f(v_z) \, dv_z \qquad\qquad (2\text{-}14\text{c})$$

Let a further restriction be imposed by seeking the number of molecules $d^2N_{v_x v_y}$ that not only have velocities with $x$ components in the range $v_x \pm \frac{1}{2}dv_x$, but also have $y$ components in the range $v_y \pm \frac{1}{2}dv_y$. The velocity vector of these molecules lies in the prism in velocity space as shown in Fig. 2-8. It is seen that $v_z$ can take on any value. Consistency with the functional form previously used for Eq. (2-14) leads to the velocity distribution of this subgroup of

$$\frac{d^2N_{v_x v_y}}{dN_{v_x}} = f(v_y) \, dv_y$$

that, to be true, must be the same as that of the original group of size $N$. Use of Eq. (2-14a) in this relationship gives

$$d^2N_{v_x v_y} = Nf(v_x) \, f(v_y) \, dv_x \, dv_y \qquad\qquad (2\text{-}15)$$

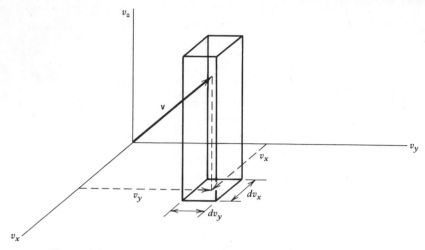

**Figure 2-8**  Molecular velocity with two components specified.

By the same reasoning, it is concluded that the even smaller number of molecules $d^3N_{v_xv_yv_z}$ whose velocities are further constrained to have $z$ components in the range of $v_z \pm \frac{1}{2}dv_z$, lying in the cube in velocity space shown in Fig. 2-9, must be related to $d^2N_{v_xv_y}$ by

$$\frac{d^3N_{v_xv_yv_z}}{d^2N_{v_xv_y}} = f(v_z)\,dv_z$$

This relation again requires the subgroup to have the same velocity distribution

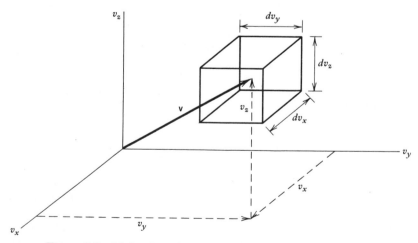

**Figure 2-9**  Molecular velocity with three components specified.

as the larger original group. Employment of Eq. (2-15) in this relationship yields

$$d^3 N_{v_x v_y v_z} = N f(v_x) f(v_y) f(v_z) \, dv_x \, dv_y \, dv_z \qquad (2\text{-}16)$$

A quantity of great utility in the next developments is the density of points in velocity space $\rho$ that is obtained from Eq. (2-16) as

$$\rho = \frac{d^3 N_{v_x v_y v_z}}{dv_x \, dv_y \, dv_z} = N f(v_x) f(v_y) f(v_z) \qquad (2\text{-}17)$$

The velocity distribution has been assumed to have all directions of motion equally probable. In other words, the velocity distribution is isotropic. Therefore, $\rho$ is constant as long as the velocity vector lies in a thin shell of radius $v$ and thickness $dv$ in velocity space. Figure 2-10 shows such a shell.

Consider now a second shell-like volume element in velocity space that is adjacent to the one previously considered. The possible difference in $\rho$ values in these thin and adjacent shells is accurately given by the first terms of the Taylor series expansion

$$d\rho = \frac{\partial \rho}{\partial v_x} dv_x + \frac{\partial \rho}{\partial v_y} dv_y + \frac{\partial \rho}{\partial v_z} dv_z \qquad (2\text{-}18)$$

From Eq. (2-17) the derivatives in Eq. (2-18) can be evaluated, allowing Eq.

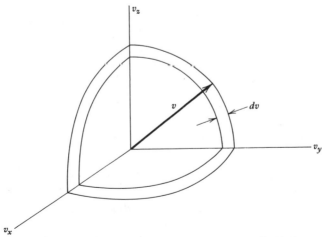

**Figure 2-10**  Molecular velocity with specified speed.

(2-18) to be recast as

$$\frac{d\rho}{\rho} = \frac{f'(v_x)}{f(v_x)}dv_x + \frac{f'(v_y)}{f(v_y)}dv_y + \frac{f'(v_z)}{f(v_z)}dv_z \tag{2-19}$$

in which, for example, $f'(v_x) = df(v_x)/dv_x$.

At this point the special case is considered in which $dv_{x,y,z}$ are selected in such a way as to make the second shell-like volume coincident with the original one; then $d\rho/\rho = 0$ in Eq. (2-19). Additionally, if the second volume coincides with the original one, the constraint that

$$v^2 = v_x^2 + v_y^2 + v_z^2 = \text{const} \tag{2-20}$$

requires that

$$v\,dv = v_x\,dv_x + v_y\,dv_y + v_z\,dv_z = 0 \tag{2-21}$$

also be satisfied. The determination of the unknown velocity distribution function $f$ now has the classical form of the problem of extremizing a function $\rho$ subject to the equality constraint in Eq. (2-21).

The method of Lagrangian multipliers will be employed for this task. Equation (2-21) is multiplied by a constant, the Lagrangian multiplier here denoted by the symbol $\lambda$, and added to Eq. (2-19), which is recalled as being required to equal zero. This procedure gives

$$\left[\frac{f'(v_x)}{f(v_x)} + \lambda v_x\right]dv_x + \left[\frac{f'(v_y)}{f(v_y)} + \lambda v_y\right]dv_y + \left[\frac{f'(v_z)}{f(v_z)} + \lambda v_z\right]dv_z = 0$$

$$\tag{2-22}$$

It can now be said that $dv_{x,y,z}$ are independent in Eq. (2-22) since there is only one equation, and it must apply to arbitrary variations of $dv_{x,y,z}$ as the velocity vector is allowed to move in velocity space around the shell in Fig. 2-10. This is a substantial simplification in view of the previous interdependence of $dv_{x,y,z}$ because the only way Eq. (2-22) can always be satisfied is for each coefficient to equal zero. Therefore, for example,

$$\frac{df(v_x)}{dv_x} = -\lambda v_x f(v_x)$$

from which it is found that

$$f(v_x) = \alpha e^{-\beta^2 v_x^2} \tag{2-23}$$

where $\alpha$ is a constant of integration, $\beta^2 = \lambda/2$ and $f(v_y)$ and $f(v_z)$ are observed to have the same functional form. Assemblage of these results in Eq. (2-17) in order to express the density of points in velocity space gives

$$\rho = N\alpha^3 e^{-\beta^2(v_x^2 + v_y^2 + v_z^2)}$$

or, in view of Eq. (2-20),

$$\rho = N\alpha^3 e^{-\beta^2 v^2} \tag{2-24}$$

In order for the density of points in velocity space as expressed by Eq. (2-24) to really be useful, the parameters $\alpha$ and $\beta$ must be evaluated in terms of easily measurable quantities. For this purpose, it will first be noted that all molecules must have speeds between zero and infinity, $0 \leqslant v \leqslant \infty$. Recall that $\rho \, dv_x \, dv_y \, dv_z$ is the number of molecules having speeds in the range of $v_{x, y, z} \pm dv_{x, y, z}/2$. Since $\rho$ depends only on the speed (velocity magnitude) and is isotropic, the number of molecules $dN_v$ having speeds in the range of $v \pm dv/2$ is computed by multiplying $\rho$ by the volume of the spherical shell in velocity space shown in Fig. 2-10; thus

$$dN_v = \rho 4\pi v^2 \, dv$$

which becomes, with the aid of Eq. (2-24),

$$dN_v = 4\pi N\alpha^3 v^2 e^{-\beta^2 v^2} \, dv \tag{2-25}$$

The requirement that all molecules must have some speed is expressed as

$$N = \int_{v=0}^{\infty} dN_v$$

With the use of Eq. (2-25), this simple relationship becomes

$$N = 4\pi N\alpha^3 \int_{v=0}^{\infty} v^2 e^{-\beta^2 v^2} \, dv$$

whose integral is evaluated to give

$$N = 4\pi N\alpha^3 \frac{\pi^{1/2}}{4\beta^3}$$

from which it is found that

$$\alpha = \frac{\beta}{\pi^{1/2}} \tag{2-26}$$

A second requirement that can also be used to relate $\alpha$ and $\beta$ is given by Eq. (2-13) as

$$\overline{v^2} = \frac{3KT}{m}$$

By the definition of $\overline{v^2}$, this relationship becomes

$$\frac{1}{n}\int_{v=0}^{\infty} v^2 \, dn_v = \frac{3KT}{m}$$

Equation (2-25) gives $dN_v$ and allows restatement as

$$N^{-1}\int_{v=0}^{\infty} v^2\left(4\pi N\alpha^3 e^{-\beta^2 v^2} \, dv\right) = \frac{3KT}{m}$$

Use of Eq. (2-26) to eliminate $\alpha$ in favor of $\beta$ then yields

$$\frac{4\beta^3}{\pi^{1/2}}\int_{v=0}^{\infty} v^4 e^{-\beta^2 v^2} \, dv = \frac{3KT}{m}$$

whose integral is evaluated to give

$$\frac{4\beta^3}{\pi^{1/2}}\frac{3\pi^{1/2}}{8\beta^5} = \frac{3KT}{m}$$

from which it is found that

$$\beta^2 = \frac{m}{2KT} \tag{2-27}$$

With this result in hand, Eq. (2-26) allows $\alpha$ to be given as

$$\alpha = \left(\frac{m}{2\pi KT}\right)^{1/2} \tag{2-28}$$

Finally, one has from Eqs. (2-23) and (2-24) that

$$f(v_x) = \left(\frac{m}{2\pi KT}\right)^{1/2} e^{-mv_x^2/2KT} \tag{2-29}$$

and

$$\rho = N\left(\frac{m}{2\pi KT}\right)^{3/2} e^{-mv^2/2KT} \tag{2-30}$$

Equation (2-30) is the Maxwell distribution function of molecular *velocities*.

A useful generalization is possible at this point by rewriting Eq. (2-14a) for the fraction of molecules that have $x$ components of velocity within $\pm dv_x/2$ of $v_x$. Use of Eq. (2-29) in Eq. (2-14a) gives

$$\frac{dN_{v_x}}{N} = \left(\frac{m}{2\pi KT}\right)^{1/2} e^{-mv_x^2/2KT}\, dv_x$$

which can be interpreted as

$$\frac{dN_{v_x}}{N} = ce^{-(\text{K.E. of }x\text{-direction motion})/KT}$$

This relationship suggests the general principle that the fraction of molecules that have a particular kind of energy near a specified value will vary as

$$e^{-(\text{energy kind})/KT} \qquad\qquad (2\text{-}30a)$$

The distribution of gases in a planetary atmosphere, the subject of Problem 2-13, is suggested by this general result to vary with height as

$$n = n(z = 0)e^{-mgz/KT} \qquad\qquad (2\text{-}30b)$$

and the kind of energy appropriately considered is potential energy. Detailed analysis confirms this result and others [10], strengthening the use of the general principle to describe the many natural phenomena in which energy must be borrowed from somewhere; the number of particles per unit volume with a specified energy is proportional, again, to $e^{-(\text{energy kind})/KT}$. Equation (2-30a) is also known as *Boltzmann's law* and states, in general terms, that the probability that molecules will be in a given spatial arrangement varies exponentially with the negative of the potential energy of that arrangement.

*Example 2-1:* Determine the manner in which the number of molecules having speeds in the range of $v \pm dv/2$ varies with $v$. From Eq. (2-25) the number of molecules is given by

$$dN_v = 4\pi N\alpha^3 v^2 e^{-\beta^2 v^2}\, dv$$

which can be rearranged with the help of Eqs. (2-27) and (2-28) into

$$N^{-1}\frac{dN_v}{dv} = \left(\frac{8m}{\pi KT}\right)^{1/2}\frac{mv^2}{2KT}e^{-mv^2/2kT} \qquad\qquad (2\text{-}31)$$

which is the Maxwell distribution function of molecular *speeds*.

The Maxwell distribution function of molecular speeds given by Eq. (2-31) predicts, as shown in Fig. 2-11, that there will be some molecules whose speeds

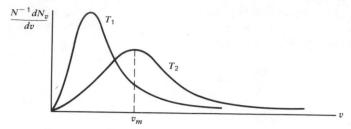

**Figure 2-11** Maxwell distribution for speed $T_2 > T_1$.

are near any specified value. There is a most probable speed $v_m$ that occurs at the peak of the curve in Fig. 2-11 and that is the speed possessed by more molecules than any other. However, a substantial fraction of the total molecules possess quite different speeds. The value of $v_m$ is obtained by setting the first derivative of Eq. (2-31) to zero and solving to find

$$v_m = \left( \frac{2KT}{m} \right)^{1/2} \tag{2-32}$$

It is noted in Fig. 2-11 that as temperature increases, the distribution of speeds broadens and the most probable speed $v_m$ itself increases.

*Example 2-2:* Determine the relation between the average speed $\bar{v}$ and absolute temperature. According to Eq. (2-7),

$$\bar{v} = n^{-1} \int_{v=0}^{\infty} v \, dn_v$$

Insertion of Eq. (2-31) into this formulation gives

$$\bar{v} = \left( \frac{8m}{\pi T} \right)^{1/2} \int_{v=0}^{\infty} v \frac{mv^2}{2KT} e^{-mv^2/2KT} \, dv$$

which can be rearranged into

$$\bar{v} = \left( \frac{8KT}{\pi m} \right)^{1/2} \int_{x=0}^{\infty} xe^{-x} \, dx$$

where $x = mv^2/2KT$. Evaluating the integral then yields

$$\bar{v} = \left( \frac{8KT}{\pi m} \right)^{1/2} \tag{2-32a}$$

*Example 2-3:* Determine the fraction of molecules that have speeds between $v_1$ and $v_2$. From Eq. (2-31) the number of molecules having speeds in the range of $v \pm \frac{1}{2} dv$ is

$$dN_v = N\left(\frac{8m}{\pi KT}\right)^{1/2} \frac{mv^2}{2KT} e^{-mv^2/2KT} \, dv \qquad (2\text{-}33)$$

The desired fraction is computed from

$$\frac{N_{v_1-v_2}}{N} = \int_{v_1}^{v_2} \frac{dN_v}{N}$$

Equation (2-33) introduced into this relationship gives

$$\frac{N_{v_1-v_2}}{N} = \left(\frac{8m}{\pi KT}\right)^{1/2} \int_{v_1}^{v_2} \frac{mv^2}{2KT} e^{-mv^2/2KT} \, dv$$

which assumes the form

$$\frac{N_{v_1-v_2}}{N} = 4\pi^{-1/2} \int_{x_1}^{x_2} x^2 e^{-x^2} \, dx$$

when $x = v(m/2KT)^{1/2}$. This integral can be evaluated by parts to achieve

$$\frac{N_{v_1-v_2}}{N} = 2\pi^{-1/2}\left(x_1 e^{-x_1^2} - x_2 e^{-x_2^2}\right) + \mathrm{erf}(x_2) - \mathrm{erf}(x_1) \qquad (2\text{-}34)$$

Note here that the error function $\mathrm{erf}(x)$ is defined as

$$\mathrm{erf}(x) = \frac{2}{\pi^{1/2}} \int_0^x e^{-\eta^2} \, d\eta \qquad (2\text{-}35)$$

and is a tabulated function readily available in such references as the *Handbook of Mathematical Functions* [5]. The variation of the error function is shown in Fig. 2-12.

***Example 2-4:*** Determine the fraction of molecules whose $x$ component of velocity lies between $v_{x_1}$ and $v_{x_2}$. According to Eq. (2-14a), the fraction near $v_x$ is

$$\frac{dN_{vx}}{N} = f(v_x) \, dv_x$$

In addition, the fraction in the specified range is given by

$$\frac{N_{v_{x_1}-v_{x_2}}}{N} = \int_{v_{x_1}}^{v_{x_2}} \frac{dN_{vx}}{N}$$

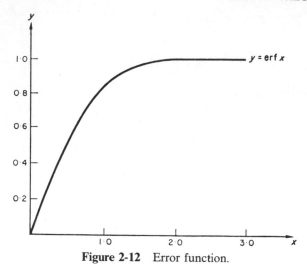

**Figure 2-12**   Error function.

Introduction of Eq. (2-29) for $f(v_x)$ then gives

$$\frac{N_{v_{x_1} - v_{x_2}}}{N} = \left( \frac{m}{2\pi KT} \right)^{1/2} \int_{v_{x_1}}^{v_{x_2}} e^{-mv_x^2/2KT} \, dv_x$$

This can be written more simply as

$$\frac{N_{v_{x_1} - v_{x_2}}}{N} = \pi^{-1/2} \int_{y_1}^{y_2} e^{-y^2} \, dy$$

by letting $y = v_x (m/2kT)^{1/2}$. This integral can be rearranged as

$$\frac{N_{v_{x_1} - v_{x_2}}}{N} = \pi^{-1/2} \left( \int_0^{y_2} e^{-y^2} \, dy - \int_0^{y_1} e^{-y^2} \, dy \right)$$

in which form the relation of the two integrals to the error function defined by Eq. (2-35) is evident. So

$$\frac{N_{vx_1 - vx_2}}{N} = \frac{\mathrm{erf}(y_2) - \mathrm{erf}(y_1)}{2} \qquad (2\text{-}36)$$

Note that the fraction of molecules moving in the positive direction is $\frac{1}{2}$—for $v_{x_1} = 0$ and $v_{x_2} = \infty$, one has $y_1 = 0$ and $y_2 = \infty$, for which case the error function behaviour of Fig. 2-12 leads Eq. (2-36) to the $\frac{1}{2}$ result. In an equilibrium situation, half the molecules should move in the positive direction and half in the opposite direction.

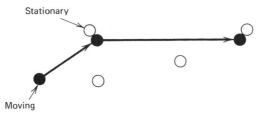

**Figure 2-13** Simplified view of molecular path with intermolecular collisions.

## 2.5 MEAN FREE PATH

The transport properties to be discussed in Chapter 3 are found to depend on the distance a molecule travels before colliding with another molecule as well as on the molecule speed. To estimate the typical distance traveled between intermolecular collisions, also called the *mean free path*, a simplified situation is assumed in which all molecules but one are assumed to be frozen in space as Fig. 2-13 illustrates; the one moving molecule travels at a relative speed $v_r$. In this way insight into the physical phenomenon is more easily achieved than would be the case for a view that accounted in detail for Maxwell velocity distributions for all molecules.

All molecules are assumed to be identical and to be perfectly elastic spheres of diameter $\sigma$. Since molecules are not really billiard balls, $\sigma$ is best interpreted as giving the collision cross section $\sigma_c = \pi\sigma^2$. As Fig. 2-14a demonstrates, on collision the center-to-center distance is $\sigma$. The same result is obtained if the moving molecule is considered to be of diameter $2\sigma$ and the stationary ones are considered to be points as shown in Fig. 2-14b.

In a time interval $t$ the moving molecule travels a relative distance given by $v_r t$ that is zigzag as a result of collisions for appreciable elapsed times. The volume it sweeps out is

$$\frac{v_r t (2\sigma)^2 \pi}{4} = \pi v_r t \sigma^2$$

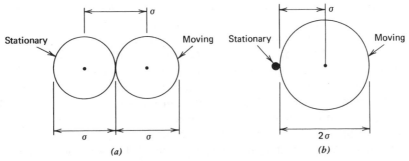

**Figure 2-14** Simplified view of a collision of two molecules.

Since all molecules are assumed to be stationary, all molecules in this swept-out volume will be struck by the moving one. There being $n$ molecules per unit volume, the number of collisions in time $t$ is

$$\pi n v_r t \sigma^2$$

and the collision frequency $z$ is obtained by dividing the number of collisions by the elapsed time to obtain

$$z = \pi n v_r \sigma^2 \qquad (2\text{-}37)$$

The mean free path $\lambda$ is merely the actual distance traveled divided by the number of collisions experienced. Thus

$$\lambda = \frac{\bar{v} t}{\pi n v_r t \sigma^2}$$

which reduces to

$$\lambda = \frac{1}{\pi n \sigma^2} \qquad (2\text{-}38)$$

for the simplistic case in which $v_r = \bar{v}$.

Since $\lambda$ depends inversely on $n$ in Eq. (2-38), it is seen that the mean free path for a perfect gas is directly proportional to absolute temperature and inversely proportional to absolute pressure. As a result, at low pressures the mean free path can be quite large. In such a case it is possible for the mean free path to approach the size $L$ of some solid object immersed in the gas. Then the transport of energy in the vicinity of the solid object is not affected by random collisions and so the mean free path does not exert exclusive control over the transport rate. As a result, the fluid can seem to slip at the solid surface, and a temperature jump can also occur there. Discontinuities in velocity and temperature that are so small when $\lambda \ll L$ that they are accurately assumed not to exist become increasingly important as $\lambda$ approaches $L$. When $\lambda \gg L$, the mean free path is unimportant; instead, the transport rate is controlled by parameters that describe the effectiveness with which properties are exchanged between a surface and the impinging molecules. Under normal conditions the following orders of magnitude are typical: mean free path $\sim 10^{-7}$ m, molecular separation $\sim 10^{-9}$ m, molecular diameter $\sim 10^{-10}$ m, and collision frequency $\sim 10^{10}$ collisions/s. At pressures so high as to render the mean free path comparable in size to molecular dimensions, there may also be unusual effects; this if often made evident by a pressure dependence of viscosity, for example, that is absent under normal conditions.

Note that Eq. (2-37) was obtained on the basis that all molecules but one were stationary. Actually, molecules not originally in the swept-out volume can move into it and be struck; those originally in it can move out and avoid collision. If all molecules are allowed to possess the Maxwell velocity distribution, it can be shown [6] that the correct relative velocity is

$$v_r = 2^{1/2}\bar{v}$$

which gives refined estimates for the collision frequency and mean free path as

$$z = 2^{1/2}\pi n\bar{v}\sigma^2 \tag{2-39}$$

$$\lambda = \frac{1}{2^{1/2}\pi n\sigma^2} \tag{2-40}$$

It must be kept in mind that the mean free path of Eq. (2-39) is only an average. Some molecules travel shorter distances before a collision, whereas others travel a longer distance. To show this, let us follow a number of molecules $N_i$ as they proceed in the $x$ direction. Let the distance traveled be $x$ and the number of molecules that have not suffered collision be $N$. The symbol $dN$ represents the number of molecules removed from $N$ as a result of collisions after traveling an additional distance $dx$. Then

$$dN = -P_c N \, dX$$

where $P_c$ is a proportionality constant called a *collision probability*. Solution yields

$$\frac{N}{N_i} = e^{-P_c x}$$

which is the same extinction law, Bougher's law, which describes the intensity of thermal radiation or visible light in its passage through an absorbing medium. The collision probability is analogous to an extinction coefficient arising from absorption and scattering of photons. Although molecules are not absorbed, they are scattered. It is clear that a large number of molecules can travel a short distance without a collision. The average, or mean, distance that can be traveled without a collision is evaluated as

$$\bar{x} = N_i^{-1}\int_{N=0}^{N_i} x \, dN$$

$$= \int_{x=0}^{\infty} x\left(-P_c e^{-P_c x} \, dx\right)$$

$$= P_c^{-1}$$

in which $\bar{x}$ is recognized to be identical to $\lambda$ by definition. The mean free path

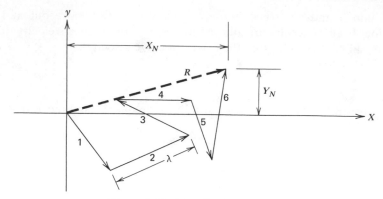

**Figure 2-15** Drunkard's walk.

is seen to be merely an expression of the probability of collision while traversing a distance $dx$—a large value of $\lambda$ corresponds to a small collision probability, and vice versa. The fraction of molecules traveling at least a distance $x$ before collision can then be written as

$$\frac{N}{N_i} = e^{-x/\lambda} \qquad\qquad (2\text{-}40a)$$

## 2.6  DRUNKARD'S WALK

Useful insight into the net results of a molecule's zigzag motion due to collisions is provided by a simple model called the *drunkard's walk*. In this model a drunkard steps in a straight line for a fixed distance $\lambda$ and then randomly chooses a new direction and again steps in a straight line for a fixed distance $\lambda$. Figure 2-15 illustrates these events.

    The average distance $R$ from the origin at which the drunkard will be located after $N$ steps can be estimated in the following way. It is first realized that

$$R^2 = X_N^2 + Y_N^2$$

where $X_N$ and $Y_N$ are the two components of $R$. Then, since each component of the net displacement is the algebraic sum of the corresponding components of the individual steps,

$$R^2 = (X_1 + X_2 + \cdots + X_N)^2 + (Y_1 + Y_2 + \cdots + Y_N)^2$$

$$= (X_1^2 + Y_1^2) + (X_2^2 + Y_2^2) + \cdots + (X_N^2 + Y_N^2)$$

$$+ 2(X_1 Y_1 + X_1 Y_2 + X_2 Y_1 + \cdots + X_N Y_N)$$

On the average there will be as many positively as negatively directed moves, so the cross-product terms should average to zero. And for each step,

$$\lambda^2 = X_i^2 + Y_i^2$$

Therefore,

$$R^2 = N\lambda^2$$

The average root-mean-square distance staggered from the starting point increases as the square root of the number of steps according to

$$R = \lambda N^{1/2} \tag{2-41}$$

A randomly chosen direction after each step reduces the net distance traveled below the magnitude it would have with unimpeded, straight-line progress, but it still increases as more steps are taken.

Naturally, the drunkard corresponds to a molecule, the step size corresponds to a mean free path, and the changes of direction correspond to molecular collisions. Since the molecule undergoes net motion, on the average, it can be appreciated that any property it carries will be transported by this diffusive process. In terms of previously derived quantities, Eq. (2-41) becomes

$$R = \left( \frac{\bar{v}}{2^{1/2}\pi n\sigma^2} \right)^{1/2} t^{1/2} \tag{2-42}$$

since the number of steps is merely the product of collision frequency from Eq. (2-39) and time, with Eq. (2-40) giving $\lambda$. Equation (2-42) reveals that the average diffused distance increases as the square root of the elapsed time. Therefore, it is reasonable to find that the equations that describe diffusive phenomena do have solutions that involve $t^{1/2}$. A more detailed exposition is given by Present [6].

## PROBLEMS

2-1 Show that an analogy exists between the rate at which molecules leave an elemental area at an angle of $\theta$ from the surface normal and Lambert's cosine law for the rate of emission of thermal radiation at the same angle. Lambert's cosine law gives

$$q(\theta) = I \cos\theta \, d\omega$$

where $I = q_{total}/\pi$, with $q_{total} = \sigma T^4$ for a blackbody, is the radiation intensity and $d\omega$ is the elemental solid angle defined by the ratio of an elemental area on a spherical surface to the square of the sphere radius. Thus, with reference to Fig. 2-3,

$$dw = \frac{dA_s}{r^2} = r^2 \sin\theta \frac{d\theta \, d\phi}{r^2}$$

*Hint*: Recast Eq. (2-5) into the form

$$\frac{d^2 N_{\theta\phi}}{dA \, dt} = \frac{\dot{N}}{\pi} \cos\theta \, dw$$

and note that equal numbers of molecules must go in the other direction for equilibrium.

**2-2** On the basis of the analogy developed in Problem 2-1, estimate the numerical value of the fraction of molecules leaving a small circular area of radius $r$ that impinge on a parallel and directly opposite disk of the same size a distance $2r$ away. *Hint*: Assume no scattering by collisions in the intervening space and use the appropriate geometric shape factor for thermal radiation.

**2-3** Equation (2-6) can be applied to flow from a thin aperture of radius $r$ sufficiently small that no molecular collisions occur near the hole since then $\lambda \gg r$. Such a flow is termed *effusion*.

   **(a)** Show that if two containers are connected by a small and thin aperture, the net molecular flux by effusion is given by

$$\dot{N}_{net} = N_0 \left( \frac{P_f}{T_f^{1/2}} - \frac{P_b}{T_b^{1/2}} \right) \frac{(8/\pi RM)^{1/2}}{4}$$

   where the subscripts $f$ and $b$ refer to conditions on the front and back sides of the aperture, respectively.

   **(b)** Show that a pressure differential can exist when there is no net molecular flux for the conditions of part a. Evaluate the numerical magnitude of this pressure difference for a typical set of temperatures.

   **(c)** Show that a net molecular flux can exist when there is no pressure difference if a temperature difference is imposed for the conditions of part a.

   **(d)** Compare the net molecular flux by effusion at constant temperature through a thin aperture from part a with that predicted for free-molecule (Knudsen) flow by [6]

$$\dot{N}_{net} = \frac{2r}{3L} \left( \frac{8}{\pi MRT} \right)^{1/2} N_A (P_f - P_b)$$

   which applies to long tubes of length $L$ and radius $r$ with $\lambda \gg r$.

Note that effusion can be viewed as a special case of Knudsen flow in which $L \ll r$. Knudsen's derivation contained errors that cancel only for a circular tube [7].

2-4 (a) Show that the equation giving the maximum rate of evaporation of a liquid (into a vacuum) derived from Eq. (2-6) is

$$\dot{m} = P\left(\frac{M}{2\pi RT}\right)^{1/2}$$

where $P$ is vapor pressure at absolute temperature $T$, $M$ is molecular weight, $R$ is the universal gas constant, and $\dot{m}$ is mass flux.

(b) Use the result of part a to plot numerical values of $\dot{m}$ versus $T$ for water in the range of 5–100°C.

(c) Comment on the degree to which the evaporation rate in commercial water cooling towers approaches the maximum rate of part a.

(d) The finding reported in the literature [8,9] is that a multiplicative constant near unity should be used to correct the result of part a. Comment on the physical reasons behind the need for a corrective multiplier and cite one of its typical numerical values.

2-5 Show that the mean speed $\bar{v}$ of Eq. (2-32a) can also be expressed as

$$\bar{v} = \left(\frac{8RT}{\pi M}\right)^{1/2}$$

where $M$ is the molecular weight.

2-6 (a) Show that the relative magnitudes of the molecular speeds are

$$v_{\text{acoustic}} : v_m : \bar{v} : v_{\text{rms}} = \left(\frac{\gamma}{2}\right)^{1/2} : 1 : 1.128 : 1.224$$

where $\gamma$ is the ratio of specific heats.

(b) Determine whether the speed at which signals can be propagated in a gas—the acoustic speed—is less than the other three speeds of part a.

(c) Determine the fraction of molecules whose speed is less than the acoustic speed. Evaluate the numerical value of this fraction for nitrogen and for hydrogen at a temperature of 20°C. Are the fractions different for these two gases? (*Answer*: 0.3 for both.)

2-7 Derive Eq. (2-30b). First make a force balance on a horizontal slice of air of thickness $dz$ to show that the pressure must increase as

$$dP = -mng\,dz$$

and then use the perfect gas equation of state for isothermal conditions.

**2-8** Determine the fraction of molecules whose $x$ component of velocity is less than the acoustic speed. Evaluate the numerical value of this fraction for nitrogen and for hydrogen at a temperature of 20°C. Are the fractions different from those found for Problem 2-6c?

**2-9** Derive the equation predicting the mean free path of an electron moving among molecules of gas

$$\lambda_e = \frac{4}{\sigma n}$$

which has application in gas discharge tubes. *Hint*: Recognize that an electron moves so much faster than the gas molecules that the latter can accurately be assumed to be stationary and that the electron diameter is so small that the collision cross section is only $\pi\sigma^2/4$.

**2-10** A spherical bulb of 10 cm radius is maintained at 27°C except for 1 cm$^2$, which is kept at a very low temperature. The bulb contains water vapor originally at a pressure of 10 mm Hg. Assuming that every water molecule striking the cold area condenses and sticks to the surface, how long a time is required for the pressure to decrease to $10^{-4}$mm Hg?

**2-11** The mean free path in a certain gas is 10 cm. If there are $10^4$ free paths, how many are longer than

(a) 10 cm?

(b) 20 cm?

(c) 50 cm?

(d) How many are longer than 5 cm but shorter than 10 cm?

(e) How many are between 9.5 and 10.5 cm in length?

(f) How many are between 9.9 and 10.1 cm in length?

(g) How many are exactly 10 cm in length?

**2-12** To what pressure, in millimeters of mercury, must a cathode ray tube be evacuated for 90% of the electrons leaving the cathode to reach the anode, 20 cm away, without experiencing a collision? The diameter of an air molecule can be taken to be $4 \times 10^{-10}$ m.

**2-13** The ability of a planet to retain an atmosphere has been examined in detail (J. H. Jeans, *The Dynamical Theory of Gases*, 4th ed., Dover, New York, 1925, p. 343). A simple estimate can be achieved, however, in the following way for a constant temperature atmosphere:

(a) Show that the flux of molecules crossing a horizontal plane with speeds exceeding the planetary escape velocity $V_e$ is

$$\dot{N} = n_0\left(\frac{KT}{2\pi m}\right)^{1/2}\left(1 + \frac{MV_e^2}{2KT}\right)e^{-mV_e^2/2KT}$$

where $V_e^2 = 2g_0r_0$, $g_0$ is the planet gravitational acceleration, $r_0$ is the planet radius, and $n_0$ is the number of molecules per volume at the planet surface. *Hint*: Review the derivation of Eq. (2-6).

**(b)** Show that for an exponential atmosphere of infinite thickness, the total number of molecules per unit area is

$$N = \frac{n_0 KT}{mg_0}$$

**(c)** Use the results of parts a and b to show that the rate of loss of molecules from the planetary atmosphere is dependent on a time constant $t_c$ according to

$$\frac{dN}{dt} = \frac{N}{t_c}$$

or

$$\frac{N}{N_0} = e^{-t/t_c}$$

where the time constant is given by

$$t_c = \left(\frac{2\pi r_0}{g_0}\right)^{1/2} \frac{B^{-1/2}e^B}{1+B}$$

with

$$B = \frac{mg_0 r_0}{KT}$$

**(d)** Presuming that five time constants give a completed decay process, estimate the time required for an atmosphere to be lost from (a) Earth, (b) Earth's moon, and (c) Mars. Assume reasonable temperatures for each case.

**2-14** It has been observed that a hot solid immersed in a stagnant dust-laden gas is enveloped by a dust-free film [W. Cawood, The movements of dust or smoke particles in a temperature gradient, *Transact. Faraday Soc.* **32**, 1068–1073 (1936); H. H. Watson, The dust-free space surrounding hot bodies, *Transact. Faraday Soc.* **32**, 1073–1084 (1936)]. The forces on a particle due to differential molecular bombardment in a steep temperature gradient near a hot solid are balanced against bulk convective flow toward the body, producing the observed dust-free zone.

Other forces can act on a spherical particle [B. Otterman, Motion of particles injected from the surface into stagnation point flow, *AIAA J.* **10**, 1079–108 (1972)]: (1) Stokes viscous drag; (2) transverse force due to slip and shear; and (3) lift force due to sphere rotation. Briefly outline a technological application of these "boundary-layer filtration" phenomena. A review of the complete articles referenced might be helpful.

**2-15**  If the absolute temperature of a gas is doubled, by what factor is the root-mean-square distance traveled in a specified time interval by a molecule increased?

**2-16**  The rate at which molecules cross an area from one side can be computed in a different manner from that leading up to Eq. (2-7). In the alternative derivation, it is considered that the number of molecules in unit volume having velocity components within a specified range is $nf(v_x)f(v_y)f(v_z)\,dv_x\,dv_y\,dv_z$. The number of these that cross unit area of a surface in unit time is the number in a volume based on unit area with a height $v_y$. Therefore, the rate at which molecules cross unit area in unit time is given by

$$\dot{N} = n\int_{v_x=-\infty}^{\infty}\int_{v_y=0}^{\infty}\int_{v_z=-\infty}^{\infty} v_y f(v_x)f(v_y)f(v_z)\,dv_x\,dv_y\,dv_z$$

Show that the preceding formulation leads to

$$\dot{N} = \frac{n\bar{v}}{4} \qquad\qquad (2\text{-}7)$$

where $\bar{v}$ is the average speed.

## REFERENCES

1   J. F. Lee, F. W. Sears, and D. L. Turcotte, *Statistical Thermodynamics*, Addison-Wesley, Reading, MA, 1963.

2   R. B. Bird, W. E. Stewart, and E. N. Lightfoot, *Transport Phenomena*, Wiley, New York, 1960, pp. 21–23.

3   R. N. Schrage, *A Theoretical Study of Interphase Mass Transfer*, Columbia University Press, New York, 1953, p. 17.

4   R. P. Feynman, R. B. Leighton, and M. Sands, *The Feynman Lectures on Physics*, Vol. I, Addison-Wesley, Reading, MA, 1963.

5   M. Abramowitz and Irene A. Stegun, Eds., *Handbook of Mathematical Functions*, Dover, New York, 1965.

6   R. D. Present, *Kinetic Theory of Gases*, McGraw-Hill, New York, 1958.

7   S. L. Matson and J. A. Quinn, Knudsen diffusion through noncircular pores: Textbook errors, *AIChE J.* **23**, 768–770 (1977).

8   J. Bonacci, A. L. Myers, G. Nongbri, and L. C. Eagleton, The evaporation and condensation coefficient of water, ice, and carbon tetrachloride, *Chem. Eng. Sci.* **31**, 609–617 (1976).

9   J. R. Maa, Evaporation coefficient of liquids, *Indust. Eng. Chem. Fund.* **6**, 504–518 (1967).

10  A. D. Wilson, Tungsten filament life under constant-current heating, *J. Appl. Phys.* **40**, 1956–1964 (1969).

11  C. Truesdell and R. G. Muncaster, *Fundamentals of Maxwell's Kinetic Theory of A Simple Monatomic Gas, Treated As A Branch of Rational Mechanics*, Academic Press, New York, 1980.

# 3
# TRANSPORT PROPERTIES

The transport properties of primary interest are viscosity, thermal conductivity, and mass diffusivity. As explained in Chapter 1, the transport properties are properties of a substance and serve as coefficients in the rate equations that relate the flux of a quantity to the gradient of that quantity. As might be expected, therefore, it is only in nonequilibrium situations that the transport properties become important.

It is desirable that each transport property be experimentally determined; in fact, for most fluids and substances of engineering interest, measured values are reported in the literature [13, 24, 25, 26]. However, correlation and explanation of data are also needed; thus the physical phenomena involved in the transport process must be considered on a microscopic level.

For gases, kinetic theory has been appealed to successfully for insight into the parameters that influence transport properties. A certain amount of surprise must be registered at such success for kinetic theory postulates an equilibrium situation, but it must be applied to the nonequilibrium situation in which there is some net transport. It seems, though, that the essential features of kinetic theory closely approximate actual events in the case of dilute gases. For the more complicated liquids, kinetic theory is much less successful.

In Chapter 2 the drunkard's walk illustrated that random molecular motion causes a net flow of any property possessed by the molecules. In the case of bulk momentum, it is clear that momentum will be exchanged on collision, as molecular collisions have been assumed to be elastic; the bulk momentum carried by a molecule is the momentum acquired at its last collision. In general, it will be assumed in the kinetic theory developments to follow that the transported property was acquired at the last collision.

## 3.1 AVERAGE COLLISION HEIGHT [1]

The average height above a surface, illustrated in Fig. 3-1 by the elemental area $dA$, at which a molecule last underwent collision before crossing the surface

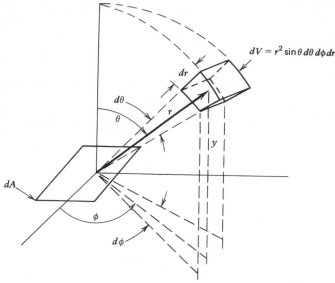

$$dV = r^2 \sin\theta \, d\theta \, d\phi \, dr$$

**Figure 3-1** Location at which a molecule last underwent intermolecular collision before crossing area $dA$.

must be determined. For this purpose, attention is focused on the events in a small volume element on one side of the area in question, keeping in mind that similar events occur on the other side of the plane.

Within the small volume element $dV$ of Fig. 3-1 many collisions occur. We are interested only in the molecules that pass through $dA$ without further collisions. Since $n$ is the number of molecules per unit volume and $z$ is the collision frequency, there are

$$\tfrac{1}{2} z \, dt \, n \, dV$$

collisions in the volume element in the time interval $dt$—the $\tfrac{1}{2}$ factor prevents collisions from being counted twice because there are two molecules per collision. At each collision two new molecular paths originate, $z \, dt \, n \, dV$ in number, of which the fraction headed toward $dA$ is the projected elemental area divided by the area of a sphere

$$\cos\theta \, \frac{dA}{4\pi r^2}$$

which can also be regarded as, with the solid angle being $d\omega = \cos\theta \, dA / r^2$,

$$\frac{d\omega}{4\pi}$$

The fraction of molecules headed toward $dA$ that experience no further collision is $e^{-r/\lambda}$, according to Eq. (2-40a). The number of molecules that cross

$dA$ that have had their last collision in $dV$ is then

$$z \, dt \, n \, dV \cos \theta \, \frac{dA}{4\pi r^2} e^{-r/\lambda}$$

The average height $\bar{y}$ above the area element at which molecules passing through had their last collision is

$$\bar{y} = \Sigma \, \frac{\text{height}}{\text{molecule}}$$

$\times$ (number of molecules of that height crossing $dA$ without collision)

$\times$ (number of molecules crossing $dA$ without collision)$^{-1}$

$$= \left[ \int_{\theta=0}^{\pi/2} \int_{\phi=0}^{2\pi} \int_{r=0}^{\infty} (r \cos \theta) \left( zn \cos\theta \, e^{-r/\lambda} \, dt \, dV \frac{dA}{4\pi r^2} \right) \right]$$

$$\times \left[ \int_{\theta=0}^{\pi/2} \int_{\phi=0}^{2\pi} \int_{r=0}^{\infty} zn \cos \theta \, e^{-r/\lambda} dt \, dV \frac{dA}{4\pi r^2} \right]^{1}$$

$$\bar{y} = \frac{2\lambda}{3} \tag{3-1}$$

in which $dV = r^2 \sin \theta \, d\theta \, d\phi \, dr$ has been used.

Equation (3-1) shows that on the average, each molecule crossing $dA$ had its last collision $2\lambda/3$ above the plane of $dA$. Of course, some molecules traveled farther and some traveled shorter distances.


## 3.2  PROPERTY FLUX

Now consider the $y$ flux $p_y$ of any property $P$ per molecule being transferred by random molecular motion across an elemental area in the $S$–$S$ plane shown in Fig. 3-2$a$. The flux of $P$ in the positive $y$ direction across $dA$ is given as

$p_y$ = (number of molecules crossing from below with last collision at $-\bar{y}$)

$\times$ ($P$ value carried)

$-$ (number of molecules crossing from above with last collision at $+\bar{y}$)

$\times$ ($P$ value carried)

$$= \left( \frac{n\bar{v}}{4} \right)^{-} P^{-} - \left( \frac{n\bar{v}}{4} \right)^{+} P^{+}$$

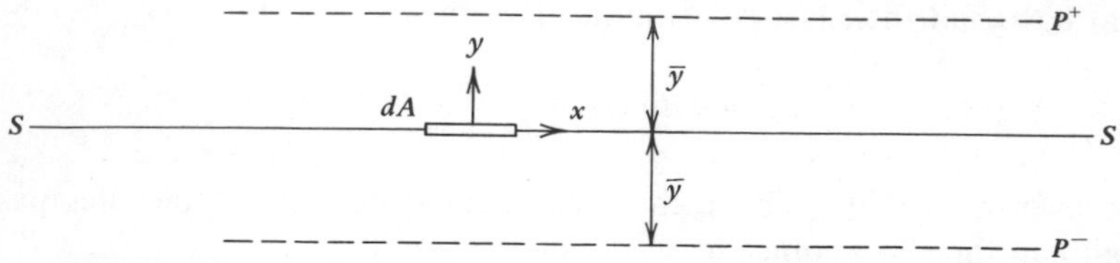


**Figure 3-2** ($a$) One-dimensional transport by molecular motion.

At this point it must be recalled that application of kinetic theory presumes equilibrium. Equal numbers of molecules then pass in the upward and downward directions. As a result, $n\bar{v}/4$ can be treated as a constant to give

$$p_y = \frac{n\bar{v}}{4}(P^- - P^+)$$

Expanding the property $P$ in a Taylor series then gives

$$p_y = \frac{n\bar{v}}{4}\left[\left(P_0 - \bar{y}\frac{dP}{dy} + \cdots\right) - \left(P_0 + \bar{y}\frac{dP}{dy} + \cdots\right)\right]$$

$$= -\left(\frac{n\bar{v}}{2}\right)\bar{y}\frac{dP}{dy}$$

Use of Eq. (3-1) then gives

$$p_y = -\frac{\bar{v}n\lambda}{3}\frac{dP}{dy} \tag{3-2}$$

Note that consideration of the physical details of diffusion has led to the result that the flux of a property is proportional to the gradient of that property, a relation that is of the same form as the experimentally verified rate equations introduced in Chapter 1.

It is as if two opposing groups of children stood on either side of a fence, each group standing $2\lambda/3$ from the fence and throwing snowballs back and forth at an equal rate. In each snowball are packed some stones, but the snowballs of the first group have slightly more stones than do those of the second group. The net flux of stones would be from the first to the second group.

## 3.3 GAS VISCOSITY

When bulk momentum is the transported quantity, it is assumed for the moment that bulk motion is parallel to the $x$ axis in Fig. 3-2$a$, $p_y$ is a $y$ flux of bulk $x$ momentum that manifests itself as a shear stress so that $p_y = -\tau_{yx}$. Further, the property per molecule is bulk $x$ momentum so that $P = mu$, where $u$ is the $x$ component of the bulk velocity and $m$ is the molecular mass. Substitution of these relations into Eq. (3-2) gives

$$\tau_{yx} = \left( \frac{\bar{v} n \lambda}{3} \right) \frac{d(mu)}{dy} \tag{3-3}$$

Comparison with the rate equation, Eq. (1-9), gives the viscosity as

$$\mu = \frac{\bar{v} \lambda m n}{3} \tag{3-4}$$

Into this result $\bar{v}$ from Eq. (2-32a) and $\lambda$ from Eq. (2-40) are introduced to yield

$$\mu = \frac{2}{3\pi^{3/2}} \frac{(KTm)^{1/2}}{\sigma^2} \tag{3-5}$$

Equation (3-5) predicts, and experiment confirms, that dilute gas viscosity is unaffected by pressure.

As an aside, it is now possible to relate the translational relaxation time of a gas to its viscosity. As discussed in Section 4.1, a gas requires a number of collisions to adjust to changed conditions. The translational energy is affected by the first few collisions, with the rotational and vibrational modes of energy storage being affected after larger and larger numbers of collisions. As a result, the order of magnitude of the translation relaxation time is proportional to the inverse of the collision frequency $1/z$. From Eqs. (2-39) and (2-40) it is evident that

$$\lambda = \frac{\bar{v}}{z}$$

Insertion of this result into the relationship for viscosity [Eq. (3-4),] gives

$$\mu = \frac{(\bar{v})^2 mn}{3z}$$

The perfect gas equations of state, $p = nKT$, and Eq. (2-32a) for $\bar{v}$ allow this result to be recast as

$$\frac{1}{z} = \frac{3\pi}{8} \frac{\mu}{p} \approx \frac{\mu}{p} \tag{3-6}$$

Equation (3-6) is the desired estimate of the translational relaxation time in terms of more easily measured properties [2, 5].

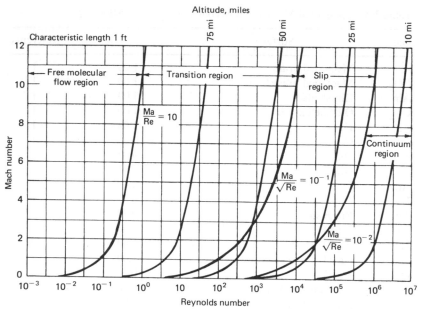

**Figure 3-2** (*b*) Defined flow regimes of gas dynamics [by permission from H. S. Tsien, Superaerodynamics, mechanics of rarefied gases, *J. Aeronaut. Sci.* **13**, 653–664 (1946)].

regimes based on the ratio of mean free path to body dimension or boundary layer thickness summarized by Eckert and Drake [3] as suggested by Tsien [32] is illustrated in Fig. 3-2*b*.

A modification of the preceding kinetic theory predictions of viscosity and slip velocity, which retains the billiard ball model of a molecule, concerns an effect that is referred to as the *persistence of velocities* as explained by Present [2]. This effect is due to the fact that when two rigid elastic spheres collide, the resultant scattering is isotropic in the center-of-mass frame of reference, but not in the fixed laboratory frame of reference in which the angular distribution after collision favors the direction of motion before collision. Because of this effect, the average distance above a plane at which molecules acquire bulk *x*-direction momentum is found to be

$$\bar{y} = 0.998\lambda$$

$$\approx \lambda \tag{3-12}$$

according to Chapman and Cowling [4]. With this modification it is slightly more accurately found that viscosity is described by

$$\mu = \frac{\bar{v}\lambda mn}{2}$$

$$= \frac{\pi^{-3/2}(KTm)^{1/2}}{\sigma^2} \tag{3-13}$$

and slip velocity is described by

$$u_0 - u_w = \frac{2 - f_s}{f_s} \lambda \frac{du(y = 0)}{dy} \tag{3-14}$$

The simplest kinetic theory application that led to Eq. (3-5) predicts that

$$\mu \sim (\text{const}) \frac{T^{1/2} M^{1/2}}{\sigma^2}$$

which is in substantial agreement with experiment. It can, therefore, be said that the billiard ball view of the kinetic theory of gases has given a qualitatively correct view of the physical events that give rise to the property of viscosity. Further improvement must come from a consideration of the forces acting between molecules and a consequent gradual abandonment of the mean-free-path concept. Sutherland treated molecules as impenetrable smooth elastic spheres but considered them to also possess a weak attractive force when separated as explained by Present [2] and Chapman and Cowling [4]. The net effect is to shorten the mean free path since two molecules that would have had only a close approach would now collide or would still exchange some momentum if they do finally only experience a close approach. As a result,

$$\frac{\lambda(\text{Maxwell})}{\lambda(\text{Sutherland})} = 1 + \frac{S}{T} \tag{3-14a}$$

in which the Sutherland constant $S$ is dependent on the gas under consideration. According to Reid and Sherwood [5], the resultant predictive equation for viscosity that has been found to be very reliable is expressed as

$$\mu = \frac{bT^{3/2}}{S + T} \tag{3-15}$$

A rearranged form

$$\frac{\mu}{\mu_0} = \left(\frac{T}{T_0}\right)^{1/2} \frac{1 + S/T_0}{1 + S/T}$$

is used to extend measured viscosity values. In Eq. (3-15) a plot of $T^{3/2}/\mu$ versus $T$ should give a slope of $1/b$ and an intercept of $S/b$. However, Eq. (3-15) should not be extended to either very low or very high temperatures.

The temperature dependence of $S$ can be observed in the success of another often used expression for the temperature dependence of viscosity

$$\frac{\mu}{\mu_0} = \left(\frac{T}{T_0}\right)^n \tag{3-15a}$$

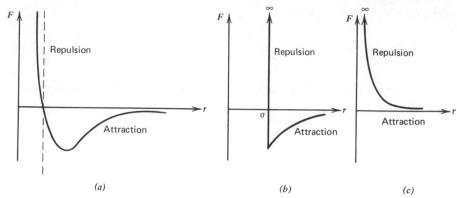

*(a)*            *(b)*            *(c)*

**Figure 3-3**  Intermolecular force models: (*a*) Leonard-Jones (6-12); (*b*) Sutherland weakly attracting spheres; (*c*) point center of repulsion.

which can be shown to result from molecules that are point centers of repulsive force whose magnitude $F$ as a function of the distance between molecular centers $r$ is

$$F = \frac{\text{const}}{r^{\nu}} \tag{3-16}$$

from which relation it follows [4] that $n = \frac{1}{2} + 2/(\nu - 1)$. This model allows molecules to interpenetrate one another on collision, but it does not account for attractive forces.

Success has been enjoyed by accounting for the possibility of both attractive and repulsive forces between molecules. Of these various models (Sutherland's is a special case) the greatest success has attended that named after J. E. Leonard-Jones, the spherically symmetric Leonard-Jones (6-12) potential

$$\phi = 4E\left[\left(\frac{d}{r}\right)^{12} - \left(\frac{d}{r}\right)^{6}\right]$$

where $E$ is the maximum energy of attraction, $r$ is the distance from the molecule center, and $d$ is a collision diameter with the dimension of length and that is related to molecular diameter in an indirect way. The intermolecular repulsive force is given by

$$F = -\frac{d\phi}{dr}$$

Figure 3-3 illustrates the variation of intermolecular forces predicted by several models. In general terms, it has been shown by Chapman and Enskog that the resulting viscosity prediction is of the same form as that provided by the simplest kinetic theory. Specifically,

$$\mu = 2.669 \times 10^{-3}\frac{(MT)^{1/2}}{d^{2}\Omega_{v}} \tag{3-16a}$$

where $\mu$ is viscosity in units of g/cm s, $T$ is absolute temperature in Kelvin units, $d$ is collision diameter in units of $10^{-10}$ m, $M$ is molecular weight, and $\Omega_v$ is the collision integral. The collision integral can be viewed as a measure of the importance of intermolecular forces; a value near unity signifies that a billiard ball closely approximates a molecule and means that intermolecular forces are unimportant until actual collision occurs [3-6]. Tabulations can be found in Table 3-6. Best results are found with nonpolar gases; polar gases such as water vapor are less accurately treated. A modification of the theoretical method credited to Bromley and Wilke is reported by Reid and Sherwood [5] as being able to predict viscosity with 5% accuracy for nonpolar gases. The modified equation is

$$\mu = 3.33 \times 10^{-3} \frac{(MT_c)^{1/2} f_1(1.33 T_r)}{V_c^{2/3}} \tag{3-17}$$

where $\mu$ is the viscosity, $cP$; $M$ is the molecular weight; $T_c$ is the critical temperature, $K$; $V_c$ is the critical volume, cm$^3$/gmol, $T_r$ is the reduced temperature, $T_r = T/T_c$; and $f_1(1.33 T_r)$ is a function in Table 3-2.

The viscosities of gas mixtures at low pressure are important in applications such as deep-sea diving. It is not uncommon for the mixture viscosity to be larger than the viscosity of either component. Apparently, such maxima are most likely in a mixture of polar and nonpolar gases of greatly different molecular weights when the viscosities of the pure components are about equal. The method developed by Wilke is reported by Reid and Sherwood [5] to give accurate predictions and is, for an $n$-component mixture,

$$\mu_m = \sum_{i=1}^{n} \frac{\mu_i}{1 + \sum_{\substack{j=1 \\ j \neq i}}^{n} \phi_{ij}(y_j/y_i)} \tag{3-18}$$

Table 3-2   Values of $f_1(1.33 T_r)$ for Viscosity [5]

| $1.33 T_r$ | $f_1(1.33 T_r)$ | $1.33 T_r$ | $f_1(1.33 T_r)$ |
|------------|-----------------|------------|-----------------|
| 0.3 | 0.1969 | 2.00 | 1.2048 |
| 0.4 | 0.2540 | 2.50 | 1.4501 |
| 0.5 | 0.3134 | 3.00 | 1.6728 |
| 0.6 | 0.3751 | 3.50 | 1.8789 |
| 0.7 | 0.4384 | 4.00 | 2.0719 |
| 0.8 | 0.5025 | 5.00 | 2.4264 |
| 0.9 | 0.5666 | 6.00 | 2.751 |
| 1.0 | 0.6302 | 8.00 | 3.337 |
| 1.2 | 0.7544 | 10.00 | 3.866 |
| 1.4 | 0.8744 | 20.00 | 6.063 |
| 1.6 | 0.9894 | 30.00 | 7.88 |
| 1.8 | 1.0999 | 40.00 | 9.488 |

with

$$\phi_{ij} = \frac{1}{\sqrt{8}} \frac{\left[1 + (\mu_i/\mu_j)^{1/2}(M_j/M_i)^{1/4}\right]^2}{(1 + M_i/M_j)^{1/2}}$$

where $\mu_m$ is the mixture viscosity in centipoise, $\mu_i$ the component viscosity, $M_i$ the component molecular weight, and $y$ the mole fraction. For a binary system ($n = 2$), these relations simplify to

$$\mu_m = \frac{\mu_1}{1 + (y_2/y_1)\phi_{12}} + \frac{\mu_2}{1 + (y_1/y_2)\phi_{21}} \qquad (3\text{-}19)$$

with

$$\phi_{12} = \frac{1}{\sqrt{8}} \frac{\left[1 + (\mu_1/\mu_2)^{1/2}(M_2/M_1)^{1/4}\right]^2}{(1 + M_1/M_2)^{1/2}}$$

$$\phi_{21} = \frac{1}{\sqrt{8}} \frac{\left[1 + (\mu_2/\mu_1)^{1/2}(M_1/M_2)^{1/4}\right]^2}{(1 + M_2/M_1)^{1/2}}$$

For pressures of up to few times normal atmospheric, an estimate of viscosity can be made accurately by the foregoing methods. At high pressures, intermolecular spacings begin to be small enough that simple kinetic theory assumptions begin to be noticeably inaccurate and different estimation methods must be employed. A common dense-gas viscosity correlation assumes that the ratios $\mu/\mu_0$ or $\mu/\mu_c$ depend only on the reduced temperature and pressure. Here, $\mu_0$ is viscosity at atmospheric pressure and the temperature of interest whereas $\mu_c$ is the viscosity at the critical point. It is possible to estimate $\mu_c$ from

$$\mu_c = \frac{7.7 \times 10^{-4} M^{1/2} P_c^{2/3}}{T_c^{1/6}}$$

where $\mu_c$ is the dense-gas viscosity in centipoise, $M$ the molecular weight, $P_c$ the critical pressure in atm, and $T_c$ the critical temperature in kelvins. If either $\mu_0$ or $\mu_c$ is known, the desired value of $\mu$ can be read from Figs. 3-4 and 3-5a, respectively, at the appropriate reduced temperature and pressure. A detailed discussion is given by Reid and Sherwood [5] and a useful brief exposition is presented by Bird et al. [7].

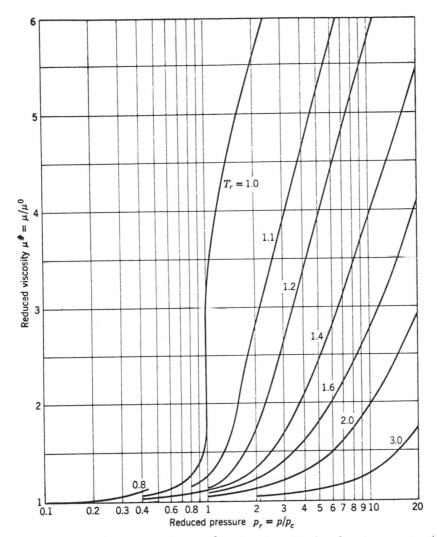

**Figure 3-4** Reduced viscosity $\mu^{\#} = \mu/\mu^0$ as function of reduced pressure $p_r = p/p_c$ and reduced temperature $T_r = T/T_c$. [By permission from N. L. Carr, R. Kobayashi, and D. B. Burroughs, Viscosity of hydrocarbon gases under pressure, *Transact. Am. Inst. Min. Met. Eng.* (*Petroleum Branch*) **201**, 264–272 (1954).]

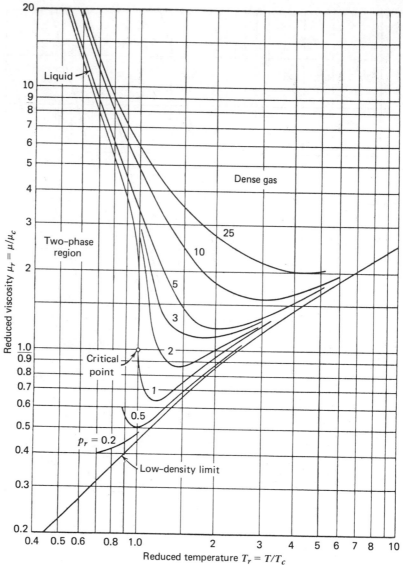

**Figure 3-5** (*a*) Reduced viscosity $\mu_r = \mu/\mu_c$ as a function of temperature for several values of the reduced pressure $p_r = p/p_c$ [by permission from O. A. Hougen, K. M. Watson, and R. A. Ragatz, *Chemical Process Principles Charts*, 3rd ed., Wiley, New York, 1964, Fig. A]. (*b*) Temperature dependence of thermal accommodation coefficient of the noble gases on different tungsten surfaces [by permission from J. Bierhals, G. Grosse, and G. Messer, Measurements of thermal accommodation coefficients of noble gases on tungsten with different surface structures between 300 and 370 K, *Proc. Sixth Int. Vacuum Congr., 1974, Kyoto, Japan Jap. J. Appl. Phys.*, Suppl. 2, Pt. 1, pp. 335–338, (1974)].

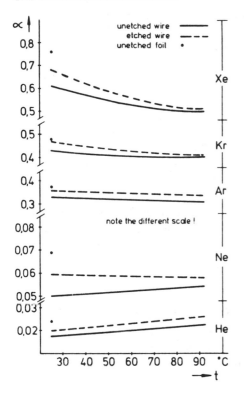

Figure 3-5   (*Continued*)

## 3.4   GAS THERMAL CONDUCTIVITY

When energy is the transported quantity, the property flux development of Section 3.2 has $p$ equal to the energy flux $q_y$. The property per molecule $P$ is energy $e$, which can be made up of translational, rotational, and vibrational energy for a polyatomic molecule. These substitutions in Eq. (3-2) give

$$q_y = -\frac{\bar{v}n\lambda}{3}\frac{de}{dy}$$

$$= -\frac{\bar{v}n\lambda}{3}\frac{de}{dT}\frac{dT}{dy} \qquad (3\text{-}20)$$

Comparison with the rate equation, Eq. (1-5), shows that the thermal conductivity is

$$k = \frac{\bar{v}n\lambda(de/dT)}{3} \qquad (3\text{-}21)$$

For the billiard ball molecule of the simplest kinetic theory, a molecule can possess only translational kinetic energy that is given by Eq. (2-13) as

$$e = \frac{3KT}{2}$$

This relationship, together with $\bar{v}$ from Eq. (2-32a) and $\lambda$ from Eq. (2-40), then gives

$$k = \frac{(1/\pi^{3/2})(K^3 T/m)^{1/2}}{\sigma^2} \tag{3-22}$$

Equation (3-22) predicts, and experiment confirms, that dilute-gas thermal conductivity is unaffected by pressure.

The consequence of having a solid wall immersed in the gas is considered next. The essential physical elements are emphasized by employing the billiard ball molecule model. In simple terms, molecules rebounding from a collision with the wall would be expected to have an energy $e_r$ different from that of impinging molecules $e_i$. Because equilibrium requires an equal number of impinging and reflecting molecules, the fluid average energy $e_0$ is reasonably taken to be

$$e_0 = \frac{e_i + e_r}{2} \tag{3-23}$$

Some of the impinging molecules will come into thermal equilibrium with the wall before rebounding; on rebounding, they will have an energy $e_w$ that corresponds to the wall temperature. These molecules might have been ad-sorbed or temporarily trapped in surface pockets on the wall [3]. The other molecules will rebound without accommodating to the wall temperature. If the fraction of molecules that accommodate their energy to the wall conditions is denoted by $\alpha_s$, the thermal accommodation coefficient, the energy of the rebounding molecules is

$$e_r = \alpha_s e_w + (1 - \alpha_s) e_i \tag{3-24}$$

which can also be rearranged to give a definition of $\alpha_s$ as

$$\alpha_s = \frac{e_i - e_r}{e_i - e_w} \tag{3-24a}$$

The impinging molecules have an average energy equal to that at the wall increased by $\bar{y}\, de/dy$; thus

$$e_i = e_0 + \bar{y}\frac{de}{dy} \tag{3-25}$$

Use of Eqs. (3-24) and (3-25) in Eq. (3-23) gives the apparent energy jump at the wall as

$$e_0 - e_w = \frac{2 - \alpha_s}{\alpha_s} \bar{y}\frac{de}{dy}\bigg|_w \tag{3-26}$$

which is readily rearranged into

$$e_0 - e_w = \frac{2 - \alpha_s}{\alpha_s} \bar{y} \frac{de}{dT} \frac{dT(y=0)}{dy} \tag{3-27}$$

Equation (3-21) relates $de/dT$ to $k$ so that

$$e_0 - e_w = \frac{2 - \alpha_s}{\alpha_s} \frac{2k}{\bar{v}n} \frac{dT}{dy}\bigg|_w \tag{3-28}$$

The simplest billiard ball kinetic theory has $\bar{y} = 2\lambda/3$ by Eq. (3-1), for which $\mu = \bar{v}mn\lambda/3$ by Eq. (3-4); it also allows only translational kinetic energy, $e = 3KT/2$ for a molecule. Because a molecule can undergo only a constant-volume process

$$mc_v = \frac{\Delta e}{\Delta T} \tag{3-28a}$$

So Eq. (3-28) then gives the temperature jump as

$$T_0 - T_w = \frac{2 - \alpha_s}{\alpha_s} \frac{\gamma}{\mathrm{Pr}} \frac{2\lambda}{3} \frac{dT(y=0)}{dy} \tag{3-29}$$

where $\gamma = C_p/C_v$ and $\mathrm{Pr} = \mu C_p/k$ is the Prandtl number. The discussion advanced for velocity slip also applies to the temperature jump whose existence was demonstrated earlier. Table 3-3 [3] displays typical numerical values of the thermal accommodation coefficient $\alpha_s$. It can be seen that $\alpha_s$ is much more likely than the tangential momentum accommodation coefficient $f_s$ to substantially differ from unity. Light gases on dirty walls are most likely to have $\alpha_s$ near unity. Variation of $\alpha_s$ with temperature is shown in Fig. 3-5b.

The success of the billiard ball model of the simplest kinetic theory is primarily the clarification of the physical phenomena involved in thermal conduction. For example, the prediction that thermal conductivity of a dilute gas is independent of pressure is confirmed by measurement, and the predicted thermal conductivity variation with temperature is in generally satisfactory agreement with experiment (although substantial improvement could be hoped for). The coefficient of Eq. (3-22) that was arrived at from simple kinetic theory seems to be a factor of greater error than any other. This can be clarified by a computation of the dimensionless Prandtl number. By definition

$$\mathrm{Pr} = \frac{\mu C_p}{k}$$

and from Eqs. (3-4), (3-21), and (3-28a) it is found that

$$\frac{\mathrm{Pr}}{\gamma} = \frac{\mu C_v}{k} = 1$$

**Table 3-3 Thermal Accommodation Coefficients $\alpha_s{}^a$**

| Gas | Surface | Adsorbed Gas | Temperature °C | $\alpha_s$ | Reference |
|-----|---------|--------------|----------------|-----------|-----------|
| $H_2$ | Pt, bright | | | 0.32 | 1 |
| | Pt, black | | | 0.74 | 1 |
| $O_2$ | Pt, bright | | | 0.81 | 1 |
| | Pt, black | | | 0.93 | 1 |
| $N_2$ | Pt | | | 0.50 | 2 |
| | W | | | 0.35 | 2 |
| Air | Flat lacquer on bronze | | | 0.88–0.89 | 3 |
| | Polished bronze | | | 0.91–0.94 | 3 |
| | Machined bronze | | | 0.89–0.93 | 3 |
| | Etched bronze | | | 0.93–0.95 | 3 |
| | Polished cast iron | | | 0.87–0.93 | 3 |
| | Machined cast iron | | | 0.87–0.88 | 3 |
| | Etched cast iron | | | 0.87–0.96 | 3 |
| | Polished aluminum | | | 0.87–0.95 | 3 |
| | Machined aluminum | | | 0.95–0.97 | 3 |
| | Etched aluminum | | | 0.89–0.97 | 3 |
| He | W | | | 0.025–0.057 | 4 |
| | Ni, not flashed | | | 0.20 | 4 |
| | Ni, flashed | | | 0.085 | 4 |
| | W, flashed | | | 0.17 | 5 |
| | W, flashed | | | 0.12 | 2 |
| | W, not flashed | | | 0.53 | 5 |
| A | W, flashed | | | 0.82 | 5 |
| | W, flashed | | | 0.46 | 5 |
| | W, not flashed | | | 1.00 | 5 |
| He | $H_2$ adsorbed on W | | | 0.041 | 6 |
| | $N_2$ adsorbed on W | | | 0.064 | 6 |
| | W clean system | | | 0.02 | 6 |
| | W | None | 30 | 0.0164 | 7 |
| | W | None | −120 | 0.0130 | 7 |
| | W | None | −196 | 0.0109 | 7 |
| Ne | W | None | 30 | 0.0412 | 7 |
| | W | None | −120 | 0.0426 | 7 |
| | W | None | −196 | 0.0495 | 7 |
| Ar | W | None | 30 | 0.271 | 7 |
| | W | None | −120 | 0.300 | 7 |
| | W | None | −196 | 0.549 | 7 |
| Xe | W | None | 30 | 0.773 | 7 |
| | W | None | −120 | 0.878 | 7 |
| | W | None | −196 | 0.942 | 7 |
| He | K | None | | 0.083 | 8 |
| Ar | K | None | | 0.444 | 8 |
| He | K | None | | 0.042 | 8 |
| Ar | K | None | | 0.41 | 8 |
| He | W | 0 (composite film) | 32 | 0.107 | 6 |
| Ne | W | 0 (composite film) | 32 | 0.204 | 6 |

**Table 3-3** (*Continued*)

| Gas | Surface | Adsorbed Gas | Temperature °C | $\alpha_s$ | Reference |
|-----|---------|--------------|----------------|------------|-----------|
| $H_2$ | W | 0 | 32 | 0.201 | 6 |
| | W | None | 32 | 0.105 | 6 |
| He | W | N (monolayer) | 32 | 0.040 | 6 |
| Ne | W | N | 32 | 0.117 | 6 |
| $H_2$ | W | N | 32 | 0.642 | 6 |
| $CO_2$ | W | $CO_2$ | 32 | 0.990 | 6 |
| He | Glass | None | 29.3 | 0.31 | 7 |

*Source*: By permission from E. R. G. Eckert and R. M. Drake, *Analysis of Heat and Mass Transfer*, McGraw-Hill, New York, 1972.
[a]Compiled by F. C. Hurlbut; data from the following references: (1) M. Knudsen, *The Kinetic Theory of Gases*, Methuen, London, 1934; (2) R. N. Oliver and M. Farber, Experimental determination of accommodation coefficients as a function of temperature for several metals and gases, Ph.D. dissertation, California Institute of Technology, 1950; (3) M. L. Wiedemann and P. R. Trumpler, Thermal accommodation coefficient, *Transact. ASME*, **68**, 57–64 (1946); (4) J. K. Roberts, The exchange of energy between gas atoms and solid surfaces, *Proc. Roy. Soc. Lond. Ser. A*, **129**, 146–151 (1930); **135**, 192 (1932); **142**, 518–524 (1933); (5) W. C. Michels, Accommodation coefficients of helium and argon against tungsten, *Phys. Rev.*, **40**, 472 (1932); (6) H. Y. Wachman, The thermal accommodation coefficient and adsorption on tungsten, Ph.D. dissertation, University of Missouri, 1957; (7) W. L. Silvernail, The thermal accommodation of helium, neon, and argon on clean tungsten from 77° to 303°K, Ph.D. dissertation, University of Missouri, 1954; (8) H. L. Petersen, The accommodation coefficients of helium, neon and argon on clean potassium surfaces from 77°K to 208°K, Ph.D. dissertation, University of Missouri, 1958.

where $\gamma = C_p/C_v$. Table 3-4 shows some measured values of $Pr/\gamma$; they depart significantly from unity, although they are of that order of magnitude. It is clear that molecular complexity exceeds that of a billiard ball.

Further improvement in the predictive accuracy of the billiard ball model of simple kinetic theory can be accomplished in the same manner as was possible for viscosity. In the case of thermal conductivity, the fact that a billard ball's energy is entirely translatory and is strongly correlated with velocity means that the fast and energetic molecules have longer mean free paths. As a consequence, the average distance from a plane at which molecules acquired their property, $\bar{y}$ in Eq. (3-1), is more accurately given for thermal conduction as

$$\bar{y} \approx \frac{5\lambda}{2} \tag{3-30}$$

according to Chapman and Cowling [4]. This correction yields

$$k = \frac{15}{4\pi^{3/2}} \left( \frac{K^3 T}{m} \right)^{1/2} \sigma^{-2} \tag{3-31}$$

**Table 3-4   Pr/γ for Various Gases**

| Gas | $Pr/\gamma = \mu C_v/k$ |
|-----|-------------------------|
| He | 0.41 |
| Ne | 0.40 |
| A | 0.413 |
| $H_2$ | 0.485 |
| $N_2$ | 0.524 |
| $O_2$ | 0.521 |
| $CO_2$ | 0.61 |
| $NH_3$ | 0.714 |
| $CH_4$ | 0.571 |

for thermal conductivity. The corresponding corrected expression for temperature jump is found to be, with $\mu = \bar{v}mn\lambda/2$,

$$T_0 - T_w = \frac{2 - \alpha_s}{\alpha_s}\frac{\gamma}{Pr}\lambda\frac{dT(y = 0)}{dy}$$

A further minor correction to the temperature jump can be made by accounting for the fact that the mean translational kinetic energy of the molecules crossing a surface is $\frac{4}{3}$ that of those present locally. As explained by Schrage [8], the result of this fact (the subject of Problem 3-9) is that

$$e_0 - e_w = (T_0 - T_w)\frac{m(C_p + C_v)}{2}$$

which leads finally to

$$T_0 - T_w = \frac{2 - \alpha_s}{\alpha_s}\frac{2\gamma}{(\gamma + 1)Pr}\lambda\frac{dT(y = 0)}{dy} \tag{3-32}$$

The benefit of these modifications to the kinetic theory predictions can again be ascertained by computation of the Prandtl number. On the basis of Eqs. (3-31) and (3-13), it is found that

$$\frac{Pr}{\gamma} = \frac{2}{5} \tag{3-33}$$

Comparison of this result with the values in Table 3-4 shows a distinct improvement in the agreement with measurement. Agreement is good for monatomic gases, although there is still a substantial difference between measurement and prediction for polyatomic gases.

The qualitative reason for the difference observed previously is that a polyatomic molecule possesses energy not only in the form of translational kinetic energy, but also in other storage modes. This additional energy is an internal energy that is transported with the molecule, and it has not been accounted for in the foregoing developments. Eucken proposed a correction as described by Reid and Sherwood [5]. Rearrangement of the relation between thermal conductivity and specific heat expressed by Eq. (3-33) shows that a monatomic gas has

$$\frac{k}{\mu} = f C_v \qquad (3\text{-}34)$$

with $f = \frac{5}{2}$.

Now the total specific heat $C_v$ should be the sum of that due to translational energy exchange $C_{v,\text{tr}}$ and internal energy exchange $C_{v,\text{int}}$. The weighted sum suggested by the form of Eq. (3-34) is then

$$f C_v = f_{\text{tr}} C_{v,\text{tr}} + f_{\text{int}} C_{v,\text{int}}$$

The value of $f_{\text{tr}}$ is set equal to $\frac{5}{2}$ to force the result for a monatomic gas to reduce to Eq. (3-33); since a monatomic gas has only translational kinetic energy, $C_v$ then is equal to $C_{v,\text{tr}}$ which, in turn, equals the classical value of $3R/2M$. For a polyatomic gas, one then has $C_{v,\text{int}} = C_v - 3R/2M$. Equation (3-34) then becomes

$$\frac{k}{\mu} = \frac{5}{2}\left(\frac{3R}{2M}\right) + f_{\text{int}}\left(C_v - \frac{3R}{2M}\right)$$

Although it has not been assumed so by all, Eucken took $f_{\text{int}} = 1$, which then gives

$$\frac{k}{\mu} = \frac{9R}{4M} + C_v$$

A perfect gas has $C_p - C_v = R/M$; hence the preceding relation can be rephrased as

$$\frac{k}{\mu C_p} = \frac{9\gamma - 5}{4\gamma} \qquad (3\text{-}35)$$

Equation (3-35) allows the thermal conductivity to be estimated with about 10% error once the viscosity, specific heat, and ratio of specific heats have been determined. Equation (3-35) also predicts that

$$\text{Pr} = \frac{4\gamma}{9\gamma - 5}$$

which is in substantial agreement with measurement, predicting for air (with $\gamma = \frac{7}{5}$) that

$$Pr = 0.737$$

This agrees well with the measured value at room temperature of 0.72.

It should be noted in passing that a semiempirical procedure has been suggested to account for the effect of temperature on the thermal conductivity of a dilute gas. This procedure employs the Sutherland correction of Eq. (3-14a) for the mean free path and the empirical observation that $fc_v$ in Eq. (3-34) varies somewhat linearly with temperature as $fC_v \sim 1 + GT$. Equation (3-34) together with the form of Eq. (3-15) then yields [9]

$$k = AT^{1/2} \frac{1 + GT}{1 + S/T}$$

Here, the three constants of $A$, $G$, and $S$ depend on the particular gas being considered. If these constants are evaluated, a sometimes troublesome task, about 5% accuracy can be achieved. In general, the effect of various models of the force exerted by one molecule on another (of which the Sutherland model is only one) seems to be best incorporated into an estimation of viscosity; if the viscosity is known, Eq. (3-35) can be used to determine thermal conductivity.

The thermal conductivity of a dilute mixture of gases is usually a nonlinear function of composition, just as is the case for viscosity. Although its accuracy is dependent on whether polar or nonpolar gases are under consideration, the correlation due to Lindsay and Bromley is reported by Reid and Sherwood [5] to be the most generally applicable of the several methods available. Their predictive equation for an $n$-component mixture is

$$k_m = \sum_{i=1}^{n} \frac{k_i}{1 + \sum_{j \neq i}^{n} A_{ij}(y_j/y_i)} \tag{3-36}$$

where $y$ is the mole fraction of a component and

$$A_{ij} = \frac{1}{4} \left\{ 1 + \left[ \left( \frac{\mu_i}{\mu_j} \right) \left( \frac{M_j}{M_i} \right)^{3/4} \frac{1 + S_i/T}{1 + S_j/T} \right]^{1/2} \right\}^2$$

$$\times \left( \frac{1 + S_{ij}/T}{1 + S_i/T} \right) \tag{3-37}$$

Here $k_{i,j}$ and $\mu_{i,j}$ are the properties for the pure components and $T$ is the absolute temperature. The Sutherland constant $S_{i,j}$ is that used in the viscosity relation of Eq. (3-15) and may be evaluated from viscosity data or estimated from the empirical rule

$$S_i = 1.5 T_{B_i}$$

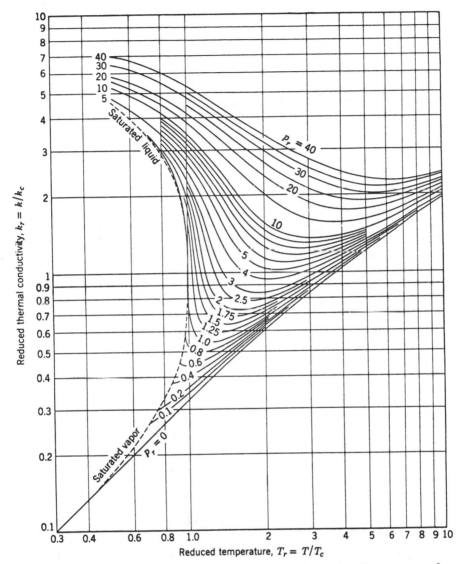

**Figure 3-6** Reduced thermal conductivity. (*a*) For monatomic substances as a function of reduced temperature and pressure [by permission from E. J. Owens and G. Thodos, Thermal-conductivity-reduced-state correlation for the inert gases, *AIChE J.* **3**, 454–461 (1957)]. A larger version of this chart is in O. A. Hougen, K. M. Watson, and R. A. Ragatz, *Chemical Process Principles Charts*, 3rd ed., Wiley, New York 1964. (*b*) The diatomic gases [by permission from C. A. Schaefer and G. Thodos, Thermal conductivity of diatomic gases: Liquid and gaseous states, *AIChE J.* **5**, 367–372 (1959)].

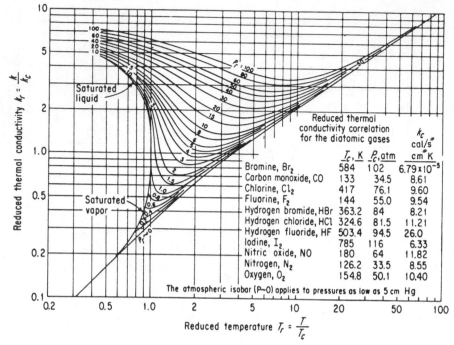

**Figure 3-6** (*Continued*)

where $T_{B_i}$ is the absolute boiling temperature at one atmosphere of pressure for the $i$th component—for $H_e$, $H_2$, and $N_e$ it is best to use $S = 79$ K. The interaction Sutherland constant $S_{ij}$ is estimated as

$$S_{ij} = (S_i S_j)^{1/2}$$

unless one of the gases is polar (e.g., water vapor), in which case a multiplying factor of 0.733 should be introduced. The Brokaw prediction is reported by Reid and Sherwood [5] to be applicable to binary mixtures of nonpolar gases. Its simple form makes it attractive and is

$$k_m = \bar{a} k_{sm} + (1 - \bar{a}) k_{rm} \tag{3-38}$$

where $x_i$ is the mole fraction, $k_{sm} = \Sigma x_i k_i$ (a simple molar mixing rule), and $1/k_{rm} = \Sigma x_i / k_i$ (a reciprocal mixing rule). The coefficient $\bar{a}$ varies from 0.3 to 0.8, depending on the mole fraction of the lightest component in a binary mixture, which can be roughly approximated as $\frac{1}{2}$. This rule has been tested only for binary mixtures but might be applicable to multicomponent mixtures as well.

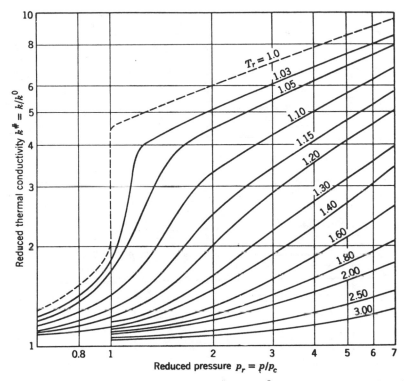

**Figure 3-7** Reduced thermal conductivity $k^\# = k/k^0$ as a function of reduced temperature and reduced pressure. [By permission from J. M. Lenoir, W. A. Junk, and E. W. Comings, Measurement and correlation of thermal conductivities of gases at high pressure, *Chem. Eng. Prog.* **49**, 539–542 (1953).]

At greater than several atmospheres of pressure, the thermal conductivity of gases is pressure dependent. It is unfortunate that the principle of corresponding states does not give accurate correlations since, as Figs. 3-6 and 3-7 suggest, a concise representation of data would then result. Figure 3-6 presents the ratio of $k/k_c = k_r$ versus reduced temperature $T_r = T/T_c$, with reduced pressure $P_r = P/P_c$ as a parameter for monatomic and diatomic gases that may also be used for polyatomic gas estimations. Although the curves in Fig. 3-6 are smooth, many irregularities have been eliminated and are not shown. An experimental value of $k_c$ is seldom available. However, it can be estimated by taking a measured value of $k$ together with a value of $k_r$ from Fig. 3-6 at the same conditions and then solving for $k_c$ from $k_c = k/k_r$. A similar use of the principal of corresponding states is employed in the correlation of $k/k_0$ versus reduced pressure with reduced temperature as a parameter as shown in Fig. 3-7. Here $k_0$ is the thermal conductivity at the temperature in question but at

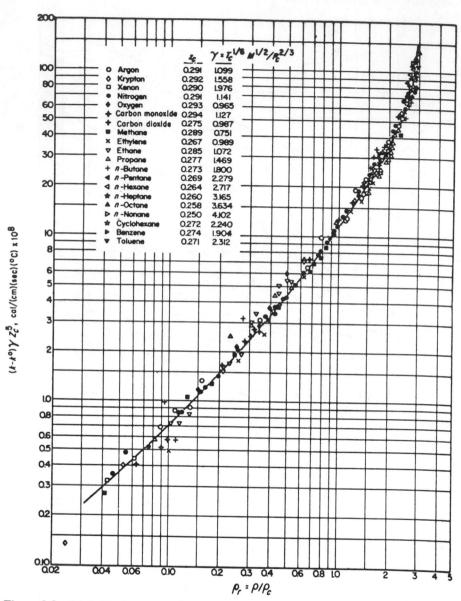

Figure 3-8   Stiel–Thodos correlation for dense-gas thermal conductivities. [By permission from L. I. Stiel and G. Thodos, The thermal conductivities of nonpolar substances in the dense gaseous and liquid regions, *AIChE J.* **10**, 26–30 (1964).]

low pressure. The best generalized correlation available at present is reported by Reid and Sherwood [5] to be one that relates the residual thermal conductivity $k - k_0$ to the reduced density $\rho_r = \rho/\rho_c$ for nonpolar gases. The correlation is shown in Fig. 3-8 with the definitions of

$$\gamma = \frac{T_c^{1/6} M^{1/2}}{P_c^{2/3}}$$

$$Z_c = \frac{P_c M}{R \rho_c T_c}$$

Errors of about 10% are possible and Fig. 3-8 should not be used for polar gases or for either hydrogen or helium.

It should be noted that the often useful density and specific heat of a mixture vary linearly with the mass fraction of the mixture components. For example, the specific heat of a binary mixture is given in terms of component properties as

$$C_{p_m} = \omega_1 C_{p_1} + \omega_2 C_{p_2}$$

in which $\omega$ is the mass fraction and $\omega_1 = \rho_1/(\rho_1 + \rho_2)$. This linear behavior contrasts strongly with the nonlinearity that characterizes the dependence of transport properties on mixture fractions.

## 3.5   GAS MASS DIFFUSIVITY

When the transported quantity is mass of species 1 molecules and the transport occurs under stagnant and isothermal conditions with only species 2 molecules additionally present, $p_y$ in Eq. (3-2) is the rate of flow of species 1 mass per unit area and time, and $P$ in Eq. (3-2) is the fraction of an average molecule that has the molecular mass of species 1. In mathematical terms, $p_y = \dot{m}_{1, y}$ and $P = m_1 n_1/(n_1 + n_2)$. Equation (3-2) becomes, with these definitions for $p$ and $P$,

$$\dot{m}_{1, y} = - \frac{\bar{v} n \lambda}{3} \frac{d(m_1 n_1/n)}{dy} \tag{3-39}$$

with $n = n_1 + n_2$. The mass of an average molecule $m$ can be taken to be constant within the framework of the assumptions that underly Eq. (3-39) in contrast to the total number of molecules per unit volume $n$ that could vary rather easily. Hence Eq. (3-39) can be rearranged by appropriate multiplication and division by $m$ into the form

$$\dot{m}_{1,y} = -\frac{\bar{v}nm\lambda}{3}\frac{d(m_1n_1/mn)}{dy}$$

In this form it is seen that $m_1n_1/mn = \rho_1/\rho = \omega_1$; the mass fraction is an appropriate potential for driving mass diffusion. As a result,

$$\dot{m}_{1,y} = -\frac{\bar{v}nm\lambda}{3}\frac{d\omega_1}{dy} \qquad (3\text{-}40)$$

Comparison of Eq. (3-40) with the Fick's law rate equation of Eq. (1-7) reveals that the mass diffusivity is related to molecular quantities by the preceding simplest kinetic theory approach as

$$\rho D_{12} = \frac{\bar{v}nm\lambda}{3} \qquad (3\text{-}41)$$

Or

$$D_{12} = \frac{2}{3\pi^{3/2}}\frac{(KT/m)^{1/2}}{n\sigma^2}$$

Note that the mass diffusivity is given as

$$D_{12} = \frac{\bar{v}\lambda}{3}$$

just as in the diffusivity of momentum where kinematic viscosity $\nu$ is given by Eq. (3-5) as

$$\nu = \frac{\bar{v}\lambda}{3}$$

It is seen that the simplest kinetic theory results give the dimensionless Schmidt number as

$$Sc = \frac{\nu}{D_{12}} = 1$$

Comparison with the data of Table 3-5 [3] shows that the simplest kinetic theory predicts the correct order of magnitude for the Schmidt number. Insertion of $\bar{v}$ from Eq. (2-32a) and $\lambda$ from Eq. (2-40) into Eq. (3-41) gives

$$\rho D_{12} = \frac{2}{3\pi^{3/2}}\frac{(KTm)^{1/2}}{\sigma^2} \qquad (3\text{-}42)$$

It is seen that the density–mass diffusivity product is independent of pressure

**Table 3-5  Coefficients of Diffusion $D$ [3]**

| Diffusing Material | Medium of diffusion | Temperature, C | Diffusion Coefficient, $m^2/s$ | Schmidt No. $Sc = \nu/D$ |
|---|---|---|---|---|
| *Solids* | | | | |
| Cu | Al | 462.2 | $8.45 \times 10^{-10}$ | |
| Hg | Pb | 177.2 | $1.95 \times 10^{-10}$ | |
| | Pb | 197.2 | $5.02 \times 10^{-10}$ | |
| *Liquids* | | | | |
| HCl | $H_2O$ | 0.0 | $2.23 \times 10^{-6}$ | 0.81 |
| $NH_3$ | $H_2O$ | 4.0 | $1.21 \times 10^{-6}$ | 1.5 |
| *Gases* | | | | |
| $NH_3$ | Air | 0.0 | $2.165 \times 10^{-5}$ | 0.634 |
| $CO_2$ | Air | 0.0 | $1.198 \times 10^{-5}$ | 1.14 |
| | $H_2$ | 18.0 | $6.048 \times 10^{-5}$ | 0.158 |
| Hg | $N_2$ | 18.4 | $325. \times 10^{-5}$ | 0.00424 |
| $O_2$ | Air | 0.0 | $1.533 \times 10^{-5}$ | 0.895 |
| | $N_2$ | 11.8 | $2.025 \times 10^{-5}$ | 0.681 |
| $H_2$ | Air | 0.0 | $5.472 \times 10^{-5}$ | 0.250 |
| | $O_2$ | 14.0 | $7.748 \times 10^{-5}$ | 0.182 |
| $H_2$ | $N_2$ | 12.5 | $7.376 \times 10^{-5}$ | 0.187 |
| $H_2O$ | Air | 8.0 | $2.062 \times 10^{-5}$ | 0.615 |
| | Air | 16.0 | $2.815 \times 10^{-5}$ | 0.488 |
| $C_6H_6$ | Air | 0.0 | $0.7497 \times 10^{-5}$ | 1.83 |
| | $CO_2$ | 0.0 | $0.5268 \times 10^{-5}$ | 1.37 |
| | $H_2$ | 0.0 | $2.936 \times 10^{-5}$ | 3.26 |
| $CS_2$ | Air | 19.9 | $0.8807 \times 10^{-5}$ | 1.68 |
| Ether | Air | 19.9 | $0.7692 \times 10^{-5}$ | 1.93 |
| Ethyl alcohol | Air | 0.0 | $1.013 \times 10^{-5}$ | 1.36 |
| | Air | 40.4 | $1.180 \times 10^{-5}$ | 1.45 |

*Source*: *Handbook of Chemistry and Physics*, 39th ed., Chemical Rubber Publishing Company, Cleveland, Ohio, 1957–1958; and other sources.
[a]Water, $H_2O$, in air: $D$, $(m^2/s) = 2.302\, p_0/p(T/T_0)^{1.81} \times 10^{-5}$; ammonia, $NH_3$, in air: $D(m^2/s) = 1.969\, p_0/p\,(T/T_0)^{1.81} \times 10^{-5}$; $p_0 = 0.98 \times 10^5\ N/m^2$; $T_0 = 256\ K$.

as was also seen to be the case for the transport properties of thermal conductivity and viscosity.

Equation (3-42) has not accounted for possible differences between the two kinds of molecules, a complexity that arose for viscosity and thermal conductivity only in consideration of mixtures. The simplest kinetic theory leading to Eq. (3-42) does give a good appreciation for the basic physical events, but the manner in which the effective molecular mass $m$ and diameter $\sigma$ are to be determined is an important matter that should be noted here.

Following Present [3], it will first be recognized that if the diffusing molecule has diameter $d_1$ and the stationary molecule has diameter $d_2$, the temporary consideration that the diffusing molecule is moving at a relative velocity of $v_r$ and the molecule of species 2 is a point as suggested by Fig. 2-14 leads to an effective molecular diameter of

$$2\sigma = \sigma_1 + \sigma_2 \qquad (3\text{-}43)$$

The number of collisions in unit time between a single moving molecule of species 1 and species 2 molecules is then

$$n_2 \pi \sigma^2 v_r$$

where $n_2$ is the number of species 2 molecules per unit volume. If there are $n_1$ species 1 molecules per unit volume, the collision frequency must be

$$n_1 n_2 \pi \sigma^2 v_r$$

It can be shown for a Maxwell velocity distribution that $v_r^2 = (\bar{v}_1)^2 + (\bar{v}_2)^2$, which reduces properly to $v_r = 2^{1/2}\bar{v}$ for equal molecules. With this result, the collision frequency between unlike molecules is

$$n_1 n_2 \pi \sigma^2 \left( \bar{v}_1^2 + \bar{v}_2^2 \right)^{1/2}$$

It is now necessary to abandon the mean-free-path approach. This is a result of the fact that collisions between like molecules do not affect the transport of a species of mass since properties are merely exchanged then. Instead, only encounters between unlike molecules are of interest (the frequency of such collisions is available in the immediately preceeding expression), and the momentum-transfer method exploits this fact. The existence of a gradient in the number of species 1 molecules per unit volume implies a gradient in the partial pressure $p_1$ of that gas since $p_1 = n_1 KT$ and the total pressure $p$ and temperature $T$ are uniform. Thus a slice of species 1 gas of thickness $dy$ has a net force per unit area of $dp_1$ exerted on it. There is no possibility of walls exerting forces (as they are far away), there is no bulk motion to exert shear stress, and collisions between species 1 molecules can exchange properties but cannot create a gradient. Therefore, there must be a net rate of $y$-momentum exchange between the two gases that supports the observed net force $dp_1$. In mathematical terms

$$dp_1 = M_{12}\, dy$$

$$KT\, dn_1 = \qquad (3\text{-}44)$$

where $M_{12}$ is the rate per unit volume and time at which species 1 and 2 gases are exchanging $y$ momentum. Now if species 1 molecules move in the $y$ direction at an average diffusive speed of $u_1$ whereas species 2 molecules possess $u_2$, the center of mass of the two molecules has a mean velocity $u_c$ given by

$$u_c = \frac{m_1 u_1 + m_2 u_2}{m_1 + m_2}$$

The $y$ momentum lost by the species 1 molecule (since the average velocity of a molecule after collision is that of the center of mass after all directions of rebound are considered) is given on the average for one unlike collision by

$$m_1(u_c - u_1) = (u_2 - u_1)\frac{m_1 m_2}{m_1 + m_2}$$

$$= (u_2 - u_1)m^*$$

where $m^* = m_1 m_2/(m_1 + m_2)$ is the reduced mass. Inclusion of the effect of the frequency of unlike collisions gives

$$M_{12} = (u_2 - u_1)m^* n_1 n_2 \pi \sigma^2 (\bar{v}_1^2 + \bar{v}_2^2)^{1/2}$$

Consider also the fact that the $y$ transport of species 1 and 2 molecules is, respectively,

$$\dot{N}_{1y} = n_1 u_1$$

and

$$\dot{N}_{2y} = n_2 u_2$$

To have total pressure remain constant, equal numbers of molecules must move in each direction so that

$$\dot{N}_{1y} = -\dot{N}_{2y}$$

Hence

$$M_{12} = -n\dot{N}_{1y} m^* \pi \sigma^2 (\bar{v}_1^2 + \bar{v}_2^2)^{1/2}$$

which, when put into Eq. (3-44), gives

$$\dot{N}_{1y} = \left[ \frac{KT}{m^* n \pi \sigma^2 (\bar{v}_1^2 + \bar{v}_2^2)^{1/2}} \right] \frac{dn_1}{dy}$$

This can be put into the standard form

$$\dot{m}_{1y} = -\left[ \frac{mKT}{m^*\pi\sigma^2(\bar{v}_1^2 + \bar{v}_2^2)^{1/2}} \right] \frac{d(n_1 m_1/mn)}{dy}$$

Recognizing that $n_1 m_1/nm = \omega_1$, the mass fraction, and comparing with Fick's law in Eq. (1-7), we see that

$$\rho D_{12} = \frac{mKT}{m^*\pi\sigma^2(\bar{v}_1^2 + \bar{v}_2^2)^{1/2}}$$

or

$$D_{12} = \frac{1}{2\pi^{1/2}} \frac{\left[KT(1/m_1 + 1/m_2)/2\right]^{1/2}}{n\sigma^2}$$

More exact calculations [4] change the numerical coefficient so that finally

$$D_{12} = \frac{3}{8\pi^{1/2}} \frac{\left[KT(1/m_1 + 1/m_2)/2\right]^{1/2}}{n\sigma^2} \tag{3-45}$$

The dimensionless Schmidt number formed from Eqs. (3-45) and (3-13) is

$$Sc = \frac{\nu}{D_{12}} = \frac{8}{3\pi}\left( \frac{2m_1}{m_1 + m_2} \right)^{1/2} \tag{3-46}$$

provided $\sigma_1 \approx \sigma_2$ and $n_1 \ll n_2$.

The case of self-diffusion $(m_1 \equiv m_2)$ gives $Sc = 0.85$. When the gases differ noticeably in molecular weight and there is only a small amount of species 1 $(n \approx 0)$, then

$$Sc = \frac{\nu}{D_{12}} = \frac{1.2}{(1 + M_2/M_1)^{1/2}} \tag{3-47}$$

When the two species are identical, as is nearly realized when species 1 is an isotope of species 2, the diffusive process is termed *self-diffusion*. It can be seen from Eq. (3-45), for instance, that it is generally true that

$$D_{12} = D_{21} \tag{3-48}$$

For spherical, nonpolar molecules, the refined calculations performed by Chapman and Enskog that use the Leonard-Jones (6-12) potential for intermolecular forces yields, as it similarly did for viscosity,

$$D_{12} = 1.858 \times 10^{-3} T^{3/2} \frac{(1/M_1 + 1/M_2)^{1/2}}{P\sigma_{12}^2 \Omega_D} \tag{3-49}$$

where $D_{12}$ is in $cm^2/s$, $P$ is in atmospheres, and $T$ is in kelvins. The collision integral for diffusion is presented in Table 3-6 as a function of reduced temperature, with use of the approximation that $KT/\varepsilon_0 = 1.33T_r$. the collision cross section for the gas pair can be estimated from

$$\sigma_{12} = \frac{\sigma_1 + \sigma_2}{2}$$

with $\sigma_1 = 5V_c^{1/3}/6$, where $V_c$ is molal volume at the critical point in $cm^3/gmol$. According to Reid and Sherwood [5], Eq. (3-49) gives mass diffusivity coefficients as accurately as any existing correlation for less than about 20 atm of pressure and gives useful estimates even for polar gases.

**Table 3-6  Values of the Collision Integral $\Omega_D$ Based on the Leonard-Jones Potential**

| $KT/\varepsilon_0{}^a$ | $\Omega_D{}^a$ | $KT/\varepsilon_0$ | $\Omega_D$ | $KT/\varepsilon_0$ | $\Omega_D$ |
|---|---|---|---|---|---|
| 0.30 | 2.662 | 1.65 | 1.153 | 4.0 | 0.8836 |
| 0.35 | 2.476 | 1.70 | 1.140 | 4.1 | 0.8788 |
| 0.40 | 2.318 | 1.75 | 1.128 | 4.2 | 0.8740 |
| 0.45 | 2.184 | 1.80 | 1.116 | 4.3 | 0.8694 |
| 0.50 | 2.066 | 1.85 | 1.105 | 4.4 | 0.8652 |
| 0.55 | 1.966 | 1.90 | 1.094 | 4.5 | 0.8610 |
| 0.60 | 1.877 | 1.95 | 1.084 | 4.6 | 0.8568 |
| 0.65 | 1.798 | 2.00 | 1.075 | 4.7 | 0.8530 |
| 0.70 | 1.729 | 2.1 | 1.057 | 4.8 | 0.8492 |
| 0.75 | 1.667 | 2.2 | 1.041 | 4.9 | 0.8456 |
| 0.80 | 1.612 | 2.3 | 1.026 | 5.0 | 0.8422 |
| 0.85 | 1.562 | 2.4 | 1.012 | 6 | 0.8124 |
| 0.90 | 1.517 | 2.5 | 0.9996 | 7 | 0.7896 |
| 0.95 | 1.476 | 2.6 | 0.9878 | 8 | 0.7712 |
| 1.00 | 1.439 | 2.7 | 0.9770 | 9 | 0.7556 |
| 1.05 | 1.406 | 2.8 | 0.9672 | 10 | 0.7424 |
| 1.10 | 1.375 | 2.9 | 0.9576 | 20 | 0.6640 |
| 1.15 | 1.346 | 3.0 | 0.9490 | 30 | 0.6232 |
| 1.20 | 1.320 | 3.1 | 0.9406 | 40 | 0.5960 |
| 1.25 | 1.296 | 3.2 | 0.9328 | 50 | 0.5756 |
| 1.30 | 1.273 | 3.3 | 0.9256 | 60 | 0.5596 |
| 1.35 | 1.253 | 3.4 | 0.9186 | 70 | 0.5464 |
| 1.40 | 1.233 | 3.5 | 0.9120 | 80 | 0.5352 |
| 1.45 | 1.215 | 3.6 | 0.9058 | 90 | 0.5256 |
| 1.50 | 1.198 | 3.7 | 0.8998 | 100 | 0.5130 |
| 1.55 | 1.182 | 3.8 | 0.8942 | 200 | 0.4644 |
| 1.60 | 1.167 | 3.9 | 0.8888 | 400 | 0.4170 |

*Source*: By permission from J. O. Hirschfelder, C. F. Curtiss, and R. B. Bird, *Molecular Theory of Gases and Liquids*, Wiley, New York, 1966.
[a]Hirschfelder uses the symbols $T^*$ for $KT/\varepsilon_0$ and $\Omega^{(1,1)\star}$ in place of $\Omega_D$.

An alternative correlation for mass diffusivity in $cm/s^2$ of dilute binary gases is that due to Slattery and Bird as reported by Reid and Sherwood [5]

$$D_{12} = 2.74 \times 10^{-4} \left( \frac{1}{M_1} + \frac{1}{M_2} \right)^{1/2} \left( P_{c_1} P_{c_2} \right)^{1/3} \left( T_{c_1} T_{c_2} \right)^{-0.495} \frac{T^{1.823}}{P} \quad (3\text{-}50)$$

where $M$, $P_c$, and $T_c$ are the molecular weights, critical pressures in atmospheres, and critical temperatures in kelvins of the two gases. The total pressure $P$ is in atmospheres of pressure and the temperature is in kelvins. Equation (3-50) does not work well for mixtures containing hydrogen, helium, or water vapor. If one component is water vapor and the other is a nonpolar gas, it is recommended that Eq. (3-50) be changed to

$$D_{12} = 3.64 \times 10^{-4} \left( \frac{1}{M_1} + \frac{1}{M_2} \right)^{1/2} \left( P_{c_1} P_{c_2} \right)^{1/3} \left( T_{c_1} T_{c_2} \right)^{-0.75} \frac{T^{2.334}}{P} \quad (3\text{-}51)$$

It appears that there are very few data for binary diffusion at high pressures. As a result, estimation techniques are poorly developed. For purposes of revealing trends, however, Fig. 3-9 shows the dimensionless mass diffusivity for self-diffusion at high density as a function of reduced temperatures and pressure. Here $(PD)^0$ refers to values computed at low pressures from Eq. (3-49). Figure 3-9 is unreliable for mixtures of two different gases, partly because the critical temperature and pressure of a mixture are somewhat nebulous quantities and their estimation on a rational basis is uncertain. Bird, et al. [7] suggest that for a mixture,

$$P_c = \sum_{i=1}^{n} y_i P_{c_i} \qquad \text{and} \qquad T_c = \sum_{i=1}^{n} y_i T_{c_i}$$

For a binary dilute gas mixture, it is found both by experiment and by rigorous analysis that the mass diffusivity does not depend appreciably on the concentrations of the individual components. The case for a multicomponent system is substantially more complex, however, as is shown in detail by Hirschfelder, et al. [6] and by Curtis and Bird [10]. Basically, the diffusivity of species 1 through the mixture depends on the concentration $y_i$ of each component as well as the binary mass diffusivity $D_{1_i}$ of species 1 through each component. The simplest case, when species 1 is the only diffusing component in an $n$-component mixture, has a mass diffusivity of species 1 through the mixture given by

$$D_{1m} = \frac{1 - y_1}{\sum_{i=2}^{n} (y_i / D_{1_i})} \quad (3\text{-}51a)$$

where $y_1$ is the mole fraction of species 1, $y_i$ is the mole fraction of component $i$, and $D_{1_i}$ is the binary mass diffusivity of species of 1 through species i.

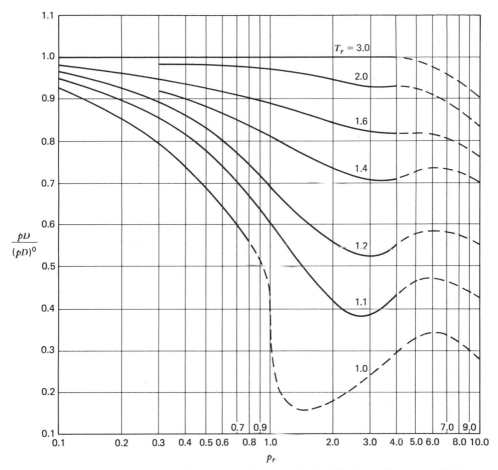

**Figure 3-9**  Generalized chart for the coefficient of self-diffusion of gases at high densities. [By permission from J. C. Slattery and R. B. Bird, Calculation of the diffusion coefficient of dilute gases and of the self-diffusion coefficient of dense gases, *AIChE J.* **4**, 137–142 (1958).]

Condensation from a multicomponent vapor requires consideration of such complexities as is discussed in Section 14.9, where pertinent literature is cited.

## 3.6  LIQUID VISCOSITY

The viscosities of liquids differ from the viscosities of gases in important ways. Liquid viscosities typically exceed gas viscosities by about two orders of magnitude. Further, liquid viscosities tend to decrease with increasing temperature, whereas gas viscosities tend to increase with increasing temperature.

In a qualitative way, these important differences are due largely to the vast difference in molecular spacing between the two states. In a gas, molecular spacings are so large that bulk momentum is transported by molecules moving long distances randomly. Intermolecular forces are not greatly important in gases, as is apparent from the predictive success enjoyed by a kinetic theory that postulates billiard ball molecules; gas molecules are most often out of the range of action of intermolecular forces, a fact that is reflected in the near unity value of the collision integrals $\Omega$ in Tables 3-2 and 3-6. A liquid is nearly three orders of magnitude more dense than a gas (e.g., $\rho_{water}/\rho_{air} \approx 900$) and has molecular spacings about $\frac{1}{10}$ as large. The intermolecular forces are quite important in liquids, as a result. It would be likely, in such a case, that momentum transport in a liquid is achieved by accompanying small displacements (of the order of the molecular spacing) and is greatly influenced by intermolecular forces.

From the standpoint of estimating viscosity, the greatest difference between the liquid and gaseous state is that there is no workable fundamental theory that is generally applicable for liquids, although many models have been formulated [5] and a hard-sphere model [33] shows promise. The kinetic theory that works so well for dilute gases has no counterpart for liquids. It is perplexing, therefore, that the simple relationship

$$\mu = Ae^{B/T} \tag{3-52}$$

where $A$ and $B$ are constants that are peculiar to the particular liquid under consideration, should so generally represent the temperature dependence of viscosity for liquids ranging from glass to water. Such a generality makes it unlikely that liquid viscosity should depend in a substantial way on the details of intermolecular forces that must surely vary greatly from one liquid to another as was previously remarked to also be the case with dilute gases. As Problem 3-15 illustrates, Eq. (3-52) does not provide an accurate correlation for all temperature ranges, however, and so it is necessary to recognize that the accuracy of Eq. (3-52) can be low if extrapolations over large temperature intervals are attempted.

The form of Eq. (3-52) was first suggested by de Guzman in 1913 without theoretical basis other than its resemblence to the Clausius–Clapeyron relation between vapor pressure, temperature, and heat of vaporization

$$\ln p = -\frac{\Delta H_{vap}}{RT} + \text{const}$$

A more physical appreciation for the applicability of this functional form can be obtained by considering the vibration theory suggested by Andrade in 1930 and subsequently refined [11]. In the vibration theory it is presumed that the liquid is monatomic with a solidlike structure in which a molecule vibrates about an equilibrium position that moves, but slowly, a presumption most likely to be correct for a monatomic metal that has just melted. As the molecule vibrates, it exchanges momentum with its neighbors at the extreme point of its displacement. A molecule below a plane conveys momentum across the plane twice in a complete oscillation, as does a molecule above the plane. Also, one-third of the molecules vibrate in the direction normal to the plane. The rate of momentum transport across a unit area of the plane is then

$$\frac{4}{3}\frac{1}{\delta^2} fm \, \Delta u$$

where $1/\delta^2$ gives the number of molecules per unit area, with $\delta$ as the molecular spacing; $f$ the oscillation frequency, $m$ the molecular mass, and $\Delta u$ the change in bulk velocity parallel to the plane experienced in moving a distance $\delta$. With $\Delta u = \delta \, du/dy$, the flux of momentum becomes

$$\frac{4}{3}\frac{1}{\delta^2} fm\delta\frac{du}{dy}$$

from which it is seen that the viscosity at the melting point is

$$\mu_m = \frac{4fm}{3\delta}$$

The frequency $f$ can be seen to be related to an average velocity divided by distance traveled. From kinetic theory principles it would then be expected that, with $v$ representing a velocity,

$$f \sim \frac{v}{\delta} \sim \frac{(T_m/m)^{1/2}}{\delta}$$

This, together with the facts that $m = M/N_A$ and $\delta = (V_m/N_A)^{1/3}$, where $V_m$ is the volume occupied by a gram mole at melting, leads to

$$\mu_m = 5.1 \times 10^{-4} \frac{(MT_m)^{1/2}}{V_m^{2/3}} \qquad (3\text{-}52a)$$

where $V_m$ is the molar volume, the absolute melting temperature is measured in kelvins, and the constant $5.1 \times 10^{-4}$ is largely taken from data pertaining to molecular oscillation frequencies in solids. The melting viscosity has units of g/cm s. The form of viscosity's temperature dependence is arrived at by viewing the just described momentum transport by "orderly" vibration to be progressively interfered with by the random agitation of molecules that accompanies a temperature increase. Restated in more general terms, as temperature increases above the melting point the liquid begins to act more like a gas and less like the crystalline solid that it originally was. If it is assumed that momentum exchange requires a molecule to have a certain energy relative to its neighbor, the number of molecules possessing this energy must be proportional to

$$e^{E/KT}$$

according to Boltzmann's law described following Eq. (2-30b). Note that the energy must be negative since momentum transfer is more likely when molecules attract, as is also suggested in Fig. 3-3. The viscosity at temperature $T$ is related to that at temperature $T_m$ by

$$\frac{\mu}{\mu_m} = e^{-E/KT_m} e^{E/KT}$$

which is of the exponential form sought. Note that the energy term has not been evaluated; only the physical basis for a functional form has been suggested. One additional temperature correction can be made to account for the fact that molecular spacing $\delta$ varies with temperature. The formula for $\mu_m$ suggests that the general form ought to be

$$\mu = \frac{\text{const}}{V^{2/3}} e^{B/T} \tag{3-53}$$

An alternative physical explanation for the exponential form of Eq. (3-52) is the hole theory due to Eyring as described by Hirshfelder et al. [6] and Bird et al. [7]. The hole theory regards the liquid as a lattice with some unoccupied, vacant sites. Viscous flow is accomplished by molecules jumping back and forth from one vacant site to another, with the number jumping in the bulk flow direction exceeding the number jumping the opposite direction. As depicted in Fig. 3-10, in order to squeeze between its nearest neighbors into a nearby vacant hole, a molecule must possess a certain minimum energy increment $\Delta E$. The shear stress $\tau_{yx}$ does work on the molecule as it moves from

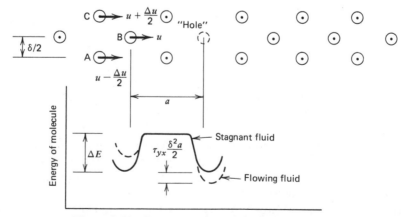

**Figure 3-10**  Representation of the hole theory.

location $B$ to the midpoint of its jump in the amount of $\tau_{yx}\delta^2 a$, which is a positive reduction of the necessary energy increment when the molecule moves in the bulk-flow (forward) direction and negative otherwise. The net rate at which molecules jump forward is given by the Eyring rate equation as

$$r_f = \frac{KT}{h}\exp\left(\frac{-\Delta E + \tau_{yx}\delta^2 a/2}{KT}\right)$$

and the net rate of jumping backward is, similarly,

$$r_b = \frac{KT}{h}\exp\left(\frac{-\Delta E - \tau_{yx}\delta^2 a/2}{KT}\right)$$

where $K$ is Boltzmann's constant, $h$ is Planck's constant, and $\Delta E$ may be regarded as an activation energy in chemical terminology. From the standpoint of molecule B at the midplane, molecule A is jumping backward while molecule C is jumping forward. The bulk velocity difference is then the distance jumped times the difference in jumping frequency, giving

$$\Delta u = a(r_f - r_b)$$

It is permissible to consider the shear stress $\tau_{yx}$ constant for the short distances involved. Then, straightforward substitution gives

$$\Delta u = a\frac{KT}{h}e^{-\Delta E/KT}\left(e^{\tau_{yx}\delta^2 a/3KT} - e^{-\tau_{yx}\delta^2 a/2KT}\right)$$

Rearrangement with $\delta\, du/dy = \Delta u$ gives

$$\frac{du}{dy} = 2\frac{a}{\delta}\frac{KT}{h}e^{-\Delta E/KT}\sin h\left(\tau_{yx}\delta^3\frac{a}{\delta}2KT\right)$$

which is mildly remarkable for predicting non-Newtonian flow in which shear stress is nonlinearly related to the gradient of the bulk velocity. Since such non-Newtonian behavior is experimentally observed in some liquids (see Fig. 4-5 and Appendix C), the above-described hole theory behaves well in a qualitative sense. For small shear stresses, linearization of the preceding relationship yields

$$\tau_{yx} = \left(\frac{\delta}{a}\right)^2\frac{h}{\delta^3}e^{\Delta E/KT}\frac{du}{dy} \tag{3-53a}$$

Setting $\delta^3 = V/N_A$, where $V$ is the volume occupied by a mole of liquid, and additionally assuming that $\delta = a$, one arrives at

$$\mu = \frac{hN_A}{V}e^{\Delta E/KT}$$

It is argued next that the energy $\Delta E$ required to form a hole is of the same order of magnitude as that required for vaporization, but without the need to include any work term for pushing back the environment so that

$$\Delta E = (\text{const})(\Delta U_{\text{vap}})$$

It is found experimentally that

$$\Delta E = 0.408 \, \Delta U_{\text{vap}}$$

and that $\Delta U_{\text{vap}}$ can be related by Trouton's rule to the boiling temperature at 1 atm of pressure $T_b$ as

$$\Delta U_{\text{vap}} = 9.4 K T_b$$

As a result

$$\mu = \frac{hN_A}{V} e^{3.8T_b/T} \tag{3-54}$$

This result is of the same form as Eq. (3-52). The physical phenomena giving rise to viscosity are somewhat different from those supposed by Andrade, but the final result is very much the same, as a comparison of Eqs. (3-53) and (3-54) reveals.

Attempts to predict the constants of Eq. (3-52) with sufficient accuracy have been unsuccessful to date. It is recommended that Eq. (3-52) be fitted to at least two known viscosity values, following which necessary extrapolations or interpolations can be accomplished. Special graph paper, such as the American

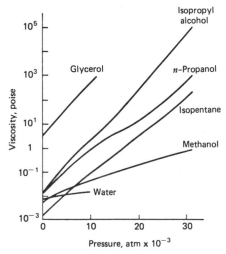

**Figure 3-11** Approximate variation of liquid viscosity with pressure at room temperature. (By permission from D. C. Munro, *High Pressure Physics and Chemistry*, Vol. 1, R. S. Bradley, Ed., Academic Press, London, 1963, p. 18; copyright by Academic Press Inc. (London) Ltd.)

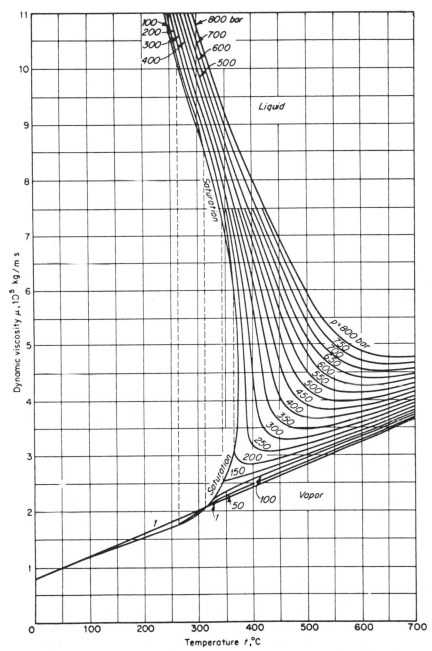

**Figure 3-12** Dynamic viscosity $\mu$ of water and water vapor as a function of pressure and temperature. [By permission from U. Grigull, F. Mayinger, and J. Bach, Viskosität, Wärmeleitfähigkeit und Prandtl-Zahl von Wasser und Wasserdampf, *Wärme-und Stoffübertragung* **1**, 15–34 (1968).]

Society for Testing and Materials (ASTM) viscosity graph paper used to represent oil viscosity versus temperature, uses distorted coordinates to force a linear $\mu$–$T$ relation [12] and can be used to find viscosity at another temperature when only one viscosity value is known. Alternatively, alignment charts can be convenient for the same purpose [13, 14].

The effect of pressure on liquid viscosity is not great, except at the most elevated pressures, and can usually be ignored. Figure 3-11 illustrates the general case of several liquids, whereas Fig. 3-12 gives particular information for water.

There is apparently no reliable method for estimation of the viscosity of liquid mixtures. For information on such methods as are available on this and related topics, the survey of Reid and Sherwood [5] should be consulted. A simple relation, reported by Hirschfelder et al. [6] to be useful for approximations, is

$$\log \mu_{mix} = y_1 \log \mu_1 + y_2 \log \mu_2$$

where $\mu_{1,2}$ are the viscosities of the pure substances and $y_{1,2}$ are the mole fractions of the constituents.

## 3.7  LIQUID THERMAL CONDUCTIVITY

The liquid thermal conductivity, like the liquid viscosity previously discussed, is not well understood on a theoretical basis. Although many correlations can be shown to have a physical basis, they are mostly empirical in nature.

A liquid thermal conductivity typically exceeds that of a gas by about two orders of magnitude, indicating a substantial difference in the physical mechanisms of energy transport. As in the case of viscosity, it is possible to view liquid thermal conductivity $k$ as being the sum of a contribution $k_{vibr}$ from the small-amplitude vibration of a molecule about an equilibrium position in a lattice of the size of molecular spacing and a contribution $k_{conv}$ from the convection that is descriptive of molecular motion over a distance much larger than the molecular spacing. In mathematical terms,

$$k = k_{vibr} + k_{conv}$$

The $k_{conv}$ contribution is rather small under usual conditions, as is evidenced by the low values of mass diffusivity that are observed in liquids. Since mass must move convectively and not just oscillate about a fixed equilibrium position in order to be transported, the $k_{conv}$ contribution, although small, is greater than zero.

Considerable success has attended correlations that are suggested by the form of the equations that result from viewing energy transport in liquids as being caused by molecules constrained to vibrate inside the cage formed by their neighbors and at the sonic velocity. In this view molecules exchange energy upon colliding at the cage boundaries. As discussed by Jakob [9], this viewpoint was first put forth by Paschki in 1915 and later (independently) by Bridgman in 1923. The formal derivation proceeds by building on the kinetic theory of gases result

$$k_{gas} = \frac{mnc_v\bar{v}\lambda}{3} \tag{3-55}$$

which, since $\rho = mn$ and $\bar{v} = v_{acoustic}^{gas}(8/\pi\gamma)^{1/2}$, can be rewritten as

$$k_{gas} = \rho c_v \left(\frac{8}{\pi\gamma}\right)^{1/2} \frac{v_{acoustic}^{gas}\lambda}{3}$$

For a liquid, the mean distance traveled between collisions is the molecular spacing; $\delta = (V/N_A)^{1/3} = (M/\rho N_A)^{1/3}$, where $V$ is the volume of a mole. Thus, by analogy

$$k_{liq} = \rho c_v \left(\frac{8}{\pi\gamma}\right)^{1/2} \frac{v_{acoustic}^{liq}(M/\rho N_A)^{1/3}}{3}$$

A monatomic liquid molecule will have half its energy in kinetic form and the other half in potential energy form because of the forces exerted on it by its nearby neighbors. So, its total energy is $3KT$, which is twice the kinetic energy of a gas molecule, since the energy in each degree of freedom is $KT/2$. Then $c_v = 3K/m$ that, on substitution into this equation, gives

$$k_{liq} = \left(\frac{\rho N_A}{M}\right)^{2/3} K \left(\frac{8}{\pi\gamma}\right)^{1/2} v_{acoustic}^{liq}$$

Equation (3-35) suggests that the additional modes of energy storage in a polyatomic molecule can be accounted for by the factor $(9\gamma - 5)/4$. Multiplication by this factor gives

$$k_{liq} = \left(\frac{\rho N_A}{M}\right)^{2/3} K \left(\frac{8}{\pi\gamma}\right)^{1/2} v_{acoustic}^{liq} \frac{9\gamma - 5}{4}$$

Comparison of this equation with data suggests that, if a liquid molecule is to be considered in the same light as a gas molecule, only translational and rotational molecular energy are effectively transported. For a gas,

$$\gamma = 1 + \frac{2}{f}$$

where $f$ is the number of degrees of freedom possessed by a molecule. Rotation and translation each require three coordinates for a complete description, so $f = 6$ if only their energies are to be considered. With this in mind, $\gamma = \frac{4}{3}$ and

$$k_{\text{liq}} = \left(\frac{7}{4}\right)\left(\frac{8}{\pi}\right)^{1/2} K\left(\frac{\rho N_A}{M}\right)^{2/3} \frac{v_{\text{acoustic}}^{\text{liq}}}{\gamma^{1/2}}$$

$$k_{\text{liq}} = 2.8K\left(\frac{\rho N_A}{M}\right)^{2/3} \frac{v_{\text{acoustic}}^{\text{liq}}}{\gamma^{1/2}} \tag{3-56}$$

in which it is usually the case that the $\gamma$ is nearly unity. Equation (3-56) provides estimates within about 10% for many liquids, including water. It gives the correct temperature dependence, $k_{\text{liq}}$ increasing with temperature for water but decreasing for most other liquids. It correctly predicts that thermal conductivity increases with pressure, but over estimates the effect by a factor of about 2.

The success of the "acoustic" view of thermal conduction lends credence to Andrade's similar view for viscosity. On the other hand, the rapid decrease of viscosity with increasing temperature considered together with the relative constancy of thermal conductivity suggests that the physical mechanism of transport are still somewhat different for energy and for momentum. The Prandtl number for many liquids is substantially different from unity, suggesting again a possible difference in transport mechanisms.

The acoustic velocity in a liquid must either be available from measurements or be estimated from auxiliary data for Eq. (3-56) to be useful. One estimation of liquid acoustic velocity used liquid compressibility data in

$$v_{\text{acoustic}}^{\text{liq}} = \left[\gamma \frac{\partial p(T = \text{const})}{\partial \rho}\right]^{1/2} \tag{3-57}$$

Another uses information from the vapor equation of state as follows. The free space available to a liquid molecule is smaller than the molecular spacing $\delta$ by the molecular radius $\sigma/2$. If energy is instantaneously transferred across a molecule on collision, it must be that

$$v_{\text{acoustic}}^{\text{liq}} = \frac{\delta}{\delta - \sigma} v_{\text{acoustic}}^{\text{gas}} \tag{3-58}$$

Here $v_{\text{acoustic}}^{\text{gas}}$ is evaluated at the temperature of interest. It is known that $\delta = (M/\rho N_A)^{1/3}$. And, the molecular diameter can be estimated from the van der Waals equation of state for the vapor [6]

$$\left(\frac{p - a}{V^2}\right) = \frac{RT}{V - b}$$

where $V$ is the volume per mole of vapor and $b$ is four times the volume occupied by a mole of vapor molecules. From this it follows that $\sigma = (3b/2\pi N_A)^{1/3}$. It can be shown that $b$ is related to the critical temperature and pressure by

$$b = \frac{RT_c}{8P_c}$$

where $R$ is the universal gas constant.

Pressure does not affect thermal conductivity noticeably until about 30 atm of pressure is achieved, except near the critical point. Figure 3-13 illustrates

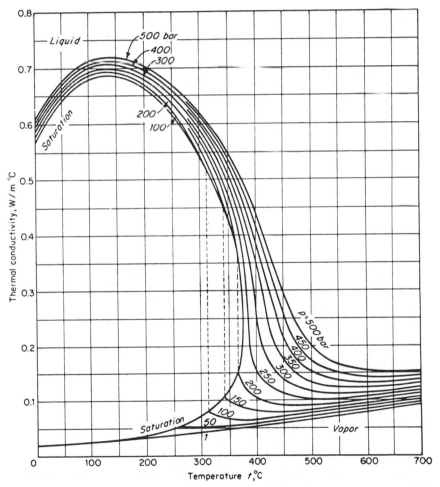

**Figure 3-13** Thermal conductivity $k$ of water and water vapor as a function of pressure and temperature. [By permission from U. Grigull, F. Mayinger, and J. Bach, *Viskosität, Wärmeleitfähigkeit und Prandtl-Zahl von Wasser und Wasserdampf, Wärme-und Stoffübertragung* **1**, 15–34 (1968).]

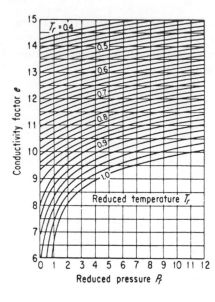

**Figure 3-14**  Effect of pressure on liquid thermal conductivities. [By permission from J. M. Lenoir, Effect of pressure on liquid thermal conductivities, *Pet. Ref.* **36**, (8), 162–164 (1957).]

this fact for water. A general correlation for the effect of pressure on thermal conductivity is

$$\frac{k_{p_2}}{k_{p_1}} = \frac{e_2}{e_1} \qquad (3\text{-}59)$$

where the conductivity factor $e$ is given in Fig. 3-14 as a function of reduced pressure and temperature.

The thermal conductivity of liquid mixtures can be estimated by the Filippov rule devised in 1955

$$k_m = k_1 x_1 + k_2 x_2 - 0.72 \, | \, k_2 - k_1 \, | \, x_1 x_2 \qquad (3\text{-}60)$$

in which $x_{1,2}$ are the weight fractions of the components. Too few data on mixture properties are available to assess the accuracy of Eq. (3-60).


## 3.8   LIQUID METAL THERMAL CONDUCTIVITY

In liquid metals the drift of free electrons provides a mechanism of energy transport that is negligible in most other liquids. The relationship between this part of thermal conductivity and the associated electrical conductivity is given by the Wiedemann–Franz–Lorenz law [9], which can be derived by considering the free electrons (each of electrical charge $e$) inside metal to be a gas. When an electric field of strength $E = dV/dx$ is imposed across the gas, each electron experiences a force $F$ of

$$F = e \frac{dV}{dx}$$

which results in an average drift velocity $\bar{v}_d$ in the direction of the imposed force. Here $V$ represents voltage. The drift velocity is quite small relative to the random thermal velocity $\bar{v}$. Newton's second law gives the electron acceleration as

$$m_e \frac{dv_d}{dt} = F$$

It is then found that

$$v_d = \frac{F}{m_e} t \qquad (3\text{-}61)$$

if the electron starts from rest. It is assumed that, on the average, an electron accelerates between collisions for a time given by $t = \lambda_e/\bar{v}$ and that the drift velocity is reduced to zero by the collision. The average drift velocity is then

$$\bar{v}_d = \frac{e}{m_e} \frac{dV}{dx} \frac{\lambda_e/\bar{v}}{2} \qquad (3\text{-}62)$$

The electrical current is given by Ohm's law as

$$i = -k_e \frac{dV}{dx} \qquad (1\text{-}11)$$

Alternatively, the electrical current can be computed by

$$i = -ne\bar{v}_d$$

Comparison of these two formulations for $i$ reveals that

$$k_e = \frac{ne\bar{v}_d}{dV/dx} \qquad (3\text{-}63)$$

Recall next from the simplest kinetic theory result of Eq. (3-21) for thermal conductivity that

$$k = \frac{n\bar{v}\lambda K}{2}$$

Combination of Eqs. (3-62), (3-63), and (3-21) then gives the electron gas thermal-to-electrical conductivity ratio as

$$\frac{k}{k_e} = K m_e \left( \frac{\bar{v}}{e} \right)^2$$

The thermal average speed is found from Eq. (2-32a) so that, finally,

$$\frac{k}{k_e T} = \frac{8}{\pi} \left( \frac{K}{e} \right)^2$$

More refined analysis gives [18]

$$\frac{k}{k_e T} = \frac{\pi^2}{3} \left( \frac{K}{e} \right)^2$$

$$= L_0 = 2.45 \times 10^{-8} \; W\,\Omega/K^2 \tag{3-64}$$

where $L_0$ is the Lorenz number.

Of course, the total thermal conductivity in a metal is the sum of the electronic transport and the transport caused by acoustic vibrations and convection of molecules (in a liquid metal). The difference between a measured value of thermal conductivity and that predicted by Eq. (3-64) is often small and is ascribed to the latter two effects. The utility of the Wiedemann–Franz–Lorenz law is that it relates thermal conductivity to electrical conductivity, about which a great deal more is known. The form of Eq. (3-64) was derived by Lorenz in 1872 as an improvement over the proposal by Wiedemann and Franz in 1853 that $k/k_e = $ const.

## 3.9  LIQUID MASS DIFFUSIVITY

Mass diffusivities are much lower in liquids than in gases. As evidence of this consider oxygen diffusing through water at 25°C and one atmosphere of pressure for which

$$D_{12} = 2.41 \times 10^{-5} \; \text{cm}^2/\text{s}$$

when the water is in liquid form and

$$D_{12} = 0.352 \; \text{cm}^2/\text{s}$$

when the water is in vapor form. The mass diffusivities in liquid and in gas typically stand in the ratio

$$\frac{D_{\text{liq}}}{D_{\text{gas}}} \sim 10^{-4}$$

It is convenient to call the diffusing substance the *solute*; the liquid through which diffusion takes place is then called the *solvent*. In a gas, diffusing molecules are relatively free to move about; but in a liquid, diffusing molecules have much less freedom of large-scale motion. As discussed earlier in the cases of viscosity and thermal conductivity, most liquid molecules undergo acoustic vibrations about an equilibrium position with an amplitude of the order of molecular spacing; very few molecules are able to experience the motion over distances large compared to molecular spacing, which is necessary for diffusion. Despite a low mass diffusivity, however, the rate of diffusion in a liquid is not necessarily lower than in a gas because a liquid can have very large gradients of the driving potential.

The general form of a relationship between mass diffusivity and viscosity can be deduced from principles already developed. Although the results are not necessarily immediately useful for accurate predictions, they do play a valuable role in suggesting likely functional forms for empirical correlations. On physical grounds, the solute molecule can be regarded as being bombarded by the random thermal motion of the solvent molecules. As a result of this bombardment, a net displacement in a random walk fashion occurs that is often called *Brownian motion*.

The desired general relationship has as its starting place the observation that a particle acted on by a force $F$ for an average time $t$ achieves a drift velocity $\bar{v}_d$ of

$$\bar{v}_d = \frac{t_{av}}{m} F \tag{3-65}$$

which may be rewritten as

$$\bar{v}_d = \eta F$$

where $\eta$ is the mobility and $\eta = t_{av}/m$. Now it is possible to evaluate the mobility $\eta$ if one imagines the diffusing molecule to be a solid sphere of radius $r$ moving steadily through a continuous fluid of viscosity $\mu$ under the influence of a force $F$. Stokes' relation for the drag force on the sphere is that

$$F_{drag} = 6\pi\mu r \bar{v}_d$$

Thus the mobility is established by comparison as

$$\eta = \frac{1}{6\pi\mu r} \tag{3-66}$$

The flux of solute molecules resulting from a drift velocity is

$$\dot{N}_1\big|_{drift} = n_1 \bar{v}_d$$

$$= n_1 \eta F$$

The force $F$ is presumed to be adjustable to just cause the drift flux to exactly cancel the diffusive flux. Then

$$\dot{N}_{1_{\text{drift}}} + \dot{N}_{1_{\text{diffuse}}} = 0$$

$$n_1 \eta F - D_{12} \frac{dn_1}{dx} = 0$$

At this equilibrium condition

$$\frac{dn_1}{dx} = \frac{n_1 \eta F}{D_{12}} \tag{3-67}$$

Of course, the origin of the randomly directed force $F$ is not explicitly described, although it must be the random bombardment of solvent molecules as explained earlier. Boltzmann's law from kinetic theory tells that the fraction of molecules occupying a particular position is dependent on the potential energy (P.E.) associated with that location. Thus the number of solute molecules at a displacement of $x$ is

$$n_1(x) = n_1(0)e^{-\text{P.E.}/KT}$$

The potential energy is just $-Fx$. This gives

$$n_1(x) = n_1(0)e^{Fx/KT}$$

Taking the derivative of this relation, one obtains

$$\frac{dn_1}{dx} = \frac{F}{KT}n_1(0)e^{Fx/KT}$$

$$= \frac{n_1 F}{KT} \tag{3-68}$$

Comparison of Eqs. (3-67) and (3-68) reveals the general result attributed to Einstein that

$$D_{12} = KT\eta \tag{3-69}$$

This brief development is parallel to that given by Feynman [15].

In the so-called hydrodynamic theory the mobility is attributed to viscous drag. Use of Eq. (3-66) in Eq. (3-69) then yields

$$D_{12} = \frac{KT}{6\pi\mu_2 r_1} \tag{3-70}$$

The hydrodynamic theory is said [16] to have started with Wiedemann's observation in 1858 that $D \sim 1/\mu$. Sutherland and Einstein in 1905 then independently determined Eq. (3-70). Equation (3-70) is satisfactorily accurate for large colloidal particles and even for very large molecules. Its accuracy becomes increasingly unsatisfactory as particle size decreases, which is a plausible trend since the solvent then is less accurately represented as a continuum.

An alternative derivation of the form of Eq. (3-70) can be made from the hole theory due to Eyring [6] previously used for liquid viscosity. It is presumed that the rate at which solute molecules are able to pass from one hole to another depends on the product of areal density $\delta n_1$ of solute molecules and the rate $r_0$ at which molecules are able to jump past the potential barrier of the neighboring solvent molecules. Figure 3-10 illustrates this situation, showing that $\delta$ is the molecular spacing, although it is not clear whether of the solvent or the solute. As in the case of liquid viscosity, the Eyring rate equation gives

$$ r_0 = \frac{KT}{h} e^{-\Delta E_d / KT} $$

Equation (3-53a) allows $r_0$ to be expressed in terms of viscosity as

$$ r_0 = \left[ \left( \frac{\delta}{a} \right)^2 e^{-(\Delta E_d - \Delta E)/KT} \right] \frac{KT}{\delta^3 \mu_2} $$

The flux of solvent molecules is then

$$ \dot{N}_1 = \delta n_1 r_0 - \delta r_0 \left( n_1 + \delta \frac{dn_1}{dy} \right) $$

$$ = -\delta^2 r_0 \frac{dn_1}{dy} \tag{3-71} $$

Comparison of Eq. (3-67) with Fick's law shows that

$$ D_{12} = \delta^2 r_0 $$

$$ D_{12} = \left[ \left( \frac{\delta}{a} \right)^2 e^{-(\Delta E_d - E)/KT} \right] \frac{KT}{\delta \mu_2} \tag{3-72} $$

Although it is not possible to evaluate the bracketed coefficient of Eq. (3-72), it is of the same form as the hydrodynamic theory result [Eq. (3-70)] if the activation energies $\Delta E_d$ and $\Delta E$ are equal for diffusion and viscosity.

It is important, when discussing the evaluation of mass diffusivities for liquids, to be aware that they often vary greatly with concentration. For gases,

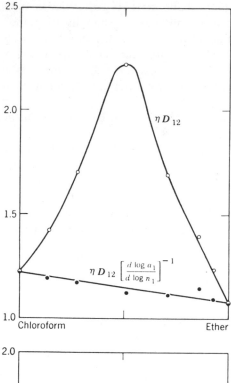

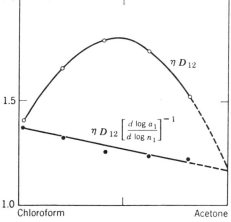

**Figure 3-15** Effect of activity on the product of the coefficient of diffusion and viscosity. [Reprinted with permission from R. E. Powell, W. E. Roseveare, and H. Eyring, Diffusion, thermal conductivity and viscous flow of liquids, *Ind. Eng. Chem.* **33**, 430–435 (1941); copyright 1941, American Chemical Society.]

mass diffusivity depends on concentration only if there are more than two components. In liquids this complication often arises in the binary case. Figure 3-15 illustrates the situation for two ideal cases. In these two cases $\mu D_{12}$ is made to vary linearly with concentration in a binary system by multiplying by a coefficient, which involves the activity [17] $a_1$ of the solute in the mixture. Unfortunately, this correction is not always dependable as the acetone–water system illustrated in Fig. 3-16 shows.

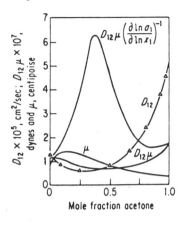

$y$-axis: $D_{12} \times 10^5$, cm$^2$/sec; $D_{12}\,\mu \times 10^7$, dynes and $\mu$, centipoise

$x$-axis: Mole fraction acetone

$D_{12}\mu\left(\dfrac{\partial \ln a_1}{\partial \ln x_1}\right)^{-1}$

$D_{12}$

$\mu$

$D_{12}\mu$

**Figure 3-16** Mutual diffusion, viscosity, uncorrected and activity-corrected $D_{12}\,\mu$ product for acetone–water system at 25.15°C. [Reprinted with permission from D. K. Anderson, J. R. Hall, and A. L. Babb, Mutual diffusion in non-ideal binary liquid mixtures, *J. Phys. Chem.* **62**, 404–409 (1958); copyright 1958, American Chemical Society.]

The estimation of $D_{12}$ for water in organic solvents can be obtained from the Wilke–Chang correlation, according to Reid and Sherwood [5], after dividing by 2.3. The Wilke–Chang correlation is

$$D_{12} = 7.4 \times 10^{-8}(\phi M_2)^{1/2}\frac{T}{\mu_2 V_1^{0.6}} \tag{3-73}$$

where $D_{12}$ is in cm$^2$/s, $M_2$ is the solvent molecular weight, $T$ is temperature in kelvins, $\mu_2$ is the solvent viscosity in centipoise, $V_1$ is the solute molar volume in cm$^3$/gmol at its boiling temperature under one atmosphere of pressure, and $\phi$ is an "association" parameter of the solvent. The values of $\phi$ are: water, 2.6; methanol, 1.9; ethanol, 1.5; benzene, 1.0; ether, 1.0; heptane, 1.0; and other unassociated solvents, 1.0. Reid and Sherwood [5] recommend that for non-aqueous solvents the Scheibel equation be used in which

$$D_{12} = \frac{KT}{\mu_2 V_1^{1/3}} \tag{3-74}$$

where, generally speaking,

$$K = 8.2 \times 10^{-8}\left[1 + \left(\frac{3V_2}{V_1}\right)^{2/3}\right]$$

In particular solvents, better values of $K$ are: water with $V_1 < V_2$, $25.2 \times 10^{-8}$; benzene with $V_1 < 2V_2$, $18.9 \times 10^{-8}$; and other solvents with $V_1 < 2.5V_2$, $17.5 \times 10^{-8}$. Here $V_2$ is the solvent molar volume in cm$^3$/gmol at its boiling temperature under one atmosphere of pressure. Equations (3-73) and (3-74) are of the general form suggested by Eq. (3-70), although the exponent of $V_1$ in Eq. (3-73) deviates. Roughly 20% accuracy can be achieved. To estimate $D_{12}$ for very low concentrations of the solute in water, Reid and Sherwood [5] recommend use of the Othmer–Thakar correlation

$$D_{12} = 1.4 \times 10^{-4} V_1^{-0.6} \mu_{\text{water}}^{-1.1} \tag{3-75}$$

where $D_{12}$ is in $cm^2/s$, $V_1$ is the volume per mole of the solute at its boiling point for one atmosphere of pressure in units of $cm^3/gmol$ and $\mu_w$ is the viscosity of water in centipoise at the temperature of interest. Errors of about 15% can be expected in the 10–30°C temperature range. The form of Eq. (3-75) is as suggested by Eq. (3-70) but differs in that the temperature dependence is expressed in terms of the solvent viscosity.

## 3.10   TWO-PHASE PARTICULATE SYSTEMS

The effective thermal conductivity and viscosity of a two-phase particulate system is important to a number of technological applications. The variety of possibilities is large, and it is almost impossible to make general statements that are applicable to all cases.

The case that is most amenable to analysis is that in which a single continuous phase has distributed within it spherical particles of concentration sufficiently low that they do not touch. The effective properties are deduced by solving the describing equations for each phase and subjecting the solutions to appropriate boundary conditions. Unfortunately, the effective property values depend somewhat on the nature of the imposed boundary conditions.

If the continuous phase has thermal conductivity $k_c$ and is a Newtonian fluid of viscosity $\mu_c$ while the dispersed phase has thermal conductivity $k_d$ and is a Newtonian fluid of viscosity $\mu_d$ with a volumetric concentration $\phi$, it is found that the effective thermal conductivity is given by

$$\frac{k_{\text{eff}}}{k_c} = \frac{1 + 2y + 2(1-y)\phi}{1 + 2y - (1-y)\phi} \tag{3-76}$$

with $y = k_c/k_d$ for a slab geometry with an imposed temperature difference. The effective viscosity is given for Couette flow on the basis of equal energy dissipation by

$$\frac{\mu_{\text{eff}}}{\mu_c} = 1 + 5.5\phi \frac{4\phi^{7/3} + 10 - \frac{84}{11}\phi^{2/3} + 4\beta(1 - \phi^{7/3})}{10(1 - \phi^{10/7}) - 25\phi(1 - \phi^{4/3}) + 10\beta(1 - \phi^{7/3})(1 - \phi)} \tag{3-77}$$

where

$$\beta = \frac{\mu_c}{\mu_d + \bar{\gamma}}$$

The $\bar{\gamma}$ parameter is the interfacial retardation viscosity due to interfacial contamination by surfactant impurities that imparts a certain rigidity to the interface. Equations (3-76) and (3-77) are intended for use at low concentrations $\phi < 1$ and were developed by Yaron [19]. Equation (3-76) agrees with the results obtained by Maxwell [20, 21] and is accurate for $\phi \leqslant 0.1$; for $\phi > 0.1$, the Bruggeman equation is preferred [31]. Equation (3-77) differs slightly from the result due to Einstein for very low concentration [22], $\mu_{eff}/\mu_c = 1 + 2.5\phi$ and is in substantial agreement with the results of Sather and Lee [23] for rigid spheres.

Further general information on a broad range of related topics is available in the literature survey by Soo [23] and by Depew and Kramer [24].

## 3.11   WATER VISCOSITY AND THERMAL CONDUCTIVITY

The properties of water are of particular interest because of its common occurance and great utility. Most water properties are tabulated [27], but new measurements have recently been incorporated into more accurate interpolating equations.

For the viscosity of water and steam, the recommended [28] interpolating equation for dynamic viscosity $\mu$ in terms of density $\rho$, absolute temperature $T$, and numerical constants is

$$\frac{\mu}{\mu_0} = \exp\left[\frac{\rho}{\rho^*} \sum_{i=0}^{5} \sum_{j=0}^{4} b_{ij}\left(-1 + \frac{T^*}{T}\right)^i \left(-1 + \frac{\rho}{\rho^*}\right)^j\right] \qquad (3\text{-}78a)$$

Here

$$\mu_0 = \frac{(T/T^*)^{1/2}}{\sum_{k=0}^{3} a_k (T^*/T)^k} \qquad (3\text{-}78b)$$

which gives $\mu_0$ in units of micropascal seconds. The constants in Eq. (3-78) have the following numerical values and those given in Table 3-7 for $b_{ij}$:

$$T^* = 647.27 \ K$$

$$\rho^* = 317.763 \ \text{kg/m}^3$$

$$a_0 = 0.0181583$$

$$a_1 = 0.0177624$$

$$a_2 = 0.0105287$$

$$a_3 = -0.0036744$$

**Table 3-7   Numerical Values of $b_{ij}$**

| $i =$ | 0 | 1 | 2 | 3 | 4 | 5 |
|---|---|---|---|---|---|---|
| $j = 0$ | 0.501938 | 0.162888 | $-0.130356$ | 0.907919 | $-0.551119$ | 0.146543 |
| 1 | 0.235622 | 0.789393 | 0.673665 | 1.207552 | 0.0670665 | $-0.0843370$ |
| 2 | $-0.274637$ | $-0.743539$ | $-0.959456$ | $-0.687343$ | $-0.497089$ | 0.195286 |
| 3 | 0.145831 | 0.263129 | 0.347247 | 0.213486 | 0.100754 | $-0.032932$ |
| 4 | $-0.0270448$ | $-0.0253093$ | $-0.0267758$ | $-0.0822904$ | 0.0602253 | $-0.0202595$ |

Equation (3-75) is accurate in the temperature range $0 < T < 800°C$ and the density range $0 < \rho < 1050$ kg/m$^3$, corresponding to an approximate pressure range of $0 < P < 100$ MPa. The domain of accuracy is extendable to $P = 1000$ MPa for $0 < T < 100°C$ and to $P = 350$ MPa for $100°C < T < 560°C$. Tabulated viscosity values are given by Nagashima [28].

The thermal conductivity of water and steam is also representable by interpolating equations. Unlike the case for dynamic viscosity, the surface representing thermal conductivity as a function of density and temperature is complicated by the nonanalytic character of the critical point whose effects cannot be ignored. Two interpolating equations are available [29], one for industrial use and one for scientific use. The one for scientific use gives a good representation of the anomaly that thermal conductivity becomes infinite at the critical point; the interpolating equation is relatively complex and requires use of the isothermal compressibility as well as the dynamic viscosity. The simpler interpolating equation for industrial use [Eq. (3-79)] is a two-variable fit to the thermal conductivity surface that excludes a rectangular region about the critical point bounded by $T = T_{\text{critical}} \pm 1.5°C$ and $\rho = \rho_{\text{critical}} \pm 100$ kg/m$^3$. Its uncertainty ranges from $\pm 2\%$ at a pressure of 0.1 MPa to $\pm 11\%$ at 100 MPa. This industrial-use interpolating equation yields a finite value at the critical point instead of the theoretically justified infinite value and is

$$k = k_0 + \bar{k} + \Delta k \tag{3-79a}$$

where

$$k_0 = \left(\frac{T}{T^*}\right)^{1/2} \sum_{i=0}^{3} a_i \left(\frac{T}{T^*}\right)^i \tag{3-79b}$$

$$\bar{k} = b_0 + b_1\left(\frac{\rho}{\rho^*}\right) + b_2 \exp\left[B_1\left(B_2 + \frac{\rho}{\rho^*}\right)\right]^2 \tag{3-79c}$$

$$\Delta k = \left[d_1\left(\frac{T^*}{T}\right)^{10} + d_2\right]\left(\frac{\rho}{\rho^*}\right)^{1.8} \exp\left\{C_1\left[1 - \left(\frac{\rho}{\rho^*}\right)^{2.8}\right]\right\}$$

$$+ d_3 S \left( \frac{\rho}{\rho^*} \right)^Q \exp \left\{ \frac{Q}{R} \left[ 1 - \left( \frac{\rho}{\rho^*} \right)^R \right] \right\}$$

$$+ d_4 \exp \left[ C_2 \left( \frac{T}{T^*} \right)^{1.5} + C_3 \left( \frac{\rho^*}{\rho} \right)^5 \right] \qquad (3\text{-}79\text{d})$$

$$Q = 2.0 + C_5 (\Delta T^*)^{-0.6} \qquad (3\text{-}79\text{e})$$

$$R = Q + 1.0 \qquad (3\text{-}79\text{f})$$

$$S = \begin{array}{ll} (\Delta T^*)^{-1.0} & \text{for} \quad \dfrac{T}{T^*} \geqslant 1 \\[2mm] C_6 (\Delta T^*)^{-0.6} & \text{for} \quad \dfrac{T}{T^*} \leqslant 1 \end{array} \qquad (3\text{-}79\text{g})$$

$$\Delta T^* = \left| \frac{T}{T^*} - 1.0 \right| + C_4 \qquad (3\text{-}79\text{h})$$

Numerical values of constants for Eq. (3-79) are:

$$T^* = 647.3 \ K$$

$$a_0 = 1.02811 \times 10^{-2} \ \text{W/K m}$$

$$a_1 = 2.99621 \times 10^{-2} \ \text{W/K m}$$

$$a_2 = 1.56146 \times 10^{-2} \ \text{W/K m}$$

$$a_3 = -4.22464 \times 10^{-3} \ \text{W/K m}$$

$$b_0 = -3.97070 \times 10^{-1} \ \text{W/K m}$$

$$b_1 = 4.00302 \times 10^{-1} \ \text{W/K m}$$

$$b_2 = 1.06000 \ \text{W/K m}$$

$$B_1 = -1.71587 \times 10^{-1}$$

$$B_2 = 2.39219$$

$$\rho^* = 317.7 \ \text{kg/m}^3$$

$$d_1 = 7.01309 \times 10^{-2} \ \text{W/K m}$$

$$d_2 = 1.18520 \times 10^{-2} \ \text{W/K m}$$

Table 3-8  Short Table of the Thermal Conductivity of Water Substance, $k$ in mW/K m [29]

| P, in MPa \ T, in °C | 0 | 25 | 50 | 75 | 100 | 150 | 200 | 250 | 300 | 350 | 375 | 400 | 450 | 500 | 550 | 600 | 700 | 800 |
|---|---|---|---|---|---|---|---|---|---|---|---|---|---|---|---|---|---|---|
| 0.1 | 561 | 607 | 644 | 667 | 25.1 | 28.9 | 33.3 | 38.2 | 43.4 | 49.0 | 51.8 | 54.8 | 60.8 | 67.0 | 73.4 | 79.9 | 93.4 | 107 |
| 0.5 | 561 | 607 | 644 | 667 | 679 | 682 | 34.9 | 39.2 | 44.1 | 49.4 | 52.3 | 55.1 | 61.1 | 67.2 | 73.6 | 80.1 | 93.6 | 108 |
| 1.0 | 562 | 608 | 644 | 667 | 679 | 682 | 37.2 | 40.5 | 45.0 | 50.1 | 52.8 | 55.6 | 61.5 | 67.6 | 73.9 | 80.4 | 93.9 | 108 |
| 2.5 | 562 | 608 | 645 | 668 | 680 | 683 | 664 | 45.2 | 47.8 | 52.1 | 54.5 | 57.1 | 62.7 | 68.7 | 74.9 | 81.4 | 94.8 | 108 |
| 5.0 | 564 | 609 | 646 | 669 | 682 | 685 | 666 | 623 | 53.9 | 56.0 | 57.9 | 60.1 | 65.1 | 70.7 | 76.8 | 83.1 | 96.3 | 110 |
| 7.5 | 565 | 611 | 647 | 670 | 683 | 687 | 668 | 626 | 63.1 | 61.0 | 62.0 | 63.6 | 67.8 | 73.0 | 78.8 | 85.0 | 98.1 | 112 |
| 10.0 | 566 | 612 | 648 | 672 | 684 | 688 | 671 | 629 | 551 | 68.1 | 67.3 | 67.9 | 71.0 | 75.6 | 81.1 | 87.1 | 100 | 113 |
| 15.0 | 569 | 614 | 651 | 674 | 687 | 692 | 675 | 635 | 562 | 101 | 85.5 | 80.7 | 79.2 | 81.9 | 86.4 | 91.9 | 104 | 117 |
| 20.0 | 572 | 616 | 653 | 677 | 690 | 695 | 679 | 641 | 572 | 463 | 141 | 105 | 91.0 | 89.9 | 92.8 | 97.6 | 109 | 121 |
| 25.0 | 575 | 618 | 655 | 679 | 693 | 698 | 684 | 647 | 581 | 481 | 412 | 168 | 109 | 100 | 101 | 104 | 115 | 126 |
| 30.0 | 577 | 621 | 658 | 682 | 695 | 702 | 688 | 652 | 589 | 496 | 438 | 331 | 136 | 114 | 110 | 112 | 121 | 131 |
| 40.0 | 583 | 625 | 662 | 687 | 701 | 708 | 696 | 662 | 605 | 520 | 473 | 414 | 227 | 152 | 134 | 130 | 135 | 143 |
| 50.0 | 588 | 630 | 667 | 692 | 706 | 715 | 704 | 672 | 619 | 541 | 498 | 451 | 315 | 203 | 164 | 152 | 151 | 156 |
| 60.0 | 593 | 634 | 671 | 696 | 712 | 721 | 711 | 682 | 631 | 559 | 519 | 477 | 371 | 256 | 198 | 177 | 168 | 170 |
| 70.0 | 598 | 639 | 676 | 701 | 717 | 728 | 719 | 691 | 643 | 576 | 537 | 498 | 408 | 301 | 233 | 203 | 186 | 185 |
| 80.0 | 603 | 643 | 680 | 706 | 722 | 734 | 726 | 700 | 654 | 591 | 554 | 516 | 435 | 339 | 265 | 228 | 204 | 199 |
| 90.0 | 608 | 647 | 685 | 711 | 727 | 740 | 734 | 708 | 665 | 604 | 569 | 533 | 457 | 369 | 294 | 251 | 221 | 213 |
| 100.0 | 612 | 652 | 689 | 715 | 732 | 746 | 741 | 717 | 675 | 617 | 583 | 548 | 476 | 395 | 319 | 272 | 236 | 215 |

100

$$d_3 = 1.69937 \times 10^{-3} \text{ W/K m}$$

$$d_4 = -1.02000 \text{ W/K m}$$

$$C_1 = 6.42857 \times 10^{-1}$$

$$C_2 = -4.11717$$

$$C_3 = -6.17937$$

$$C_4 = 3.08976 \times 10^{-3}$$

$$C_5 = 8.22994 \times 10^{-2}$$

$$C_6 = 1.00932 \times 10^{1}$$

Equation (3-79) is subject to exponential underflows when programmed for solution on a digital computer. It is preferred that density be computed from the 1967 IFC Formulation for Industrial Use. Information on the effects of temperature and pressure (as well as isotopic composition and dissolved gases) on the density of liquid water is given by Kell [30]. A brief compilation of the new thermal conductivity values predictable by Eq. (3-79) is given in Table 3-8. Its indicated pressure effect can be compared against the earlier results of Lenoir in Eq. (3-59) and Fig. 3-14. New values of the Prandtl number of water and steam are also available [34]; the procedure that gives the values also gives the specific heat.


## PROBLEMS

**3-1** (a) Estimate the molecular diameters of hydrogen and nitrogen, using measured viscosities in conjunction with a kinetic-theory-based relationship between viscosity and molecular properties. (*Answer*: $\sigma_{H_2} = 2.2 \times 10^{-10}$ m, $\sigma_{N_2} = 3 \times 10^{-10}$ m.)

(b) Use the results of part a to calculate the corresponding molecular spacing, mean free path, collision frequency, and translational relaxation time. (*Answer*: $\delta = 3 \times 10^{-9}$ m; $\lambda_{H_2} = 1.9 \times 10^{-7}$ m, $Z_{H_2} = 9 \times 10^{9}$ s$^{-1}$.)

**3-2** Kinetic theory predicts that gas viscosity varies as the square root of the molecular weight, all else being equal. Plot $\mu/\mu_{air}$ at the same absolute temperature against $M/M_{air}$ for various gases on log–log coordinates to ascertain the accuracy of the kinetic theory prediction.

**3-3** In view of the fact that experiment shows most gas molecules to reflect diffusely from a collision with a wall as evidenced by the fact that $f_s \sim 1$, comment on the accuracy of Eq. (2-13), whose derivation assumed elastic (specular) reflection.

**3-4** As illustrated in Fig. 1-3, two parallel plates have vertical velocities of 300 m/s and 0 m/s, respectively. The plate separation is 1 mm, and air fills the intervening space.

(a) Estimate the magnitude of the slip velocity at a plate at normal atmospheric conditions.

(b) At what pressure will the slip velocity become 1% of the imposed velocity difference if temperature does not change?

**3-5** The slip velocity at a solid wall can be influenced by a temperature gradient along the wall as given by [2]:

$$u_0 - u_w = \frac{2 - f_s}{f_s} \lambda \frac{du}{dy}\bigg|_w + 3\left(\frac{RT}{8\pi M}\right)^{1/2} \frac{\lambda}{T} \frac{dT}{dx}\bigg|_w$$

where the second term on the right relates to the phenomenon known as *thermal creep* that is observed in high-vacuum technology.

(a) Estimate the thermal creep term without rigor by showing that a bulk flow of molecules could be caused by a temperature difference (as in Problem 2.3) as

$$n(u_0 - u_w) = \frac{n_1 \bar{v}_1 - n_2 \bar{v}_2}{8}$$

which can be manipulated to yield

$$u_0 - u_w = \left(\frac{KT}{8\pi m}\right)^{1/2} \frac{\Delta T}{T}$$

where $T_1 = T - \Delta T$ and $T_2 = T + \Delta T$.

(b) Presuming that a molecule acquired its energy at an average distance $\bar{x}$ away from the central plane of interest so that $\Delta T = \bar{x}(dT/dx)$, show that if $\bar{x} = 5\lambda/2$, as it does in precise computation of thermal conductivity, a factor of $\frac{5}{2}$ appears in the thermal creep term (instead of 3), whereas if $\bar{x} = 2\lambda/3$, as it does in the simplest kinetic theory, a factor of $\frac{2}{3}$ appears.

(c) Compare the results of part b and those initially stated with the carefully derived result (G. N. Patterson, *Molecular Flow of Gases*, Wiley, New York, 1956, p. 124) for the case of a unity accomodation coefficient

$$u_0 - u_w = \frac{5\pi\lambda}{16}\left[\frac{du}{dy}\bigg|_w + 3\left(\frac{RT}{8\pi}\right)^{1/2} \frac{1}{T} \frac{dT}{dx}\bigg|_w\right]$$

where the $5\pi/16$ factor may be taken as a minor difference in definition of the mean free path.

(d) For the conditions in Problem 3-4, determine the numerical value of $dT/dx$ that would make thermal creep about one-half the total slip velocity.

**3-6** Determine the Sutherland constants for air, water vapor, and hydrogen from three plots of viscosity data.

**3-7** Determine the best value of $n$ in Eq. (3-15) for air from a plot of viscosity versus absolute temperature on log–log coordinates. Is $n = \frac{1}{2}$ as simple kinetic theory predicts? If not, explain briefly what is responsible for the deviation.

**3-8** (a) Estimate the numerical value of the viscosity of a gas mixture that is 97.5% helium—2.5% oxygen by weight at a total pressure of 11 atm and a temperature of 20°C.

(b) Estimate the thermal conductivity of this mixture.

(c) Compare the effect of the oxygen on the mixture for parts a and b. Is thermal conductivity or viscosity affected more?

**3-9** It is to be shown that the temperature jump and the energy jump at a solid wall are related by

$$e_0 - e_w = (T_0 - T_w)\frac{m(C_p + C_v)}{2}$$

(a) First, verify that the average translational kinetic energy per molecule crossing a plane is $\frac{4}{3}$ that of local molecules. In other words,

$$\frac{\text{Average K.E.}}{\text{crossing molecule}} = \frac{\text{K.E. flow rate}}{\text{molecular flow rate}}$$

$$= \frac{\int_{v=0}^{\infty}\int_{\theta=0}^{\pi/2}\int_{\phi=0}^{2\pi}(mv^2/2)\sin\theta\cos\theta\,d\theta\,d\phi\,v\,dn_v/4\pi}{\int_{v=0}^{\infty}\int_{\theta=0}^{\pi/2}\int_{\phi=0}^{2\pi}\sin\theta\cos\theta\,d\theta\,d\phi\,v\,dn_v/4\pi}$$

$$= 2KT$$

$$= \frac{3KT}{2}\frac{4}{3}$$

$$= \frac{4}{3}\ (\text{local translational K.E.})$$

(b) Show that since the energy jump at a solid wall is composed of the sum of translational and internal (rotational plus vibrational) energy

$$e_0 - e_w = \Delta(\text{translational K.E.}) + \Delta(\text{internal energy})$$

$$= 2K(T_0 - T_w) + \Delta(\text{internal energy})$$

and since specific heat at constant volume is defined as

$$mC_v \Delta T = \Delta(\text{local translational K.E.} + \text{internal energy})$$

$$= \Delta(3KT/2) + \Delta(\text{internal energy})$$

with

$$C_p - C_v = \frac{R}{M}$$

$$= \frac{R}{MN_A}$$

for a perfect gas, in addition, it follows that the lead-off equation is a correct relationship.

**3-10** Two horizontal parallel plates are separated by a distance $L$ and are maintained at a constant temperature difference $\Delta T$ with the hotter plate uppermost. The space between the plates is filled with a gas at pressure $p$.

**(a)** Show that the conductive heat flux between the plates is given by

$$q = \frac{k\Delta T/L}{1 + 2\xi/L}$$

where

$$\xi = \frac{2 - \alpha_s}{\alpha_s} \frac{2\gamma}{(\gamma + 1)P_r}\lambda$$

**(b)** From the result of part a, show that the apparent conductivity $k_{\text{eff}}$ of the intervening gas is given by

$$\frac{k_{\text{eff}}}{k} = \frac{1}{1 + 2\xi/L}$$

**(c)** Plot $k_{\text{eff}}/k$ against pressure for air, assuming that $L = 10\lambda_1$, where $\lambda_1$ is the mean free path at a pressure of 1 atm. Also plot $k_{\text{eff}}/k$ against $L/\lambda$.

**(d)** Comment on the use by Smoluchowski of granular powders to produce insulators better than air (M. Jakob, *Heat Transfer*, Vol. 1, Wiley, New York, 1949, p. 90).

**3-11** The carefully derived result for temperature jump (G. N. Patterson, *Molecular Flow of Gases*, Wiley, New York, 1956, p. 125) for a mona-tomic gas and unity accommodation coefficients is

$$T_0 - T_w = \frac{75\pi}{128}\lambda \frac{dT}{dy}\bigg|_w - \frac{5}{24}\frac{\lambda}{\bar{v}}\frac{du}{dx}\bigg|_w$$

Explain on physical grounds how it is that an $x$-direction variation of $u$ can give rise to a $y$-direction temperature jump.

**3-12** Show that the thermal conductivity for a dilute polyatomic gas is related to the value predicted by a monatomic model as

$$k_{poly} = k_{mon} \frac{9\gamma - 5}{10}$$

where $k_{mon}$ is given by Eq. (3-34).

**3-13** Compare values of Schmidt number ($Sc = \nu/D_{12}$) predicted by Eq. (3-47) against measured values from Table 3-5.

**3-14** Compare mass diffusivities for $CO_2$ through air estimated by taking air to be a (a) 1-component gas and (b) 2-component gas made up of nitrogen and oxygen. Explain the probable source of any observed difference.

**3-15** (a) Plot liquid water viscosity versus inverse absolute temperature, using semilog coordinates, with data ranging from the melting point up to the critical temperature.

    (1) Determine the range in which Eq. (3-52) applies as evidenced by a linear curve—mark the melting, boiling at 1 atm, and critical temperature on the plot,

    (2) Determine the accuracy of $B \approx 3.8\ T_b$.

  (b) Repeat part a, but plot viscosity against the internal energy of vaporization divided by absolute temperature, again using semilog coordinates. Does this plot yield a straighter line than that of part a? Explain why this did or did not happen.

**3-16** (a) Show analytically that for all liquids near their melting temperature there is reason to expect that, with $M$ being molecular weight,

$$\frac{Mk}{\mu} = \text{const}$$

  (b) Compare the prediction of part (a) with data for several different liquids and evaluate the constant.

**3-17** Estimate the thermal conductivity of liquid water at 20°C by making use of Eq. (3-56) with

  (a) Acoustic velocity from Eq. (3-57).

  (b) Acoustic velocity from Eq. (3-58). (Answer: $k = 2.4$ W/m K.)

**3-18** Estimate the percentage change in thermal conductivity of water that accompanies a pressure change from 1 to 500 atm. Use Eqs. (3-56) and (3-59) and compare the results against Fig. 3-13.

**3-19**  Estimate the thermal conductivity of a 50% by weight mixture of liquid water and $n$-butyl alcohol at 20°C:

| Liquid | Thermal Conductivity | Viscosity, $lb_m$/ft sec |
|---|---|---|
| Water | 0.347 Btu/hr ft °F | $0.658 \times 10^{-3}$ |
| $n$-Butyl alcohol | 0.096 Btu/hr ft °F | $2.26 \times 10^{-3}$ |

**3-20**  Estimate the viscosity of a 50% mole fraction mixture of liquid water and $n$-butyl alcohol at 20 °C.

**3-21**  (a) Predict the thermal conductivities of aluminum and liquid mercury from values of their electrical conductivities by the Wiedemann–Franz–Lorenz law.

(b) Compare the results with data and state the fraction of the total thermal conductivity accounted for by electron transport.

**3-22**  Estimate the mass diffusivity of oxygen gas through both liquid water and water vapor. Evaluate the ratio of the mass diffusivity through liquid water to that through water vapor.

**3-23**  Estimate the mass diffusivity of blood through water.

**3-24**  The value of $D_{12}$ for a dilute solution of methanol in water at 15°C is reported to be $1.28 \times 10^{-5}$ cm$^2$/s.

(a) Is the reported value of $D_{12}$ reasonable?

(b) Estimate, taking whichever is the best result for part a as a datum, the value of $D_{12}$ at 100°C.

**3-25**  Estimate the activation energy for diffusion $\Delta E_d$ from a comparison of Eqs. (3-72) and (3-73).

**3-26**  Compare the numerical value of $K$ in the Scheibel Eq. (3-74) for the case in which the solvent can be considered to be a continuum (i.e., $V_2 = 0$) with the numerical value that is predicted by the hydrodynamic theory of Eq. (3-70).

**3-27**  Estimate the numerical value of the mass diffusivity of small, spherical particles through air if the particle radius is $10^{-4}$ cm. How much larger would it be for a particle whose radius is $10^{-5}$ cm? Would these particles diffuse more rapidly?

**3-28**  Explain on physical grounds why mass diffusivity and viscosity are directly proportional for gases but are inversely proportional for liquids.

**3-29**  Plot $k_{eff}/k_c$ against $\phi$ from Eq. (3-65). Use numerical values for $y = k_c/k_d$ of 10 and $\frac{1}{10}$.

# REFERENCES

1   J. F. Lee, F. W. Sears, and D. L. Turcotte, *Statistical Thermodynamics*, Addison-Wesley, Reading, MA, 1963, p. 81.

2   R. D. Present, *Kinetic Theory of Gases*, McGraw-Hill, New York, 1958.

3   E. R. G. Eckert and R. M. Drake, *Analysis of Heat and Mass Transfer*, McGraw-Hill, New York, 1972.

4   S. Chapman and T. G. Cowling, *The Mathematical Theory of Non-Uniform Gases*, University Press, Cambridge, 1958.

5   R. C. Reid and T. K. Sherwood, *The Properties of Gases and Liquids*, 2nd ed., McGraw-Hill, New York, 1966.

6   J. O. Hirschfelder, C. F. Curtis, and R. B. Bird, *Molecular Theory of Gases and Liquids*, Wiley, New York, 1966.

7   R. B. Bird, W. E. Stewart, and E. N. Lightfoot, *Transport Phenomena*, Wiley, New York, 1965.

8   R. W. Schrage, *Interphase Mass Transfer*, Columbia University Press, New York, 1953, pp. 20–24.

9   M. Jakob, *Heat Transfer*, Vol. I, Wiley, New York, 1949, pp. 77, 80, 112.

10  C. F. Curtis and J. O. Bird, Transport properties of multicomponent gas mixtures, *J. Chem. Phys.* **17**, 550–555 (1949).

11  E. N. da C. Andrade, The viscosity of liquids, *Endeavor* **13**, 117–127 (July 1954); A theory of the viscosity of liquids—Part I, *Phil. Mag.* **17**, 497–511 (1934); A theory of the viscosity of liquids—Part II, *Phil. Mag.* **17**, 698–732, (1934).

12  K. M. Watson, F. R. Wien, and G. B. Murphy, High temperature viscosities of liquid petroleum fractions, *Ind. Eng. Chem.* **28**, 605–609 (1936).

13  R. H. Perry and C. H. Chilton, Eds., *Chemical Engineers' Handbook*, 5th ed., McGraw-Hill, New York, 1973.

14  T. Baumeister and L. S. Marks, Eds., *Standard Handbook for Mechanical Engineers*, 7th ed., McGraw-Hill, New York, 1967.

15  R. P. Feynman, R. B. Leighton, and M. Sands, *The Feynman Lectures on Physics*, Vol. I, Addison-Wesley, Reading, MA, 1963.

16  P. A. Johnson and A. L. Babb, Liquid diffusion of non-electrolytes, *Chem. Rev.* **56**, 387–453 (1956).

17  E. F. Obert, *Concepts of Thermodynamics*, McGraw-Hill, New York, 1960, p. 378.

18  C. L. Tien and J. H. Lienhard, *Statistical Thermodynamics*, Holt, Rinehart, and Winston, New York, 1971, p. 328.

19  I. Yaron, Cell model approach to transport phenomena in porous media, in *Workshop on Heat and Mass Transfer in Porous Media* (at Case Western Reserve University, Cleveland, Ohio and sponsored by Engineering Division, Heat Transfer Program, National Science Foundation), A. Dybbs (Department of Fluid, Thermal, and Aerospace Sciences at Case Western Reserve University) October 14–15, 1974, pp. 40–87; Regular expansion solutions for heat or mass transfer in concentrated two-phase particulate solutions at small Peclet and Reynolds numbers, *Internatl. J. Heat Mass Transf.* **19**, 61–69 (1976).

20  J. C. Maxwell, *A Treatise on Electricity and Magnetism*, 3rd ed., Vol. 1, Dover, New York, 1954, p. 440.

21  H. S. Carslaw and J. C. Jaeger, *Conduction of Heat in Solids*, 2nd ed., Clarendon Press, Oxford, 1959, p. 428.

22  A. Einstein, Eine neue Bestimmung der Molekül-dimensionen, *Annalen der Physik*, **19**, 289–306 (1906); Berichtigung zu meiner Arbeit: Eine neue Bestimmung der Molekül-dimensionen, *Annalen der Physik*, **34** 12, 591 (1911).

23  S. L. Soo, *Fluid Dynamics of Multiphase Systems*, Blaisdell, Waltham, MA, 1967.

24  C. A. Depew and T. J. Kramer, Heat transfer to flowing gas–solid mixtures, *Adv. Heat Transf.* **9**, 113–180 (1973).

25  Y. S. Touloukinan, Ed., *Thermophysical Properties of Matter*, IFI/Plenum, New York, 1970.

26  *International Critical Tables*, McGraw-Hill, New York, 1933.

27  J. H. Keenan, F. G. Keyes, P. G. Hill, and J. G. Moore, *Steam Tables*, Wiley, New York, 1978; *ASME Steam Tables*, 4th ed. ASME, United Engineering Center, New York 1979.

28  A. Nagashima, Viscosity of water substance—New international formulation and its background, *J. Phys. Chem. Ref. Data*, **6**, 1133–1166 (1977) [see also *Mech. Eng.* **98**, (7), 79 (1976).

29  J. Kestin, Thermal conductivity of water and steam, *Mech. Eng.* **100** (7), 46–48 (1978) (see also J. V. Sengers et al., to be published in *J. Phys. Chem. Ref. Data*).

30  G. S. Kell, Effects of isotopic composition, temperature, pressure, and dissolved gases on the density of liquid water, *J. Phys. Chem. Ref. Data*, **6**, 1109–1131 (1977).

31  T. Y. R. Lee and R. E. Taylor, Thermal diffusivity of dispersed materials, *Transact. ASME, J. Heat Transf.* **100**, 720–724 (1978).

32  H. S. Tsien, Superaerodynamics, mechanics of rarefied gas, *J. Aeronaut. Sci.* **13**, 653–664 (1946).

33  J. A. Barker and D. Henderson, The fluid phases of matter, *Sci. Am.* **245**, 130–138 (1981).

34  J. V. Fengerf, R. S. Basu, B. Kamgar-Parsi, and J. Kestin, Problems with the Prandtl number of steam, *Mechanical Engineering*, volume 104, number 5, 1982, pp. 60–63.

# 4

# EQUATIONS OF CONTINUITY, MOTION, ENERGY, AND MASS DIFFUSION

Detailed consideration of the motion of gas molecules leads, as has been seen in Chapter 3, to insight into the manner in which properties are transported through a fluid. The resulting information concerning the manner in which transport properties vary with temperature, pressure, and molecular weight is valuable not only for prediction of property values, but also for correlation of experimental measurements. Very often, however, transport properties can be taken to be known quantities.

Most commonly, the desire is to determine the distributions of bulk velocity, temperature, or mass fraction. These, when inserted into the appropriate rate equation (e.g., Fourier's law), then yield the quantity of major engineering interest that is a surface shear stress, heat flux, or mass flux. Naturally, the greater the detail with which this information is obtained, the more accurate the final results. The determination of every detail of bulk velocity distribution that comes from accounting for random molecular motion in gases, for example, leads to the Boltzmann integrodifferential equation. The Boltzmann equation represents the most detailed description of transport, but it is so difficult that it has been solved neither exactly nor numerically for many cases [1, 17]. The first terms of a series solution of the Boltzmann equation verify that (1) fluid bulk velocity and so forth can usually be accurately obtained from simpler equations that consider the fluid to be a continuous substance of known properties and (2), as long as extremely steep gradients and rapid changes are not encountered, the rate equations of Chapter 1 are accurate.

For these reasons, the equations describing bulk velocity, temperature, and mass fraction distributions are derived for a continuous fluid. Conservation principles are applied to a convenient control volume, and the rate equations of Chapter 1 are used to relate fluxes to gradients.

## 4.1   GENERAL ASSUMPTIONS

Information concerning property values is assumed to be available. Transport properties such as thermal conductivity and viscosity are often found in tabular form, but their representation by accurate equations is also frequent. When large temperature excursions are encountered, it is presumed that the variation of properties with temperature can be evaluated and appropriately taken into account. Equations of state are assumed to be known to relate density to temperature and pressure; the equation of state $\rho = p/RT$ for a perfect gas is a simple example. It is usually assumed that the materials encountered are isotropic—no property depends on direction.

The continuous nature of matter is assumed. This is a particular convenience because the derivation of differential equations involves taking the limit as a control volume becomes of infinitesimal size. Although extremely rarefied gases may violate this assumption somewhat when the mean free path of molecules is of the same order of magnitude as the system dimensions, it can be appreciated that at room conditions even a gas can be considered to be a continuous medium. Liquids and solids, which are substantially denser than gases, are even better regarded as continuous materials; hence this continuous approximation is readily seen to be accurate for most substances of engineering interest.

It is assumed that pure substances in the thermodynamic sense are under consideration. A pure substance [2] is homogeneous in composition and invariable in chemical aggregation. For example, water existing as a vapor, a liquid, or a solid or a combination of these states is a pure substance. A mixture of hydrogen and oxygen gases is a pure substance as long as no liquid or solid phase appears since then one phase will be richer in hydrogen than the other. If part of the system is combined to form water, this system is not a pure substance because it is not homogeneous in chemical composition. This assumption is useful because it, when taken together with the local equilibrium assumption, allows quantities such as enthalpy and internal energy to be related to such easily measured quantities as temperature and pressure.

Local equilibrium is assumed. This is somewhat contradictory since if there really is equilibrium, even if only locally, there can be no transport by diffusion—the rate equations reveal that there must be a variation in the diffusing quantity to have a nonzero flux. The essence of this assumption really is that the departure from equilibrium is sufficiently slight that the rate equations are accurate. This requires that in the immediate vicinity of a point there is locally uniform velocity and so forth. Without equilibrium, the concept of temperature loses much of its meaning. The temperature of a gas, for instance, is strictly defined only if all energy storage modes are in a state of thermal equilibrium. Since individual gas molecules store energy in translational, rotational, and electronic modes of motion, complete equilibrium is attained only when all the modes of energy storage have population distributions corresponding to the same temperature [3]. A rapid change such as might be experienced in a shock wave, rapid decompression, or rapid chemical reaction may result in a measurable delay before equilibrium is again attained among the translational and other modes of energy storage. In a monatomic gas only three translation modes exist for energy storage, with rotation seemingly unimportant at normal temperatures, and equilibrium is achieved rather rapidly. A diatomic or polyatomic gas, on the other hand, first stores energy in translational motion. Then, after a number of collisions with other molecules, energy is stored in the rotational mode of motion; after a larger number of collisions, vibrational modes of motion store energy. Thus the effective vibrational temperature of a gas may be quite different from the translational temperature in a nonequilibrium situation. The number of collisions required for a gas to reachieve true equilibrium appears to range from a few in the case of air to several thousand in the case of carbon dioxide [4, 5]. The new equilibrium condition is asymptotically approached in an exponential manner [6] with a time constant, or relaxation time, proportional to $\mu/p$ for the translational mode [4] [see Eq. (3-6) and its derivation].

## 4.2  EQUATION OF CONTINUITY

The conservation of mass principle is applied first to the general control volume shown in Fig. 4.1. This principle merely states that the mass of a system of fixed identity is constant. Let the system of fixed identity be bounded by the dotted line; as the fluid flows and distorts the system, the dotted line will also move and distort in such a manner as to always enclose the same fluid particles. The solid line denotes an arbitrary control volume within which the system is initially contained. At time $t$ the system occupies regions 1 and 3; at a later time $t + \Delta t$ it occupies regions 1 and 2. It is presumed that the elapsed time $\Delta t$ is sufficiently small that region 1 occupies some part of the control volume. Since the system mass is constant,

$$m_1(t) + m_3(t) = m_1(t + \Delta t) + m_2(t + \Delta t)$$

in which $m$ represents the mass in the identified region of space. Rearrangement gives

$$m_1(t + \Delta t) - m_1(t) = m_3(t) - m_2(t + \Delta t)$$

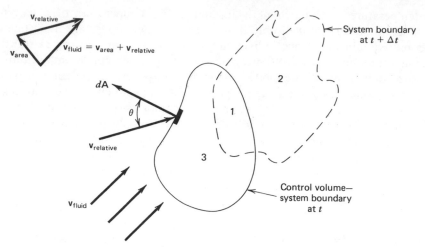

**Figure 4-1**   General control volume through which fluid flows.

Division by $\Delta t$ then yields

$$\frac{m_1(t + \Delta t) - m_1(t)}{\Delta t} = \frac{m_3(t)}{\Delta t} - \frac{m_2(t + t)}{\Delta t} \tag{4-1}$$

In the limit as $\Delta t \to 0$, region 1 coincides with the control volume. It is, therefore, permissible to intepret the parts of Eq. (4-1) as follows. The left-hand side represents the rate at which mass is stored in the control volume. The first term on the right-hand side represents the rate at which mass enters the control volume since the mass $m_3(t)$ initially occupying region 3 must be replaced if the fluid is continuous. The second term on the right-hand side represents the rate at which mass leaves the control volume since the mass $m_2(t + \Delta t)$ in region 2 was initially inside the control volume. The conservation of mass principle applied to a control volume is then stated as

$$\Sigma \dot{m}_{stored} = \Sigma \dot{m}_{in} - \Sigma \dot{m}_{out} \tag{4-2}$$

The dot notation signifies a rate with respect to time and $m$ signifies mass. For many engineering problems involving finite control volumes, Eq. (4-2) is the most useful statement of the conservation of mass principle.

Equation (4-2) can be generally reformulated in mathematical terms as

$$\frac{\partial \left( \int_V \rho \, dV \right)}{\partial t} = -\int_A \rho \left( \mathbf{V}_{rel} \cdot d\mathbf{A} \right) \tag{4-3}$$

since the mass in any infinitesimal volume $dV$ is $\rho \, dV$ and the net rate at which mass flows across an infinitesimal area $d\mathbf{A}$ is

$$\rho(V_{\text{fluid}} - V_{\text{area}})\cos\theta\, dA$$

or

$$\rho\mathbf{V}_{\text{rel}} \cdot d\mathbf{A}$$

in which it is assumed that the surface normal of an area element points outward. Equation (4-3) can be manipulated into a more convenient form by appealing to the divergence theorem, which relates volume integrals to area integrals according to

$$\int_A \mathbf{X} \cdot d\mathbf{A} = \int_V \text{div}(\mathbf{X})\, dV$$

in which $A$ is the area that encloses the volume $V$. Recognizing that $\mathbf{X} = \rho\mathbf{V}_{\text{rel}}$ in this case allows Eq. (4-3) to be rewritten as

$$\frac{\partial\left(\int_V \rho\, dV\right)}{\partial t} + \int_V \text{div}(\rho\mathbf{V}_{\text{rel}})\, dV = 0 \qquad (4\text{-}4)$$

Equation (4-4) applies to a control volume that can distort as time passes. A more special circumstance of particular interest occurs when the control volume is of fixed shape. Then the time derivative can be taken inside the integral to obtain

$$\int_V \left[\frac{\partial\rho}{\partial t} + \text{div}(\rho\mathbf{V}_{\text{rel}})\right] dV = 0 \qquad (4\text{-}5)$$

Focusing attention on an infinitesimal control volume (over which no property experiences appreciable variation), Eq. (4-5) is well approximated by

$$\left[\frac{\partial\rho}{\partial t} + \text{div}(\rho\mathbf{V}_{\text{rel}})\right] dV = 0$$

which is satisfied only by

$$\frac{\partial\rho}{\partial t} + \text{div}(\rho\mathbf{V}_{\text{rel}}) = 0 \qquad (4\text{-}6)$$

which can be called a *conservation-law form*. If the control volume is additionally constrained to be motionless, $\mathbf{V}_{\text{area}} = 0$ and so $\mathbf{V}_{\text{rel}} = \mathbf{V}_{\text{fluid}}$. If subscripts are dropped and $\mathbf{V}_{\text{fluid}}$ is denoted by $\mathbf{V}$, Eq. (4-6) then is

$$\frac{\partial\rho}{\partial t} + \text{div}(\rho\mathbf{V}) = 0 \qquad (4\text{-}7)$$

Equation (4-7) is called the *continuity equation* (because the divergence theorem employed at one step in its derivation requires density and velocity to be continuous functions) and is the form of the conservation of mass principle that is most useful for infinitesimal control volumes.

Equation (4-7) has been derived without specification of a coordinate system. It is, therefore, applicable to any coordinate system. In rectangular coordinates, for which $\mathbf{V} = u\hat{i} + v\hat{j} + w\hat{k}$, the continuity equation takes the form

$$\frac{\partial \rho}{\partial t} + \frac{\partial(\rho u)}{\partial x} + \frac{\partial(\rho v)}{\partial y} + \frac{\partial(\rho w)}{\partial z} = 0 \qquad (4\text{-}8)$$

In steady state $\partial \rho / \partial t = 0$; and for a constant-density fluid, $\rho = \text{const}$, it is always the case that

$$\frac{\partial u}{\partial x} + \frac{\partial v}{\partial y} + \frac{\partial w}{\partial z} = 0 \qquad (4\text{-}9)$$

The continuity equation has already been derived in a general way, following which specific assumptions were made to arrive at Eq. (4-8). It is instructive to first make the specific assumptions and then to apply the conservation principle. To this end, it is assumed from the beginning that the control volume is a cube fixed in space as shown in Fig. 4-2. A fluid, having velocity components, $u$, $v$, $w$ flows through the space occupied by the cube from time $t$ to time $t + \Delta t$. The cube faces are of length $\Delta x$ in the $x$ direction, $\Delta y$ in the $y$ direction, and $\Delta z$ in the $z$ direction. As shown in Fig. 4-2, the rate at which total mass flows through a face is evaluated at the average time $t < t_{av} < t + \Delta t$ as density $\times$ velocity perpendicular to area at area center and average time $\times$ area. The conservation principle expressed by Eq. (4-2) is, in terms of the mathematical quantities shown in Fig. 4-2,

$$\frac{\rho \, \Delta x \, \Delta y \, \Delta z \, |_{x, y, z, t+\Delta t} - \rho \, \Delta x \, \Delta y \, \Delta z \, |_{x, y, z, t}}{\Delta t}$$

$$= \rho u \, \Delta y \, \Delta z \, |_{x-\Delta x/2, y, z, t_{av}} - \rho u \, \Delta y \, \Delta z \, |_{x+\Delta x/2, y, z, t_{av}}$$

$$+ \rho v \, \Delta x \, \Delta z \, |_{x, y-\Delta y/2, z, t_{av}} - \rho v \, \Delta x \, \Delta z \, |_{x, y+\Delta y/2, z, t_{av}}$$

$$+ \rho w \, \Delta x \, \Delta y \, |_{x, y, z-\Delta z/2, t_{av}} - \rho w \, \Delta x \, \Delta y \, |_{x, y, z+\Delta z/2, t_{av}}$$

Dividing this equation by $\Delta x \, \Delta y \, \Delta z$ and taking the limit as all differential quantities approach zero, one has, by the definition of a partial derivative,

$$\frac{\partial \rho}{\partial t} + \frac{\partial(\rho u)}{\partial x} + \frac{\partial(\rho v)}{\partial y} + \frac{\partial(\rho w)}{\partial z} = 0 \qquad (4\text{-}9a)$$

which is a conservation-law form. This is exactly Eq. (4-8) as it should be. It is important to fully comprehend this second manner of derivation.

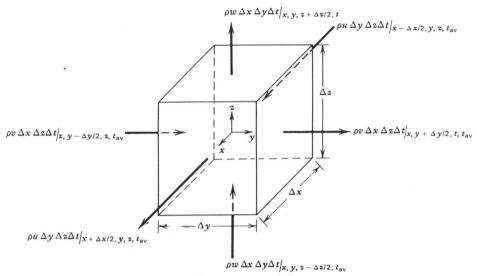

**Figure 4-2** Bulk mass flows across the surfaces of a stationary control volume in rectangular coordinates.

Note that the continuity equation has dimensions of mass per unit volume per unit time. The density is the bulk (average) density of the possibly multicomponent fluid, and the velocity is the bulk (average) macroscopic velocity that gives the bulk mass flow rate—$\mathbf{V} = \Sigma_i \rho_i \mathbf{V}_i / \rho$, where $\rho = \Sigma_i \rho_i$. The continuity equation for cylindrical and spherical coordinates is given in Appendix D.

## 4.3  SUBSTANTIAL DERIVATIVE

The continuity equation for an infinitesimal control volume fixed in space, Eq. (4-7) in general or Eq. (4-8) for rectangular coordinates, can be manipulated into an alternative form that will be convenient later. To see the nature of the manipulation, Eq. (4-8) is considered first. Taking the derivatives of products in Eq. (4-8), one obtains

$$\left( \frac{\partial \rho}{\partial t} + u\frac{\partial \rho}{\partial x} + v\frac{\partial \rho}{\partial y} + w\frac{\partial \rho}{\partial z} \right) + \rho\left( \frac{\partial u}{\partial x} + \frac{\partial v}{\partial y} + \frac{\partial w}{\partial z} \right) = 0$$

This result can be more compactly written as

$$\frac{D\rho}{Dt} + \rho\left( \frac{\partial u}{\partial x} + \frac{\partial v}{\partial y} + \frac{\partial w}{\partial z} \right) = 0 \qquad (4\text{-}10)$$

where the operator

$$\frac{D(\ )}{Dt} = \frac{\partial(\ )}{\partial t} + u\frac{\partial(\ )}{\partial x} + v\frac{\partial(\ )}{\partial y} + w\frac{\partial(\ )}{\partial z} \qquad (4\text{-}11)$$

is the *substantial derivative* in rectangular coordinates.

The general form of the substantial derivative $D/Dt$ can be established by executing manipulations on the general form of the continuity equation [Eq. (4-6)] that parallel those used previously. Equation (4-6) is

$$\frac{\partial \rho}{\partial t} + \nabla \cdot (\rho \mathbf{V}_{rel}) = 0 \qquad (4\text{-}6)$$

whose second term can be separated into two parts as

$$\left( \frac{\partial \rho}{\partial t} + \mathbf{V}_{rel} \cdot \nabla \rho \right) + \rho \nabla \cdot \mathbf{V}_{rel} = 0$$

This result is more compactly written as

$$\frac{D\rho}{Dt} + \rho \operatorname{div}(\mathbf{V}_{rel}) = 0 \qquad (4\text{-}12)$$

which is, of course, still the continuity equation. Here, the general form of the substantial derivative is seen to be

$$\frac{D(\ )}{Dt} = \frac{\partial(\ )}{\partial t} + \mathbf{V}_{rel} \cdot \nabla(\ ) \qquad (4\text{-}13)$$

whose compactness explains its frequent use in convective heat transfer and fluid mechanics. Because the words of Davis [7] are particularly well chosen, they are quoted here:*

Substantial differentiation, which is sometimes called differentiation by pursuit, is a type of differentiation commonly used in the continuum approach to fluid mechanics. However, its common usage does not necessarily mean that it is commonly understood. The inspiration to write this note has stemmed from seeing the substantial derivative derived rather carelessly or vaguely in some textbooks in the field.

To properly understand the derivation of the substantial derivative, one must first have a clear understanding of the two different kinematic descriptions of particle motion due to Leonhard Euler (1707–1783) and Joseph Louis Lagrange (1738–1813). In fact, both descriptions are due to Euler but one was explored so much further by Lagrange that it is now usually attributed to him.

*By permission from P. K. Davis, *Mechanical Engineering News* **2**, 29 (1965).

In the *Eulerian description*, the velocity field is given by,

$$V = V(x, y, z) \tag{4-14}$$

in which the coordinates $x$, $y$, and $z$ are independent of time. In words; if attention is focused on any point $(x, y, z)$ in the flow field, Eq. (4-14) defines the velocity of the various particles passing through the point as time passes. For example, if the probe of an "ideal" hot-wire anemometer were placed at the point $(x, y, z)$, it would read out the velocity according to the Eulerian description of the various particles passing the point. There is no need to permanently identify the particles in this description.

In the *Lagrangian description*, the fluid particles are labeled at an initial instant and then their subsequent histories are followed. The locus of the path of a particle located at $x = a$, $y = b$, and $z = c$ at $t_0$ is subsequently given by

$$X = X(a, b, c, t)$$

$$Y = Y(a, b, c, t)$$

$$Z = Z(a, b, c, t) \tag{4-15}$$

The material coordinates $a$, $b$, and $c$ serve to identify the particle. Once the particle has been chosen and labeled $X$, $Y$, and $Z$ are functions of time alone. The velocity components are given by

$$u = \frac{dX}{dt}, \qquad v = \frac{dY}{dt}, \qquad w = \frac{dZ}{dt} \tag{4-16}$$

The Lagrangian description is commonly employed in solid particle dynamics, but it is difficult to use in fluid mechanics because fluid mechanics problems deal with an infinite number of particles in a flow field and it is simply not practical to attempt to permanently identify them. However, in order to determine the time rate of change of any of the various fluid or flow variables at a point in Eulerian space it is necessary to pursue the particle for a short interval of time. Thus, it is necessary to use both modes of thinking in the derivation of the substantial derivative and this is the crux.

Let $P(x, y, z)$ represent any one of the various flow or fluid variables described in Eulerian space; then let it be required to determine the time rate of change of $P$. The error that is commonly made is to merely take the total derivative according to the chain rule, i.e.

$$\frac{dP(x, y, z, t)}{dt} = \frac{\partial P}{\partial t}\frac{dt}{dt} + \frac{\partial P}{\partial x}\frac{dx}{dt} + \frac{\partial P}{\partial y}\frac{dy}{dt} + \frac{\partial P}{\partial z}\frac{dz}{dt} \tag{4-17}$$

The $dx/dt$, $dy/dt$, and $dz/dt$ are identified as the velocity components and the result is the substantial derivative of $P$. However $x$, $y$, and $z$ are Eulerian coordinates which are independent of time; thus, the time derivatives of these coordinates are meaningless! Such a careless derivation serves only to confuse the student.

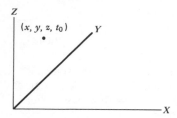

$-X$          **Figure 4-3**   Eulerian space frozen at time $t_0$.

Let us focus our attention on the particle that happens to be located at the point $(x, y, z)$ in Eulerian space at time $t_0$ as shown in Fig. 4-3. Let us now correctly determine the time rate of change of $P$ at the point. Now that the point (particle) is chosen, it is necessary to identify it permanently; but we must identify it for only a very short interval of time and the Eulerian coordinates serve this purpose. We now switch our thinking to Lagrangian space in order to pursue the particle a short distance as shown in Fig. 4-4. By virtue of the continuum, the Taylor series expansion gives a first order approximation of the change in $P$ from time $t_0$ to $t_0 + \Delta t$, i.e.

$$\Delta P = \frac{\partial P}{\partial t}\Delta t + \frac{\partial P}{\partial x}\Delta X + \frac{\partial P}{\partial y}\Delta Y + \frac{\partial P}{\partial z}\Delta Z \qquad (4\text{-}18)$$

Dividing by $\Delta t$ and taking the limit as $\Delta t$ approaches zero yields

$$\lim_{\Delta t \to 0} \frac{\Delta P}{\Delta t} = \frac{\partial P}{\partial t} + u\frac{\partial P}{\partial x} + v\frac{\partial P}{\partial y} + w\frac{\partial P}{\partial z}$$

which is called the substantial derivative of $P$ and is usually written $DP/Dt$. The first term, $\partial P/\partial t$, represents a "local" change, i.e., this would be the total change in $P$ that the particle would experience if it were not moving at all. The remaining three terms represent the "convective" change, i.e. the change in $P$ due to the particle's change in position.

One can see that conceptually it is necessary to distinguish between the Eulerian and Lagrangian coordinates. Writers who neglect to do this, or who look upon this as something too basic to bother with, tend to confuse the student who is just beginning to study the theory of fluid mechanics and, consequently, cause him to look upon this fundamental operation as something difficult to understand. The basic operations and the basic equations should not be treated

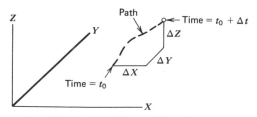

**Figure 4-4**   Lagrangian space.

hastily in fear of oversimplifying because students who have not mastered the basics will later find themselves solving a system of differential equations which they really do not understand and herein lies the danger.

An alternative discussion of the substantial derivative is given by Thorpe [8].


## 4.4   EQUATIONS OF MOTION

The motion of a fluid particle must be influenced by the external forces acting on it, a fact that can be taken into account by use of Newton's second law. Newton's second law for an inertial coordinate system and a system of fixed identity is

$$\mathbf{F} = \frac{m}{g_c} \frac{d\mathbf{V}}{dt} \tag{4-19}$$

where $\mathbf{F}$ is the vector sum of all external forces, $m$ is the total mass of the system of fixed identity, and $\mathbf{V}$ is the center of mass velocity referred to inertial coordinates. The constant of proportionality $1/g_c$ has been explicitly written in Eq. (4-19) as a reminder it must always be included at some stage to make the equation dimensionally consistent. In the English system of units $g_c = 32.2 \text{ lb}_m$ ft/lb$_f$ sec$^2$, and serious numerical errors (in addition to dimensional inconsistency) can result from its omission. In the SI system of units $g_c = 1 \text{ kg m/N}$ s$^2$, and no numerical error results from its omission. To emphasize the importance of momentum changes, Newton's second law can be rewritten in a slightly different form from that of Eq. (4-19) as

$$\mathbf{F} = \frac{d(m\mathbf{V} g_c)}{dt} \tag{4-20}$$

The momentum of the system $m\mathbf{V}/g_c$ is shown by Eq. (4-20) to change with time by the action of the external force. Inasmuch as the change in momentum is stored in the system, it could be said that

$$\mathbf{F} = \overset{\bullet}{\mathbf{momentum}}_{\text{stored}}$$

for a system of fixed identity and that $\mathbf{F}$ acts like a momentum generation term.

Because it is impractical in a moving fluid to permanently identify a system of fixed identity, Newton's second law in the form of Eq. (4-20) is not directly applicable. Rather, it is desired to have a form that is appropriate for a control volume through whose surfaces fluid flows. To this end, a system of fixed identity is identified and followed for only a short time $\Delta t$. Referring to Fig.

4-1, it is seen that, in general, the system momentum is initially equal to

$$\frac{m_1V_1(t)}{g_c} + \frac{m_3V_3(t)}{g_c}$$

and is finally equal to

$$\frac{m_1V_1(t + \Delta t)}{g_c} + \frac{m_2V_2(t + \Delta t)}{g_c}$$

The change in system momentum is, accordingly,

$$\Delta\frac{mV}{g_c} = \frac{[m_1V_1(t + \Delta t) - m_1V_1(t)]}{g_c} + \frac{m_2V_2(t + \Delta t)}{g_c} - \frac{m_3V_3(t)}{g_c}$$

Division by the elapsed time $\Delta t$—assumed to be sufficiently small that region 1 lies within the control volume—followed by substitution into Eq. (4-20) shows that

$$F = \frac{m_1V_1(t + \Delta t) - m_1V(t)}{g_c\,\Delta t} + \frac{m_2V_2(t + \Delta t)}{g_c\,\Delta t} - \frac{m_3V_3(t)}{g_c\,\Delta t}$$

In the limit as $\Delta t \to 0$, region 1 coincides with the control volume. The terms on the right-hand side of Eq. (4-21) are identical in form with those of Eq. (4-1) in the continuity equation derivation and can be similarly interpreted. The only difference is that momentum takes the place of mass. Accordingly, Eq. (4-21) can be stated as

$$\textbf{Force} + \overset{\textbf{.}}{\textbf{momentum}}_{\text{in}} - \overset{\textbf{.}}{\textbf{momentum}}_{\text{out}} = \overset{\textbf{.}}{\textbf{momentum}}_{\text{stored}} \tag{4-22}$$

Again, the dot notation signifies a rate with respect to time. Equation (4-22) is the most convenient general form of Newton's second law for problems involving finite control volumes. It should be remembered that it is a vector equation, having three components.

Equation (4-22) can be stated generally as

$$F - \int_A \frac{V}{g_c}\rho V_{rel} \cdot d\mathbf{A} = \frac{\partial\left[\int_V (\rho V/g_c)\,dV\right]}{\partial t} \tag{4-23}$$

The second term on the left-hand side is the net efflux of momentum, which is seen to be momentum per unit mass × mass flow rate. The vector identity

$$\int_A \mathbf{y}(\mathbf{x} \cdot d\mathbf{A}) = \int_V [\mathbf{y}\,\text{div}(\mathbf{x}) + (\mathbf{x} \cdot \nabla)\mathbf{y}]\,dV$$

allows a surface integral to be expressed as an integral over the volume enclosed by the surface. Its use allows Eq. (4-23) to be rewritten as

$$\mathbf{F} = \frac{\partial\left[\int_V (\rho\mathbf{V}/g_c)\, dV\right]}{\partial t} + \int_V \frac{(\mathbf{V}/g_c)\,\mathrm{div}(\rho\mathbf{V}_{\mathrm{rel}}) + (\rho\mathbf{V}_{\mathrm{rel}} \cdot \nabla)\mathbf{V}}{g_c}\, dV \quad (4\text{-}24)$$

If the control volume is of fixed shape, the order of differentiation and integration can be interchanged to give

$$\mathbf{F} = \int_V [\partial(\rho\mathbf{V})\partial t + \mathbf{V}\,\mathrm{div}(\rho\mathbf{V}_{\mathrm{rel}}) + (\rho\mathbf{V}_{\mathrm{rel}} \cdot \nabla)\mathbf{V}]\frac{dV}{g_c}$$

If the control volume is of infinitesimal size, experiencing no appreciable property variation over its extent, this relation is well approximated by

$$\frac{g_c\mathbf{F}}{dV} = \frac{\partial(\rho\mathbf{V})}{\partial t} + \mathbf{V}\,\mathrm{div}(\rho\mathbf{V}_{\mathrm{rel}}) + (\rho\mathbf{V}_{\mathrm{rel}} \cdot \nabla)\mathbf{V}$$

which is a conservation-law form. This equation can be rearranged as

$$\frac{g_c\mathbf{F}}{dV} = \mathbf{V}\left[\frac{\partial\rho}{\partial t} + \mathrm{div}(\rho\mathbf{V}_{\mathrm{rel}})\right] + \rho\left[\frac{\partial\mathbf{V}}{\partial t} + (\mathbf{V}_{\mathrm{rel}} \cdot \nabla)\mathbf{V}\right]$$

The first bracketed term on the right-hand side is identically zero from the continuity equation [Eq. (4-6)]. Thus Newton's second law applied to an infinitesimal control volume of fixed shape is

$$\frac{\mathbf{F}}{dV} = \rho\left[\frac{\partial(\mathbf{V}/g_c)}{\partial t} + (\mathbf{V}_{\mathrm{rel}} \cdot \nabla)\frac{\mathbf{V}}{g_c}\right] \quad (4\text{-}25)$$

The substantial derivative notation, generally given by Eq. (4-13), gives a more compact form of Eq. (4-25) as

$$\frac{\mathbf{F}}{dV} = \rho\frac{D(\mathbf{V}/g_c)}{Dt} \quad (4\text{-}26)$$

Equation (4-26) is identical in form with Newton's second law for a system of fixed identity, as comparison with Eq. (4-19) shows. Because of the demonstrated similarity some derivations of the equations of motion merely adopt Eq. (4-19) and identify the derivative as the "derivative following the motion," termed the *substantial derivative* here. Such a derivation, although it has speed and brevity to commend it, is not really correct, as was discussed in Section 4.2. Both Eqs. (4-25) and (4-26) suggest that force/volume $\mathbf{F}/dV$ is required to affect the transported quantity, which is momentum/mass $\mathbf{V}/g_c$. Having been derived without specification of a coordinate system, Eq. (4-25) is applicable to any coordinate system.

In the case for which the control volume is stationary, $\mathbf{V}_{area} = 0$; thus $\mathbf{V}_{rel} = \mathbf{V}_{fluid}$. If subscripts are dropped so that $\mathbf{V}$ represents only the fluid velocity, Eq. (4-25) becomes

$$\frac{\mathbf{F}}{dV} = \rho \frac{\partial \mathbf{V}/\partial t + (\mathbf{V} \cdot \nabla)\mathbf{V}}{g_c} \tag{4-27}$$

In rectangular coordinates for which $\mathbf{F} = F_x \hat{\mathbf{i}} + F_y \hat{\mathbf{j}} + F_z \hat{\mathbf{k}}$ and $\mathbf{V} = u\hat{\mathbf{i}} + v\hat{\mathbf{j}} + w\hat{\mathbf{k}}$, Eq. (4-27) becomes

$$\frac{F_x}{dV} = \rho \frac{\partial u/\partial t + u\,\partial u/\partial x + v\,\partial u/\partial y + w\,\partial u/\partial z}{g_c} \tag{4-28a}$$

$$\frac{F_y}{dV} = \rho \frac{\partial v/\partial t + u\,\partial v/\partial x + v\,\partial v/\partial y + w\,\partial v/\partial z}{g_c} \tag{4-28b}$$

$$\frac{F_z}{dV} = \rho \frac{\partial w/\partial t + u\,\partial w/\partial x + v\,\partial w/\partial y + w\,\partial w/\partial z}{g_c} \tag{4-28c}$$

As in the derivation of the continuity equation, it is instructive to first make all assumptions and then to apply the conservation principle. It is, therefore, assumed from the beginning that the control volume is a cube fixed in space with rectangular coordinates as shown in Fig. 4-5. The rate at which a momentum component flows through one of the cube faces is evaluated at the average time $t \leqslant t_{av} \leqslant t + \Delta t$ as

Momentum component/mass × mass flux at area center and average time × area

where the flow is observed for the time interval between $t$ and $t + \Delta t$. With the $x, y$, and $z$ velocity components being $u, v$, and $w$, respectively, and focusing on the $x$ component of momentum whose flow through each face is shown in Fig. 4-5, application of the momentum principle expressed by Eq. (4-22) gives

$$F_x = \frac{\left[(u/g_c)\rho\,\Delta x\,\Delta y\,\Delta z\,|_{x,y,z,t+\Delta t} - (u/g_c)\rho\,\Delta x\,\Delta y\,\Delta z\,|_{x,y,z,t}\right]}{\Delta t}$$

$$+ \frac{u}{g_c}\rho u\,\Delta y\,\Delta z\,|_{x+\Delta x/2,y,z,t_{av}} - \frac{u}{g_c}\rho u\,\Delta y\,\Delta z\,|_{x-\Delta x/2,y,z,t_{av}}$$

$$+ \frac{u}{g_c}\rho v\,\Delta x\,\Delta z\,|_{x,y+\Delta y/2,z,t_{av}} - \frac{u}{g_c}\rho v\,\Delta x\,\Delta z\,|_{x,y-\Delta y/2,z,t_{av}}$$

$$+ \frac{u}{g_c}\rho w\,\Delta x\,\Delta y\,|_{x,y,z+\Delta z/2,t_{av}} - \frac{u}{g_c}\rho w\,\Delta x\,\Delta y\,|_{x,y,z-\Delta z/2,t_{av}}$$

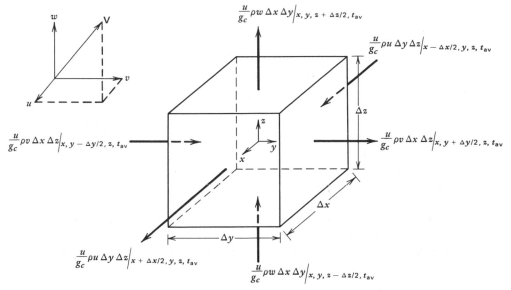

**Figure 4-5** x-Momentum flows across the surfaces of a stationary control volume in rectangular coordinates.

Dividing by $\Delta x \, \Delta y \, \Delta z$ and taking the limit as all differential quantities approach zero one obtains, by the definition of a partial derivative,

$$\frac{g_c F_x}{dx \, dy \, dz} = \frac{\partial(\rho u)}{\partial t} + \frac{\partial(\rho u u)}{\partial x} + \frac{\partial(\rho u v)}{\partial y} + \frac{\partial(\rho u w)}{\partial z}$$

Taking the derivatives of the products, one then has

$$\frac{g_c F_x}{dx \, dy \, dz} = u\left[\frac{\partial \rho}{\partial t} + \frac{\partial(\rho u)}{\partial x} + \frac{\partial(\rho v)}{\partial y} + \frac{\partial(\rho w)}{\partial z}\right]$$

$$+ \rho\left[\frac{\partial u}{\partial t} + u\frac{\partial u}{\partial x} + v\frac{\partial u}{\partial y} + w\frac{\partial u}{\partial z}\right]$$

The first bracketed term on the right-hand side of this equation is identically zero according to the continuity equation [Eq. (4-9a)]. Hence the result is

$$\frac{F_x}{dx \, dy \, dz} = \frac{\rho(\partial u/\partial t + u\,\partial u/\partial x + v\,\partial u/\partial y + w\,\partial u/\partial z)}{g_c}$$

which is the same as Eq. (4-28a), as it should be.

The external forces acting on the control volume greatly influence the fluid motion and must be dealt with in a detailed fashion. These forces either act on the surface of the control volume and are called *surface forces*, or they act on all the distributed mass in the control volume and are called *body forces*. Shear and pressure give rise to surface forces, for example, whereas gravity gives rise to a body force. Because of the somewhat complex manner in which these external forces act on a control volume, a general treatment is not undertaken in this first discussion. For clarity, attention is instead first directed to a stationary cubical control volume in rectangular coordinates. As illustrated in Fig. 4-6, the surface force acting on a face is composed of three components as is also the case for the body force. A surface force is given by the product of the force/area, denoted by $\tau$, and area. The stress on a face is denoted by $\tau_{ij}$ in which the convention is adopted that the first subscript $i$ designates the face on which the stress acts and the second subscript $j$ designates the direction in which the stress acts. On a face whose outward directed normal points in a positive coordinate direction, the positive stress points in a positive coordinate direction; if the outward directed surface normal points in a negative coordinate direction, so does the positive stress. The body force is given by the product of the force/volume, denoted by $B$, and volume. The net $x$ component of the external force is then

$$F_x = \left(\tau_{xx}\big|_{x+\Delta x/2,\, y,\, z,\, t_{av}} - \tau_{xx}\big|_{x-\Delta x/2,\, y,\, z,\, t_{av}}\right)\Delta y \Delta z$$
$$+ \left(\tau_{yx}\big|_{x,\, y+\Delta y/2,\, z,\, t_{av}} - \tau_{yx}\big|_{x,\, y-\Delta y/2,\, z,\, t_{av}}\right)\Delta x \Delta z$$
$$+ \left(\tau_{zx}\big|_{x,\, y,\, z+\Delta z/2,\, t_{av}} - \tau_{zx}\big|_{x,\, y,\, z-\Delta z/2,\, t_{av}}\right)\Delta x \Delta y$$
$$+ B_x\big|_{x,\, y,\, z,\, t_{av}}\Delta x \Delta y \Delta z$$

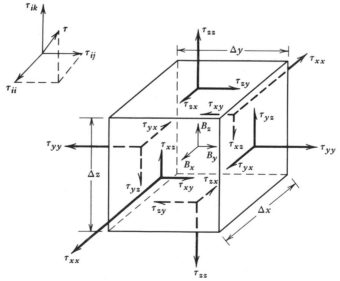

**Figure 4-6** Surface and body-force components acting on a stationary control volume in rectangular coordinates.

in which the possibility that quantities are time dependent is taken into account by evaluating them at an average time $t \leqslant t_{av} \leqslant t + \Delta t$, which lies between the start of the observation period $t$ and its end $t + \Delta t$. Dividing by $\Delta x \, \Delta y \, \Delta z$ and taking the limit as all differential quantities approach zero, one obtains

$$\frac{F_x}{dV} = B_x + \frac{\partial \tau_{xx}}{\partial x} + \frac{\partial \tau_{yx}}{\partial y} + \frac{\partial \tau_{zx}}{\partial z} \qquad (4\text{-}29)$$

where $dV = dx \, dy \, dz$. Substitution of Eq. (4-19) into the $x$ equation of motion [Eq. (4-28a)] gives

$$\rho \frac{D(u/g_c)}{Dt} = B_x + \frac{\partial \tau_{xx}}{\partial x} + \frac{\partial \tau_{yx}}{\partial y} + \frac{\partial \tau_{zx}}{\partial z} \qquad (4\text{-}30a)$$

In similar fashion it is found for the $y$ and $z$ equations of motion in rectangular coordinates that

$$\rho \frac{D(v/g_c)}{Dt} = B_y + \frac{\partial \tau_{xy}}{\partial x} + \frac{\partial \tau_{yy}}{\partial y} + \frac{\partial \tau_{zy}}{\partial z} \qquad (4\text{-}30b)$$

$$\rho \frac{D(w/g_c)}{Dt} = B_z + \frac{\partial \tau_{xz}}{\partial x} + \frac{\partial \tau_{yz}}{\partial y} + \frac{\partial \tau_{zz}}{\partial z} \qquad (4\text{-}30c)$$

for which the substantial derivative notation is defined by Eq. (4-11).

The number of unknowns in Eqs. (4-30) must be reduced if a solution is to eventually be achieved, and information from an independent source must be provided for that purpose. It is observed that stresses in solids are related to displacements, whereas stresses in fluids are related to velocities (the rates at which displacements occur). This difference prevents the simple development of equations of motion that are applicable to both solids and fluids. Because interest here is centered on the convection that is peculiar to fluids, attention is directed to fluids.

For most important fluids, such as air and water, there is a linear relationship between stresses and velocities; such fluids are called *Newtonian fluids*. Fluids having a nonlinear relation between stresses and velocities are called *non-Newtonian fluids*. The major types of relationship between stresses and velocities are illustrated in Fig. 4-7 for a simple flow situation. Examples of shear-thickening (dilatant) fluids are water–starch solutions and quicksand. Some shear-thinning (pseudoplastic) fluids are mayonnaise and many printer's inks. Toothpaste and drilling mud are examples of Bingham plastic fluids. In yet another class are shear-thinning fluids in which the viscosity depends on the length of time shear has been applied; these are called *thixotropic fluids* and are exemplified by margarine and catsup. A rheopectic fluid is one that exhibits negative thixotropy; its viscosity increases with time as shear is applied. A further class of non-Newtonian fluids displays elastic as well as

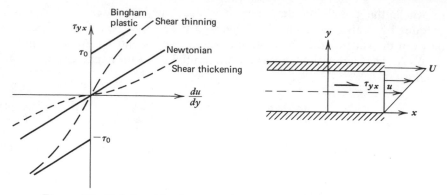

**Figure 4-7** Relationships between shear stress and velocity gradient.

viscous properties; examples are the thick portion of egg white, STP Oil Treatment, and some condensed soups. Additional discussion is given in Appendix C.

For a Newtonian fluid, the linear relation between stresses and velocities (shown in detail in Appendix C) in rectangular coordinates is

$$\tau_{xy} = \mu\left(\frac{\partial u}{\partial y} + \frac{\partial v}{\partial x}\right) = \tau_{yx} \tag{4-31a}$$

$$\tau_{yz} = \mu\left(\frac{\partial v}{\partial z} + \frac{\partial w}{\partial y}\right) = \tau_{zy} \tag{4-31b}$$

$$\tau_{xz} = \mu\left(\frac{\partial w}{\partial x} + \frac{\partial u}{\partial z}\right) = \tau_{zx} \tag{4-31c}$$

$$\tau_{xx} = -p + \mu\left(2\frac{\partial u}{\partial x} - \frac{2}{3}\operatorname{div}\mathbf{V}\right) \tag{4-31d}$$

$$\tau_{yy} = -p + \mu\left(2\frac{\partial v}{\partial y} - \frac{2}{3}\operatorname{div}\mathbf{V}\right) \tag{4-31e}$$

$$\tau_{zz} = -p + \mu\left(2\frac{\partial w}{\partial z} - \frac{2}{3}\operatorname{div}\mathbf{V}\right) \tag{4-31f}$$

The viscosity $\mu$ appearing in these relations is a property of the fluid and originates in random molecular motion. The stress normal to a surface $\tau_{ii}$ is composed of the sum of pressure stress $p$ and viscous stress (which is not always a shear stress) as

$$\tau_{ii} = -p + \tau_{ii_{\text{viscous}}}$$

If the fluid is inviscid, $\mu \equiv 0$, so that there cannot be any viscous stress

$$\tau_{ii} = -p$$

as would be expected. Equations (4-31d–f) added show that for a viscous fluid pressure can be defined as

$$p = -\frac{(\tau_{xx} + \tau_{yy} + \tau_{zz})}{3}$$

which conforms to the expectation one has for inviscid fluids.

Introducing the experimentally verified relations of Eqs. (4-31) into the equations of motion that follow from Newton's second law [Eqs. (4-30)], one obtains

$$\frac{\rho}{g_c}\frac{Du}{Dt} = B_x - \frac{\partial p}{\partial x} + \frac{\partial}{\partial x}\left[\mu\left(2\frac{\partial u}{\partial x} - \frac{2}{3}\operatorname{div}\mathbf{V}\right)\right] + \frac{\partial}{\partial y}\left[\mu\left(\frac{\partial u}{\partial y} + \frac{\partial v}{\partial x}\right)\right]$$

$$+ \frac{\partial}{\partial z}\left[\mu\left(\frac{\partial w}{\partial x} + \frac{\partial u}{\partial z}\right)\right] \tag{4-32a}$$

$$\frac{\rho}{g_c}\frac{Dv}{Dt} = B_y - \frac{\partial p}{\partial y} + \frac{\partial}{\partial x}\left[\mu\left(\frac{\partial u}{\partial y} + \frac{\partial v}{\partial x}\right)\right] + \frac{\partial}{\partial y}\left[\mu\left(2\frac{\partial v}{\partial y} - \frac{2}{3}\operatorname{div}\mathbf{V}\right)\right]$$

$$+ \frac{\partial}{\partial z}\left[\mu\left(\frac{\partial v}{\partial z} + \frac{\partial w}{\partial y}\right)\right] \tag{4-32b}$$

$$\frac{\rho}{g_c}\frac{Dw}{Dt} = B_z - \frac{\partial p}{\partial z} + \frac{\partial}{\partial x}\left[\mu\left(\frac{\partial w}{\partial x} + \frac{\partial u}{\partial z}\right)\right] + \frac{\partial}{\partial y}\left[\mu\left(\frac{\partial v}{\partial z} + \frac{\partial w}{\partial y}\right)\right]$$

$$+ \frac{\partial}{\partial z}\left[\mu\left(2\frac{\partial w}{\partial z} - \frac{2}{3}\operatorname{div}\mathbf{V}\right)\right] \tag{4-32c}$$

for rectangular coordinates.

When viscosity is constant, Eqs. (4-32) assume a simpler form as can be shown through the example of the $x$ direction equation of motion. Expansion of the derivatives in Eq. (4-32a) with $\mu$ constant gives

$$\frac{\rho}{g_c}\frac{Du}{Dt} = B_x - \frac{\partial p}{\partial x} + \mu\left\{\left[\frac{\partial(\partial u/\partial x)}{\partial x} + \frac{\partial(\partial u/\partial y)}{\partial y} + \frac{\partial(\partial u/\partial z)}{\partial z}\right]\right.$$

$$\left. + \left[\frac{\partial(\partial u/\partial x)}{\partial x} + \frac{\partial(\partial v/\partial x)}{\partial y} + \frac{\partial(\partial w/\partial x)}{\partial z}\right] - \frac{2}{3}\frac{\partial\operatorname{div}(\mathbf{V})}{\partial x}\right\}$$

The second bracketed term inside the braces is recognized as being just $\operatorname{div}(V)$

if the order of differentiation is interchanged for each term. Then the $x$-direction equation of motion is

$$\frac{\rho}{g_c}\frac{Du}{Dt} = B_x - \frac{\partial p}{\partial x} + \mu\left[\frac{\partial^2 u}{\partial x^2} + \frac{\partial^2 u}{\partial y^2} + \frac{\partial^2 u}{\partial z^2} + \frac{1}{3}\frac{\partial\,\mathrm{div}\,(\mathbf{V})}{\partial x}\right] \qquad (4\text{-}33)$$

The equations of motion for the other two directions are of the same form. In general the constant viscosity case can be expressed as

$$\frac{\rho}{g_c}\frac{D\mathbf{V}}{Dt} = \mathbf{B} - \nabla p + \mu\left[\nabla^2\mathbf{V} + \frac{1}{3}\nabla(\nabla\cdot\mathbf{V})\right]$$

If density is also constant, the continuity equation [Eq. (4-7)] shows that $\mathrm{div}(\mathbf{V}) = 0$. Then the equations of motion in rectangular coordinates are

$$\frac{\rho}{g_c}\frac{Du}{Dt} = B_x - \frac{\partial p}{\partial x} + \mu\left(\frac{\partial^2 u}{\partial x^2} + \frac{\partial^2 u}{\partial y^2} + \frac{\partial^2 u}{\partial z^2}\right) \qquad (4\text{-}34a)$$

$$\frac{\rho}{g_c}\frac{Dv}{Dt} = B_y - \frac{\partial p}{\partial y} + \mu\left(\frac{\partial^2 v}{\partial x^2} + \frac{\partial^2 v}{\partial y^2} + \frac{\partial^2 v}{\partial z^2}\right) \qquad (4\text{-}34b)$$

$$\frac{\rho}{g_c}\frac{Dw}{Dt} = B_z - \frac{\partial p}{\partial z} + \mu\left(\frac{\partial^2 w}{\partial x^2} + \frac{\partial^2 w}{\partial y^2} + \frac{\partial^2 w}{\partial z^2}\right) \qquad (4\text{-}34c)$$

This set of equations describing the motion of a constant density and viscosity fluid can be more compactly written in vector notation as

$$\frac{\rho}{g_c}\frac{D\mathbf{V}}{Dt} = \mathbf{B} - \nabla p + \mu\nabla^2\mathbf{V} \qquad (4\text{-}35)$$

Equations of motion for cylindrical and spherical coordinate systems are given in Appendix D. It is usually most convenient to obtain the equations of motion by a coordinate transformation of the equations for rectangular coordinates, rather than by applying conservation principles directly to an often awkwardly shaped control volume in nonrectangular coordinates.

## 4.5   ENERGY EQUATION

The conservation of energy principle is most familiar in the form that applies to a system of fixed identity (i.e., a system always composed of the same particles of mass). Also known as the *first law of thermodynamics*, it is expressed in mathematical terms as

$$Q_{\mathrm{in}} = \Delta E + W_{\mathrm{out}} \qquad (4\text{-}36)$$

where $Q_{in}$ is the heat transferred into the system from its surroundings, $W_{out}$ is the work done by the system on its surroundings, and $\Delta E$ is the change in the system internal energy. It must be remembered that the mass of the system can store energy internally in a number of forms; thus the internal energy is

$$E = E_{kinetic} + E_{\substack{potential \\ due\ to\ body\ force}} + E_{surface\ tension} + E_{electromagnetic} + E_{thermal}$$

For the developments of this chapter, the possible contributions of surface tension, electromagnetic, and other energy forms are taken as unchanged by the physical processes encountered; since they do not change, they can be neglected. Also, the potential energy due to location in a body force field is considered to be the result of work done by the body force, a permissible though arbitrary decision. With these choices in mind, $E = E_{kinetic} + E_{thermal}$ in the remainder of this chapter.

To cast the conservation of energy principle into the form needed for application to a control volume through which fluid flows, Eq. (4-36) is applied to the system of fixed identity illustrated in Fig. 4-1 for a brief time $\Delta t$. During this time interval the internal energy change of the system is

$$\Delta E = [E_1(t + \Delta t) + E_2(t + \Delta t)] - [E_1(t) + E_3(t)]$$

Insertion of this result into Eq. (4-36) and division by $\Delta t$ gives

$$\frac{Q_{in}}{\Delta t} - \frac{W_{out}}{\Delta t} = \frac{E_1(t + \Delta t) - E_1(t)}{\Delta t} + \frac{E_2(t + \Delta t)}{\Delta t} - \frac{E_3(t)}{\Delta t}$$

In the limit as $\Delta t$ approaches zero the terms on the right-hand side of this equation can be interpreted as the rate of internal energy leaving and entering, respectively, as explained in detail during the continuity equation derivation. Thus the conservation of energy principle for a control volume can be expressed as

$$\dot{Q}_{in} - \dot{W}_{out} = \text{internal energy}_{stored} - \text{internal energy}_{in} + \text{internal energy}_{out}$$

$$(4\text{-}37)$$

The dot notation signifies a rate with respect to time. Equation (4-37) is the most useful form of the conservation of energy principle for problems involving finite control volumes.

Equation (4-37) can be stated generally in mathematical terms as

$$\dot{Q}_{in} - \dot{W}_{out} = \frac{d\left(\int_V e\rho\, dV\right)}{dt} + \int_A e(\rho \mathbf{V}_{rel} \cdot d\mathbf{A})$$

in which $e$ represents internal energy per unit mass. The divergence theorem permits the area integral to be cast as a volume integral, giving

$$\dot{Q}_{in} - \dot{W}_{out} = \frac{d\left(\int_V e\rho \, dV\right)}{dt} + \int_V \text{div}(e\rho \mathbf{V}_{rel}) \, dV$$

For a control volume of fixed shape, the order of integration and differentiation can be interchanged in the preceding equation to obtain

$$\dot{Q}_{in} - \dot{W}_{out} = \int_V \left[\frac{\partial(e\rho)}{\partial t} + \text{div}(e\rho \mathbf{V}_{rel})\right] dV$$

For an infinitesimal control volume, this relation becomes

$$\frac{\partial(e\rho)}{\partial t} + \text{div}(e\rho \mathbf{V}_{rel}) = \frac{\dot{Q}_{in}}{dV} - \frac{\dot{W}_{out}}{dV} \tag{4-38}$$

which is a conservation-law form. Equation (4-38) can be rearranged into

$$e\left[\frac{\partial \rho}{\partial t} + \text{div}(\rho \mathbf{V}_{rel})\right] + \rho\left[\frac{\partial e}{\partial t} + \mathbf{V}_{rel} \cdot \nabla e\right] = \frac{\dot{Q}_{in}}{dV} - \frac{\dot{W}_{out}}{dV}$$

The first bracketed term identically equals zero by the continuity equation [Eq. (4-6)], whereas the second bracketed term can be written as $De/Dt$ by use of the substantial derivative defined by Eq. (4-13). With these facts employed, Eq. (4-38) becomes

$$\rho \frac{De}{Dt} = \frac{\dot{Q}_{in}}{dV} - \frac{\dot{W}_{out}}{dV} \tag{4-39}$$

The two terms on the right-side of Eq. (4-39) must now be evaluated. The heat term $\dot{Q}_{in}$ is treated first. A general treatment without specification of a coordinate system is parallel to the energy equation derivation presented thus far. Heat enters the control volume by diffusion relative to the bulk flow due to random molecular motion. It is also possible for some energy that entered the control volume in chemical, atomic, or electromagnetic forms (which were assumed to be negligible earlier) to be converted into a thermal form of energy by way of what acts like an internally distributed heat source—it really just accounts for things not specifically enumerated. Hence

$$\dot{Q}_{in} = -\int_A \mathbf{q} \cdot d\mathbf{A} + \int_V q''' \, dV$$

in which $\mathbf{q}$ is the diffusive flux of heat and $q'''$ is the rate at which thermal energy is liberated per unit volume. Appealing once again to the divergence theorem, we see that

$$\dot{Q}_{in} = \int_V [q''' - \text{div}(\mathbf{q})] \, dV$$

For an infinitesimal control volume, this relationship becomes

$$\frac{\dot{Q}_{in}}{dV} = q''' - \text{div}(\mathbf{q})$$

Insertion of this result into Eq. (4-39) gives the energy equation for a nondeforming infinitesimal control volume as

$$\rho \frac{De}{Dt} + \text{div}(\mathbf{q}) - q''' = \frac{-\dot{W}_{out}}{dV} \tag{4-40}$$

In rectangular coordinates, Eq. (4-40) would be written in full as

$$\rho \left( \frac{\partial e}{\partial t} + u\frac{\partial e}{\partial x} + v\frac{\partial e}{\partial y} + w\frac{\partial e}{\partial z} \right) + \left( \frac{\partial q_x}{\partial x} + \frac{\partial q_y}{\partial y} + \frac{\partial q_z}{\partial z} \right) - q''' = \frac{-\dot{W}_{out}}{dx \, dy \, dz}$$

$$\tag{4-41}$$

Although the foregoing derivation of Eq. (4-40) is quite general and is independent of any coordinate system, it does tend to obscure the important point that

Total energy flux = bulk convection + diffusion relative to bulk motion

To amplify this point, the foregoing development is repeated for a stationary control volume in rectangular coordinates. The energy flows across surfaces of the control volume are shown in Fig. 4-8. The energy conservation principle of Eq. (4-37) accepts the terms partially enumerated in Fig. 4-8 to show

$$\left( e\rho|_{x,y,z,t+\Delta t} - e\rho|_{x,y,z,t} \right) \frac{\Delta x \, \Delta y \, \Delta z}{\Delta t}$$
$$+ \left[ (e\rho u + q_x)|_{x+\Delta x/2, y, z, t_{av}} - (e\rho u + q_x)|_{x-\Delta x/2, y, z, t_{av}} \right] \Delta y \, \Delta z$$
$$+ \left[ (e\rho v + q_y)|_{x, y+\Delta y/2, z, t_{av}} - (e\rho v + q_y)|_{x, y-\Delta y/2, z, t_{av}} \right] \Delta x \, \Delta z$$
$$+ \left[ (e\rho w + q_z)|_{x, y, z+\Delta z/2, t_{av}} - (e\rho w + q_z)|_{x, y, z-\Delta z/2, t_{av}} \right] \Delta x \, \Delta y$$
$$- q'''|_{x, y, z, t_{av}} \Delta x \, \Delta y \, \Delta z = -\dot{W}_{out}$$

in which $t \leq t_{av} \leq t + \Delta t$. With division by $\Delta x \, \Delta y \Delta z$, and if the limit is taken

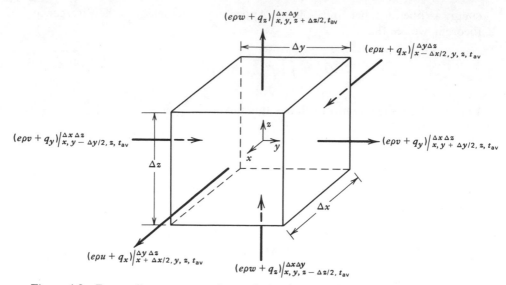

**Figure 4-8**   Energy flows across surfaces of a stationary control volume in rectangular coordinates.

as all differential quantities approach zero, the above relationship is put into the form

$$\frac{\partial(e\rho)}{\partial t} + \frac{\partial(e\rho u + q_x)}{\partial x} + \frac{\partial(e\rho v + q_y)}{\partial y} + \frac{\partial(e\rho w + q_z)}{\partial z} - q''' = \frac{-\dot{W}_{out}}{dx\,dy\,dz}$$

This result can be rearranged, by taking derivatives of products and sums, into

$$\rho\left(\frac{\partial e}{\partial t} + u\frac{\partial e}{\partial x} + v\frac{\partial e}{\partial x} + w\frac{\partial e}{\partial z}\right) + \left(\frac{\partial q_x}{\partial x} + \frac{\partial q_y}{\partial y} + \frac{\partial q_z}{\partial z}\right)$$

$$+ e\left(\frac{\partial\rho}{\partial t} + \frac{\partial\rho u}{\partial x} + \frac{\partial\rho v}{\partial y} + \frac{\partial\rho w}{\partial z}\right) - q''' = \frac{-\dot{W}_{out}}{dx\,dy\,dz}$$

The third term in parentheses on the right-hand side is identically zero, according to the continuity equation [Eq. (4-8)], so that Eq. (4-41) is duplicated as it should be. Although this derivation is of less generality than that leading to Eq. (4-40), it desirably emphasizes the fact that diffusion provides a flux relative to bulk convection.

The work term $-\dot{W}_{out}$ on the right-hand side of Eq. (4-40) is evaluated next. To do this without specifying a coordinate system, it is necessary to accept the fact that the stresses acting on the control volume surface (as illustrated in Fig. 4-6) are tensors of the second order. (A scalar is a tensor of zeroth order, and a vector is a tensor of first order.) It is necessary to become familiar with tensor notation, which requires more time and effort to describe than is appropriate here. Clarity of insight into the physical processes involved is most easily achieved by considering a stationary control volume in rectangular coordinates. Recognize that

$$ -\dot{W}_{out} = \dot{W}_{in} $$

which is the work done by the environment on the mass inside the control volume. This work is the scalar product of the force exerted by the environment and the velocity of the particle exerting the force. Since the particle exerting the force on the control volume is itself touching the control volume, its velocity is the fluid velocity at the control volume surface. Figure 4-9 shows the $\dot{W}_{in}$ at two control volume surfaces (note that the rate of work input is negative when force and velocity are oppositely directed); similar terms appear for the other four faces. The body force is shown as doing work, also; this is permissible on physical grounds, although it means that the potential energy due to position in the force field cannot then also be considered to be a part of internal energy of the mass in the control volume. Collecting terms, dividing by $\Delta x \Delta y \Delta z$, and taking the limit as all differential quantities approach zero (again, $t \leqslant t_{av} \leqslant t + \Delta t$), one obtains

$$
\begin{aligned}
&+ \frac{\partial(\tau_{xx}u)}{\partial x} + \frac{\partial(\tau_{yx}u)}{\partial y} + \frac{\partial(\tau_{zx}u)}{\partial z} + B_x u \\
-\dot{W}_{out}/dV = &+ \frac{\partial(\tau_{xy}v)}{\partial x} + \frac{\partial(\tau_{yy}v)}{\partial y} + \frac{\partial(\tau_{zy}v)}{\partial z} + B_y v \qquad (4\text{-}42) \\
&+ \frac{\partial(\tau_{xz}w)}{\partial x} + \frac{\partial(\tau_{yz}w)}{\partial y} + \frac{\partial(\tau_{zz}w)}{\partial z} + B_z w
\end{aligned}
$$

which is a conservation law form. Expansion and collection of terms gives,

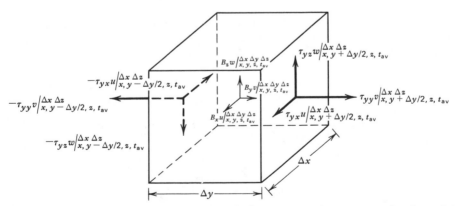

**Figure 4-9** Work done by surface and body forces on stationary control volume in rectangular coordinates.

with Eq. (4-30) taken into account,

$$
\begin{aligned}
-\frac{\dot{W}_{out}}{dV} = \quad & u\left(\frac{\partial \tau_{xx}}{\partial x} + \frac{\partial \tau_{yx}}{\partial y} + \frac{\partial \tau_{zx}}{\partial z} + B_x\right)\frac{\rho}{g_c}\frac{Du}{Dt} \\[2mm]
& + v\left(\frac{\partial \tau_{xy}}{\partial x} + \frac{\partial \tau_{yy}}{\partial y} + \frac{\partial \tau_{zy}}{\partial z} + B_y\right)\frac{\rho}{g_c}\frac{Dv}{Dt} \\[2mm]
& + w\left(\frac{\partial \tau_{xz}}{\partial x} + \frac{\partial \tau_{yz}}{\partial y} + \frac{\partial \tau_{zz}}{\partial z} + B_z\right)\frac{\rho}{g_c}\frac{Dw}{Dt} \\[2mm]
& + \tau_{xx}\frac{\partial u}{\partial x} + \tau_{yx}\frac{\partial u}{\partial y} + \tau_{zx}\frac{\partial u}{\partial z} \\[2mm]
& + \tau_{xy}\frac{\partial v}{\partial x} + \tau_{yy}\frac{\partial v}{\partial y} + \tau_{zy}\frac{\partial v}{\partial z} \\[2mm]
& + \tau_{xz}\frac{\partial w}{\partial x} + \tau_{yz}\frac{\partial w}{\partial y} + \tau_{zz}\frac{\partial w}{\partial z}
\end{aligned}
\tag{4-43}
$$

The first three terms of Eq. (4-43) constitute $\rho D(\text{kinetic energy}/\text{mass})/Dt$ as can be seen by summing them as

$$
\frac{\rho}{g_c}\left(u\frac{Du}{Dt} + v\frac{Dv}{Dt} + w\frac{Dw}{Dt}\right) = \frac{\rho}{2g_c}\left(\frac{Du^2}{Dt} + \frac{Dv^2}{Dt} + \frac{Dw^2}{Dt}\right)
$$

$$
= \rho\frac{D}{Dt}\left(\frac{u^2 + v^2 + w^2}{2g_c}\right)
$$

$$
= \rho\frac{D}{Dt}\left(\frac{|\mathbf{V}|^2}{2g_c}\right)
$$

Introducing the preceding expression for $-\dot{W}/dV$ into Eq. (4-40) and letting the internal energy per unit mass due to temperature be $I = e - $ kinetic energy per unit mass $= e - |\mathbf{V}|^2/2g_c$, one finds that

$$
\begin{aligned}
\rho\frac{DI}{Dt} + \text{div}(\mathbf{q}) - q''' = \quad & \tau_{xx}\frac{\partial u}{\partial x} + \tau_{yx}\frac{\partial u}{\partial y} + \tau_{zx}\frac{\partial u}{\partial z} \\[2mm]
& + \tau_{xy}\frac{\partial v}{\partial x} + \tau_{yy}\frac{\partial v}{\partial y} + \tau_{zy}\frac{\partial v}{\partial z} \\[2mm]
& + \tau_{xz}\frac{\partial w}{\partial x} + \tau_{yz}\frac{\partial w}{\partial y} + \tau_{zz}\frac{\partial w}{\partial z}
\end{aligned}
\tag{4-44}
$$

Note that both kinetic and potential forms of energy have been considered and have been shown to be absent from the "energy" equation [Eq. (4-44)]. As it stands, Eq. (4-44) is applicable to all fluids and solids, but it is no longer truly an energy equation.

Surface stresses $\tau_{ij}$ are rarely specified directly (largely because it is almost impossible to measure a stress in the interior of a fluid), so it is necessary to relate them to the velocities that are more commonly specified (largely because they are more easily measured). Although necessary, this step introduces a restriction on the generality of Eq. (4-44). For a Newtonian fluid, the most common sort, Eqs. (4-31) provide the needed relationship for a stationary control volume in rectangular coordinates. Their introduction into Eq. (4-44) results in

$$
\rho\frac{DI}{Dt} + \mathrm{div}(\mathbf{q}) - q''' = 
\begin{aligned}
&-p\left(\frac{\partial u}{\partial x} + \frac{\partial v}{\partial y} + \frac{\partial w}{\partial z}\right) + \mu\left\{\left[2\left(\frac{\partial u}{\partial x}\right)^2 - \frac{2}{3}\frac{\partial u}{\partial x}\mathrm{div}(\mathbf{V})\right]\right.\\
&+\left[\frac{\partial u}{\partial y} + \frac{\partial v}{\partial x}\right]\frac{\partial u}{\partial y} + \left[\frac{\partial w}{\partial x} + \frac{\partial u}{\partial z}\right]\frac{\partial u}{\partial z}\\
&+\left[\frac{\partial u}{\partial y} + \frac{\partial v}{\partial x}\right]\frac{\partial v}{\partial x} + \left[2\left(\frac{\partial v}{\partial y}\right)^2 - \frac{2}{3}\frac{\partial v}{\partial y}\mathrm{div}(\mathbf{V})\right]\\
&+\left[\frac{\partial v}{\partial z} + \frac{\partial w}{\partial y}\right]\frac{\partial v}{\partial z} + \left[\frac{\partial w}{\partial x} + \frac{\partial u}{\partial z}\right]\frac{\partial w}{\partial z} + \left[\frac{\partial v}{\partial z} + \frac{\partial w}{\partial y}\right]\frac{\partial w}{\partial y}\\
&+\left.\left[2\left(\frac{\partial w}{\partial z}\right)^2 - \frac{2}{3}\frac{\partial w}{\partial z}\mathrm{div}(\mathbf{V})\right]\right\}
\end{aligned}
$$

This lengthy equation can be more compactly written in the general form

$$
\rho\frac{DI}{Dt} + \mathrm{div}(\mathbf{q}) - q''' = -p\,\mathrm{div}(\mathbf{V}) + \mu\Phi \tag{4-45}
$$

Here, $\Phi$ is defined in rectangular coordinates as

$$
\Phi = 2\left[\left(\frac{\partial u}{\partial x}\right)^2 + \left(\frac{\partial v}{\partial y}\right)^2 + \left(\frac{\partial w}{\partial z}\right)^2\right] - \frac{2}{3}[\mathrm{div}(\mathbf{V})]^2
$$

$$
+ \left(\frac{\partial u}{\partial y} + \frac{\partial v}{\partial x}\right)^2 + \left(\frac{\partial v}{\partial z} + \frac{\partial w}{\partial y}\right)^2 + \left(\frac{\partial w}{\partial x} + \frac{\partial u}{\partial z}\right)^2 \tag{4-46}
$$

and is called the *dissipation function* since it represents the irreversible conversion of mechanical forms of energy to a thermal form. The squared terms show the irreversibility since $\Phi \geqslant 0$ and *always* acts as a *source* of thermal energy. In contrast, $p\,\mathrm{div}(\mathbf{V})$ represents the reversible work done on the environment by the expanding mass inside the control volume.

The "energy" equation in the form of Eq. (4-45) has thermal energy forms on the left-hand side and mechanical energy forms on the right-hand side.

Although this is correct, it is often more convenient to have only thermal forms of energy appear explicitly. This desire can be largely satisfied by introducing enthalpy per unit mass $H$ as a replacement for thermal internal energy per unit mass $I$. By definition,

$$H = I + \frac{p}{\rho}$$

Thus

$$\frac{DI}{Dt} = \frac{DH}{Dt} - \rho^{-1}\frac{Dp}{Dt} + p\rho^{-2}\frac{D\rho}{Dt} \qquad (4\text{-}47)$$

The continuity equation [Eq. (4-12)] shows that the last term on the right-hand side of this relation equals $-p\rho^{-1}\,\mathrm{div}(\mathbf{V})$. Incorporation of this fact into Eq. (4-47) and substitution of that result into Eq. (4-45) gives

$$\rho\frac{DH}{Dt} + \mathrm{div}(\mathbf{q}) - q''' = \frac{Dp}{Dt} + \mu\Phi \qquad (4\text{-}48)$$

The $Dp/Dt$ term in Eq. (4-48) is simpler than the $p\,\mathrm{div}(\mathbf{V})$ term in Eq. (4-45) because it is more common for a fluid to undergo a nearly constant pressure process (for which $Dp/Dt \sim 0$) than a nearly constant volume process (for which $\mathrm{div}\,\mathbf{V} \sim 0$). It is apparent that the "energy" equation can be manipulated into a variety of forms, some of which are more convenient for a given purpose than others. In fact, if attention is confined to purely mechanical forms of energy, there is no need for the developments of this section that have added a new unknown as well as a new equation. Instead, a "mechanical" energy equation can be formed [9] by merely taking the scalar product of the velocity vector with the vector equation of motion [Eq. (4-35)]—an interest in thermal forms of energy makes the "mechanical" energy equation of too narrow a scope, however.

Equation (4-48) is occasionally a convenient form for such applications as gases whose specific heat varies significantly or that undergo dissociation or ionization [10]. Usually, however, a further manipulation of Eq. (4-48) is required so that boundary conditions given in terms of temperature can be met by the solution. Accordingly, recourse to thermodynamics is next.

For a pure substance in the absence of motion, surface tension, and electromagnetic effects, there are only two independent properties [11]—$H = H(p, T)$, where $T$ is thermodynamic equilibrium temperature, for example. Thus

$$DH = \frac{\partial H}{\partial T}\bigg|_{p=\text{const}} DT + \frac{\partial H}{\partial p}\bigg|_{T=\text{const}} Dp \qquad (4\text{-}49)$$

From thermodynamics, with $s = $ entropy,

$$DH = T\,Ds + \frac{Dp}{\rho}$$

which can be rephrased as

$$\frac{DH}{Dp} = T\frac{Ds}{Dp} + \frac{1}{\rho} \tag{4-50}$$

This last relation may be applied to an infinitesimal process at thermodynamic equilibrium, assumed to be locally achieved. Since thermodynamic equilibrium implies constant temperature, Eq. (4-50) gives

$$\frac{\partial H}{\partial p}\bigg|_{T=\text{const}} = T\left(\frac{\partial s}{\partial p}\right)\bigg|_{T=\text{const}} + \frac{1}{\rho} \tag{4-51}$$

The Maxwell relations of thermodynamics [10] reveal the additional information that*

$$\frac{\partial s}{\partial p}\bigg|_{T=\text{const}} = \rho^{-2}\frac{\partial \rho}{\partial T}\bigg|_{p=\text{const}}$$

This relationship, when substituted into Eq. (4-51), gives

$$\left(\frac{\partial H}{\partial p}\right)\bigg|_{T=\text{const}} = T\rho^{-2}\frac{\partial \rho}{\partial T}\bigg|_{p=\text{const}} + \frac{1}{\rho}$$

$$= \rho^{-1}\left[1 + T\rho^{-1}\frac{\partial \rho}{\partial T}\bigg|_{p=\text{const}}\right] \tag{4-52}$$

Recall that the specific heat at constant pressure is defined as

$$C_p = \frac{\partial H}{\partial T}\bigg|_{p=\text{const}} \tag{4-53}$$

---

*Gibb's free energy is $G = I + p/\rho - Ts$. Its change is given by

$$dG = \left[dI + p\left(d\rho^{-1}\right) - T\,ds\right] + \rho^{-1}\,dp - s\,dT$$

The bracketed term is zero according to the first law of thermodynamics. Thus

$$dG = \rho^{-1}\,dp - s\,dT$$

Comparison with the chain rule $dG = (\partial G/\partial p)\,dp + (\partial G/\partial T)\,dT$ shows

$$\frac{\partial G}{\partial p} = \rho^{-1} \quad \text{and} \quad \frac{\partial G}{\partial T} = -s$$

Taking of second derivatives yields

$$\frac{\partial G^2}{\partial T\,\partial p} = -\rho^{-2}\frac{\partial \rho}{\partial T}\bigg|_{p=\text{const}} \quad \text{and} \quad \frac{\partial^2 G}{\partial p\,\partial T} = -\frac{\partial s}{\partial p}\bigg|_{T=\text{const}}$$

and the coefficient of thermal expansion is defined as

$$\beta = -\rho^{-1}\left(\frac{\partial \rho}{\partial T}\right)_{p=\text{const}} \tag{4-54}$$

Insertion of Eqs. (4-52)–(4-54) into Eq. (4-49) shows that

$$DH = C_p\,DT + \rho^{-1}(1 - \beta T)Dp \tag{4-55}$$

The energy equation, on substitution of Eq. (4-55) into Eq. (4-48), then takes the form

$$\rho C_p \frac{DT}{Dt} = -\text{div}(\mathbf{q}) + q''' + \beta T \frac{Dp}{Dt} + \mu \Phi \tag{4-56}$$

Relation of conductive heat flux to temperature is accomplished by Fourier's law $\mathbf{q} = -k\nabla T$. The energy equation then is

$$\rho C_p \frac{DT}{Dt} = \text{div}(k\nabla T) + q''' + \beta T \frac{Dp}{Dt} + \mu \Phi \tag{4-57}$$

For a constant-density fluid, $\beta = 0$. If viscous dissipation is neglected, $\Phi = 0$. Then, for the case of constant thermal conductivity, Eq. (4-57) in rectangular coordinates for a stationary control volume is

$$\rho C_p \left(\frac{\partial T}{\partial t} + u \frac{\partial T}{\partial x} + v \frac{\partial T}{\partial y} + w \frac{\partial T}{\partial z}\right) = k \left(\frac{\partial^2 T}{\partial x^2} + \frac{\partial^2 T}{\partial y^2} + \frac{\partial^2 T}{\partial z^2}\right) \tag{4-57a}$$

## 4.6  BINARY MASS DIFFUSION EQUATION

The conservation of mass principle that led to the continuity equation makes no distinction among the different kinds (species) of mass that can comprise a fluid. If a fluid is made up of several kinds of mass and if one of these kinds of mass is unusually abundant in a particular region, it is known that the fluid will tend to diffuse in such a way as to render its abundance uniform. But the continuity equation gives no hint of such a smoothing process; it cannot, for the density and velocity it uses are bulk (mass-averaged) values. To obtain the desired behavior, a single species of mass must be tracked, using the conservation of a species of mass principle.

The conservation of a species of mass principle, differing slightly from the conservation of mass principle in that it is possible to create or destroy a species of mass by a chemical reaction, is given for species $i$ as

$$\overset{\cdot}{\text{mass}}_{i\,\text{in}} - \overset{\cdot}{\text{mass}}_{i\,\text{out}} + \overset{\cdot}{\text{mass}}_{i\,\text{generated}} = \overset{\cdot}{\text{mass}}_{i\,\text{stored}} \tag{4-58}$$

This principle can be applied to the general control volume illustrated in Fig.

4-1. In doing so, it must be remembered that each species moves with its own velocity $V_i$, has its own mass density $\rho_i$, and might be created at the rate per unit volume $r_i$. The procedure is parallel to that which led to the continuity equation [Eq. (4-6)] and gives the mathematical equivalent of Eq. (4-58) as

$$\frac{\partial \rho_i}{\partial t} + \mathrm{div}(\rho_i V_{i_{rel}}) = r_i''' \tag{4-59}$$

For each component of the fluid, there will be one equation of the preceding form. These equations can be summed to give

$$\frac{\partial(\Sigma_i \rho_i)}{\partial t} + \mathrm{div}\left(\sum_i \rho_i V_{i_{rel}}\right) = \sum_i r_i''' \tag{4-60}$$

It is reasonable to define the bulk density as

$$\rho = \sum_i \rho_i$$

The bulk mass flow across the control volume surface is also reasonably defined as

$$\rho V_{rel} = \sum_i \rho_i V_{i_{rel}}$$

from which it is seen that the bulk velocity is a mass-averaged velocity

$$V_{rel} = \rho^{-1} \sum_i \rho_i V_{i_{rel}}$$

If it is additionally realized that $\Sigma_i r_i''' = 0$ since bulk mass must be conserved —as much species mass must be destroyed as is created if bulk mass is to be conserved—Eq. (4-60) is recognized to just be the continuity equation for the bulk fluid

$$\frac{\partial \rho}{\partial t} + \mathrm{div}(\rho V_{rel}) = 0 \tag{4-6}$$

Thus only $n - 1$ components of an $n$-component fluid need be tracked if the bulk continuity equation is included.

As Eq. (4-59) stands, two new unknowns have been introduced—the species mass density $\rho_i$ and the species mass velocity $V_{i_{rel}}$. To reduce the number of new unknowns to just one, an idea used for the derivation of the energy equation is employed. The flux of species $i$ mass across a surface $\rho_i V_{i_{rel}}$ will be set equal to the sum of that convected by the bulk flow $\rho_i V_{rel}$ and that which diffuses relative to the bulk convection $\dot{m}_i$. Thus

$$\rho_i V_{i_{rel}} = \rho_i V_{rel} + \dot{m}_i \tag{4-61}$$

Introduction of the relation of Eq. (4-61) into Eq. (4-59) gives

$$\frac{\partial \rho_i}{\partial t} + \text{div}(\rho_i \mathbf{V}_{\text{rel}}) = r_i''' - \text{div}(\dot{\mathbf{m}}_i) \tag{4-62}$$

Let the mass fraction of species $i$ $\omega_i$ be defined as

$$\omega_i = \frac{\rho_i}{\rho} \tag{4-63}$$

Then Eq. (4-62) becomes

$$\frac{\partial(\rho \omega_i)}{\partial t} + \text{div}(\omega_i \rho \mathbf{V}_{\text{rel}}) = r_i''' - \text{div}(\dot{\mathbf{m}}_i)$$

which can be expanded into

$$\omega_i \left[ \frac{\partial \rho}{\partial t} + \text{div}(\rho \mathbf{V}_{\text{rel}}) \right] + \rho \left[ \frac{\partial \omega_i}{\partial t} + (\overline{\mathbf{V}}_{\text{rel}} \cdot \nabla) \omega_i \right] = r_i''' - \text{div}(\dot{\mathbf{m}}_i)$$

The first bracketed term identically equals zero by the bulk continuity equation [Eq. (4-6)]. The second bracketed term can be written as $D\omega_i/Dt$ by use of the substantial derivation defined by Eq. (4-13). Thus the conservation of species principle gives

$$\rho \frac{D\omega_i}{Dt} = r_i''' - \text{div}(\dot{\mathbf{m}}_i) \tag{4-64}$$

For a binary fluid, Fick's law relates the mass fraction to the diffusive flux of species 1 relative to the bulk convection as

$$\dot{\mathbf{m}}_1 = -\rho D_{12} \nabla \omega_1 \tag{4-65}$$

Although Eq. (4-64) applies to a multicomponent fluid, Eq. (4-65) applies only to a binary fluid, and its use in Eq. (4-64) results in the specialized form for a binary fluid

$$\rho \frac{D\omega_1}{Dt} = r_1''' + \text{div}(\rho D_{12} \nabla \omega_1) \tag{4-66}$$

which is called the *binary mass-diffusion equation*. For a binary fluid, no corresponding equation need be written for the second component if the bulk continuity equation is used to obtain the bulk density since then

$$\rho_2 = \rho - \rho_1 = (1 - \omega_1)\rho$$

or

$$\frac{\rho_2}{\rho} = \omega_2 = 1 - \omega_1 \tag{4-67}$$

For a constant-property fluid such as would result if there were vanishingly small amounts of species 1 in the fluid, Eq. (4-66) becomes

$$\frac{D\omega_1}{Dt} = D_{12}\nabla^2\omega_1 + \frac{r_1'''}{\rho}$$ (4-66a)

In rectangular coordinates for a stationary control volume and for a constant property fluid, Eq. (4-66a) is

$$\frac{\partial\omega_1}{\partial t} + u\frac{\partial\omega_1}{\partial x} + v\frac{\partial\omega_1}{\partial y} + w\frac{\partial\omega_1}{\partial z} = \left(\frac{\partial^2\omega_1}{\partial x^2} + \frac{\partial^2\omega_1}{\partial y^2} + \frac{\partial^2\omega_1}{\partial z^2}\right)D_{12} + \frac{r_1'''}{\rho}$$

(4-68)

These ideas can be more concretely developed by applying the conservation of species of mass principle to a cubical control volume fixed in a rectangular coordinate system. It is assumed that a binary fluid flows. The rate at which species 1 mass flows across two control volume surfaces is shown in Fig. 4-10. In particular, the flux of species 1 mass across a surface in an $x$–$z$ plane, for example, is either $\rho_1 v_1$ or $\rho_1 v + \dot{m}_{1y}$, where $v_1$ is the velocity of species 1 and $v$ is the bulk mass-average fluid velocity. In other words,

Species 1 mass flux = bulk convection + diffusion relative to bulk convection

or

$$\rho_1 v_1 = \rho_1 v + \dot{m}_{1y}$$ (4-69)

Application of the conservation of species principle expressed by Eq. (4-58)

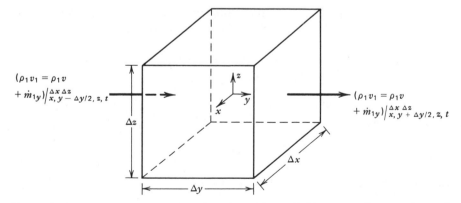

**Figure 4-10** Flow of one species of mass in a binary fluid across the $x$ surface of a stationary control volume in rectangular coordinates.

then results in

$$
\begin{aligned}
&\left(\rho_1 u_1\big|_{x-\Delta y/2,\,y,\,z,\,t_{av}} - \rho_1 u_1\big|_{x+\Delta x/2,\,y,\,z,\,t_{av}}\right)\Delta y\,\Delta z \\
&+ \left(\rho_1 v_1\big|_{x,\,y-\Delta y/2,\,z,\,t_{av}} - \rho_1 v_1\big|_{x,\,y+\Delta y/2,\,z,\,t_{av}}\right)\Delta x\,\Delta z \\
&+ \left(\rho_1 w_1\big|_{x,\,y,\,z-\Delta z/2,\,t_{av}} - \rho_1 w_1\big|_{x,\,y,\,z+\Delta z/2,\,t_{av}}\right)\Delta x\,\Delta y + r_1'''\big|_{x,\,y,\,z,\,t_{av}}\Delta x\,\Delta y\,\Delta z \\
&= \left(\rho_1\big|_{x,\,y,\,z,\,t+\Delta t} - \rho_1\big|_{x,\,y,\,z,\,t}\right)\frac{\Delta x\,\Delta y\,\Delta z}{\Delta t}
\end{aligned}
$$

Dividing by $\Delta x\,\Delta y\,\Delta z$ and taking the limit as all differential quantities approach zero, one finds that this equation becomes

$$
\frac{\partial \rho_1}{\partial t} + \frac{\partial(\rho_1 u_1)}{\partial x} + \frac{\partial(\rho_1 v_1)}{\partial y} + \frac{\partial(\rho_1 w_1)}{\partial z} = r_1''' \tag{4-70}
$$

For the second fluid, one finds by similar means that

$$
\frac{\partial \rho_2}{\partial t} + \frac{\partial(\rho_2 u_2)}{\partial x} + \frac{\partial(\rho_2 v_2)}{\partial y} + \frac{\partial(\rho_2 w_2)}{\partial z} = r_2''' \tag{4-71}
$$

When the two equations for the two components [Eqs. (4-70) and (4-71)] are added, the result is

$$
\frac{\partial(\rho_1 + \rho_2)}{\partial t} + \frac{\partial(\rho_1 u_1 + \rho_2 u_2)}{\partial x} + \frac{\partial(\rho_1 v_1 + \rho_2 v_2)}{\partial y} + \frac{\partial(\rho_1 w_1 + \rho_2 w_2)}{\partial z}
$$

$$
= r_1''' + r_2'''
$$

which is just the bulk continuity equation

$$
\frac{\partial \rho}{\partial t} + \frac{\partial(\rho u)}{\partial x} + \frac{\partial(\rho v)}{\partial y} + \frac{\partial(\rho u)}{\partial z} = 0 \tag{4-8}
$$

since $\rho = \rho_1 + \rho_2$, $r_1''' + r_2''' = 0$, $\rho u = \rho_1 u_1 + \rho_2 u_2$, and so forth. Again, only one component need be tracked if the bulk continuity equation is taken into account.

Focusing on species 1, only Eq. (4-70) is considered in further detail. With use of the idea in Eq. (4-69), Eq. (4-70) becomes

$$
\frac{\partial \rho_1}{\partial t} + \frac{\partial(\rho_1 u)}{\partial x} + \frac{\partial(\rho_1 v)}{\partial y} + \frac{\partial(\rho_1 w)}{\partial z} = r_1''' - \left(\frac{\partial \dot{m}_{1x}}{\partial x} + \frac{\partial \dot{m}_y}{\partial_y} + \frac{\partial \dot{m}_{1z}}{\partial z}\right)
$$

$$
\tag{4-72}
$$

The mass fraction $\omega_1$ of species 1 is defined as

$$
\omega_1 = \frac{\rho_1}{\rho} = \frac{\rho_1}{\rho_1 + \rho_2}
$$

With $\omega_1$ employed in Eq. (4-72), one has

$$\frac{\partial(\rho\omega_1)}{\partial t} + \frac{\partial(\rho u \omega_1)}{\partial x} + \frac{\partial(\rho v \omega_1)}{\partial y} + \frac{\partial(\rho w \omega_1)}{\partial z}$$

$$= r_1''' - \left( \frac{\partial \dot{m}_{1x}}{\partial x} + \frac{\partial \dot{m}_{1y}}{\partial y} + \frac{\partial \dot{m}_{1z}}{\partial z} \right)$$

Expansion of the left-hand side of this equation gives

$$\omega_1 \left[ \frac{\partial \rho}{\partial t} + \frac{\partial(\rho u)}{\partial x} + \frac{\partial(\rho v)}{\partial y} + \frac{\partial(\rho w)}{\partial z} \right]$$

$$+ \rho \left[ \frac{\partial \omega_1}{\partial t} + u \frac{\partial \omega_1}{\partial x} + v \frac{\partial \omega_1}{\partial y} + w \frac{\partial \omega_1}{\partial z} \right]$$

The first bracketed term is zero by the bulk continuity equation [Eq. (4-8)]. Use of Fick's law additionally as

$$\dot{m}_{1x} = -\rho D_{12} \frac{\partial \omega_1}{\partial x}$$

with similar expressions for the $y$ and $z$ fluxes then results in, assuming constant $\rho$,

$$\frac{\partial \omega_1}{\partial t} + u \frac{\partial \omega_1}{\partial x} + v \frac{\partial \omega_1}{\partial y} + w \frac{\partial \omega_1}{\partial z} = D_{12} \left( \frac{\partial^2 \omega_1}{\partial x^2} + \frac{\partial^2 \omega_1}{\partial y^2} + \frac{\partial^2 \omega_1}{\partial z^2} \right) + \frac{r_1'''}{\rho}$$

which is identical to Eq. (4-68), as it should be.

A balanced summary of multicomponent diffusion is provided by Cussler [12], which can be consulted if other than binary diffusion is encountered. The important anomalous departures from pseudobinary behavior are discussed, and simple molecular-level hypotheses are advanced in explanation.

## 4.7  ENTROPY EQUATION

The irreversible diffusion of heat and momentum that is caused by nonuniformities of temperature and velocity in a fluid tends to establish equilibrium conditions in the fluid. This tendency toward uniform temperature and velocity must be accompanied by a diffusion and volumetric generation of entropy. Because it is occasionally useful to be able to quantitatively evaluate entropy changes, the equation that describes the entropy production associated with heat conduction and viscous friction is developed.

A review of the generalized development of the mass diffusion equation [Eq. (4-64)] shows that if $P$ is a property per unit mass of the bulk fluid and $\dot{P}$ is its diffusive flux relative to bulk convection and $P'''$ is its volumetric rate of generation, the describing equation is

$$\rho \frac{DP}{Dt} = -\mathrm{div}(\dot{\mathbf{P}}) + P'''$$

Let the property per unit mass of the bulk fluid be entropy $s$, the diffusive flux of entropy relative to bulk convection be $\dot{s}$, and the volumetric rate of generation of entropy per unit mass be $s'''$. Then

$$\rho \frac{Ds}{Dt} = -\mathrm{div}(\dot{s}) + s''' \tag{4-73}$$

For a reversible process, it is known from thermodynamics that

$$T\,Ds = DI + p\,D\rho^{-1} \tag{4-74}$$

It is assumed that local equilibrium is present, even though a nonequilibrium situation is under consideration, so that Eq. (4-74) can be applied with substantial accuracy. The energy equation provides the needed relation between internal energy per unit bulk mass $I$ and other quantities in Eq. (4-45). Its use in Eq. (4-74) gives

$$T\frac{Ds}{Dt} = \frac{-\mathrm{div}(\mathbf{q}) + q''' - p\,\mathrm{div}(\mathbf{V}) + \mu\Phi}{\rho} - \frac{p}{\rho^2}\frac{D\rho}{Dt}$$

The continuity equation [Eq. (4-12)] allows the last term to be put into the form

$$\frac{p}{\rho}\,\mathrm{div}(\mathbf{V})$$

Combination of terms then gives

$$\frac{Ds}{Dt} = \frac{-\mathrm{div}(\mathbf{q})}{T} + \frac{q''' + \mu\Phi}{T}$$

Recall that the divergence of a scalar times a vector is

$$\mathrm{div}\left(\frac{\mathbf{q}}{T}\right) = \frac{\mathrm{div}(\mathbf{q})}{T} + \mathbf{q}\cdot\nabla\left(\frac{1}{T}\right)$$

$$= \frac{\mathrm{div}(\mathbf{q})}{T} - \frac{\mathbf{q}\cdot\nabla T}{T^2}$$

Therefore, the substantial derivative of entropy is

$$\rho\frac{Ds}{Dt} = -\mathrm{div}\left(\frac{\mathbf{q}}{T}\right) + \left(\frac{-\mathbf{q}\cdot\nabla T}{T^2} + \frac{q''' + \mu\Phi}{T}\right) \tag{4-75}$$

Comparison of Eqs. (4-75) and (4-73) shows that the diffusive flux of entropy relative to bulk convection is given by

$$\dot{s} = \frac{q}{T} \qquad (4\text{-}76)$$

and the volumetric rate of generation of entropy per unit mass is given by

$$s''' = -\frac{q \cdot \nabla T}{T^2} + \frac{q''' + \mu \Phi}{T} \qquad (4\text{-}77)$$

where the dissipation function $\Phi$ is defined in Eq. (4-46).

These two relations are subject to the same limitations as the energy equation that was used. A Newtonian fluid was assumed and the effects of mass diffusion (e.g., in a binary fluid) were not taken into account. A fuller discussion and treatment is given by Hirschfelder et al. [13]. If Fourier's law of heat conduction is employed, then

$$\dot{s} = -\frac{k \nabla T}{T} \qquad (4\text{-}78)$$

and

$$s''' = \frac{k |\nabla T|^2}{T^2} + \frac{q''' + \mu \Phi}{T} \qquad (4\text{-}79)$$

The viscous dissipation function $\Phi$ is always positive, as it is made up of squared velocity derivatives, and so viscous friction always generates entropy as befits an irreversible process. The flow of heat is also seen to always generate entropy.

Bejan [15, 16] has applied relations similar to Eq. (4-79) to the design of heat exchangers that minimize the production of entropy.

## PROBLEMS

**4-1**  Derive the equation of continuity for cylindrical coordinates by:

(a) Applying the conservation of mass principle to the stationary cylindrical control volume indicated in Fig. 4P-1 for which $dV = r\,\Delta\theta\,\Delta r\,\Delta z$.

(b) Mathematically transforming the continuity equation from rectangular coordinates through the use of the relationships that

| | | |
|---|---|---|
| $x = r\cos\theta$ | $r = (x^2 + y^2)^{1/2}$ | $v_x = v_r \cos\theta - v_\theta \sin\theta$ |
| $y = r\sin\theta$ | $\theta = \arctan\dfrac{y}{x}$ | $v_y = v_r \sin\theta + v_\theta \cos\theta$ |
| $z = z$ | $z = z$ | $v_z = v_z$ |

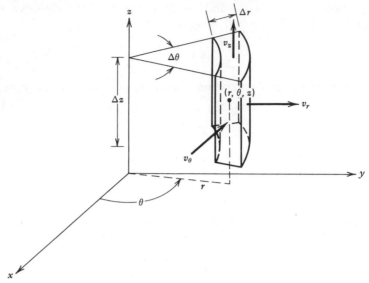

**Figure 4P-1**

**4-2**  Derive the equation of continuity for spherical coordinates by applying the conservation of mass principle to the stationary spherical control volume indicated in Fig. 4P-2 for which $dV = r^2 \sin \theta \Delta \theta \Delta \phi \Delta r$.

**4-3**  Consider the constant property, steady laminar flow illustrated in Fig. 4P-3. A fluid is contained between an infinitely large horizontal stationary bottom plate and an infinitely large top plate that moves parallel to the bottom plate at a constant velocity $U$ in response to a constant external force per unit area $F/A$ applied to the upper plate. Because of the infinite extent, it can be assumed that no quantity varies in the $x$ and $z$ directions.

  **(a)** Apply the conservation of mass principle to the dotted control volume (after first making all assumptions) to derive the specialized form of the continuity equation appropriate to this physical situation.

  **(b)** Apply Newton's second law for control volumes to the dotted control volume (after first making all assumptions) to derive the specialized forms of the $x$- and $y$-direction equations of motion in terms of pressures and stresses appropriate to this physical situation. Show all nonzero surface and body forces acting on the control volume; note that since $v = 0$, it is reasonable to assume $\tau_{yy} = -p$ is the only stress acting on the control volume top and bottom other than $\tau_{yx}$.

  **(c)** Relate $\tau_{yx}$ to $du/dy$ in the equations of motion of part b by applying Newton's viscosity law.

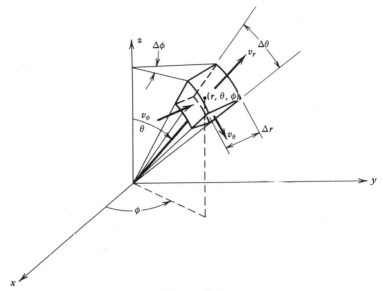

**Figure 4P-2**

(d) Draw a free-body diagram of the upper plate and relate the external force per unit area $F/A$ to the shear stress $\tau_{yx}$ in the fluid.

**4-4**  Consider the constant property, steady laminar flow illustrated in Fig. 4P-4. A fluid flows through a very long tube in response to an externally imposed pressure gradient along the tube length. Because of the infinite length, it can be assumed that no quantity (other than pressure) varies in the $\theta$ and $z$ directions. Also, under these conditions it can be experimentally observed that a fluid particle moves solely in the $z$ direction (i.e., $v_\theta = 0 = v_r$).

(a) Apply the conservation of mass principle to the dotted control volume (after first making all assumptions) to derive the specialized

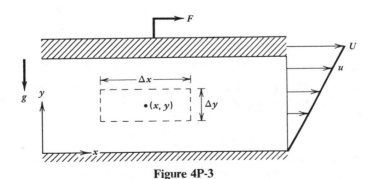

**Figure 4P-3**

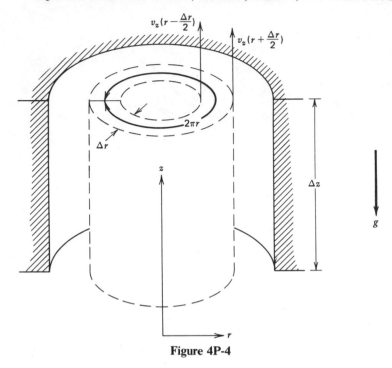

**Figure 4P-4**

form of the continuity equation appropriate to this physical situation.

**(b)** Apply Newton's second law for control volumes to the dotted control volume (after first making all assumptions) to derive the specialized $z$-direction equation of motion in terms of pressures and stresses. Show all nonzero surface and body forces acting on the control volume—note that, since $v_\theta = 0 = v_r$, it is reasonable to assume that only $\tau_{rz}$ is acting on the sides of the annular control volume and that, since $v_z$ does not vary in the $z$ direction, $\tau_{zz} = -p$ is the only stress acting on the ends of the annular control volume.

**(c)** Relate $\tau_{rz}$ to $dv_z/dr$ in the equation of motion of part b by considering the annulus to be so thin that it can be flattened to a flat plate of thickness $\Delta r$ for which Newton's viscosity law is directly applicable.

**4-5** Show that Newtonian fluid of constant viscosity but variable density has $y$- and $z$-direction equations of motion that are of the same form as Eq. (4-33).

**4-6** On the three positive faces of the cylindrical and spherical elemental control volumes illustrated in Figs. 4P-1 and 4P-2, draw all the stresses acting and label them with appropriate subscripts.

**4-7**  Manipulate the energy equation given by Eq. (4-45) into a form that involves $T$ and $C_v$.

**4-8**  Show that $\beta = 1/T$ for a perfect gas.

**4-9**  Show that for a reversible adiabatic process (viscous dissipation and thermal conduction are negligible), the solution to the energy equation [Eq. (4-57)] for a perfect gas is $p/\rho^\gamma = \text{const}$, where $\gamma$ is the ratio of specific heats $\gamma = C_p/C_v$.

**4-10**  Plot the coefficient of thermal expansion–absolute temperature product $\beta T$ against absolute temperature for water and for air. Comment on the magnitude of $\beta T$ at high and low temperatures.

**4-11**  Derive the energy equation in rectangular coordinates for the situation described in Problem 4-3 by first making all assumptions and then applying the conservation of energy principle to the indicated control volume.

**4-12**  Derive the energy equation in cylindrical coordinates for the situation described in Problem 4-4 by first making all assumptions and then applying the conservation of energy principle to the indicated control volume.

**4-13**  To explore the different technique used to derive boundary conditions, consider a stationary plane interface between two phases of a fluid. As illustrated in Fig. 4P-13, the fluid flows from left to right in a one-dimensional manner. It is assumed that properties are constant in each phase.

   **(a)** Apply the conservation of mass principle to the control volume shown with dashed lines to show that, in the limit as $\Delta x \to 0$,

$$\rho_1 u_1 = \rho_2 u_2$$

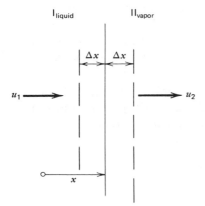

**Figure 4P-13**

(b) Apply Newton's second law for a control volume to show that, in the limit as $\Delta x \rightarrow 0$,

$$\tau_{2_{xx}} - \tau_{1_{xx}} + \frac{\rho_1 u_1^2}{g_c} - \frac{\rho_2 u_2^2}{g_c} = 0$$

or

$$P_1 - P_2 = \frac{\rho_2 u_2^2 (1 - \rho_2/\rho_1)}{g_c}$$

(c) Apply the conservation of energy principle for control volumes to show that, in the limit as $\Delta x \rightarrow 0$,

$$q_1 - q_2 + \tau_{2_{xx}} u_2 - \tau_{1_{xx}} u_1 = e_2 \rho_2 u_2 - e_1 \rho_1 u_1$$

or

$$k_2 \frac{dT}{dx} - k_1 \frac{dT}{dx} = \rho_2 u_2 \left[ \Lambda + \left( 1 - \frac{\rho_2^2}{\rho_1^2} \right) \frac{u_2^2}{2 g_c} \right]$$

where $\Lambda$ is the latent heat of vaporization $\Lambda = H_2 - H_1$.

4-14  Show for a binary fluid that, if $\rho_i \mathbf{V}_i = \rho_i \mathbf{V} + \dot{m}_i$ for mass diffusion, the definition of mass average velocity as

$$\rho \mathbf{V} = \sum_i \rho_i \mathbf{V}_i$$

requires that

$$\dot{m}_1 = -\dot{m}_2 \quad \text{and} \quad \dot{m}_i = \rho_i (\mathbf{V}_i - \mathbf{V})$$

In other words, for a binary fluid, the two species of mass diffuse at equal rates in opposite directions.

4-15  The consequences of having a multicomponent fluid rather than a pure fluid can be approximately explored by considering the one-dimensional flow of a binary fluid (see Appendix B for the results of detailed analysis). Consider the control volume shown with dashed lines in Fig. 4P-15. The body force per unit volume acting on the $i$th species has an $x$ component of $B_{ix} \rho_i / \rho$.

(a) Apply the conservation of species principle to the control volume. With bulk density and velocity defined as $\rho = \rho_1 + \rho_2$ and $u(\rho_1 u_1 + \rho_2 u_2)/\rho$, respectively, show that the bulk continuity equation is

$$\frac{\partial \rho}{\partial t} + \frac{\partial (\rho u)}{\partial x} = 0$$

for one-dimensional flow.

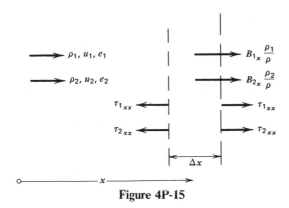

**Figure 4P-15**

(b) Apply Newton's second law for control volumes to the control volume of Fig. 4P-15. With bulk stress defined as

$$\tau_{xx} = \tau_{1_{xx}} + \tau_{2_{xx}}$$

and with species velocity defined as

$$u_i = u + \frac{\dot{m}_{ix}}{\rho_i}$$

show that the relation from Newton's second law, which is

$$\frac{\rho_1 B_{1_x} + \rho_2 B_{2_x}}{\rho} + \frac{\partial(\tau_{1_{xx}} + \tau_{2_{xx}})}{\partial x} - \frac{\partial(\rho_1 u_1 u_1/g_c + \rho_2 u_2 u_2/g_c)}{\partial x}$$

$$= \frac{\partial(\rho_1 u_1/g_c + \rho_2 u_2/g_c)}{\partial t}$$

can be cast into the form

$$\frac{\rho_1 B_{1_x} + \rho_2 B_{2_x}}{\rho} + \frac{\partial \tau_{a_{xx}}}{\partial x} = \frac{\rho}{g_c} \frac{Du}{Dt}$$

where $\tau_{a_{xx}}$ is the apparent normal surface stress that is related to the stress computed using bulk properties by

$$\tau_{a_{xx}} = \tau_{xx} - \frac{\rho}{g_c} \left[ \omega_1 \left( \frac{\dot{m}_1}{\rho_1} \right)^2 + \omega_2 \left( \frac{\dot{m}_2}{\rho_2} \right)^2 \right]$$

(c) Apply the conservation of energy principle for control volumes to the control volume in Fig. 4P-15. With bulk energy (internal plus kinetic) defined as

$$e = \frac{e_1 \rho_1 + e_2 \rho_2}{\rho}$$

show that the conservation of energy principle gives

$$q''' - \frac{\partial q_x}{\partial x} + \frac{\partial(\tau_{1_{xx}} u_1 + \tau_{2_{xx}} u_2)}{\partial x} + \frac{\rho_1 B_{1_x} u_1 + \rho_2 B_{2_x} u_2}{\rho}$$

$$\frac{-\partial(e_1 \rho_1 u_1 + e_2 \rho_2 u_2)}{\partial x} = \frac{\partial(e_1 \rho_1 + e_2 \rho_2)}{\partial t}$$

which can be cast into the form

$$q''' + \frac{B_{1_x} \dot{m}_1 + B_{2_x} \dot{m}_2}{\rho} - \frac{\partial(q_{ax})}{\partial x} + \tau_{a_{xx}} \frac{\partial u}{\partial x} = \rho \, DI/Dt$$

Here $q_{ax}$ is the apparent conductive heat flux that is related to the conductive heat flux computed using bulk properties by

$$q_{ax} = q_x + \dot{m}_1(H_1 - H_2)$$

with the additive higher-order kinetic term $\dot{m}_1[(\dot{m}_1/\rho_1)^2 - (\dot{m}_2/\rho_2)^2]/2g_c$ neglected on the right-hand side of the definition of $q_{ax}$.

(d) For the case in which the body forces $B_{1_x}$ and $B_{2_x}$ are the same for the two species (they would be equal for gravitational fields, but might be different for electrical fields), show that these equations describing one-dimensional convection in a binary fluid are identical to those describing a single component fluid if the energy flow relative to bulk motion is given by $q_{ax}$ and the normal stress is given by $\tau_{a_{xx}}$.

4-16 To explore the manner in which a coordinate transformation can be employed to take an equation from one coordinate system over to another, consider the one-dimensional form of the energy equation for a constant-property fluid without dissipation or volume sources in stationary coordinates (see Fig. 4P-16)

$$\frac{\partial T}{\partial t} + u \frac{\partial T}{\partial x} = \alpha \frac{\partial^2 T}{\partial x^2} \qquad (4\text{-}57a)$$

This equation is to be transformed to a coordinate system moving to the right at steady velocity $u'$. The coordinates are to be transformed according to

$$t' = t \qquad \text{and} \qquad x' = x - u't$$

Figure 4P-16

The chain rule gives

$$\frac{\partial T}{\partial t} = \frac{\partial T}{\partial t'}\frac{\partial t'}{\partial t} + \frac{\partial T}{\partial x'}\frac{\partial x'}{\partial t} = \frac{\partial T}{\partial t'} - u'\frac{\partial T}{\partial x'}$$

and

$$\frac{\partial T}{\partial x} = \frac{\partial T}{\partial t'}\frac{\partial t'}{\partial x} + \frac{\partial T}{\partial x'}\frac{\partial x'}{\partial x} = \frac{\partial T}{\partial x'}$$

from which it is also found that

$$\frac{\partial^2 T}{\partial x^2} = \frac{\partial T}{\partial t'}\frac{\partial^2 t'}{\partial x^2} + \frac{\partial t'}{\partial x}\left(\frac{\partial^2 T}{\partial t'^2}\frac{\partial t'}{\partial x} + \frac{\partial^2 T}{\partial x'\partial t'}\frac{\partial x'}{\partial x}\right)$$

$$+ \frac{\partial T}{\partial x'}\frac{\partial^2 x'}{\partial x^2} + \frac{\partial x'}{\partial x}\left(\frac{\partial^2 T}{\partial t'\partial x'}\frac{\partial t'}{\partial x} + \frac{\partial^2 T}{\partial x^2}\frac{\partial x'}{\partial x}\right)$$

$$= \frac{\partial^2 T}{\partial' x^2}$$

Substitute these chain rule results into Eq. (4-57a) to show that in $x'$, $t'$ coordinates the energy equation is

$$\frac{\partial T}{\partial t'} + (u - u')\frac{\partial T}{\partial x'} = \alpha\frac{\partial^2 T}{\partial x'^2}$$

*Note:* Detailed examples of coordinate transformations for a stagnant medium are given in reference 14.

**4-17** The boundary conditions at the moving interface between two phases of a fluid can be obtained by applying the several conservation principles to a control volume that moves with the interface. Consider the spherical vapor bubble in an infinite liquid as sketched in Fig. 4P-17.

**(a)** Apply the conservation of mass principle to the spherical annulus control volume suggested in Fig. 4P-17*b* and show that, since the mass flux across a surface is $\rho v_{rel}$,

$$\rho_1(v_1 - \dot{R}) = \rho_2(v_2 - \dot{R}) \quad \text{at} \quad r = R(t)$$

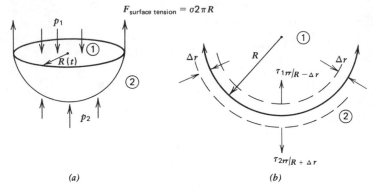

$$F_{\text{surface tension}} = \sigma 2\pi R$$

(a)                                                        (b)

**Figure 4P-17**

(b) Apply Newton's second law for a control volume to the hemispherical annulus, and show that

$$P_1 - P_2 = \frac{2\sigma}{R} + \rho_1(u_1 - \dot{R})(u_2 - u_1) + \frac{4}{3}\mu_1\left(\frac{\partial u_1}{\partial r} - \frac{u_1}{r}\right)$$

$$-\frac{4}{3}\mu_2\left(\frac{\partial u_2}{\partial r} - \frac{u_2}{r}\right)$$

at $r = R(t)$.

In view of the continuity equation for radial flow, for an incompressible fluid, the $\frac{4}{3}\mu(\partial u/\partial r - u/r)$ term can also be written as $2\mu\,\partial u/\partial r$.

(c) Apply the conservation of energy principle for control volumes to the spherical annulus, and show that

$$k_2\frac{\partial T_2}{\partial r} - k_1\frac{\partial T_1}{\partial r} = \rho_1(\dot{R} - u_1)\left[\Lambda + \frac{(u_1 - \dot{R})^2}{2g_c} - \frac{(u_2 - \dot{R})^2}{2g_c}\right.$$

$$\left. + \frac{4}{3}\frac{\mu_2}{\rho_2}\left(\frac{\partial u_2}{\partial r} - \frac{u_2}{r}\right) - \frac{4}{3}\frac{\mu_1}{\rho_1}\left(\frac{\partial u_1}{\partial r} - \frac{u_1}{r}\right)\right] \quad \text{at} \quad r = R(t)$$

For detailed discussion, see G. M. Cho and R. A. Seban, On some aspects of steam bubble collapse, *Transact. ASME, J. Heat Transf.* **91** 537–542 (1969) and D. Y. Hsieh, Some analytical aspects of bubble dynamics, *Transact. ASME, J. Basic Eng.* **87**, 991–1005 (1965).

4-18 The boundary at the interface between a fluid and a porous solid is shown in Fig. 4P-18. The porous solid is assumed to consist of cylindri-

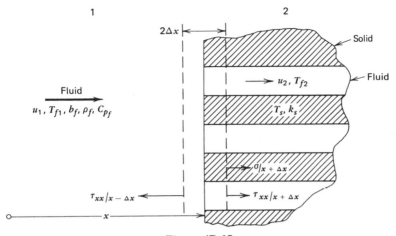

**Figure 4P-18**

cal capillary pores so that the fluid cross-sectional area is only a fraction $P$ of the total. The fluid flows steadily and horizontally into the porous solid and all properties are constant.

**(a)** Apply the conservation of mass principle to the control volume at the interface shown in dashed lines, and show that at the interface

$$u_2 = \frac{u_1}{P}$$

**(b)** Apply Newton's second law for control volumes to the control volume at the interface, and show that at the interface

$$\tau_{2_{xx}} P + \sigma(1 - P) - \tau_{1_{xx}} = \frac{\rho_f}{g_c}(u_1^2 - Pu_2^2)$$

**(c)** Apply the conservation of energy principle for control volumes to the control volume at the interface to show that at the interface

$$-k_f \frac{\partial T_{f_1}}{\partial x} + Pk_f \frac{\partial T_{f_2}}{\partial x} + (1 - P)k_s \frac{\partial T_s}{\partial x}$$

$$-\rho_f u_1 \left[ \left( e_2 - \frac{\tau_{2_{xx}}}{\rho_f} \right) - \left( e_1 - \frac{\tau_{1_{xx}}}{\rho_f} \right) \right] = 0$$

Realizing that $\tau_{xx} = -p$ for constant density and that, with $H = e_2 + p/\rho_f$, $H_2 = H_1$ if there is no fluid temperature jump at the

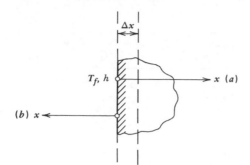

Figure 4P-19

interface, one can rearrange this relationship into

$$k_f \frac{\partial T_{f_1}}{\partial x} = (1 - P)k_s\frac{\partial T_s}{\partial x} + Pk_f\frac{\partial T_{f_2}}{\partial x} - \frac{\rho_f u_1^3(P^{-2} - 1)}{2g_c}$$

Next, it can reasonably be assumed that kinetic energy is negligible and that most conduction occurs in the solid state. Then

$$k_f \frac{\partial T_{f_1}}{\partial x} = (1 - P)k_s\frac{\partial T_s}{\partial x}$$

For a more detailed discussion, see D. A. Nealy and P. W. McFadden, Comment on an investigation of porous wall cooling *Transact. ASME, J. Heat Transf.* **91**, 284–285 (1969).

**4-19**  Consider a solid that contacts a fluid with heat exchange occurring by convection. The fluid temperature away from the solid is $T_f$, and the heat transfer coefficient is $h$. Apply the conservation of energy to the control volume shown in Fig. 4P-19 to show that at the interface
(a) $k_s\, \partial T_s/\partial x = h(T_s - T_f)$ and (b) $-k_s\, \partial T_s/\partial x = h(T_s - T_f)$.

## REFERENCES

1  S. Harris, *An Introduction To The Theory Of The Boltzmann Equation*, Holt, Rinehart, and Winston, New York, 1971.

2  J. H. Keenan, *Thermodynamics*, Wiley, New York, 1941, p. 18.

3  D. A. Russell, Gas dynamic lasers, *Astronaut. Aeronaut.* **13**, 50–55 (1975).

4  E. R. G. Eckert and R. M. Drake, *Analysis Of Heat And Mass Transfer*, McGraw-Hill, New York, 1972, p. 479.

5  J. O. Hirschfelder, C. F. Curtis, and R. B. Bird, *Molecular Theory Of Gases And Liquids*, Wiley, New York, 1966, p. 21.

6  S. Harris, *Introduction To The Theory Of The Boltzmann Equation*, Holt, Rinehart, and Winston, New York, 1971, p. 125.

7  P. K. Davis, On the substantial derivative, *Mech. Eng. News*, **2**, (1965).

8  J. F. Thorpe, On the substantial derivative, *Mech. Eng. News* **3**, 19–20 (1966).

9  J. O. Hinze, *Turbulence*, 2nd ed., McGraw-Hill, New York, 1975, p. 68.

10  E. R. G. Eckert and R. M. Drake, *Analysis Of Heat And Mass Transfer*, McGraw-Hill, New York, 1972, p. 433.

11  E. F. Obert, *Concepts of Thermodynamics*, McGraw-Hill, New York, 1960, p. 368, p. 266.

12  E. L. Cussler, *Multicomponent Diffusion*, American Elsevier, New York, 1976. (A three-page Errata is available from the publisher.)

13  J. O. Hirschfelder, C. F. Curtis, and R. B. Bird, *Molecular Theory Of Gases and Liquids*, Wiley, New York, 1966, p. 700.

14  L. M. K. Boelter, V. H. Cherry, H. A. Johnson, and R. C. Martinelli, *Heat Transfer Notes*, McGraw-Hill, New York, 1965.

15  A. Bejan, A study of entropy generation in fundamental convective heat transfer, *Transact. ASME, J. Heat Transf.* **101**, 718–725 (1979).

16  A. Bejan, The concept of irreversibility in heat exchanger design: Counterflow heat exchangers for gas-to-gas application, *Transact. ASME, J. Heat Transf.* **99**, 374–380 (1977).

17  C. Truesdell and R. G. Muncaster, *Fundamentals of Maxwell's Kinetic Theory of A Simple Monatomic Gas, Treated As A Branch of Rational Mechanics*, Academic Press, New York, 1980.

# 5

# ONE-DIMENSIONAL SOLUTIONS

The major points discernable from the describing equations in Chapter 4 can be emphasized by solution of well-chosen one-dimensional problems. Although the solutions themselves are of restricted applicability, the conclusions drawn from them are much broader.

As a preliminary, note that if all properties are constant, the energy and diffusion equation do not affect either the equations of motion or the continuity equation. In other words, the velocity distribution can be determined first, although by way of nonlinear equations. With velocities known, the temperature and mass fraction distributions can subsequently be determined. The energy and diffusion equations are then linear so that superposition may be employed if desired; the only complication introduced by velocity distributions then is that the energy and diffusion equations have variable coefficients. This simplification is generally characteristic of the problems discussed in this chapter.

Before solving specific problems, let us first determine whether solutions are possible on basic grounds. To be solvable, there should be as many equations as there are unknowns. The listing of Table 5-1 shows that a determinate system exists.

## 5.1 COUETTE FLOW

Couette flow consists of a fluid contained between two parallel plates, the upper one moving with velocity $U$ while the lower one is stationary as illustrated in Fig. 5-1. Additional simplifying assumptions are customarily

**Table 5-1  Equations and Unknowns**

| Equations | Unknowns |
|---|---|
| 1  continuity (plus $n - 1$ diffusion equations for an $n$-component fluid) | 1  density (plus $n - 1$ mass fractions for an $n$-component fluid) |
| 3  equations of motion | 3  velocity components |
| 1  energy equation | 1  temperature |
| 1  $\mu = \mu(p, T)$ | 1  viscosity |
| 1  $k = k(p, T)$ | 1  thermal conductivity |
| $[n - 1\ D_{ab} = D_{ab}(p, T)$ for $n$-component fluid] | $(n - 1$ mass diffusivities for $n$-component fluid) |
| 1  equation of state ($n$ for an $n$-component fluid) | 1  pressure (plus $n - 1$ partial pressures for an $n$-component fluid) |
| 8 ($5 + 3n$ for $n$-component fluid) | 8 ($5 + 3n$ for $n$-component fluid) |

made and appear later. The simplicity of this flow situation is attractive particularly because it allows exact solutions of the complete describing equations. The effect of the imposed flow on the heat transfer resulting from different temperatures at the top and bottom plates is ascertained. Assumptions for the following developments are (1) steady conditions, (2) laminar flow, (3) constant properties, (4) no pressure gradient in the $x$ direction, (5) no edge effects ($\partial/\partial z = 0$), (6) no end effects ($\partial/\partial x = 0$), (7) Newtonian fluid, and (8) accuracy of Fourier law. The equations of continuity, motion, and energy then reduce to

Continuity
$$\frac{dv}{dy} = 0 \qquad\qquad (5\text{-}1a)$$

$x$ Motion
$$\frac{d^2 u}{dy^2} = 0 \qquad\qquad (5\text{-}1b)$$

$y$ Motion
$$\frac{\rho g}{g_c} + \frac{dp}{dy} = 0 \qquad\qquad (5\text{-}1c)$$

$z$ Motion
$$\frac{d^2 w}{dy^2} = 0 \qquad\qquad (5\text{-}1d)$$

Energy
$$\frac{d^2 T}{dy^2} + \left(\frac{du}{dy}\right)^2 \frac{\mu}{k} = 0 \qquad\qquad (5\text{-}1e)$$

Boundary conditions to be imposed on the solutions to these equations are

| At $y = 0$ | $u = 0 = w$ | no slip | (5-2a) |
| | $v = 0$ | impermeable wall | (5-2b) |
| | $T = T_0$ | no temperature jump | (5-2c) |
| At $y = L$ | $u = U,$    $w = 0$ | no slip | (5-2d) |
| | $v = 0$ | impermeable wall | (5-2e) |
| | $T = T_1$ | no temperature jump | (5-2f) |

Note that the boundary conditions and the basic assumptions have been used to simplify the continuity, motion, and energy equations. The full continuity equation, for example, for constant properties is simplified as

$$\underset{\substack{\text{Assumption} \\ 0}}{\frac{\partial u}{\partial x}} + \frac{\partial v}{\partial y} + \underset{\substack{(5) \\ 0}}{\frac{\partial w}{\partial z}} = 0$$

Assumption        (6)                (5)

The first and last terms vanish by assumptions 6 and 5, giving Eq. (5-1a), which in turn shows that $v = \text{const}$. But since $v = 0$ at the two plates, $v = 0$ everywhere. The constant-property $z$-motion equation is simplified as

0 continuity                                              0 none imposed

$$\frac{\partial w}{\partial t} + u\frac{\partial w}{\partial x} + v\frac{\partial w}{\partial y} + w\frac{\partial w}{\partial z} = \frac{g_c}{\rho}\left(B_z - \frac{\partial p}{\partial z}\right)$$

Assumption   (1)        (6)                (5)              (5)
            0          0                0                0

$$+ g_c\nu\left(\frac{\partial^2 w}{\partial x^2} + \frac{\partial^2 w}{\partial y^2} + \frac{\partial^2 w}{\partial z^2}\right)$$

                                              0          0
                                             (6)        (5)

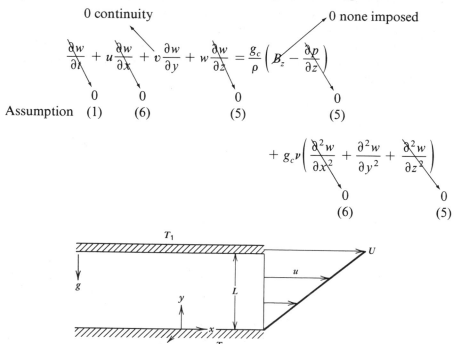

**Figure 5-1**   Geometry and coordinate system for Couette flow.

giving Eq. (5-1d). The constant-property $y$-motion equation is easily simplified, since it has already been shown that $v = 0$, as

$$\frac{\partial v}{\partial t} + u\frac{\partial v}{\partial x} + v\frac{\partial v}{\partial y} + w\frac{\partial v}{\partial z} = \frac{g_c}{\rho}\left(B_y - \frac{\partial p}{\partial y}\right) + g_c\nu\left(\frac{\partial^2 v}{\partial x^2} + \frac{\partial^2 v}{\partial y^2} + \frac{\partial^2 v}{\partial z^2}\right)$$

_____
                    0 continuity                          $-\dfrac{\rho g}{g_c}$                          0 continuity

giving Eq. (5-1c). The constant-property $x$-motion equation is simplified as

0 continuity                                                        0 none imposed

$$\frac{\partial u}{\partial t} + u\frac{\partial u}{\partial x} + v\frac{\partial u}{\partial y} + w\frac{\partial w}{\partial z} = \frac{g_c(B_x - \partial p/\partial x)}{\rho}$$

0                  0                                0                                0
Assumption   (1)        (6)                          (5)                              (6)

$$+ g_c\nu\left(\frac{\partial^2 u}{\partial x^2} + \frac{\partial^2 u}{\partial y^2} + \frac{\partial u^2}{\partial z^2}\right)$$

0                        0
(6)                      (5)

giving Eq. (5-1b). The constant-property energy equation is simplified as

0 continuity

$$\frac{\partial T}{\partial t} + u\frac{\partial T}{\partial x} + v\frac{\partial T}{\partial y} + w\frac{\partial T}{\partial z} = \alpha\left(\frac{\partial^2 T}{\partial x^2} + \frac{\partial^2 T}{\partial y^2} + \frac{\partial^2 T}{\partial z^2}\right)$$

0                  0                       0                   0                        0
Assumption   (1)        (6)                  (5)                 (6)                       (5)

0 none imposed

$$+ \frac{q'''}{\rho C_p} + \frac{\mu}{\rho C_p}\Phi$$

$$\Phi = 2\left[\left(\frac{\partial u}{\partial x}\right)^2 + \left(\frac{\partial v}{\partial y}\right)^2 + \left(\frac{\partial w}{\partial z}\right)^2\right] - \frac{2}{3}(\text{div } \mathbf{V})^2 + \left(\frac{\partial u}{\partial y} + \frac{\partial v}{\partial x}\right)^2$$

0 continuity      0 continuity

Assumption    (6)       0       (5)      0       (6)    0

0 z-motion and boundary conditions

$$+ \left(\frac{\partial v}{\partial z} + \frac{\partial w}{\partial y}\right)^2 + \left(\frac{\partial w}{\partial x} + \frac{\partial u}{\partial z}\right)^2$$

0      0    0
(6)     (5)

giving Eq. (5-1e).

The $y$-motion equation [Eq. (5-1c)] shows that pressure undergoes a hydro-static variation in the $y$ direction as

$$p = p_0 - \rho\left(\frac{g}{g_c}\right)y$$

The $z$-motion equation [Eq. (5-1d)] and its boundary conditions show that $w = 0$. From the $x$-motion equation it is found that

$$\frac{u}{U} = \frac{y}{L} \tag{5-3}$$

which represents a linear variation of velocity. The energy equation can be solved now that velocity is known. With the use of $du/dy = U/L$, the energy equation is

$$\frac{d^2T}{dy^2} = -\frac{\mu}{k}\frac{U^2}{L^2} \tag{5-4}$$

$$\frac{T - T_0}{T_1 - T_0} = \frac{y}{L} + \frac{\text{E Pr}}{2}\frac{y}{L}\left(1 - \frac{y}{L}\right) \tag{5-5}$$

where E = Eckert number = $U^2/C_p(T_1 - T_0)$, representing a measure of the importance of viscous dissipation relative to the imposed temperature dif-ference, and Pr = Prandtl number = $\mu C_p/k$. The heat flow at the lower surface is obtained from the Fourier law applied to Eq. (5-5) as

$$q_w = -k\frac{dT(y=0)}{dy} = -\frac{k(T_1 - T_0)(1 + \mathrm{E\,Pr}/2)}{L}$$

which can be rearranged into

$$q_w = \frac{k\left[T_0 - \left(T_1 + \mathrm{Pr}\,U^2/2C_p\right)\right]}{L} \tag{5-6}$$

The simple Couette flow situation here considered is important not only as an approximation to conditions in a hydrodynamically lubricated bearing, but also because it resembles boundary-layer flow over a flat plate far from the leading edge as shown in Fig. 5-2. The outer edge of the boundary layer is represented by the moving upper plate in the Couette flow model. The disturbing aspect of the heat flow found in Eq. (5-6) is that more than a temperature difference drives the heat flow, and it is expected that the same conclusion applies to boundary layer flow. Of course, the physical mechanism responsible for this inconvenient state of affairs is that some kinetic energy is dissipated near the solid surface into a thermal form of energy.

In an attempt to recover the idea that heat flows only in response to a temperature difference, consider $\mathrm{Pr}\,U^2/2C_p$ to be a recovery or adiabatic wall temperature. To show that this is a reasonable thing to do, the case where the lower surface is insulated is solved next and the temperature it attains under the influence of viscous dissipation determined. The velocity distribution is unchanged from Eq. (5-3), so that specification of an insulated lower surface requires as the only change that $dT(y=0)/dy = 0$. As a result, the solution to Eq. (5-4) is

$$T - T_1 = \frac{\mathrm{Pr}\,U^2}{2C_p}\left(\frac{1 - y^2}{L^2}\right) \tag{5-7}$$

At the lower surface it is then found that

$$T(y=0) = T_{aw} = \frac{\mathrm{Pr}\,U^2}{2C_p} + T_1 \tag{5-8}$$

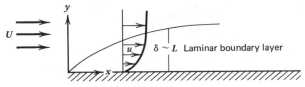

**Figure 5-2**   Laminar boundary-layer flow resembles Couette flow.

where $T_{aw}$ is the temperature an adiabatic surface assumes under the influence of viscous dissipation. In the previous expression for heat flow at the lower surface [Eq. (5-6)], one can review $T_0$ as $T_{wall}$ and $T_1 + \Pr(U^2/2Cp)$ as $T_{aw}$. Then the interpretation of

$$q_w = \frac{k(T_{wall} - T_{aw})}{L} \tag{5-9}$$

holds true. The conclusion is that the driving temperature difference for convection is really $T_{wall} - T_{aw}$, with the possibility yet remaining that $T_{aw}$ may vary for different geometries and flow conditions. Heat-transfer coefficients measured or calculated at low speeds, where viscous dissipation is negligible can then be used in high-speed situations where $U^2/C_p$, dissipation, is large provided that fluid properties are evaluated at the proper reference temperature, and no further correlations of experimental data are necessary.

The convenience offered by basing the wall heat flux on the difference between the wall temperature and the adiabatic wall temperature as suggested by Eq. (5-9) can be appreciated by considering the temperature distribution given by Eq. (5-5) in more detail. Figure 5-3 shows that at $E \Pr > 2$ the wall heat flux actually is into the upper plate even though the imposed temperature difference would initially have been considered to cause the wall heat flux to be out of the upper plate. The heat-transfer coefficient $h$ at the lower plate based on the imposed temperature difference $T_0 - T_1$ is

$$h = \frac{q_w}{T_0 - T_1} = \frac{k}{L}\left(1 + \frac{E \Pr}{2}\right) = \frac{k}{L}\left(1 + \frac{U^2\mu/k}{T_1 - T_0}\right)$$

according to Eq. (5-6). If the heat-transfer coefficient is based on $T_0 - T_1$, the result is a coefficient that depends on both the imposed velocity and the imposed temperature difference; occasionally, this will yield a negative $h$ since the driving potential difference for local heat transfer is incorrect—the im-

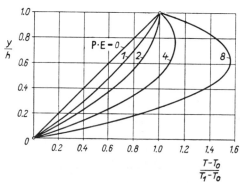

**Figure 5-3**   Dimensionless temperature variation in Couette flow.

posed temperature difference is *not* proportional to the temperature gradient at the wall. If the heat-transfer coefficient is based on the difference between the imposed and the adiabatic wall temperatures, $T_w - T_{aw}$, then

$$h = \frac{q_w}{T_w - T_{aw}} = \frac{k}{L} \qquad (5\text{-}10)$$

according to Eq. (5-9), which leads to a Nusselt number Nu of $\text{Nu} = hL/k = 1$. This result is more convenient due to its constancy at the value for negligible viscous dissipation $T_{aw} \approx T_1$. Because the driving potential difference for local heat transfer is correctly ascertained, $h$ will never be negative.

Equation (5-8) provides another item of information that is particularly useful in the boundary layer flows, which Couette flow resembles. Equation (5-8) demonstrates that the kinetic energy of the fluid at the upper plate $U^2/2$ is partially converted into thermal form. If the lower plate is insulated, some of the converted energy shows up there and is manifested by a temperature rise $T_{aw} - T_1$. The ratio of this "recovered" thermal energy at the lower wall to the fluid kinetic energy at the upper plate is called the *recovery factor r*. Thus, for Couette flow

$$r = C_p \frac{T_{aw} - T_1}{U^2/2} = \text{Pr} \qquad (5\text{-}11)$$

For gases such as air, $\text{Pr} < 1$ and $r$ is less than unity. For liquids such as water, $\text{Pr} > 1$ and $r$ exceeds unity.

## 5.2 POISEUILLE FLOW

Poiseuille flow is much like Couette flow, except that both surfaces are usually considered to be stationary. Flow is caused by maintaining a pressure differential down the flow channel. Figure 5-4 and the following list of assumptions

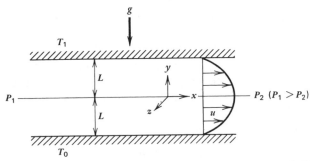

**Figure 5-4**  Geometry and coordinate system for Poiseuille flow.

describe the physical problem to be treated. Assumptions for the following developments are (1) steady conditions, (2) laminar flow, (3) constant properties, (4) constant $x$-direction pressure gradient, (5) no edge effects ($\partial/\partial z = 0$), (6) no end effects ($\partial/\partial x = 0$ except for pressure), (7) Newtonian fluid, and (8) accuracy of Fourier law.

The equations of continuity, motion, and energy then reduce to

$$\text{Continuity} \qquad \frac{dv}{dy} = 0 \qquad\qquad (5\text{-}12a)$$

$$x \text{ Motion} \qquad \frac{d^2u}{dy^2} = \frac{1}{\mu}\frac{dp}{dx} \qquad\qquad (5\text{-}12b)$$

$$y \text{ Motion} \qquad \frac{dp}{dy} = -\frac{\rho g}{g_c} \qquad\qquad (5\text{-}12c)$$

$$z \text{ Motion} \qquad \frac{d^2w}{dy^2} = 0 \qquad\qquad (5\text{-}12d)$$

$$\text{Energy} \qquad \frac{d^2T}{dy^2} + \frac{\mu}{k}\left(\frac{du}{dy}\right)^2 = 0 \qquad\qquad (5\text{-}12e)$$

Boundary conditions to be imposed are

| | | | |
|---|---|---|---|
| At $y = -L$ | $u = 0 = w$ | no slip | (5-13a) |
| | $v = 0$ | impermeable wall | (5-13b) |
| | $T = T_0$ | no temperature jump | (5-13c) |
| At $y = L$ | $u = 0 = w$ | no slip | (5-13d) |
| | $v = 0$ | impermeable wall | (5-13e) |
| | $T = T_1$ | no temperature jump | (5-13f) |

The initial assumptions and boundary conditions have been used to simplify the full continuity, motion, and energy equations in the same manner as for the Couette flow case.

The continuity equation [Eq. (5-12a)] and its boundary conditions [Eqs. (5-13b and (5-13e)] yield $v = 0$. Similarly, the $z$-motion equation [Eq. (5-12d)] and its boundary conditions [Eqs. (5-13a) and (5-13d)] give $w = 0$. The pressure variation in the vertical direction is found from the $y$-motion equation to be purely hydrostatic as

$$p = p_0 - \rho\frac{g}{g_c}y$$

From the $x$-motion equation and its boundary conditions it is found that

$$u = \frac{L^2}{2\mu}\left(-\frac{dp}{dx}\right)\left(1 - \frac{y^2}{L^2}\right) \tag{5-14}$$

The maximum velocity occurs at the center line where $y = 0$ and is

$$u_m = \frac{L^2}{2}\left(-\frac{dp}{dx}\right) = \frac{3}{2}u_{av} \tag{5-15}$$

The average velocity is found from its definition to be

$$u_{av} = \frac{1}{2L}\int_L^L u\,dy$$

$$= \frac{L^2}{3}\left(-\frac{dp}{dx}\right)$$

Thus

$$u = u_m\left(1 - \frac{y^2}{L^2}\right) \tag{5-16}$$

Or

$$u = \frac{3}{2}u_{av}\left(1 - \frac{y^2}{L^2}\right)$$

Note that it was again possible to solve the equations of motion without consideration of temperature distributions because of the constancy of properties. If velocities are known, the energy equation, [Eq. (5-12e)] can next be attacked to find temperature as

$$T = T_0 + \frac{T_1 - T_0}{2}\left(1 + \frac{y}{L}\right) + \frac{\mu}{3k}u_m^2\left(1 - \frac{y^4}{L^4}\right) \tag{5-17}$$

The temperature distribution predicted by Eq. (5-17) is shown in Fig. 5-5, where it is seen that for $E\,Pr > \frac{3}{8}$, heat flows into the upper plate even though the upper plate might be warmer than the bottom one. The heat flow at the lower surface $(y = -L)$ is obtained from the Fourier law as

$$q_w = -k\frac{dT(y = -L)}{dy} = -\frac{k}{2L}(T_1 - T_0)\left(1 + \frac{8}{3}Pr\,E\right)$$

$$q_w = \frac{k}{2L}\left(T_0 - T_1 - \frac{8}{3}\frac{Pr\,u_m^2}{C_p}\right) \tag{5-18}$$

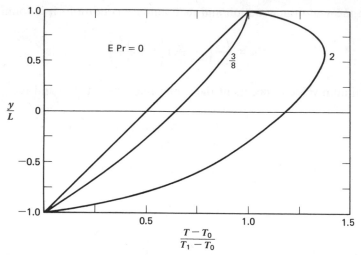

**Figure 5-5**  Dimensionless temperature variation in Poiseuille flow.

where $E =$ Eckert number $= u_m^2/C_p(T_1 - T_0)$ and $Pr =$ Prandtl number $= \mu C_p/k$.

It is presumed that the preceding relationship [(Eq. (5-18)] for lower surface heat flow can be interpreted in the same manner as for the Couette flow case $T_1 + \frac{8}{3} Pr\, u_m^2/C_p = T_{aw}$. To determine with certainty that this can be done, let the lower surface be insulated and adiabatic. The boundary condition at the lower surface is restated as

$$\frac{dT(y = -L)}{dy} = 0$$

The velocity distribution being unchanged, the resultant temperature is

$$T - T_1 = \frac{\mu}{3k} u_m^2 \left(5 - \frac{y^4}{L^4} - 4\frac{y}{L}\right)$$

The adiabatic temperature of the lower surface is then

$$T_{aw} = T(y = -L) = T_1 + \frac{8}{3}\frac{Pr\, u_m^2}{C_p}$$

$$= T_1 + 6\frac{Pr\, u_{av}^2}{C_p} \qquad\qquad (5\text{-}19)$$

The truth of the interpretation

$$q_w = \frac{k}{2L}(T_{wall} - T_{aw})$$

is again seen as is the convenience of having the corresponding heat transfer coefficient $h = k/2L$ apply to both low-speed and high-speed flow (in which viscous dissipation is important). The Nusselt number Nu is seen to be $\text{Nu} = 2hL/k = 1$. For this slightly different flow condition, Eq. (5-19) shows that $T_{aw}$ must be computed from a formula slightly different than in Eq. (5-8) for Couette flow. As a practical matter, accurate computation of $T_{aw}$ requires that properties be evaluated at some suitable reference temperature to account for their temperature dependence.

The moral again is that, when accounting for the effects of viscous dissipation, the proper driving temperature difference for convective heat transfer is $T_{\text{wall}} - T_{aw}$. Inasmuch as the only major assumption is the linearity of the energy equation with respect to $T$, the same conclusion is expected to hold for more complicated flow conditions. The only difficulty is accurate computation of $T_{aw}$.

The Poiseuille flow case just treated resembles flow down a duct, where the the duct is quite long and wide. In contrast to Couette flow whose resemblence to boundary layer flow makes the imposed velocity a natural choice for a reference velocity, the Poiseuille flow has several plausible reference velocities (e.g., $u_m$ or $u_{av}$), but none is obvious by reason of being imposed. In duct flow it is usually the average velocity that is known, and it is related to the maximum velocity by Eq. (5-15). Accordingly, the adiabatic wall temperature is most conveniently given by

$$T_{aw} = T_1 + 6\frac{\text{Pr } u_{av}^2}{C_p}$$

The recovery factor for this Poiseuille flow is then

$$r = \frac{C_p(T_{aw} - T_1)}{u_{av}^2/2} = 12\text{Pr}$$

which is of the same form as Eq. (5-11) for Couette flow, but with a different coefficient.

The cases of temperature dependent viscosity in Poiseuille flow between two planes and in a circular duct are reported by Schlichting [4], as is flow between convergent planes.

## 5.3  STEFAN'S DIFFUSION PROBLEM

In the two foregoing examples the fluid was considered to be a single-component fluid. Exploration of the consequences of allowing the fluid to have more than one component is undertaken by examining a two-component (binary) convection problem referred to as *Stefan's diffusion problem*. The problem is

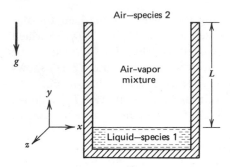

**Figure 5-6**  Geometry and coordinate system for Stefan's diffusion problem.

sufficiently simple physically to substantiate any necessary assumptions so that the ensuing solutions can be taken as exact solutions of the complete describing equations.

In Stefan's problem a container partially filled with a volatile liquid is exposed to an atmosphere free of the vapor. The rate of evaporation is to be related to the saturation vapor pressure specified by the liquid temperature. Figure 5-6 shows the physical arrangement. Before going into the details of a mathematical solution, the physical events are briefly discussed. As the liquid evaporates, its vapor moves from a region of high concentration at the bottom to a region of low concentration at the top. As the top is open, the vapor moves out of the container. In like fashion the air will try to move from the top to the bottom of the container; but if the liquid is impermeable to air or does not absorb air, the air is stagnant and cannot experience any net motion downward. Therefore, the diffusion of air downward must be countered by an upward bulk flow. Assumptions for this situation are (1) steady state, (2) stagnant conditions (laminar flow), (3) liquid impermeable to air, (4) constant total pressure and temperature, (5) liquid and vapor in equilibrium at their interface, (6) air and vapor as ideal gases, and (6) variations in the $y$ direction only.

Since it is assumed that $u = 0 = w$ (this can be inaccurate [14]) the $x$- and $z$-motion equations need not be considered. The assumed constancy of temperature renders the energy equation irrelevant (since its solution has been assumed). Likewise, the $y$-motion equation is unneeded because of the assumed constancy of total pressure. The two equations of continuity and diffusion are all that are needed for determination of the vertical bulk velocity $v$ and the mass fraction $\omega_1$ of species 1. These reduce to

$$\text{Continuity} \qquad \frac{d(\rho v)}{dy} = 0 \qquad\qquad (5\text{-}20a)$$

$$\text{Diffusion} \qquad \rho v \frac{d\omega_1}{dy} = D_{12} \frac{d(\rho\, d\omega_1/dy)}{dy} \qquad (5\text{-}20b)$$

The applicable boundary conditions are

At $y = 0^+$ $\qquad \omega_1 = \omega_{10}$ or $\rho_1 = \rho_{10}$ $\qquad$ vapor in equilibrium $\qquad$ (5-21a)
with liquid

$\qquad\qquad\qquad \rho_2 v_2 = 0$ $\qquad\qquad\qquad\qquad\qquad$ liquid impermeable $\qquad$ (5-21b)
to air

At $y = L$ $\qquad \omega_1 = \omega_{1L}$ or $\rho_1 = \rho_{1L}$ $\qquad$ some vapor in $\qquad\qquad$ (5-21c)
ambient air

Note that bulk density $\rho$ has not been assumed to be constant.

An important conclusion can be reached without solving either the diffusion or the continuity equation. The mass flow rate of air (species 2) is given by

$$\rho_2 v_2 = \rho_2 v - \rho D_{12} \frac{d\omega_2}{dy} \qquad (5\text{-}22)$$

But there can be no net flow of air since the liquid is impermeable to air; thus it is everywhere true that $\rho_2 v_2 = 0$. This fact gives the bulk velocity from Eq. (5-22) as

$$v = \frac{\rho}{\rho_2} D_{12} \frac{d\omega_2}{dy}$$

By definition $\omega_2 = \rho_2/\rho$, so this relationship becomes

$$v = \frac{D_{12}}{\omega_2} \frac{d\omega_2}{dy}$$

Recall also that by definition $\rho = \rho_1 + \rho_2$, so that $1 = \omega_1 + \omega_2$. Incorporation of these relations into the equation for $v$ gives

$$v = -\frac{D_{12}}{1 - \omega_1} \frac{d\omega_1}{dy} \qquad (5\text{-}23)$$

Somewhat surprisingly, there is an upward bulk flow, a "blowing," which exactly counteracts the downward diffusion of air. The mass flow rate of vapor (species 1) must be constant so that

$$\dot{m}_1 = \rho_1 v - \rho D_{12} \frac{d\omega_1}{dy} \qquad (5\text{-}24)$$

If Eq. (5-23) is introduced into Eq. (5-24), the vapor mass flow rate becomes

$$\dot{m}_1 = -\frac{\rho}{1 - \omega_1} D_{12} \frac{d\omega_1}{dy} \qquad (5\text{-}25)$$

Note that because $\dot{m}_1$ is constant, Eq. (5-23) compared with Eq. (5-25) reveals that $\rho v$ is also constant. Evaluation of all quantities at the liquid–vapor interface puts Eq. (5-25) in the form

$$\dot{m}_1 = - \frac{(\rho D_{12} \, d\omega_1/dy) \, |_{y-0}}{1 - \omega_1 |_{y=0}} \tag{5-26}$$

The point to be observed here is that Fick's law $\dot{m}_1 = -[\rho D_{12} \, d\omega_1/dy] |_{y=0}$, represents diffusive flow, which is also given in terms of a mass transfer film coefficient $h_D$ as $\dot{m}_{1_y} = \rho h_d \, \Delta\omega_1$; the effect of bulk convection induced by the mass transfer is important in some cases and is incorporated in Eq. (5-26) by $(1 - \omega|_{y=0})$ in the denominator. In general, then, the effect of bulk convection can be approximately taken into account in mass transfer in which one surface is impermeable to species 2 of a binary mixture by use of

$$\dot{m}_1 = \frac{\rho h_D \Delta\omega_1}{1 - \omega_{10}} \tag{5-27}$$

Do the continuity and diffusion equations yield the same result? Their integration will show. The continuity equation [Eq. (5-20a)] gives, after one integration,

$$\rho v = C$$

which agrees with the earlier deduction. The diffusion equation [Eq. (5-20b)] gives, after one integration,

$$C_1 + \rho v \omega_1 = \rho D_{12} \frac{d\omega_1}{dy} \tag{5-28}$$

Equation (5-23) has already established that

$$\rho v = - \frac{\rho D_{12}}{1 - \omega_1} \frac{d\omega_1}{dy}$$

which, when inserted into Eq. (5-28), gives

$$C_1 = \frac{\rho D_{12}}{1 - \omega_1} \frac{d\omega_1}{dy} \tag{5-29}$$

Equation (5-26) compared with Eq. (5-29) immediately reveals that $\dot{m}_1 = -C_1$; $\dot{m}_1$ is shown by the diffusion and continuity equations to be constant, just as previously argued on physical grounds.

The distribution of species 1 mass fraction $\omega_1$ or its partial pressure $p_1$ can be determined by a second integration of the diffusion equation. This requires that Eq. (5-29) be integrated once. If total density $\rho$ is assumed constant, it can be immediately integrated. However, accuracy and a complete solution are sought, so variable total density is considered. Relations pertinent to a binary mixture of ideal gases now are useful and are developed next.

Because the equilibrium vapor pressure of a liquid at a specified temperature is most commonly given, Eq. (5-19) is recast into a form that displays partial pressures instead of species 1 mass fraction. Recall that for a binary mixture of isothermal perfect gases, the total pressure is the sum of the partial pressures

$$p = p_1 + p_2 \tag{5-30}$$

Also, the equation of state for a perfect gas requires $p = \rho RT/M$ for the mixture and $p_i = \rho_i RT/M_i$ for each component. Insertion of these equations of state into Eq. (5-30) gives

$$\frac{1}{M} = \frac{\rho_1}{\rho}\frac{1}{M_1} + \frac{\rho_2}{\rho}\frac{1}{M_2} \tag{5-31}$$

Use of the equation of state in the definition of mass fraction gives

$$\omega_1 = \frac{\rho_1}{\rho} = \frac{P_1 M_1}{RT}\frac{RT}{pM} = \frac{p_1}{p}\frac{M_1}{M}$$

Substitution of Eq. (5-31) into this relation for $\omega_1$ gives

$$\omega_1 = \frac{p_1}{p}\left(\frac{\rho_1}{\rho} + \frac{\rho_2}{\rho}\frac{M_1}{M_2}\right)$$

which can be rearranged, in view of the definition of $\omega_i = \rho_i/\rho$ and the fact that $1 = \omega_1 + \omega_2$, to give

$$\omega_1 = \frac{p_1(M_1/M_2)}{p - p_1(1 - M_1/M_2)} \tag{5-32}$$

It is further found that the bulk density is given by

$$\rho = \frac{pM}{RT} = \frac{M_2}{RT}\left[p - p_1\left(1 - \frac{M_1}{M_2}\right)\right] \tag{5-33}$$

Equation (5-29) becomes, after incorporation of the relations for $\omega_1$ and $\rho$ provided by Eqs. (5-32) and (5-33),

$$C_1 = \frac{pM_1}{RT}\frac{D_{12}}{p - p_1}\frac{dp_1}{dy}$$

whose solution is

$$C_2 - C_1 \frac{RT}{pM_1 D_{12}} y = \ln(p - p_1) \tag{5-34}$$

Evaluation of $C_1$ and $C_2$ is accomplished by consideration of the boundary conditions expressed by Eqs. (5-21a) and (5-21c) with the result that

$$\ln\left(\frac{p - p_1}{p - p_{10}}\right) = \frac{y}{L} \ln\left(\frac{p - p_{1L}}{p - p_{10}}\right)$$

$$= \ln\left[\left(\frac{p - p_{1L}}{p - p_{10}}\right)^{y/L}\right]$$

which can be rewritten as either

$$\frac{p - p_1}{p - p_{10}} = \left(\frac{p - p_{1L}}{p - p_{10}}\right)^{y/L} \tag{5-35a}$$

or

$$\frac{p_2}{p_{20}} = \left(\frac{p_{2L}}{p_{20}}\right)^{y/L} \tag{5-35b}$$

As illustrated in Fig. 5-7, the partial pressure distribution of Eq. (5-35) is nonlinear. Additionally, it is recalled that $C_1 = -\dot{m}_1$ so that the evaporation rate of species 1 is given by

$$\dot{m}_1 = \frac{pM_1 D_{12}}{RTL} \ln\left(\frac{p - p_{1L}}{p - p_{10}}\right) \tag{5-36}$$

At this point it is appropriate to use the exact solution expressed by Eq. (5-36) to test the suggestion of Eq. (5-27). To do this, Eq. (5-36) is first recast to employ mass fractions with the aid of Eq. (5-32) to obtain

$$\dot{m}_1 = \frac{pM_1 D_{12}}{RTL} \ln\left\{ 1 + \frac{\omega_{10} - \omega_{1L}}{(1 - \omega_{10})\left[\frac{M_1}{M_2} + \omega_{1L}\left(1 - \frac{M_1}{M_2}\right)\right]} \right\} \tag{5-37}$$

or,

$$\dot{m}_1 = \frac{pM_1 D_{12}}{RTL} \ln\left(1 + \frac{p_{10} - p_{1L}}{p - p_{10}}\right)$$

The series expansion of $\ln(1 + x) = x + \cdots$, accurate for small values of $x$, in

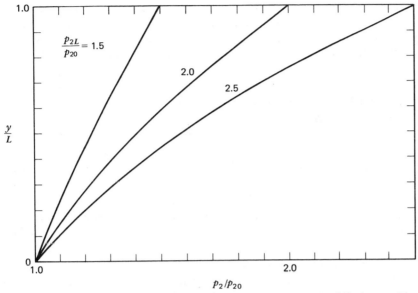

**Figure 5-7**  Dimensionless partial pressure variation in Stefan's diffusion problem.

Eq. (5-37) shows that for small values of $\omega_1$ (low rates of species 1 evaporation)

$$\dot{m}_1 \approx \frac{pM_2 D_{12}}{RTL} \frac{\omega_{10} - \omega_{1L}}{1 - \omega_{10}} \qquad (5\text{-}38)$$

or

$$\dot{m}_1 \approx \frac{pM_1 D_{12}}{RTL} \frac{p_{10} - p_{1L}}{p - p_{10}}$$

Comparison with Eq. (5-27) shows that for this particular case the mass transfer coefficient $h_D$ at low transport rates is

$$\dot{m}_1 \frac{1 - \omega_{10}}{\rho(\omega_{10} - \omega_{10})} = h_D \approx \frac{pM_2 D_{12}}{\rho RTL} \approx \frac{D_{12}}{L}$$

which is a constant independent of the species 1 mass fraction. From this result, the Sherwood number Sh is found to be

$$\text{Sh} = \frac{h_D L}{D_{12}} = 1$$

Note that the "blowing" associated with mass transfer in which one surface is impermeable is accounted for by the $1 - \omega_{10}$ term in the denominator of Eq. (5-27), but only approximately. Also, note that $\dot{m}_1$ is more simply expressed in terms of partial pressure than in terms of mass fraction. At large mass transfer rates, the full solutions to the describing differential equations [e.g., Eqs. (5-34) or (5-35) for Stefan's diffusion problems] must be used for accuracy—the alternative is to consider the mass-transfer coefficient to be dependent on mass fraction, which would be awkward since its greatest utility is as a constant unaffected by the driving potential.

The rate at which a volatile liquid evaporates into an infinitely long tube has been analytically determined by Arnold [5]. This result is useful in estimating the time required to achieve steady-state conditions in a tube of finite length.

Not every binary diffusion problem has one surface impermeable to the second species. Another important simple case is that of equimolar counterdiffusion in which the molar transport rates of the two species are constrained to be equal in magnitude and oppositely directed. This case is the subject of Problem 5-11.

Vapor pressure–temperature data can be obtained from such sources as the *International Critical Tables* [6], as well as the books by Nesmeyanov [7] and Jordan [8].

## 5.4   TRANSIENT CHANGE OF PHASE

Up to this point no change of phase of the fluid has been taken into account. Applications such as quenching during heat treatment, boiling heat exchangers, and ablative heat shields for reentry vehicles rely on the fact that a large amount of energy must be absorbed by a fluid in order to change phase. Consequently, large amounts of heat must be transferred to accomplish a very small temperature change; thus heat-transfer coefficients become quite large.

Illustration of this point is initiated by treating a transient vaporization problem in which a large body of stagnant liquid is initially at the boiling temperature corresponding to the imposed system pressure. A large flat plate rests horizontally in the pool. Suddenly, the plate attains a high temperature that is substantially above the liquid boiling temperature. As a result, a film of vapor forms between the plate and the liquid. As time passes, the vapor film increases in thickness as a result of heat transfer through the film into the liquid–vapor interface, where more liquid is vaporized (see Fig. 5-8). This is also referred to as *Stefan's problem*, a classical discussion of which is given by Carslaw and Jaeger [3].

Assumptions applied to this problem are: (1) only $x$-direction variations are important; (2) pressure is constant; (3) the plate is at constant temperature; (4) the liquid is everywhere at its saturation temperature; (5) the density–thermal conductivity product is constant; (6) viscous dissipation is negligible; and (7) thermal radiation is negligible. With these simplifications, the describing

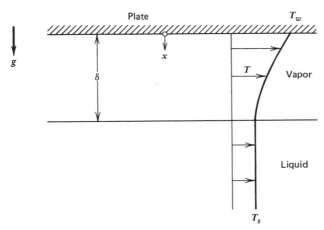

**Figure 5-8** Geometry and coordinate system for the problem of one-dimensional transient change of phase.

equations are, for $0 \leqslant x \leqslant \delta$,

$$\text{Continuity} \qquad \frac{\partial \rho}{\partial t} + \frac{\partial \rho u}{\partial x} = 0 \qquad\qquad (5\text{-}39)$$

$$\text{Energy} \qquad \rho C_p \left( \frac{\partial T}{\partial t} + u \frac{\partial T}{\partial x} \right) = \frac{\partial}{\partial x} \left( k \frac{\partial T}{\partial x} \right) \qquad (5\text{-}40)$$

subject to the boundary conditions that

At $x = 0$       $T = T_w$       no temperature jump       (5-41a)

                  $u = 0$       wall impermeable       (5-41b)

At $x = \delta$       $T = T_s$       no temperature jump       (5-41c)

$$-k \frac{\partial T}{\partial x} = \Lambda \frac{dM}{dt}$$

      heat conducted into liquid–vapor interface all used to vaporize liquid       (5-41d)

where $\delta$ is vapor film thickness, $\Lambda$ is latent heat of vaporization, and $M$ is the vapor mass in the film per unit area.

A coordinate transformation is now undertaken that will place the continuity and energy equations of Eqs. (5-37) and (5-38) in a Lagrangian form. This transformation, referred to as a *Lagrangian transformation* because it removes convective terms, is quite useful in what are basically one-dimensional problems. To begin, integrate the continuity equation of Eq. (5-39) from $x = 0$ to obtain

$$\int_0^x \frac{\partial \rho}{\partial t} dx + \int_0^x \frac{\partial \rho u}{\partial x} dx = 0$$

$$\int_0^x \frac{\partial \rho}{\partial t} dx + (\rho u)|_{x,t} - (\rho u)|_{x=0,t}^0 = 0 \quad \text{by Eq. (5-41b)}$$

Since $x$ is independent of $t$, allowing the order of integration and differentiation to be interchanged, one has

$$\frac{\partial}{\partial t} \left( \int_0^x \rho \, dx \right) = -\rho u$$

or

$$\frac{\partial m}{\partial t} = -\rho u \qquad (5\text{-}42)$$

where $m = \int_0^x \rho(x, t) \, dx$ is the mass of vapor per unit area contained between
the plate and a parallel plane a distance $x$ away. Although $m$ depends on both $t$
and $x$, $m$ can be used as a replacement for $x$ as a spatial position indicator in
the energy equation since at any fixed time $m$ is directly related to $x$. Next, the
energy equation is transformed from $x, t$ to $m, t'$ as independent variables. By
the chain rule

$$\frac{\partial}{\partial t} = \frac{\partial}{\partial t'} \frac{\partial t'}{\partial t} + \frac{\partial}{\partial m} \frac{\partial m}{\partial t}$$

$$\frac{\partial}{\partial x} = \frac{\partial}{\partial t'} \frac{\partial t'}{\partial x} + \frac{\partial}{\partial m} \frac{\partial m}{\partial x}$$

It has already been specified that $m = \int_0^x \rho \, dx$, and we additionally arbitrarily
set $t' = t$. As a result,

$$\frac{\partial}{\partial t} = \frac{\partial}{\partial t} + \frac{\partial}{\partial m} \overset{\rho u}{\frac{\partial m}{\partial t}} \quad \text{and} \quad \frac{\partial}{\partial x} = \frac{\partial}{\partial m} \overset{\rho}{\frac{\partial m}{\partial x}}$$

The energy equation in terms of $m$ and $t$ is then

$$\rho C_p \left[ \left( \frac{\partial T}{\partial t} - \rho u \frac{\partial T}{\partial m} \right) + \rho u \frac{\partial T}{\partial m} \right] = \frac{\partial}{\partial x} \left( k\rho \frac{\partial T}{\partial m} \right)$$

Note on the right-hand side that $k\rho$ can accurately be assumed to be constant.
Its removal from the derivative leaves only a function of $m$ and $t$ to be
differentiated with respect to $x$. The chain rule has divulged the relation

between an $x$ and an $m$ derivative so that the energy equation reduces to, for $0 \leqslant m \leqslant M$,

$$C_p \frac{\partial T}{\partial t} = (k\rho) \frac{\partial^2 T}{\partial m^2} \tag{5-43}$$

with the boundary conditions of $T(m = 0, t) = T_w$, $T(m = M, t) = T_s$, and

$$- (k\rho) \frac{\partial T(m = M, t)}{\partial m} = \Lambda \frac{dM}{dt} \tag{5-44}$$

The transformed energy equation of Eq. (5-43) certainly has a Lagrangian aspect—all convective terms have disappeared, and it has the form of a conduction problem. Indeed, all specific problems treated so far contained very little convection, and when there was significant convection, the equations to be solved were linear. In the present case the apparent linearity is illusory. Although it is true that the partial differential equation of Eq. (5-43) is linear, the boundary condition of Eq. (5-44) is not. This is characteristic of problems in which the location of a surface is initially unknown and must be determined in the course of the problem solution. To bring this point into sharper focus, consider that at the liquid–vapor interface $dT = (\partial T/\partial m)\, dm + (\partial T/\partial t)\, dt = 0$ since there $T = T_s = \text{const}$; or

$$\frac{\partial T}{\partial m} dm + \frac{k\rho}{C_p} \frac{\partial^2 T}{\partial m^2} dt = 0$$

So

$$\frac{dM}{dt} = \frac{-k\rho}{C_p} \frac{\partial^2 T/\partial m^2}{\partial T/\partial m}$$

which is a nonlinear relation.

As things now stand the unknown maximum value of $m$—$M$—is bound up in a boundary condition. This awkwardness can be partially overcome by another coordinate transformation which fixes the liquid–vapor interface at unity distance from the plate by using $m/M$ as a measure of spatial position. This puts $M$ into the differential equation while removing it from the boundary conditions. Let $t'' = t$ and $y = m/M$. By the chain rule

$$\frac{\partial}{\partial t} = \frac{\partial}{\partial t''} \frac{\partial t''}{\partial t} + \frac{\partial}{\partial y} \frac{\partial y}{\partial t}$$

$$\underset{1}{\Big\downarrow} \qquad \qquad -\frac{m}{M^2} \frac{dM}{dt} = \frac{-y}{M} \frac{dM}{dt} = \frac{-y}{M^2} M \frac{dM}{dt} = -\frac{y}{2M^2} \frac{dM^2}{dt}$$

$$\frac{\partial}{\partial m} = \frac{\partial}{\partial t''} \frac{\partial t''}{\partial m} + \frac{\partial}{\partial y} \frac{\partial y}{\partial m}$$

$$\underset{0}{\Big\downarrow} \qquad \qquad \underset{\dfrac{1}{M}}{\Big\downarrow}$$

The new form of the energy equation, as a consequence of this transformation, is then, for $0 \leqslant y \leqslant 1$,

$$\frac{\partial T}{\partial t} - \frac{y}{2M^2} \frac{dM^2}{dt} \frac{\partial T}{\partial y} = \frac{1}{M^2} \frac{(k\rho)}{C_p} \frac{\partial^2 T}{\partial y^2}$$

Rearrangement gives

$$M^2 \frac{\partial T}{\partial t} = \frac{(k\rho)}{C_p} \frac{\partial^2 T}{\partial y^2} + \frac{y}{2} \frac{dM^2}{dt} \frac{\partial T}{\partial y} \tag{5-45}$$

with boundary conditions of

$$\text{At } y = 0 \quad T = T_w \qquad \text{at } y = 1 \quad T = T_s \tag{5-46}$$

$$-2\frac{(k\rho)}{C_p} \frac{C_p}{\Lambda} \frac{\partial T}{\partial y} = \frac{dM^2}{dt}$$

The effects of the transformations are depicted in Fig. 5-9. The idea to be pursued next is that it is possible that all temperature profiles will be of similar appearance in $y$, $t$ coordinates as time progresses. In other words, it is possible that $T$ depends only on $y$ and not at all on $t$. In order for this "similarity solution" to be valid, it must be that $dM^2/dt = 4B^2(k\rho/C_p) = \text{const.}$ With a dimensionless temperature defined as $\theta = (T - T_s)/(T_w - T_s)$, the dimensionless form of the energy equation and boundary conditions is, for $0 \leqslant y \leqslant 1$,

$$0 = \frac{d^2\theta}{dy^2} + 2B^2 y \frac{d\theta}{dy} \tag{5-47}$$

with boundary conditions of

$$\theta(y = 0) = 1 \tag{5-48a}$$

$$\theta(y = 1) = 0 \tag{5-48b}$$

$$\frac{-C_p(T_w - T_s)}{\Lambda} \frac{d\theta(y = 1)}{dy} = 2B^2 \tag{5-48c}$$

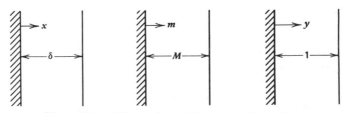

**Figure 5-9**   Effects of coordinate transformations.

Integration of the transformed and dimensionless energy equation of Eq. (5-47) twice results in

$$\frac{T - T_s}{T_w - T_s} = \theta = 1 - \frac{\mathrm{erf}(By)}{\mathrm{erf}(B)} \tag{5-49}$$

where the error function is defined as

$$\mathrm{erf}(z) = 2\pi^{-1/2} \int_0^z e^{-u^2}\, d\eta$$

The constant $B$ is evaluated from Eq. (5-48c) in terms of the relationship

$$\pi^{1/2} Be^{B^2}\, \mathrm{erf}(B) = \frac{C_p(T_w - T_s)}{\Lambda} \tag{5-50}$$

For small values of $C_p(T_w - T_s)/\Lambda$, Eq. (5-50) is closely approximated by

$$2B^2 \approx \frac{C_p(T_w - T_s)}{\Lambda} \tag{5-51}$$

which compares to the exact result of Eq. (5-50) as shown in Fig. 5-10. The "blowing" directed toward the plate caused by vaporization at the liquid–vapor interface is responsible for the nonlinear film temperature distribution given by Eq. (5-49) and illustrated in Fig. 5-10.

If the dimensionless parameter $C_p(T_w - T_s)/\Lambda$ is known, the rate of evaporation is available since

$$\frac{dM^2}{dt} = \frac{4B^2 k\rho}{C_p}$$

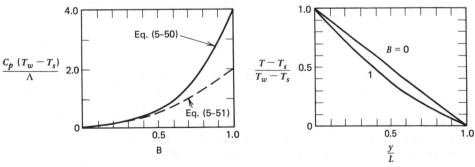

**Figure 5-10** Parameter $B$ dependence on $C_p(T_w - T_0)/\Lambda$ and dimensionless temperature variation.

One integration gives

$$M^2 = \frac{4B^2 k\rho}{C_p} t + \text{const}$$

If there is initially no vapor film, the constant of integration must vanish. Then

$$M = 2B\left(\frac{k\rho}{C_p}\right)^{1/2} t^{1/2} \tag{5-52}$$

Provided vapor density is reasonably constant, $M \approx \rho\delta$, and Eq. (5-52) shows that the film thickness also varies as the square root of elapsed time as

$$\delta = 2B\alpha^{1/2} t^{1/2} \tag{5-53}$$

The finding that $\delta \sim t^{1/2}$ agrees with the expectation provided by the random walk discussion in Section 2.6. Also, Eq. (5-53) suggests that the Fourier number Fo is the constant

$$\text{Fo} = \frac{\alpha t}{\delta^2} = \frac{1}{4B^2}$$

The infinitely large heat flux this solution predicts at the beginning when $\delta = 0$ cannot be entirely correct as is discussed in Appendix B.

The vaporization problem that has been solved is in some ways a boundary-layer problem. The vapor film is a layer of thermally affected fluid near a boundary. Although the boundary layer in this case has a definite and real thickness that is marked by the liquid–vapor interface, it is similar to the boundary layer of less physically definite thickness that forms as air flows over an airfoil. It would be expected, then, that the boundary layer thickness on an airfoil immersed in an airstream flowing at velocity $V$ would increase in the manner suggested by Eq. (5-53). In other words, since the elapsed time for which a fluid particle moving with the free stream has been exposed to disturbance is $t = x/V$, where $x$ is the distance from the leading edge of the airfoil, the boundary layer thickness should vary with distance as

$$\delta \sim x^{1/2}$$

The effect of thermal radiation on transient vaporization of a saturated liquid at a constant temperature plate was included in a study by Limpiyakorn and Burmeister [9]. In problems of this type there must be a pressure excursion since some time must elapse before the initially stagnant liquid can be displaced to make room for the newly formed vapor. Studies [10–12] of this effect show that the pressure excursion is usually moderate and does not markedly affect the rate of vapor formation, although the film thickness is often substantially affected.

## PROBLEMS

**5-1** (a) Show from Eq. (5-5) that in Couette flow with both plate temperatures equal, the maximum temperature in the fluid occurs at $y/L$
$= \frac{1}{2}$ and equals

$$T_{max} - T_0 = \frac{\mu U^2}{8k}$$

which is independent of the plate separation. The temperature
distribution is then as illustrated in Fig. 5P-1.

(b) Show from the energy equation for Couette flow that the viscous
dissipation rate is uniform at the value $\mu U^2/L^2$ regardless of the
boundary conditions imposed and acting as a uniformly distributed
heat source.

(c) Show from Eq. (5-3) that the shear stress $\tau_{yx}$ is

$$\tau_{yx} = \frac{\mu U}{L}$$

which is constant throughout the space between the plates.

**5-2** (a) The maximum oil temperature in a nominally 60°F journal bearing
that rotates at 3000 rpm and has an inner diameter of 2 in. and a
radial clearance of 0.002 in. is to be estimated. The oil properties are
$\mu = 5.82 \times 10^{-2}$ lb$_m$/ft sec, $k = 0.077$ Btu/hr ft °F. Show that the
maximum temperature rise in the oil is 10°F, assuming that both
bearing surfaces are maintained at 60°F. *Note*: At 100°F, $\mu = 1.53$
$\times 10^{-2}$ lb$_m$/ft sec so that $A = 9726°$R in the viscosity–temperature
relation $\mu/\mu_0 = \exp[A(1/T - 1/T_0)]$.

(b) Show that the shear stress in the fluid is

$$\tau_{yx} = \frac{\mu U}{L} = 284 \text{ lb}_f/\text{ft}^2$$

and that the torque per unit length required to turn the bearing is

$$\text{Torque} = \tau_{yx} \frac{\pi D^2}{2} = 12.3 \text{ ft lb}_f/\text{ft}$$

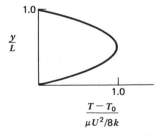

$$\frac{T - T_0}{\mu U^2/8k}$$

**Figure 5P-1**

which yields a power requirement to overcome bearing friction of

$$P = \text{torque} \times \text{angular velocity} = 7 \text{ hp}$$

5-3  Couette flow with temperature dependent properties has been studied
[1, 2], and it is found that the velocity distribution ceases to be linear.
For the case in which both the upper and lower plates of Fig. 5-1 are
maintained at $T_0$ the velocity distribution in the fluid is to be estimated
for viscosity which varies with temperature as

$$\frac{\mu}{\mu_0} = e^{A(1/T - 1/T_0)}$$

The describing equations are

$$x \text{ Motion} \qquad \frac{d(\mu \, du/dy)}{dy} = 0$$

$$\text{Energy} \qquad \frac{d^2T}{dy^2} + \frac{\mu}{k}\left(\frac{du}{dy}\right)^2 = 0$$

with boundary conditions of $u(0) = 0$, $u(L) = U$, $T(0) = T_0 = T(L)$. A
first integration of the x-motion equation gives

$$\frac{du'}{d\eta} = \frac{c\mu_0}{\mu}$$

$$= c \exp\left(\frac{A}{T_0}\frac{\theta}{\theta + 1}\right) \qquad\qquad\qquad \text{(3P-1)}$$

Substitution of this result into the energy equation then gives

$$\frac{d^2\theta}{d\eta^2} + BC^2 \exp\left(\frac{A}{T_0}\frac{\theta}{\theta + 1}\right) = 0, \qquad \theta(0) = 0 = \theta(1) \quad \text{(3P-2)}$$

Here, $\theta = (T - T_0)/T_0$, $u' = u/U$, $\eta = y/L$, and $B = \mu_0 U^2/T_0 k$
(a)  The viscosity–temperature relation is next expanded in series to
obtain

$$\frac{\mu_0}{\mu} = 1 + \beta_1\theta + \beta_2\theta^2 + \cdots$$

With this series inserted into the dimensionless energy equation, one
then has

$$\frac{d^2\theta}{d\eta^2} + BC_1^2\left(1 + \beta_1\theta + \beta_2\theta^2 + \cdots\right) = 0$$

subject to

$$\theta(0) = 0 = \theta(1)$$

Next a series solution in terms of the source $B$ is assumed in the form

$$\theta = F_0(\eta) + BF_1(\eta) + B^2F_2(\eta) + \cdots$$

$$C_1 = C_0 + BC_1 + B^2C_2 + \cdots$$

$$u' = f_0(\eta) + Bf_1(\eta) + B^2f_2(\eta) + \cdots$$

Introducing these expressions into the energy and motion equations, equating coefficients of the various powers of $B$, and imposing the boundary conditions gives

$$\frac{d^2F_0}{d\eta^2} = 0, \qquad F_0(0) = 0 = F_0(1)$$

$$\frac{d^2F_1}{d\eta^2} + C_0^2\left(1 + \beta_1 F_0 + \beta_2 F_0^2 + \beta_3 F_0^3 + \cdots\right) = 0,$$

$$F_1(0) = 0 = F_1(1)$$

$$+$$

$$\vdots$$

$$\frac{df_0}{d\eta} = C_0\left(1 + \beta_1 F_0 + \beta_2 F_0^2 + \beta_3 F_0^3 + \cdots\right), \qquad f_0(0) = 0, f_0(1) = 1$$

$$\frac{df_1}{d\eta} = C_0 + \left(\beta_1 F_1 + 2\beta_2 F_0 F_1 + 3\beta_3 F_0^2 F_1 + \cdots\right)$$

$$+ C_1\left(1 + \beta_1 F_0 + \beta_2 F_0^2 + \beta_3 F_0^3 + \cdots\right) \qquad f_1(0) = 0 = f_1(1)$$

$$+$$

$$\vdots$$

Solution of these equations gives the answer to any desired accuracy by increasing the number of terms evaluated. Show that the first few terms of the solution are

$$\theta = B\frac{1-\eta}{2} + \cdots$$

$$u' = \eta\left[1 - B\beta_1\frac{1 - 3\eta + 2\eta^2}{12} + \cdots\right]$$

**(b)** An alternative solution procedure is to adopt the approximation that Eqs. (3P-1) and (3P-2) are closely given by

$$\frac{du'}{d\eta} = Ce^{A\theta/T_0}, \qquad u'(0) = 0, \qquad u'(1) = 1 \qquad \text{(3P-3)}$$

$$\frac{d^2\theta}{d\eta^2} + BC^2e^{A\theta/T_0} = 0, \qquad \theta(0) = 0 = \theta(1) \qquad \text{(3P-4)}$$

if $\theta \ll 1$, as is often true. Then a result pertaining to the steady temperature in a slab with heat generation at a rate that is an exponential function of temperature from Carslaw and Jaeger [3] can be employed as follows. Equation (3P-4) is recast into

$$\frac{d\theta}{d\eta} \frac{d(d\theta/d\eta)}{d\theta} = -BC^2e^{A\theta/T_0}$$

and integrated from $\eta = \frac{1}{2}$ (by symmetry, the presumed location of the maximum, at which location $d\theta/d\eta = 0$ and $\theta = \theta_m$) to $\eta$ to obtain

$$\frac{d\theta}{d\eta} = -\left[\frac{2T_0BC^2}{A}\left(e^{A\theta_m/T_0} - e^{A\theta/T_0}\right)\right]^{1/2}$$

Integration again from $\eta = \frac{1}{2}$ to $\eta$ gives

$$\theta = \theta_m - 2\frac{T_0}{A}\ln\left\{\cosh\left[2\left(\eta - \frac{1}{2}\right)\left(\frac{ABC^2}{2T_0}e^{A\theta_m/T_0}\right)^{1/2}\right]\right\}$$

Imposition of the condition that $\theta(1) = 0$ on this result requires that

$$e^{A\theta_m/2T_0} = \cosh\left[\left(\frac{ABC^2}{2T_0}e^{A\theta_m/T_0}\right)^{1/2}\right]$$

or with

$$\xi = \left(\frac{ABC^2}{2T_0}e^{A\theta_m/T_0}\right)^{1/2}$$

one has

$$\cosh(\xi) = \xi\left(\frac{2T_0}{ABC^2}\right)^{1/2} \qquad \text{(3P-5)}$$

If $0 < 2T_0/ABC_1^2 < 0.88 \cdots$, this transcendental equation has two roots, corresponding to two possible steady-state solutions. If $\beta > 0.88 \cdots$, no steady solution exists (in such a case the approximation that $\theta \ll 1$ loses accuracy, and a steady-state solution is still predicted by the full equations). Show that this temperature distribution inserted into Eq. (3P-3) results in the velocity distribution

$$u' = \frac{1}{2} + \frac{1}{2} \frac{\left[ x\left( e^{A\theta_m/T_0} - \frac{1}{2} \right) - \cosh(x)\sinh(x)/2\xi \right]}{\left[ e^{A\theta_m/T_0} - \frac{1}{2} - \dfrac{\cosh\xi\sinh\xi}{2\xi} \right]} \qquad (3P\text{-}6)$$

where $x = 2(\eta - 1/2)\xi$ and $\xi$ is as defined before.

A trial-and-error procedure is necessary since $u'$ is not really known until $C$ and $\theta_m$ have been numerically evaluated for the specified conditions. Since

$$C = 1 / \left[ e^{A\theta_m/T_0} - \frac{1}{2} - \frac{\cosh\xi\sinh\xi}{2\xi} \right] \qquad (3P\text{-}7)$$

which is the equivalent (due to the definition of $\xi$) of

$$\xi = \frac{\left[ (AB/2T_0) e^{A\theta_m/T_0} \right]^{1/2}}{e^{A\theta_m/T_0} - \frac{1}{2} - \dfrac{\cosh\xi\sinh\xi}{2\xi}} \qquad (3P\text{-}8)$$

the procedure is to first guess a value of $\theta_m$. Then $\xi$ is evaluated from Eq. (3P-8) and $C$ is evaluated from Eq. (3P-7). This value of $C$ is inserted into Eq. (3P-5), and a new value of $\xi$ is computed. From this new $\xi$ value, a new $\theta_m$ value is determined. This procedure is repeated until convergence occurs; then $u'$ can be plotted.

(c) Plot the velocity and temperature profiles according to the results of parts a and b for the conditions in Problem 5-2.

5-4 The steady laminar flow of a constant property Newtonian fluid down a circular pipe of radius $r_0$ and constant temperature $T_0$ in response to an imposed constant pressure drop $dp/dz$ is to be determined.

(a) Show that the motion and energy equations reduce to

$$z \text{ Motion} \qquad \frac{d(r\,dv_z/dr)}{dr} = \frac{r(dp/dz)}{\mu}$$

$$\text{Energy} \qquad \frac{d(r\,dT/dr)}{dr} = -\frac{\mu r}{k} \left( \frac{dv_z}{dr} \right)^2$$

(b) Show that the boundary conditions to be imposed are

$$\text{At } r = r_0 \qquad v_z = 0 = v_\theta \qquad \text{at } r = 0 \qquad v_z \text{ finite}$$

$$v_r = 0 \qquad\qquad\qquad T \text{ finite}$$

$$T = T_0$$

(c) On the basis of parts a and b, show that the velocity distribution is given by

$$v = \frac{r_0^2}{4\mu}\left(-\frac{dp}{dz}\right)\left(1 - \frac{r^2}{r_0^2}\right)$$

with a maximum velocity given by

$$v_m = \frac{r_0^2}{4\mu}\left(-\frac{dp}{dz}\right)$$

and an average velocity given by

$$v_{av} = \frac{r_0^2}{8\mu}\left(-\frac{dp}{dz}\right) = \frac{v_m}{2}$$

Thus the velocity distribution can be written alternatively as

$$v = 2v_{av}\left(1 - \frac{r^2}{r_0^2}\right)$$

For a pipe of length $\Delta L$, the volumetric flow rate is then given by

$$Q = \frac{\pi r_0^4}{8\mu}\frac{\Delta P}{\Delta L}$$

(d) On the basis of parts a, b, and c, show that the temperature distribution is given by

$$T = T_0 + \frac{\mu}{k}v_{av}^2\left(1 - \frac{r^4}{r_0^4}\right)$$

and the maximum temperature (at the center line) is

$$T_m = T_0 + \frac{\mu}{k}v_{av}^2$$

**5-5** For oil at 60°F ($\mu = 5.82 \times 10^{-2}$ lb$_m$/ft sec, $k = 0.077$ Btu/hr ft F°, $\rho = 57$ lb$_m$/ft$^3$) flowing down a pipe of 12 in. diameter, determine the maximum oil temperature. Assume that the oil is flowing at a Reynolds number of Re $= V_{av}D\rho/\mu = 1000$. Determine the pressure gradient down the pipe and the volumetric flow rate as well.

**5-6** For the Couette flow in Problem 5-1, evaluate the entropy flow rate at each plate and evaluate the rate at which entropy is generated per unit volume in the fluid.

**5-7** Show that if bulk density $\rho$ is constant the rate of mass transfer of species 1 in Stefan's diffusion problem is immediately obtainable from Eq. (5-29) as

$$\dot{m}_1 = \frac{\rho D_{12}}{L} \frac{\omega_{10} - \omega_{1L}}{1 - \omega_{10}}$$

**5-8** Show that in Stefan's diffusion problem the "blowing" represented by the upward bulk velocity $v$ becomes large at large rates of species 1 mass transfer. Comment on the importance of this effect in view of the predicted rate of mass transfer when the liquid is at its nominal boiling temperature. Show that $v = \dot{m}_1/\rho$ and is always directed upward, always aiding the diffusive transport.

**5-9** The solutions to Stefan's diffusion problem have been used in experimental determination of gas diffusivities [C. Y. Lee and C. R. Wilke, Measurements of vapor diffusion coefficient, *Indust. Eng. Chem.* **46**, 2381–2387 (1954)]. In one such experiment the diffusivity of gaseous carbon tetrachloride through oxygen was measured. The distance between the CCl$_4$ liquid and the tube top was 17.1 cm, the tube cross-sectional area was 0.82 cm$^2$, the total pressure was 775 mm Hg and the saturation pressure of the CCl$_4$ at the ambient temperature of 0°C was 33 mm Hg. If 0.0208 cm$^3$ of liquid CCl$_4$ evaporated 10 h after steady state was reached (liquid CCl$_4$ density at OC is 1.59 g/cm$^3$), show that the diffusivity is $D_{12} = 0.0636$ cm$^2$/s.

**5-10** Perform a simple experiment to test the accuracy of the analytical solution to Stefan's diffusion problem.

(a) Partially fill a tall cylinder with water and maintain the water at a constant temperature of above 150°F so that evaporation proceeds at a reasonably rapid rate.

(b) Record the time required for about 1 cm of water to evaporate, and compare this experimental result with that predicted by Eq. (5-36) by use of a reported air–water vapor mass diffusivity.

(c) Comment on the importance of such sources of discrepancy between measurement and prediction as the cooling effect of evaporation, free convection effects arising from the fact that the evaporating water is less dense than air, and two-dimensional flow patterns.

**5-11** Consider the long tube shown in Fig. 5P-11 through which two gases diffuse steadily. At the left end species 1 exists at partial pressure $p_{10}$ and corresponding mass fraction $\omega_{10}$. At the right end species 1 exists at partial pressure $p_{1L}$ and corresponding mass fraction $\omega_{2L}$. The total pressure $p$ and temperature $T$ are uniform and constant. The rate at which moles of species 1 are transported to the right is equal in magnitude and opposite in direction to the rate at which moles of species 2 are transported to the left, perhaps because of some chemical reactions occuring outside the tube. This is an equimolar counterdiffusion problem in which $\dot{m}_1(x)/M_1 = -\dot{m}_2(x)/M_2$, where $M_{1,2}$ are the molecular weights of species 1, 2.

**(a)** Show that the pertinent describing differential equations are

$$\text{Continuity} \qquad \frac{d(\rho u)}{dx} = 0$$

$$\text{Diffusion} \qquad \rho u \frac{d\omega_1}{dx} = D_{12} \frac{d(\rho \, d\omega_1/dx)}{dx}$$

and that the boundary conditions are

$$\omega_1(x = 0) = \omega_{10} \qquad \text{or} \qquad p_1(x = 0) = p_{10}$$

$$\omega_1(x = L) = \omega_{1L} \qquad \text{or} \qquad p_1(x = L) = p_{1L}$$

**(b)** Integrate the continuity equation to find

$$\rho u = C_1$$

then integrate the diffusion equation to find

$$C_1 \omega_1 - C_2 = \rho D_{12} \frac{d\omega_1}{dx}$$

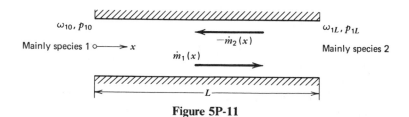

Figure 5P-11

(c) Show that the constraint of equimolar counterdiffusion leads to the result that there is bulk motion—the gas mixture is not truly stagnant and there is "blowing." *Hint*: Consider

$$\frac{\dot{m}_1(x)}{M_1} = -\frac{\dot{m}_2(x)}{M_2}$$

$$\frac{(\rho_1 u - \rho D_{12}\, d\omega_1/dx)}{M_1} = -\frac{(\rho_2 u - \rho D_{12}\, d\omega_2/dx)}{M_2}$$

$$\rho u = -\frac{\rho D_{12}(1 - M_1/M_2)}{\omega_1(1 - M_1/M_2) + M_1/M_2}\frac{d\omega_1}{dx} = C_1$$

(d) Taking into account the result of part c, show that

$$\dot{m}_1(x) = \frac{\rho D_{12}(M_1/M_2)}{\omega_1(1 - M_1/M_2) + M_1/M_2}\frac{d\omega_1}{dx}$$

and show that the first integral of the diffusion equation in part b can be expressed as

$$C_2 = -\frac{\rho D_{12}(M_1/M_2)}{\omega_1(1 - M_1/M_2) + M_1/M_2}\frac{d\omega_1}{dx} = \dot{m}_1(x)$$

a result that is also expressible in terms of partial pressure as

$$C_2 = -\frac{D_{12}M_1}{RT}\frac{dp_1}{dx}$$

(e) From the last relation of part d, show that the partial pressure distribution is linear, as expressed by

$$\frac{p_1 - p_{10}}{p_{1L} - p_{10}} = \frac{x}{L}$$

and that the mass transport rate of species 1 is given by

$$\dot{m}_1 = \frac{D_{12}M_1}{RT}\frac{(p_{10} - p_{1L})}{L}$$

or

$$\dot{m}_1 = \frac{D_{12}M_2 p}{RTL}\frac{\omega_{10} - \omega_{1L}}{[1 + \omega_{10}(M_2/M_1 - 1)][1 + \omega_{1L}(M_2/M_1 - 1)]}$$

Note the simpler expression for $\dot{m}_1$ in terms of partial pressure. Note also that the correction for "blowing" in the denominator of the expression involving mass fractions is different from the case in which one surface is impermeable to one species.

(f) Show that the "blowing" is related to the rate of species 1 mass transport by

$$u = \left(1 - \frac{M_2}{M_1}\right)\frac{\dot{m}_1}{\rho}$$

a bulk velocity that is not constant since bulk density $\rho$ varies along the tube length. Depending on the ratio of molecular weights, the "blowing" can either help or hinder diffusive transport.

Show also that the fraction of the rate of species 1 mass transport caused by blowing $\rho_1 u / \dot{m}_1$ is

$$\frac{\rho_1 u}{\dot{m}_1} = \omega_1\left(1 - \frac{M_2}{M_1}\right)$$

which is small at low mass transport rates.

(g) Comment on the possible need to relax the assumption that total pressure is constant in cases where the mass transport rate is sufficiently high to make "blowing" large. Would the $x$-motion equation provide the needed information concerning total pressure variation? Comment on the possible importance of natural convection when molecular weights differ appreciably for the case of a horizontal tube where the gases might stratify (light gas in upper half of tube and heavy gas in lower half of tube), a vertical tube with light gas transport predicted upward, and a vertical tube with heavy gas transport predicted upward.

(h) With a mass transfer coefficient $h_D$ defined as

$$\dot{m}_1 = \frac{\rho h_D(\omega_{10} - \omega_{1L})}{\left[1 + \omega_{10}\left(\frac{M_2}{M_1} - 1\right)\right]\left[1 + \omega_{1L}\left(\frac{M_2}{M_1} - 1\right)\right]}$$

show that the result of part e leads to

$$h_D = \frac{D_{12}M_2 p}{\rho RTL} \approx \frac{D_{12}}{L}$$

from which the Sherwood number Sh is found to be

$$\text{Sh} = \frac{h_D L}{D_{12}} = 1$$

**5-12** Consider a sphere of radius $r_0$ immersed in a very large fluid body of uniform temperature $T_f$. The sphere is just slightly warmer than the fluid, so there is no motion in the fluid and is uniformly at temperature $T_s$.

(a) Show that the energy equation in spherical coordinates describing the temperature distribution in the fluid is

$$\frac{1}{r^2}\frac{d(r^2 dT/dr)}{dr} = 0$$

with boundary conditions of

$$T(r = r_0) = T_s$$

$$T(r = \infty) = T_f$$

(b) Show from the result of part a that the temperature distribution is

$$\frac{T - T_f}{T_s - T_f} = \frac{r_0}{r}$$

(c) Show from the result of part b that the heat transfer rate from the sphere to the fluid is

$$q_w = -k\frac{dT(r = r_0)}{dr}4\pi r_0^2$$

$$= 4\pi k r_0(T_s - T_f)$$

(d) Show from the result of part c that the heat transfer coefficient is

$$h = \frac{q_w}{4\pi r_0^2(T_s - T_f)}$$

$$= \frac{k}{r_0}$$

and that the Nusselt number based on sphere diameter is

$$Nu = \frac{hD}{k} = 2$$

Internal circulation caused by an electric field can increase this result by an order of magnitude [13].

**5-13**  Consider a sphere of radius $r_0$ immersed in a very large body of fluid. The sphere has an excess of mass of species 1 on its surface, but it is sufficiently slight that there is no buoyancy-induced motion in the fluid at a partial pressure of $p_{10}$ and corresponding mass fraction of $\omega_{10}$. Far away from the sphere, species 1 partial pressure is $p_{1f}$ and its corresponding mass fraction is $\omega_{1f}$. The rate at which moles of species 1 are transported outward is equal and opposite to the rate at which moles of species 2 are transported inward. This is equimolar counterdiffusion from a sphere.

(a) Show that the pertinent describing differential equations are

$$\text{Continuity} \qquad \frac{d(r^2 \rho v_r)}{dr} = 0$$

$$\text{Diffusion} \qquad \rho v_r \frac{d\omega_1}{dr} = \frac{D_{12}}{r^2} \frac{d(\rho r^2 d\omega_1/dr)}{dr}$$

with boundary conditions of

$$\omega_1(r = r_0) = \omega_{10} \qquad \text{or} \qquad p_1(r = r_0) = p_{10}$$

$$\omega_1(r = \infty) = \omega_{1f} \qquad \text{or} \qquad p_1(r = \infty) = p_{1f}$$

(b) Show that the first integrals of the continuity and diffusion equations are, respectively,

$$r^2 \rho v_r = C_1$$

and

$$C_1 \omega_1 - C_2 = \rho D_{12} r^2 \frac{d\omega_1}{dr}$$

(c) Show that the equimolar counterdiffusion requirement

$$\frac{\dot{m}_1(r)}{M_1} = -\frac{\dot{m}_2(r)}{M_2}$$

$$\frac{\rho_1 v_r - \rho D_{12} \, d\omega_1/dr}{M_1} = -\frac{(\rho_2 v_r - \rho D_{12} \, d\omega_2/dr)}{M_2}$$

leads to the relationship that

$$\rho v_r = \frac{\rho D_{12}(1 - M_1/M_2)}{\omega_1(1 - M_1/M_2) + M_1/M_2} \frac{d\omega_1}{dr}$$

from which it follows that

$$\dot{m}_1(r) = -\frac{M_1}{M_2}\frac{\rho D_{12}}{\omega_1(1 - M_1/M_2) + M_1/M_2}\frac{d\omega_1}{dr}$$

(d) By use of the result of part c, show that the first integral of the diffusion equation in part b can be written as

$$-\frac{C_2}{r^2} = \frac{\rho D_{12}(M_1/M_2)}{\omega_1(1 - M_1/M_2) + M_1/M_2}\frac{d\omega_1}{dr}$$

(e) Employ the relations for a binary mixture of perfect gases to express the result of part d in terms of partial pressure as

$$-\frac{C_2}{r^2} = \frac{D_{12}M_1}{RT}\frac{dp_1}{dr}$$

(f) Integrate the result of part e and apply the boundary conditions to show that

$$\frac{p_1 - p_{1L}}{p_{10} - p_{1L}} = \frac{r_0}{r}$$

(g) Show that $\dot{m}_1(r)$ is given by the results of parts c and f as

$$\dot{m}_1(r) = -\frac{D_{12}M_1}{RT}\frac{dp_1}{dr}$$

$$= \frac{D_{12}M_1}{RT}\frac{(p_{10} - p_{1f})r_0}{r^2}$$

from which it follows that the total rate of species 1 mass transfer from the sphere is

$$4\pi\frac{D_{12}M_1}{RT}r_0(p_{10} - p_{1f})$$

(h) Show that the mass-transfer rate of part g can also be expressed as

$$\dot{m}_{1,\text{total}} = \frac{4\pi D_{12}pM_2r_0}{RT}$$

$$\times\frac{(\omega_{10} - \omega_{1f})}{[1 + \omega_{10}(M_2/M_1 - 1)][1 + \omega_{1f}(M_2/M_1 - 1)]}$$

(i) For this equimolar counterdiffusion from a sphere case, show from the result of part h that defining a mass transfer coefficient $h_D$ as

$$\dot{m}_{1,\,\text{total}} = \frac{A\rho h_D(\omega_{10} - \omega_{1f})}{[1 + \omega_{10}(M_2/M_1 - 1)][1 + \omega_{1f}(M_2/M_1 - 1)]}$$

leads to

$$h_D = \frac{D_{12} M_2 p}{RTr\rho_0} \approx \frac{D_{12}}{r_0}$$

from which the Sherwood number Sh is found to be

$$\text{Sh} = \frac{Dh_D}{D_{12}} = 2$$

**5-14** A carbon particle of diameter $D = 0.1$ in. burns in an oxygen atmosphere. Temperature is uniform at $1800R$, and the total pressure is 1 atm. At the carbon surface the reaction $C + O_2 \rightarrow CO_2$ occurs, so this is equimolar counterdiffusion about a sphere. Assume that at the carbon surface $p_{CO_2} = 1$ atm while far from the carbon surface $p_{CO_2} = 0$ atm.
(a) Verify that $D_{CO_2-O_2} \approx 4 \text{ ft}^2/\text{hr}$.
(b) Verify that the rate at which carbon burns, assuming the combustion to be limited by the rate at which oxygen can be transported to the carbon surface, is about $0.9 \times 10^{-6} \text{ lb}_m/\text{hr}$.

**5-15** Consider a sphere of radius $r_0$ immersed in a very large body of fluid. The sphere has an excess of species 1 on its surface, but it is sufficiently slight that there is no buoyancy-induced motion in the fluid, at a partial pressure of $p_{10}$ and corresponding mass fraction of $\omega_{10}$. Far away from the sphere, species 1 partial pressure is $p_{1f}$ and its corresponding mass fraction is $\omega_{1f}$. The sphere surface is impermeable to species 2.
(a) Show that the describing equations are

Continuity                $$\frac{d(r^2 \rho v_r)}{dr} = 0$$

Diffusion                $$\rho v_r \frac{d\omega_1}{dr} = \frac{D_{12}}{r^2} \frac{d(\rho r^2 \, d\omega_1/dr)}{dr}$$

with boundary conditions of

$$\omega_1(r = r_0) = \omega_{10} \quad \text{or} \quad p_1(r = r_0) = p_{10}$$

$$\omega_1(r = \infty) = \omega_{1f} \quad \text{or} \quad p_1(r = \infty) = p_{1f}$$

**(b)** Show that the first integrals of the continuity and diffusion equations are

$$r^2 \rho v_r = C_1 \quad \text{and} \quad C_1 \omega_1 - C_2 = \rho D_{12} r^2 \frac{d\omega_1}{dr}$$

**(c)** Show that the requirement of the sphere surface being impermeable to species 2

$$\dot{m}_2(r) = 0 \quad \text{or} \quad \rho_2 v_r - \rho D_{12} \frac{d\omega_2}{dr} = 0$$

leads to the relationship that

$$\rho v_r = -\frac{\rho D_{12}}{1 - \omega_1} \frac{d\omega_1}{dr}$$

from which it follows that

$$\dot{m}_1(r) = -\frac{\rho D_{12}}{1 - \omega_1} \frac{d\omega_1}{dr}$$

**(d)** Making use of the result of part c, show that the first integral of the diffusion equation in part b can be written as

$$-\frac{C_2}{r^2} = \frac{\rho D_{12}}{1 - \omega_1} \frac{d\omega_1}{dr}$$

**(e)** Employ the relations for a binary mixture of perfect gases to express the result of part d in terms of partial pressure as

$$-\frac{C_2}{r^2} = \frac{D_{12} M_1 p}{RT} \frac{1}{p - p_1} \frac{dp_1}{dr}$$

**(f)** Integrate the result of part e and apply the boundary conditions to show that

$$\frac{p - p_1}{p - p_{1f}} = \frac{p - p_{10}}{p - p_{1f}} \frac{r_0}{r}$$

**(g)** Show that $\dot{m}_1(r)$ is given by the results of parts c and f as

$$\dot{m}_1(r) = -\frac{D_{12} M_1 p}{RT} \frac{1}{p - p_1} \frac{dp_1}{dr}$$

$$= \frac{D_{12} M_1 p r_0}{RT r^2} \ln\left(\frac{p - p_{1f}}{p - p_{10}}\right)$$

from which it follows that the total rate of species 1 mass transfer from the sphere is

$$\frac{4\pi D_{12} M_1 \, p r_0}{RT} \ln\left(\frac{p - p_{1f}}{p - p_{10}}\right)$$

**(h)** Show that the mass-transfer rate of part g can also be expressed in terms of mass fractions as

$$\dot{m}_{1,\text{total}} = \frac{4\pi D_{12} M_1 \, p r_0}{RT}$$

$$\times \ln\left\{ 1 + \frac{\omega_{10} - \omega_{1f}}{(1 - \omega_{10})\left[M_1/M_2 + \omega_{1f}(1 - M_1/M_2)\right]} \right\}$$

**(i)** For this case of diffusion from a semipermeable sphere, show from the result of part h that defining a mass transfer coefficient $h_D$ as

$$\dot{m}_{1,\text{total}} = A\rho h_D \frac{\omega_{10} - \omega_{1f}}{1 - \omega_{10}}$$

for small mass transfer rate ($\omega_{1f} \approx 0$ and $\omega_{10} \approx 0$) gives

$$h_D \approx \frac{D_{12} M_2 \, p}{\rho R T r_0} \approx \frac{D_{12}}{r_0}$$

from which the Sherwood number Sh is found to be

$$\text{Sh} = \frac{D h_D}{D_{12}} = 2$$

**5-16** A droplet of water initially of 0.1 in. diameter is maintained at 70°F in a 70°F dry stagnant air atmosphere.

**(a)** Estimate the water evaporation rate per unit area from this sphere. Is "blowing" significant? What fraction of the maximum possible evaporation is this estimate?

**(b)** Derive the equation

$$\frac{d\left[(r/r_0)^2\right]}{dt} = \frac{\rho_{\text{air}} D_{12} \text{Sh}}{\rho_{\text{liq H}_2\text{O}} r_0^2} \frac{\omega_{10} - \omega_{1f}}{1 - \omega_{10}}$$

where $\text{Sh} = 2 r_0 h_D / D_{12}$ to describe the rate at which the droplet radius decreases. Estimate the time required for the droplet to completely evaporate.

**5-17** A droplet of water is placed in a dry stagnant atmosphere of constant temperature $T_f$. The droplet evaporates and cools to an equilibrium temperature $T_w$.

(a) To estimate $T_w$, derive the equation

$$k_{air} \text{Nu}(T_w - T_f) = \Lambda D_{12} \rho_{air} \text{Sh} \frac{\omega_{10} - \omega_{1f}}{1 - \omega_{10}}$$

where $\Lambda$ is the water latent heat of vaporization $\text{Nu} = hD/k$, and $\text{Sh} = Dh_D/D_{12}$, assuming thermal radiation to be negligible. (It is likely to require consideration in a practical case since convection is so low in a stagnant situation.)

(b) Show that for the case of a stagnant atmosphere, the result of part a can be reduced to

$$\frac{C_{p_{air}}(T_w - T_f)}{\Lambda} = \text{Le} \frac{\omega_{10} - \omega_{1f}}{1 - \omega_{10}}$$

where the Lewis number Le is defined as $\text{Le} = D_{12}/\alpha$ and $\alpha = k/\rho C_p$.

(c) For the case of a stagnant dry atmosphere ($\omega_{1f} = 0$) at 70°F, determine the amount by which the droplet is cooled below the atmosphere. Note that an iterative (trial-and-error) solution procedure is required since $\omega_{10}$ depends on the unknown $T_w$. How long will a 0.1-in.-diameter droplet last if evaporating at this equilibrium rate?

**5-18** Consider steady Couette flow of a constant-property fluid between large parallel and horizontal plates that are separated by a distance $L$, with the lower plate stationary and the upper plate moving at velocity $U$ as illustrated in Fig. 5-1. The two plates are porous and fluid enters the interplate space through the bottom plate and leaves at the same steady and uniform rate through the top plate—the fluid vertical velocity $v$ is a nonzero constant as a result.

(a) Show that the x-motion equation is

$$v \frac{du}{dy} = \nu \frac{d^2 u}{dy^2}$$

subject to the boundary conditions that

$$u(y = 0) = 0 \quad \text{and} \quad u(y = L) = U$$

from which it follows that the velocity distribution is

$$\frac{u}{U} = \frac{e^{vy/\nu} - 1}{e^{vL/\nu} - 1}$$

**(b)** Plot $u/U$ against $y/L$ for several values of the dimensionless Reynolds number $vL/v$, and demonstrate the effect the imposed "blowing" exerts on the velocity profile.

Show that the viscous drag at the bottom plate $\tau_w$ is also influenced by the "blowing" and is

$$\tau_w = \mu \frac{du(y=0)}{dy}$$

$$= \frac{\mu U}{L} \frac{vL/v}{e^{vL/v} - 1}$$

When $v$ is positive, the upward "blowing" is representative of that which occurs in evaporation or transpiration cooling of a turbine blade. When $v$ is negative, the downward "suction" is representative of that which occurs in condensation or boundary-layer suction to control flow over an airfoil.

**(c)** Show that if the bottom and top plates are maintained at temperatures $T_0$ and $T_1$, respectively, the energy equation is (neglecting viscous dissipation)

$$v \frac{dT}{dy} = \alpha \frac{d^2 T}{dy^2}$$

with the boundary conditions of

$$T(y=0) = T_0 \quad \text{and} \quad T(y=L) = T_1$$

from which it follows that the temperature distribution is

$$\frac{T - T_0}{T_1 - T_0} = \frac{e^{\Pr vy/v} - 1}{e^{\Pr vL/v} - 1}$$

**(d)** Plot $(T - T_0)/(T_1 - T_0)$ against $y/L$ and comment as in part b. Show also that the diffusive heat flux at the bottom plate is also affected by the imposed "blowing" as

$$q_w = -k \frac{dT(y=0)}{dy}$$

$$= \frac{(T_0 - T_1)k}{L} \frac{\Pr vL/v}{e^{\Pr vL/v} - 1}$$

Demonstrate that as "blowing" increases the wall heat flux decreases, an effect called *transpiration cooling* that can be used to protect surfaces from excessively hot atmospheres.

**5-19** An application of transpiration cooling to a spherical geometry is illustrated in Fig. 5P-19, in which two concentric porous spheres of radii $r_1$ and $r_2$ are maintained at temperatures $T_1$ and $T_2$, respectively. A gas moves radially outward at a total mass flow rate of $w$. Assuming steady state, laminar flow, and constant properties and neglecting viscous dissipation, determine the rate of heat gain by the inner sphere as a function of the mass flow rate of the gas.

(a) Show that the describing equations are

$$\text{Continuity} \qquad \frac{d(r^2 v_r)}{dr} = 0$$

$$\text{Energy} \qquad r^2 v_r \frac{dT}{dr} = \alpha \frac{d(r^2\, dT/dr)}{dr}$$

with boundary conditions of

$$r^2 v_r = \frac{w}{4\pi\rho} = \text{const}$$

$$T(r_1) = T_1$$

$$T(r_2) = T_2$$

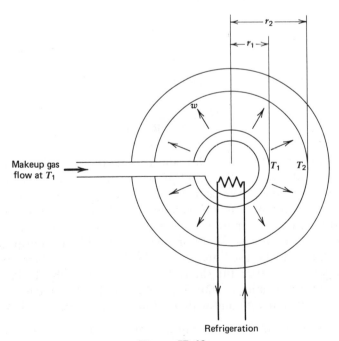

Figure 5P-19

**(b)** Show that the temperature distribution is given by

$$\frac{T - T_2}{T_1 - T_2} = \frac{e^{-\operatorname{Re}\operatorname{Pr} r_2/r} - e^{-\operatorname{Re}\operatorname{Pr}}}{e^{-\operatorname{Re}\operatorname{Pr} r_2/r_1} - e^{-\operatorname{Re}\operatorname{Pr}}}$$

and that the rate of heat flow into the inner sphere is

$$Q = 4\pi r_1^2 k \frac{dT(r_1)}{dr} = 4\pi r_2 k (T_2 - T_1) \frac{\operatorname{Re}\operatorname{Pr}}{e^{\operatorname{Re}\operatorname{Pr}(r_2/r_1 - 1)} - 1}$$

**(c)** Demonstrate that in the absence of a radial gas flow, heat flows into the inner sphere at a rate given by

$$Q_0 = 4\pi k r_1^2 \frac{dT(r_1)}{dr} = 4\pi r_2 k \frac{T_2 - T_1}{r_2/r_1 - 1}$$

since the temperature distribution is then given by

$$\frac{T - T_2}{T_1 - T_2} = \frac{r_2/r - 1}{r_2/r_1 - 1}$$

**(d)** Show from the results of parts b and c that the transpiration reduces the heat flow into the inner sphere as

$$\frac{Q}{Q_0} = \frac{\operatorname{Re}\operatorname{Pr}(r_2/r_1 - 1)}{e^{\operatorname{Re}\operatorname{Pr}(r_2/r_1 - 1)} - 1}$$

Show that for small "blowing" rates

$$\frac{Q}{Q_0} \approx \frac{1}{1 + \operatorname{Re}\operatorname{Pr}(r_2/r_1 - 1)/2}$$

**5-20** It is proposed to reduce the rate of evaporation of liquified oxygen in small containers by taking advantage of the transpiration the evaporation can produce. The geometry is that of Fig. 5P-19, but the makeup tube is absent as is the refrigeration coil. Estimate the rate of heat gain and the rate of evaporation for an inner sphere radius of 6 in. and an outer sphere radius of 12 in. with and without transpiration. The conditions to be used are $T_1 = -297°F$ and $T_2 = 30°F$ with oxygen properties $k = 0.02$ Btu/hr ft°F, $C_p = 0.22$ Btu/lb$_m$°F, Pr $= 0.7$, and heat of vaporization $= 92$B/lb$_m$. (*Answer:* Heat gain equals 61 Btu/hr and 82 Btu/hr with and without transpiration, respectively.)

**5-21** Consider the adiabatic flow of an ideal gas through a one-dimensional standing shock wave in a duct as illustrated in Fig. 5P-21. The velocity,

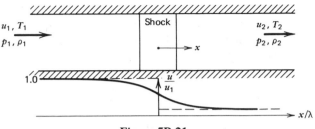

**Figure 5P-21**

temperature, and pressure distributions in and near the standing shock wave are to be determined. Assume all properties save density are constant.

**(a)** Show that the describing equations are

Continuity $\quad \dfrac{d(\rho u)}{dx} = 0$

$x$ Motion $\quad \rho u \dfrac{du}{dx} = -\dfrac{dp}{dx} + \dfrac{4}{3}\dfrac{d(\mu \, du/dx)}{dx}$

Energy $\quad C_p \rho u \dfrac{dT}{dx} = \dfrac{d(k \, dT/dx)}{dx} + u\dfrac{dp}{dx} + \dfrac{4}{3}\mu\left(\dfrac{du}{dx}\right)^2$

**(b)** Integrate the continuity equation to find that

$$\rho u = \text{const} = \rho_1 u_1$$

**(c)** Eliminate $dp/dx$ in the energy equation by making use of the $x$-motion equation and incorporate the result of part b to put the energy equation in the form

$$\rho_1 u_1\left(C_p \dfrac{dT}{dx} + u\dfrac{du}{dx}\right) = \dfrac{4}{3}\mu\dfrac{d(u \, du/dx)}{dx} + \dfrac{d(k \, dT/dx)}{dx}$$

Integrate once to find

$$\dfrac{C_p \rho_1 u_1}{k}\left(C_p T + \dfrac{u^2}{2}\right) = \dfrac{4}{3}\operatorname{Pr}\dfrac{d(u^2/2)}{dx} + C_p\dfrac{dT}{dx} + C_1$$

where $C_1$ is an integration constant. Assume $\operatorname{Pr} = \tfrac{3}{4}$ and integrate again to find

$$C_p T + \dfrac{u^2}{2} = C_1 + C_2 e^{\rho_1 u_1 C_p x/k}$$

(d) Evaluate the constants of integration by the following considerations. Since $C_pT + u^2/2$ cannot increase indefinitely in the downstream direction, $C_2 = 0$. Far upstream it must be that $C_pT_1 + u_1^2/2 = C_1$. So

$$C_pT + \frac{u^2}{2} = C_pT_1 + \frac{u_1^2}{2}$$

(e) Substitute the result of part b into the x-motion equation and integrate once to obtain

$$\rho_1 u_1 u = -p + \frac{4}{3}\mu\frac{du}{dx} + C_3$$

Note that evaluation of $C_3$ from upstream conditions where $du/dx = 0$ gives

$$C_3 = p_1 + \rho_1 u_1^2 = \rho_1\left(u_1^2 + \frac{RT_1}{M}\right)$$

(f) Eliminate $p$ from the result of part e by making use of the ideal gas equation of state ($\rho = pM/RT$) and the energy equation of part d to obtain

$$\frac{4}{3}\frac{\mu}{\rho_1 u_1}u\frac{du}{dx} - \frac{\gamma+1}{2\gamma}u^2 + \frac{C_3}{\rho_1 u_1}u = \frac{\gamma-1}{\gamma}C_1$$

where $\gamma = C_p/C_v$. Rearrange this result into the form

$$\phi\frac{d\phi}{d\xi} = \beta Ma_1(\phi - 1)(\phi - \alpha)$$

where

$$\phi = \frac{u}{u_1}, \qquad \xi = \frac{x}{\lambda_1}, \qquad Ma = u_1\left(\frac{m}{\gamma RT_1}\right)^{1/2} = \text{Mach number}$$

$$\alpha = \frac{\gamma-1}{\gamma+1} + \frac{2}{\gamma+1}\frac{1}{Ma_1^2}, \qquad \beta = \frac{9}{8}(\gamma+1)\left(\frac{\pi}{8\gamma}\right)^{1/2}$$

$$\lambda_1 = 3\frac{\mu_1}{\rho_1}\left(\frac{\pi M}{8RT_1}\right)^{1/2} = \text{mean free path}$$

(g) Integrate the result of part f to obtain

$$\frac{1 - \phi}{(\phi - \alpha)^\alpha} = e^{\beta(1-\alpha)Ma_1(\xi - \xi_0)}$$

where $\xi_0$ is an integration constant. With this result, temperature and pressure distributions can be found from the results of parts d and e.

(h) Note from the result of part g that $\phi$ must approach 1 as $\xi \rightarrow -\infty$, which requires $\alpha < 1$. This requirement can be met only if $Ma_1 > 1$, the upstream flow is supersonic. Note also that as $\xi \rightarrow \infty$, $\phi \rightarrow \alpha$. For $Ma_1 = 2$ and $T_1 = 530R$ and air, plot $\phi$ versus $(x - x_0)/\lambda_1$ as illustrated in Fig. 5P-21 to demonstrate that the standing shock wave is only a few mean free paths thick. [M. Morduchow and P. A. Libby, On a complete solution of the one-dimensional flow equations of a viscous, heat-conducting, compressible gas, *J. Aeronaut. Sci.* **16** 674–684 (1949); R. von Mises, on the thickness of a steady shock wave, *J. Aeronaut. Sci.* **17**, 551–554 (1950); S. Sherman, A low-density wind-tunnel study of shock wave structure and relaxation phenomena in gases, NACA TN 3298, 1955.

**5-22**  An approximate solution to the vaporization problem treated in Section 5-4 and illustrated in Fig. 5-8 can be made by assuming the temperature distribution in the film to be linear. Then the requirement that heat conducted into the interface vaporize liquid can be expressed as

$$k\frac{T_w - T_s}{\delta} = \Lambda\rho_{vap}\frac{d\delta}{dt}$$

(a) Show that the solution to the preceding equation is

$$\delta^2 = 2\frac{k\Delta T}{\Lambda\rho}t$$

for an initially zero film thickness.

(b) Show that the result of part a is the same as that predicted by the approximation of Eq. (5-51).

**5-23**  A stagnant pond of water that is uniformly at its freezing temperature of 32°F is suddenly exposed to 22°F air. Estimate the time required for 4 in. of ice to form. A linear temperature distribution in the ice may be assumed if desired. How long a time is required to double the ice thickness?

**5-24**  Plot the numerical value of the heat flux from a piece of metal at 812°F exposed to liquid water at 212°F against time. The heat flux at the metal surface can be obtained from the derivative of Eq. (5-49).

**5-25**  For the transient vaporization problem of Section 5-4, show that the velocity in the liquid $u_L$ is given by

$$\rho_L\left(\frac{d\delta}{dt} - u_L\right) = \frac{dM}{dt}$$

from which, for the case in which the vapor density is constant, it is found that the velocity in the liquid varies with time as

$$u_L = B\left(1 - \frac{\rho_{\text{vapor}}}{\rho_L}\right)\left(\frac{\alpha_{\text{vapor}}}{t}\right)^{1/2}$$

Note that unless the two phases are of equal density, the liquid velocity is initially large and is never zero. Plot $u_L$ against $t$ for the conditions of Problem 5-24; comment on the accelerating force required to achieve this result with an initially stagnant liquid.

**5-26** The heat transfer in a piston–cylinder system such as is important to a Stirling-cycle heat engine has been analyzed as a one-dimensional problem and the results have been compared to measurements [15]. Prepare a brief report outlining the salient features of the analysis and the results.

## REFERENCES

1 R. B. Bird, W. E. Stewart, and E. N. Lightfoot, *Transport Phenomena*, Wiley, New York, 1960, pp. 272, 306.

2 R. Nahme, Beiträge zur hydrodynamischen Theorie der Lagerreibung, *Ing.-Arch.* **11**, 191–209 (1940).

3 H. S. Carslaw and J. C. Jaeger, *Conduction of Heat in Solids*, 2nd ed., Clarendon Press, Oxford, 1959, pp. 405–406.

4 H. Schlichting, *Boundary-Layer Theory* (translated by J. Kestin), McGraw-Hill, New York, 1968, pp. 277–278.

5 J. H. Arnold, Studies in diffusion: III. Unsteady state vaporization and absorption, *Transact. AIChE* **40**, 361–378 (1944).

6 *International Critical Tables*, McGraw-Hill, New York, 1933.

7 A. N. Nesmeyanov, *Vapor Pressure of the Elements*, Infosearch Ltd., London, 1963.

8 T. E. Jordan, *Vapor Pressure of Organic Compounds*, Interscience, New York, 1954.

9 C. Limpiyakorn and L. Burmeister, The effect of thermal radiation on transient vaporization of a saturated liquid at a constant temperature plate, *Transact. ASME, J. Heat Transf.* **94**, 415–418 (1972).

10 M. Rooney and L. Burmeister, Pressure excursions in transient film boiling from a sphere, *Internatl. J. Heat Mass Transf.* **18**, 671–675 (1975).

11 S. Kibbee, J. A. Orozco, and L. C. Burmeister, Influence of liquid heat conduction on maximum pressure during transient film boiling from a sphere to a saturated liquid, *Internatl. J. Heat Mass Transf.* **20**, 1069–1075 (1977).

12 L. C. Burmeister, Pressure Excursions in transient film boiling from a sphere to a subcooled liquid, *Internatl. J. Heat Mass Transf.* **21**, 1411–1420 (1978).

13 S. K. Griffiths and F. A. Morrison, Low Peclet number heat and mass transfer from a drop in an electric field, *Transact. ASME, J. Heat Transf.* **101**, 484–488 (1979).

14 J. P. Meyer and M. D. Kostin, Circulation phenomena in Stefan diffusion, *Internatl. J. Heat Mass Transf.* **18**, 1293–1297 (1975).

15 J. Polman, Heat transfer in a piston–cylinder system, *Internatl. J. Heat Mass Transf.* **24**, 184–187 (1981).

# 6

# LAMINAR HEAT TRANSFER IN DUCTS

Convective transport of heat or mass can be generally classified as forced or natural. In forced convection the fluid motion is primarily caused by some mechanism not connected with the heat or mass transfer at the location of interest; usually, a pump or fan at a distant location "forces" the fluid into motion. In natural convection the fluid motion depends on the heat or mass transfer because of a body force acting on density differences induced by the local heat or mass transfer; the fluid flows "naturally."

A second classification of convective heat or mass transfer is as either bounded or unbounded. A somewhat more common term is *internal* or *external flow*. In bounded or internal cases the flow is constrained on all sides by solid boundaries, as in flow through a pipe. In unbounded or external flow the fluid has at least one side extending to infinity without encountering a solid surface. An airplane flying through the air presents an external flow example insofar as the atmospheric air is concerned. Also, a flat plate suspended in air puts the fluid in an unbounded or external flow condition.

These two classification schemes are really independent. An internal flow can be either forced or natural convection, and an external flow can likewise be either forced or natural convection.

## 6.1 MIXING-CUP TEMPERATURE

As seen in Sections 5.1 and 5.2 when accounting for the effects of viscous dissipation, a careful choice of the driving temperature difference greatly simplifies heat-transfer calculations. Although mass transfer does not readily admit an analog to viscous dissipation, it is safe to expect that the similarity of

207

the describing equations will enable the benefits of a careful choice of the driving potential difference to apply to mass transfer as well. The most useful driving potential difference depends somewhat on the physical situation under consideration.

In an external flow the temperature of the fluid far removed from the heat- or mass-transfer surface is usually known (and may even be a constant). However, in an internal flow there is usually no well-defined temperature except, perhaps, for a wall or an inlet temperature. In the case of heat transfer, there will of necessity be a temperature variation perpendicular to the solid surface, and there will probably be a temperature variation in the direction of fluid flow. Thus there are several choices available for the driving temperature difference.

To facilitate discussion, attention is focused in what follows on the common engineering situation of flow down a duct. Such a flow is illustrated in Fig. 6-1. In the choice of a temperature for the fluid, the fact that a designer is most interested in the rate of flow of total mass and energy down the duct should be recognized. The rate of mass flow down the duct $\dot{m}$ is obtained from the velocity distribution as

$$\dot{m} = \int_A \rho v \, dA = \rho_{av} v_{av} A$$

from which it follows that the average velocity is given by

$$v_{av} = \frac{1}{A} \int_A \frac{\rho}{\rho_{av}} V \, dA \tag{6-1}$$

Energy flow rate down the duct $\dot{E}$ is obtained from the velocity and temperature distribution by

$$\dot{E} = \int_A \rho C_p T v \, dA = \rho_{av} C_{p_{av}} v_{av} A T_m$$

from which it is found that the mixing-cup temperature $T_m$ is given by

$$T_m = \frac{1}{A} \int_A \frac{\rho}{\rho_{av}} \frac{C_p}{C_{p_{av}}} \frac{v}{v_{av}} T \, dA \tag{6-2}$$

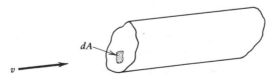

**Figure 6-1**   Flow down a duct.

For a circular duct of radius $R$ when velocity and temperature depend only on radius, $dA = 2\pi r\,dr$ so that Eqs. (6-1) and (6-2) become

$$v_{av} = 2\int_0^1 \frac{\rho}{\rho_{av}} vr'\,dr'$$

and

$$T_m = 2\int_0^1 \frac{\rho}{\rho_{av}} \frac{C_p}{C_{av}} \frac{v}{v_{av}} Tr'\,dr'$$

with $r' = r/R$.

The average, mixed temperature defined by Eq. (6-2) is usually called the *mixing-cup temperature* since it represents the temperature that a cup of fluid, scooped out of the stream, would attain. The mixing-cup temperature can, of course, vary along the length of the duct. Because the mixing-cup temperature is likely to be known, it is anticipated that heat flux at a local wall position can be calculated most simply from

$$q_w = h(T_{wall} - T_m) \qquad (6\text{-}3)$$

## 6.2  ENTRANCE LENGTH

Before any detailed solutions of particular cases are attempted, a brief review of salient features is in order. First, consider the expected behavior of the velocity profile in a round duct of constant cross section. Figure 6-2 shows the change in axial velocity profile at successive stations down the duct. As flow proceeds down the duct, the effect of the wall diffuses into the main stream. A boundary layer builds up and eventually occupies the entire tube, with $v_{av}$ remaining constant. After this entrance region $L_e$ the flow is fully developed and no longer changes. If the flow is laminar (Re $<$ 2300) a characteristic parabolic velocity profile exists; if the flow is turbulent (Re $>$ 2300), a much flatter velocity profile results. It is important to be able to judge whether flow is fully developed. For this purpose the rules of thumb

$$\frac{L_e}{D} \gtrsim 0.05\,\text{Re} \qquad \text{laminar flow}$$

$$\frac{L_e}{D} \gtrsim 40 - 100 \qquad \text{turbulent flow} \qquad (6\text{-}4)$$

specify the minimum entrance length required to achieve fully developed velocity profiles.

A similar behavior is observed for thermal cases. A new complication is that a heated section may not occur at the beginning of the duct; there may be an

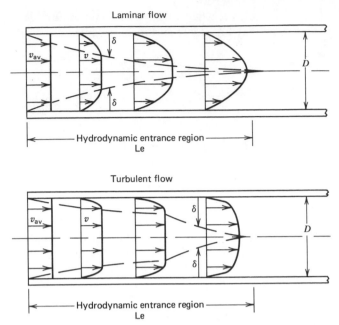

**Figure 6-2** Laminar and turbulent velocity profiles in the entrance region of a circular duct.

unheated starting section in which velocity profiles can develop before a temperature profile can even start. Figure 6-3 shows the general trend of events. Note the thermal boundary-layer growth. Evidently a thermal entrance region is to be expected, and constant heat-transfer coefficients will be achieved only after some distance down the duct. If the unheated section is very small ($L_e/D \ll 0.05$ Re for laminar flow, $L_e/D \ll 40$–$100$ for turbulent flow), the velocity may be assumed constant (slug flow). At the other extreme, if the unheated section is much longer than $L_e$, the velocity profile is fully developed and depends only on radial position (assuming that properties are not temperature dependent). In either of these two extreme cases there is no appreciable radial component of velocity. For intermediate lengths of the unheated starting section, the velocity profile is still developing and there may be appreciable radial velocity. Note that a thermal effect on velocity profiles is not taken into account in the preceding discussion. As shown in Fig. 6-4 [1] for laminar flow, a temperature-dependent viscosity can result in departure from the usually expected parabolic velocity profile (curve $a$). If heat flows into a liquid, the resultant velocity profile is flattened (curve $b$) because the viscosity near the wall is less than in the core of the tube. If heat flows out of the fluid, the velocity profile is peaked (curve $c$) because the viscosity in the core of the tube is less than near the wall. Of course, for a gas the effects are opposite from those for a liquid since gas viscosity increases with increasing temperature.

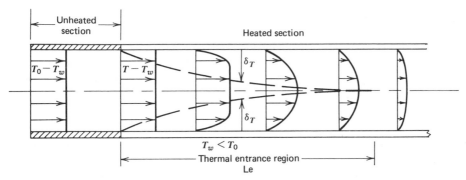

**Figure 6-3** Temperature profiles in the entrance region of a circular duct.

The mass diffusion case has a behavior that is similar to the thermal case discussed previously. The only additional complication is a possible "blowing" at the wall.

In the inlet region of a duct where velocity and temperature profiles are developing simultaneously, the portion of the duct very near the inlet is filled with fluid that, for the most part, is as yet unaffected by the conditions at the wall. In this region the fluid in the essentially undisturbed core can be regarded as an external flow and the fluid very near the duct wall, as a boundary layer. Hence the boundary layer concepts of Chapter 7 can be applied to estimate wall friction, heat transfer, and so forth there.

A complete discussion and survey of laminar heat transfer in ducts by Shah and London [2] can be consulted for greater detail and variety of method and result than is afforded by the present chapter. Various cross-section shapes and boundary conditions are treated as are fully developed and entrance region cases.

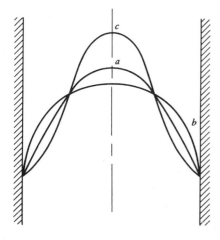

**Figure 6-4** Temperature-dependent liquid viscosity effects on fully developed laminar velocity profiles: (a) constant viscosity; (b) heat flow into liquid; (c) heat flow out of liquid.

## 6.3  CIRCULAR DUCT

The manner in which heat-transfer coefficients vary along the length of a duct can be determined from knowledge of the temperature distribution in the fluid. To illustrate the procedure employed, consider flow of a Newtonian fluid down a circular duct of radius $R$.

Assumptions to be generally imposed are (1) steady laminar flow, (2) constant properties, (3) velocity distribution known and dependent only on radial position, (4) fluid temperature dependent only on axial and radial position in duct, and (5) inlet temperature known. As a consequence of assuming the axial velocity radial distribution to be known and unchanging, the continuity and motion equations need not be considered. Only the energy equation is of concern and has the form

$$V\frac{\partial T}{\partial z} = \alpha\frac{1}{r}\frac{\partial(r\,\partial T/\partial r)}{\partial r} + \frac{\partial^2 T}{\partial z^2} + \frac{\mu}{\rho C_p}\left(\frac{\partial V}{\partial r}\right)^2 \tag{6-5}$$

with boundary conditions of

$$T(r,0) = T_0 \tag{6-6a}$$

$$T(0,z)\text{ finite} \tag{6-6b}$$

$$T(r,z_{max}) = ? \tag{6-6c}$$

$$T(R,z) = T_w, \qquad \text{specified wall temperature} \tag{6-6d}$$

or

$$k\frac{\partial T(R,z)}{\partial r} = q_w, \qquad \text{specified heat flux into tube}$$

where $V$ is the known axial velocity that can vary with radius.

Clues as to the importance of various terms can be obtained by examining the dimensionless forms of Eqs. (6-5) and (6-6), which are

$$v\frac{\partial\theta}{\partial(4/Gz)} = \frac{1}{r'}\partial\frac{(r'\,\partial\theta/\partial r')}{\partial r'} + \frac{4}{P_e^2}\frac{\partial^2\theta}{\partial(4/Gz)^2} + \mathrm{E}\,P_r\left(\frac{\partial v}{\partial r'}\right)^2 \tag{6-7}$$

$$\theta(r',0) = \frac{T_0 - T_r}{\Delta T} \tag{6-8a}$$

$$\theta(0,4/Gz)\text{ finite} \tag{6-8b}$$

$$\theta(r',4/Gz_{max}) = ? \tag{6-8c}$$

$$\theta(1,4/Gz) = \frac{T_w - T_r}{\Delta T}, \qquad \text{specified wall temperature} \tag{6-8d}$$

or

$$\frac{\partial\theta(1,4/\text{Gz})}{\partial r'} = \frac{q_w R}{k\,\Delta T}, \qquad \text{specified heat flux into tube}$$

The dimensionless parameters are defined in conventional form as

$Z^+ = $ dimensionless axial distance $= 4(z/D)/\text{Re Pr} = 4\alpha z/D^2 V_{av}$

$\text{Gz} = $ Graetz number $= \text{Pr Re } D/z = D^2 V_{av}/\alpha z = 4/Z^+$

$\text{Pr} = $ Prandtl number $= \nu/\alpha = \mu C_p/k$

$\text{Re} = $ Reynolds number $= DV_{av}/\nu$

$\text{Pe} = $ Peclet number $= DV_{av}/\alpha = \text{Re Pr}$

$\text{E} = $ Eckert number $= V_{av}^2/C_p\,\Delta T$

$r' = r/R$

$\theta = (T - T_r)/\Delta T$

$T_r = $ reference temperature

$\Delta T = $ reference temperature difference

$v = V/V_{av}$

Inspection of Eq. (6-7) reveals several possible simplifications. If Pe is large (Pe $\gtrsim$ 100 [23]), axial conduction (the next-to-last term) is negligible. This can occur if Re is large, Pr is large, or both. If Pe is small, axial conduction will be of importance, but in such a case the major source of axial conduction will be down the tube walls [29]. Basically, the Peclet number is a measure of the relative importance of axial convection to axial conduction. As a practical matter, Pe is small only if Pr is small as is the case for liquid metals. Clearly, if E Pr is small, viscous dissipation (the last term) can be discarded. This can occur if Pr is small, E is small, or both. For the low velocities associated with the assumed laminar flow, it is likely that E is small; a possible exception might be encountered should the reference temperature difference be even smaller than $V_{av}^2/C_p$ [27, 30].

How does one use a solution to Eq. (6-7) to evaluate a heat-transfer coefficient? Presume that viscous dissipation is neglected so that $T_w - T_m$ drives the heat transfer. Then the local heat flux into the tube is given in terms of the local heat-transfer coefficient $h$ by

$$q_w = h(T_w - T_m) = k\frac{\partial T(R, z)}{\partial r}$$

which gives, in terms of the previously defined dimensionless parameters,

$$\frac{hR}{k}(\theta_w - \theta_m) = \frac{\partial\theta(1,4/\text{Gz})}{\partial r'}$$

From this it is found that the local Nusselt number Nu is

$$Nu = \frac{hD}{k} = \frac{2}{\theta_w - \theta_m} \frac{\partial\theta(1, 4/\mathrm{Gz})}{\partial r'} \tag{6-9a}$$

where

$$\theta_w\left(\frac{4}{\mathrm{Gz}}\right) = \theta\left(1, \frac{4}{\mathrm{Gz}}\right)$$

and, from Eq. (6-2),

$$\theta_m\left(\frac{4}{\mathrm{Gz}}\right) = 2\int_0^1 v(r')\theta\left(r', \frac{4}{\mathrm{Gz}}\right) r'\, dr'$$

If the temperature distribution in the fluid is known, the local Nusselt number can be obtained in principle from Eq. (6-9a).

How does one use a solution to Eq. (6-7) to evaluate an average heat-transfer coefficient? The most useful procedure for ducts of specified wall temperature is to define the average coefficient in a manner that is convenient for calculation of the mixing-cup temperature of the fluid at the outlet. This leads to an effective temperature difference between the fluid and the wall, which is referred to as the *log-mean-temperature* difference (LMTD). To see this, consider the energy balance for the crosshatched control volume shown in Fig. 6-5, which is

$$V_{av}\rho C_p \pi R^2\, dT_m = h2\pi R(T_w - T_m)\, dz$$

$$T_m(z = 0) = T_0$$

In dimensionless terms this equation is

$$d\theta_m = \mathrm{Nu}(\theta_w - \theta_m)d\left(\frac{4}{\mathrm{Gz}}\right)$$

$$\theta_m\left(\frac{4}{\mathrm{Gz}} = 0\right) = \theta_{m0}$$

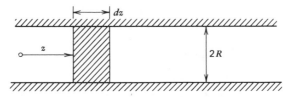

**Figure 6-5**  Control volume for derivation of average heat-transfer coefficient.

If $T_r = T_w$ and $\Delta T = T_0 - T_w$, $\theta_w = 0$, and one then has

$$-\frac{d\theta_m}{\theta_m} = \text{Nu}\, d\left(\frac{4}{\text{Gz}}\right)$$

$$\theta_m\left(\frac{4}{\text{Gz}} = 0\right) = 1$$

whose solution is

$$\theta_m = \frac{T_m - T_w}{T_0 - T_w} = \exp\left[-\int_0^{4/\text{Gz}}\text{Nu}\, d\left(\frac{4}{\text{Gz}}\right)\right] \qquad (6\text{-}9\text{b})$$

In other words, the difference between the fluid and wall temperatures decays exponentially. Naturally, Eq. (6-9b) would be more easily used to compute the outlet mixing-cup temperature $T_m$ if the exponential were given in terms of an average Nusselt number $\overline{\text{Nu}} - \bar{h}D/k$ defined in terms of the local Nusselt number as

$$\overline{\text{Nu}} = \frac{1}{4/\text{Gz}}\int_0^{4/\text{Gz}}\text{Nu}\, d\left(\frac{4}{\text{Gz}}\right) \qquad (6\text{-}9\text{c})$$

It is then possible to employ Eqs. (6-9b) and (6-9c) to obtain

$$\overline{\text{Nu}} = -\frac{\ln(\theta_m)}{4/\text{Gz}} \qquad (6\text{-}9\text{d})$$

The total heat transferred is obtained as

$$\rho C_p V_{\text{av}}\pi R^2 (T_{m,\text{out}} - T_{m,\text{in}}) = \bar{h}2\pi Rz\,\Delta T_{\text{av}} \qquad (6\text{-}9\text{e})$$

In dimensionless terms, this relationship is

$$\Delta T(\theta_m - 1) = \overline{\text{Nu}}\frac{4}{\text{Gz}}\Delta T_{\text{av}}$$

Introduction of Eq. (6-9d) into this result leads to

$$\Delta T_{\text{av}} = \frac{\Delta T(-1 + \theta_m)}{\ln(1/\theta_m)}$$

$$= \frac{(T_w - T_0) - (T_w - T_m)}{\ln[(T_w - T_0)/(T_w - T_m)]}$$

a result that is often written as

$$\text{LMTD} = \frac{\Delta T_{\text{max}} - \Delta T_{\text{min}}}{\ln(\Delta T_{\text{max}}/\Delta T_{\text{min}})} \qquad (6\text{-}9\text{f})$$

The average heat flux $\bar{q}_w$ is directly obtained from Eq. (6-9e) as

$$\bar{q}_w = \bar{h}\,\Delta T_{av} = \bar{h}\,\text{LMTD}$$

with LMTD from Eq. (6-9f).

## 6.4   LAMINAR SLUG FLOW IN A CIRCULAR DUCT

When the unheated entrance section is very short and the diffusivity of
momentum is much less than the diffusivity of heat ($\text{Pr} = \nu/\alpha \ll 1$), temperature profiles develop much more quickly than do velocity profiles. This case
corresponds to flow entering a tube from a larger reservoir. It is then accurate
to assume the axial velocity to be constant ($v = V/V_{av} = 1$). Naturally, the
results of such an assumption cannot be extended to great distances down the
duct since nonuniform velocity profiles will evolve eventually. Note also that a
uniform velocity has no viscous dissipation.

**Constant Wall Temperature**

The simplified form of Eq. (6-7) is to be solved for the case of a constant tube
wall temperature $T_w$ and a uniform inlet temperature $T_0$. The equation and
boundary conditions of interest are

$$\frac{\partial\theta}{\partial(4/\text{Gz})} = \frac{1}{r'}\frac{\partial(r'\,\partial\theta/\partial r')}{\partial r'} \qquad (6\text{-}10)$$

$$\theta(r',0) = 1 \qquad (6\text{-}11a)$$

$$\theta\left(0,\frac{4}{\text{Gz}}\right) \text{ finite} \qquad (6\text{-}11b)$$

$$\theta\left(1,\frac{4}{\text{Gz}}\right) = 0 \qquad (6\text{-}11c)$$

with $T_r = T_w$ and $\Delta T = T_0 - T_w$. Equations (6-10) and (6-11) also describe the
transient temperature in a cylinder of initially uniform temperature whose
periphery is suddenly reduced to zero temperature. The $4/\text{Gz}$ parameter plays
the role of elapsed time—it can be seen that $4/\text{Gz} == \alpha t/R^2$, where $t = z/V_{av}$.
The solution technique to be employed here is separation of variables in which
a product solution is assumed of the form

$$\theta\left(r',\frac{4}{\text{Gz}}\right) = R_1(r')Z_1\left(\frac{4}{\text{Gz}}\right)$$

Insertion of this assumed form into Eq. (6-10) and division by $R_1Z$ gives

$$\frac{1}{r'R_1}\frac{d(r'\,dR_1/dr')}{dr'} = \frac{1}{Z_1}\frac{dZ_1}{d(4/\text{Gz})} = -\lambda^2$$

or

$$\text{Function of } r' = \text{function of } 4/\text{Gz}$$

Here the variables have been separated, inasmuch as the left-hand side is solely a function of $r'$ whereas the right-hand side is solely a function of $4/\text{Gz}$. These two functions must be equal for all values of $r'$ and $4/\text{Gz}$ a requirement that can be met only if they both equal a constant $-\lambda^2$ (the constant must be negative to avoid predicting a temperature which increases exponentially down the duct). Two separate ordinary differential equations must now be solved:

$$\frac{dZ_1}{d(4/\text{Gz})} = -\lambda^2 Z_1 \qquad r'^2 \frac{d^2 R_1}{dr'^2} + r' \frac{dR_1}{dr} + \lambda^2 r'^2 R_1 = 0$$

whose solutions are

$$R_1 = C_1 J_0(\lambda r) + C_2 Y_0(\lambda r)$$

$$Z_1 = C_3 e^{-\lambda^2(4/\text{Gz})}$$

Here $J_0$ is the zero-order Bessel function of the first kind and $Y_0$ is the zero-order Bessel function of the second kind with behavior as illustrated in Fig. 6-6 [3]. The boundary condition of Eq. (6-11b) requires, since $Y_0(0) = -\infty$,

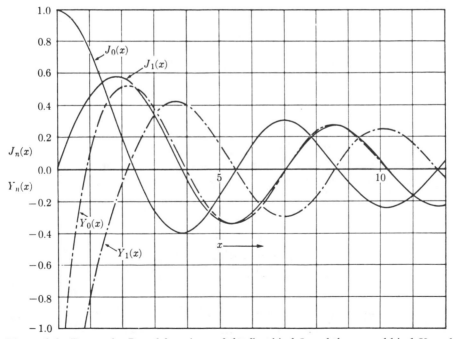

**Figure 6-6**   Zero-order Bessel functions of the first kind $J_0$ and the second kind $Y_0$ and first-order Bessel functions of the first kind $J_1$ and the second kind $Y_1$.

that $C_2 = 0$. Hence

$$\theta = C \exp\left(\frac{-4\lambda^2}{Gz}\right) J_0(\lambda r')$$

The boundary condition of Eq. (6-11c) requires that $J_0(\lambda) = 0$, from which it is found that

$$\lambda = 2.4048, 5.5201, 8.6537, 11.7915, 14.93309, \ldots \qquad (6\text{-}12)$$

Each of the infinite number of $\lambda$ values satisfies the problem conditions thus far, and each must be considered. Therefore

$$\theta = \sum_{n=1}^{\infty} C_n \exp\left(\frac{-4\lambda_n^2}{Gz}\right) J_0(\lambda_n r')$$

with $J_0(\lambda_n) = 0$. The boundary condition of Eq. (6-11a) demands

$$\sum_{n=1}^{\infty} C_n J_0(\lambda_n r') = 1$$

which is a requirement to express a constant in terms of a Bessel function over the range $0 \leqslant r' \leqslant 1$. It is found [4] that the individual constants $C_n$ are

$$C_n = \frac{2}{\lambda_n J_1(\lambda_n)}$$

The temperature distribution is then finally found to be

$$\theta = 2 \sum_{n=1}^{\infty} \exp\left(\frac{-4\lambda_n^2}{Gz}\right) \frac{J_0(\lambda_n r')}{\lambda_n J_1(\lambda_n)} \qquad (6\text{-}13)$$

with $\lambda_n$ as given by Eq. (6-12)

The mixing-cup temperature is needed next and is obtained by insertion of Eq. (6-13) into Eq. (6-2) as

$$\theta_m = 2 \int_0^1 \theta r' \, dr'$$

$$= 4 \int_0^1 \exp\left(\frac{-4\lambda_n^2}{Gz}\right) \frac{r' J_0(\lambda_n r') \, dr'}{\lambda_n J_1(\lambda_n)}$$

$$\theta_m = 4 \sum_{n=1}^{\infty} \frac{\exp(-4\lambda_n^2/Gz)}{\lambda_n^2} \qquad (6\text{-}14)$$

The next needed item is the gradient of the wall. Realization that $dJ_0(\lambda_n r')/dr'$

$= -\lambda_n J_1(\lambda_n r')$ gives

$$\theta \frac{\partial(1,4/\mathrm{Gz})}{\partial r'} = -2 \sum_{n=1}^{\infty} \exp\left(\frac{-4\lambda_n^2}{\mathrm{Gz}}\right) \tag{6-15}$$

Introduction of Eqs. (6-14) and (6-15) into Eq. (6-9a) finally yields the local Nusselt number as

$$\mathrm{Nu} = \frac{hD}{k} = \frac{\sum_{n=1}^{\infty} \exp\left(-4\lambda_n^2/\mathrm{Gz}\right)}{\sum_{n=1}^{\infty} \exp\left(-4\lambda_n^2/\mathrm{Gz}\right)/\lambda_n^2} \tag{6-16}$$

Although the limit as $4/\mathrm{Gz} \rightarrow \infty$ has no real importance for slug flow as discussed earlier, Eq. (6-16) shows that the local Nusselt number is then

$$\lim_{4/\mathrm{Gz} \rightarrow \infty} \mathrm{Nu} = \lambda_1^2 = 5.7831 \tag{6-17}$$

The average coefficient based on log–mean temperature difference is found from substitution of Eqs. (6-14) and (6-15) into Eq. (6-9d). Realization that $\theta_w = 0$ and $\theta_{m_{inlet}} = 1$ for this case then gives

$$\overline{\mathrm{Nu}} = \frac{\bar{h}D}{k} = \frac{-\ln(\theta_m)}{4/\mathrm{Gz}} \tag{6-18}$$

with $\theta_m$ given by Eq. (6-14). Taking the limit as $4/\mathrm{Gz} \rightarrow \infty$, one obtains, as for the local Nusselt number,

$$\lim_{4/\mathrm{Gz} \rightarrow \infty} \overline{\mathrm{Nu}} = 5.7831$$

Table 6-1 shows the manner in which Nu, $\overline{\mathrm{Nu}}$, and $\theta_m$ asymptotically approach their limiting values far down the duct.

### Constant Wall Heat Flux into Fluid

The case of a specified constant heat flux $q_w$ into the fluid leads to slightly different results for the heat transfer coefficients. The only changes in Eqs. (6-10) and (6-11) required to formulate the problem are to redefine $T_r = T_0$ and $\Delta T = q_w R/k$ and to change the boundary condition at $r' = 1$. Then the equations to be solved for the temperature distribution in the fluid are

$$\frac{\partial \theta}{\partial(4/\mathrm{Gz})} = \frac{1}{r} \frac{\partial(r' \partial\theta/\partial r')}{\partial r'} \tag{6-19a}$$

$$\theta(r',0) = 0 \tag{6-19b}$$

$$\theta\left(0, \frac{4}{\mathrm{Gz}}\right) \text{ finite} \tag{6-19c}$$

$$\frac{\partial\theta(1,4/\mathrm{Gz})}{\partial r'} = 1 \tag{6-19d}$$

**Table 6-1   Solutions for Slug Flow in a Circular Duct;
Constant Wall Temperature**

| $1/Gz$ | Nu | $\overline{Nu}$ | $\theta_m$ |
|--------|------|------|------------|
| 0 | $\infty$ | $\infty$ | 1 |
| 0.001 | 19.7 | 38.9 | 0.856 |
| 0.0025 | 13.1 | 24.3 | 0.784 |
| 0.005 | 9.92 | 17.7 | 0.701 |
| 0.01 | 7.75 | 13.2 | 0.59 |
| 0.025 | 6.18 | 9.31 | 0.394 |
| 0.05 | 5.81 | 7.62 | 0.218 |
| 0.1 | 5.79 | 6.71 | 0.0684 |
| 0.25 | 5.783 | 6.15 | 0.0021 |
| 0.5 | 5.783 | 5.97 | $6.5 \times 10^{-6}$ |
| 1.0 | 5.783 | 5.88 | $6.1 \times 10^{-11}$ |
| $\infty$ | 5.783 | 5.783 | 0 |

Note that Eqs. (6-19) also describe the temperature in a cylinder of initially zero temperature to whose periphery a constant heat flux is suddenly applied. Again, a separation of variables procedure is employed with the assumed solution form

$$\theta\left(r',\frac{4}{Gz}\right) = \underbrace{R_1(r')Z_1\left(\frac{4}{Gz}\right)}_{\substack{\text{decaying} \\ \text{initial} \\ \text{transient}}} + \underbrace{Z_2\left(\frac{4}{Gz}\right)}_{\substack{\text{axial temperature} \\ \text{rise due to} \\ \text{accumulated} \\ \text{wall flux}}} + \underbrace{R_2(r')}_{\substack{\text{radial temperature} \\ \text{variation to let} \\ \text{wall flux into} \\ \text{fluid}}}$$

Insertion of this assumed form into Eq. (6-19a) gives

$$R_1\frac{dZ_1}{d(4/Gz)} + \frac{dZ_2}{d(4/Gz)} = Z_1\frac{1}{r'}\frac{d(r'\,dR_1/dr')}{dr'} + \frac{1}{r'}\frac{d(r'\,dR_2/dr')}{dr'}$$

The next step is to set

$$\frac{dZ_2}{d(4/Gz)} = \frac{1}{r'}\frac{d(r'\,dR_2/dr')}{dr'} \qquad (6\text{-}20)$$

and then to separate variables in the remainder of the equation to obtain, after dividing by $R_1Z_1$,

$$\underbrace{\frac{1}{r'R_1}\frac{d(r'dR_1/dr')}{dr'}}_{\text{function of } r'} = \underbrace{\frac{1}{Z_1}\frac{dZ_1}{d(4/Gz)}}_{\text{function of } 4/Gz} = -\lambda^2$$

As before, requiring a function of $r'$ to always equal a function of $4/Gz$ when the variables are independent requires that both functions equal a separation constant $-\lambda^2$. Or

$$r'^2\frac{d^2R_1}{dr'^2} + r'\frac{dR_1}{dr'} + \lambda^2 r'^2 R_1 = 0 \qquad \text{and} \qquad \frac{dZ_1}{d(4/Gz)} = -\lambda^2 Z_1$$

The solutions for $R_1$ and $Z_1$ are, as before,

$$R_1 = C_1 J_0(\lambda r') + C_2 Y_0(\lambda r')$$

$$Z_1 = C_3 e^{-\lambda^2 4/Gz}$$

The boundary condition of Eq. (6-19c) requires that $C_2 = 0$. The same arguments for a separation constant apply to Eq. (6-20), so that

$$\frac{dZ_2}{d(4/Gz)} = C_4 = \frac{1}{r'}\frac{d(r'\,dR_2/dr')}{dr'}$$

from which it is found that the solutions for $Z_2$ and $R_2$ are

$$Z_2 = C_4\left(\frac{4}{Gz}\right) + C_5 \qquad \text{and} \qquad R_2 = \frac{C_4 r'^2}{4} + C_5 \ln(r') + C_6$$

The boundary condition of Eq. (6-19c) requires that $C_5 = 0$. The boundary conditions of Eqs. (6-19b) and (6-19d), respectively, then require the solution for $\theta$ to satisfy

$$C_1 C_3 J_0(\lambda r') + C_5 + C_6 + \frac{C_4 r'^2}{4} = 0$$

and

$$-\lambda C_1 C_3 e^{-\lambda^2 4/Gz} J_1(\lambda) + \frac{C_4}{2} = 1$$

Because there are no additional conditions to satisfy, it is permissible to arbitrarily set $C_5 + C_6 = 0$ and $C_4 = 2$. Then

$$C_1 C_3 J_0(\lambda r') = -\frac{r'^2}{2}$$

$$J_1(\lambda) = 0$$

The last of these two requirements gives an infinite number of roots of

$J_1(\lambda) = 0$ as

$$\lambda = 0, 3.8317, 7.0156, 10.1735, 13.3237, 16.4706, \ldots \qquad (6\text{-}21)$$

Each value of $\lambda$ is admissible, and so the solution is a linear combination of all of the possibilities; the first of these two requirements is then

$$\sum_{n=0}^{\infty} C_n J_0(\lambda_n r') = -\frac{r'^2}{2}$$

which requires that $-r'^2/2$ be expressed in a series of Bessel functions over the interval $0 \leqslant r' \leqslant 1$. It is found [4] that

$$C_n = \frac{-2}{\lambda_n^2 J_0(\lambda_n)} \qquad n \neq 0$$

and

$$C_0 = -\tfrac{1}{4}$$

The final result for dimensionless temperature is

$$\theta = +2\frac{4}{\text{Gz}} + \frac{r'^2}{2} - \frac{1}{4} - 2\sum_{n=1}^{\infty} \exp\left(\frac{-4\lambda_n^2}{\text{Gz}}\right) \frac{J_0(\lambda_n r')}{\lambda_n^2 J_0(\lambda_n)} \qquad (6\text{-}22)$$

with $J_1(\lambda_n) = 0$ as given by Eq. (6-21).

From Eq. (6-22) the mixing-cup temperature is found by insertion into Eq. (6-2) to be

$$\theta_m = 2\frac{4}{\text{Gz}} \qquad (6\text{-}23)$$

This shows, as expected on physical grounds, that the average fluid temperature increases linearly down the duct in response to the need to absorb an amount of heat that is proportional to the distance traveled. This is like the temporal variation of a cylinder's average temperature with a constant peripheral heat flux—it must increase linearly with time. The wall temperature is evaluated from Eq. (6-22) as

$$\theta_w = \theta\left(1, \frac{4}{\text{Gz}}\right) = 2\frac{4}{\text{Gz}} + \frac{1}{4} - 2\sum_{n=1}^{\infty} \frac{\exp\left(-4\lambda_n^2/\text{Gz}\right)}{\lambda_n^2} \qquad (6\text{-}24)$$

Introduction of Eqs. (6-23) and (6-24) into Eq. (6-9a) finally yields the local Nusselt number as

$$\text{Nu} = \frac{hD}{k} = \frac{8}{1 - 8\sum_{n=1}^{\infty} \exp\left(-4\lambda_n^2/\text{Gz}\right)/\lambda_n^2} \qquad (6\text{-}25)$$

The difference between wall and mixing cup temperature is

$$\theta_w - \theta_m = \frac{1}{4} - 2 \sum_{n=1}^{\infty} \frac{\exp(-4\lambda_n^2/Gz)}{\lambda_n^2} \qquad (6\text{-}26)$$

Although the limit as $4/Gz \to \infty$ has no real meaning, Eq. (6-25) shows that the local Nusselt number is then

$$\lim_{4/Gz \to \infty} Nu = 8 \qquad (6\text{-}27)$$

Also, Eq. (6-26) gives the limiting value

$$\lim_{4/Gz \to \infty} (\theta_w - \theta_m) = \tfrac{1}{4} \qquad (6\text{-}28)$$

The average fluid temperature at any position along the duct is already known in a specified wall heat flux case. Thus knowledge of the heat-transfer coefficient is of most utility in determining the wall temperature, an important quantity when fluid properties degrade at high temperature.

The average Nusselt number $\overline{Nu}$ is usually based on the average difference between the wall and mixing-cup temperatures

$$\overline{T}_w - \overline{T}_m = \frac{1}{4/Gz} \int_0^{4/Gz} (T_w - T_m) d\left(\frac{4}{Gz}\right)$$

so that the average heat transfer coefficient $\bar{h}$ is defined as

$$q_w = \bar{h}\left(\overline{T}_w - \overline{T}_m\right)$$

From this definition of $\bar{h}$, the average Nusselt number is

$$\overline{Nu} = \frac{\bar{h}D}{k} = \frac{2}{\overline{\theta}_w - \overline{\theta}_m}$$

$$\overline{Nu} = \frac{8}{1 - 8\sum_{n=1}^{\infty}\{[1 - \exp(-4\lambda_n^2/Gz)]/(\lambda_n^4 4/Gz)\}} \qquad (6\text{-}28a)$$

Table 6-2 shows the manner in which Nu, $\overline{Nu}$, and $\theta_m$ asymptotically approach their limiting values far down the duct.

Note that the results for constant wall heat flux are, far from the entrance, the same as for the cases of linearly increasing wall temperature and of a constant-temperature difference between the wall and the fluid. The latter case is pertinent to a counterflow heat exchanger in which the heat capacities of the two streams are equal.

**Table 6-2   Solutions for Slug Flow in a Circular Duct; Constant Wall Heat Flux**

| $1/Gz$ | Nu | $\overline{Nu}$ | $\theta_w - \theta_m$ |
|---|---|---|---|
| 0 | $\infty$ | $\infty$ | 0 |
| 0.001 | 30.6 | 41.5 | 0.0655 |
| 0.0025 | 20.4 | 28.6 | 0.0981 |
| 0.005 | 15.3 | 21.5 | 0.131 |
| 0.01 | 11.9 | 16.3 | 0.168 |
| 0.025 | 9.16 | 11.9 | 0.218 |
| 0.05 | 8.24 | 9.95 | 0.243 |
| 0.1 | 8.01 | 8.92 | 0.2496 |
| 0.25 | 8 | 8.34 | 0.25 |
| 0.5 | 8 | 8.17 | 0.25 |
| 1.0 | 8 | 8.08 | 0.25 |
| $\infty$ | 8 | 8 | 0.25 |

## 6.5   FULLY DEVELOPED LAMINAR FLOW IN A CIRCULAR DUCT

When the unheated entrance region is very long, velocity profiles develop fully before a fluid particle enters the heated region. This case corresponds to flow down a long pipe, only the latter portion of which is heated.

Fully developed velocity profiles for a power-law fluid (with Newtonian fluids a special case) supply a measure of generality. These are derived by consideration of a momentum balance on the control volume sketched in Fig. 6-7, which yields (either by derivation from first principles or from Appendix D) the $z$-motion equation

$$\frac{d(r\tau_{rz})}{dr} = r\frac{dp}{dz}$$

for the assumed case of flow parallel to the tube walls. One integration gives

$$\tau_{rz} = \frac{r}{2}\frac{dp}{dz} + \frac{c_1}{r} \tag{6-29}$$

where a constant axial pressure gradient has been assumed. The requirement that shear stress be finite at the center line requires $c_1 = 0$; it is then seen that for any fluid, shear stress varies linearly in the tube. In terms of the shear stress at the wall $\tau_w$,

$$\tau = \frac{\tau_w r}{R}$$

For a power-law fluid, $\tau = \mu\,|\,dv/dr\,|^{n-1}dv/dr$. Expecting that velocity de-

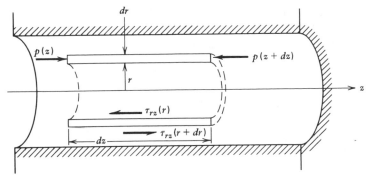

**Figure 6-7** Control volume for deriving velocity profile in fully developed laminar flow in a circular duct.

creases with increasing radius $dv/dr < 0$, set

$$\tau = -\mu \left| -\frac{dv}{dr} \right|^{n-1} \left( -\frac{dv}{dr} \right) = -\mu \left( -\frac{dv}{dr} \right)^{n} \tag{6-30}$$

to avoid possible ambiguity in exponentiating a negative number. Introduction of Eq. (6-30) into Eq. (6-29) results in

$$-\mu \left( -\frac{dv}{dr} \right)^{n} = \frac{r}{2} \frac{dp}{dz}$$

from which it is found (incorporating the no-slip condition at the wall, $V(r = R) = 0$) that the velocity distribution is

$$v = \left( -\frac{1}{2\mu} \frac{dp}{dz} \right)^{1/n} \frac{n}{n+1} R^{(n+1)/n} \left[ 1 - \left( \frac{r}{R} \right)^{(n+1)/n} \right]$$

The average velocity $v_{av}$ is found from Eq. (6-1) to be

$$v_{av} = \left[ \left( -\frac{1}{2\mu} \frac{dp}{dz} \right)^{1/n} \frac{n}{n+1} R^{(n+1)/n} \right] \frac{n+1}{3n+1} \tag{6-30a}$$

and it is then seen that the local velocity can be simply expressed in terms of the average velocity as

$$\frac{v}{v_{av}} = \frac{3n+1}{n+1} \left[ 1 - \left( \frac{r}{R} \right)^{(n+1)/n} \right] \tag{6-31}$$

With a fully developed velocity profile now in hand, consideration of heat transfer aspects can proceed. The energy equation should not be taken directly from Eq. (6-5), which is for a Newtonian fluid; instead, either by derivation from first principles or from Appendix D, it is found that for a general fluid in this flow situation the energy equation is

$$v\frac{\partial T}{\partial z} = \frac{\alpha}{r}\frac{\partial(r\partial T/\partial r)}{\partial r} + \frac{\tau_{rz}}{\rho C_p}\frac{dv}{dr}$$

with the last term representing viscous dissipation. Introduction of the stress relation for a power-law fluid of Eq. (6-30) gives the energy equation as

$$v\frac{\partial T}{\partial z} = \frac{\alpha}{r}\frac{\partial(r\partial T/\partial r)}{\partial r} + \frac{\mu}{\rho C_p}\left(-\frac{dv}{dr}\right)^{n+1}$$

With inclusion of viscous dissipation, the dimensionless energy equation is then

$$\frac{3n+1}{n+1}[1 - r'^{(n+1)/n}]\frac{\partial\theta}{\partial(4/\mathrm{Gz})}$$

$$= \frac{1}{r'}\frac{\partial(r'\partial\theta/\partial r')}{\partial r'} + \Pr\mathrm{E}\left(\frac{3n+1}{n}\right)^{n+1}r'^{(n+1)/n} \qquad (6\text{-}32)$$

where all quantities are as defined following Eq. (6-7) except for the Eckert number, which is defined here for a power-law fluid as $\mathrm{E} = v_{av}^{n+1}R^{1-n}/C_p\Delta T$. The boundary conditions are as written in Eq. (6-8). Naturally, the viscous dissipation term can be neglected if $\mathrm{E}\Pr$ is small as it probably would be for the low velocities of laminar flow (unless $\Delta T \to 0$).

### Constant Wall Heat Flux into Fluid—Limit

The case of a specified and constant wall heat flux into the fluid $q_w$ with uniform inlet temperature $T_0$ is considered first.

If viscous dissipation is neglected, the describing equations are

$$\frac{3n+1}{n+1}[1 - r'^{(n+1)/n}]\frac{\partial\theta}{\partial(4/\mathrm{Gz})} = \frac{1}{r'}\frac{\partial(r'\partial\theta/\partial r')}{\partial r'} \qquad (6\text{-}33a)$$

$$\theta(r',0) = 0 \qquad (6\text{-}33b)$$

$$\theta\left(0, \frac{4}{\mathrm{Gz}}\right) \text{ finite} \qquad (6\text{-}33c)$$

$$\frac{\partial\theta(1,4/\mathrm{Gz})}{\partial r'} = 1 \qquad (6\text{-}33d)$$

where $T_r = T_0$ and $\Delta T = q_w R/k$.

A separation of variables solution is assumed of the form

$$\theta\left(r', \frac{4}{Gz}\right) = \underbrace{R_1(r')Z_1\left(\frac{4}{Gz}\right)}_{\substack{\text{decaying initial} \\ \text{transient}}} + \underbrace{Z_2\left(\frac{4}{Gz}\right)}_{\substack{\text{axial} \\ \text{temperature rise} \\ \text{due to} \\ \text{accumulated} \\ \text{wall flux}}} + \underbrace{R_2(r')}_{\substack{\text{radial temperature} \\ \text{variation to let wall} \\ \text{flux into fluid}}}$$

Difficulties are imminent in the determination of the decaying initial transient, however. The separation of variables idea is sound, but the functions that are determined are not common, although they are tabulated, as is discussed later. Therefore, focus on the idea that the initial transients will eventually decay to insignificance, leaving only $\theta = R_2 + Z_2$ and allowing only a determination of the asymptotic Nusselt number. Inclusion of $\theta = R_2 + Z_2$ into the differential equation and separation of variables gives

$$\frac{dZ_2}{d(4/Gz)} = \frac{n+1}{3n+1} \frac{1}{r'[1 - r'^{(n+1)/n}]} \frac{d(r'\, dR_2/dr')}{dr'} = \lambda$$

in which $\lambda$ is the separation constant. From the two resulting ordinary differential equations it is found that

$$Z_2 = \lambda \frac{4}{Gz} + C_1$$

$$R_2 = \frac{3n+1}{n+1}\lambda\left[\frac{r'^2}{4} - \left(\frac{n}{3n+1}\right)^2 r'^{(3n+1)/n}\right] + C_2 \ln r' + C_3$$

Equation (6-33c) requires that $C_2 = 0$. Subjection of $R_2$ to the condition of Eq. (6-33d) leads to $\lambda = 2$. These results give the local temperature far down the duct as

$$\theta = C + 2\frac{4}{Gz} + 2\frac{3n+1}{n+1}\left[\frac{r'^2}{4} - \left(\frac{n}{3n+1}\right)^2 r'^{(3n+1)/n}\right] \qquad (6\text{-}34)$$

To evaluate the remaining constant of integration $C$, the mixing-cup temperature is examined. Insertion of Eq. (6-34) into Eq. (6-2) yields

$$\theta_m = C + 2\frac{4}{Gz} + \frac{(3n+1)^3 - 8n^3}{4(n+1)(3n+1)(5n+1)}$$

Since Eq. (6-33b) requires

$$\theta_m(4/Gz = 0) = 0, \qquad C = -\frac{(3n+1)^3 - 8n^3}{4(n+1)(3n+1)(5n+1)}$$

Then the asymptotic mixing-cup temperature is

$$\theta_m = 2\frac{4}{\text{Gz}} \tag{6-35}$$

Insertion of the determined $C$ into Eq. (6-34) results in an asymptotic difference between wall and mixing-cup temperatures of

$$\theta_w - \theta_m = \frac{1 + 12n + 31n^2}{4(3n + 1)(5n + 1)} \tag{6-36}$$

Insertion of Eq. (6-63) into Eq. (6-9) gives the asymptotic Nusselt number as

$$\text{Nu}_\infty = \frac{hD}{k} = \frac{8(3n + 1)(5n + 1)}{1 + 12n + 31n^2} \tag{6-37}$$

The manner in which the asymptotic Nusselt number is affected by the velocity distribution can be ascertained from Eq. (6-37). Table 6-3 shows that the effect is noticeable but is not large; Nusselt number varies by a factor of only 2 as $n$ varies from 0 (slug flow) to $\infty$ (linear velocity distribution). The slug flow result shown in Tale 6-3 agrees perfectly with the previously found $\text{Nu}_\infty = 8$ result of Eq. (6-27). Note that Table 6-3 reveals that for a Newtonian fluid ($n = 1$) in fully developed flow with a constant wall heat flux $\text{Nu}_\infty = \frac{48}{11} = 4.36$.

An alternative relation is often used to determine the limiting Nusselt numbers and temperature distributions far downstream. A fully developed temperature profile exists when $(T_w - T)/(T_w - T_m)$ is dependent solely on radial position as expressed by

$$\frac{T_w - T}{T_w - T_m} = f(r)$$

**Table 6-3  Limiting Nusselt Numbers for a Circular Duct; Constant Wall Heat Flux**

|  | $n$ | $\text{Nu}_\infty$ |
|---|---|---|
| Slug flow, fully developed | 0 | 8 |
|  | 1/10 | 6.22 |
|  | 1/2 | 4.75 |
| Newtonian fluid, fully developed | 1 | 4.36 |
|  | 2 | 4.14 |
|  | 5 | 3.98 |
| Linear profile, fully developed | $\infty$ | 3.87 |

This means that

$$\frac{\partial[(T_w - T)/(T_w - T_m)]}{\partial z} = 0 \tag{6-37a}$$

Expansion of the derivative in Eq. (6-37a) gives

$$\frac{\partial T_w}{\partial z} - \frac{\partial T}{\partial z} - \left(\frac{T_w - T}{T_w - T_m}\right)\left(\frac{\partial T_w}{\partial z} - \frac{\partial T_m}{\partial z}\right) = 0 \tag{6-37b}$$

For the case of $q_w = h(T_w - T_m) = $ const, one has $T_w - T_m = $ const, so that Eq. (6-37b) yields

$$\frac{\partial T}{\partial z} = \frac{\partial T_w}{\partial z} = \frac{\partial T_m}{\partial z} = \text{const} \tag{6-37c}$$

Here the value of $\partial T_w/\partial z$ need not be specified to directly obtain the radial distribution of temperature and the resulting Nusselt number. For the case of $T_w = $ const, one has $\partial T_w/\partial z = 0$; thus Eq. (6-37b) yields

$$\frac{\partial T}{\partial z} = \frac{T_w - T}{T_w - T_m}\frac{\partial T_m}{\partial z} = f(r)\frac{\partial T_m}{\partial z} \tag{6-37d}$$

Here a trial-and-error method is utilized to obtain the radial distribution of temperature and the resulting Nussult number. Usually two or three iterations of an initial assumed radial temperature distribution are sufficient.

## Constant Wall Temperature—Limit

The case of a specified and constant wall temperature is considered next. As in the constant wall heat flux case, the limiting value asymptotically approached by the Nusselt number is ascertained first.

Neglecting viscous dissipation, the describing equations are

$$\frac{3n + 1}{n + 1}(1 - r'^{(n+1)/n})\frac{\partial\theta}{\partial(4/Gz)} = \frac{1}{r'}\frac{\partial(r'\partial\theta/\partial r')}{\partial r'} \tag{6-38a}$$

$$\theta(r', 0) = 1 \tag{6-38b}$$

$$\theta\left(0, \frac{4}{Gz}\right) \text{ finite} \tag{6-38c}$$

$$\theta\left(1, \frac{4}{Gz}\right) = 0 \tag{6-38d}$$

where $T_r = T_w$ and $\Delta T = T_0 - T_w$.

A separation of variables solution of the form

$$\theta\left(r', \frac{4}{Gz}\right) = R(r')Z\left(\frac{4}{Gz}\right)$$

is substituted into Eq. (6-38a), giving

$$\frac{dZ}{d(4/Gz)} = -\lambda^2 Z$$

and

$$\frac{d(r'\, dR/dr')}{dr'} + \frac{3n+1}{n+1}\lambda^2(r' - r'^{(2n+1)/n})R = 0 \qquad (6\text{-}38e)$$

The equation for $Z$ can be easily solved to find

$$Z = C e^{-\lambda^2 4/Gz}$$

but the equation for $R$ offers substantial difficulties. There are undoubtedly an infinite number of $\lambda$ and $C$ values, and they are unknown at this point. Far down the duct, however, only the first term in the equation for $Z$ persists, representing the fact that the fluid temperature has closely approached the wall temperature. Then it is accurate to say that

$$\theta \approx C e^{-\lambda_0^2 4/Gz} R(r')$$

Substitution of this accurate approximation for $\theta$ into Eq. (6-38a) gives a relation involving only $R(r)$ that is

$$\frac{d(r'\, dR/dr')}{dr'} = -\lambda_0^2 \frac{3n+1}{n+1}(r' - r'^{(2n+1)/n})R \qquad (6\text{-}39a)$$

$$R(0) \quad \text{finite} \qquad (6\text{-}39b)$$

$$R(1) = 0 \qquad (6\text{-}39c)$$

An iterative solution to Eq. (6-39) can be accomplished by relying on the smoothing process of integration. An initial assumed temperature profile $R_0$ is set into the left-hand side of Eq. (6-39), and an improved temperature profile $R_1$ is found by integration and satisfaction of the boundary conditions. This profile is used in place of $R_0$ to generate a second improvement $R_2$. By proceeding in this fashion, any desired accuracy can be attained. The Nusselt number for each temperature profile (requiring evaluation of the mixing cup temperature and gradient at the wall) can be calculated; the iterative procedure is terminated when the resulting Nusselt number approaches a final and steady value.

In the case for which $n = 1$ (a Newtonian fluid with parabolic velocity distribution), the assumption of a polynomial for $R_{i-1}$ as

$$R_{i-1} = \sum_{m=0}^{\infty} C_m r'^m$$

when substituted into Eq. (6-39) leads to

$$R_i = -4 \sum_{m=0}^{\infty} C_m \frac{m+3}{(m+2)^2(m+4)^2} + \sum_{m=0}^{\infty} C_m \left[ \frac{r'^{m+2}}{(m+2)^2} - \frac{r'^{m+4}}{(m+4)^2} \right]$$

After the mixing-cup temperature and gradient at the wall are determined the Nusselt number that corresponds to the $i$th iteration is

$$\mathrm{Nu}_{\infty i} = \frac{\sum_{m=0}^{\infty} [C_m/(m+2)(m+4)]}{\sum_{m=0}^{\infty} C_m[(m+11)/(m+2)(m+4)(m+6)(m+8)]}$$

This iterative procedure converges fairly rapidly. Use of the temperature profile for a constant wall flux from Eq. (6-34) $R_0 = \frac{3}{4} - r'^2 + r'^4/4$ gives rise to the Nusselt number sequence: $\mathrm{Nu}_{\infty 1} = 3.729$; $\mathrm{Nu}_{\infty 2} = 3.667$; $\mathrm{Nu}_{\infty 3} = 3.6585$. The limiting value that is asymptotically approached far down the duct is

$$\mathrm{Nu}_{\infty} = 3.658 \tag{6-40}$$

In the case for which $n = \infty$ (a fluid with linear velocity profile), the previous assumption of a polynomial for $R_{i-1}$ when substituted into Eq. (6-40) leads to

$$R_i = -\sum_{m=0}^{\infty} C_m \frac{2m+5}{(m+2)^2(m+3)^2} + \sum_{m=0}^{\infty} C_m \left[ \frac{r^{m+2}}{(m+2)^2} - \frac{r^{m+3}}{(m+3)^2} \right]$$

The Nusselt number corresponding to the $i$th iteration is

$$\mathrm{Nu}_{\infty i} = \frac{2\sum_{m=0}^{\infty} [C_m/(m+2)(m+3_i)]}{\sum_{m=0}^{\infty} C_m[(2m^2+25m+62)/(m+2)(m+3)(m+4)(m+5)(m+6)]}$$

Use of the temperature profile for a constant wall flux from Eq. (6-34), $R_0 = \frac{5}{9} - r'^2 + 4r'^3/9$, gives rise to the sequence: $\mathrm{Nu}_{\infty 1} = 3.324$; $\mathrm{Nu}_{\infty 2} = 3.272$; $\mathrm{Nu}_{\infty 3} = 3.265$; $\mathrm{Nu}_{\infty 4} = 3.2641$. The limiting value that is approached far down the duct is

$$\mathrm{Nu}_{\infty} = 3.264 \tag{6-41}$$

Table 6-4   Limiting Nusselt Numbers for a Circular
Duct; Constant Wall Temperature

|  | $n$ | $\mathrm{Nu}_\infty$ |
|---|---|---|
| Slug flow, fully developed | 0 | 5.783 |
| Newtonian fluid, fully developed | 1 | 3.658 |
| Linear profile, fully developed | $\infty$ | 3.264 |

The manner in which the asymptotic Nusselt number is affected by the fully
developed velocity distribution is revealed by the results summarized in Table
6-4. As was found for the specified wall heat flux case, roughly a factor of 2
variation occurs in Nusselt number as $n$ varies from zero (slug flow) to infinity
(linear velocity distribution). A similar $\mathrm{Nu}_\infty$ variation with velocity distribution was found by Schenk and van Laar [5] for an Eyring hole theory fluid
model and by others as reported in the survey by Skelland [6].

Also of interest is the information provided by Tables 6-3 and 6-4 regarding
the influence of the nature of boundary conditions on Nusselt number. The
two boundary conditions considered represent extreme conditions; for a Newtonian fluid ($n = 1$), it is seen that $\mathrm{Nu}_\infty$ for constant wall temperature is only
16% lower than for constant wall heat flux. For slug flow ($n = 0$), the constant
wall temperature $\mathrm{Nu}_\infty$ is 25% lower; for a linear velocity profile ($n = \infty$), it is
only 2% lower.

## Constant Wall Temperature—Entrance Region

Equations (6-38) have been solved for the entrance region of a constant wall
temperature round duct by Whiteman and Drake [7] and Lyche and Bird [8]
for the case of fully developed flow of a power-law fluid. The study by Sellars
et al. [9] for a Newtonian fluid is more accurately done, however, and is of
primary interest. Their results are briefly summarized below for fully developed flow.

As previously discussed following Eq. (6-38), the temperature distribution in
the entrance region can be obtained as

$$\theta\left(r', \frac{4}{\mathrm{Gz}}\right) = \sum_{n=0}^{\infty} C_n R_n(r') \exp\left(\frac{-4\lambda_n^2}{\mathrm{Gz}}\right) \tag{6-42}$$

Here the $C_n$ are constants of integration, the $\lambda_n$ are separation constants, and
the $R_n$ are solutions to Eq. (6-38e). From Eq. (6-42) the local wall heat flux
into the fluid is obtained as

$$q_w = \frac{2k(T_w - T_0)}{R} \sum_{n=0}^{\infty} G_n \exp\left(\frac{-4\lambda_n^2}{\mathrm{Gz}}\right)$$

where $G_n = -0.5 C_n \, dR_n(1)/dr'$. Also, the mixing-cup temperature is found to be

$$\theta_m = 4 \sum_{n=0}^{\infty} G_n \frac{\exp(-4\lambda_n^2/Gz)}{\lambda_n^2}$$

The local and average (based on LMTD) Nusselt numbers are then evaluated as

$$Nu = \frac{hD}{k} = \frac{\sum_{n=0}^{\infty} G_n \exp(-4\lambda_n^2/Gz)}{4 \sum_{n=0}^{\infty} G_n \exp(-4\lambda_n^2/Gz)/\lambda_n^2}$$

and

$$\overline{Nu} = \frac{\bar{h}D}{k} = -\frac{Gz}{4} \ln\left[\frac{4\sum_{n=0}^{\infty} G_n \exp(-4\lambda_n^2/Gz)}{\lambda_n^2}\right]$$

Table 6-5 displays the first five values appropriate to a determination of heat-transfer coefficients. From them it is found that the Nusselt numbers and mixing-cup temperature vary with distance down the duct as displayed in Table 6-6.

The local Nusselt number is seen to have essentially achieved its limiting value of 3.66 (a value predicted in the preceding section) at $1/Gz = (Z/D)/Re \, Pr \approx 0.05$. Although at this Graetz value the local Nusselt number is within 2% of its limiting value, the average Nusselt number approaches that limit more slowly, as it is about 25% greater at $1/Gz = 0.05$. Nevertheless, it is reasonable to take the entrance region as being the interval

$$0 \leqslant \frac{1}{Gz} \leqslant 0.05 \qquad\qquad (6\text{-}43)$$

Table 6-5    Eigenvalues and Eigenfunctions for Entrance Region of a Circular Duct; Fully Developed Laminar Flow; Constant Wall Temperature [9]

| $n$ | $\lambda_n^2$ | $G_n$ |
|---|---|---|
| 0 | 3.656 | 0.749 |
| 1 | 22.31 | 0.544 |
| 2 | 56.9 | 0.463 |
| 3 | 107.6 | 0.414 |
| 4 | 174.25 | 0.382 |

**Table 6-6  Solutions for Entrance Region of a Circular Duct; Fully Developed Laminar Flow; Constant Wall Temperature [9]**

| $1/Gz$ | Nu | $\overline{Nu}$ | $\theta_m$ |
|--------|------|------|--------|
| 0 | $\infty$ | $\infty$ | 1 |
| 0.0005 | 12.86 | 22.96 | 0.956 |
| 0.002 | 7.91 | 12.59 | 0.904 |
| 0.005 | 5.99 | 8.99 | 0.836 |
| 0.02 | 4.18 | 5.87 | 0.626 |
| 0.04 | 3.79 | 4.89 | 0.457 |
| 0.05 | 3.71 | 4.66 | 0.394 |
| 0.1 | 3.66 | 4.16 | 0.1895 |
| $\infty$ | 3.66 | 3.66 | 0 |

or

$$0 \leqslant \frac{z_{\text{entrance}}}{D} \leqslant 0.05 \, \text{Re Pr}$$

It should be noted that Nusselt numbers in the entrance region can substantially exceed the limiting value since "short" ducts are commonly encountered.

These results for a constant wall temperature can be employed to ascertain the consequences of an axially varying wall temperature. The linearity of the describing equations is employed in a superposition scheme. First consider a situation in which $T_w = T_0$ until a position $Z^*$ is attained, after which point $T_w - T_0 = \Delta T_w$. The temperature distribution is then found from Eq. (6-42) to be

$$T(r', z) - T_0 = \Delta(T_w - T_0) - \Delta(T_w - T_0)$$

$$\times \left\{ \sum_{n=0}^{\infty} C_n R_n(r') \exp\left[ \frac{-4\lambda_n^2(z - z^*)}{D \, \text{Re Pr}} \right] \right\} \quad (6\text{-}43a)$$

for $z \geqslant z^*$. As shown in Fig. 6-8, an arbitrary wall temperature variation can be considered to be a sequence of such differential steps. Summation of their effects gives

$$T(r', z) - T_0 = (T_{w0} - T_0)[1 - f(z)] + \Delta_1(T_w - T_0)[1 - f(z - z_1^*)]$$

$$+ \Delta_2(T_w - T_0)[1 - f(z - z_2^*)] + \cdots$$

where $f$ is the term in braces in Eq. (6-43a). In the limit as the incremental increase in wall temperature approaches zero, this sum can be represented by

$$T(r', z) - T_0 = (T_{w0} - T_0)[1 - f(z)] + \int_{T_{w0} - T_0}^{T_w - T_0} [1 - f(z - z^*)] \, d(T_w - T_0)$$

$$(6\text{-}43b)$$

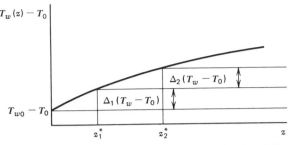

**Figure 6-8** Representation of wall temperature variation by differential steps.

Integration by parts with $u = [1 - f(z - z^*)]$ and $dv = dT_w$ puts this relation into the form

$$T(r', z) - T_0 = (T_{w0} - T_0)[1 \quad f(z)] + (T_w - T_0)[1 - f(z - z^*)]\Big|_{T_{w0}-T_0}^{T_w-T_0}$$

$$- \int_0^z \frac{d[1 - f(z - z^*)]}{dz^*}(T_w - T_0)\, dz^*$$

or

$$T(r', z) - T_0 = \int_0^z \frac{df(z - z^*)}{dz^*}(T_w - T_0)\, dz^*$$

From Eq. (6-42) it is then finally found that

$$T(r', z) - T_0 = -\int_0^z \left\{ \sum_{n=0}^{\infty} C_n R_n(r') \frac{4\lambda_n^2}{D\,\mathrm{Re}\,\mathrm{Pr}} \exp\left[\frac{-4\lambda_n^2(z - z^*)}{D\,\mathrm{Re}\,\mathrm{Pr}}\right]\right\}$$

$$\times (T_w - T_0)\, dz^* \tag{6-43c}$$

The result embodied in Eq. (6-43c) is one form of Duhamel's theorem and may be expressed in other forms by various manipulations. For example, the form of Eq. (6-43b) can be used directly. In any case, system response to a step change in forcing function is utilized.

The local wall heat flux and Nusselt number can be determined from Eq. (6-43c). The local wall heat flux is obtained from

$$q_w(z) = \frac{k}{R}\frac{\partial T(1, z)}{\partial r'}$$

Taking the indicated derivative of Eq. (6-42c), one obtains

$$\frac{q_w(z)R}{k} = 2\int_0^z \left\{ \sum_{n=0}^{\infty} G_n \frac{4\lambda_n^2}{D\,\mathrm{Re}\,\mathrm{Pr}} \exp\left[\frac{-4\lambda_n^2(z-z^*)}{D\,\mathrm{Re}\,\mathrm{Pr}}\right] \right\} (T_w - T_0)\,dz^*$$

(6-43d)

where $G_n = -0.5\, C_n\, dR_n(1)/dr'$ and allows the local wall heat flux to be evaluated for an axially varying wall temperature since the $G_n$ and the $\lambda_n$ are given in Table 6-5. The local mixing-cup temperature is evaluated from

$$2\pi R \int_0^z q_w(z)\,dz = \pi R^2 v_{av} \rho C_p \left[T_m(z) - T_0\right]$$

which recognizes that all the heat entering the fluid must be convected downstream. Rearrangement of this relation yields

$$T_m(z) - T_0 = 8\int_0^z \frac{q_w R}{k} \frac{dz}{D\,\mathrm{Re}\,\mathrm{Pr}}$$

(6-43e)

The local Nusselt number can now be obtained since

$$q_w = h(T_w - T_m)$$

giving

$$\mathrm{Nu} = \frac{hD}{k} = \frac{2(q_w R/k)}{(T_w - T_0) - (T_m - T_0)}$$

(6-43f)

Equations (6-43d–f) are too inconvenient to use without a specific wall temperature variation to consider. Note from Eq. (6-43f), however, that it is possible to have the denominator approach zero from above or below to give a local Nusselt number that approaches $\pm\infty$. Such a case can arise when $T_w$ starts out above $T_0$ and decreases axially to a value below $T_0$. The strange Nusselt behavior observed in such a case means only that the local wall heat flux is based on an inconvenient driving temperature difference $T_w - T_m$, which happens to not be directly proportional to the temperature gradient at the wall, which is the real motivator of the heat flux.

### Constant Wall Heat Flux into Fluid—Entrance Region

Equations (6-33) have been solved for the entrance region of a round duct for a Newtonian fluid ($n = 1$) with constant wall heat flux into the fluid by Siegel et al. [10]. Their results are briefly summarized in the paragraphs that follow.

The dimensionless temperature distribution is assumed to be of the form

$$\theta = \theta^+\left(r', \frac{4}{Gz}\right) + 2\frac{4}{Gz} + r'^2 - \frac{r'^4}{4} - \frac{7}{24}$$

where $\theta^+ = R_1(r')Z_1(4/\text{Gz})$ as discussed earlier following Eqs. (6-33). Introduction of this assumed form into Eqs. (6-33) and separation of variables leads to

$$\theta^+\left(r',\frac{4}{\text{Gz}}\right) = \sum_{n=1}^{\infty} C_n R_{1n} \exp\left(\frac{-2\lambda_n^2}{\text{Gz}}\right)$$

Here the $\lambda_n^2$ are the separation constants and the $R_{1n}$ are the solutions to

$$\frac{d(r'\,dR_{1n}/dr')}{dr'} + \lambda_n^2 r'(1 - r'^2)R_{1n} = 0$$

with the boundary conditions $dR_{1n}/dr = 0$ at $r' = 0$ and $r' = 1$. After the $\lambda_n$ and $R_{1n}$ were determined by numerical methods, the $C_n$ were obtained from the requirement that

$$\theta^+(r',0) = \sum_{n=1}^{\infty} C_n R_{1n} = -\left(r'^2 - \frac{r'^4}{4} - \frac{7}{24}\right)$$

to be given by

$$C_n = -\frac{\int_0^1 r'^3(1 - r'^2)(1 - r'^2/4)R_{1n}\,dr'}{\int_0^1 r'(1 - r'^2)R_{1n}^2\,dr'}$$

The results for the first seven values of $n$ are shown in Table 6-7. From these results the local Nusselt number is obtained as

$$\text{Nu} = \frac{48/11}{1 + (24/11)\sum_{n=1}^{\infty} C_n \exp\left(-2\lambda_n^2/\text{Gz}\right)R_{1n}(1)} \tag{6-44}$$

which varies as shown in Table 6-8. It can be seen there that the entrance

Table 6-7  Eigenvalues and Eigenfunctions for Entrance Region of a Circular Duct; Fully Developed Laminar Flow; Constant Wall Heat Flux [10]

| $n$ | $\lambda_n^2$ | $R_{1n}(1)$ | $C_n$ |
|---|---|---|---|
| 1 | 25.6796 | −0.492517 | 0.403483 |
| 2 | 83.8618 | 0.395508 | −0.175111 |
| 3 | 174.167 | −0.345872 | 0.105594 |
| 4 | 296.536 | 0.314047 | −0.0732804 |
| 5 | 450.947 | −0.291252 | 0.0550357 |
| 6 | 637.387 | 0.273808 | −0.043483 |
| 7 | 855.850 | −0.259852 | 0.035597 |

**Table 6-8  Solutions for Entrance Region of a
Circular Duct; Constant Wall Heat Flux [10]**

| 1/Gz | Nu |
|------|-----|
| 0 | $\infty$ |
| 0.0013 | 11.5 |
| 0.0025 | 9.0 |
| 0.005 | 7.5 |
| 0.01 | 6.1 |
| 0.025 | 5.0 |
| 0.05 | 4.5 |
| 0.1 | 4.364 |
| $\infty$ | 4.364 |

region, based on the closeness of the local Nusselt number to its limiting value, is reasonably taken to be

$$0 < 1/\text{Gz} \leqslant 0.05$$

or

$$\frac{z}{D} \leqslant 0.05 \text{Re Pr}$$

just as for the constant wall temperature case.

A result of interest is the longitudinal variation of wall temperature which accompanies these results. This is found to be

$$\theta_w = \frac{T_w - T_0}{q_w R/k} = 2(4/\text{Gz}) + \frac{11}{24} + \sum_{n=1}^{\infty} C_n \exp\left(\frac{-2\lambda_n^2}{\text{Gz}}\right) R_{1n}(1) \quad (6\text{-}45)$$

This result can be employed to ascertain the wall temperature variation that accompanies axially varying wall heat flux into the fluid. Such a circumstance occurs in a nuclear reactor, for example. The linearity of the describing equations is employed in a superposition technique. First consider a process in which $q_w$ is zero up to a position $z^*$ after which point $q_w = \Delta q$. The wall temperature is then found from Eq. (6-45) to be

$$\frac{T_w - T_0}{R/k} = \left\{ \frac{8(z - z^*)}{D \text{ Re Pr}} + \frac{11}{24} + \sum_{n=1}^{\infty} C_n R_{1n}(1) \exp\frac{[-2\lambda_n^2(z - z^*)]}{D \text{ Re Pr}} \right\} \Delta q$$

$$(6\text{-}46)$$

for $z \geqslant z^*$. As shown in Fig. 6-9, an arbitrary wall heat-flux variation can be

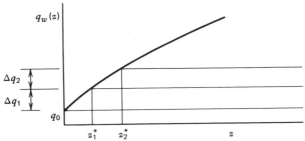

**Figure 6-9** Representation of wall heat-flux variation by differential steps.

considered as a sequence of such differential steps. Summation of their effects gives

$$\frac{T_w(z) - T_0}{R/k} = q_0 f(z) + \Delta q_1 f(z - z_1^*) + \Delta q_2 f(z - z_2^*) \cdots$$

where $f$ is the term in braces of Eq. (6-46). In the limit as the incremental increase in wall heat flux approaches zero, the sum can be replaced by the integral

$$\frac{T_w(Z) - T_0}{R/k} = q_0 f(z) + \int_{q_0}^{q} (z - z^*) \, dq$$

Integration by parts with $u = f(z - z^*)$ and $dv = dq$ in the general relation $\int u \, dv = uv - \int v \, du$ puts this relation in the form

$$\frac{T_w(z) - T_0}{R/k} = q_0 f(z) + q f(z - z^*) \left.\right|_{q_1}^{q} - \int_{Z^*=0}^{z} q \frac{df(z - z^*)}{dz^*} dz^*$$

or

$$[T_w(z) - T_0]k = \int_{z^*=0}^{0} \left\{ \frac{4}{\text{Re Pr}} - \sum_{n=1}^{\infty} C_n R_{1n}(1) \frac{\lambda_n^2}{\text{Re Pr}} \right.$$

$$\left. \times \exp\left[ \frac{-2\lambda_n^2 (z - z^*)}{D \, \text{Re Pr}} \right] \right\} q_w(z^*) \, dz^* \qquad (6\text{-}47)$$

Since Table 6-7 gives all needed numerical values, the wall temperature variation accompanying a specified wall heat flux can be ascertained. The heat-transfer coefficient is not constant in this situation [31].

## 6.6   FULLY DEVELOPED FLOW IN NONCIRCULAR DUCTS

Although circular ducts represent the most common geometry, other cross sections are encountered often enough to generate interest in heat-transfer coefficients for them. Common noncircular cross-sections are the rectangle, the triangle, and the annulus.

Just as for the circular duct, it is assumed that the duct flows full. In other words, unlike the case of a storm sewer that may be only half-filled with flowing water that wets only a portion of the sewer perimeter, the only cases to be considered are those in which the duct perimeter is completely wetted. The characteristic length dimension to be used is the hydraulic diameter $D_h$. The hydraulic diameter is defined as

$$D_h = 4\frac{\text{flow area}}{\text{wetted perimeter}}$$

which for a duct flowing full is

$$D_h = 4\frac{\text{duct area}}{\text{duct perimeter}} \tag{6-48}$$

The hydraulic diameter is used in place of a circular duct diameter in the definitions of the Nusselt, Graetz, and Reynolds numbers so that

$$\text{Nu} = \frac{hD_h}{K}, \qquad \frac{1}{\text{Gz}} = \frac{(z/D_h)}{\text{Re Pr}}, \qquad \text{and} \qquad \text{Re} = \frac{v_{av}D_h}{\nu}$$

### Nusselt Number—Limit

The limiting Nusselt number results for specified wall heat flux and specified wall temperature are displayed in Table 6-9 as reported by Shah and London [11] and are not greatly different from the circular duct results, varying by a factor of about 2. These results demonstrate that the adjustments provided by use of the hydraulic diameter as the characteristic length are insufficient to allow circular duct results to be applied to noncircular ducts unless a substantial error can be tolerated. In Table 6-9 the characteristic length is the hydraulic diameter, $f$ is the shear-stress coefficient ($\tau_w = f v_{av}^2/2g_c$), $H_1$ refers to a constant axial wall heat flux with constant peripheral wall temperature at a given cross section, $H_2$ refers to constant axial wall heat flux with uniform peripheral wall heat flux at a given cross section, and $T$ refers to constant wall temperature axially and peripherally. Axial conduction in both the fluid and the duct wall has been studied (see reference 11 for citations).

Table 6-9   Solutions for Heat Transfer and Friction for Fully
Developed Laminar Flow Through Specified Ducts[11]

| GEOMETRY ($L/D_h > 100$) | $Nu_{H1}$ | $Nu_{H2}$ | $Nu_T$ | fRe |
|---|---|---|---|---|
| $\frac{2b}{2a} = \frac{\sqrt{3}}{2}$ | 3.014 | 1.474 | 2.39* | 12.630 |
| $60°$  $\frac{2b}{2a} = \frac{\sqrt{3}}{2}$ | 3.111 | 1.892 | 2.47 | 13.333 |
| $\frac{2b}{2a} = 1$ | 3.608 | 3.091 | 2.976 | 14.227 |
| (hexagon) | 4.002 | 3.862 | 3.34* | 15.054 |
| $\frac{2b}{2a} = \frac{1}{2}$ | 4.123 | 3.017 | 3.391 | 15.548 |
| (circle) | 4.364 | 4.364 | 3.657 | 16.000 |
| $\frac{2b}{2a} = .9$ | 5.099 | 4.35* | 3.66 | 18.700 |
| $\frac{2b}{2a} = \frac{1}{4}$ | 5.331 | 2.930 | 4.439 | 18.233 |
| $\frac{2b}{2a} = \frac{1}{8}$ | 6.490 | 2.904 | 5.597 | 20.585 |
| $\frac{2b}{2a} = 0$ | 8.235 | 8.235 | 7.541 | 24.000 |
| $\frac{b}{a} = 0$  insulated | 5.385 | - | 4.861 | 24.000 |

*Interpolated values.

Results for the concentric-circular annulus geometry are discussed by Kays
and by Lundberg et al. [12]. They are of interest not only in their own right,
but also because the circular tube and the gap between parallel-plane geometries are two limiting cases of the annulus.

Some possible peculiarities are illustrated by the results found by Eckert
et al. [13] for a duct whose cross section is a sector of a circle with fully

developed velocity and temperature distributions and with the wall heat flux into the fluid a constant per unit of length down the duct. The two cases of a constant peripheral wall temperature at any axial location and a constant wall heat flux into the fluid at any axial location were considered, and the limiting Nusselt number results shown in Fig. 6-10 were found with $q_w = h(T_{w,\,av} - T_m)$. Again, note that the hydraulic diameter does not allow accurate application of circular results to noncircular cross sections. The friction factor for use in calculating fully developed pressure drop is also given in Fig. 6-10, $-dp/dz = f\rho V_{av}^2/2g_c$. The local heat-transfer coefficient approaches a zero value in the corners; a similar occurrence accompanies most geometries with sharp corners. Note that the results for a sector of a circle approximate those for an isosceles triangle (see Table 6-9).

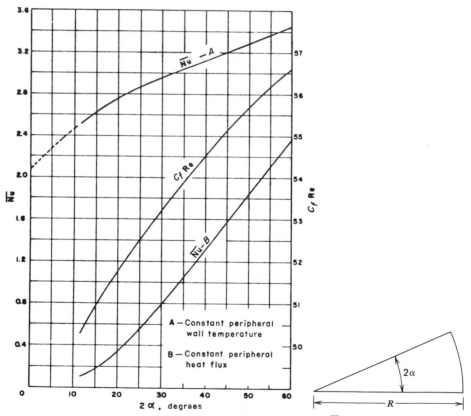

**Figure 6-10** Friction factor $C_f$ and average Nusselt numbers $\overline{Nu}$ for hydraulically and thermally developed laminar flow through a duct whose cross section is a sector of a circle: $\alpha$, half-angle of sector; Re, Reynolds number based on average velocity and hydraulic diameter; $\overline{Nu}$, Nusselt number based on hydraulic diameter and difference between bulk temperature and average wall temperature. [From E. R. G. Eckert, T. F. Irvine, and J. T. Yen, Local laminar heat transfer in wedge-shaped passages, *Transact. ASME* **80**, 1433–1438 (1958) [13].]

**Table 6-10   Eigenfunctions for Entrance Region of a Rectangular Duct; Fully Developed Laminar Flow; Constant Wall Heat Flux [14]**

| Eigenvalue | Short Side/Long Side | | | | | |
|---|---|---|---|---|---|---|
| | 1 | $\frac{2}{3}$ | $\frac{1}{2}$ | $\frac{1}{4}$ | $\frac{1}{8}$ | 0 |
| $\lambda_0^2$ | 5.96 | 6.25 | 6.78 | 8.88 | 11.19 | 15.09 |
| $\lambda_1^2$ | 35.54 | | | | | 171.3 |
| $\lambda_2^2$ | 78.9 | | | | | 498 |
| $G_0$ | 0.598 | 0.627 | 0.669 | 0.839 | 1.030 | 1.717 |
| $G_1$ | 0.462 | | | | | 1.139 |
| $G_2$ | 0.138 | | | | | 0.952 |
| $Nu_\infty$ | 2.98 | 3.125 | 3.39 | 4.44 | 5.595 | 7.545 |

## Nusselt Number—Entrance Region

The Nusselt number results for noncircular cross sections are needed for instances in which the duct is sufficiently short that temperature profiles are not fully developed.

For the rectangular duct with constant wall temperature, Table 6-10 gives the reliable eigenvalues and associated constants obtained by Dennis et al. [14] which are needed for a Nusselt number calculation. The Nusselt number based on LMTD $\overline{Nu}$ is achievable from these results by substitution into

$$\overline{Nu} = -\left(\frac{4}{Gz}\right)^{-1} \ln\left[8\sum_{n=0}^{\infty} \frac{G_n \exp(-2\lambda_n^2/Gz)}{\lambda_n^2}\right]$$

When only the first eigenvalue $\lambda_0$ is known, $\overline{Nu}$ can be evaluated from this equation for only $1/Gz \gtrsim 0.05$, which is somewhat unfortunate since the thermal entrance region lies in the range $0 \lesssim 1/Gz \lesssim 0.05$. In such cases information can still be extracted from the preceding equation since only the first term in the summation need be retained when $1/Gz$ becomes large. Then

$$Nu \approx \frac{\lambda_0^2}{2} + \frac{\ln(\lambda_0^2/8G_0)}{4/Gz}, \qquad \frac{1}{Gz} \gtrsim 0.05 \qquad (6\text{-}48a)$$

which is a general limiting form for all geometries, of course.

For the concentric–circular annulus with constant wall heat flux, the local Nusselt numbers are presented in Table 6-11 (see reference 12 for a full discussion). In Table 6-11 $Nu_{ii}$ and $Nu_{oo}$ refer to heat applied only at the inner and outer surface, respectively; these results can be used with influence coefficients [12] for any combination of inner and outer heat fluxes to find the resultant Nusselt number.

**Table 6-11   Limiting Nusselt Numbers for Laminar Flow in a
Concentric-Circular Annulus; Constant Wall Heat Flux [12]**

| $r^* = (r_1/r_0)$ | 2/Gz | $Nu_{ii}$ | $Nu_{00}$ |
|---|---|---|---|
| 0.05 | 0.001 | 33.2 | 13.4 |
| | 0.005 | 24.2 | 7.99 |
| | 0.01 | 21.5 | 6.58 |
| | 0.05 | 18.1 | 4.92 |
| | 0.10 | 17.8 | 4.80 |
| | $\infty$ | 17.8 | 4.79 |
| 0.10 | 0.001 | 25.1 | 13.5 |
| | 0.005 | 17.1 | 8.08 |
| | 0.01 | 14.9 | 6.65 |
| | 0.05 | 12.1 | 4.96 |
| | 0.10 | 11.9 | 4.84 |
| | $\infty$ | 11.9 | 4.83 |
| 0.25 | 0.001 | 18.9 | 13.8 |
| | 0.005 | 12.1 | 8.28 |
| | 0.01 | 10.2 | 6.80 |
| | 0.05 | 7.94 | 5.04 |
| | 0.10 | 7.76 | 4.91 |
| | $\infty$ | 7.75 | 4.90 |
| 0.50 | 0.001 | 16.4 | 14.2 |
| | 0.005 | 10.1 | 8.55 |
| | 0.01 | 8.43 | 7.03 |
| | 0.05 | 6.35 | 5.19 |
| | 0.10 | 6.19 | 5.05 |
| | $\infty$ | 6.18 | 5.04 |
| 1.00 | 0.0005 | 23.5 | 23.5 |
| (parallel | 0.005 | 11.2 | 11.2 |
| planes) | 0.01 | 7.49 | 7.49 |
| | 0.05 | 5.55 | 5.55 |
| | 0.125 | 5.39 | 5.39 |
| | $\infty$ | 5.38 | 5.38 |

Fully developed laminar flow between parallel planes with specified heat
flux has been studied by Cess and Shaffer [25], who give the local Nusselt
number as

$$Nu = \frac{4}{17/35 + \sum_{n=1}^{\infty} c_n y_n(1) \exp(-8\beta_n^2 z/3a\,Pe)}$$

where $Pe = 4v_{av}a/\alpha$ and $2a$ is the plane spacing (consult reference 25 for
further details). The limiting value is $Nu = 140/17 = 8.235$, which agrees
exactly with the limiting value in Table 6-9.

**Table 6-12  Solutions for Simultaneous Velocity and Temperature Development in a Circular Duct [15]**

| $1/\text{Gz} = (z/D)/\text{Re Pr}$ | $T_w = \text{const, Pr} = 0.7$ Developing Velocity | | $\text{Pr} = 0.7, q_w = \text{const}$ Developing Velocity | $\text{Pr} = 0.7, T_w - T_m = \text{const}$ Developing Velocity | |
|---|---|---|---|---|---|
| | $\overline{\text{Nu}}$ | Local Nu | Local Nu | $\overline{\text{Nu}}$ | Nu |
| 0.001 | | 18.46 | 21.62 | | 18.8 |
| 0.0025 | 17.44 | 11.31 | 14.53 | 17.8 | 11.8 |
| 0.005 | 13.36 | 7.9 | 10.54 | 13.76 | 8.36 |
| 0.01 | 10.00 | 5.82 | 7.8 | 10.48 | 6.4 |
| 0.025 | 6.91 | 4.29 | 5.64 | 7.48 | 4.94 |
| 0.05 | 5.43 | 3.77 | 4.78 | 6.06 | 4.48 |
| 0.01 | 5.43 | 3.67 | 4.54 | 5.24 | 4.41 |
| 0.2 | 4.11 | 3.66 | 4.45 | 4.82 | 4.39 |
| $\infty$ | 3.66 | 3.66 | 4.364 | 4.364 | 4.364 |

## 6.7  SIMULTANEOUS VELOCITY AND TEMPERATURE DEVELOPMENT

In many applications the velocity distribution is not fully developed by the time a fluid particle enters the heated section. This situation can occur, for example, when the fluid enters a duct from a much larger reservoir. Although the start of the heating section does not always coincide with the duct entrance, it is the most common situation and the only one to receive analytical and experimental attention.

The criterion for the thermal entrance region for laminar flow is found to be reasonably expressed by

$$0 \leqslant \frac{z/D_h}{\text{Re Pr}} \lesssim 0.05$$

The criterion for the velocity entrance region in laminar flow is, similarly,

$$0 \leqslant \frac{z}{D_h} \text{Re} \leqslant 0.05$$

Comparison of these two criteria reveals that when $\text{Pr} \gg 1$, as is the case for oils, the temperature profile takes the longer distance to develop—a fully developed velocity profile may be appropriate if the duct is sufficiently long to permit its achievement over a major portion of its length. When $\text{Pr} \approx 1$, as is the case for air and water, the temperature and velocity profiles develop at about the same rate. When $\text{Pr} \ll 1$, as is the case for liquid metals, the temperature profile takes the shorter distance to develop—a slug flow velocity profile is appropriate.

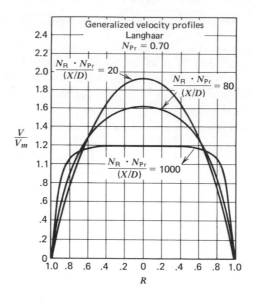

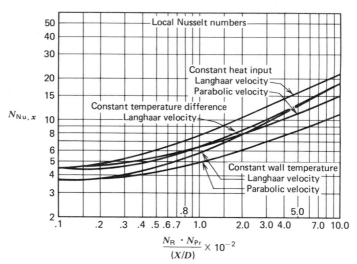

**Figure 6-11** Laminar velocity and temperature profiles and local and average Nusselt numbers in the entrance region of a circular duct for constant wall temperature, constant wall heat flux, and constant fluid–wall temperature difference. [From W. M. Kays, *Transact. ASME* **77**, 1265–1274 (1955) [15].]

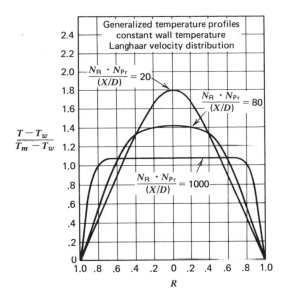

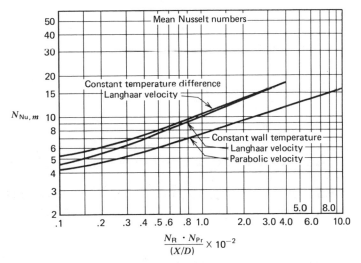

**Figure 6-11** (*Continued.*)

Insight into the nature of the results that can be expected can be obtained by a comparison of Nusselt number results for a circular duct with fully developed and slug flow. For a Newtonian fluid and a constant wall temperature, Table 6-6 shows $\overline{Nu}(Gz^{-1} = 0.005) = 8.99$ for a fully developed, parabolic velocity profile, and Table 6-1 shows $\overline{Nu}(Gz^{-1} = 0.005) = 17.7$ for slug flow. It is evident that the Nusselt number is greater with a uniform velocity profile. Then the uniform velocity at the duct inlet during simultaneous development of velocity and temperature profiles should result in higher Nusselt numbers than in the fully developed velocity case.

This expected trend was quantitatively determined by Kays [15] for constant wall temperature, constant wall heat flux, and constant fluid–wall temperature difference. His results are summarized in Table 6-12 and Fig. 6-11 and are only for $Pr = 0.7$ (the calculations were extended to other Prandtl numbers by Goldberg [16]). These results for the average Nusselt number are encapsulated in the correlating equation, whose form was apparently first proposed by Hausen [17]:

$$\overline{Nu} = \overline{Nu}_{\infty} + \frac{K_1(Re\,Pr\,D/x)}{1 + K_2(Re\,Pr\,D/x)^n} \tag{6-49}$$

for $T_w = $ const. For $q_w = $ const, the left-hand side must be replaced by the local Nusselt number Nu. The constants to use in Eq. (6-49) are shown in Table 6-13. In developing these results, the velocity profiles in the entrance region found by Langhaar [18]—who solved a linearized form of the radial-motion equation—were used. Then the energy equation

$$\frac{1}{r}\frac{\partial(r\,\partial T/\partial r)}{\partial r} = \left(u\frac{\partial T}{\partial z} + v_r\frac{\partial T}{\partial r}\right)\frac{Pr}{\nu}$$

was solved, neglecting the $v_r$ term, which is important only very near the inlet. For the assumed constant properties, the velocity profile is independent of the Prandtl number, but the temperature profile (and the Nusselt number) are dependent on the Prandtl number.

The constant wall heat flux condition was studied by Heaton et al. [19] for simultaneously developing velocity and temperature profiles for the con-

**Table 6-13   Constants $K_1$ and $K_2$ for Eq. (6-49)**

| Wall Condition | Inlet Velocity | Pr | $\overline{Nu}_{\infty}$ | $K_1$ | $K_2$ | $n$ |
|---|---|---|---|---|---|---|
| $T_w = $ const | Developing | 0.7 | 3.66 | 0.104 | 0.016 | 0.8 |
| $T_w = $ const | Parabolic | Any | 3.66 | 0.0668 | 0.04 | $\frac{2}{3}$ |
| $T_w - T_m = $ const | Developing | 0.7 | 4.36 | 0.1 | 0.016 | 0.8 |
| $q_w = $ const | Developing | 0.7 | 4.36 | 0.036 | 0.0011 | 1 |
| $q_w = $ const | Parabolic | Any | 4.36 | 0.023 | 0.0012 | 1 |

**Table 6-14  Nusselt Numbers and Influence Coefficients for the
Circular-Tube-Annulus Family; Constant Heat Rate;
Combined Thermal and Hydrodynamic Entry Length [19]**

| Pr | $1/Gz$ | Circular Tube Nu | Parallel Planes $Nu_{ii}$ | $\theta_1^*$ | Circular-Tube Annulus, $r^* = 0.50$ $Nu_{ii}$ | $Nu_{oo}$ |
|---|---|---|---|---|---|---|
| 0.01 | 0.001 | 24.2 | 24.2 | 0.048 | ... | 24.2 |
|  | 0.005 | 12.2 | 11.7 | 0.117 | ... | 11.8 |
|  | 0.010 | 9.10 | 8.80 | 0.176 | 9.43 | 8.90 |
|  | 0.05 | 6.08 | 5.77 | 0.378 | 6.40 | 5.88 |
|  | 0.10 | 5.73 | 5.53 | 0.376 | 6.22 | 5.60 |
|  | $\infty$ | 4.36 | 5.39 | 0.346 | 6.18 | 5.04 |
| 0.70 | 0.001 | 17.8 | 18.5 | 0.037 | 19.22 | 18.30 |
|  | 0.005 | 9.12 | 9.62 | 0.096 | 10.47 | 9.45 |
|  | 0.010 | 7.14 | 7.68 | 0.154 | 8.52 | 7.50 |
|  | 0.05 | 4.72 | 5.55 | 0.327 | 6.35 | 5.27 |
|  | 0.10 | 4.41 | 5.40 | 0.345 | 6.19 | 5.06 |
|  | $\infty$ | 4.36 | 5.39 | 0.346 | 6.18 | 5.04 |
| 10.0 | 0.001 | 14.3 | 15.6 | 0.0311 | 16.86 | 15.14 |
|  | 0.005 | 7.87 | 9.20 | 0.092 | 10.20 | 8.75 |
|  | 0.010 | 6.32 | 7.49 | 0.149 | 8.43 | 7.09 |
|  | 0.05 | 4.51 | 5.55 | 0.327 | 6.35 | 5.20 |
|  | 0.10 | 4.38 | 5.40 | 0.345 | 6.19 | 5.05 |
|  | $\infty$ | 4.36 | 5.39 | 0.346 | 6.18 | 5.04 |

centric–circular annulus family. Especially noteworthy in this work, summarized in Table 6-14, is the effect of the Prandtl number. As can be seen the local Nusselt number varies inversely with Prandtl number; from comparison with Tables 6-8 and 6-11 it is seen that results for $Pr = 10$ are very close to those for $Pr = \infty$ (fully developed flow), whereas from comparison with Table 6-2 the results for $Pr = 0.01$ are close to those for $Pr = 0$ (slug flow) at small values of $(z/D_h)/Re\,Pr$.

## 6.8  PRESSURE DROP IN LAMINAR FLOW THROUGH DUCTS

The pressure drop required to achieve the flow that leads to the preceding heat-transfer coefficients must be considered, for the high heat-transfer rates that can be achieved with forced convection are purchased at the cost of pumping power. Even though only laminar flow has been considered in detail and it is often considered to be accompanied by small pressure drops, the drops are not negligible. Flow is usually considered to be laminar if $Re < 2300$, based on the duct hydraulic diameter.

## Straight Circular Duct—Limit

A brief review of the pressure drop that is required for fully developed laminar flow of a Newtonian fluid in a round duct as embodied in Eqs. (6-30a) and (6-31) with $n = 1$ reveals that

$$P_0 - p = 8v_{av}\frac{\mu L}{R^2}$$

This can be rearranged into the form that is most useful in connection with turbulent flow of

$$P_0 - p = f\frac{L}{D}\frac{\rho v_{av}^2}{2g_c} \tag{6-50}$$

By comparison it is then seen that the friction factor $f$ for laminar flow in a round duct is

$$f = \frac{64}{\text{Re}} \quad \text{or} \quad f\,\text{Re} = 64 \tag{6-51}$$

where $\text{Re} = v_{av}D\rho/\mu$. The friction factor depends strongly on the Reynolds number, unlike the turbulent flow case.

Equation (6-51) can be generalized to the case of a power-law non-Newtonian fluid by suitably redefining the Reynolds number. It has been seen that for a Newtonian fluid

$$\mu = \frac{1}{32}\frac{(P_0 - p)D^2}{Lv_{av}} \tag{6-52}$$

An effective viscosity $\mu_{eff}$ for use in Reynolds number is defined as that viscosity that makes Eq. (6-52) fit the power-law case [6]. From Eq. (6-30a) one has

$$\mu_{eff} = \frac{1}{32}\frac{(P_0 - p)D^2}{Lv_{av}} = \mu\left(\frac{3n+1}{4n}\right)^n\left(\frac{8v_{av}}{D}\right)^{n-1}$$

With this definition the Reynolds number for a power-law fluid is

$$\text{Re} = v_{av}\frac{D\rho}{\mu_{eff}} = \left(v_{av}^{2-n}\frac{D^n\rho}{\mu}\right)\frac{[4n/(3n+1)]^n}{8^{n-1}}$$

Then the preceding relations for a Newtonian fluid can be used directly. For noncircular ducts, consult Skelland [6].

## Straight Circular Ducts—Entrance

The shape of a duct inlet can have a considerable effect; in the following it is assumed to be sharp edged. In a round duct that receives fluid from a much

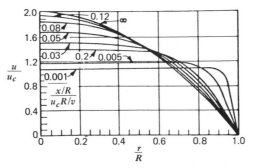

**Figure 6-12**  Laminar velocity profiles in the inlet section of a tube. [By permission from E. Sparrow, S. H. Lin, and T. Lundgren, Flow development in the hydrodynamic entrance region of tubes and ducts, *Physi. Fluids* **7**, 338–347 (1964) [20].]

larger reservoir, the velocity distribution develops from a rather uniform one at the inlet to the fully developed profile far downstream. Figure 6-12 shows the calculations by Sparrow et al. [20] for a Newtonian fluid, by use of an integral method. As can be seen there, a fully developed velocity profile is achieved at

$$\frac{z/D}{\mathrm{Re}} \gtrsim 0.05$$

which is the conventional criterion. In this development process the flow is accelerated, causing a pressure drop that is greater than that due to wall shear

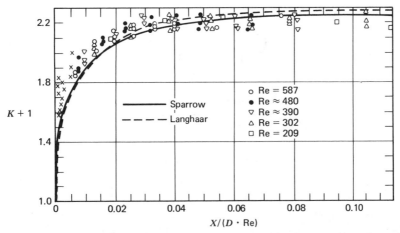

**Figure 6-13**  Coefficient $K$ for the pressure drop caused by laminar flow through the inlet section of a tube. [From S. T. McComas and E. R. G. Eckert, *Transact. ASME, J. Appl. Mech.* **32**, 765–770 (1965) [21].]

Table 6-15   Limiting $K$ Values for Eq. (6-53): Rectangular Ducts

| Aspect Ratio | $K_\infty$ | $f$Re |
|---|---|---|
| 0 | 0.6857 | 96 |
| $\frac{1}{20}$ | 0.7613 | 89.91 |
| $\frac{1}{10}$ | 0.8392 | 84.675 |
| $\frac{1}{8}$ | 0.8788 | 82.34 |
| $\frac{1}{6}$ | 0.9451 | 78.81 |
| $\frac{1}{4}$ | 1.076 | 72.931 |
| $\frac{2}{5}$ | 1.282 | 65.472 |
| $\frac{1}{2}$ | 1.383 | 62.19 |
| $\frac{3}{4}$ | 1.5203 | 57.89 |

stresses alone. This effect is accounted for by a modification of Eq. (6-50) for fully developed flow. The modified pressure-drop formula, whose form is justified on theoretical grounds, is

$$P_0 - p = \left( \frac{fz}{D} + K \right) \frac{v_{av}^2}{2g_c} \qquad (6\text{-}53)$$

with $f$ from Eq. (6-51). The entrance-effect parameter $K$ is initially quite small and itself approaches a limit of 1.24 far downstream from the inlet. The experimental study of McComas et al. [21] confirms the calculations, as Fig. 6-13 shows.

## Straight Noncircular Ducts

Pressure drop calculations for the laminar flow of Newtonian fluids in noncircular ducts is accomplished by insertion of appropriate $f$ and $K$ values in Eq.

Table 6-16   Isosceles Triangle Ducts

| Included Angle, Deg | $K_\infty$ | $f$Re |
|---|---|---|
| 10 | 2.418 | 49.9 |
| 20 | 2.128 | 51.29 |
| 30 | 1.966 | 52.26 |
| 40 | 1.876 | 52.88 |
| 50 | 1.831 | 53.23 |
| 60 | 1.818 | 53.32 |
| 70 | 1.829 | 53.24 |
| 80 | 1.86 | 52.99 |

Table 6-17 Annular Ducts

| $r_i/r_0$ | $K_\infty$ | $f\mathrm{Re}$ |
|-----------|-----------|----------------|
| 0.0001 | 1.1312 | 71.781 |
| 0.001 | 1.0727 | 74.683 |
| 0.01 | 0.9733 | 80.113 |
| 0.05 | 0.8644 | 86.27 |
| 0.1 | 0.8087 | 89.372 |
| 0.2 | 0.7542 | 92.352 |
| 0.4 | 0.71 | 94.713 |
| 0.8 | 0.6872 | 95.92 |
| 1.0 | 0.6857 | 96 |

(6-53). In general,

$$f\mathrm{Re} = C$$

where Re is based on hydraulic diameter and the constant $C$ depends on the geometry of the duct cross section. The $K$ parameter approaches a limit $K_\infty$ that is also geometry dependent. Lundgren et al. [22] give Tables 6-15–6-17 for three common geometries; they also give the fully developed velocity distributions for these geometries and the ellipse. Results for the annular sector geometry are given by Sparrow et al. [23] for $f\mathrm{Re}$, $K$, and Nusselt numbers under constant axial heat-flux conditions.

## 6.9  TEMPERATURE-DEPENDENT PROPERTIES

Heretofore no consideration has been usefully directed toward the effect of temperature-dependent fluid properties on heat-transfer coefficients and pressure drop. Because substantial temperature variations are often experienced in both the flow and transverse-flow directions and because the accompanying property variations are often substantial, it is possible that the heat-transfer coefficient differs appreciably from its constant property value. This can occur not only because $h = k\,\mathrm{Nu}/L$ with $k$ being temperature dependent, but also because a temperature-dependent viscosity affects the velocity profile, even in the case of fully developed flow. Also, a temperature-dependent density gives rise to a larger and longer-lasting radial velocity component that lengthens the entrance region.

The properties of different fluids exhibit different temperature dependencies. For a gas, specific heat and Prandtl number vary only slightly with temperature, but viscosity and thermal conductivity vary roughly with the 0.8 power of absolute temperature, whereas density varies inversely with absolute temperature. For a liquid, the specific heat, density, and thermal conductivity

vary only slightly with temperature, but viscosity varies strongly (often exponentially) with temperature. Naturally, if property variations are moderate,
simple corrections to the constant-property results will yield satisfactory accuracy.

Two schemes for temperature correction of constant-property results are the
reference-temperature and the property-ratio schemes. In the reference-
temperature scheme a characteristic temperature is selected at which *all*
properties are evaluated. The characteristic temperature can be the mixing-cup
temperature, suitably averaged between inlet and exit values, the surface
temperature, or a combination of these. In the property-ratio scheme, all
properties are evaluated at the characteristic temperature described previously,
and all effects of property variation transverse to the flow are then expressed
as ratios of properties evaluated at the characteristic and surface temperatures.

For liquids, where the viscosity variation in the direction transverse to the
flow can be substantial, the corrections for temperature-dependent properties
conventionally take the form, according to Kays and London [24], of

$$\frac{\text{Nu}}{\text{Nu}(T_m)} = \left( \frac{\mu_w}{\mu_m} \right)^n \tag{6-54}$$

$$\frac{f}{f(T_m)} = \left( \frac{\mu_w}{\mu_m} \right)^m \tag{6-55}$$

where $\text{Nu}(T_m)$, $f(T_m)$, and $\mu_m$ are the Nusselt number, friction factor, and
viscosity evaluated at the mixing-cup temperature, which is the arithmetic
average (satisfactory for gases if the absolute temperature varies by a factor of
less than 2 from inlet to outlet) of its inlet and outlet values. For gases, all
property variations can be represented by the absolute temperature; thus, the
correction for temperature dependent properties is

$$\frac{\text{Nu}}{\text{Nu}(T_m)} = \left( \frac{T_w}{T_m} \right)^n \tag{6-56}$$

$$\frac{f}{f(T_m)} = \left( \frac{T_w}{T_m} \right)^m \tag{6-57}$$

For gases, the exponents to be used in conjunction with Eqs. (6-56) and
(6-57) are displayed in Table 6-18. These values are for circular tubes but may
also be used for noncircular geometries in the absence of other alternatives.

For liquids, the exponents to be used in conjunction with Eqs. (6-54) and
(6-55) are displayed in Table 6-19 and can also be applied to liquid metals if it
is presumed that their low Prandtl values give the same *m*, *n* values as are
found for laminar flow. Results for power-law non-Newtonian liquids have
also been reported [27, 28].

**Table 6-18   Fully Developed Exponents for $(T_w/T_m)$, Gases [12, 24]**

|  | $n$ | $m$ |
|---|---|---|
| *Laminar Boundary Layer on Flat Plate* | | |
| Gas heating | −0.08 | −0.08 |
| Gas cooling | −0.045 | −0.045 |
| *Flow Normal to Circular Tube or Bank of Circular Tubes* | | |
| Gas heating | 0.0 | 0.0 |
| Gas cooling | 0.0 | 0.0 |
| *Fully Developed Laminar Flow in Circular Tube* | | |
| Gas heating | 0.0 | 1.0 |
| Gas cooling | 0.0 | 1.0 |
| *Fully Developed Turbulent Flow in Circular Tube* | | |
| Gas heating | −0.5 | 0.1 |
| Gas cooling | 0.0 | −0.1 |

## 6.10   NATURAL-CONVECTION EFFECTS

The low velocities that characterize laminar flow are accompanied by low values of viscous forces. Since this is the case, it is possible for the buoyant body forces to approach the magnitude of the viscous forces. The density gradients caused by the interaction of temperature gradients on the temperature-dependent density lead to a combination of forced and natural convection, which can then result in velocity and temperature distributions differing substantially from those that occur in the complete absence of natural convection.

**Table 6-19   Fully Developed Exponents for $(\mu_w/\mu_m)$, Liquids [24]**

| | $n$ | | $m$ | |
|---|---|---|---|---|
| Pr | Heating | Cooling | Heating | Cooling |
| Laminar | −0.14 | −0.14 | 0.58 | 0.50 |
| 1 | −0.20 | −0.19 | 0.09 | 0.12 |
| 3 | −0.27 | −0.21 | 0.06 | 0.09 |
| 10 | −0.36 | −0.22 | 0.03 | 0.05 |
| 30 | −0.39 | −0.21 | 0.00 | 0.03 |
| 100 | −0.42 | −0.20 | −0.04 | 0.01 |
| 1000 | −0.46 | −0.20 | −0.12 | −0.02 |

Natural-convection effects on laminar flow in ducts are often unimportant, but they can be appreciable in solar collectors, for example (consult Section 12.6 for additional information).

## 6.11 CURVED DUCTS

Information concerning the effect of a curved duct on heat transfer and pressure drop is given in Section 11.9.

## 6.12 TWISTED-TAPE INSERTS

Devices for establishment of fluid swirl are often used to increase heat transfer in duct flows. In many cases the increase in pressure drop is less than the increase in heat transfer. Twisted-tape inserts have been studied by Hong and Bergles [32], who found that in laminar flow heat transfer could be increased by a factor of as much as 9 whereas pressure drop increased by a factor of less than 4 over empty tube values—correlations of measurements are given. In turbulent flow twisted-tape inserts can improve heat transfer by a factor of 2 over empty tube values, but the pressure drop is increased by several orders of magnitude.

## PROBLEMS

**6-1** On the basis of the solution for the difference between wall and fluid temperatures provided by Eq. (6-26) for laminar slug flow in a round duct with specified wall heat flux, estimate the Graetz number value required for entrance effects to have diminished by a factor of 100. Sketch the variation of $\theta_w - \theta_m$ versus Gz.

**6-2** On the basis of the solution for the mixing-cup temperature provided by Eq. (6-14) for laminar slug flow in a round duct with specified wall temperature, estimate the Graetz number value required for mixing-cup temperature to have diminished by a factor of 100. Sketch the variation of $\theta_m$ versus Gz.

**6-3** Plot $V/V_{av}$ against $r/R$ for a power-law fluid for $n = 0$, $\frac{1}{10}$, 1, 5, and $\infty$ in fully developed flow through a round duct.

**6-4** A Newtonian fluid enters a circular duct in fully developed laminar flow. The heat flux into the fluid at the wall is specified and varies sinusoidally along the tube length $L$ as $q_w = q_{w_{max}} \sin(\pi z/L)$—as in a nuclear reactor where heat flux is actually more likely to vary as $A + B \sin(\pi z/L)$.

(a) Show, on the basis of Eq. (6-47), that the wall temperature along the tube length varies as

$$\frac{T_w(z) - T_0}{\left(q_{w_{max}} R/k\right)} = \frac{8L(1 - \cos bz)}{\pi D \operatorname{Re} \operatorname{Pr}}$$

$$- \sum_{n=1}^{\infty} C_n R_{1n}^{(1)}\left(\frac{a}{b}\right) \frac{e^{-az} + \left[(a/b)\sin bz - \cos bz\right]}{(a/b)^2 + 1}$$

where $a = 2\lambda^2/\operatorname{Re}\operatorname{Pr} D$ and $b = \pi/L$.

(b) Compare the result for part a with that obtained by using the result for a constant wall heat flux that equals the average heat flux—$q_w = 2q_{w_{max}}/\pi$. For the purpose of this comparison, set $L/D \operatorname{Re} \operatorname{Pr} = 0.05$ and plot dimensionless wall temperature against $z/L$.

(c) Comment on the agreement of the predictions of part b with the more exact predictions of part a.

**6-5** Repeat Problem 6-4 for the case of slug flow.

**6-6** A gas-cooled nuclear reactor (perhaps intended for a coal gasification or a nuclear rocket application) is to be constructed of a core with $\frac{1}{8}$-in.-diameter holes through it. Each flow passage is 4 ft long. The wall heat flux will vary along the length of each flow channel as

$$q_w = 300 \text{ Btu/hr ft}^2 + 800 \text{ Btu/hr ft}^2 \sin\frac{\pi z}{400 \text{ ft}}$$

where $z$ is distance from the inlet. For the purposes of simplicity, air will be the fluid (hydrogen or helium might be considered for the illustrative applications described previously) that enters at 200°F and 100 psig at a mass velocity of 5500 $\text{lb}_m/\text{hr ft}^2$. Plot wall heat flux, air-mixing-cup temperature, and wall temperature against distance along the flow passage.

**6-7** Consider a circular duct through which a Newtonian fluid flows steadily with a fully developed laminar velocity profile. The entrance temperature is uniform at $T_0$; the wall temperature starts out above $T_0$ and then decreases linearly along the duct, finally reaching $T_0$ at $\operatorname{Gz}^{-1} = (z/D)/\operatorname{Re}\operatorname{Pr} = 0.1$. In other words

$$T_w - T_0 = (1 - 10\operatorname{Gz}^{-1})(T_{w0} - T_0)$$

(a) Calculate and plot against $\operatorname{Gz}^{-1}$ the exact local values of $(q_w R/k)/(T_{w0} - T_0)$, $(T_m - T_0)/(T_{w0} - T_0)$, and Nu for $0 \leqslant \operatorname{Gz}^{-1} \leqslant 0.1$.

(b) Calculate and show in the plot of part a the approximate local values of the same quantities based on an average value of $T_w - T_0 = (T_{w0} - T_0)/2$.

**6-8** The hydraulic diameter for various cross-section geometries is defined by Eq. (6-48). Show that for (a) a rectangle of sides $a$ and $b$, $D_h = 2a/(1 + a/b)$, (b) an annulus of inner diameter $D_i$ and outer diameter $D_o$, $D_h = D_o - D_i$, (c) an equilateral triangle of side $a$, $D_h = a/4(3)^{1/2}$, (d) a circle of diameter $D$, $D_h = D$, (e) a sector of a circle of radius $D/2$ and included angle $\theta$, $D_h = D\theta/(2 + \theta)$, and (f) a gap of infinite width and spacing $a$, $D_h = 2a$.

**6-9** A fluid enters a gap bounded by two infinite parallel planes, both of whose wall temperatures are equal to $T_w$ as shown in Fig. 6P-9. The fluid experiences slug flow and has a uniform entrance temperature of $T_0$. The temperature distribution in the fluid is described by

$$\rho C_p V \frac{\partial T}{\partial z} = k \frac{\partial^2 T}{\partial x^2}, \qquad -a < x < a, \quad z > 0$$

$$T(a, z) = T_w = T(-a, z)$$

$$T(x, 0) = T_0$$

**(a)** Show that the dimensionless form of the describing equations is

$$\frac{\partial \theta}{\partial (16\mathrm{Gz}^{-1})} = \frac{\partial^2 \theta}{\partial x'^2}$$

$$\theta(1, 16\mathrm{Gz}^{-1}) = 0 = \theta(-1, 16\mathrm{Gz}^{-1})$$

$$\theta(x', 0) = 1$$

where $\theta = (T - T_w)/(T_0 - T_w)$, $x' = x/a$, $\mathrm{Gz}^{-1} = (z/D_h)/\mathrm{Re}\,\mathrm{Pr}$, $D_h = 4a$, $\mathrm{Re} = VD_h/\nu$.

**(b)** Show that the solution for the temperature of part a is

$$\theta = \frac{4}{\pi} \sum_{n=0}^{\infty} \frac{(-1)^n}{(2n + 1)} \cos\left[(2n + 1)\frac{\pi}{2}x'\right] \exp\left[-(2n + 1)^2 \frac{4\pi^2}{\mathrm{Gz}}\right]$$

from which the mixing cup temperature is found to be

$$\theta_m = \frac{8}{\pi^2} \sum_{n=0}^{\infty} \frac{1}{(2n + 1)^2} \exp\left[-(2n + 1)^2 \frac{4\pi^2}{\mathrm{Gz}}\right]$$

and the heat flux into the fluid at the upper surface is found to be

$$\frac{q_w a}{(T_w - T_0)k} = 2 \sum_{n=0}^{\infty} \exp\left[-(2n + 1)^2 \frac{4\pi^2}{\mathrm{Gz}}\right]$$

*Hint:* Consult H. S. Carslaw and J. C. Jaeger, *Conduction of Heat In Solids*, 2nd ed., Clarendon Press, Oxford, 1959, p. 97.

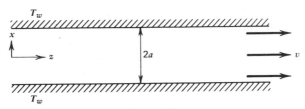

**Figure 6P-9**

(c) From the results of part b show that if the local heat transfer coefficient is based on the heat flux into the fluid from one surface, $q_w = h(T_w - T_m)$, the local Nusselt number is given by

$$\text{Nu} = \frac{hD_h}{k} = \frac{\pi^2 \Sigma_{n=0}^{\infty} \exp\left[-(2n+1)^2 4\pi^2/\text{Gz}\right]}{\Sigma_{n=0}^{\infty}\left[1/(2n+1)^2\right]\exp\left[-(2n+1)^2 4\pi^2/\text{Gz}\right]}$$

(d) From the result of part c, show that the limiting local Nusselt number is

$$\text{Nu}_{\infty} = \pi^2$$

(e) Show that the average Nusselt number based on LMTD is obtained from substitution of the mixing-cup temperature into

$$\overline{\text{Nu}} = \frac{\bar{h}D_h}{k} = -\frac{1}{4\text{Gz}^{-1}}\ln(\theta_m)$$

(f) Plot $\overline{\text{Nu}}$, Nu, and $\theta_m$ against $\text{Gz}^{-1}$ and determine a thermal entrance length criterion from that plot.

**6-10** Repeat Problem 6-9 for the case in which a constant heat flux $q_w$ flows into the fluid at each surface. *Hint*: Consult p. 112 of the reference cited for Problem 6-9.

(a) Show that the temperature distribution is

$$\frac{T - T_0}{q_w a/k} = 16\text{Gz}^{-1} + \frac{3x'^2 - 1}{6}$$

$$- \frac{2}{\pi^2}\sum_{n=1}^{\infty} \frac{(-1)^n}{n^2} \exp\left(\frac{-16n^2\pi^2}{\text{Gz}}\right)\cos(n\pi x')$$

from which the mixing-cup and wall temperatures are found to be

$$\frac{T_m - T_0}{q_w a/k} = 16\text{Gz}^{-1}$$

$$\frac{T_w - T_0}{q_w a/k} = 16\text{Gz}^{-1} + \frac{1}{3} - \frac{2}{\pi^2}\sum_{n=1}^{\infty}\frac{1}{n^2}\exp\left(\frac{-16n^2\pi^2}{\text{Gz}}\right)$$

and the limiting temperature distribution is

$$\lim_{\mathrm{Gz}^{-1}\to\infty}\frac{T-T_0}{q_w a/k}=16\mathrm{Gz}^{-1}+\frac{3x'^2-1}{6}$$

$$\lim_{\mathrm{Gz}^{-1}\to\infty}\frac{T_w-T_m}{q_w a/k}=\frac{1}{3}$$

**(b)** From the results of part a show that the local Nusselt number based on the heat flux from one surface is

$$\mathrm{Nu}=\frac{hD_h}{k}=\frac{12}{1-(6/\pi^2)\Sigma_{n=1}^{\infty}(1/n^2)\exp(-16n^2\pi^2/\mathrm{Gz})}$$

from which it is found that the limiting Nusselt number is

$$\mathrm{Nu}_{\infty}=12$$

**(c)** Compare $\mathrm{Nu}_{\infty}$ for the constant $q_w$ and constant $T_w$ cases.

**6-11** A power-law fluid enters a gap bounded by two infinite parallel planes, both of whose wall heat fluxes are constant and equal to $q_w$. The fluid experiences fully developed flow and has a uniform entrance temperature of $T_0$. Figure 6P-9 illustrates the geometry.

**(a)** Show that the velocity distribution is described by

$$\frac{d(\mu|\,dv/dx\,|^{n-1}\,dv/dx)}{dx}=\frac{-dp}{dz}=\mathrm{const}$$

$$v(x=a)=0=v(x=-a)$$

from which it is found that

$$\frac{v}{v_{\mathrm{av}}}=\frac{2n+1}{n+1}(1-x'^{(n+1)/n})$$

where $x'=x/a$.

**(b)** Show that the temperature distribution far downstream from the entrance is described by

$$\frac{(T-T_0)}{q_w a/k}=-\frac{(2n+1)(24n^2+13n+2)}{6(3n+1)(4n+1)(5n+2)}+16\mathrm{Gz}^{-1}$$

$$+\frac{2n+1}{n+1}\left[\frac{x'^2}{2}-\frac{n^2}{(2n+1)(3n+1)}x'^{(3n+1)/n}\right]$$

(c) From the result of part b show that the limiting value of the Nusselt number is

$$\mathrm{Nu}_\infty = \frac{hD_h}{k} = 12\frac{(4n+1)(5n+2)}{32n^2 + 17n + 2}$$

where $D_h = 4a$ is the hydraulic diameter, $\mathrm{Gz}^{-1} = (z/D_h)/\mathrm{Re}\,\mathrm{Pr}$, and $\mathrm{Re} = v_{av}D_h/\nu$.

(d) Compare the results of part c against the value reported for a Newtonian fluid and develop a table to show how $\mathrm{Nu}_\infty$ varies with $n$ (identify the values of $n$ that correspond to slug and linear velocity profiles.)

**6-12** Reconsider Problem 6-11 for the case in which both surfaces are maintained at the same constant temperature $T_w$. Far downstream from the entrance the temperature distribution can be determined by iteratively solving the energy equation, as was done to find $\mathrm{Nu}_\infty$ for fully developed flow in a circular duct. Use this technique to determine $\mathrm{Nu}_\infty = hD_h/k$ for fully developed slug, parabolic, and linear velocity profiles. Compare the result for a Newtonian fluid with the value reported in the text. The temperature distribution of $1 - x'^2$ can be used to begin the iterations. (*Answer:* $n = 0$; $\mathrm{Nu}_\infty = 9.87$; $n = 1$, $\mathrm{Nu}_\infty = 7.54$; $n = \infty$, $\mathrm{Nu}_\infty = 6.96$.)

**6-13** Repeat Problem 6-11 to find $\mathrm{Nu}_\infty$ for the case in which one surface is adiabatic while the other surface delivers a constant heat flux $q_w$ into the fluid. This approximates the situation in a wide and long solar collector with the adiabatic surface representing the glazing through which the insolation comes, the constant-heat-flux surface representing the absorber plate, and the heat flux representing the absorbed insolation.

(a) Determine $\mathrm{Nu}_\infty$ for slug, parabolic, and linear velocity profiles.

(b) Determine Nu in the entrance region, using the assumption that the dimensionless temperature has the form

$$\theta = \underbrace{X_1(x)Z_1(z)}_{\substack{\text{decaying}\\\text{transient}}} + \underbrace{X_2(x)}_{\substack{\text{axial temperature}\\\text{rise due to}\\\text{accumulated}\\\text{wall flux}}} + \underbrace{Z_2(z)}_{\substack{\text{transverse temperature}\\\text{variation to let}\\\text{wall flux into fluid}}}$$

Confine efforts to a Newtonian fluid until the solution procedure is grasped.

**6-14** Consider a $\frac{1}{4}$-in.-i.d. (inner diameter), 4-ft-long circular duct that receives a constant wall heat flux from an electric resistance heating element wrapped around the duct. An organic liquid flows through the duct at 10 $\mathrm{lb}_m/\mathrm{hr}$ and is to be heated from its inlet temperature of 50°F to an outlet temperature of $150°F$. The fluid properties are $Pr = 10$, $\rho = 47$ $lb_m/ft^3$, $C_p = 0.5$ $Btu/lb_m$ $°F$, $k = 0.079$ $Btu/hr$ $ft$ $°F$, and $\mu(Newtonian) = 1.6$ $lb_m/hr$ $ft$.

(a) Determine whether the duct can be considered to be infinitely long.

(b) Determine the maximum fluid temperature, the maximum wall temperature, and the wall heat flux.

6-15 Consider fully developed laminar flow of a constant property Newtonian fluid through a circular duct. The wall temperature is constant. There is uniformly distributed heat generated in the fluid (due, perhaps, to chemical or nuclear reaction) at a rate $S$, $J/s$ $m^3$. Determine the temperature distribution far from the inlet and thus the limiting value of the Nusselt number $Nu_\infty = h_\infty D/k$.

6-16 Consider the fully developed laminar flow of a Newtonian fluid in a circular duct. The duct wall temperature is constant at $T_w$, and the fluid entrance temperature is uniform at $T_0$. Fluid properties are constant, the duct is of radius $R$, and the fluid liberates thermal energy uniformly at the rate $q'''$, $J/s$ $m^3$.

(a) Show that the describing energy equation is

$$2(1 - r'^2)\frac{\partial\theta}{\partial(4/Gz)} = \frac{1}{r'}\frac{\partial(r'\partial\theta/\partial r')}{\partial r'} + E\,Pr\,16r'^2 + Q$$

$$\theta\left(r' = 1, \frac{4}{Gz}\right) = 0, \qquad \theta(r',0) = 1, \qquad \frac{\partial\theta(r' = 0, 4/Gz)}{\partial r'} \quad \text{finite}$$

where $r' = r/R$, $\theta = (T - T_w)/(T_0 - T_w)$, $E = v_{av}^2/C_p(T_0 - T_w)$, $Pr = \nu/\alpha$, and $Q = q'''R^2/k(T_0 - T_w)$.

(b) Show that far downstream from the beginning of the heated section where $\partial\theta/\partial(4/Gz) = 0$,

$$\theta(r', 4/Gz \to \infty) = E\,Pr(1 - r'^4) + \frac{Q(1 - r'^2)}{4}$$

(c) From the result of part b show that the mixing-cup temperature is

$$\theta_m\left(r, \frac{4}{Gz} \to \infty\right) = \frac{5}{6}E\,Pr + \frac{Q}{6}$$

(d) From the result of part c show that the limiting Nusselt number, based on a driving temperature difference of $T_w - T_m$, is

$$Nu_\infty = \frac{hD}{k} = \frac{48E\,Pr + 6Q}{5E\,Pr + Q}$$

(e) Discuss the result of part d with particular attention to the $Nu_\infty$ that is found when $E\,Pr = 0 = Q$ and to the appropriateness of $T_w - T_m$ as a driving temperature difference. *Note*: Consult reference 26 for entrance region details.

**6-17** Compare the constant heat flux and constant wall temperature limiting Nusselt number values for an equilateral triangle from Table 6-9 with those for a sector of a circle with a 60° included angle from Fig. 6-10. For which condition is there the closer agreement? A fluid can flow laminarly through either a circular duct or a rectangular duct whose cross section has a length/width ratio of 8. Both ducts have equal cross-sectional area, so the average velocity is the same. Because the circular duct has less surface area, it will have the lesser heat transfer rate per unit axial length for constant wall temperature conditions unless the rectangular duct has a substantially smaller heat transfer coefficient. Determine the ratio of the heat-transfer coefficient–surface area product for these two geometries.

**6-18** Show, for a 2:1 rectangle with constant wall temperature and fully developed laminar flow of a Newtonian fluid, that the Nusselt number based on LMTD is

$$\overline{Nu} = 3.39 + 0.06 Re\,Pr\frac{D_h}{x}, \qquad \frac{x/D_h}{Re\,Pr} \gtrsim 0.05$$

and determine the value of $x/D_h$ at which Nu exceeds its limiting value by 10%.

**6-19** On the basis of the criterion for thermal entrance region for laminar flow of $0 \leqslant (x/D_h)/Re\,Pr \lesssim 0.05$, estimate the length of the thermal entrance region when $Re = 500$ for (a) an oil ($Pr = 500$), (b) air ($Pr = 0.7$), and (c) liquid metal ($Pr = 0.01$). If the duct length is $100\,D_h$, determine whether either the velocity or the temperature profiles are fully developed over a major portion of the tube length.

**6-20** The Nusselt number for a Newtonian fluid based on LMTD very near the inlet of a circular duct of constant wall temperature can be obtained from Eq. (6-49). Show that for $Re\,Pr\,D/x$ small,

(a)

$$\overline{Nu} \approx 3.66 + 6.5\left(\frac{Re\,Pr\,D}{x}\right)^{1/5}$$

for simultaneously developing velocity and temperature profiles.

(b)

$$\overline{Nu} = 3.66 + 1.67\left(\frac{Re\,Pr\,D}{x}\right)^{1/3}$$

for a fully developed parabolic velocity profile.

(c) Compare by plotting a curve of $\overline{\text{Nu}}$ against $\text{Re Pr } D/x$ the accuracy of the result of part b with the full correlating equation [Eq. (6-49)] and with the empirical correlation

$$\overline{\text{Nu}} = 1.86\left(\frac{\text{Re Pr } D}{x}\right)^{1/3}\left(\frac{\mu_m}{\mu_w}\right)^{0.14}$$

due to E. N. Seider and G. E. Tate, Heat transfer and pressure drop of liquids in tubes, *Ind. Eng. Chem.* **28**, 1429–1435 (1936).

(d) On the basis of the results of parts a and b, show that near the inlet the Nusselt number with a developing velocity profile is four times greater than the Nusselt number with a fully developed velocity. Is this result limited to $\text{Pr} = 0.7$?

**6-21** Show that for a circular duct, $\text{Re Pr } D/x = (4/\pi)C_p\dot{M}/kx$, where $\dot{M}$ is the mass flow rate.

**6-22** Show that for a duct of any cross section, $\text{Re Pr } D_h/x = (16A/C^2)C_p\dot{M}/kx$, where $A$ is the cross-sectional area and $C$ is the cross-sectional circumference.

**6-23** From the results of Table 6-12 or Fig. 6-11, show that $0 \leqslant (x/D)/\text{Re Pr} \lesssim 0.05$ is a reasonable criterion for the extent of the thermal entrance region for $\text{Pr} \approx 0.7$ with simultaneous development of velocity and temperature profiles.

**6-24** Calculate the heat-transfer coefficient for laminar flow of a Newtonian fluid ($k = 0.1$ Btu/hr ft °F) inside a $\frac{1}{4}$-in. i.d. tube in the hydrodynamically and thermally developed region with a uniform wall temperature. Also, determine the heat transfer rate between the tube wall and the fluid if the inlet temperature is 300°F and the wall temperature is 100°F. [*Hint:* $h_\infty = 17.6$ Btu/hr ft$^2$ °F, $Q = \rho C_p v_{av}\pi D^2/4(200°F) =$ indeterminant without flow rate.]

**6-25** Compare the limiting values of heat-transfer coefficients for a Newtonian fluid flowing laminarly through (a) a circular duct of diameter $D$, (b) a square duct of side $D$, and (c) and equilateral-triangle duct of side $D$.

**6-26** Oil at 70°F with a mean inlet velocity of 2 ft/sec enters a 1/2-in.-i.d., 5-ft-long tube with a constant wall temperature of 150°F. Determine the pressure drop and outlet temperature as accurately as possible, accounting for viscosity variation and entrance effects, and the length of the entrance region. Repeat the outlet temperature and pressure drop calculations, first neglecting the viscosity variation and then neglecting all entrance effects. Discuss the importance of these two factors; in particular, compare their relative importance for air and oil. Fluid properties are: at 70°F, $\mu = 0.015$ $\text{lb}_m/\text{ft}$ sec; at 150°F, $\mu = 0.0055$ $\text{lb}_m/\text{ft}$ sec, $\rho = 55$ $\text{lb}_m/\text{ft}^3$, $C_p = 0.45$ $\text{Btu}/\text{lb}_m$ °F, $k = 0.1$ $\text{Btu}/\text{hr ft}$ °F.

**6-27** Air at 70°F with a mean inlet velocity of 2 ft/sec enters a $\frac{1}{2}$-in.-id., 1-ft-long tube with a constant wall temperature of 150°F. Determine the air outlet temperature, pressure drop, and the length of the entrance region as accurately as possible, accounting for property variation and entrance effects. Repeat the outlet temperature and pressure drop calculations, first neglecting the property variations and then neglecting all entrance effects. Discuss the importance of these two factors; in particular, discuss their relative importance for air and oil (do not repeat the calculation for oil).

**6-28** Water at 100°F from a large header enters a 1-cm-i.d. straight circular tube of 2 m length (whose wall temperature is constant at 150°F) that is part of a flat-plate solar collector. The water flow is forced at the rate of 0.003 kg/s, which is a typical value near the optimum for flat-plate solar collectors. Estimate the outlet temperature, pressure drop, the thermal and hydraulic entrance lengths, and the value of the film coefficient. Judge whether the Nusselt number correlation (J. A. Duffie and W. A. Beckman, *Solar Energy Thermal Processes*, Wiley-Interscience, New York, 1974, p. 83)

$$\text{Nu} = \frac{-\text{Re Pr } D}{4L} \ln\left[1 - \frac{2.654}{\text{Pr}^{0.167}(\text{Re Pr } D/L)^{0.5}}\right], \quad \frac{L}{D} < 0.0048\text{Re}$$

is of good accuracy. If a better correlation exists, state it.

**6-29** Air at 60°F from a large header enters a flat-plate solar collector that is approximated by two parallel planes, one side insulated to represent the collector's glazing and the other side at 200°F constant temperature (or constant absorbed solar heat flux, depending on one's point of view).

The flow channel is 4 m long (and 1 m × 0.01 m in cross section), and the flow rate is 40 kg/h. Estimate the outlet temperature, the pressure drop, the thermal and hydraulic entrance lengths, and the value of the heat-transfer coefficient.

**6-30** Determine the delivered solar power : pumping power ratio for water (using the conditions of Problem 6-28) and for air (using the conditions of Problem 6-29). On the basis of equal delivered solar power, judge which of these two methods of cooling a solar collector consumes the least pumping power.

# REFERENCES

1  C. S. Keevil and M. M. McAdams, How heat transmission affects fluid friction in pipes, *Chem. Metallurg. Eng.* **36**, 464–467 (1929).

2  R. K. Shah and A. L. London, Supplement 1—Laminar flow forced convection in ducts, in *Advances in Heat Transfer*, T. F. Lrvine and J. P. Hartnett, Eds., Academic Press, New York, 1978.

3  M. Abramowitz and I. A. Stegun, *Handbook Of Mathematical Functions*, National Bureau of Standards, Applied Mathematics Series 55, 1965.

4  H. S. Carslaw and J. C. Jaeger, *Conduction of Heat in Solids*, 2nd ed., Clarendon Press, Oxford, 1959, pp. 199, 203.

5  J. Schenk and J. van Laar, Heat transfer in non-Newtonian laminar flow in tubes, *Appl. Sci. Res.* **7**, 449–462 (1958).

6  A. H. P. Skelland, *Non-Newtonian Flow And Heat Transfer*, Wiley, New York, 1967, pp. 353–383.

7  I. R. Whiteman and W. B. Drake, Heat transfer to flow in a round tube with arbitrary velocity distribution, *Transact. ASME* **80**, 728–732 (1980).

8  B. C. Lyche and R. B. Bird, The Graetz–Nusselt problem for a power-law non-Newtonian fluid, *Chem. Eng. Sci.* **6**, 35–41 (1956).

9  J. A. Sellars, M. Tribus, and J. S. Klein, Heat transfer to laminar flow in a round tube or flat conduit—the Graetz problem extended, *Transact. ASME* **78**, 441–448 (1956).

10  R. Siegel, E. M. Sparrow, and T. M. Hallman, Steady laminar heat transfer in a circular tube with prescribed wall heat flux, *Appl. Sci. Res.* **7A**, 386–392 (1958).

11  R. K. Shah and A. L. London, Thermal boundary conditions and some solutions for laminar duct flow forced convection, *Transact. ASME, J. Heat Transf.* **96**, 159–165 (1974).

12  W. M. Kays, *Convective Heat And Mass Transfer*, McGraw-Hill, 1966, pp. 112–116, 128–133; R. E. Lundberg, W. C. Reynolds, and W. M. Kays, Heat transfer with laminar flow in concentric annuli with constant and variable wall temperature and heat flux, NASA TN D-1972, Washington, DC, August 1963.

13  E. R. G. Eckert, T. F. Irvine, Jr., and J. T. Yen, Local laminar heat transfer in wedge-shaped passages, *Transact. ASME* **80**, 1433–1438 (1958).

14  S. C. R. Dennis, A. McD. Mercer, and G. Poots, Forced heat convection in laminar flow through rectangular ducts, *Quart. Appl. Math.* **17**, 285–297 (1959).

15  W. M. Kays, Numerical solutions for laminar-flow heat transfer in circular tubes, *Transact. ASME* **77**, 1265–1274 (1955).

16  P. Goldberg, M. S. Thesis, Mechanical Engineering Department, Massachusetts Institute of Technology, Cambridge, MA, January 1958.

17  H. Hausen, *Z. Ver. deutsch. Ing., Beih. Verfahrenstech.* **4**, 91–98 (1943).

18  H. L. Langhaar, Steady flow in the transition length of a straight tube, *Transact. ASME, J. Appl. Mech.* **64**, A55–A58 (1942).

19  H. S. Heaton, W. C. Reynolds, and W. M. Kays, Heat transfer in annular passages. Simultaneous development of velocity and temperature fields in laminar flow, *Internatl. J. Heat Mass Transf.* **7**, 763–781 (1964).

20  E. Sparrow, S. H. Lin, and T. Lundgren, Flow developments in the hydrodynamic entrance region of tubes and ducts, *Phys. Fluids* **7**, 338–347 (1964).

21  S. T. McComas and E. R. G. Eckert, Laminar pressure drop associated with the continuum entrance region and for slip flow in a circular tube, *Transact. ASME, J. Appl. Mech.* **32**, 765–770 (1965).

22  T. S. Lundgren, E. M. Sparrow, and J. B. Starr, Pressure drop due to the entrance region in ducts of arbitrary cross-section, *Transact. ASME, J. Basic Eng.* **86**, 620–626 (1964).

23  E. M. Sparrow, T. S. Chen, and V. K. Jonsson, Laminar flow and pressure drop in internally finned annular ducts, *Internatl. J. Heat Mass Transf.* **7**, 583–585 (1964); E. M. Sparrow and A. Haji-Sheikh, Flow and heat transfer in ducts of arbitrary shape with arbitrary thermal boundary conditions, *Transact. ASME, J. Heat Transf.* **88**, 351–358 (1966).

24  W. M. Kays and A. L. London, *Compact Heat Exchangers*, McGraw-Hill, New York, 1964, pp. 86–91.

25  R. D. Cess and E. C. Shaffer, Heat transfer to laminar flow between parallel plates with a prescribed wall heat flux, *Appl. Sci. Res.* **8A**, 339–344 (1958).

26  S. N. Singh, Heat transfer by laminar flow in a cylindrical tube, *Appl. Sci. Res.* **7A**, 325–340 (1958).

27  S. H. Lin and W. K. Ksu, Heat transfer to power-law non-Newtonian flow between parallel plates, *Transact. ASME, J. Heat Transf.* **102**, 382–384 (1980).

28  S. D. Joshi and A. E. Bergles, Experimental study of laminar heat transfer to in-tube flow of non-Newtonian fluids, *Transact. ASME, J. Heat Transf.* **102**, 397–401 (1980).

29  M. Faghri and E. M. Sparrow, Simultaneous wall and fluid axial conduction in laminar pipe-flow heat transfer, *Transact. ASME, J. Heat Transf.* **102**, 58–63 (1980).

30  A. B. Metzner, Heat transfer in non-Newtonian fluids, *Adv. Heat Transf.* **2**, 376 (1965).

31  S. Golos, When Is It Allowed To Treat The Heat Transfer Coefficient $\alpha$ as constant?, *Internatl. J. Heat Transf.* **18**, 1467–1471 (1975).

32  S. W. Hong and A. E. Bergles, Augmentation of laminar flow heat transfer in tubes by means of twisted-tape inserts, *Transact. ASME, J. Heat Transf.* **98**, 251–256 (1976).

# 7

## LAMINAR
## BOUNDARY LAYERS

As can be appreciated, the complete equations of continuity, motion, energy, and diffusion are very difficult to solve in most important cases. Although there is little difficulty for such simple cases as Couette flow, they can generally be solved analytically only with considerable effort and ingenuity and then often only with numerical methods. Accordingly, an order of magnitude analysis is necessary to reveal insignificant terms. Hopefully, discarding these terms will permit a more easily solved system of equations to be taken as the describing equations without an appreciable sacrifice of accuracy.

### 7.1   LAMINAR BOUNDARY-LAYER EQUATIONS

The basic ideas underlying the approximations that yield the boundary layer equations were developed by Prandtl about 1904. The essential idea is to divide a flow into two major parts. The larger part concerns a free stream of fluid, far from any solid surface, which is accurately considered to be inviscid. The smaller part is a thin layer next to a solid surface in which the effects of molecular transport (viscosity, thermal conductivity, and mass diffusivity) are considered at the expense of some approximations.

To gain insight, note the experimentally observed distributions sketched in Fig. 7-1. Note that the major part of all variations occur in a thin layer adjacent to the solid boundary. As Fig. 7-1 indicates, the main, or free, stream is steadily flowing parallel to a smooth flat plate. The plate velocity, tempera-

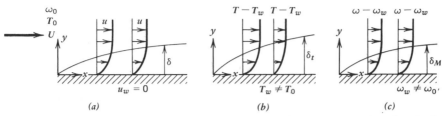

**Figure 7-1** Profiles in a laminar boundary layer: ($a$) horizontal velocity ($\delta$, velocity boundary-layer thickness); ($b$) temperature ($\delta_T$, thermal boundary-layer thickness); ($c$) mass fraction ($\delta_M$, diffusion boundary-layer thickness).

ture, and mass fraction are constant and the fluid properties are also constant in this relatively simple case. Because of the stabilizing effect of viscosity, tending to make the fluid next to the plate acquire the plate velocity, the flow in the boundary layer is laminar. (Actually, it might only be laminar in the vicinity of the leading edge; at a position farther downstream, transition from laminar to turbulent flow might occur in the boundary layer.) In some respects it is noteworthy that a boundary layer could be laminar in view of the adjacent probably turbulent free stream; on the other hand, since the boundary layer is also adjacent to the solid boundary, its possible laminar state of flow is not surprising, either.

As discussed by Schlichting [1], neglect of viscous effects in the free stream is accurate and greatly simplifies the Navier–Stokes equations, which describe the velocity distribution. An advantageous result is that inviscid incompressible flow is irrotational since then $V = \text{grad } \Phi$, where $\Phi$ is a potential satisfying Laplace's equation $\nabla^2\Phi = 0$, from which it follows that $\text{grad}(\nabla^2\Phi) = \nabla^2(\text{grad } \Phi) = \nabla^2 V = 0$. A disadvantageous result is the vanishing of the second order terms for which viscosity is a coefficient, rendering impossible the simultaneous satisfaction of the conditions of no slip and impermeability at a solid boundary. If the impermeability condition is satisfied, as it usually is since it most characterizes a solid surface, the potential flow solution shows a velocity slip at the solid boundary. Analytically and experimentally it is found that even in the limit as viscosity effects approach zero, the no-slip condition applies at a solid boundary (provided the fluid can be considered to be continuous, of course); the smaller the viscosity is, the closer to the solid boundary one must be before the potential flow solution loses accuracy. In this thin fluid layer near the boundary, viscous effects must not be ignored (as they successfully are in the free stream). In mathematical terms, inclusion of viscous effects in the Navier–Stokes equations provides two orders in the direction normal to the solid surface and enables both the no-slip and the impermeability conditions to be satisfied there.

Historically, the velocity boundary layer was of first concern and is here, too. Referring to Fig. 7-1 for the coordinate system, the describing equations are

Continuity $\quad \dfrac{\partial u}{\partial x} + \dfrac{\partial v}{\partial y} = 0$

$x$ Motion $\quad u\dfrac{\partial u}{\partial x} + v\dfrac{\partial u}{\partial y} = \dfrac{B_x - \dfrac{\partial p}{\partial x} + \mu(\partial^2 u/\partial x^2 + \partial^2 u/\partial y^2)}{\rho}$

$y$ Motion $\quad u\dfrac{\partial v}{\partial x} + v\dfrac{\partial v}{\partial y} = \dfrac{B_y - \dfrac{\partial p}{\partial y} + \mu(\partial^2 v/\partial x^2 + \partial^2 v/\partial y^2)}{\rho}$

with $u = 0 = v$ at $y = 0$ and $u \rightarrow U$ at $y \rightarrow \infty$. These equations are made dimensionless by dividing through by the free stream velocity $U$, and a characteristic length $L$ (perhaps the length of the plate). Let $u' = u/U$, $x' = x/L$, and so forth. Consider the continuity equation first and put it in the dimensionless form

$$\frac{\partial u'}{\partial x'} + \frac{\partial v'}{\partial y'} = 0$$

To ascertain whether the continuity equation contains negligible terms, this dimensionless form is subjected to an order of magnitude analysis. Its reasoning is that one is interested principally in locations removed from the leading edge, so that $x' = x/L \sim O(1)$. The symbol $O(\ )$ is used to denote "order of." In most of the boundary layer $u' = u/U \sim O(1)$. Furthermore, the values of $y$ of interest are only in the boundary layer. So $y' = y/L \sim O(\delta' = \delta/L)$. Substitution of these orders of magnitude into the dimensionless continuity equation gives

$$\frac{O(1)}{O(1)} + \frac{v'}{O(\delta')} \sim 0 \tag{7-1}$$

which can be true only if

$$v' \sim O(\delta') \tag{7-2}$$

which predicts that the velocity normal to the solid surface is quite small. One of the essential assumptions here is that $\delta/L$ is small—the boundary layer is thin. On physical grounds Eq. (7-2) is plausible since the plate is impermeable and the nearby free stream flows parallel to the plate. Turning next to the $x$-motion equation, one finds its dimensionless form to be

$$u'\frac{\partial u'}{\partial x'} + v'\frac{\partial u'}{\mu\, y'} = \left( B'_x - \frac{\partial p'}{\partial x'} \right) + \frac{\nu}{UL} \left( \frac{\partial^2 u'}{\partial x'^2} + \frac{\partial^2 u'}{\partial y'^2} \right)$$

An order of magnitude analysis, with the use of Eq. (7-12) as well as the previously explained ideas regarding orders of magnitude, shows the dimensionless $x$-motion equation to have terms whose orders of magnitude are, with $\text{Re} = UL/\nu$,

$$O(1)\frac{O(1)}{O(1)} + O(\delta')\frac{O(1)}{O(\delta')} \sim \frac{1}{\text{Re}}\left[\frac{O(1)}{\cancel{O(1^2)}}\text{moderate} + \frac{O(1)}{\cancel{O(\delta'^2)}}\text{large}\right] \quad (7\text{-}3)$$

Consolidation of the terms of the left-hand side and retaining only the large term in parentheses on the right-hand side leads to

$$O(1) \sim \frac{1}{\text{Re}}\frac{1}{O(\delta'^2)}$$

This relationship can be true only if

$$\text{Re} \sim \frac{1}{O(\delta'^2)} \quad (7\text{-}4)$$

Note that none of the convective terms on the left-hand side of the $x$-motion equation can be neglected. Most importantly, it is seen that $\partial^2 u/\partial x^2$ can be dropped from the viscous term, changing the equation from an elliptic to a parabolic form, but that viscous effects are retained since $\partial^2 u/\partial y^2$ must be kept. Also note that it is necessary for $\text{Re} \sim O(1/\delta'^2)$—the Reynolds number must be very large for the boundary layer ideas and approximations to be accurate. For the case of parallel flow over a flat plate of this discussion, it is to be understood that $B'_x - dp'/dx = (B_x - dp/dx)L/\rho U^2 = 0$ since outside the boundary layer in the free stream

$$u'\underset{0}{\frac{\partial \cancel{u'}}{\cancel{\partial x'}}} + v'\underset{0}{\frac{\partial \cancel{u'}}{\cancel{\partial y'}}} = \left(B'_x - \frac{dp'}{dx'}\right) + \frac{1}{\text{Re}}\left(\underset{0}{\frac{\partial^2 \cancel{u'}}{\cancel{\partial x'^2}}} + \underset{0}{\frac{\partial^2 \cancel{u'}}{\cancel{\partial y'^2}}}\right)$$

inasmuch as $u$ is there constant. Similarly, subjection of the dimensionless form of the $y$-motion equation to an order of magnitude analysis gives

$$u'\frac{\partial v'}{\partial x'} + v'\frac{\partial v'}{\partial y'} = B'_y - \frac{\partial p'}{\partial y'} + \frac{1}{\text{Re}}\left(\frac{\partial^2 v'}{\partial x'^2} + \frac{\partial^2 v'}{\partial y'^2}\right)$$

$$O(1)\frac{O(\delta')}{O(1)} + O(\delta')\frac{O(\delta')}{O(\delta')} \sim B'_y - \frac{\partial p'}{\partial y'} + O(\delta'^2)\left[\frac{O(\delta')}{O(1)} + \frac{O(\delta')}{O(\delta'^2)}\right]$$

$$(7\text{-}5)$$

$$O(\delta') \sim B'_y - \frac{\partial p'}{\partial y'} + O(\delta')$$

$$O(\delta') \sim B'_y - \frac{\partial p'}{\partial y'} \quad (7\text{-}6)$$

Equation (7-6) shows that pressure does not vary much across the boundary layer; thus the boundary layer pressure at any $x$-position is imposed by the free stream and represents the main channel of communication between the boundary layer and the free stream.

The thermal boundary layer is subjected to the same order of magnitude analysis. The energy equation in its dimensionless form is

$$u'\frac{\partial\theta}{\partial x'} + v\frac{\partial\theta}{\partial y'} = \frac{1}{Re\,Pr}\left(\frac{\partial^2\theta}{\partial x'^2} + \frac{\partial^2\theta}{\partial y'^2}\right) + \frac{E}{Re}\Phi' + \beta TE\left(u'\frac{\partial p'}{\partial x'} + v'\frac{\partial p'}{\partial y'}\right)$$

where the dimensionless temperature is $\theta = (T - T_{ref})/\Delta T_{ref}$. With incorporation of the information provided by Eqs. (7-2) and (7-4), it is found that the dimensionless energy equation terms have the orders of magnitude of

$$O(1)\frac{O(1)}{O(1)} + O(\delta')\frac{O(1)}{O(\delta'_T)} \sim \frac{O(\delta'^2)}{Pr}\left[\frac{O(1)}{O(1^2)}\ \text{moderate} + \frac{O(1)}{O(\delta'^2_T)}\ \text{large}\right]$$

$$+ E\frac{O(\delta'^2)}{O(\delta'^2)} + \beta TE[O(1)O(0) + O(\delta')O(\delta')]$$

$$(7\text{-}7)$$

This can be consolidated, assuming $O(\delta')/O(\delta'_T) \sim O(1)$ on the left-hand side, to

$$O(1) \sim \frac{1}{Pr}O\left(\frac{\delta'^2}{\delta'^2_T}\right) + E + \beta TE\,O(\delta'^2) \qquad (7\text{-}8)$$

From this it can be deduced that $Pr \sim O(\delta'^2/\delta'^2_T)$ from which a preliminary inference of $\delta'_T/\delta' = 1/Pr^{1/2}$ can be made, that $E \sim O(1)$ if viscous dissipation is to be important, and that $\beta TE \sim O(1/\delta'^2)$ if compressive work is to be important. Dimensionless viscous dissipation is

$$\Phi' = 2\left[\left(\frac{\partial u'}{\partial x'}\right)^2 + \left(\frac{\partial v'}{\partial y'}\right)^2\right] + \left[\frac{\partial v'}{\partial x'} + \frac{\partial u'}{\partial y'}\right]^2$$

$$\sim \left[\frac{O(1)}{O(1)} + \frac{O(\delta')}{O(\delta')}\right]^2 + \left[\frac{O(\delta')}{O(1)} + \frac{O(1)}{O(\delta')}\right]^2$$

$$\sim \frac{1}{O(\delta'^2)}$$

It is important to note that $\partial^2 T/\partial x^2$ is negligibly small compared to $\partial^2 T/\partial y^2$

in the diffusive heat conduction term of the energy equation. This is as plausible a result on physical grounds as the analogous finding for the x-motion equation $(\partial^2 u/\partial x^2 < \partial^2 u/\partial y^2)$ since gradients are greatest in the $y$ direction (for a thin boundary layer).

The mass diffusion equation can be similarly treated with the result that the second derivative with respect to $x$ in the molecular transport term is negligible.

The laminar boundary-layer equations, finally, are

Continuity $\quad \dfrac{\partial \rho}{\partial t} + \dfrac{\partial \rho u}{\partial x} + \dfrac{\partial \rho v}{\partial y} = 0$

$x$ Motion $\quad \rho \left( \dfrac{\partial u}{\partial t} + u\dfrac{\partial u}{\partial x} + v\dfrac{\partial u}{\partial y} \right) = B_x - \dfrac{\partial p}{\partial x} + \dfrac{\partial}{\partial y}\left( \mu \dfrac{\partial u}{\partial y} \right)$

$y$ Motion $\quad B_y - \dfrac{\partial p}{\partial y} = 0$

$(7\text{-}9)$

Energy $\quad \rho C_p \left( \dfrac{\partial T}{\partial t} + u\dfrac{\partial T}{\partial x} + v\dfrac{\partial T}{\partial y} \right) = \dfrac{\partial}{\partial y}\left( k\dfrac{\partial T}{\partial y} \right) + \mu \left( \dfrac{\partial u}{\partial y} \right)^2$

$\qquad\qquad\qquad\qquad\qquad\qquad + \beta T \left( \dfrac{\partial p}{\partial t} + u\dfrac{\partial p}{\partial x} \right)$

Diffusion $\quad \rho \left( \dfrac{\partial \omega_a}{\partial t} + u\dfrac{\partial \omega_a}{\partial lx} = v\dfrac{\partial \omega_a}{\partial y} \right) = \dfrac{\partial}{\partial y}\left( D_{ab}\rho \dfrac{\partial \omega_a}{\partial y} \right) + r_a'''$

Although the boundary-layer equations [Eqs. 7–9)] are written for a Newtonian fluid and seem specialized, they are still quite general and bear simplification when specific problems are attacked. The incorporation of time derivatives is justified on the grounds that they must be present if unsteady problems are to be solved—very rapid transients, such as occur in blasts and explosions, may not be well handled by boundary-layer equations. Note that the possibility of variable properties has been included. This is a rational thing to do, even in light of the constant-property fluid postulated during the order of magnitude arguments, since the property variation is seldom large. If a non-Newtonian fluid is encountered, express the boundary-layer equations [Eqs. (7–9)] in terms of shear stress with order of magnitude arguments of the nature previously discussed.

As was mentioned earlier, the free stream communicates mainly with the boundary layer through the fact that

$$\rho \left( \dfrac{\partial U}{\partial t} + U\dfrac{\partial U}{\partial x} \right) = B_x - \dfrac{\partial p}{\partial x} \qquad (7\text{-}10)$$

where all terms are to be taken as just outside the boundary layer. Because, the free stream velocity $U$ is either known or is determined from the potential flow (often also referred to as inviscid or irrotational flow) solution, $\partial p/\partial x$ can also be regarded as a known forcing function in the boundary layer equations.

## 7.2   SIMILARITY SOLUTION FOR PARALLEL FLOW OVER A FLAT PLATE—VELOCITY DISTRIBUTION

The sketches in Fig. 7-1, representing measurements, lead one to suspect that all profiles are basically similar to one another if the coordinates are properly stretched. If such a similarity actually exists, a mathematical transformation of coordinates can be made to reflect this fact. Of course, not all boundary layers have "similar" profiles, but many do. The boundary-layer equations are observed to be nonlinear partial-differential equations; a similarity transformation results in nonlinear equations, still, but they are ordinary differential equations—a big advantage.

To introduce similarity solutions, consider a flat plate over which a constant-property fluid steadily flows parallel to the plate. Initial interest is in the velocity distribution in the boundary layer. Note that $\partial p/\partial x = 0$ since $U$ is constant. The applicable boundary-layer equations are

$$\frac{\partial u}{\partial x} + \frac{\partial v}{\partial y} = 0$$

$$u\frac{\partial u}{\partial x} + v\frac{\partial u}{\partial y} = \nu\frac{\partial^2 u}{\partial y^2}$$

with the boundary conditions that

$$u(x,0) = 0 = v(x,0) \qquad \text{and} \qquad u(x,\infty) = U$$

A clue as to the form of the coordinate transformation that will give a similarity solution can be obtained from the results of the transient vaporization study of Section 5.4. There it was found that the vapor film thickness always increased as $t^{1/2}$, and the random-walk exercise placed this result on firmer physical ground, showing that the distance moved by a random walking particle always increases as $t^{1/2}$. If one views $y$ as the vertical distance from the plate the effect of the plate has penetrated, one has some reason to suspect that $y \sim t^{1/2}$; the time elapsed for an affected fluid particle $t$ can be viewed as $x/U$. Therefore, a variable $y/t^{1/2} \sim y/(x/U)^{1/2}$ may well be the one sought. A rigorous analysis (see Appendix E) shows that the variable sought is

$$\eta = y\left(\frac{U}{x}\right)^{1/2} \tag{7-11}$$

The salient point is that a rigorous analysis (see Hansen [2] or Schlichting [1]) cannot start without a general idea as to the form of the coordinate transformation.

As is usual in boundary-layer work, the continuity equation is first satisfied by assuming the existence of a stream function $\Psi$ such that $u = \partial\Psi/\partial y$ and

$v = -\partial\Psi/\partial x$. The fact that the continuity equation is satisfied is shown by direct substitution to obtain

$$\frac{\partial u}{\partial x} + \frac{\partial v}{\partial y} = 0$$

$$\frac{\partial^2\Psi}{\partial x\,\partial y} - \frac{\partial^2\Psi}{\partial y\,\partial x} = 0$$

If the order of differentiation can be interchanged, requiring only that $\partial\Psi/\partial x$ and $\partial\Psi/\partial y$ exist as well as $\partial^2\Psi/\partial x\,\partial y$ be continuous, it is seen that the continuity equation is indeed satisfied regardless of what the stream function may turn out to be.

The stream function is given here without proof (see Appendix E for details) as

$$\Psi = \sqrt{\nu U x}\, F\left( \eta = y\sqrt{\frac{U}{\nu x}} \right) \tag{7-12}$$

Equation (7-12) displays a separation-of-variables form inasmuch as $\Psi = H(x)F(\eta)$ is the product of two functions that separately depend on independent variables. Having the stream function, the velocity components are found from

$$u = \frac{\partial\Psi}{\partial y} = \frac{\partial\Psi}{\partial x}\frac{\overset{0}{\cancel{\partial x}}}{\partial y} + \frac{\partial\Psi}{\partial\eta}\frac{\partial\eta}{\partial y} = \sqrt{\nu U}\,\frac{dF}{d\eta}\sqrt{\frac{U}{\nu x}}$$

to be

$$\frac{u}{U} = \frac{dF}{d\eta} = F' \tag{7-13}$$

and

$$-v = \frac{\partial\Psi}{\partial x} = \frac{\partial\Psi}{\partial x}\frac{\overset{1}{\cancel{\partial x}}}{\partial x} + \frac{\partial\Psi}{\partial\eta}\frac{\partial\eta}{\partial x}$$

$$= \frac{1}{2}\sqrt{\frac{\nu U}{x}}\,F + \left( \sqrt{\nu U x}\,\frac{dF}{d\eta} \right)\left( -\frac{1}{2}\frac{y}{x}\sqrt{\frac{U}{\nu x}} \right)$$

or

$$\frac{v}{U}\sqrt{\frac{Ux}{\nu}} = \frac{\eta F' - F}{2} \tag{7-14}$$

The convention that $F' = dF/d\eta$ is employed. Equations (7-13) and (7-14) are substituted into the $x$-motion boundary-layer equation for our flat-plate problem to achieve

$$F''' + \tfrac{1}{2}FF'' = 0 \tag{7-15a}$$

which, although nonlinear, is an ordinary differential equation. The requirement that the plate be impermeable and that the no-slip condition be observed at the plate surface gives

$$F(0) = 0 = F'(0) \tag{7-15b}$$

and that far from the plate the free stream velocity must be approached gives

$$F'(\infty) = 1 \tag{7-15c}$$

Verification of Eq. (7-15) is the subject of Problem 7-2.

Equation (7-15), often referred to as the *Blasius equation* in honor of H. Blasius, who first solved it, can be solved numerically to obtain the solution for $F$. It is a boundary value problem—some conditions are given at the plate surface and others at infinity, and the difficulty really is in finding $F''(0)$. The results are shown in Fig. 7-2 and Table 7-1 as numerically determined by Howarth [3] and can be viewed as exact.

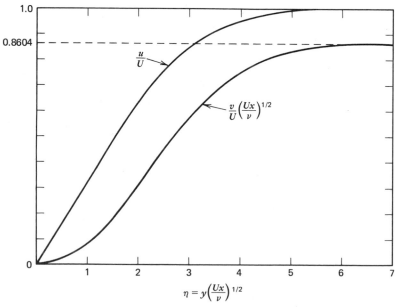

**Figure 7-2**   Velocity profiles in a laminar boundary layer on a flat plate.

**Table 7-1   Function $F(\eta)$ for Boundary Layer Along a Flat Plate at Zero Incidence**

| $\eta = y\sqrt{U_\infty/\nu x}$ | $F$ | $F' = u/U_\infty$ | $F''$ |
|---|---|---|---|
| 0 | 0 | 0 | 0.33206 |
| 0.2 | 0.00664 | 0.06641 | 0.33199 |
| 0.4 | 0.02656 | 0.13277 | 0.33147 |
| 0.6 | 0.05974 | 0.19894 | 0.33008 |
| 0.8 | 0.10611 | 0.26471 | 0.32739 |
| 1.0 | 0.16557 | 0.32979 | 0.32301 |
| 1.2 | 0.23795 | 0.39378 | 0.31659 |
| 1.4 | 0.32298 | 0.45627 | 0.30787 |
| 1.6 | 0.42032 | 0.51676 | 0.29667 |
| 1.8 | 0.52952 | 0.57477 | 0.28293 |
| 2.0 | 0.65003 | 0.62977 | 0.26675 |
| 2.2 | 0.78120 | 0.68132 | 0.24835 |
| 2.4 | 0.92230 | 0.72899 | 0.22809 |
| 2.6 | 1.07252 | 0.77246 | 0.20646 |
| 2.8 | 1.23099 | 0.81152 | 0.18401 |
| 3.0 | 1.39682 | 0.84605 | 0.16136 |
| 3.2 | 1.56911 | 0.87609 | 0.13913 |
| 3.4 | 1.74696 | 0.90177 | 0.11788 |
| 3.6 | 1.92954 | 0.92333 | 0.09809 |
| 3.8 | 2.11605 | 0.94112 | 0.08013 |
| 4.0 | 2.30576 | 0.95552 | 0.06424 |
| 4.2 | 2.49806 | 0.96696 | 0.05052 |
| 4.4 | 2.69238 | 0.97587 | 0.03897 |
| 4.6 | 2.88826 | 0.98269 | 0.02948 |
| 4.8 | 3.08534 | 0.98779 | 0.02187 |
| 5.0 | 3.28329 | 0.99155 | 0.01591 |
| 5.2 | 3.48189 | 0.99425 | 0.01134 |
| 5.4 | 3.68094 | 0.99616 | 0.00793 |
| 5.6 | 3.88031 | 0.99748 | 0.00543 |
| 5.8 | 4.07990 | 0.99838 | 0.00365 |
| 6.0 | 4.27964 | 0.99898 | 0.00240 |
| 6.2 | 4.47948 | 0.99937 | 0.00155 |
| 6.4 | 4.67938 | 0.99961 | 0.00098 |
| 6.6 | 4.87931 | 0.99977 | 0.00061 |
| 6.8 | 5.07928 | 0.99987 | 0.00037 |
| 7.0 | 5.27926 | 0.99992 | 0.00022 |
| 7.2 | 5.47925 | 0.99996 | 0.00013 |
| 7.4 | 5.67924 | 0.99998 | 0.00007 |
| 7.6 | 5.87924 | 0.99999 | 0.00004 |
| 7.8 | 6.07923 | 1.00000 | 0.00002 |
| 8.0 | 6.27923 | 1.00000 | 0.00001 |
| 8.2 | 6.47923 | 1.00000 | 0.00001 |
| 8.4 | 6.67923 | 1.00000 | 0.00000 |
| 8.6 | 6.87923 | 1.00000 | 0.00000 |
| 8.8 | 7.07923 | 1.00000 | 0.00000 |

After L. Howarth [3].

A numerical solution procedure often adopted to solve Eq. (7-15) is a so-called shooting method, in which a value of the missing $F''(0)$ is guessed and the profile (trajectory) is generated out to a large $\eta$, say, $\eta = 20$; if $F'(20) \neq 1$, an adjustment is made to $F''(0)$ and a new profile is generated. This procedure is repeated until $F'(20)$ is as close to unity as desired; at the termination the last $F''(0)$ value used is the one sought.

It should be noted in Fig. 7-1 and Eq. (7-14) that the vertical component of velocity $v$ is not zero at the edge of the boundary layer ($\eta \approx 5$), even though the free stream flows parallel to the plate. Equation (7-2) suggests that $v/U$ must be small for the boundary-layer approximation to be accurate. Thus accuracy would not be expected near the plate leading edge, where $x$ is small and, correspondingly, $v/U$ is large.

The boundary-layer thickness $\delta$ is now available since it is observed that $u/U = 0.99$ at $\eta = 5$. Realizing that $y/x = \eta/(UX/\nu)^{1/2}$, one has

$$\frac{\delta}{x} = \frac{5}{\sqrt{\text{Re}_x}} \tag{7-16}$$

The conventional practice of saying that edge of the boundary layer occurs where $u/U = 0.99$ has been followed. Clearly this is arbitrary—the 1% deviation from free stream conditions is not magical. For this case the boundary layer is slightly more a point of view than a tangible thing. In order to really verify the present analysis, which agrees well with pitot tube velocity measurements as Fig. 7-3 shows, and put things on a solid basis, something more

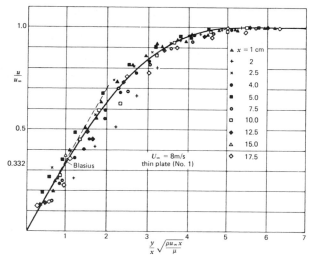

**Figure 7-3** Comparison of the velocity profile in a laminar boundary layer predicted by Blasius with measurements of Hansen. (From M. Hansen, Velocity distribution in the boundary layer of a submerged plate, NACA TM 585, 1930.)

tangible and more definitely measureable than a boundary-layer thickness must be predicted. Compute viscous drag at the plate surface. If $\tau_w$ is the shear stress, then

$$\tau_w = \mu \frac{\partial u(y=0)}{\partial y} = \frac{\rho U^2 C_f}{2g_c}$$

$$\frac{C_f}{2} = \sqrt{\frac{\nu}{UX}} F''(\eta = 0) = \frac{F''(\eta = 0)}{\sqrt{\text{Re}_x}}$$

$$\frac{C_f}{2} = \frac{0.33206}{\sqrt{\text{Re}_x}} \tag{7-17}$$

where $\text{Re}_x = UX/\nu$.

Only local values of shear stress $\tau_w$ and friction coefficient $C_f$ are given here, and the value of $F''(0) = 0.33206$ is only for this flat-plate problem with constant free-stream velocity. The inapplicability of the boundary-layer equations at the leading edge of the plate is shown by the tendency of the skin friction coefficient $C_f$ to approach infinity as $x$ approaches zero. Recall that $v(x \to 0)$ behaves similarly.

For purposes of comparison with experimental measurement, it is more convenient to have the average value of $C_f$, denoted by $\overline{C}_f$. This is obtained from the drag on one side of the plate as

$$\text{Drag} = x\tau_w = \int_0^x \tau_w \, dx = \tfrac{1}{2}\rho U^2 \overline{C}_f x$$

$$\frac{1}{x} \int_0^x \rho U^2 \frac{C_f}{2} dx = \rho U^2 \frac{\overline{C}_f}{2}$$

$$\frac{1}{x} \int_0^x \sqrt{\frac{\nu}{UX}} F''(0) dx = \frac{\overline{C}_f}{2}$$

$$2\sqrt{\frac{\nu}{UX}} F''(0) = \frac{\overline{C}_f}{2}$$

$$2\frac{C_f}{2} = \frac{\overline{C}_f}{2} \tag{7-18}$$

In other words, the average friction coefficient is twice the local friction coefficient. Thus $\overline{C}_f/2 = 0.66412/\text{Re}_x^{1/2}$.

The viscous drag that can act on the edges, also parallel to the external flow, of a flat plate not only increases the drag coefficient above the prediction of

Eq. (7-18) in its own right, but the secondary flow perturbations produced by these edge boundary layers cause yet a further increase in total drag. For references dealing with these considerations, consult Schlichting [1, p. 247].

The decrease in fluid flow through the boundary-layer region due to the viscous influence exerted by the plate is

$$\int_0^\infty \rho U \, dy - \int_0^\infty \rho u \, dy$$

Because of this decrease, the streamlines of the external flow field are displaced a distance $\delta_1$ away from the plate since the flow decrease in the boundary layer must be compensated elsewhere. Thus

$$U\delta_1 = \int_0^\infty (U - u) \, dy \qquad \text{or} \qquad \delta_1 = \int_0^\infty \left( \frac{1 - u}{U} \right) dy$$

From Eqs. (7-11) and (7-13) it follows that the displacement thickness $\delta_1$ is given by

$$\delta_1 \left( \frac{\nu x}{U} \right)^{-1/2} = \int_0^\infty (1 - F') \, d\eta = (\eta - F) \big|_{\eta = \infty}$$

From Table 7.1 it is then found that

$$\delta_1 \left( \frac{\nu X}{U} \right)^{-1/2} = 1.7208 \qquad (7\text{-}19)$$

The displacement thickness is related to the boundary-layer thickness according to Eq. (7-16) as

$$\frac{\delta_1}{\delta} = 0.3442 \approx \frac{1}{3} \qquad (7\text{-}20)$$

The decrease in $x$ momentum flowing through the boundary-layer region due to the viscous influence exerted by the plate is

$$\int_0^\infty \frac{\rho U U}{g_c} \, dy - \int^\infty \frac{\rho u u}{g_c} \, dy - \frac{\rho v(\eta = \infty)}{g_c} U$$

where, by conservation of mass, $\rho v(\eta = \infty) = \int_0^\infty (\rho U - \rho u) \, dy$. On physical grounds, this decrement in $x$-momentum flow must be equal to the drag force on the plate as Problem 7-9 discusses. The thickness, which possesses the free-stream velocity, would accommodate this decrement, is called the *momentum thickness* $\delta_2$ and is seen to be

$$U U \delta_2 = \int_0^\infty u(U - u) \, dy$$

or

$$\delta_2 = \int_0^\infty \frac{u}{U}\left(\frac{1-u}{U}\right) dy$$

This relationship can be put into the form

$$\delta_2 \left(\frac{\nu x}{U}\right)^{-1/2} = \int_0^\infty F'(1 - F') \, d\eta = F \Big|_0^\infty - \int_0^\infty (F')^2 \, d\eta$$

An integration by parts gives

$$\delta_2 \left(\frac{\nu x}{U}\right)^{-1/2} = F(1 - F') \Big|_0^\infty + \int_0^\infty FF'' \, d\eta$$

Equation (7-15) provides further information that permits this relation to be expressed as

$$\delta_2 \left(\frac{\nu x}{U}\right)^{-1/2} = -2\int_0^\infty F''' \, d\eta = -2F'' \Big|_0^\infty = 0.66412 \qquad (7\text{-}21)$$

The momentum thickness is related to the *boundary-layer thickness* according to Eq. (7-16) as

$$\frac{\delta_2}{\delta} = 0.1328 \approx \frac{1}{7} \qquad (7\text{-}22)$$

## 7.3  SIMILARITY SOLUTION FOR PARALLEL FLOW OVER A FLAT PLATE—TEMPERATURE DISTRIBUTION

The similarity solutions introduced previously to determine the velocity distribution in the laminar boundary layer for a constant-property fluid in steady flow over a flat plate can be extended to determine the temperature distribution as well. The free-stream velocity $U$ is constant (termed *parallel flow*) so that $dp/dx = 0$. The assumption of constant properties, as has been seen before, uncouples the equations of motion and energy, allowing the velocity profiles to be determined first and the temperature profiles second. The same problem as before is solved, but with the additional stipulation that the free-stream temperature and the wall temperature are both constant. These restrictions are not necessary to achieve a similarity solution, but their simplicity allows one to focus on the essential features of the solution procedure. Particularly recall that the temperature profiles are assumed to be similar as sketched in Fig. 7-1b and that the velocity components are now all known.

The energy equation for the described problem, from Eq. (7-9) with viscous dissipation ignored, is

$$u\frac{\partial T}{\partial x} + v\frac{\partial T}{\partial y} = \alpha\frac{\partial^2 T}{\partial y^2} \qquad\qquad (7\text{-}23a)$$

with boundary conditions of

$$T(x,0) = T_w \qquad \text{and} \qquad T(x,\infty) = T_\infty \qquad\qquad (7\text{-}23b)$$

With definition of a dimensionless temperature as $\theta = (T - T_w)/(T_\infty - T_w)$, Eq. (7-23) becomes

$$u\frac{\partial \theta}{\partial x} + v\frac{\partial \theta}{\partial y} = \frac{1}{\text{Pr}}\frac{\partial^2 \theta}{\partial y^2} \qquad\qquad (7\text{-}24a)$$

with

$$\theta(x,0) = 0 \qquad \text{and} \qquad \theta(x,\infty) = 1 \qquad\qquad (7\text{-}24b)$$

Note that $\theta$ occupies the same position as did $u/U$ in the $x$-motion boundary-layer equation. The boundary conditions are the same as for $u/U$, and the differential equation is also the same if $\text{Pr} = 1$. For $\text{Pr} = 1$, the solution for $\theta$ would be identical to that for $u/U$. The plate drag and heat transfer would also be expected to be closely related. The close relation expected can be seen by the parallel developments of

$$\tau_w = \mu\frac{\partial u}{\partial y}\bigg|_{y=0} \qquad q_w = -k\frac{\partial T}{\partial y}\bigg|_{y=0}$$

$$\frac{\tau_w}{\mu U} = \frac{\partial u/U}{\partial y}\bigg|_{y=0} \qquad \frac{q_w}{k(T_w - T_\infty)} = \frac{\partial\big[(T - T_w)/(T_\infty - T_w)\big]}{\partial y}\bigg|_{y=0} = \frac{\partial \theta}{\partial y}\bigg|_{y=0}$$

If $\text{Pr} = 1$, the two derivatives at the wall are equal and

$$\frac{\tau_w}{\mu U} = \frac{q_w}{k(T_w - T_\infty)}$$

or

$$\frac{U^2 C_f/2}{\mu U} = \frac{h(T_w - T_\infty)}{k(T_w - T_\infty)}$$

which can be rearranged into

$$\frac{C_f}{2} = \frac{hL}{k}\frac{1}{UL/\nu} = \frac{\text{Nu}}{\text{Re}} \qquad\qquad (7\text{-}25)$$

This relationship between the skin friction coefficient and the heat-transfer coefficient [Eq. (7-25)] is Reynolds analogy, in which $L$ is a characteristic length.

If $\Pr \neq 1$, the simple Reynolds analogy of Eq. (7-25) needs correction. For this purpose, the boundary-layer energy equation is next considered in detail. With $u/U = F'$, $(v/U)(Ux/\nu)^{1/2} = (\eta F' - F)/2$, $\eta = y(U/\nu x)^{1/2}$, and a prime denoting $d/d\eta$ Eq. (7-24) acquires the similarity form

$$\theta'' + \frac{\Pr}{2} F\theta' = 0 \tag{7-26a}$$

$$\theta(0) = 0 \tag{7-26b}$$

$$\theta(\infty) - 1 \tag{7-26c}$$

Numerical solutions to Eq. (7-26) show $\theta$ to vary with $\eta$ as depicted in Fig. 7-4. As expected from the order of magnitude analysis whose result is Eq. (7-8), the ratio of the thermal boundary-layer thickness to the velocity boundary-layer thickness varies inversely with the Prandtl number. An approximate relation for flow parallel to a flat plate is

$$\frac{\delta_T}{\delta} = \Pr^{-1/3} \qquad 0.6 \leqslant \Pr \leqslant 10 \tag{7-27}$$

The important quantity $d\theta(0)/d\eta$ depends on $\Pr$ as illustrated in Fig. 7-5. Approximate relations are

$$\frac{d\theta(0)}{d\eta} = 0.564\,\Pr^{1/2} \qquad \Pr < 0.05 \tag{7-28a}$$

$$= 0.33206\,\Pr^{1/3} \qquad 0.6 \leqslant \Pr \leqslant 10 \tag{7-28b}$$

$$= 0.339\,\Pr^{1/3} \qquad \Pr > 10 \tag{7-28c}$$

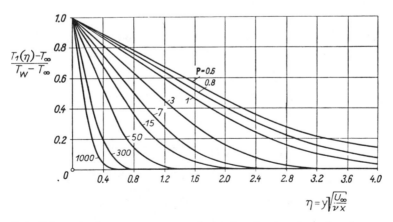

**Figure 7-4** Dimensionless temperature variation in a laminar boundary layer on a flat plate.

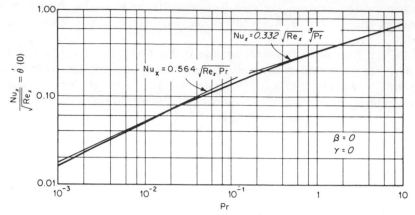

**Figure 7-5**  Local Nusselt number for a laminar boundary layer on a flat plate. (By permission from E. R. G. Eckert and R. M. Drake, *Analysis Of Heat And Mass Transfer*, McGraw-Hill, New York, 1972 [6].)

From this information the local heat-transfer coefficient $h$ is obtainable since

$$q_w = h(T_w - T_\infty) = -k\frac{\partial T(x,0)}{\partial y}$$

This gives

$$h = k\frac{\partial\theta(x,0)}{\partial y} = k\left(\frac{U}{\nu x}\right)^{1/2}\frac{d\theta(0)}{d\eta}$$

Note that unboundedly large heat flux is predicted near the plate leading edge for the same reasons that the drag is predicted to be large there—the boundary layer assumptions lose accuracy.

When the information of Eq. (7-28) is used in the foregoing development of Reynolds analogy, the needed correction for $\text{Pr} \neq 1$ is found to result in

$$\frac{\text{Nu}}{\text{Re}\,\text{Pr}^{1/2}} = 1.7\frac{C_f}{2} \qquad \text{Pr} < 0.05 \tag{7-29a}$$

$$\frac{\text{Nu}}{\text{Re}\,\text{Pr}^{1/3}} = \frac{C_f}{2} \qquad 0.6 \leqslant \text{Pr} \leqslant 10 \tag{7-29b}$$

$$\frac{\text{Nu}}{\text{Re}\,\text{Pr}^{1/3}} = 1.021\frac{C_f}{2} \qquad \text{Pr} > 10 \tag{7-29c}$$

Now that the salient features of the temperature distribution are known some common methods for solving Eq. (7-26), a boundary-value problem, are described. In the first method, it is recognized that the primary difficulty is that $d\theta(0)/d\eta$ is not specified and must be determined in the course of the solution. However, the equation is linear in $\theta$. Therefore, any two solutions to the equation can be added together in such a way as to satisfy the given boundary conditions. So, let

$$\theta = \theta_1 + C\theta_2$$

where

$$\theta_1'' + \frac{\text{Pr}}{2} F\theta_1' = 0 \qquad \theta_2'' + \frac{\text{Pr}}{2} F\theta_2' = 0$$

$$\theta_1(0) = 0 \quad \text{and} \quad \theta_1'(0) = 1 \qquad \theta_2(0) = 0 \quad \text{and} \quad \theta_2'(0) = -1$$

Note that determination of $\theta_1$ and $\theta_2$ requires only the relatively simple solution of two initial-condition problems. To satisfy the condition that $\theta(\infty) = 1$, $C$ must be given by

$$1 = \theta_1(\infty) + C\theta_2(\infty)$$

or

$$C = \frac{1 - \theta_1(\infty)}{\theta_2(\infty)}$$

The equations for $\theta_1$ and $\theta_2$ are solved numerically so that $C$ can be numerically evaluated. The missing value of $d\theta(0)/d\eta$ is then found from

$$\frac{d\theta(0)}{d\eta} = \frac{d\theta_1(0)}{d\eta} + C\frac{d\theta_2(0)}{d\eta}$$

to be

$$\frac{d\theta(0)}{d\eta} = 1 - C = \frac{\theta_1(\infty) + \theta_2(\infty) - 1}{\theta_2(\infty)} .$$

A second method makes use of the fact that $F(\eta)$ is known from a prior solution for the velocity distribution and was first used by Pohlhausen [4]. Equation (7-26) can be written as

$$\frac{1}{\theta'} \frac{d\theta'}{d\eta} = -\frac{\text{Pr}}{2} F$$

and integrated once with respect to $\eta$ from $\eta = 0$ to $\eta$ to find

$$\theta'(\eta) = \theta'(0) \exp\left(-\frac{\Pr}{2} \int_0^\eta F \, d\eta\right)$$

A second integration with respect to $\eta$ from $\eta = 0$ to $\eta$ gives

$$\theta(\eta) - \theta(\emptyset)^0 = \theta'(0) \int_0^\eta \exp\left(-\frac{\Pr}{2} \int_0^\eta F \, d\eta\right) d\eta$$

From Eq. (7-15) it is found that $-F/2 = F'''/F''$, thus allowing the simplification that

$$-\int_0^\eta \frac{F}{2} \, d\eta = \int_0^\eta \frac{F'''}{F''} \, d\eta = \ln\left[\frac{F''}{F''(0)}\right]$$

Use of this relation in that for $\theta(\eta)$ then gives

$$\theta(\eta) = \theta'(0) \int_0^\eta \left(\frac{F''}{0.33206}\right)^{\Pr} d\eta \tag{7-30}$$

It remains only to evaluate the missing $\theta'(0)$. From the condition that $\theta(\infty) = 1$ it follows that

$$\theta'(0) = 1 \bigg/ \int_0^\infty \left(\frac{F''}{0.33206}\right)^{\Pr} d\eta = a_1(\Pr) \tag{7-31}$$

It is often convenient to have the average heat-transfer coefficient $\bar{h}$ for the total heat transfer. This is obtained from the total heat transfer from one side of the plate as

$$q_w = \bar{h}x(T_w - T_\infty) = -k \int_0^x \frac{\partial T(x', 0)}{\partial y} \, dx'$$

This is rearranged into the form

$$\bar{h} = \frac{k}{x} \int_{x'=0}^x \left(\frac{U}{\nu x'}\right)^{1/2} \frac{d\theta(0)}{d\eta} \, dx'$$

$$= 2k\left(\frac{U}{\nu x}\right)^{1/2} \frac{d\theta(0)}{d\eta}$$

$$\bar{h} = 2h$$

Thus the average heat-transfer coefficient is twice the local one, just as the average drag coefficient is twice the local drag coefficient. Equation (7-29) also applies to the average heat-transfer coefficient as

$$\frac{\overline{Nu}}{Re\,Pr^{1/2}} = 1.7\frac{\overline{C}_f}{2} \qquad Pr < 0.05 \qquad \text{(7-32a)}$$

$$\frac{\overline{Nu}}{Re\,Pr^{1/3}} = \frac{\overline{C}_f}{2} \qquad 0.6 \leqslant Pr \leqslant 10 \qquad \text{(7-32b)}$$

$$\frac{\overline{Nu}}{Re\,Pr^{1/3}} = 1.021\frac{\overline{C}_f}{2} \qquad Pr > 10 \qquad \text{(7-32c)}$$

where $\overline{Nu} = \bar{h}x/k$ and $Re - Ux/\nu$. Experiment confirms the accuracy of these relations.

## 7.4  SIMILARITY SOLUTION FOR PARALLEL FLOW OVER A FLAT PLATE—VISCOUS DISSIPATION EFFECTS ON TEMPERATURE DISTRIBUTION

Beyond introductory discussion given in Sections 5.1 and 5.2 for Couette and Poiseuille flows, no investigation of the effects of viscous dissipation has heretofore been undertaken. The resultant simplicity was desirable in order to clarify ideas of analysis. Recall that the influence of viscous dissipation on heat transfer was suggested to be embodied in $q_w = h(T_w - T_{aw})$, where $T_{aw}$ is the temperature achieved by an adiabatic wall. This relationship must be verified and the relation between $T_{aw}$ and $T_\infty$ ascertained for steady parallel flow of a constant-property fluid over a flat plate.

The describing equations, including viscous dissipation, are

$$\frac{\partial u}{\partial x} + \frac{\partial v}{\partial y} = 0$$

$$u\frac{\partial u}{\partial x} + v\frac{\partial u}{\partial y} = \nu\frac{\partial^2 u}{\partial y^2}$$

$$u\frac{\partial T}{\partial x} + v\frac{\partial T}{\partial y} = \alpha\frac{\partial^2 T}{\partial y^2} + \frac{\nu}{C_p}\left(\frac{\partial u}{\partial y}\right)^2$$

with the boundary conditions that

$$u(y = 0) = 0 = v(y = 0)$$

$$T(y = 0) = T_w \quad \text{for specified wall temperature}$$

$$\frac{\partial T(y = 0)}{\partial y} = 0 \quad \text{for adiabatic wall}$$

$$u(y \to \infty) = U$$

$$T(y \to \infty) = T_\infty$$

At $y \to \infty$, $u \to U = \text{const}$ and $T \to T_\infty$.

Restricting the problem to one which admits a similarity solution, set $\eta = y\sqrt{U/\nu x}$ and $\psi = \sqrt{\nu x U}\, F(\eta)$ so that

$$u = UF'(\eta)$$

$$v = \frac{1}{2}\sqrt{\frac{U\nu}{x}}\,[\eta F' - F]$$

The energy equation then reduces to

$$\frac{d^2 T}{d\eta^2} + \frac{\text{Pr}}{2} F \frac{dT}{d\eta} = \underbrace{-\text{Pr}\frac{U^2}{C_p}(F'')^2}_{\substack{\text{forcing function due} \\ \text{to viscous dissipation}}} \qquad (7\text{-}33)$$

$$T(\eta = 0) = T_w \quad \text{for specified wall temperature}$$

$$\frac{dT(\eta = 0)}{d\eta} = 0 \quad \text{for adiabatic wall}$$

$$T(\eta \to \infty) = T_\infty$$

Equation (7-33) is linear in $T$; thus a solution for particular boundary conditions can be formed by properly adding solutions that individually satisfy equation (7-33). Therefore, let

$$T - T_\infty = C \underbrace{\theta_1(\eta)}_{\substack{\text{complementary} \\ \text{solution}}} + \frac{U^2}{2C_p}\underbrace{\theta_2(\eta)}_{\substack{\text{particular} \\ \text{solution}}}$$

where

$$\theta_1(\eta = 0) = 1 \qquad \theta_2'(\eta = 0) = 0$$

$$\theta_1(\eta \to \infty) = 0 \qquad \theta_2(\eta \to \infty) = 0$$

and

$$\theta_1'' + \frac{Pr}{2} F\theta_1' = 0 \qquad \theta_2'' + \frac{Pr}{2} F\theta_2' = -2Pr(F'')^2 \qquad (7\text{-}34)$$

Note that $\theta_1$ is the solution for $(T - T_\infty)/(T_w - T_\infty) = 1 - (T - T_w)/(T_\infty - T_w)$ without viscous dissipation and $\theta_2$ is the contribution of viscous dissipation. Recall from Eq. (7-30) that

$$\theta_1(\eta) = 1 - \theta = \frac{\int_\eta^\infty (F'')^{Pr} \, d\eta}{\int_0^\infty (F'')^{Pr} \, d\eta}$$

To express the solution for $\theta_2$ in similar fashion, recall that the Blasius equation has $-F'''/F'' = F/2$. Introduction of this relationship into the $\theta_2$ equation gives

$$\frac{d\theta_2'}{d\eta} - Pr\left(\frac{F'''}{F''}\right)\theta_2' = -2Pr(F'')^2$$

An integrating factor is $(F'')^{-Pr}$; hence a first integration yields

$$\theta_2' = -2Pr(F'')^{Pr} \int_0^\eta (F'')^{2-Pr} \, d\eta$$

A second integration then gives

$$\theta_1(\eta) = \theta_2(0) - 2Pr \int_0^\eta [F''(\xi)]^{Pr} \left\{ \int_0^\xi [F''(\tau)]^{2-Pr} \, d\tau \right\} d\xi$$

The value of $\theta_2(0)$ is obtained from the condition that $\theta_2(\infty) = 0$ as

$$\theta_2(0) = 2Pr \int_0^\infty [F''(\xi)]^{Pr} \left\{ \int_0^\xi [F''(\tau)]^{2-Pr} \, d\tau \right\} d\xi$$

Hence

$$\theta_2(\eta) = 2Pr \int_\eta^\infty [F''(\xi)]^{Pr} \left\{ \int_0^\xi [F''(\tau)]^{2-Pr} \, d\tau \right\} d\xi \qquad (7\text{-}35)$$

The important result is $\theta_2(0)$, which is displayed in Fig. 7-6. Approximations are

$$\theta_2(0) = Pr^{1/2} \qquad 0.5 \leqslant Pr < 47$$

$$= 1.9 \, Pr^{1/3} \qquad Pr \geqslant 47 \qquad (7\text{-}36)$$

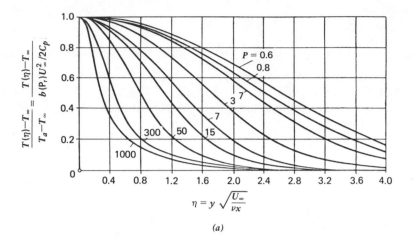

(a)

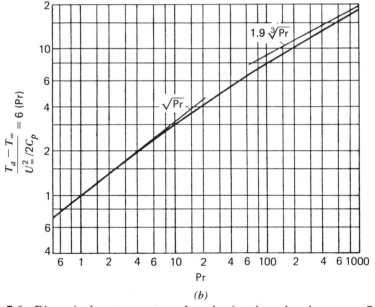

Pr

(b)

**Figure 7-6** Dimensionless temperatures for a laminar boundary layer on a flat plate: (a) temperature excess $\theta_2$ variation on an adiabatic plate; (b) adiabatic wall temperature $T_{aw}$ dependence on Prandtl number. (By permission from H. Schlichting, *Boundary-Layer Theory*, McGraw-Hill, New York, 1972 [1].)

Now focus on the use of the solutions to solve a few particular cases. First, consider the plate to be adiabatic so that at $y = 0$, $dT/dy = 0$. Now

$$T - T_\infty = C\theta_1(\eta) + \frac{U^2}{2C_p}\theta_2(\eta)$$

disregard

Here $\theta_1$ is disregarded since it does not allow the imposed boundary condition to be satisfied. Then

$$T(y = 0) - T_\infty = \frac{U^2}{2C_p}\theta_2(0)$$

(with $T_{aw}$ labeling $T(y=0)$)

or

$$\frac{T_{aw} - T_\infty}{U^2/2C_p} = \theta_2(0) = b(\mathrm{Pr}) = r \qquad (7\text{-}37)$$

Here $r$ is called the *recovery factor* and represents the fraction of the kinetic that is "recovered" at the wall. As Eq. (7-36) indicates, $r = \mathrm{Pr}^{1/2}$ for moderate Prandtl numbers. Note that $r$ depends on the velocity distribution since $r = \mathrm{Pr}$ for Couette flow from Section 5.1. Second, consider the plate temperature to be specified as $T_w$. Then the energy equation and boundary conditions are satisfied by

$$T - T_\infty = [(T_w - T_\infty) - (T_{aw} - T_\infty)]\theta_1(\eta) + \frac{U^2}{2C_p}\theta_2(\eta)$$

since at $\eta = 0$

$$T_w - T_\infty = [T_w - T_\infty - (T_{aw} - T_\infty)][1] + \frac{U^2}{2C_p}\overset{T_{aw}-T_\infty}{\theta_2(0)}$$

Now the heat-transfer coefficient is ascertained by

$$q_w = -k\frac{\partial T}{\partial y}\Big|_{y=0}$$

$$= \underbrace{k\left(\frac{U}{\nu x}\right)^{1/2}[-\theta_1'(0)]}_{h}(T_w - T_{aw})$$

$$= \underbrace{\frac{1}{\mathrm{Pr}}\left(\frac{U\rho\mu}{x}\right)^{1/2}[-\theta_1'(0)]C_p}_{h}(T_w - T_{aw}) \qquad (7\text{-}37a)$$

This is exactly the same form for $h$ as when viscous dissipation was neglected. The moral remains as it was suggested earlier; use low-speed results to evaluate $h$ [e.g., (7-29)], but use a driving temperature difference of $T_w - T_{aw}$, where $T_{aw} - T_\infty = (U^2/2C_p)r$.

The variation of properties with temperature for most gases can be accurately accounted for, according to Eckert [5], by evaluating all properties at the reference temperature.

$$T^* = 0.22(T_{aw} - T_\infty) + \frac{T_w + T_\infty}{2} \tag{7-38}$$

Equation (7-38) applies as well for turbulent boundary layers (for which $r = Pr^{1/3}$).

## 7.5  VARIABLE-PROPERTY EFFECT ON BOUNDARY LAYER FOR PARALLEL FLOW OVER A FLAT PLATE

The manner in which temperature and velocity distributions are affected by variable properties is explored for the case of steady parallel flow of a fluid over a flat plate. Although these effects are of interest for general fluids, they have been most intensively studied for gases, with which most of the high-temperature-difference applications have been found to cause large variations in properties.

The boundary-layer equations describing steady flow of a variable property fluid over a flat plate are taken from Eq. (7-9) as the continuity equation

$$\frac{\partial(\rho u)}{\partial x} + \frac{\partial(\rho v)}{\partial y} = 0 \tag{7-39a}$$

and the $x$-motion equation

$$\rho\left(u\frac{\partial u}{\partial x} + v\frac{\partial u}{\partial y}\right) = \frac{\partial(\mu\partial u/\partial y)}{\partial y} \tag{7-39b}$$

The energy equation is most generally handled by a form that involves stagnation enthalpy $H_0 = H + u^2/2$ instead of temperature. Reference to Eqs. (4-40) and (4-42) gives the energy equation as

$$\rho\left(u\frac{\partial H_0}{\partial x} + v\frac{\partial H_0}{\partial y}\right) = \frac{\partial(k\partial T/\partial y)}{\partial y} + \frac{\partial(\mu u\,\partial u/\partial y)}{\partial y} + uB_x$$

Equation (4-55) gives $DH = C_p DT$ if pressure is constant and allows the diffusive heat flux to be related to the enthalpy gradient. Introduction of this relationship into the energy equation for the case in which body forces are negligible ($B_x = 0$) results in

$$\rho\left(u\frac{\partial H_0}{\partial x} + v\frac{\partial H_0}{\partial y}\right) = \frac{\partial(\mu/\text{Pr})\partial H_0/\partial y)}{\partial y} + \frac{\partial[(1 - 1/\text{Pr})\mu u\,\partial u/\partial y]}{\partial y}$$

$$(7\text{-}39c)$$

For an impermeable plate at whose surface there is no slip and no thermal jump, the boundary-conditions imposed on Eq. (7-39) are

$$\text{At } y = 0 \quad u = 0 = v \quad \text{and} \quad H_0 = H_w \qquad (7\text{-}40a)$$

$$\text{At } y \to \infty \quad u \to U \quad \text{and} \quad H_0 = H_{0\infty} \qquad (7\text{-}40b)$$

A similarity solution is begun by defining as stream function $\psi$ as

$$\frac{\rho u}{\rho_\infty} = \frac{\partial\psi}{\partial y} \quad \text{and} \quad \frac{\rho v}{\rho_\infty} = -\frac{\partial\psi}{\partial x}$$

which automatically satisfies the continuity equation. Then, parallel to the idea utilized in Eq. (5-40) for transient change of phase with variable properties, a coordinate transformation is effected by defining

$$\eta = \left(\frac{U}{\nu_\infty x}\right)^{1/2}\int_0^y \frac{\rho}{\rho_\infty}\,dy \qquad (7\text{-}41)$$

which is also similar to the relation of Eq. (7-11) used for the constant property case and is often referred to as a *Dorodnitzyn–Stewartson* [7, 8] transformation. Additionally, the separation-of-variables idea is again employed for the stream function as

$$\psi = (\nu_\infty Ux)^{1/2}F(\eta) \qquad (7\text{-}42)$$

With the definitions of Eqs. (7-41) and (7-42), it is found that

$$u = UF'$$

$$\frac{\rho}{\rho_\infty}\frac{v}{U}\left(\frac{Ux}{\nu_\infty}\right)^{1/2} = \frac{1}{2}\left\{\left[\eta - 2\left(\frac{Ux}{\nu_\infty}\right)^{1/2}\frac{\partial}{\partial x}\int_0^y \frac{\rho}{\rho_\infty}dy\right]F' - F\right\}$$

with a prime used to denote differentiation with respect to $\eta$. Then the $x$-motion equation [Eq. (7-39b)] becomes

$$\frac{d(CF'')}{d\eta} + \frac{1}{2}FF'' = 0 \qquad (7\text{-}43)$$

where $C = \mu\rho/\mu_\infty\rho_\infty$.

Similarly, the energy equation [Eq. (7-39c)], becomes

$$\frac{d\left[C(1/\text{Pr})dH_0/d\eta\right]}{d\eta} + \frac{1}{2}F\frac{dH_0}{d\eta} = -\frac{U^2}{2}\frac{d\left[(1 - 1/\text{Pr})Cd(F')^2/d\eta\right]}{d\eta}$$

$$(7\text{-}44)$$

The boundary conditions imposed on Eqs. (7-43) and (7-44) are

$$F(0) = 0 = F'(0) \qquad \text{and} \qquad H_0(0) = H_w \qquad (7\text{-}45a)$$

$$F'(\infty) = 1 \qquad \text{and} \qquad H_0(\infty) = H_{0\infty} \qquad (7\text{-}45b)$$

Use of the enthalpy $H$ removes variable $C_p$ from the energy equation. Since enthalpy is related to temperature in a known way, it is possible to relate all other temperature-dependent properties, such as $\mu$ and Pr, to enthalpy. Beyond this point, use of the total enthalpy $H_0$ renders the effects of shear work negligible if $\text{Pr} \approx 1$. As Eqs. (7-43) and (7-44) stand, they are coupled and must be solved simultaneously.

For gases, Pr varies less with temperature than does $C$, which experiences a variation with absolute temperature closely described by $T^n$ as mentioned in Chapter 3. For liquids, Pr varies sharply with temperature, as does $\mu$. Fortunately, numerous detailed numerical solutions of the describing equations in similarity form show that less than 5% error results if $C = 1$ and Pr constant is assumed in Eqs. (7-43) and (7-44), provided all properties are evaluated at the reference enthalpy $H^*$ corresponding to the reference temperature $T^*$ of Eq. (7-38):

$$H^* = 0.22(H_{aw} - H_\infty) + \frac{H_w + H_\infty}{2} \qquad (7\text{-}46)$$

This has been demonstrated for air by Eckert [5], for nitrogen and carbon dioxide by Simon et al. [9], for the extreme conditions of ionized and dissociated gases as well as of plasmas as surveyed by Eckert and Pfender [10], and by the numerical solutions of Reshotko and Cohen [11, 12] and Levy [13]. The study by Poots and Raggett [14] for water suggests a reference temperature of

$$T^* = \begin{cases} T_w + 0.6(T_\infty - T_w) & \text{heated wall} \\ T_w + 0.69(T_\infty - T_w) & \text{cooled wall} \end{cases}$$

The solution to Eq. (7-43) with $C = 1$ is the same as for the constant-property case described by Eq. (7-15a)—the energy and $x$-motion equations are uncoupled. The energy equation [Eq. (7-44)] solution with Pr constant is achieved by a linear combination of the homogeneous and the particular solutions, in both of which $F$ is known. The homogeneous problem is taken as

$$\theta_3'' + \frac{\text{Pr}}{2}F\theta_3' = 0$$

$$\theta_3(0) = 1$$

$$\theta_3(\infty) = 0$$

whose solution is given by Eq. (7-30) subtracted from unity as

$$\theta_3 = \frac{\int_\eta^\infty (F'')^{\text{Pr}}\, d\eta}{\int_0^\infty (F'')^{\text{Pr}}\, d\eta} = \theta_1(\eta) \tag{7-47}$$

The particular solution is that derived from the inhomogeneous problem

$$\theta_4'' + \frac{\text{Pr}}{2}F\theta_4' = -(\text{Pr} - 1)\frac{d^2(F')^2}{d\eta^2}$$

$$\theta_4'(0) = 0$$

$$\theta_4(\infty) = 0$$

This can be rearranged into

$$\frac{d^2\left[\theta_4 - (F')^2\right]}{d\eta^2} + \frac{\text{Pr}}{2}F\frac{d\left[\theta_4 - (F')^2\right]}{d\eta} = -\text{Pr}\left[\frac{d^2(F')^2}{d\eta^2} + \frac{F}{2}\frac{d(F')^2}{d\eta}\right]$$

$$= -2\text{Pr}(F'')^2$$

whose solution is

$$\theta_4(\eta) = [F'(\eta)]^2 - 1 + 2\text{Pr}\int_\eta^\infty [F(\xi)]^{\text{Pr}}\left\{\int_0^\xi [F(\tau)]^{2-\text{Pr}}\, d\tau\right\} d\xi$$

$$= \left(\frac{u}{U}\right)^2 - 1 + \theta_2(\eta) \tag{7-48}$$

from Eq. (7-35).

For the case in which the wall enthalpy $H_w$ is specified, the solution can be constructed as the appropriate sum of Eqs. (7-47) and (7-48) as

$$H_0 - H_\infty = [(H_{0w} - H_{0\infty}) - (H_{0aw} - H_{0\infty})]\theta_3(\eta) + \frac{U^2}{2}\theta_4(\eta) \tag{7-49}$$

Here, the stagnation enthalpy at an insulated surface $H_{0aw}$ is defined as

$$H_{0aw} - H_{0\infty} = \frac{U^2}{2}\theta_4(0) \tag{7-50}$$

Reference to Eqs. (7-35)–(7-37) shows that $\theta_4(0) = -1 + r_H$, where $r_H \approx$ $\mathrm{Pr}^{1/2}$ is the recovery factor. Hence Eq. (7-50) gives the enthalpy at an insulated wall as

$$H_{aw} - H_\infty - \frac{U^2}{2} = \frac{U^2}{2}(r_H - 1)$$

or

$$\frac{H_{aw} - H_\infty}{U^2/2} = r_H \tag{7-51}$$

The heat flux at the flat-plate surface is found as

$$q_w = -k_w \frac{\partial T(x,0)}{\partial y}$$

$$= -\frac{k_w}{C_{p_w}} \frac{\partial H_0(x,0)}{\partial y}$$

$$= -\frac{\mu_w}{\mathrm{Pr}_w} \frac{dH_0(0)}{d\eta} \frac{\rho_w}{\rho_\infty} \left(\frac{U}{\nu_\infty x}\right)^{1/2}$$

Drawing on Eq. (7-49) to evaluate $dH_0(0)/d\eta$, one finds that this relationship for $q_w$ becomes

$$q_w = \underbrace{\frac{C_w}{\mathrm{Pr}_w}\left(\frac{U\rho_\infty\mu_\infty}{x}\right)^{1/2}[-\theta_3'(0)]}_{h_H}(H_w - H_{aw})$$

with, again, $C = \mu\rho/\mu_\infty\rho_\infty$. For $C = 1$, as it has been taken in the foregoing discussion, the severity of the variable property problem is greatly reduced if constant-property results are taken over directly with enthalpy used in place of temperature as the primary variable. Then one has

$$q_w = h_H(H_w - H_{aw}) \tag{7-52}$$

with the local heat-transfer coefficient based on enthalpy defined as

$$h_H = \frac{0.33206(U\rho_\infty\mu_\infty/x)^{1/2}}{\mathrm{Pr}^{2/3}} \tag{7-53}$$

or

$$\mathrm{Nu} = \frac{h_H x}{k} = \frac{0.33206 \mathrm{Re}^{1/2} \mathrm{Pr}^{1/3}}{C_p}$$

The average Nusselt number is

$$\overline{\mathrm{Nu}} = \frac{h_H X}{k} = \frac{0.6641 \mathrm{Re}^{1/2} \mathrm{Pr}^{1/3}}{C_p}$$

Note that when specific heat is constant, $h_H = h/C_p$ and $r_H = r$.


## 7.6  SIMILARITY SOLUTION FOR FLOW OVER A WEDGE

The boundary-layer equation describing the velocity distribution for steady parallel flow of a constant-property fluid over a flat plate was shown by Prandtl [15] in 1904 to be capable of being transformed into a single ordinary differential equation. Blasius [16] then obtained the first similarity solution, following this lead, in 1908. Because of the complexity of the flow situations for which boundary-layer applications are desired, similarity solutions for the case of constant free-stream velocity do not have the desired generality. Accordingly, extensive effort has been devoted to finding the conditions of surface geometry and free-stream velocity that admit a similarity solution. In the following discussion, constant properties are assumed.

As remarked by Hansen [17], the two-dimensional steady laminar boundary-layer equations seem to have similarity solutions only when there is no characteristic dimension— at least one dimension must extend to infinity. The wedge geometry shown in Fig. 7-7 has this necessary characteristic and is also one that occurs more commonly in practice than does a flat plate. Potential (inviscid) flow theory shows that the free-stream velocity at the wedge surface varies with distance from the tip as

$$U = Cx^m \tag{7-54}$$

where the exponent $m$ is related to the wedge angle $\beta\pi$ by

$$m = \frac{\beta}{2 - \beta} \quad \text{or} \quad \beta = \frac{2m}{1 + m}$$

When $\beta$ is positive, the free-stream velocity increases along the wedge surface; for negative $\beta$, it decreases. It is physically impossible to construct a wedge of negative opening angle, of course, but Fig. 7-7c shows that this situation could be nearly realized by sucking the boundary layer off the leading part of the wedge. Near the leading edge of a blunt object, the free-stream velocity also

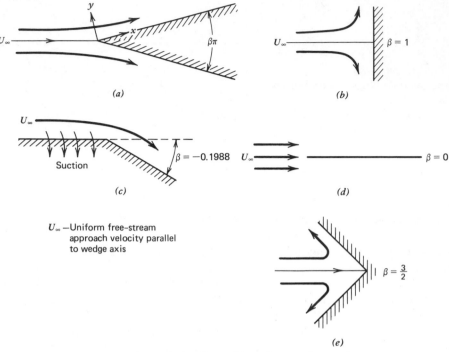

**Figure 7-7** Wedge flow with various wedge angles.

varies as $x^m$. In particular as discussed in Schlichting [1], the free-stream velocity on the surface of a cylinder and a sphere of radius $R$ is predicted by potential flow theory to vary as

$$U = 2U_\infty \sin \frac{x}{R} \approx 2U_\infty \frac{x}{R}$$

and

$$U = \frac{3}{2} U_\infty \sin \frac{x}{R} \approx \frac{3}{2} U_\infty \frac{x}{R}$$

respectively. Thus wedge flow is applicable near a stagnation point.

**Velocity Distribution**

Falkner and Skan [18] discovered the similarity transformation appropriate to wedge flow and presented numerical results in 1931. Hartree [19] performed a more detailed study and gave more accurate results in 1937. Then in 1939 Goldstein [20] investigated in detail the conditions under which a similarity transformation is possible, finding that free-stream velocity variations of the forms $U = Cx^m$ and $U = Ce^{\alpha x}(\alpha \geq 0)$ are the only allowable ones for $u/U = F(\eta)$ with $\eta = y/\phi(x)$.

In parallel with the procedure for the case of a flat plate at zero incidence, the similarity variable $\eta$ for wedge flow (where $U = Cx^m$) is defined as

$$\eta = y\left(\frac{U}{\nu x}\right)^{1/2} = y\left(\frac{C}{\nu}\right)^{1/2} x^{(m-1)/2} \tag{7-55}$$

A stream function $\psi$ that automatically satisfies the continuity equation since $u = \partial\psi/\partial y$ and $v = -\partial\psi/\partial x$ is defined by adhering to the previously used separation of variables idea as

$$\psi = (\nu U x)^{1/2} F(\eta) = (\nu C x^{m+1})^{1/2} F(\eta) \tag{7-56}$$

Appendix E discusses the considerations that lead to these definitions of $\eta$ and $\psi$ in greater detail.

The boundary-layer equations of continuity and $x$ motion are

$$\frac{\partial u}{\partial x} + \frac{\partial v}{\partial y} = 0 \tag{7-57a}$$

$$u\frac{\partial u}{\partial x} + v\frac{\partial u}{\partial y} = U\frac{dU}{dx} + \nu\frac{\partial^2 u}{\partial y^2} \tag{7-57b}$$

It is seen from Eq. (7-54) that $U\,dU/dx = mC^2 x^{2m-1}$. The boundary conditions on an impermeable wedge surface with no-slip conditions and achievement of the free-stream velocity far from the wedge surface require

$$u(y = 0) = v(y = 0) \quad \text{and} \quad u(y \to \infty) \to U = Cx^m \tag{7-57c}$$

Application of Eqs. (7-55) and (7-56) yields

$$u = \frac{\partial\psi}{\partial y} = \frac{\partial\psi}{\partial x}\frac{\partial x}{\partial y} + \frac{\partial\psi}{\partial\eta}\frac{\partial\eta}{\partial y} = UF'$$

Here again, $U = Cx^m$ and $F' = dF/d\eta$. Similarly,

$$\frac{\partial u}{\partial y} = \frac{\partial(UF')}{\partial x}\frac{\partial x}{\partial y} + \frac{\partial(UF')}{d\eta}\frac{\partial\eta}{\partial y} = U\left(\frac{U}{\nu x}\right)^{1/2} F''$$

$$\frac{\partial^2 u}{\partial y^2} = \frac{\partial(\partial u/\partial y)}{\partial x}\frac{\partial x}{\partial y} + \frac{\partial(\partial u/\partial y)}{\partial\eta}\frac{\partial\eta}{\partial y} = \frac{U^2}{\nu x}F'''$$

$$-v = \frac{\partial\psi}{\partial x} = \frac{\partial\psi}{\partial x}\frac{\partial x}{\partial x} + \frac{\partial\psi}{\partial\eta}\frac{\partial\eta}{\partial x} = U\left(\frac{Ux}{\nu}\right)^{-1/2}\left(\frac{m+1}{2}\right)\left[F - \frac{1-m}{1+m}\eta F'\right]$$

Substitution into Eqs. (7-57) finally gives the nonlinear ordinary differential equation

$$F''' + \frac{m+1}{2} FF'' + m\left[1 - (F')^2\right] = 0$$

$$F(0) = 0 = F'(0) \qquad \text{and} \qquad F'(\infty) = 1 \qquad (7\text{-}58)$$

commonly called the *Falkner–Skan equation*, after its first discoverers. The trend of the results is displayed in Fig. 7-8, where it is seen that for accelerating flows ($m$, $\beta > 0$), the boundary layer is thinner than for a flat plate at zero incidence. For decelerating flows ($m$, $\beta < 0$), the boundary layer is thicker than for a flat plate at zero incidence. At $m = -0.1988$, the velocity gradient at the wedge surface is seen to be zero, indicating that back flow and flow separation are imminent. Typical values of some useful quantities are tabulated in Table 7-2. Note that the boundary-layer thickness (really a mental thing in this case) occurs when $u/U = 0.99$, and this occurs at values of $\eta$ that depend on $m$. The variation of the boundary-layer thickness $\delta$ is available from

$$\eta_\delta = (y = \delta)\sqrt{\frac{U}{\nu x}}$$

$$\delta = \eta_\delta \sqrt{\frac{\nu x}{U}} = \eta_\delta \sqrt{\frac{\nu}{C}} x^{(1-m)/2}$$

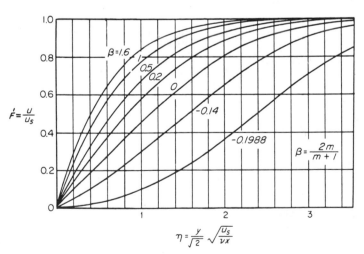

**Figure 7-8** Velocity profiles for laminar wedge-type flow based on the calculations of Hartree [19]. (By permission from E. R. G. Eckert and R. M. Drake, *Analysis of Heat and Mass Transfer*, McGraw-Hill, New York, 1972 [6].)

**Table 7-2    Laminar Wedge Flow Result [19]**

| $\beta$ | $m$ | $\eta_\delta = (Ux/\nu)^{1/2}\delta/x$ | $F''(0) = \mathrm{Re}^{1/2}C_f/2$ | |
|---|---|---|---|---|
| 2 | $\infty$ | | $\infty$ | |
| 1.6 | 5 | 1.3 | 2.6344 | |
| 1.0 | 1 | 2.4 | 1.2326 | Two-dimensional stagnation |
| 0.5 | 1/3 | 3.4 | 0.75746 | |
| 0.2 | 1/9 | 4.2 | 0.51199 | |
| 0 | 0 | 5.0 | 0.33206 | Flat plate |
| $-0.14$ | $-0.06542$ | 5.5 | 0.16372 | |
| $-0.18$ | $-0.08257$ | 5.8 | 0.08228 | |
| $-0.1988$ | $-0.09041$ | 6.5 | 0 | Separation |

For stagnation flow, $m = 1$ and $\delta = \mathrm{const.}$ A more tangible item also available is the local drag coefficient obtained from

$$\tau_w = \mu \frac{\partial u}{\partial y}\Big|_{y=0} = \frac{C_f}{2}\rho U^2$$

$$\mu \frac{C^{3/2}}{\nu^{1/2}} x^{(3m-1)/2} F''(0) = \frac{C_f}{2}\rho C^2 x^{2m}$$

$$\frac{C_f}{2} = \sqrt{\frac{\nu}{C}}\, x^{-(m+1)/2} F''(0) = \frac{F''(0)}{\sqrt{\mathrm{Re}_x}} \qquad (7\text{-}59)$$

Here again $\mathrm{Re}_x = Ux/\nu$. The total drag on one wedge surface is given by

$$\mathrm{Drag} = \int_0^x \tau_w \, dx = \mu F''(0) \int_0^x \left(\frac{U^3 x^{-1}}{\nu}\right)^{1/2} dx$$

thus allowing the average skin friction coefficient to be evaluated as

$$\bar{\tau}_w x = \mu F''(0) \frac{2}{3m+1} U\left(\frac{Ux}{\nu}\right)^{1/2}$$

$$\frac{\bar{C_f}}{2}\rho U^2 x =$$

$$\frac{\bar{C_f}}{2} = \left(\frac{2}{3m+1}\right)\frac{C_f}{2} \qquad (7\text{-}60)$$

For $-0.1988 \leqslant \beta < 0$, similarity solutions also exist (as a so-called lower branch) with negative wall shear, corresponding to the reverse flow that characterizes flows beyond the separation point as first found by Stewartson [45] and refined by Cebeci and Keller [46]. For $-0.5 < \beta < 0$, similarity solutions exist that correspond to a boundary layer bounded on one side by a free-streamline of zero shear rather than a wall—this can occur downstream from the flow separation point.

**Temperature Distribution**

The rate of convective heat transfer from a wedge surface can be determined from the laminar boundary-layer energy equation. At this point constant properties are assumed, rendering the energy equation linear in so far as temperature is concerned and allowing velocity to be taken as a known quantity, and the wedge surface temperature is considered as known.

The energy equation to be solved and its boundary conditions for a Newtonian fluid are

$$u\frac{\partial T}{\partial x} + v\frac{\partial T}{\partial y} = \alpha\frac{\partial^2 T}{\partial y} + \frac{\nu}{C_p}\left(\frac{\partial u}{\partial y}\right)^2 \tag{7-61a}$$

$$T(y = 0) = T_w \quad \text{and} \quad T(y \to \infty) \to T_\infty \tag{7-61b}$$

Inasmuch as similarity solutions are sought, a dimensionless temperature is defined as $\theta = (T - T_w)/(T_\infty - T_w)$, and the preceding energy equation is

$$u\frac{\partial \theta}{\partial x} + v\frac{\partial \theta}{\partial y} + u\theta\frac{1}{T_\infty - T_w}\frac{d(T_\infty - T_w)}{dx} + \frac{u\,dT_w/dx}{T_\infty - T_w}$$

$$= \frac{\nu}{\text{Pr}}\frac{\partial^2 \theta}{\partial y^2} + \frac{\nu}{C_p}\frac{(\partial u/\partial y)^2}{T_\infty - T_w}$$

with $\theta(y = 0) = 0$ and $\theta(y \to \infty) \to 1$. The same coordinate transformation as before $(x, y \to \xi = x, \eta = y\sqrt{U/\nu x})$ is effected; and the allowable form of $d(T_\infty - T_w)/dx$, which is consistent with a similarity solution, is determined. Then

$$\frac{\partial \theta}{\partial x} = \frac{\partial \theta}{\partial \xi}\frac{\partial \xi}{\partial x}^{\,1} + \frac{\partial \theta}{\partial \eta}\frac{\partial \eta}{\partial x} = \frac{\partial \theta}{\partial x} + \frac{\partial \theta}{\partial \eta}\left(y\sqrt{\frac{C}{\nu}}\frac{m-1}{2}x^{(m-3)/2}\right)$$

and since similarity is required, $\partial \theta/\partial x = 0$ arbitrarily. Thus

$$\frac{\partial \theta}{\partial x} = \theta y\sqrt{\frac{C}{\nu}}\frac{m-1}{2}x^{(m-3)/2} = \theta'\frac{m-1}{2}\frac{\eta}{x}$$

$$\frac{\partial \theta}{\partial y} = \frac{\partial \theta}{\partial \xi}\frac{\partial \xi}{\partial y}^{\,0} + \frac{\partial \theta}{\partial \eta}\frac{\partial \eta}{\partial y} = \theta'\sqrt{\frac{C}{\nu}}x^{(m-1)/2}$$

$$\frac{\partial^2 \theta}{\partial y^2} = \frac{\partial}{\partial y}\left(\frac{\partial \theta}{\partial y}\right) = \frac{\partial}{\partial \xi}(\quad)\frac{\partial \xi}{\partial y}^{\,0} + \frac{\partial}{\partial \eta}(\quad)\frac{\partial \eta}{\partial y} = \theta''\frac{C}{\nu}x^{m-1}$$

Also, $u = Cx^m F$ and $v = [(m+1)/2]\sqrt{\nu C}\, x^{(m-1)/2}[\{(1-m)/(1+m)\}\eta F' - F]$; thus the energy equation [Eq. (7-61)] becomes

$$
Cx^m F'\left[\theta' \frac{m-1}{2}\frac{\eta}{x}\right] + \frac{m+1}{2}\sqrt{C\nu}\, x^{(m-1)/2}\left[\frac{1-m}{1+m}\eta F' - F\right]
$$

$$
\times\left[\theta\sqrt{\frac{C}{\nu}}\, x^{(m-1)/2}\right] + Cx^m F'\frac{\theta}{T_\infty - T_w}\frac{d(T_\infty - T_w)}{dx}
$$

$$
+ Cx^m \frac{F'}{T_\infty - T_w}\frac{dT_w}{dx} = \frac{\nu}{\mathrm{Pr}}\frac{C}{\nu}x^{(2m-2)/2}\theta'' + \frac{Cx^{2m}}{C_p(T_\infty - T_w)}(F')^2
$$

Rearrangement gives

$$
\underbrace{\theta'' + \frac{m+1}{2}\mathrm{Pr}\, F'\theta}_{\mathrm{func}(\eta)}
$$

$$
= \underbrace{\frac{x}{T_\infty - T_w}\left[(\theta-1)\frac{d(T_\infty - T_w)}{dx} + \frac{dT_\infty}{dx}\right]\mathrm{Pr}\, F' - \frac{\mathrm{Pr}}{C_p}C^2(F'')^2\frac{x^{2m}}{T_\infty - T_w}}_{\text{possible func}(x)}
$$

$$\tag{7-62}$$

It is seen that Eq. (7-62) possesses the sought after similarity form in which $\theta = \theta(\eta)$ only if special conditions are satisfied by the right-hand side.

Neglecting viscous dissipation [the last term on the right-hand side of Eq. (7-62)] for the moment, it is seen that similarity is achievable if

$$
\underbrace{\frac{\theta'' + [(m+1)/2]\mathrm{Pr}\, F\theta'}{\mathrm{Pr}\, F'}}_{\mathrm{func}(\eta)} = \underbrace{\frac{x}{T_\infty - T_w}\left[(\theta-1)\frac{d(T_\infty - T_w)}{dx} + \frac{dT_\infty}{dx}\right]}_{\text{possible func}(x)}
$$

This leads to three major cases. The first case is of constant-temperature difference $T_\infty - T_w = \mathrm{const}$, which requires

$$
\frac{x}{T_\infty - T_w}\frac{dT_\infty}{dx} = \gamma
$$

where $\gamma$ is a separation constant. Or

$$
T_\infty = K_{1\infty} + (T_\infty - T_w)\gamma \ln x
$$

$$
T_w = K_{1w} + (T_\infty - T_w)\gamma \ln x
$$

The energy equation without viscous dissipation is then

$$\theta'' + \frac{m+1}{2} \Pr F\theta' - \gamma \Pr F' = 0$$

$$\theta(0) = 0 \quad \text{and} \quad \theta(\infty) = 1 \tag{7-63}$$

Similarity is not preserved if viscous dissipation is included.

The second case is of constant wedge temperature $T_w = $ const, which requires that, neglecting viscous dissipation again,

$$\frac{\theta'' + [(m+1)/2]\Pr F\theta'}{\Pr F'\theta} = \frac{x}{T_\infty - T_w}\frac{dT_\infty}{dx} = \gamma$$

where $\gamma$ is a separation constant. Or

$$T_\infty = T_w + K_2 x^\gamma$$

The energy equation with viscous dissipation is then

$$\theta'' + \frac{m+1}{2}\Pr F\theta' - \gamma \Pr F'\theta = -\Pr E(F'')^2 x^{2m-\gamma}$$

$$\theta(0) = 0 \quad \text{and} \quad \theta(\infty) = 1 \tag{7-64}$$

with $E = C^2/C_p K_2$. It is seen that a similarity solution exists when viscous dissipation is included only if $2m = \gamma$; the free-stream temperature and velocity must vary in a precise way. If $\gamma \neq 0$, the surface temperature varies along the wedge and numerical methods are required.

The third case is of constant free-stream temperature $T_\infty = $ const, which requires, neglecting viscous dissipation again, that

$$\frac{\theta'' + [(m+1)/2]\Pr F\theta'}{\Pr F'(\theta-1)} = -\frac{x}{T_\infty - T_w}\frac{dT_w}{dx} = \gamma$$

or

$$T_w = T_\infty - K_3 x^\gamma$$

The energy equation with viscous dissipation is then

$$\theta'' + \frac{m+1}{2}\Pr F\theta' - \gamma \Pr F'(\theta-1) = -\Pr E(F'')^2 x^{2m-\gamma} \tag{7-65}$$

with $E = C^2/C_p K_3$, $\theta(0) = 0$, and $\theta(\infty) = 1$. As for the case of $T_w = $ const, no similarity solution exists when viscous dissipation is included, unless $2m = \gamma$. Stojanovic [21] investigated the conditions for the existence of similarity solutions to the energy equation for not only wedge flows, but also for rotating bodies of revolution and for bodies of revolution in a rotating fluid.

The heat flow from a wedge surface is obtained from the solution to the appropriate similarity form of the energy equation according to

$$q_w = -k\frac{\partial T(y=0)}{\partial y} = k(-T_\infty + T_w)\left(\frac{Cx^{m-1}}{\nu}\right)^{1/2}[\theta'(0)] \quad (7\text{-}66)$$

For the most common second and third cases just discussed in which $T_w - T_\infty = Kx^\gamma$, it follows that $q_w \sim x^{(2\gamma+m-1)/2}$. Then if wall heat flux is specified to be a constant, it is necessary for $\gamma = (1-m)/2$ and $T_w - T_\infty = Kx^{(1-m)/2}$. For parallel flow over a flat plate where $m = 0$, a nonzero constant wall heat flux requires that $T_w - T_\infty = Kx^{1/2}$.

A local heat-transfer coefficient is found by setting $q_w$ from Eq. (7-66) equal to $h(T_w - T_\infty)$. This yields

$$h = k\left(\frac{Cx^{m-1}}{\nu}\right)^{1/2}\theta'(0)$$

Note that in two-dimensional stagnation flow ($m = 1$, $\beta = 1$) the heat-transfer coefficient is constant, indicating that the thermal boundary layer is of constant thickness as was previously remarked for the velocity boundary layer. In dimensionless terms this relation can be expressed as

$$\mathrm{Nu} = \theta'(0)\mathrm{Re}^{1/2}$$

where $\mathrm{Nu} = hx/k$ and $\mathrm{Re} = Ux/k = Cx^{m+1}/\nu$. The average heat-transfer coefficient is often more useful and is found from

$$\bar{q}_w x = \int_0^x q_w\,dx$$

$$\bar{h}(T_w - T_\infty)x = \int_0^x q_w\,dx$$

to be related to the local coefficient by

$$\bar{h} = \frac{2h}{2\gamma + m + 1}$$

From this it follows that

$$\overline{\mathrm{Nu}} = \frac{\bar{h}x}{k} = \frac{2\theta'(0)}{2\gamma + m + 1}\mathrm{Re}^{1/2} \quad (7\text{-}67)$$

The dimensionless temperature distribution for the case of constant free stream and wall temperatures, with viscous dissipation neglected, is given by Eckert [22], who solved Eq. (7-65) with $\gamma = 0$. Figure 7-9 illustrates the results, showing that the thermal boundary layer thins as the wedge angle increases in a manner similar to that depicted in Fig. 7-8 for velocities. Extensive tabulated numerical results by Evans [23] give the local Nusselt number over $Re^{1/2}$, again for constant free stream and wall temperatures and neglecting viscous dissipation, which are combined in Fig. 7-10 and are briefly

$$\lim_{Pr \to 0} \frac{Nu}{Re^{1/2}} = \frac{Pr^{1/2}}{[\pi(1 - \beta/2)]^{1/2}}$$

$$\lim_{Pr \to \infty} \frac{Nu}{Re^{1/2}} = 0.224 Pr^{1/4} \quad \text{and} \quad \beta = -0.199$$

$$\lim_{Pr \to \infty} \frac{Nu}{Re^{1/2}} \sim Pr^{1/3}* \quad \text{and} \quad \beta \neq -0.199$$

represented in Table 7-3. Levy [25] solved Eq. (7-65) for $\gamma = 0$, finding temperature distributions such as are illustrated in Fig. 7-11 for parallel flow over a flat plate ($\beta = 0$). These trends, also observed for $\beta \neq 0$, show that the thickness of the thermal boundary layer is reduced as $\gamma$ increases just as was previously found to be true for the parameter Pr. For $\gamma > -1/(2 - \beta)$, heat flows always into the wall despite the wall temperature's continual excess over

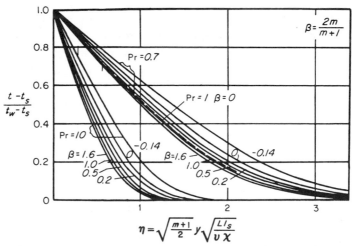

**Figure 7-9** Temperature profiles for laminar wedge-flow with constant surface temperature. [By permission from E. R. G. Eckert and R. M. Drake, *Analysis Of Heat And Mass Transfer*, McGraw-Hill, New York, 1972 (after E. Eckert, VDI-Forschungsh., p. 416, 1942) [6].]

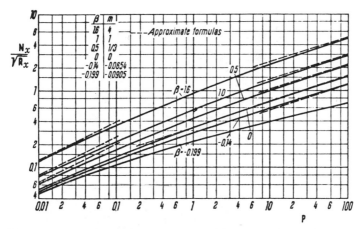

**Figure 7-10** Local Nusselt number versus Prandtl number for wedge flow in which $m = \beta/(2 - \beta)$ with constant wall temperature and without viscous dissipation:

$$\lim_{\mathrm{Pr} \to 0} \frac{\mathrm{Nu}}{\mathrm{Re}^{1/2}} = \frac{\mathrm{Pr}^{1/2}}{[\pi(1 - \beta/2)]^{1/2}}$$

$$\lim_{\mathrm{Pr} \to \infty} \frac{\mathrm{Nu}}{\mathrm{Re}^{1/2}} = 0.224\mathrm{Pr}^{1/4}, \qquad \beta = -0.199$$

$$\approx \mathrm{Pr}^{1/3}, \qquad \beta \neq -0.199 \quad (\text{*see reference 6, p. 284 or reference 24}).$$

(By permission from H. Schlichting, *Boundary-Layer Theory*, McGraw-Hill, New York, 1968 [1].)

**Table 7-3** $\mathrm{Nu}/\mathrm{Re}^{1/2}$ for Laminar Wedge Flow [23]

| $m$ | Pr | | | | |
|---|---|---|---|---|---|
| | 0.7 | 0.8 | 1.0 | 5.0 | 10.0 |
| $-0.0753$ | 0.242 | 0.253 | 0.272 | 0.457 | 0.570 |
| 0 | 0.292 | 0.307 | 0.332 | 0.585 | 0.730 |
| 0.111 | 0.331 | 0.348 | 0.378 | 0.669 | 0.851 |
| 0.333 | 0.384 | 0.403 | 0.440 | 0.792 | 1.013 |
| 1.0 | 0.496 | 0.523 | 0.570 | 1.043 | 1.344 |
| 4.0 | 0.813 | 0.858 | 0.938 | 1.736 | 2.236 |

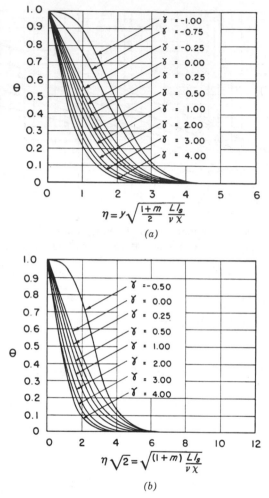

Figure 7-11 Dimensionless temperature distribution with variable surface temperature at Pr = 0.7 for (a) stagnation flow and (b) flat plate flow. [By permission from S. Levy, Heat transfer to constant-property laminar boundary-layer flows with power-function free-stream velocity and wall-temperature variations, *J. Aeronaut. Sci.* **19**, 341–348 (1952) [25].]

the free-stream temperature. This is a consequence of a fluid particle heated to nearly the wall temperature being convected downstream to a place at which wall temperature is lower. Then, since the fluid particle is warmer than the wall, heat flows into the wall. Such an occurrence results in negative heat-transfer coefficients and means only that the temperature gradient at the wall is no longer proportional to $T_w - T_\infty$. Levy suggests that the local Nusselt number results be correlated by

$$\frac{\text{Nu}}{\text{Re}^{1/2}} = B(m, \gamma)\text{Pr}^\lambda \tag{7-68}$$

where, for $-0.0904 \leqslant m \leqslant 4$,

$$B(m, \gamma) = 0.57\left(\frac{2m}{m+1} + 0.205\right)^{0.104}\left(1 + \frac{2\gamma}{m+1}\right)^{0.37+0.12m/(m+1)}\left(\frac{m+1}{2}\right)^{1/2}$$

and $\lambda \approx \frac{1}{3}$ for $\text{Pr} \approx 1$, but with variations as reported in Table 7-4 that are based on results for $0.7 < \text{Pr} \leqslant 4$: accuracy is $\pm 5\%$, generally, but deteriorates at large negative $\gamma$ values. The prediction of Eq. (7-68) is depicted in Fig. 7-12.

The work of Brun [26] can be consulted to obtain results that include the effects of viscous dissipation.

The results for wedge flow with varying wall temperature can be applied to the more general wall-temperature variation

$$T_w - T_\infty = a_0 + a_1 x + a_2 x^2 + \cdots = \sum_{\gamma=0}^{\infty} a_\gamma x^\gamma$$

as expounded by Chapman and Rubesin [27].

Advantage is taken of the linearity of the energy equation in terms of temperature. Thus the wall heat flux for $T_w - T_\infty = a_n x^n$, being given by Eq. (7-66), is

$$q_w = k(T_w - T_\infty)\theta'(0)\left(\frac{Cx^{m-1}}{\nu}\right)^{1/2}$$

$$= h(T_w - T_\infty)$$

Summation of the individual contributions then leads to

$$q_w = h_0 a_0 + h_1 a_1 x + h_2 a_2 x^2 + \cdots$$

or

$$\frac{q_w x}{k} = a_0 \text{Nu}_0 + a_1 x \text{Nu}_1 + a_2 x^2 \text{Nu}_2 + \cdots = \sum_{\gamma=0}^{\infty} a_\gamma x^\gamma \text{Nu}_\gamma$$

**Table 7-4   Values of $\gamma$ for Eq. (7-68) [25]**

| $\beta$ | 1.6 | 1.0 | 0 | $-0.199$ |
|---|---|---|---|---|
| $\lambda$ | 0.367 | 0.355 | 0.327 | 0.254 |

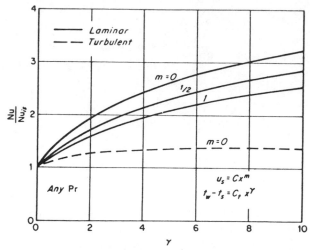

**Figure 7-12**  Ratio of Nusselt number for a surface with locally varying temperature to that for an isothermal surface for wedge flow. (By permission from E. R. G. Eckert and R. M. Drake, *Analysis Of Heat And Mass Transfer*, McGraw-Hill, New York, 1972 [6].)

where $Nu_\gamma$ is the Nusselt number which corresponds to $T_w - T_\infty = ax^\gamma$ as found, for example, from Eq. (7-68). A practical difficulty with this procedure is that the surface temperature may not be well approximated over the required range of $x$ by a finite number of terms, each of the form $a_\gamma x^\gamma$. Further detailed discussion of the effect of an arbitrary variation of surface temperature is deferred for treatment by the approximate integral methods. Surveys of the many methods that have been developed for calculating thermal boundary layers along isothermal surfaces are available in Schlichting [1, p. 295] and Eckert [6, p. 321].

**Local Similarity**

On a physical basis the difficulties encountered in calculations with $x$-dependent free-stream velocities and surface temperatures can be ascribed to the fact that local wall drag and heat flux are determined by upstream events as well as by local differences in velocities and temperatures. This difficulty is more pronounced for laminar than for turbulent flows in boundary layers and in ducts.

The concept of local similarity provides a means by which free-stream velocity and wall temperature variations that do not admit similarity solutions can be approximately taken into account. Local similarity, as its name suggests, ignores the effects of upstream events and presumes that local conditions predominantly influence local boundary-layer behavior. Spalding and Pun [28] compared 15 computational schemes applied to the prediction of Nusselt numbers on the forward part of a constant temperature cylinder in cross flow and found the best local similarity methods to give results with 5% error. Figure 7-13 shows some typical results and shows the local similarity method as described by Eckert and Livingood [29] to be an often-adequate approximation. Flow separation on the back side of the cylinder renders the boundary-layer equations (in the forms heretofore discussed) inapplicable there. A recent rapid method [44] is more accurate.

The solutions for wedge flow are used to obtain approximate heat-transfer coefficients for flow about a cylinder with normal impingement, for example, by presuming that the coefficient at any location is identical to the coefficient

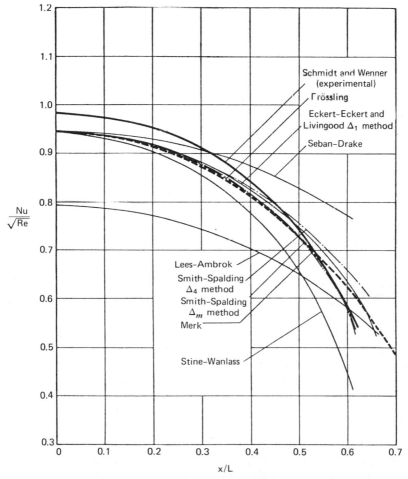

**Figure 7-13**  $\mathrm{Nu}/\mathrm{Re}^{1/2}$ versus $x/L$ for circular cylinder in cross-flow at $\mathrm{Pr} = 0.7$ as predicted by different methods; $x$ is circumferential distance from the forward stagnation point. [By permission from D. B. Spalding and W. M. Pun, A review of methods for predicting heat-transfer coefficients for laminar uniform property boundary layer flows, *Internatl. J. Heat Mass Transf.* **5**, 239–249 (1962) [28].]

on a wedge for which, at the same distance from the stagnation point, the free-stream velocity and its gradient are the same as those on the cylinder. To illustrate the procedure, consider flow normal to a cylinder for which

$$U = 2U_\infty \sin \frac{x}{R}$$

It is presumed that there is an equivalent wedge flow for which $U = Cx^m$, and $m$ is to be determined; once it is known, $h$ can be estimated from wedge flow results. Taking the derivation of the equivalent wedge velocity, one obtains

$$\frac{dU}{dx} = Cx^{m-1}m$$

or

$$x\frac{dU/dx}{U} = m$$

from which the local value of $m$ to be used to find the local $h$ can be determined. This elementary local-similarity procedure appears to give results with about 15% deviation from data [43] on flow normal to a constant-temperature cylinder. Its application is required in Problem 7-13. (See Morgan [58] for yawed cylinders.) For cross flow over cylinders, Churchill and Bernstein [43] recommend

$$\overline{Nu} = 0.3 + 0.62Re^{1/2}\,Pr^{1/3}\left[1 + \left(\frac{Re}{282,000}\right)^{5/8}\right]^{4/5}\left[1 + \left(\frac{0.4}{Pr}\right)^{2/3}\right]^{-1/4},$$

$$Re\,Pr > 0.2$$

As discussed by Eckert and Livingood [29], this procedure can be improved if consideration is taken of the flow and temperature events in the boundary layer upstream from the surface location in question. The additional requirement that the rate of increase of the thermal boundary-layer thickness be the same on the equivalent wedge and on the cylinder at the point in question leads to a substantial improvement in accuracy.

Cogent discussions of local similarity procedures are available by Kays and Crawford [40] and Schlichting [1]. A particularly convenient and yet accurate scheme based on an integral method is the subject of Problem 8-15.

## 7.7  ROTATIONALLY SYMMETRIC STAGNATION FLOW AND MANGLER'S TRANSFORMATION

The flow of a fluid past a body of revolution is of considerable importance. The reentry of a space vehicle into the Earth's atmosphere is just one example of many possible applications. For simplicity, only the case in which the fluid

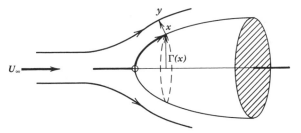

**Figure 7-14** Coordinate system for flow parallel to the axis of a body of revolution.

flows parallel to the body axis is considered, as illustrated in Fig. 7-14, and steady flow of a constant property fluid is assumed.

The laminar boundary layer that forms on the solid surface can be taken to be thin. Application of the boundary-layer assumptions to the continuity, $x$-motion, and energy equations for cylindrical coordinates from Appendix D gives the laminar boundary-layer equations as

$$\frac{\partial(ur)}{\partial x} + \frac{\partial(vr)}{\partial y} = 0 \tag{7-69a}$$

$$u\frac{\partial u}{\partial x} + v\frac{\partial u}{\partial y} = U\frac{dU}{dx} + v\frac{\partial^2 u}{\partial y^2} \tag{7-69b}$$

$$u\frac{\partial T}{\partial x} + v\frac{\partial T}{\partial y} = \alpha\frac{\partial^2 T}{\partial y^2} + \frac{\mu}{\rho C_p}\left(\frac{\partial u}{\partial y}\right)^2 \tag{7-69c}$$

Here $r$ is the radial distance from the body centerline to the surface location in question.

Only the continuity equation is changed from the two-dimensional form to which previous discussions have been confined. To see more clearly that this is so, consider the extreme rotationally symmetric stagnation case depicted in Fig. 7-15. There, the continuity equation in cylindrical coordinates is

$$\frac{\partial(rv_r)}{\partial r} + r\frac{\partial(\rho v_z)}{\partial z} = 0$$

Assuming that $r = x$, $z = y$, $v_r = u$, and $v_z = v$ then gives Eq. (7-69a) since $r$ is independent of $z$ and can be taken inside the $z$ derivative. Similarly, the $r$ equation of motion in cylindrical coordinates under the boundary-layer assumption is

$$v_r\frac{\partial v_r}{\partial r} + v_z\frac{\partial v_z}{\partial z} = \frac{-dp}{dr} + \mu\frac{\partial^2 v_r}{\partial z^2}$$

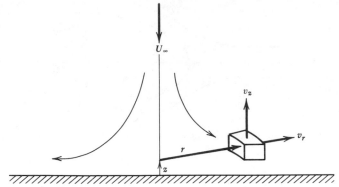

**Figure 7-15**   Coordinate system for stagnation flow perpendicular to a plane.

which becomes, with the preceding substitutions, as written in Eq. (7-69c). In like manner, the energy equation is given by Eq. (7-69c). A mass balance on a pillbox-shaped control volume centered at the origin and extending just above the boundary layer shows that the velocity just outside the boundary layer varies as $U = Cr = Cx$.

The solution to the rotationally symmetric laminar boundary-layer equations of Eq. (7-69) is more involved because, since $r$ depends on $x$, the body contour is explicitly involved. Fortunately, Mangler's coordinate transformation [30] provides a relationship between the two-dimensional boundary-layer equations and those for rotationally symmetric flow. In Mangler's transformation the equivalent two-dimensional coordinates ($\bar{x}$ and $\bar{y}$) are related to those for rotational symmetry by

$$\bar{x} = L^{-2} \int_0^x r^2(x)\, dx \qquad \text{and} \qquad \bar{y} = L^{-1} r(x) y \qquad (7\text{-}70)$$

where $L$ is an arbitrary constant with a dimension of length. The basis for these transformations can be glimpsed by referring briefly to Eq. (7-69b) in the form

$$u\frac{\partial u}{\partial x} + \cdots = \cdots + \nu \frac{\partial^2 u}{\partial y^2}$$

Application of the chain rule for $\bar{x} = \bar{x}(x, y)$ and $\bar{y} = \bar{y}(x, y)$ puts this into the form $u(\partial u/\partial \bar{x})(\partial \bar{x}/\partial x) + \cdots = \cdots + (\partial \bar{y}/\partial y)^2 \partial^2 u/\partial \bar{y}^2$. If this is to have the same form as for a two-dimensional case, $\partial \bar{x}/\partial x = (\partial \bar{y}/\partial y)^2$. Furthermore, if the boundary condition $u(y = \infty) = (L/r)\partial \psi/\partial y = (L/r)(\partial \bar{y}/\partial y)\partial \psi/\partial \bar{y}$ is to have the same form as the two-dimensional case, it is necessary that $\partial \bar{y}/\partial y = r/L$. Hence $d\bar{y} = (r/L)\, dy$ and $d\bar{x} = (r^2/L^2)\, dx$ as stated previously.

To show that Eq. (7-70) is successful in removing the influence of $r(x)$ from Eq. (7-69), a stream function to satisfy Eq. (7-69a) is first defined as

$$\frac{\partial \psi}{\partial y} = \frac{ru}{L} \quad \text{and} \quad -\frac{\partial \psi}{\partial x} = \frac{rv}{L}$$

Then Eq. (7-69b) becomes

$$\frac{L}{r}\frac{\partial \psi}{\partial y}\frac{\partial}{\partial x}\left(\frac{L}{r}\frac{\partial \psi}{\partial y}\right) - \frac{L}{r}\frac{\partial \psi}{\partial x}\frac{\partial}{\partial y}\left(\frac{L}{r}\frac{\partial \psi}{\partial y}\right) = U\frac{dU}{dx} + \nu\frac{\partial^2}{\partial y^2}\left(\frac{L}{r}\frac{\partial \psi}{\partial y}\right)$$

If the coordinate transformation of Eq. (7-70) is effected with

$$\frac{\partial}{\partial y} = \frac{\partial}{\partial \bar{x}}\frac{\partial \bar{x}}{\partial y}^{\,0} + \frac{\partial}{\partial \bar{y}}\frac{\partial \bar{y}}{\partial y}^{\,r/L} = \frac{r}{L}\frac{\partial}{\partial \bar{y}}$$

$$\frac{\partial}{\partial x} = \frac{\partial}{\partial \bar{x}}\frac{\partial \bar{x}}{\partial x}^{\,r^2/L^2} + \frac{\partial}{\partial \bar{y}}\frac{\partial \bar{y}}{\partial x}^{\,(1/L)\,dr/dx} = \left(\frac{r}{L}\right)^2\frac{\partial}{\partial \bar{x}} + \frac{r'}{L}\frac{\partial}{\partial \bar{y}}$$

then results in, with $r' = dr/dx$,

$$\frac{L}{r}\left(\frac{r}{L}\frac{\partial \psi}{\partial \bar{y}}\right)\left[\left(\frac{r}{L}\right)^2\frac{\partial^2 \psi}{\partial \bar{x}\,\partial \bar{y}} + \frac{r'}{L}\frac{\partial^2 \psi}{\partial \bar{y}^2}\right] - \frac{L}{r}\left[\left(\frac{r}{L}\right)^2\frac{\partial \psi}{\partial \bar{x}} + \frac{r'}{L}\frac{\partial \psi}{\partial \bar{y}}\right]\left[\frac{r}{L}\frac{\partial^2 \psi}{\partial \bar{y}^2}\right]$$

$$= U\left(\frac{r}{L}\right)^2\frac{dU}{dx} + \nu\left(\frac{r}{L}\right)^2\frac{\partial^3 \psi}{\partial \bar{y}^3}$$

Simplification of this equation reduces it to the form, identical with that for two-dimensional boundary layers (when $u = \partial\psi/\partial\bar{y}$ and $v = -\partial\psi/\partial\bar{x}$),

$$\frac{\partial \psi}{\partial \bar{y}}\frac{\partial^2 \psi}{\partial \bar{x}\,\partial \bar{y}} - \frac{\partial \psi}{\partial \bar{x}}\frac{\partial^2 \psi}{\partial \bar{y}^2} = U\frac{dU}{dx} + \nu\frac{\partial^3 \psi}{\partial \bar{y}^3}$$

and verifies the transformation's claimed property. It is also true that the velocities at $x$, $y$ in the rotationally symmetric case equal the velocities at $\bar{x}$, $\bar{y}$ in the two-dimensional case. This follows from

$$u(x, y) = \frac{\partial \psi}{\partial y}\frac{L}{r} = \left(\frac{r}{L}\frac{\partial \psi}{\partial \bar{y}}\right)\frac{L}{r} = \frac{\partial \psi}{\partial \bar{y}} = u(\bar{x}, \bar{y})$$

For the rotationally symmetric stagnation flow of Fig. 7-15, representing an infinite stream impinging perpendicularly on a flat plate, $r = x$ and $U = Cx$. Then Eq. (7-70) gives $\bar{x} = x^3/3L^2$ and $\bar{y} = xy/L$. Furthermore,

$$U = C(3L^2\bar{x})^{1/3}$$

Thus it is seen that the equivalent two-dimensional flow is one for which $U \sim x^m$ with $m = \frac{1}{3}$, or $\beta = \frac{1}{2}$, which is a wedge of included angle $\pi/2$. For convenience, it is desirable to have $\bar{x} = x$ at one given nonzero point (say, $x_0$), so that $L = x_0/3^{1/2}$. At that point the corresponding value of $\bar{y}$ is related to $y$ as $3^{-1/2}\bar{y} = y$, showing that the rotationally symmetric boundary layer is thinner than the equivalent two-dimensional boundary layer by the factor $3^{1/2}$. The heat transfer coefficient is obtained from the equivalent two-dimensional problem solution as

$$q_w = -k\frac{\partial T(x, y = 0)}{\partial y}$$

$$h(T_w - T_\infty) = \underbrace{\left[-k\frac{\partial T(\bar{x}, \bar{y} = 0)}{\partial \bar{y}}\right]}_{h_{2-D}\left(\bar{x},\, m = \frac{1}{3}\right)(T_w - T_\infty)}\frac{r}{L}$$

Reference to Table 7-2 at $\mathrm{Pr} = 1$ and $m = \frac{1}{3}$ gives

$$h_{2-D}\left(\bar{x},\, m = \frac{1}{3}\right) = 0.44k\left(\frac{U}{\nu\bar{x}}\right)^{1/2}$$

Hence

$$h = 0.44k\left(\frac{U}{\nu\bar{x}}\right)^{1/2}\frac{r}{L}$$

or

$$h = 0.44k(3^{1/2})\left(\frac{C}{\nu}\right)^{1/2} \tag{7-71}$$

In dimensionless form, this result for rotationally symmetric stagnation flow can be expressed as

$$\frac{\mathrm{Nu}}{\mathrm{Re}^{1/2}\mathrm{Pr}^{0.4}} = 0.76 \tag{7-72}$$

where $\mathrm{Nu} = hx/k$ and $\mathrm{Re} = Ux/\nu$ and the Prandtl number influence is approximately accounted for by a 0.4 exponent [11], a procedure that is of doubtful improvement over an exponent of $\frac{1}{3}$ but that is conventional, nevertheless. The constant value of $h$ in Eq. (7-71) is a result of a boundary layer of constant thickness. This can be seen from the definition of

$$\eta = \bar{y}\left(\frac{U}{\nu\bar{x}}\right)^{1/2}$$

which, with $\bar{x} \sim x^3$ and $\bar{y} \sim xy$, gives

$$\eta_\delta \sim \delta$$

Use of the Mangler transformation and the concept of local similarity in cases of flow over an arbitrary body in situations in which properties are variable (to the point that the gas dissociates) has been studied by Lees [31] and Eckert and Tewfik [32]. The survey by Dewey and Gross [33] provides additional references and discussion of various local-similarity schemes, in addition to extensive tables of numerical solutions to the laminar boundary-layer equations. Of particular interest is a development of the idea that the effect of slightly nonsimilar effects can be taken into account by expanding the full boundary-layer equations in terms of small parameters that describe departure from exact similarity. The resulting coefficients of the small parameters are obtained from similarity equations that need be calculated only once.

The simplification provided by a well-chosen transformation is evident. Sun [34] described the most prominent ones in his survey, which merit mention here. The Meksyn–Görtler transformation [1, p. 164] is reminiscent of local-similarity and allows accounting for nonsimilarity effects in the equation of motion, setting $\xi = \int_0^x (U/v)\,dx$ and $\eta = yU/[2v\int_0^x U\,dx]$ together with $\psi = v(2\xi)^{1/2}F(\xi, \eta)$ so that the equation of motion is $F_{\eta\eta\eta} + FF_{\eta\eta} + \beta(\xi)(1 - F_\eta^2) = 2\xi(F_\eta F_{\xi\eta} - F_\xi F_{\eta\eta})$ and the boundary conditions are $F(\eta = 0) = 0 = F_\eta(\eta = 0)$ and $F_\eta(\eta = \infty) = 1$; since $\beta = 2(dU/dx)(\int_0^x U\,dx)/U^2$, it is possible to express $F(\xi, \eta)$ and $\beta(\xi)$ as a series of powers of $\xi$ whose coefficients are given by similarity equations. The Hantzche–Wendt transformation [1, p. 319] treats compressible fluids with zero pressure gradient. The von Mises transformation [1, p. 143] uses the streamline $\psi$ as a vertical coordinate for two-dimensional boundary layers so that the convective terms vanish from the equation of motion, leaving

$$\frac{\partial(p + \rho u^2/2)}{\partial x} = vu\frac{\partial^2(p + \rho u^2/2)}{\partial \psi^2}$$

which is of the diffusion equation form and is somewhat suited for numerical solution. The Crocco transformation [1, p. 324] simplifies the equations for variable viscosity. The Mangler transformation discussed earlier was extended by Probstein and Elliot [1, p. 229] for boundary layer flow on very slender bodies of revolution. The Howarth [1, p. 324], Illingworth–Stewartson [1, p. 324], and Cope–Hartree [1, p. 358] transformations put the equations for compressible flow into the same form as for incompressible flow. Moore's transformation [1, p. 397] is one of the few available for applying similarity concepts to unsteady boundary-layer problems; it has been applied to unsteady flight velocity for the flat-plate geometry and to unsteady film boiling by Burmeister and Schoenhals [35]. It resembles the Meksyn–Görtler transformation but is more general in that the transformation for a specific problem is suggested by the form of the results which appear as the procedure is applied.

The heat-transfer coefficient between impinging gas jets and solid surfaces somewhat resembles that for the rotationally symmetric stagnation case of an infinite stream impinging on a flat surface. The round or slot nozzles, used either singly or in arrays, lead to important differences as surveyed by Martin [36]. For a single round nozzle, the average heat-transfer coefficient for perpendicular impingement is given by

$$\frac{\overline{Nu}}{Pr^{0.42}} = \frac{D}{r} \frac{1 - 1.1D/r}{1 + 0.1(H/D - 6)D/r} \left[ 2Re^{1/2} \left( 1 + \frac{Re^{0.55}}{200} \right)^{0.5} \right] \quad (7\text{-}73)$$

where $\overline{Nu} = \bar{h}D/k$, $Re = V_{\text{nozzle exit}} D/\nu$, $H = $ nozzle height above surface, $D = $ nozzle diameter, and $r = $ radius from nozzle centerline and for $2 \times 10^3 \leqslant Re \leqslant 4 \times 10^5$, $2.5 \leqslant r/D \leqslant 7.5$, and $2 \leqslant H/D \leqslant 12$. Close to the stagnation point, $r/D < 2.5$, higher heat-transfer coefficients are encountered (see Fig. 7-16). For a single-slot nozzle, the average heat-transfer coefficient is given by

$$\frac{\overline{Nu}}{Pr^{0.42}} = \frac{1.53Re^m}{x/s + H/s + 1.39}$$

$$m = 0.695 - \left[ \frac{x}{s} + \left( \frac{H}{s} \right)^{1.33} + 3.06 \right]^{-1} \quad (7\text{-}74)$$

for $3 \times 10^3 \leqslant Re \leqslant 9 \times 10^4$, $2 \leqslant H/s \leqslant 10$, and $2 \leqslant x/s \leqslant 25$ (although $x/s$

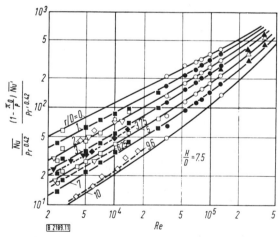

**Figure 7-16** Heat and mass transfer to impinging flow from a single round nozzle at $H/D = 7.5$ according to E. U. Schlünder and V. Gnielinski, *Chem. -Ing. -Tech.* **39**, 578 (1967); ●○ Schlünder and Gnielinski, ▲△ Petzold, ■□ Gardon and Cobonpue, ▼▽ Brdlick and Savin ($r/D = 2.34, 3.75, 6.25$) ◆◇ [By permission from H. Martin, *Adv. Heat Transf.* **13**, 17 (1977) [36]; copyright 1977, Academic Press, New York.]

$\geqslant 0$ can be used if errors of about 40% are acceptable at $x/s = 0$) and where $\mathrm{Nu} = hs/k$, $\mathrm{Re} = V_{\text{nozzle exit}}s/\nu$, $H$ = nozzle height above surface, $B$ = slot nozzle width, $x$ = distance from nozzle centerline, and $s = 2B$. When round nozzles are employed in a regular array, the average heat-transfer coefficient is given by

$$\frac{\overline{\mathrm{Nu}}}{\mathrm{Pr}^{0.42}} = K\frac{f^{1/2}(1 - 2.2f^{1/2})\mathrm{Re}^{2/3}}{1 + 0.2(H/D - 6)f^{1/2}} \tag{7-75}$$

with

$$K = \left[1 + \left(1.67\frac{H}{D}\right)^6 f^3\right]^{-0.05}$$

for $2 \times 10^3 \leqslant \mathrm{Re} \leqslant 10^5$, $2 \leqslant H/D \leqslant 12$, and $0.004 \leqslant f \leqslant 0.04$, where all quantities are as defined in Eq. (7-73) and $f$ = nozzle exit area/solid surface area served by one nozzle. For a regular array of slot nozzles, the average heat transfer coefficient is given by

$$\frac{\overline{\mathrm{Nu}}}{\mathrm{Pr}^{0.42}} = 0.67f_0^{3/4}\left(\frac{2\mathrm{Re}}{f/f_0 + f_0/f}\right)^{2/3}$$

$$f_0 = \left[60 + 4\left(\frac{H}{s} - 2\right)^2\right]^{-1/2} \tag{7-76}$$

for $1.5 \times 10^3 \leqslant \mathrm{Re} \leqslant 4 \times 10^4$, $1 \leqslant H/s \leqslant 40$, and $0.008 \leqslant f \leqslant 2f_0H/s$. Maximum average heat-transfer coefficients result for regular arrays when $D/H \approx \frac{1}{5}$ for round nozzles and $2B/H \approx \frac{1}{3}$ for slot nozzles with a nozzle-to-nozzle distance of $7H/5$. Obliquely impinging circular jets have also been studied [42].

## 7.8 TRANSPIRATION ON A FLAT PLATE—LAMINAR BOUNDARY-LAYER SIMILARITY SOLUTIONS

The behavior of a laminar boundary layer can be substantially influenced by either adding or removing fluid at the solid surface by injection or suction processes, respectively. As might be expected, suction removes decelerated fluid particles from the boundary layer before they have a chance to cause flow separation. A major benefit of suction on airfoils is to reduce drag. Injection can also reduce drag since it is possible that the injection process will add sufficient additional energy to the retarded fluid particles near the wall to prevent flow separation. The primary heat-transfer application of injection, however, is to reduce the net wall heat flux as a result of the injected fluid motion through the porous wall in a direction opposite to the heat flux from the external fluid to the wall. Another term for injection cooling is *transpiration cooling*, since the injected coolant fluid transpires through a porous wall.

## Velocity Distribution

The laminar boundary-layer equation of motion has a similarity solution only if the vertical velocity at the wall varies in a particular manner. Reference to Eq. (7-58) shows that for wedge flow with $U = Cx^m$, it is required that $v_w \sim x^{(m-1)/2}$. Although the injection, or "blowing," velocity at the wall is unlikely to vary in exactly this way in an application, such a variation is sufficiently realistic to enable one to make use of solutions based on that assumption. The $x$-motion equation for wedge flow is unchanged. Its boundary conditions still require the no-slip condition at the wall and merger with the free-stream velocity at the edge of the boundary layer. The sole change is that $v_w$ does not necessarily equal zero. Thus

$$F''' + \frac{m+1}{2} FF'' + m\left[1 - (F')^2\right] = 0$$

$$F(0) = -\frac{v_w}{U}\left(\frac{Ux}{\nu}\right)^{1/2}\left(\frac{2}{m+1}\right), \qquad F'(0) = 0, \qquad F'(\infty) = 1 \quad (7\text{-}77)$$

For a flat plate ($m = 0$), the velocity distribution is as illustrated in Fig. 7-17 [37]. The thickening of the boundary layer with injection and its thinning with suction are evident. The local skin friction coefficient displayed in Fig. 7-18 demonstrates the influence of injection (or transpiration) on the viscous stress at the wall. It should be noted that the boundary-layer assumptions do

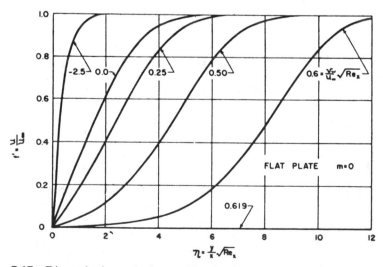

**Figure 7-17** Dimensionless velocity profiles for laminar flow over a flat plate with $(v_w/U)\mathrm{Re}_x^{1/2}$ as an injection parameter. [From J. P. Hartnett and E. R. G. Eckert, *Transact. ASME* **79**, 247–254 (1957) [37].]

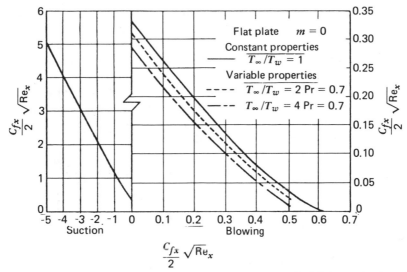

**Figure 7-18** Dimensionless local skin friction coefficient $C_f \text{Re}_x^{1/2}/2$ for laminar flow over a flat plate with $(v_w/U)\text{Re}_x^{1/2}$ as an injection parameter. [From J. P. Hartnett and E. R. G. Eckert, *Transact. ASME* **79**, 247–254 (1957) [37].]

not permit a solution of the boundary-layer equations for $(v_w/U)\sqrt{Ux/\nu}$, the "blowing" parameter, in excess of 0.619 since the boundary layer is almost literally "blown" off the plate.

A particularly simple solution for the constant suction rate case on a flat plate has been obtained by Iglisch [38]—easily obtained from Problem 5-18 (with $v_w = \text{const} < 0$ and $L = \infty$)—to the effect that far from the leading edge the velocity distribution is $u/U = 1 - e^{-v_w y/\nu}$. Iglisch's study indicates that if $v_w = \text{const} < 0$, this limit is achieved after a distance from the leading edge given by

$$\left(\frac{v_w}{U}\right)^2 \frac{Ux}{\nu} = 2$$

This solution is only a limiting case, of course, for the boundary layer must be thin near the leading edge, and will most likely undergo transition to turbulent flow before it becomes infinitely thick.

Velocity distributions for wedge flow have been obtained [39] that can be consulted if there is need for detailed results. The trend of events with transpiration is generally as just indicated for the flat plate.

### Temperature Distribution

Under the assumption that the injected fluid does not differ in properties from the fluid of the free stream (properties are constant), the energy equation is easily formulated. It is assumed that there is no difference between the wall temperature and the fluid temperature near the wall in spite of the fact that the wall is often porous so that opportunity for a temperature discrepancy does exist. This assumption is similar in spirit to the no-slip assumption of a porous wall during transpiration.

The energy equation and boundary conditions to be solved are identical with Eqs. (7-64) or (7-65), but $F$ is now given by Eq. (7-77). Specialization to the case of constant wall and free-stream temperature requires that $\gamma = 0$ and prohibits a similarity solution if viscous dissipation is included (unless $m = 0$ as it is for a flat plate). With $\theta = (T - T_w)/(T_\infty - T_w)$ the equation to be solved is, with $\gamma = 0$,

$$\theta'' + \frac{m + 1}{2} \Pr F\theta' = -\Pr E(F'')^2 x^{2m}$$

$$\theta(0) = 0 \quad \text{and} \quad \theta(\infty) = 1 \tag{7-78}$$

For the case of a flat plate ($m = 0$), the results due to Hartnett and Eckert [37] show the "blowing parameter" $(v_w/U)(Ux/\nu)^{1/2}$ to have similar effects on the velocity and temperature profiles. As might be expected, suction thins the thermal boundary layer, whereas injection thickens it (see Fig. 7-19). From

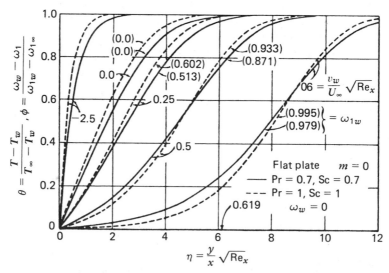

**Figure 7-19** Dimensionless temperature and mass-fraction profiles for laminar flow over a flat plate with $(v_w/U)\mathrm{Re}_x^{1/2}$ as an injection parameter for $\Pr = 0.7$ and $1.0$. [From J. P. Hartnett and E. R. G. Eckert, *Transact. ASME* **79**, 247–254 (1957) [37].]

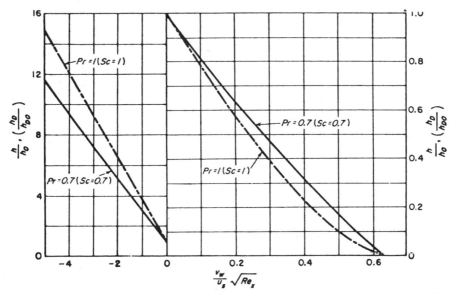

**Figure 7-20**  Heat- and mass-transfer coefficients for laminar flow over a flat plate. The subscript 0 indicates heat- and mass-transfer coefficients with vanishing values of the injection parameter $(v_w/U)\mathrm{Re}_x^{1/2}$; Pr belongs to $h$ and Sc belongs to $h_D$. [From J. P. Hartnett and E. R. G. Eckert, *Transact. ASME* **79**, 247–254 (1957) [37].]

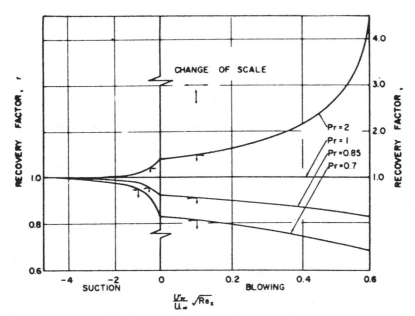

**Figure 7-21**  Recovery factor $r$ for laminar flow over a flat plate for several Prandtl numbers with $(v_w/U)\mathrm{Re}_x^{1/2}$ as an injection parameter. [From J. P. Hartnett and E. R. G. Eckert, *Transact. ASME* **79**, 247–254 (1957) [37].]

323

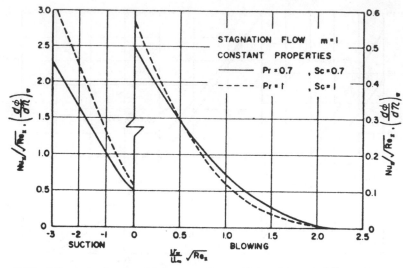

**Figure 7-22** Dimensionless local heat-transfer coefficient $\mathrm{Nu}_x/\mathrm{Re}_x^{1/2}$ and local mass-transfer coefficients $(d\Phi/d\eta)_w$ for laminar plane stagnation flow with $(v_w/U)\mathrm{Re}_x^{1/2}$ as an injection parameter. [From J. P. Hartnett and E. R. G. Eckert, *Trans. ASME* **79**, 247–254 (1957) [37].]

these results the local heat-transfer coefficient is found to vary with the "blowing parameter" as depicted in Fig. 7-20; a linear variation is a reasonable approximation. The recovery factor $r$ for a flat plate is also affected by transpiration (see Fig. 7-21).

For flow over a surface other than a flat plate, a single adiabatic wall temperature does not exist. Therefore, available similarity solutions ignore the effect of viscous dissipation. For two-dimensional stagnation flow ($m = 1$), the local heat-transfer coefficient depends on the "blowing parameter" as shown in Fig. 7-22. A substantial influence on $\mathrm{Nu}/\mathrm{Re}^{1/2}$ is exerted by the wedge angle, as Table 7-5 from Kays [40] shows for various values of $m$ and $(v_w/U)(Ux/\nu)^{1/2}$. From the results of Table 7-5 for $m = \frac{1}{3}$, Mangler's transformation has been employed by Howe and Mersman [41] to obtain the local heat-transfer coefficients for rotationally symmetric stagnation flow with injection as presented in Table 7-6. The study by Donoughe and Livingood [39] encapsulated in Table 7-7 shows the effect of variable wall temperature along with transpiration on wedge flow. The value of $\gamma$ (in $T_w - T_\infty = Kx^\gamma$) that produces a constant wall heat flux is easily determined [as explained in conjunction with Eq. (7-66), $2\gamma = 1 - m$ for such a condition] as is the corresponding local Nusselt number.

The manner in which the local heat-transfer and skin friction coefficients previously presented for transpiration are to be used deserves explicit discus-

**Table 7-5**  $\mathrm{Nu}_x\,\mathrm{Re}_x^{-1/2}$ for Various Rates of Blowing or Suction and Various Values of $m$; Laminar Constant-Property Boundary Layer $(t_\infty, t_0 = \text{const}; \mathrm{Pr} = 0.7)$

| $\dfrac{v_0}{u_\infty}\sqrt{\dfrac{\rho u_\infty x}{\mu}}$ | $-0.04175$ | $-0.0036$ | $0$ | $0.0257$ | $0.0811$ | $0.333$ | $0.500$ | $1.000$ |
|---|---|---|---|---|---|---|---|---|
| 0 | | | 0.292 | | | 0.384 | | 0.496 |
| 0.239 | 0.103 | | | | | | | |
| 0.250 | | | 0.166 | | | | | |
| 0.333 | | | | | | 0.242 | | |
| 0.375 | | | 0.107 | | | | 0.259 | |
| 0.500 | | 0.0251 | 0.0517 | | | | | 0.293 |
| 0.518 | | | | 0.087 | | | | |
| 0.558 | | | | | 0.109 | | | |
| 0.667 | | | | | | 0.131 | | |
| 1.000 | | | | | | | | 0.146 |

*Source*: By permission from W. M. Kays and M. E. Crawford, *Convective Heat and Mass Transfer*, McGraw-Hill, New York, 1980.

sion for there is an element of arbitrariness involved. With reference to Fig. 7-23, an energy balance on control volume $a$–$b$–$c$–$d$ shown by dashed lines gives, if such extraneous factors as radiative heat fluxes and horizontal conduction in the wall are ignored,

$$q_{\text{net from wall}} = -\frac{k\,\partial T(0)}{\partial y} + \rho C_p v_w T_w - \rho C_p v_w T_c$$

On rearrangement, this relation becomes

$$q_{\text{net from wall}} = \rho C_p v_w (T_w - T_c) - k\frac{\partial T(0)}{\partial y}$$

**Table 7-6**  $\mathrm{Nu}_x\,\mathrm{Re}_x^{-1/2}$ for Various Rates of Blowing for Axisymmetric Stagnation Point; Laminar Constant-Property Boundary Layer $(t_\infty, t_0 = \text{const}; \mathrm{Pr} = 0.7)$

| $\dfrac{v_0}{u_\infty}\sqrt{\dfrac{\rho u_\infty x}{\mu}}$ | $0$ | $0.567$ | $1.154$ |
|---|---|---|---|
| $\mathrm{Nu}_x\,\mathrm{Re}_x^{-1/2}$ | 0.664 | 0.419 | 0.227 |

Table 7-7  **Summary of Heat-Transfer and Friction Parameters and Boundary-Layer Thicknesses [39]**

| $\dfrac{2}{m+1}\dfrac{v_w}{U}\sqrt{Re}$ | $m$ | $\gamma$ | $\dfrac{Nu}{\sqrt{Re}}$ | $\dfrac{C_f}{2}\sqrt{Re}^{\,a}$ |
|---|---|---|---|---|
| 0 | 0 | −0.5000 | 0 | 0.3320 |
| | | 0 | 0.2927 | |
| | | 0.5000 | 0.4059 | |
| | | 1.000 | 0.4803 | |
| | 0.5 | −0.7500 | 0 | 0.89975 |
| | | 0 | 0.4162 | |
| | | 0.5000 | 0.5426 | |
| | | 1.000 | 0.6350 | |
| | 1.0 | −1.000 | 0 | 1.2326 |
| | | −0.5000 | 0.3228 | |
| | | 0 | 0.4958 | |
| | | 0.5000 | 0.6159 | |
| | | 1.000 | 0.7090 | |
| −0.5 | 0 | −0.3702 | 0 | 0.1645 |
| | | 0 | 0.1661 | |
| | | 0.5000 | 0.2611 | |
| | | 1.000 | 0.3211 | |
| | 0.5 | −0.5356 | 0 | 0.6974 |
| | | −0.5000 | 0.0272 | |
| | | 0 | 0.2594 | |
| | | 0.5000 | 0.3834 | |
| | | 1.0000 | 0.4711 | |
| | 1.0 | −0.6789 | 0 | 0.9692 |
| | | 0 | 0.2934 | |
| | | 0.5000 | 0.4132 | |
| | | 1.0000 | 0.5030 | |
| −1.0 | 0 | −0.2384 | 0 | 0.0355 |
| | | 0 | 0.0516 | |
| | | 0.5000 | 0.1052 | |
| | | 1.0000 | 0.1383 | |
| | 0.5 | −0.3585 | 0 | 0.5345 |
| | | 0 | 0.1392 | |
| | | 0.5000 | 0.2528 | |
| | | 1.0000 | 0.3314 | |
| | 1.0 | −0.4235 | 0 | 0.7565 |
| | | 0 | 0.1457 | |
| | | 0.5000 | 0.2553 | |
| | | 1.0000 | 0.3360 | |

$^{a}Re = \dfrac{Ux}{\nu}$

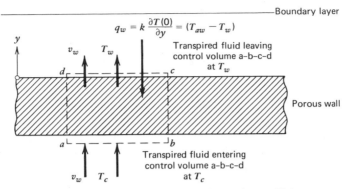

**Figure 7-23** Control volume illustrating the use of transfer coefficients at a transpiring wall.

On setting $h(T_w - T_{aw}) = -k\,\partial T(0)/\partial y$, which signifies that only the diffusive energy flow is represented by the heat-transfer coefficient, one has

$$q_{\text{net from wall}} = \rho C_p v_w(T_w - T_c) + h(T_w - T_{aw}) \qquad (7\text{-}79)$$

In steady state and in the absence of other heat-flow mechanisms, it is necessary that $q_{\text{net from wall}} = 0$, which results in

$$h(T_w - T_{aw}) = \rho C_p v_w(T_w - T_c)$$

From this it can be seen that, subject to the assumptions employed, $T_w = T_{aw}$ when $T_w = T_c$.

## 7.9 MASS TRANSFER ON A FLAT PLATE—LAMINAR BOUNDARY-LAYER SIMILARITY SOLUTION

Mass transfer occurs in as many diverse situations as does heat transfer, not all of which can be accurately described as boundary-layer problems. However, a general analogy between heat and mass transfer is likely to be common everywhere. If this is so, mass transfer rates can be calculated if heat-transfer rates can be. Looking ahead, it is possible that the transpiration caused by the small vertical velocity components at the surface from which mass leaves will distort temperature, velocity, and concentration profiles in such a way as to render low-mass-transfer-rate results substantially different from high-mass-transfer-rate results.

A simple problem, Stefan's diffusion problem, with no forced fluid motion has been examined in Section 5.3 in order to explore the effect of the flow induced by the mass transfer. One of the noteworthy results of that work was the realization that the induced velocity effects could be combined with the diffusive mass transfer by using

$$\dot{m}_1 = \underbrace{\left(\frac{1}{1-\omega_1}\right)}_{\substack{\text{induced} \\ \text{flow} \\ \text{correction}}} \underbrace{\left(-\rho D_{12}\frac{d\omega_1}{dy}\right)}_{\substack{\text{diffusion}}} \tag{7-80}$$

where species 1 is diffusing through species 2. If the rate of mass transfer is quite small as evidenced by small values of the mass fraction of species 1, the induced-flow correction is small. Anticipation of design usage of results suggests determination of a mass-transfer coefficient; realization that the diffusive flow is given by $\dot{m}_1 = -\rho D_{12}\,d\omega_1/dy$ leads naturally to

$$\dot{m}_1 = \rho h_D(\omega_{1w} - \omega_{1\infty})$$
$$\substack{\text{mass-transfer} \\ \text{coefficient,} \\ \text{neglecting} \\ \text{induced flow}}$$

Combination of these two ideas suggests a way to account for induced flow as

$$\overset{\substack{\text{transfer coefficient} \\ \text{neglecting} \\ \text{induced flow}}}{\dot{m}_1 = \rho h_D(\omega_{1w} - \omega_{1\infty})} / \underbrace{(1 - \omega_{1w})}_{\substack{\text{correction for} \\ \text{induced flow}}} \tag{7-81}$$

The form of Eq. (7-81) requires detailed verification for a specific case. For this purpose, consider the boundary-layer situation posed in the following paragraph and illustrated in Fig. 7-24.

A flat plate is exposed to a parallel-flowing free stream of constant velocity $U$, temperature $T_\infty$, and a fixed proportion of species 1 and 2 so that $\omega_{1\infty}$ is also constant. Species 1 is the diffusing substance. The plate temperature is constant at $T_w$, and the mass fraction of 1 on the plate is also constant at $\omega_{1w}$. This is a steady-state problem, the boundary layer is laminar, viscous dissipation is negligible, and bulk properties are constant. The describing boundary

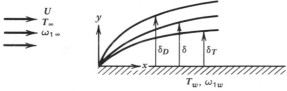

**Figure 7-24**   Simultaneous heat and mass transfer in laminar boundary-layer flows.

layer equations are

$$\text{Continuity} \quad \frac{\partial u}{\partial x} + \frac{\partial v}{\partial y} = 0$$

$$\text{Motion} \quad u\frac{\partial u}{\partial x} + v\frac{\partial u}{\partial y} = \nu\frac{\partial^2 u}{\partial y^2}$$

$$\text{Energy} \quad u\frac{\partial T}{\partial x} + v\frac{\partial T}{\partial y} = \alpha\frac{\partial^2 T}{\partial y^2} = \frac{\nu}{Pr}\frac{\partial^2 T}{\partial y^2}$$

$$\text{Diffusion} \quad u\frac{\partial \omega_1}{\partial x} + v\frac{\partial \omega_1}{\partial y} = D_{12}\frac{\partial^2 \omega_1}{\partial y^2} = \frac{\nu}{Sc}\frac{\partial^2 \omega_1}{\partial y^2} \qquad (7\text{-}82)$$

with boundary conditions of

$$\text{At } y = 0; \quad u = 0, \quad v = v_w, \quad \omega_1 = \omega_{1w}, \quad T = T_w$$

$$\underset{\substack{\text{not specified} \\ \text{initially as to} \\ \text{magnitude}}}{}$$

$$\text{At } y \to \infty; \quad u \to U, \quad \omega_1 \to \omega_{1\infty}, \quad T \to T_\infty$$

Here $Pr = \nu/\alpha$ and $Sc = \nu/D_{12}$.

The condition at $y = 0$, saying that $v = v_w$, makes this situation similar to the transpiration studies described in Section 7.8. Also, the constant-property assumption has uncoupled the energy and the diffusion equations from the x-motion equation in the sense that the velocity distribution can be found first and the other two distributions later and separately. However, the occurrence of mass transfer at the wall may cause $v_w$ to be noticeably different from zero. If the mass-transfer rate $\dot{m}_1$ is specified

$$\dot{m}_1 = \rho_1|_w v_w - \rho D_{12} \left.\frac{\partial \omega_1}{\partial y}\right|_w = \rho\left(\omega_1 v_w - D_{12} \left.\frac{\partial \omega_1}{\omega y}\right|_w\right)$$

the mass fraction at the wall $\omega_{1w}$ may not be known. On the other hand, if a particular value of $\omega_{1w}$ is forced to exist, $\dot{m}_1$ and $v_w$ may not be known. There is a possibility of $v_w = 0$ if the plate surface is permeable to both species, but the mass flow rates of the two species must be rather delicately balanced in such a case.

If $v_w$ is sufficiently small to be assumed zero, it is recognized that the diffusion equation is very similar to the energy equation. This is demonstrated by the dimensionless set

$$\theta = \frac{T - T_w}{T_\infty - T_w}, \qquad \phi = \frac{\omega_1 - \omega_{1w}}{\omega_{1\infty} - \omega_{1w}}$$

as follows:

$$\frac{\partial u}{\partial x} + \frac{\partial v}{\partial y} = 0$$

$$u\frac{\partial u}{\partial x} + v\frac{\partial u}{\partial y} = \nu\frac{\partial^2 u}{\partial y^2}$$

$$u\frac{\partial \theta}{\partial x} + v\frac{\partial \theta}{\partial y} = \frac{\nu}{\mathrm{Pr}}\frac{\partial^2 \theta}{\partial y^2}$$

$$u\frac{\partial \phi}{\partial x} + v\frac{\partial \phi}{\partial y} = \frac{\nu}{\mathrm{Sc}}\frac{\partial^2 \phi}{\partial y^2} \qquad (7\text{-}83)$$

At $y = 0$;     $u = 0$,     $v = v_w$,     $\theta = 0 = \phi$

At $y \to \infty$;     $u \to U$,     $\theta \to 1 \leftarrow \phi$

Since the differential equations and boundary conditions for $\theta$ and $\phi$ are the same in Eq. (7-83), the solutions for $\phi$ and $\theta$ will be the same with Sc and Pr interchanged as appropriate. It was found in Section 7.3 that

$$\frac{\mathrm{Nu}}{\mathrm{Re}_x \,\mathrm{Pr}^{1/3}} = \frac{C_f}{2} \qquad \text{or} \qquad \frac{hx/k}{\mathrm{Re}_x \,\mathrm{Pr}^{1/3}} = \frac{C_f}{2}$$

Therefore, it can be asserted that

$$\frac{\mathrm{Sh}}{\mathrm{Re}_x \,\mathrm{Sc}^{1/3}} = \frac{C_f}{2} \qquad \text{or} \qquad \frac{h_D x/D_{12}}{\mathrm{Re}_x \,\mathrm{Sc}^{1/3}} = \frac{C_f}{2}$$

Since both quantities equal the same thing, it must be that an extended Reynolds analogy is

$$\frac{\mathrm{Nu}}{\mathrm{Pr}^{1/3}} = \frac{\mathrm{Sh}}{\mathrm{Sc}^{1/3}} \qquad (7\text{-}84)$$

where Nu = Nusselt number and Sh = Sherwood number, relating heat- and mass-transfer coefficients with the proviso, so far, that mass-transfer-induced flow is negligible. To give a bit more background consider that (with $v_w \approx 0$) for mass and heat transfer, respectively,

$$\dot{m}_1 = -\rho D_{12}\frac{\partial \omega_1}{\partial y}(y=0) = \rho h_D(\omega_{1w} - \omega_{1\infty})$$

and

$$q_w = -k\frac{\partial T \ (y = 0)}{\partial y} = h(T_w - T_\infty)$$

In terms of dimensionless $\phi$ and $\theta$, these relationships become

$$-\rho D_{12}(\omega_{1\infty} - \omega_{1w})\frac{\partial \phi \ (y = 0)}{\partial y} = -\rho h_D(\omega_{1\infty} - \omega_{1w})$$

and

$$-k(T_\infty - T_w)\frac{\partial \theta \ (y = 0)}{\partial y} = -h(T_\infty - T_w)$$

Employment of the approximate result of Eq. (7-28) in these relationships gives

$$\frac{\partial \phi (y = 0)}{\partial y} = \frac{h_D}{D_{12}} \qquad\qquad \frac{\partial \theta (y = 0)}{\partial y} = \frac{h}{k}$$

$$\approx 0.33206 \ Sc^{1/3} \qquad\qquad \approx 0.33206 \ Pr^{1/3}$$

Equation of the preceding mass- and heat-transfer results then leads to

$$\frac{h_D}{D_{12}} Sc^{-1/3} = \frac{h}{k} Pr^{-1/3}$$

$$\frac{h_D x}{D_{12}} Sc^{-1/3} = \frac{hx}{k} Pr^{-1/3}$$

$$\frac{Sh}{Sc^{1/3}} = \frac{Nu}{Pr^{1/3}} \qquad\qquad (7\text{-}85)$$

Two points are important here: (1), the coefficients $h_D$ and $h$ in $Sh = h_D x/D_{12}$ and $Nu = hx/k$ give the diffusive transport; any convective transport must be an additional contribution; and (2) the extended Reynolds analogy of Eq. (7-85) has thus far been verified in detail only for the restricted case of laminar boundary-layer flow over a flat plate; its applicability to more general cases such as turbulent flow and other geometries is not assured, but it is likely (as experiment confirms).

With the insight so far acquired, the refinements afforded by a complete solution of the boundary-layer equations can be appreciated and used. For this purpose, a similarity solution is achieved by letting $\eta = y\sqrt{U/\nu x}$ and $\Psi = \sqrt{\nu U x} \ F(\eta)$ so that $u = UF'$ and $v = \frac{1}{2}\sqrt{\nu U/x}[\eta F' - F)$ as before. As determined previously, this leads to

$$F''' + \frac{F}{2}F'' = 0 \qquad\qquad F'(0) = 0, \qquad \theta(0) = 0 = \phi(0)$$

$$\theta'' + \frac{Pr}{2}F\theta' = 0 \qquad\qquad v_w = -\frac{1}{2}\sqrt{\frac{vU}{x}}\,F(0) = -2\frac{v_w}{U}\sqrt{Re_x}$$

$$\phi'' + \frac{Sc}{2}F\phi' = 0 \qquad\qquad F'(\infty) = 1, \qquad \theta(\infty) = 1 = \phi(\infty) \quad (7\text{-}86)$$

If $v_w = 0$, there is no doubt that everything is dependent only on $\eta$, and a similarity solution is guaranteed. If $v_w \neq 0$, a similarity solution exists only if $v_w$ varies as $x^{-1/2}$; this condition is not far from what actually happens, although it cannot be an exact representation of the "blowing" that really occurs.

Before closely examining the solutions of Eq. (7-86) the manner in which the results can be used should be determined. Recall from the Stefan diffusion problem in Section 5.3 that

$$\dot{m}_1 = \left(\rho_1 v_w - \rho D_{12}\frac{\partial \omega_1}{\partial y}\right)\bigg|_0 \quad \text{and} \quad \dot{m}_2 = \left(\rho_2 v_w - \rho D_{12}\frac{\partial \omega_2}{\partial y}\right)\bigg|_0$$

$$= \rho\omega_{1w}v_w - \rho D_{12}\frac{\partial \omega_1}{\partial y}\bigg|_0 \qquad \dot{m}_2 = \rho(1-\omega_1)\big|_0 v_w + \rho D_{12}\frac{\partial \omega_1}{\partial y}\bigg|_0$$

$$\frac{\dot{m}_2 - \rho D_{12}\,\partial\omega_1/\partial y\,|_0}{1-\omega_{1w}} = \rho v_w$$

Elimination of $\rho v_w$ from these two relationships results in

$$\dot{m}_1 = \frac{\omega_{1w}}{1-\omega_{1w}}\dot{m}_2 - \rho D_{12}\frac{\partial \omega_1}{\partial y}\bigg|_0\left[1 + \frac{\omega_{1w}}{1-\omega_{1w}}\right]$$

$$\dot{m}_1 = \frac{\omega_{1w}}{1-\omega_{1w}}\dot{m}_2 + \left(\frac{1}{1-\omega_{1w}}\right)\underbrace{\left(-\rho D_{12}\frac{\partial \omega_1}{\partial y}\bigg|_0\right)}_{\substack{\rho h_D(\omega_{1w}-\omega_{1\infty}) \\ \text{neglecting} \\ \text{induced flow}}}$$

Although certain precise values of $\dot{m}_2$ could possibly cancel the effect of $1/(1-\omega_{1w})$ given in the preceding equation, the most important case is where $\dot{m}_2 = 0$. Then

$$\dot{m}_1 = \frac{\rho h_D(\omega_{1w}-\omega_{1\infty})}{1-\omega_{1w}} = \frac{\rho}{1-\omega_{1w}}\left[-D_{12}\frac{\partial\omega_1(0)}{\partial y}\right] \quad (7\text{-}87)$$

and

$$v_w = -\frac{D_{12}}{1 - \omega_{1w}} \frac{\partial \omega_1}{\partial y}\bigg|_0 \qquad (7\text{-}88)$$

Of course, Eqs. (7-87) and (7-88) stipulate that (with $\dot{m}_2 = 0$)

$$\dot{m}_1 = \rho v_w \qquad (7\text{-}88a)$$

and reveal that specification of either the "blowing" velocity $v_w$ or the mass transfer rate of 1 $\dot{m}_1$ also specifies the other.

A further useful relation can be developed by again considering Eq. (7-88). In terms of the parameters of the similarity solution to Eq. (7-86),

$$v_w = -\frac{D_{12}}{1} \frac{}{\omega_{1w}} (\omega_{1\infty} - \omega_{1w}) \frac{d\phi(0)}{d\eta} \frac{\partial \eta}{\partial y}$$

$$= D_{12}\left(\frac{\omega_{1w} - w_{1\infty}}{1 - \omega_{1w}}\right)\phi'(0)\left(\frac{U}{\nu x}\right)^{1/2}$$

or

$$\left[\frac{v_w}{U}\left(\frac{Ux}{\nu}\right)^{1/2}\right]\frac{Sc}{\phi'(0)} = \frac{w_{1w} - \omega_{1\infty}}{1 - w_{1w}} \qquad (7\text{-}89)$$

Equation (7-89) shows that specification of $v_w$ (or $\dot{m}_1$) fixes the value of $(\omega_{1w} - \omega_{1\infty})$ and vice versa. Although it is not necessary to continue this development for the present purpose of acquiring insight into the use of solutions, Eq. (7-77) can be inserted into Eq. (7-89) to give the alternative relation

$$-Sc\frac{F(0)}{2\phi'(0)} = \frac{\omega_{1w} - \omega_{1\infty}}{1 - \omega_{1w}} \qquad (7\text{-}90)$$

The solutions to Eq. (7-86) are as shown in Fig. 7-17 for the velocity distribution; Fig. 7-18 for the skin friction coefficient; Fig. 7-19 for the temperature distribution, which is also the distribution of $(\omega_1 - \omega_{1w})/(\omega_{1\infty} - \omega_{1w})$ with Sc used in place of Pr; and in Fig. 7-20 for Nu $= hx/k$, which is also applicable for Sh $= h_D x/D_{12}$ with Sc used in place of Pr. In addition, Eq. (7-90) provides a means of relating $(\omega_{1w} - \omega_{1\infty})/(1 - \omega_{1w})$ to the "blowing parameter" $(v_w/U)(Ux/\nu)^{1/2}$ with the result shown in Fig. 7-25. Note from Fig. 7-25 that the boundary-layer assumptions fail when the "blowing parameter" exceeds 0.619, blowing the boundary layer off the plate; the value of $\omega_{1w}$ at which this occurs is seen to be $\omega_{1w} = 1$, which does not permit the boiling

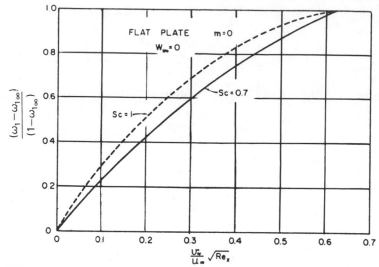

**Figure 7-25** Dimensionless surface mass fractions for laminar flow over a flat plate with $\omega_{1\infty} = 0$ with $(v_w/U)\mathrm{Re}_x^{1/2}$ as an injection parameter. [From J. P. Hartnett and E. R. G. Eckert, *Transact. ASME* **79**, pp. 247–254 (1957) [37].]

point to be closely approached unless the free stream is nearly saturated, really representing a limitation of $v_w$. The case of $m = 1$ has also been treated [37].

The situations of major interest fall into three cases. In case 1, $(\omega_{1w} - \omega_{1\infty})/(1 - \omega_{1w})$ is known but transfer coefficients and the "blowing parameter" are unknown. Figure 7-25 can be first used to obtain the "blowing parameter," and then Fig. 7-20 can be consulted for transfer coefficients. In case 2, the "blowing parameter" and $\omega_{1\infty}$ are known, but transfer coefficients and $\omega_{1w}$ are unknown. Figure 7-25 can be used to obtain $\omega_{1w}$, and Fig. 7-20 can be consulted for transfer coefficients. In case 3, $\dot{m}_1$ and $\omega_{1\infty}$ are known, but transfer coefficients as well as $\omega_{1w}$ and the "blowing parameter" are unknown. An iterative procedure must be employed: (1) assume a value of $h_D$; (2) compute the value of $\omega_{1w}$ from Eq. (7-87); (3) obtain the "blowing parameter" from Fig. 7-5; and (4) calculate an improved $h_D$ from Fig. 7-20. This procedure is repeated until $h_D$ and other unknowns converge. Once they have been evaluated, $h$ for possible heat-transfer conditions is readily available from Fig. 7-20.

It is appropriate to emphasize that the diffusive transport of mass is given by

$$\rho h_D(\omega_{1w} - \omega_{1\infty}) = -\rho D_{12}\frac{\partial \omega_1(y = 0)}{\partial y}$$

to which must be added any convective transport in order to obtain the total mass flux. Thus Eq. (7-87) must be recognized as applying only to binary diffusion with the surface impermeable to one of the species.

Application of the boundary-layer equations to multicomponent mass diffusion is discussed by Eckert [6, pp. 723–752]. The combustion of a carbon surface in an airstream is discussed in detail under both constant-property and variable-property assumptions, accounting for the complications of chemical reactions in the laminar boundary layer and transpiration at the carbon surface. Two extreme boundary-layer cases are usually considered for multicomponent diffusion with chemical reactions. In the first extreme case, called an *equilibrium* boundary layer, the chemical processes proceed very rapidly compared to the diffusion processes so that the fluid is in chemical equilibrium at every point in the boundary layer. Local composition is then dependent only on local temperature and pressure. In the second extreme case, called a *frozen* boundary layer, the chemical processes proceed very slowly compared to the diffusion processes. Unless the wall functions as a catalyst (as it sometimes does, allowing recombination at the wall), the boundary layer composition will be uniform at the free-stream conditions.

## 7.10  FINITE-DIFFERENCE SOLUTIONS

In many applications the free-stream velocity, the wall blowing velocity, or the wall temperature vary in such a manner that similarity solutions (e.g., available only for $U = Cx^m$ or $U = Ce^{\alpha x}$, $\alpha \geq 0$) do not exist for the boundary-layer equations. In such cases local similarity can often be assumed and such methods as Ambrok's (see Problem 10-19) can be employed for approximate predictions. Greatest accuracy is obtained by finite difference solution of the boundary-layer equations, however. Figure 7-26 illustrates for a flat plate the steady free stream $U(x)$ and wall blowing velocities $v_w(x)$, which might be encountered and might result in a lack of similarity.

The case of wedge flow with constant properties is discussed to illustrate the essential features of one of the several solution procedures that have been developed. Since properties are constant, the velocity distributions can be determined without regard for the temperature distribution. The velocities in the boundary layer could be directly determined by a finite-difference procedure as outlined by Schlichting [1, p. 181] and Pletcher [57]. But it is computationally more efficient to first subject the boundary-layer equations to

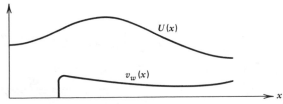

**Figure 7-26**  Free-stream $U$ and wall blowing $v_w$ velocity variations that can cause a laminar boundary layer to be nonsimilar.

a similarity like coordinate transformation since the departure from similarity is rarely more than a factor of 2 or 3. The continuity equation is automatically satisfied by a stream function $\psi$ defined in the usual way such that $u = \partial\psi/\partial y$ and $v = -\partial\psi/\partial x$. The similarity variable and stream function are defined much as before [see Eqs. (7-55) and (7-56)] as

$$\eta = y\left(\frac{U}{\nu x}\right)^{1/2} \qquad \text{and} \qquad \psi = (\nu x U)^{1/2} F(\eta, x)$$

The coordinate transformation $x, y \to x' = x, \eta$ reduces variations in the $x$ direction as illustrated in Fig. 7-27 and is effected by applying the chain rule to obtain

$$u = (\nu x U)^{1/2}\left(\frac{\partial F}{\partial x'}\overset{0}{\frac{\partial x'}{\partial y}} + \frac{\partial F}{\partial \eta}\overset{(U/\nu x)^{1/2}}{\frac{\partial \eta}{\partial y}}\right) = UF'$$

and

$$-v = \left(\frac{\nu U}{4x}\right)^{1/2}\left(1 + \frac{x}{U}\frac{dU}{dx}\right)F + (\nu x U)^{1/2}\left(\frac{\partial F}{\partial x'}\overset{1}{\frac{\partial x'}{\partial x}} + \frac{\partial F}{\partial \eta}\overset{-\eta/2x}{\frac{\partial \eta}{\partial x}}\right)$$

$$= \left(\frac{U}{4x}\right)^{1/2}\left[\left(1 + \frac{x}{U}\frac{dU}{dx}\right)F + 2x\frac{\partial F}{\partial x} - \eta F'\right]$$

where $F' = \partial F(\eta, x)/\partial\eta$. Repeated applications of the chain rule further result in

$$\frac{\partial u}{\partial x} = F'\frac{dU}{dx} + U\frac{\partial F'}{\partial x} - \frac{U\eta}{2x}F''$$

$$\frac{\partial u}{\partial y} = U\left(\frac{U}{\nu x}\right)^{1/2}F''$$

$$\frac{\partial^2 u}{\partial y^2} = \frac{U^2}{\nu x}F'''$$

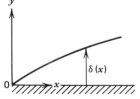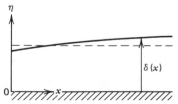

**Figure 7-27**  Boundary-layer thickness variation in the downstream direction in untransformed and transformed coordinates.

Substitution into the $x$-motion equation

$$u \frac{\partial u}{\partial x} + v \frac{\partial u}{\partial y} = U \frac{dU}{dx} + v \frac{\partial^2 u}{\partial y^2}$$

$$u(x, y = 0) = 0$$

$$v(x, y = 0) = v_w(x)$$

$$u(x, y \to \infty) \to U(x)$$

gives, with $P = (x/U) \, dU/dx$,

$$F''' + P\left[1 - (F')^2\right] + \left(\frac{1+P}{2}\right) FF'' = x\left(F' \frac{\partial F'}{\partial x} - F'' \frac{\partial F}{\partial x}\right)$$

$$F'(x, \eta = 0) = 0$$

$$F(x, \eta = 0) = -(vxU)^{-1/2} \int_0^x v_w \, dx$$

$$F'(x, \eta \to \infty) \to 1 \qquad\qquad (7\text{-}91a)$$

The boundary condition for $F(x, \eta = 0)$ is most easily obtained from the realization that the expression for $v$ can be rewritten as

$$-v = \frac{\partial\left[(vxU)^{1/2} F\right]}{\partial x} + \eta \left(\frac{vU}{4x}\right)^{1/2} F'$$

Note that the similarity form of Eq. (7-58) is recovered in cases for which $\partial F/\partial x = 0 = \partial F'/\partial x$, giving an ordinary differential equation, which occurs when $U = Cx^m$ and $v_w = a/x^{1/2}$, for example. A slight rearrangement for the purpose of controlling error in a finite difference solution is accomplished by setting $f = F - \eta$ to finally achieve

$$f''' - P\left[2f' + (f')^2\right] + \left(\frac{1+P}{2}\right)(f+\eta)f'' = x\left[(f'+1)\frac{\partial f'}{\partial x} - f'' \frac{\partial f}{\partial x}\right]$$

$$f'(x, \eta = 0) = -1$$

$$f(x, \eta = 0) = -(vxU)^{1/2} \int_0^x v_w \, dx$$

$$f'(x, \eta \to \infty) \to 0 \qquad\qquad (7\text{-}91)$$

The transformed $x$-motion equation [Eq. (7-91)] is next approximated in the $x$-direction by finite differences. For this purpose, the Taylor series for

$f(\eta, x_{n-1}) = f_{n-1}$ and $f(\eta, x_{n-2}) = f_{n-2}$ can be written in terms of $f(\eta, x_n) = f_n$ as

$$f_{n-1} = f_n + \frac{\partial f_n}{\partial x}(x_{n-1} - x_n) + \frac{\partial^2 f_n}{\partial x^2}\frac{(x_{n-1} - x_n)^2}{2!} + \cdots$$

$$f_{n-2} = f_n + \frac{\partial f_n}{\partial x}(x_{n-2} - x_n) + \frac{\partial^2 f_n}{\partial x^2}\frac{(x_{n-2} - x_n)^2}{2!} + \cdots$$

Solution for $\partial f_n/\partial x$ gives the three-point backward difference formula, with an error $O(\Delta x^2)$,

$$\frac{\partial f(\eta, x)}{\partial x} = \frac{2x_n - x_{n-1} - x_{x-2}}{(x_n - x_{x-1})(x_{n-2} - x_n)}f_n$$

$$-\frac{x_n - x_{n-2}}{(x_n - x_{n-1})(x_{n-1} - x_{n-2})}f_{n-1}$$

$$+\frac{x_n - x_{n-1}}{(x_n - x_{n-2})(x_{n-1} - x_{n-2})}f_{n-2} \qquad (7\text{-}92)$$

A less accurate two-point backward difference formula, with an error $O(\Delta x)$, is found in a similar way to be

$$\frac{\partial f(\eta, x)}{\partial x} = \frac{f_n - f_{n-1}}{x_n - x_{n-1}} \qquad (7\text{-}93)$$

The equation to be solved [Eq. (7-91)], has been discretized in the $x$ direction by finite differences. Note that the discrete locations along the axis need not be equally spaced. To illustrate the solution procedures, consider, however, the case of equal spacing $\Delta x$ in the $x$ direction. The values of $f$ at $x = 0$ can be obtained with satisfactory accuracy from similarity solutions by fitting $U(x)$ with $Cx^m$. At the next downstream location the two-point Eq. (7-93) approximation is employed in a manner similar to the following procedures. For further downstream locations, the three-point Eq. (7-92) gives Eq. (7-91) as

$$f_n''' - P\left[2f_n + (f_n')^2\right] + \frac{(1 + P)}{2}(f_n + \eta)f_n''$$

$$= \frac{x}{\Delta x}\left[(f_n' + 1)(3f_n' - 2f_{n-1}' + f_{n-2}') - f_n''(3f_n - 2f_{n-1} + f_{n-2})\right]$$

$$(7\text{-}94)$$

with $f_n' = df(\eta, x_n)/d\eta$, for example. Since $f_{n-1}$ and $f_{n-2}$ are now known, $f_n$ is

found by numerically solving (perhaps by a fourth-order Runge–Kutta procedure such as is described in the text by Chow [51]) this ordinary differential
equation in the $\eta$ direction. Equally spaced increments in the $\eta$ direction of
$\Delta\eta = 0.05$ for $0 \leqslant \eta \leqslant \eta_{max} \approx 12$ have been found to give accurate results. It is
suggested that $\Delta x \geqslant x/50$ for the x-direction spacing.

Note that the solution proceeds by marching downstream to successive $x$
locations, with the results being dependent mostly on local conditions at $x_n$.
(Conditions at $x_{n-1}$ are less strongly influential, whereas those at $x_{n-2}$ are of
much weaker influence.) This explains the success of methods that assume
local similarity. The equations being solved are parabolic (as a result of the
boundary-layer assumptions), so that downstream events do not affect upstream events; if the boundary-layer assumptions are not made, the full
equations are elliptic and downstream events do affect upstream ones (although often weakly), and numerical computation is more complicated.

Finite difference solution of the energy equation is accomplished by similar
means. Although this discussion continues to assume constant properties, the
energy equation is written in terms of the total enthalpy $H_0 = C_p T + u^2/2$, to
show its application to variable-property situations. The energy equation is

$$u\frac{\partial H_0}{\partial x} + v\frac{\partial H_0}{\partial y} = \frac{v}{\text{Pr}}\frac{\partial[\partial H_0/\partial y + (\text{Pr} - 1)u\,\partial u/\partial y]}{\partial y} \qquad (7\text{-}39c)$$

with either

$$H_0(x, y = 0) = H_w(x) \qquad \text{or} \qquad \frac{\partial H_0(x, y = 0)}{\partial y} = -\frac{C_p q_w(x)}{k}$$

$$H_0(x, y \to \infty) \to H_\infty$$

where $H_w(x) = C_{p,w} T_w$. To begin, let $\theta = (H - H_\infty)/H_\infty$ and again effect the
coordinate transformation $x, y \to x' = x, \eta = y(U/vx)^{1/2}$. Application of the
chain rule gives

$$\frac{\partial H_\theta}{\partial x} = \frac{\partial H_0}{\partial x'}\frac{\partial x'}{\partial x} + \frac{\partial H_0}{\partial \eta}\frac{\partial \eta}{\partial x}$$

$$= H_\infty\frac{\partial \theta}{\partial x} + (\theta + 1)\frac{dH_\infty}{dx} - (\eta/2x)H_\infty\frac{\partial \theta}{\partial \eta}$$

and

$$\frac{\partial H_0}{\partial y} = \frac{\partial H_0}{\partial x'}\frac{\partial x'}{\partial y} + \frac{\partial H_0}{\partial \eta}\frac{\partial \eta}{\partial y}$$

$$= H_\infty\left(\frac{U}{vx}\right)^{1/2}\frac{\partial \theta}{\partial \eta}$$

Repeated application of the chain rule also gives

$$\frac{\partial^2 H_0}{\partial y^2} = H_\infty \frac{U}{\nu x} \frac{\partial^2 \theta}{\partial \eta^2}$$

Substitution into Eq. (7-39c) finally results in

$$\frac{\partial \left[ \theta'/\mathrm{Pr} + (1 - 1/\mathrm{Pr})(U^2/H_\infty)(f' + 1)f'' \right]}{\partial \eta}$$

$$= - \left( \frac{1 + P}{2} \right)(f + \eta)\theta' + x \left[ (f' + 1) \frac{\partial \theta}{\partial x} - \theta' \frac{\partial f}{\partial x} \right] \qquad (7\text{-}95)$$

with either

$$\theta(x, \eta = 0) = \frac{H_w(x) - H_\infty}{H_\infty} \qquad \text{or} \qquad \theta'(x, \eta = 0) = -\frac{C_p q_w(x)}{k H_\infty}$$

$$\theta(x, \eta \to \infty) \to 0$$

Here it has been recognized that $H_\infty$ is constant in the free stream since viscosity and thermal conductivity are unimportant there. Also, $\theta' = \partial \theta(x, \eta/\partial \eta)$, for example. Comparison of Eq. (7-95) with the similarity form given by Eqs. (7-44) and (7-45) reveals that departure from similarity can occur because of awkward variations of $U(x)$, $v_w(x)$, $H_w(x)$, or $q_w(x)$. The finite-difference solution procedure for Eq. (7-95) is similar to that discussed for Eq. (7-91).

Complexities such as temperature-dependent properties can be numerically included in this way. The energy and $x$-motion equations then have additional terms and are coupled. For details, including axisymmetric boundary-layer applications, consult Smith and Clutter [47]. Additional useful information is given in a recent reference book by Cebeci and Bradshaw [48]; their FORTRAN programs for solving the laminar boundary-layer equations in finite-difference form are based on the Keller–Cebeci [49] box method and employ a nonuniform spacing in the $\eta$ direction as shown in Fig. 7-28. This spacing puts the nodal points close together at the wall where detail and accuracy are needed and where rapid changes are likely to occur without requiring close spacing where detail is unneeded and rapid changes do not occur. The distance between two nodes varies in a geometric progression as $\Delta\eta_j = K \Delta\eta_{j-1}$, where $K > 1$ is a parameter; the distance from the wall to the $j$th node is $\eta_j = h_1(K^j - 1)/(K - 1)$, where $h_1$ is the length of the first $\Delta\eta$ increment and the total number of points $J$ is

$$J = \frac{\ln \left[ 1 + (K - 1)(\eta_\infty/h_1) \right]}{\ln (K)}$$

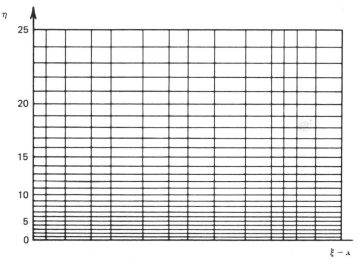

**Figure 7-28** Nonuniform grid spacing perpendicular to surface.

For $h_1 = 0.01$, $K = 1.1$, and $\eta_\infty = 10$, it is found that $J = 24$ whereas a uniform spacing ($h_1 = 0.01$, $K = 1$, $\eta_\infty = 10$) needs 1,000 points, demonstrating the utility of variable spacing. Figure 7-28, due to Cebeci et al. [50], illustrates a variable-grid system of the sort that has just been described.

Boundary-fitted coordinate systems transform the problem domain into one (often rectangular) in which numerical calculations are more easily done and eliminate boundary shapes as a complicating factor. An application to heat conduction [52], utilizing many developments due to Thompson et al. [53], discusses and illustrates the numerical generation of meshes in such transformed domains. Schemes are being developed for moving the mesh points in transient problems [54] in order to maintain accuracy in regions that contain large gradients. An overview of the rapidly developing area of numerical computation of convective transport is provided by two survey papers [55, 56].

When a conjugate problem is encountered, involving a coupling of heat conduction in a solid and convection in the adjacent fluid, a finite difference method such as that devised by Patankar [60, 61] is often convenient. This method eliminates the involved matching of boundary conditions at fluid–solid interfaces and the complications of an irregular fluid domain by formulating the problem in terms of a variable viscosity and thermal conductivity fluid over a composite domain that includes both the solid and the fluid. The resulting domain is often regular. That part of the composite domain occupied by the solid has the thermal conductivity of the solid and an infinite viscosity, whereas the part occupied by the fluid has the properties of the fluid. To illustrate the method, consider a finite-difference grid laid out so that the fluid–solid interface is between control volumes on either side with a grid point at the center of each control volume. Let $L_1$ and $L_2$ be the perpendicular distances from each grid point to the interface. Then the effective viscosity $\mu^*$ at the interface is evaluated in terms of the viscosities $\mu_1$ and $\mu_2$ at the grid points as

$$\frac{1}{\mu^*} = \frac{L_1/\mu_1 + L_2/\mu_2}{L_1 + L_2}$$

In similar fashion, the effective thermal conductivity $k^*$ at the interface is evaluated as

$$\frac{1}{k^*} = \frac{L_1/k_1 + L_2/k_2}{L_1 + L_2}$$

The basis for these representations of effective interface properties is grasped by considering one-dimensional heat conduction through the composite slab described previously in which the temperatures of the two grid points are $T_1$ and $T_2$, respectively. An effective thermal conductivity satisfies the relationship

$$\frac{k^*(T_1 - T_2)}{L_1 + L_2} = \frac{T_1 - T_2}{k_1/L_1 + k_2/L_2}$$

The previous expression for the effective thermal conductivity at the interface follows. This method has been applied to laminar flow in a square duct of finite wall thickness [60] and to ducts with internal fins [60, 62].

## PROBLEMS

7-1  (a) Show that the importance of the time-dependent terms in the continuity and $x$-motion boundary-layer equations can be ascertained by an order of magnitude analysis that gives

$$\text{Continuity} \quad \frac{L}{t_r U} \frac{\partial \rho'}{\partial t'} + \frac{\partial(p'u')}{\partial x'} + \cdots = 0$$

$$x \text{ Motion} \quad \rho'\left( \frac{L}{t_r U} \frac{\partial u'}{\partial t'} + u' \frac{\partial u'}{\partial x'} + \cdots - \right) = \cdots$$

where $t' = t/t_r$, $\rho' = \rho/\rho_r$, $u' = u/U$, and $x' = x/L$ with $t_r$ a time characteristic of that required for the contemplated change to occur (e.g., in free-stream velocity $U$). Thus, if $L/t_r U \ll 1$, unsteady effects can be neglected and the viscous drag can be determined at any instant from the corresponding instantaneous free-stream velocity.

(b) Show that for representative values of $L = 3$ ft and $U = 300$ ft/sec, if $t_r > 10^{-2}$ sec, unsteady effects have negligible effect on velocity distributions.

**7-2** (a) Show by means of an order of magnitude analysis that the boundary-layer equations for a general fluid can be expressed as

$$x \text{ Motion} \quad \rho \left( \frac{\partial u}{\partial t} + u \frac{\partial u}{\partial x} + v \frac{\partial u}{\partial y} \right) = B_x - \frac{\partial p}{\partial x} + \frac{\partial \tau_{yx}}{\partial y}$$

$$\text{Energy} \quad \rho C_p \left( \frac{\partial T}{\partial t} + u \frac{\partial T}{\partial x} + v \frac{\partial T}{\partial y} \right) = \frac{\partial (k \, \partial T/\partial y)}{\partial y} + \tau_{yx} \frac{\partial u}{\partial y}$$

$$+ q''' + \beta T \left( \frac{\partial p}{\partial t} + u \frac{\partial p}{\partial x} \right)$$

with the others remaining unchanged from their form for a Newtonian fluid.

(b) Show that for steady parallel flow of a constant property fluid over a flat plate where $U = \text{const}$, $u = \partial \psi/\partial y$, and $v = -\partial \psi/\partial x$ with the stream function defined as $\psi = (\nu U x)^{1/2} F(\eta)$ and the similarity variable defined as $\eta = y(U/\nu x)^{1/2}$, the chain rule gives

$$(1) \quad u = \frac{\partial \psi}{\partial y} = \frac{\partial \psi}{\partial x} \overset{0}{\underset{\partial y}{\cancel{\frac{\partial x}{\partial y}}}} + \frac{\partial \psi}{\partial \eta} \frac{\partial \eta}{\partial y} = UF'$$

$$(2) \quad -v = \frac{\partial \psi}{\partial x} = \frac{\partial \psi}{\partial x} \overset{1}{\underset{\partial x}{\cancel{\frac{\partial x}{\partial x}}}} + \frac{\partial \psi}{\partial \eta} \frac{\partial \eta}{\partial x} = \left( \frac{U\nu}{x} \right)^{1/2} \frac{F - \eta F'}{2}$$

$$(3) \quad \frac{\partial u}{\partial y} = \frac{\partial \, UF'}{\partial y} = \frac{\partial \, UF'}{\partial x} \overset{0}{\underset{\partial y}{\cancel{\frac{\partial x}{\partial y}}}} + \frac{\partial \, UF'}{\partial \eta} \frac{\partial \eta}{\partial y} = \left( \frac{U^3}{\nu x} \right)^{1/2} F''$$

$$(4) \quad \frac{\partial^2 u}{\partial y^2} = \frac{\partial \left[ (U^3/\nu x)^{1/2} F'' \right]}{\partial x} \overset{0}{\underset{\partial y}{\cancel{\frac{\partial x}{\partial y}}}} + \frac{\partial \left[ (U^3/\nu x)^{1/2} F'' \right]}{\partial \eta} \frac{\partial \eta}{\partial y}$$

$$= \frac{U^2}{\nu x} F'''$$

$$(5) \quad \frac{\partial u}{\partial x} = \frac{\partial \, UF'}{\partial x} \overset{1}{\underset{\partial x}{\cancel{\frac{\partial x}{\partial x}}}} + \frac{\partial \, UF'}{\partial \eta} \frac{\partial \eta}{\partial x} = -\frac{U \eta F''}{2x}$$

(c) Substitute the results of parts 1–5 into the $x$-motion laminar boundary-layer equation $u \, \partial u/\partial x + v \, \partial u/\partial y = \nu \, \partial^2 u/\partial y^2$ to verify Eq. (7-15).

**7-3** Construct an additional column for Table 7-1 to show the value of $v(Ux/\nu)^{1/2}/U$ for steady parallel flow of a constant property fluid over a flat plate.

**7-4**  An iterative procedure [J. Piercy and N. A. V. Preston, A simple solution of the flat plate problem of skin friction and heat transfer, *Phil. Mag.* **21**, 995–1005 (1936)] for approximating the velocity distribution is as follows. Starting with

$$\frac{d(F'')}{d\eta} = -\left(\frac{F}{2}\right)F'' \tag{7-15}$$

a first integration gives

$$\ln F''(\eta) - \ln F''(0) = -\int_0^\eta \frac{F}{2}d\eta$$

$$F'' = F''(0)\exp\left(-\int_0^\eta \frac{F}{2}d\eta\right)$$

A second integration gives

$$\overset{u/U}{F'(\eta)} - \overset{0}{F'(\emptyset)} = F''(0)\int_0^\eta \exp\left(-\int_0^\eta \frac{F}{2}d\eta\right)d\eta$$

Now $F''(0)$ can be evaluated since

$$F'(\infty) = 1 = F''(0)\int_0^\infty \exp\left(-\int_0^\eta \frac{F}{2}d\eta\right)d\eta$$

so that

$$F''(0) = \frac{1}{\int_0^\infty \exp\left(-\int_0^\eta \frac{F}{2}d\eta\right)d\eta}$$

One further integration gives

$$F(\eta) - \overset{0}{F(\emptyset)} = F''(0)\int_0^\eta \left[\int_0^\eta \exp\left(-\int_0^\eta \frac{F}{2}d\eta\right)d\eta\right]d\eta$$

**(a)** Starting with $F(\eta) = 2\eta$, perform at least two iterations with a new and updated function for $F(\eta)$ obtained from the last iteration.

**(b)** Compare the value of $F''(0)$ from part a after each iteration with the exact answer. Plot $u/U$ against $\eta$ from part a and compare it with the exact profile.

**7-5**  The average drag coefficient for steady parallel flow of a constant property fluid over a flat plate as given by Eq. (7-18) includes the effect of an integrable singularity at the leading edge, $x = 0$. As pointed out by Schlichting [1, p. 131], a more refined analysis that accounts for the singularity results in

$$\frac{\overline{C}_f}{2} = \frac{0.664}{Re_x^{1/2}} + 1.163/Re_x$$

For $Re_x$ of $10^3$, $10^4$, and $10^5$, evaluate the error involved in using only $\overline{C}_f/2 = 0.664/Re_x^{1/2}$ to compute the skin friction coefficient.

**7-6** A method for converting a boundary-value problem into an initial-value problem is sometimes helpful in easing the numerical solution of the laminar boundary-layer equations [(1) S. Goldstein, *Modern Develop-ments in Fluid Dynamics*, Oxford University Press, London, 1938, p. 135; (2) T. Y. Na, Further extension on transformation from boundary value of initial value problems, *SIAM Rev.* **10**, 85–87 (1968); and (3) T. Y. Na, An initial value method for the solution of a class of nonlinear equations in fluid mechanics, ASME Paper 69-WA/FE-8]. For the Blasius equation of Eq. (7-15), consider the transformations

$$\eta = A^a n \quad \text{and} \quad F = A^b f$$

where $a$ and $b$ are constants to be determined and $A$ is a parameter. Under these transformations Eq. (7-15) becomes

$$A^{b-3a}\frac{d^3f}{dn^3} + \frac{1}{2}A^{2b-2a}f\frac{d^2f}{dn^2} = 0$$

$$f(n=0) = 0 = \frac{df(n=0)}{dn}$$

$$A^{b-a}\frac{df(n=\infty)}{dn} = 1$$

For the differential equation to be invariant under the transformation,

$$b - 3a = 2b - 2a$$

Now, realizing that the unknown value of $d^2F(0)/d\eta^2$ is the difficulty in the first place, set it equal to a constant so that

$$\frac{d^2F(0)}{d\eta^2} = A$$

This transforms to

$$A^{b-2a}\frac{d^2F(0)}{dn^2} = A \qquad\qquad (P6\text{-}1)$$

which, to be invariant under the transformation, requires that

$$b - 2a = 1$$

From these two relations between $b$ and $a$ it is found that

$$-a = \tfrac{1}{3} = b$$

The value of $d^2F(0)/d\eta^2$, an item of major interest because of its connection with viscous drag, is known if $A$ is known. From the boundary condition

$$\frac{dF(\infty)}{d\eta} = 1$$

one has the transformed condition

$$A^{b-a}\, df(\infty)/dn = 1$$

Thus

$$A = \left[\frac{df(\infty)}{dn}\right]^{-3/2} \tag{P6-2}$$

where

$$\frac{d^3f}{dn^3} + \frac{1}{2}f\frac{d^2}{dn^2} = 0$$

$$f(0) = 0 = \frac{df(0)}{dn} \quad \text{and} \quad \frac{d^2f(0)}{dn^2} = 1 \tag{P6-3}$$

(a) Solve Eq. (P6-3) numerically on a computer and determine $d^2F(0)/d\eta^2$ from the result. Also, in view of the relation that $u/U = dF/d\eta = A^{2/3}\, df/dn$, determine the velocity profile. Compare both results with the exact answers.

(b) Solve the Blasius equation [Eq. (7-15)], numerically on a computer, using a shooting method.

(c) Cast Eq. (P6-3) into the form suited for an iterative solution (refer to Problem 7-4) of

$$\frac{df(n)}{dn} = \int_0^n \exp\left(-\int_0^n \frac{f}{2}\,dn\right) dn$$

$$f(n) = \int_0^n \left[\int_0^n \exp\left(-\int_0^n \frac{f}{2}\,dn\right) dn\right] dn$$

(d) Starting with $f = 2n$ and using the results of part c, perform at least two iterations with a new and updated $f$ obtained from the last equation in part c after each iteration. Compare the value of $d^2F(0)/d\eta^2$ after each iteration with the exact answer. Comment briefly on the relative ease of this iterative technique in solving the initial-value problem of Problem 7-6 and the boundary-value problem of Problem 7-4.

**7-7** (a) For the steady parallel flow of air over a flat plate at atmospheric pressure at 100°F and a free-stream velocity of 100 ft/sec, determine the magnitude of the vertical component of velocity at the outer edge of the boundary layer at distances of 0.3, 3, and 6 cm from the plate leading edge. At each point determine the boundary-layer thickness and the local viscous shear stress.

(b) Assess the probable accuracy of the numerical results of part a. How well are the boundary-layer assumptions satisfied?

(c) Determine the total drag force acting on one side of the plate up to each point of computation.

(d) If a second plate were to be brought opposite and parallel to the plate of part a, estimate the separation between them that would be required to keep the boundary layers from meeting 6 cm from the leading edge.

**7-8** To assess the importance of fluid properties, work Problem 7-7, but take the fluid to be water at 22°C.

**7-9** Consider application of the conservation principles of Chapter 4 to control volume $A-B-C-D$ for the steady laminar boundary-layer problem sketched in Fig. 7P-9.

(a) Show that the mass flow rate across $B-C$ equals

$$\int_0^y (\rho U - \rho u)\, dy = \int_{x_A}^{x_D} \rho v(y)\, dx$$

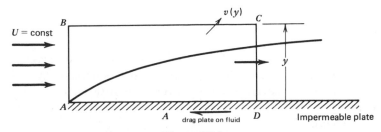

**Figure 7P-9**

**(b)** Show that the drag force exerted on the fluid by the plate is

$$\int_0^y \frac{\rho U U}{g_c} \, dy - \left[ \int_0^y \frac{\rho u u}{g_c} \, dy + \int_{x_A}^{x_B} \frac{\rho v(y) U}{g_c} \, dx \right] - F_{\substack{\text{drag} \\ \text{plate on fluid}}} = 0$$

**(c)** Use the results of parts a and b to show that, when $y \to \infty$ to avoid shear stresses that exist near the plate,

$$F_{\substack{\text{drag} \\ \text{fluid on plate}}} = \frac{\rho}{g_c} \int_0^\infty u(U - u) \, dy$$

$$= \frac{\rho U^2 \delta_2}{g_c} = x \frac{\overline{C_f}}{2} \rho U^2 g_c$$

Hence $\delta_2$ is not only related to drag, but its value is given in terms of the already evaluated $\overline{C_f}/2$. Thus

$$\delta_2 = x \frac{\overline{C_f}}{2} = 0.66412 \left( \frac{\nu x}{U} \right)^{1/2}$$

**7-10** Prepare a brief report, explaining the manner in which a small movable element mounted flush with a plate surface can be used to measure local drag. [Consult: (1) S. Dhawan, Direct measurements of skin friction, NASA Report 1121, 1953; (2) H. W. Liepmann and S. Dhawan, Direct measurements of local skin friction in low-speed and high-speed flow, *Proc. First U.S. Nat. Congr. Appl. Mech.*, 1951, p. 869; (3) J. M. Allen, Improved sensing element for skin-friction balance measurements, *AIAA J.* **18**, 1342–1345 (1980); and (4) reference [59].

**7-11** Show that the similarity solution for steady parallel flow of a constant property fluid over a flat plate results in the local Nusselt number being related to $\theta = (T - T_w)/(T_\infty - T_w)$ by

$$\text{Nu} = \frac{hx}{k} = \left( \frac{Ux}{\nu} \right)^{1/2} \theta'(0)$$

**7-12** Show that Eq. (7-31) reduces to the simple relation $\theta(\eta) = u/U$ if $\text{Pr} = 1$.

**7-13** Consider the case of a steady parallel flow of a constant property fluid over a flat plate. When $\text{Pr} \to 0$, the thermal boundary layer is much thicker than the velocity boundary layer, so that throughout the thermal boundary layer $F \approx \eta$ since then $u/U = F \approx 1$. Such is the case for liquid metals.

(a) Show that in such a case Eq. (7-26) gives

$$\frac{d\theta(0)}{d\eta} = \left(\frac{Pr}{\pi}\right)^{1/2}$$

(b) Show that the result of part a gives

$$\lim_{Pr \to 0} Nu \left(\frac{Re\,Pr}{\pi}\right)^{1/2}$$

where $Nu = hx/k$ and $Re = Ux/\nu$. Or

$$\lim_{Pr \to 0} \frac{Nu}{Re\,Pr^{1/2}} = \frac{0.33206}{\pi^{1/2}} \frac{C_f}{2} = 0.187\frac{C_f}{2}$$

**7-14**  Consider the case of a steady parallel flow of constant-property fluid over a flat plate. When $Pr \to \infty$, the thermal boundary layer is much thinner than the velocity boundary layer, so that throughout the thermal boundary layer $u/U$ is nearly linear. This case was solved by M. A. Leveque in 1928 according to Schlichting [1, p. 272], who notes that this situation can also occur if there is an unheated starting length $x_0$ at the leading edge. A velocity distribution that retains accuracy while giving the necessary linearity is $u = \tau_w y/\mu$. The continuity equation then gives

$$v = - \left(\frac{y^2}{2\mu}\right)\frac{d\tau_w}{dx}$$

(a) Show that the substitution

$$n = \frac{y(\tau_w/\mu)^{1/2}}{\left[9\alpha\int_{x_0}^{x}(\tau_w/\mu)\,dx\right]^{1/3}}$$

transforms the energy equation into the similarity form

$$\frac{d^2\theta}{dn^2} + 3n^2\frac{d\theta}{dn} = 0$$

$$\theta(0) = 0 \quad \text{and} \quad \theta(\infty) = 1$$

(b) Show that the solution to the differential equation of part a is

$$\theta = \frac{d\theta(0)}{dn}\int_0^n e^{-Z^3}\,dZ$$

and that

$$\frac{d\theta(0)}{dn} = \frac{1}{\int_0^\infty e^{-Z^3}\, dZ} = \frac{1}{\Gamma(4/3)} = \frac{1}{0.893}$$

Here $\Gamma(Z)$ is the tabulated gamma function that can be shown to be related to the integral in question as

$$\Gamma(Z+1) = \int_0^\infty \exp(-x^{1/Z})\, dx$$

(c) Show that the result of part b leads to

$$\lim_{\text{Pr} \to \infty} \text{Nu} = \frac{hx}{k} = x\left(\frac{\tau_w}{\mu}\right)^{1/2} \frac{d\theta(0)/dn}{\left[9\alpha\int_0^x(\tau_w/\mu)\, dx\right]^{1/3}}$$

for the case of a flat plate at zero incidence. From Eq. (7-17) the information relating $\tau_w$ to $x$ must now be obtained and the integral evaluated to yield

$$\lim_{\text{Pr} \to \infty} \text{Nu} = 0.339\, \text{Re}^{1/2}\, \text{Pr}^{1/3}$$

or

$$\lim_{\text{Pr} \to \infty} \frac{\text{Nu}}{\text{Re}\,\text{Pr}^{1/3}} = 1.021\frac{C_f}{2}$$

(d) Show that the results of parts a–c can also be achieved by realizing that near the plate, $d^2F/d\eta^2 = 0.33206$ so that $dF/d\eta \approx 0.33206\eta$ and $F \approx 0.33206\eta^2/2$.

**7-15** Find the value of Pr at which
(a) Eqs. (7-28a) and (7-28b) are equal.
(b) Eqs. (7-28b) and (7-28c) are equal.

**7-16** Show that if $\text{Pr} = 1$, Eq. (7-34) gives $\theta_2(\eta) = 1 - [F'(\eta)]^2 = 1 - (u/U)^2$ and the recovery factor as $r = 1$.

**7-17** Consider steady parallel flow of a constant property fluid over a flat plate with viscous dissipation. The value of $\theta_2(0)$ from Eq. (7-34) is desired for the limiting case of $\text{Pr} \to \infty$. Since the thermal boundary layer is much thinner than the velocity boundary layer, $F'' \approx 0.33206$, $F' \approx 0.33206\eta$, and $F \approx 0.33206\eta^2/2$. Thus Eq. (7-34) is approximately given by

$$\theta_2'' + \frac{\text{Pr}}{2}\frac{0.33206\eta^2}{2}\theta_2' = -2\,\text{Pr}\,(0.33206)^2$$

$$\theta_2'(0) = 0 = \theta_2(\infty)$$

(a) Show that the solution is

$$\theta_2(\eta) - \theta_2(0) = -2.4105\,\mathrm{Pr}^{1/3}\int_0^x e^{-y^3}\left[\int_0^y e^{Z^3}\,dZ\right]dy$$

This result, and that of part b, was determined by D. Meksyn, Plate thermometer, *Zeitschrift für angewandte Mathematic und Physik* **11**, 63–68 (1960).

(b) Show that the result of part a yields

$$\lim_{\mathrm{Pr}\to\infty}\theta_2(0) = 1.9\,\mathrm{Pr}^{1/3}$$

(c) Show that the Prandtl number at which the two asymptotic approximations given by Eq. (7-36) achieve equality is $\mathrm{Pr} = 47$.

**7-18** Consider a constant-property fluid in steady parallel flow over an adiabatic flat plate. Show that the total temperature in the fluid $T^\circ = T + u^2/2C_p$ differs between the plate and the free stream by

$$T_\infty^\circ - T_{aw}^\circ = \frac{(1-r)U^2}{2C_p}$$

Note particularly that if $r \equiv 1$, no energy separation occurs. Comment on the magnitude of the energy separation predicted for the case of laminar boundary layers, including pertinent points from the related presentation for energy separation in the Hilsch–Ranque tube by Eckert [6, pp. 427–430].

**7-19** The effect of variable properties described in Section 7.5 made use of the coordinate transformation $\eta = (U/\nu_\infty x)^{1/2}\int_0^y(\rho/\rho_\infty)\,dy$.

(a) Show that the incremental distance $dy$ from the solid surface is expressed according to this transformation by

$$dy = \frac{\nu_\infty x}{U}\frac{\rho_\infty}{\rho}\,d\eta$$

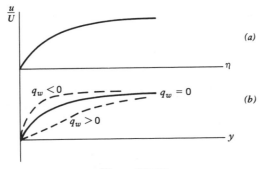

Figure 7P-19

**(b)** On the basis of the result of part a, verify that a velocity distribution as shown in Fig. 7P-19$a$ for the $\eta$ coordinate would have velocity distributions as sketched in Fig. 7P-19$b$ for the $y$ coordinate for gases.

**(c)** Repeat part b for a liquid.

**7-20** Show for wedge flow that the displacement thickness

$$\delta_1 = \int_0^\infty \left(1 - \frac{u}{U}\right) dy$$

is given by

$$\delta_1 \frac{U}{\nu x} = (\eta - F)\,|_0^\infty$$

and that the momentum thickness

$$\delta_2 = \int_0^\infty \frac{u}{U}\left(1 - \frac{u}{U}\right) dy$$

is given by

$$\delta_2 \left(\frac{U}{\nu x}\right)^{1/2} = -\frac{2\left[F''(0) + m(\eta - F)\,|_0^\infty\right]}{3m + 1}$$

Either consult the published solutions of Hartree [19], noting that his definition of $\eta$ may cause his $F$ values to differ from those in this text, or solve the Falker–Skan equation [Eq. (7-58)] numerically to construct two additional columns for $\delta_1$ and $\delta_2$ in Table 7-1.

**7-21** Compare the magnitudes of the heat-transfer coefficients for laminar boundary layer flow over a flat plate for the specified heat flux and specified temperature cases for both the local and average values.

**7-22** Determine whether Reynolds analogy $Nu/Re\,Pr^{1/3} = C_f/2$ is valid for wedge flow on both local and average bases.

**7-23** Determine the variation in local laminar heat-transfer coefficient on the forward part of a cylinder in normal crossflow by use of local similarity as discussed just before Section 7.7. Compare your results with those displayed in Fig. 7-13.

**7-24** Table 7-3 shows that for two-dimensional stagnation flow ($m = 1$), $Nu/Re^{1/2}\,Pr^{0.4} = 0.57$; and Eq. (7-72) shows that for rotationally symmetric stagnation flow, $Nu/Re^{1/2}\,Pr^{0.4} = 0.76$. Here $Nu = hx/k$ and $Re = Ux/\nu$.

**(a)** Show that for cross flow over a cylinder, $U \approx 2U_\infty x/R$. Substitute this relation into the two-dimensional correlation to show that the two-dimensional stagnation point heat-transfer coefficient varies with radius of curvature as

$$\frac{Nu_R}{Re_R^{1/2} Pr^{0.4}} = 0.81$$

or

$$h \sim R^{-1/2}$$

where $Nu_R = hR/k$ and $Re_R = U_\infty R/\nu$

(b) Repeat part a for a sphere, showing $U \approx 3U_\infty x/2R$, to find

$$\frac{Nu_R}{Re_R^{1/2} Pr^{0.4}} = 0.93$$

or

$$h \sim R^{-1/2}$$

(c) On the basis of the results of parts a and b, comment on the reduction in heat flux on entry of a space vehicle into a planetary atmosphere offered by a large radius of curvature.

**7-25** Compare the heat-transfer coefficient that results from an infinite stream impinging perpendicularly on a flat surface with that resulting from an optimal array of round nozzles under the condition that the same average fluid flow per unit plate area is the same in both cases.

**7-26** From Table 7-6 determine the ratio of the local heat-transfer coefficients for constant heat flux to constant-temperature wall conditions for several wedge angles and transpiration rates.

**7-27** Ablative cooling of a fast-moving airborne vehicle can be provided by a solid surface that vaporizes as it is heated by the process of viscous dissipation. During ablation the material surface gradually recedes, and a steady temperature gradient is established in the solid material as sketched in Fig. P-27, where control volume $a$–$b$–$c$–$d$ is attached to the slowly moving surface. The conservation of energy principle applied to control volume $a$–$b$–$c$–$d$ yields (with $H$ the surface material enthalpy corresponding to temperature $T$)

$$\dot{m}_1 H_c(T_c) - \dot{m}_1 H_w(T_w) - q_w = 0$$

The heat flux at the wall is given by

$$q_w = -k\,\frac{\partial T(y = 0)}{\partial y} = h(T_w - T_\infty)$$

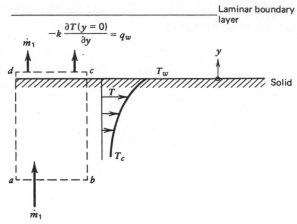

**Figure 7P-27**

and the mass flux is given for binary diffusion with the surface impermeable to air by

$$\dot{m}_1 = \rho h_D \frac{\omega_{1w} - \omega_{1\infty}}{1 - \omega_{1w}} \tag{7P-27}$$

**(a)** Combine these three equations to provide the relation between wall
temperature and mass fraction of

$$\rho h_D \left( \frac{\omega_{1w} - \omega_{1\infty}}{1 - \omega_{1w}} \right)(H_c - H_w) = h(T_w - T_\infty)$$

and rearrange it into

$$\frac{\text{Pr}}{\text{Sc}} \frac{\text{Sh}}{\text{Nu}} = \frac{C_p(T_w - T_\infty)}{H_c - H_w}$$

Here, note that as a result of the change of phase, $H_c - H_w \neq C_{p_1}(T_c - T_w)$. For the laminar boundary layer postulated, the extended
Reynolds analogy of Eq. (7-84) gives

$$\left( \frac{\text{Pr}}{\text{Sc}} \right)^{2/3} = C_p \frac{T_w - T_\infty}{H_c - H_w}$$

**(b)** With the aid of Fig. 7-20 and Eq. (7-88a), show that the effect of
transpiration on $h_D$ is approximately accounted for by, with $h_{D0}$
representing $h_D$ at $v_w = 0$,

$$h_D = h_{D0} - 0.619 v_w = h_{D0} - 0.619 \frac{\dot{m}_1}{\rho}$$

which when used in Eq. (7P-27) gives the mass transfer rate as

$$\dot{m}_1 = \rho h_{D0} \frac{(\omega_{1w} - \omega_{1\infty})/(1 - \omega_{1w})}{1 + 0.619(\omega_{1w} - \omega_{1\infty})/(1 - \omega_{1w})}$$

(c) Take the surface to be composed of liquid water at 50°F in contact with a flowing airstream of 100°F that is perfectly dry. Assume the boundary layer to be laminar, and determine the numerical values of $T_w$, $\dot{m}_1$, and $v_w$.

7-28 A large and dependable energy supply is one of our greatest needs. Although solar energy (about 100 Btu/hr ft² as a daily average in sunny regions) is free, a cheap and large collector is needed to employ such a diffuse energy source. Consider the possibility of damming the Red Sea and closing the Suez Canal. As illustrated in the sketch in Fig. 7P-28, no rivers empty into the Red Sea, and thus the solar-induced evaporation would eventually create a large difference in water level across the dam. Hydraulic turbines could then be utilized to generate electricity by then permitting a steady inflow of Indian Ocean water at a rate equal to evaporation. Explore the feasibility of this idea by estimating:

(a) The time required to create a 35-m difference in water level across the dam. *Note*: Although laminar boundary layers are not likely to be encountered, the analogies they suggest can be applied.

(b) The steady-state water flow rate across the dam that would allow this head difference to be maintained.

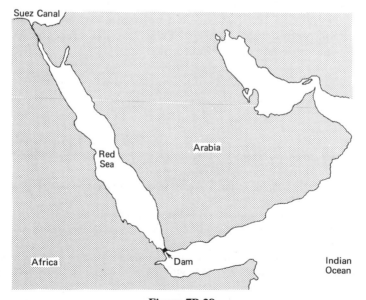

**Figure 7P-28**

(c) The electrical power output of hydraulic turbine-driven electrical generators. The Red Sea is approximately 1300 mi long and 150 mi wide.

(d) The fraction of incident solar energy converted into electrical form. References that can be consulted are: (1) Willy Ley, *Engineers' Dreams*, Viking Press, New York, 1954; (2) R. Hamil, Terraforming the Earth, *Analog Sci. Fict. Sci. Fact* **98** (7), 46–65 (1978); (3) M. A. Kettani and L. M. Gonsalves, Heliohydroelectric power generation, *Solar Energy* **14**, (September 1972); and (4) M. A. Kettani and L. M. Gonsalves, Feasibility of a heliohydroelectric power plant on the eastern shore of Saudi Arabia, in *Proceedings of the Sixth Annual Conference of the Marine Technology Society*, Washington, D.C., June 29–July 1, 1970. A similar scheme was advanced in 1928 by the Bavarian engineer, Herman Sörgel, to dam the Mediterranean Sea from the Atlantic Ocean at Gibraltar and from the Black Sea at the Dardanelles. He also proposed in 1935 that a dam be constructed across the mouth of the Congo River to form a huge lake. The 400-ft elevation difference between the Mediterranean Sea and the Qattara Depression 40 mi inland (a terrain feature near El Alemain in Egypt that forced the German general, Rommel, to disastrously concentrate his tanks when pushing toward Alexandria in World War II) has been proposed for a heliohydroelectric power station. Similarly, the 1200-ft elevation difference between the Mediterranean Sea and the Dead Sea 50 mi inland has been studied by the Israelis —a tunnel or pipeline would be required since there is an intervening mountain range. Additional possibilities occur with South Australia's Lake Eyre, which is below sea level and is near Spencer's Gulf, with the dry "Lake of Pallas" west of Gabès in Tunisia, and with the Salton Sea in California.

# REFERENCES

1   H. Schlichting, *Boundary-Layer Theory*, 6th ed., McGraw-Hill, New York, 1968.

2   A. G. Hansen, *Similarity Analyses of Boundary Value Problems In Engineering*, Prentice-Hall, Englewood Cliffs, NJ, 1964.

3   L. Howarth, On the solution of the laminar boundary layer equations, *Proc. Roy. Soc. Lond.*, **164A**, 547–579 (1938).

4   E. Pohlhausen, Der Wärmeaustausch zwischen festen Körpern und Flüssigkeiten mit kleiner Reibung und kleiner Wärmeleitung, *Zeitschrift für angewandte Mathematik und Mechanik* **1**, 115–121 (1921).

5   E. R. G. Eckert, Engineering relations for heat transfer and friction in high-velocity laminar and turbulent boundary layer flow over surfaces with constant pressure and temperature, *Transact. ASME* **78**, 1273–1284 (1956).

6   E. R. G. Eckert and R. M. Drake, *Analysis of Heat and Mass Transfer*, McGraw-Hill, New York, 1972.

7   A. Dorodnitzyn, On the boundary layer of a compressible gas, *Prikladnaia Matematika i Mekhanika* **6**, 449–485 (1942).

8   K. Stewartson, Correlated incompressible and compressible boundary layers, *Proc. Roy. Soc. Lond.*, Series A **200**, 84–100 (1950).

9   H. A. Simon, C. S. Liu, and J. P. Hartnett, The Eckert reference formulation applied to high speed laminar boundary layers of nitrogen and carbon dioxide, *Internatl. J. Heat Mass Transf.* **10**, 406–409 (1967).

10  E. R. G. Eckert and E. Pfender, Advances in plasma heat transfer, *Adv. Heat Transf.* **4**, 229–316 (1967).

11  E. Reshotko and C. B. Cohen, Heat transfer at the forward stagnation point of blunt bodies, NACA TN 3513, Washington, DC, July 1955.

12  C. B. Cohen and E. Reshotko, Similar solutions for the compressible laminar boundary layer with heat transfer and pressure gradient, NASA Technical Report 1293, 1956, pp. 919–956.

13  S. Levy, Effect of large temperature changes (including viscous heating) upon laminar boundary layers with variable free-stream velocity, *J. Aeronaut. Sci.* **21**, 459–474 (1954).

14  G. Poots and G. F. Raggett, Theoretical results for variable property laminar boundary layers in water, *Internatl. J. Heat Mass Transf.* **10**, 597–610 (1967).

15  L. Prandtl, Uber Flüssigkeits bewegung bei sehr kleiner Reibung, *Proc. 3rd Internatl. Math. Kong. Heidelberg*, 1904.

16  H. Blasius, Grenzschichten in Flüssigkeiten mit kleiner Reibung, *Zeitschrift für angewandte Mathematic und Physik* **56**, 1 (1908); see also NASA TM 1256.

17  A. G. Hansen, *Similarity Analyses of Boundary Value Problems in Engineering*, Prentice-Hall, Englewood Cliffs, NJ, 1964.

18  V. M. Falkner and S. W. Skan, "Some approximate solutions of the boundary-layer equations, *Phil. Mag.* **12**, 865–896 (1931); see also Report of the Memorial Aeronautical Research Committee, London, No. 1 314, 1930.

19  D. R. Hartree, On an equation occurring in Falkner and Skan's approximate treatment of the equations of the boundary layer, *Proc. Cambridge Phil. Soc.* **33**, Part II, 223–239 (1937).

20  S. Goldstein, A note on the boundary layer equations, *Proc. Cambridge Phil. Soc.* **35**, 338–340 (1939).

21  D. Stojanovic, Similar temperature boundary layers, *J. Aerospace Sci.* **26**, 571–574 (1959).

22  E. Eckert, Die Berechnung des Wärmeüberganges in der Laminaren Grenzschicht um strömter Körper, *VDI-Forschungsheft*, No. 416, Berlin, 1942.

23  H. L. Evans, Mass transfer through laminar boundary layers. 7. Further similar solutions to the B-equation for the case $B = 0$, *Internatl. J. Heat Mass Transf.* **5**, 35–57 (1962).

24  H. Schuh, On asymptotic solutions for the heat transfer at varying wall temperatures in a laminar boundary-layer with Hartree's velocity profiles, *J. Aeronaut. Sci.* **20**, 146–147 (1953).

25  S. Levy, Heat transfer to constant-property laminar boundary-layer flows with power-function free-stream velocity and wall-temperature variation, *J. Aeronaut. Sci.* **19**, 341–348 (1952).

26  E. A. Brun, *Selected Combustion Problems*, Vol. II, AGARD, Pergamon Press, New York, 1956, pp. 105–198.

27  D. R. Chapman and W. M. Rubesin, Temperature and velocity profiles in the compressible laminar boundary layer with arbitrary distribution of surface temperature, *J. Aeronaut. Sci.* **16**, 547–565 (1949).

28  D. B. Spalding and W. M. Pun, A review of methods for predicting heat-transfer coefficients for laminar uniform-property boundary layer flows, *Internatl. J. Heat Mass Transf.* **5**, 239–249 (1962).

29  E. R. G. Eckert and J. N. B. Livingood, Method for calculation of heat transfer in laminar region of air flow around cylinders of arbitrary cross section, NASA TN 2733, Washington, DC, 1952.

30   W. Mangler, Zusammenhang zwischen ebenen und rotationssymmetrischen Grenzschichten in kompressiblen Flüssigkeiten, *Zeitschrift für angewandte Mathematik und Mechanik* **28**, 97–103 (1948).

31   L. Lees, Laminar heat transfer over blunt-nosed bodies at hypersonic flight speeds, *Jet Propul.* **26**, 259–269 (1956).

32   E. R. G. Eckert and O. E. Tewfik, Use of reference enthalpy in specifying the laminar heat-transfer distribution around blunt bodies in dissociated air, *J. Aerospace Sci.* **27**, 464–466 (1960).

33   C. F. Dewey and J. F. Gross, Exact similar solutions of the laminar boundary-layer equations, *Adv. Heat Transf.* **4**, 317–446 (1967).

34   E. Y. C. Sun, A compilation of coordinate transformations applied to the boundary-layer equations for laminar flows, Deutsche Versuchanstalt für Luftfahrt Report No. 121, 1960.

35   L. C. Burmeister and R. G. Schoenhals, Effect of pressure fluctuations on laminar film boiling, *Prog. Heat Mass Transf.* **2**, 371–394 (1969).

36   H. Martin, Heat and mass transfer between impinging gas jets and solid surfaces, *Adv. Heat Transf.* **13**, 1–60 (1977).

37   J. P. Hartnett and E. R. G. Eckert, Mass-transfer cooling in a laminar boundary layer with constant fluid properties, *Transact. ASME* **79**, 247–254 (1957).

38   R. Iglisch, Exakte Berechnung der laminaren Reibungsschicht an der längsangeströmten ebenen Platte mit homogener Absaugung, *Schriften d. dt. Akad. d. Luftfahrtforschung* **8B** (1) (1944); see also NASA RM 1205, Washington, DC, 1949.

39   P. L. Donoughe and J. N. B. Livingood, Exact solutions of laminar-boundary-layer equations with constant property values for porous wall with variable temperature, NASA TN 3151, Washington, DC, 1954.

40   W. M. Kays and M. E. Crawford, *Convective Heat and Mass Transfer*, McGraw-Hill, New York, 1980.

41   J. T. Howe and W. A. Mersman, Solutions of the laminar compressible boundary-layer-equations with transpiration which are applicable to the stagnation regions of axisymmetric blunt bodies, NASA TN D-12, Washington, DC, 1959.

42   E. M. Sparrow and B. J. Lovell, Heat transfer characteristics of an obliquely impinging circular jet, *Transact. ASME, J. Heat Transf.* **102**, 202–209 (1980).

43   S. W. Churchill and M. Bernstein, A correlating equation for forced convection from gases and liquids to a circular cylinder in crossflow, *Transact. ASME, J. Heat Transf.* **99**, 300–306 (1977).

44   T. Kao and H. G. Elrod, Rapid calculation of heat transfer in nonsimilar laminar incompressible boundary layers, *AIAA J.* **14**, 1746–1751 (1976).

45   K. Stewartson, Further solutions of the Falkner–Skan equation, *Proc. Cambridge Phil. Soc.* **50** 454–465 (1954).

46   T. Cebeci and H. B. Keller, Shooting and parallel shooting methods for solving the Falkner–Skan boundary-layer equation, *J. Comput. Phys.* **7**, 289–300 (1971).

47   A. M. O. Smith and D. W. Clutter, Machine calculation of compressible laminar boundary layers, *AIAA J.* **3**, 639–647 (1965).

48   T. Cebeci and P. Bradshaw, *Momentum Transfer In Boundary Layers*, Hemisphere (McGraw-Hill), New York, 1977.

49   H. B. Keller and T. Cebeci, Accurate numerical methods for boundary-layer flows, Part 1, Two-dimensional laminar flows, in *Lecture Notes In Physics*, Vol. 8, *Proceedings Of The Second International Conference On Numerical Methods In Fluid Dynamics*, Springer-Verlag, New York, 1971, p. 92.

50   T. Cebeci, A. M. O. Smith, and G. Mosinskis, Solution of the incompressible turbulent boundary-layer equations with heat transfer, *Transact. ASME, J. Heat Transf.* **92**, 133–143 (1970).

**51** C. Chow, *An Introduction To Computational Fluid Mechanics*, Wiley, New York, 1979, p. 228.

**52** J. C. McWhorter and M. H. Sadd, Numerical anisotropic heat conduction solutions using boundary-fitted coordinate systems, *Transact. ASME, J. Heat Transf.* **102**, 308–311 (1980).

**53** J. F. Thompson, F. C. Thames, and C. W. Mastin, TOMCAT—a code for numerical generation of boundary-fitted curvilinear coordinate systems on fields containing any number of arbitrary two-dimensional bodies, *J. Comput. Phys.* **24**, 274–302 (1977).

**54** H. A. Dwyer, R. J. Kee, and B. R. Sanders, Adaptive grid method for problems in fluid mechanics and heat transfer, *AIAA J.* **18**, 1205–1212 (1980).

**55** D. G. Fox and J. W. Deardorff, Computer methods for simulation of multidimensional, nonlinear, subsonic, incompressible flow, *Transact. ASME, J. Heat Transf.* **94**, 337–346 (1972).

**56** A. Brandt, Multilevel adaptive computations in fluid dynamics, *AIAA J.* **18**, 1165–1172 (1980).

**57** R. H. Pletcher, On a finite-difference solution for the constant-property turbulent boundary layer, *AIAA J.* **7**, 305–311 (1969).

**58** V. T. Morgan, The overall convective heat transfer from smooth circular cylinders, *Adv. Heat Transf.* **11**, 199–264 (1975).

**59** D. J. Monson and H. Higuchi, Skin friction measurements by a dual-laser-beam interferometer technique, *AIAA J.* **19**, 739–744 (1981).

**60** S. V. Patankar, A numerical method for conduction in composite materials, flow in irregular geometries and conjugate heat transfer, in *Proceedings of the Sixth International Heat Transfer Conference, Toronto, Canada*, Vol. 3, 1978, paper CO-14, pp. 297–302.

**61** S. V. Patankar, *Numerical Heat Transfer And Fluid Flow*, Hemisphere (McGraw-Hill), New York, 1980.

**62** E. M. Sparrow and C. F. Hsu, Analytically determined fin-tip heat transfer coefficients, *Transact. ASME, J. Heat Transf.* **103**, 18–25 (1981).

# 8

# INTEGRAL METHODS

The problems treated thus far have in most cases been of interest in their own right, and the analytical techniques employed in the solution process resulted in exact solutions, if the underlying assumptions are accepted. A rather high price was paid for this exactitude, however, since the usually assumed simple geometries and, most importantly, laminar flow are restrictive.

In many instances the solution methods used previously are either not directly applicable (e.g., when flow is turbulent) or do not give answers in closed form. A solution method is desired that, in addition to overcoming these difficulties, gives accurate answers fairly easily for complex situations, even though the answers might not be exact. An integral method meets these needs, as is seen later.

## 8.1 LEIBNITZ FORMULA FOR DIFFERENTIATION OF AN INTEGRAL WITH RESPECT TO A PARAMETER

In formulating an integral description of a phenomenon, it is helpful to be able to differentiate an integral with respect to a parameter. Leibnitz' formula for this purpose is

$$\frac{\partial}{\partial t}\left[\int_{\alpha(t)}^{\beta(t)} f(x, t)\, dx\right] = \int_{\alpha(t)}^{\beta(t)} \frac{\partial f(x, t)}{\partial t}\, dx + f(x = \beta(t), t)\frac{d\beta(t)}{dt}$$

$$-f(x = \alpha(t), t)\frac{d\alpha(t)}{dt} \qquad (8\text{-}1)$$

Although this formula is proved in numerous advanced calculus texts, a graphical reminder of its truth is helpful here. Figure 8-1 shows the basic operations to be performed with the understanding that one basically seeks,

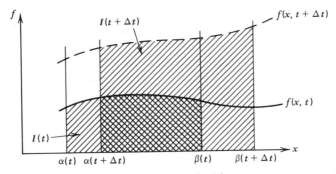

**Figure 8-1**  Differentiation of an integral with respect to a parameter.

with $I(t) = \int_{\alpha(t)}^{\beta(t)} f(x, t) \, dt$,

$$\lim_{\Delta t \to 0} \frac{I(t + \Delta t) - I(t)}{\Delta t}$$

It can be seen that

$$\lim_{\Delta t \to 0} \frac{I(t + \Delta t) - I(t)}{\Delta t}$$

$$= \lim_{\Delta t \to 0} \left\{ \left[ \int_{\alpha(t)}^{\beta(t)} f(x, t + \Delta t) \, dx + f\left( x = \beta\left( t + \frac{\Delta t}{2} \right) \right) [\beta(t + \Delta t) - \beta(t)] \right. \right.$$

$$\left. \left. -f\left( x = \alpha\left( t + \frac{\Delta t}{2} \right) \right) [\alpha(t + \Delta t) - \alpha(t)] \right] - \int_{\alpha(t)}^{\beta(t)} f(x, t) \, dx \right\} / \Delta t$$

or, which was to be proved,

$$\frac{\partial}{\partial t} \left[ \int_{\alpha(t)}^{\beta(t)} f(x, t) \, dx \right] = \int_{\alpha(t)}^{\beta(t)} \frac{\partial}{\partial t} f(x, t) \, dx$$

$$+ f(x = \beta(t), t) \frac{d\beta(t)}{dt} - f(x = \alpha(t), t) \frac{d\alpha(t)}{dt}$$

## 8.2   TRANSIENT ONE-DIMENSIONAL CONDUCTION

Although integral methods appear to have been first used to solve boundary-layer problems by von Karman [1] and Pohlhausen [2], they are most conveniently introduced by considering transient heat conduction. The simplicity of the mathematics in such an application allows concentration on the method rather than on the problem. Goodman [3–5] treats one-dimensional transient heat conduction extensively, whereas Sfeir [6] treats two-dimensional steady conduction problems; Langford [7] can be consulted for additional discussion.

Consider a semi-infinite solid that is initially at a uniform temperature $T_0$ whose surface temperature is suddenly raised to a new constant level $T_s$ as shown in Fig. 8-2. The describing equation and boundary conditions are

$$\rho C_p \frac{\partial T}{\partial t} = \frac{\partial (k \, \partial T / \partial x)}{\partial x} \tag{8-2a}$$

$$T(x, 0) = T_0 \tag{8-2b}$$

$$T(\infty, t) = T_0 \tag{8-2c}$$

$$T(0, t > 0) = T_s \tag{8-2d}$$

This is a classical problem whose solution is available from many sources, such as Carslaw and Jaeger [8]. Here we wish to convert to an integral formulation that is done by integrating the differential equation with respect to $x$. This gives

$$\int_0^x \rho C_p \frac{\partial T}{\partial t} \, dx = \int_0^x \frac{\partial}{\partial x} \left( k \frac{\partial T}{\partial x} \right) dx$$

The right-hand side is immediately integrable to put the preceding relation into the form

$$\int_0^x \rho C_p \frac{\partial T}{\partial t} \, dx = \left[ -k \frac{\partial T(0, t)}{\partial x} \right] - \left[ -k \frac{\partial T(x, t)}{\partial x} \right]$$

$$= q_w(t) - q_x(t) \tag{8-3}$$

where $q_w$ is the heat flux into the wall at $x = 0$ and $q_x$ is the heat flux in the $x$ direction at $x$. Next it is assumed that the thermal disturbance penetrates into

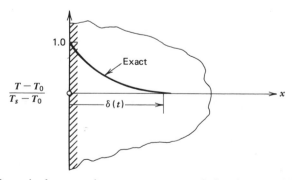

**Figure 8-2** Dimensionless transient temperature variation in a semi-infinite solid of initially uniform temperature subjected to a step change in surface temperature.

the wall only a distance $\delta(t)$ that can also be regarded as a boundary-layer thickness—an approximation is hereby introduced since the exact solution shows that the disturbance is nonzero, although often small, everywhere for $t > 0$—beyond which nothing of interest occurs. Therefore, the upper limit of the integral in Eq. (8-3) is set equal to $\delta(t)$, giving

$$\int_0^{\delta(t)} \rho C_p \frac{\partial T}{\partial t} \, dx = q_w(t) - q_\delta(t)$$

In accordance with this assumption, $q_\delta = 0$ hereafter, so no heat is conducted out of the penetration thickness. Assuming that $\rho$ and $C_p$ are constant and applying the Leibnitz formula [Eq. (8-1)] leads to

$$\frac{\partial}{\partial t} \left( \int_0^\delta T \, dx \right) - T_\delta \frac{d\delta}{dt} + T_0 \frac{d0}{dt} = \frac{q_w}{\rho C_p}$$

But according to the earlier assumption, $T_\delta = T_0$, so the preceding relation simplifies to

$$\rho C_p \frac{\partial}{\partial t} \left[ \int_0^{\delta(t)} (T - T_0) \, dx \right] = q_w \qquad (8\text{-}4)$$

The integral formulation of Eq. (8-4) expresses the conservation of energy principle, stating that the difference between the rate of energy flow in and out must equal the rate of energy storage, for a very specific finite control volume somewhat more clearly than does the partial differential formulation of Eq. (8-2) for an infinitesimal control volume. Also, the steps followed in achieving the integral formulation from the partial differential formulation have reversed those followed in deriving the partial differential formulation from an integral one such as were discussed in Chapter 4. The advantage of Eq. (8-4) is that a reasonable temperature profile can be assumed for insertion into the temperature integral, following which $\delta(t)$ can be evaluated from the resulting first-order differential equation that will be ordinary rather than partial, although nonlinear.

Any temperature profile must satisfy $T(x = 0, t) = T_s$ and $T(x = \delta, t) = T_0$, which corresponds to the two boundary conditions of Eqs. (8-2c) and (8-2d). The simplest profile satisfying these conditions is

$$T - T_0 = (T_s - T_0)\left(1 - \frac{x}{\delta}\right), \qquad x \leqslant \delta$$

$$= 0, \qquad x > \delta$$

From this it is found that

$$\int_0^\delta (T - T_0) \, dx = \delta \int_0^1 (T - T_0) d(x/\delta) = \delta(T_s - T_0)\frac{1}{2}$$

$$q_w = -k_w \frac{\partial T(0, t)}{\partial x} = k_w \frac{T_s - T_0}{\delta}$$

Introduction of these results into the integral formulation of Eq. (8-4) gives

$$\rho C_p \frac{\partial}{\partial t} \frac{(T_s - T_0)\delta}{2} = k_w \frac{T_s - T_0}{\delta}$$

Note that $q_\delta = 0$, even though a linear temperature profile suggests $q_{cond}$ is constant, since $T = T_0$ for $x \geqslant \delta$ as sketched in Fig. 8-3. This differential equation can be simplified to

$$\frac{d\delta^2}{dt} = 4\alpha_w$$

where $\alpha_w = k_w/\rho C_p$. Assumption that $\delta(t = 0) = 0$ gives

$$\delta = (4\alpha_w t)^{1/2} \qquad (8-5)$$

showing that, as expected from the drunkard's walk view of diffusion in Section 2.6, the penetration depth of a disturbance varies as the square root of elapsed time. Knowing $\delta$, a somewhat fictitious quantity, one can now evaluate the wall heat flux, a real quantity, as

$$q_w = k_w \frac{T_s - T_0}{\delta} = (T_s - T_0)\left(k_w \frac{\rho C_p}{4t}\right)^{1/2} \qquad (8-6)$$

Comparison of the result of Eq. (8-6) with the exact answer for the case of constant properties $q_{w,\,exact} = (T_s - T_0)(k\rho C_p/\pi t)^{1/2}$ gives

$$\frac{q_w}{q_{w,\,exact}} = \left(\frac{\pi}{4}\right)^{1/2}$$

Thus the integral method with a linear profile gives a wall heat flux prediction that is only 12% low, which is surprisingly good agreement in view of the coarseness of the assumed temperature profile.

Review of the effort just expended reveals that the major approximation occurs in using an assumed profile in the integral. Note that integration over space eliminated the most difficult part of the problem—it is second order in space but only first order in time. Reliance on the smoothing effect of integration makes use of an assumed profile reasonably accurate since the

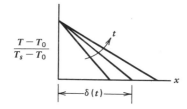

$$\frac{T - T_0}{T_s - T_0}$$

**Figure 8-3** Linear approximation to the transient variation of dimensionless temperature in a semi-infinite solid of initially uniform temperature subjected to a step change in surface temperature.

conservation of energy principle is satisfied on the average. The conservation of energy principle is unlikely to be satisfied at a particular interior position, however; collocation methods [5] have been devised to improve local accuracy.

Recall that the assumed profile used before had a linear characteristic, which is unlikely to lead to the best possible accuracy, and was a two-parameter curve composed of the first two terms of a polynomial—$T - T_0 = a + bx + cx^2 + \cdots$. If a two-parameter profile is to be used, there is no really compelling reason for it to be a polynomial, of course. It could be a collection of functions that more nearly fit the true shape of the profile such as $T - T_0 = [1 - \sin(\frac{1}{2}\pi x/\delta)](T_s - T_0)$. The primary advantage of a polynomial profile is that it is easy to integrate and differentiate.

Acquisition of a more accurate temperature profile and thus increase of the integral methods's accuracy requires consideration of the various conditions that can be imposed. Although not necessarily exhaustive, these are

$$\text{At } x = 0 \quad T - T_0 = T_s - T_0 \xleftarrow{\substack{\text{from} \\ \text{boundary conditions} \\ (must \text{ be satisfied})}} T - T_0 = 0 \quad \text{at } x = \delta$$

$$\xrightarrow{\substack{\text{physical insight} \\ \text{since } q = 0 \text{ at } x = \delta}} \frac{\partial T}{\partial x} = 0$$

$$\frac{\partial^2 T}{\partial x^2} = 0 \xleftarrow{\substack{\text{from D. E. (conservation of energy} \\ \text{at the two boundaries) since}}} \frac{\partial^2 T}{\partial x^2} = 0$$

$$\frac{\partial T}{\partial t} = 0 \quad \text{at} \quad x = 0, \delta$$

The last two conditions from the energy equation presume that $k$ is constant and guarantee that conservation of energy will be satisfied at the two boundaries. Assuming a polynomial profile $T - T_0 = a + bx + cx^2 + \cdots$ and keeping in mind that the lowest-order terms of the polynominal are the most important (as are the lowest-order derivatives of the preceding conditions to be imposed) [4], one can evaluate the coefficients from the five algebraic equations:

$$T(0) - T_0 = T_s - T_0 = a$$

$$T(\delta) - T_0 = 0 = a + b\delta + c\delta^2 + d\delta^3 + e\delta^4$$

$$\frac{\partial T(\delta)}{\partial x} = 0 = b + 2c\delta + 3d\delta^2 + 4e\delta^3$$

$$\frac{\partial^2 T(0)}{\partial x^2} = 0 = 2c$$

$$\frac{\partial^2 T(\delta)}{\partial x^2} = 0 = 2c + 6d\delta + 12e\delta^2$$

The resulting temperature profile is the quartic

$$\frac{T - T_0}{T_s - T_0} = 1 - 2\left(\frac{x}{\delta}\right) + 2\left(\frac{x}{\delta}\right)^3 - \left(\frac{x}{\delta}\right)^4 \qquad (8\text{-}7)$$

Note that this profile displays a similarity form inasmuch as $x/\delta$ is the only parameter and all coefficients are constant. If the coefficients of the assumed polynomical do not turn out to be constants, the problem being described is a *nonsimilar* one; the coefficients must then be evaluated in the course of the problem solution, usually by way of the solution of a differential equation. The integral method is well suited for nonsimilar problems. From Eq. (8-7) it follows that

$$\int_0^\delta (T - T_0)\,dx = \delta \int_0^1 (T - T_0)\,d(x/\delta) = (T_s - T_0)\delta\frac{3}{10}$$

$$q_w = -k\frac{\partial T(0)}{\partial x} = 2k\frac{T_s - T_0}{\delta}$$

Introduction of these results into the integral formulation [Eq. (8-4)] gives

$$\rho C_p \frac{d}{dt}\left[(T_s - T_0)\delta\frac{3}{10}\right] = 2k\frac{T_s - T_0}{\delta}$$

which simplifies to

$$\frac{d\delta^2}{dt} = \frac{40\alpha t}{3}$$

The solution for the penetration depth is, assuming $\delta(t = 0) = 0$,

$$\delta = \left(\frac{40\alpha t}{3}\right)^{1/2} \qquad (8\text{-}8)$$

This is nearly twice the value predicted for a linear profile in Eq. (8-5), but both results agree that $\delta^2/\alpha t = \text{const}$. The wall heat flux is available, now that $\delta$ is known, as

$$q_w = 2k\frac{T_s - T_0}{\delta} = (T_s - T_0)\left(3k\frac{\rho C_p}{10t}\right)^{1/2} \qquad (8\text{-}9)$$

Comparison with the exact answer gives

$$q_w/q_{w,\text{exact}} = \left(\frac{3\pi}{10}\right)^{1/2}$$

which is only 3% low.

This success in wall heat flux prediction attests to the potential accuracy of the integral method.

## 8.3 LAMINAR BOUNDARY LAYERS BY THE INTEGRAL METHOD

Enough laminar boundary-layer problems have been solved to provide a feel for proper trends and to enable the accuracy of approximate answers to be judged. Because turbulent boundary layers possess peculiar uncertainties, one wishes to be confident of the integral method before applying it to them. The following discussion is primarily aimed at supplying that confidence.

In the interests of simplicity, only those geometries already familiar are treated. Specifically, a flat plate is always assumed as shown in Fig. 8-4. At the outset it will not be specified whether the flow is forced or natural.

### Motion Equations

The boundary-layer equations and their boundary conditions are (for constant properties, steady state, and laminar flow):

$$\frac{\partial u}{\partial x} + \frac{\partial v}{\partial y} = 0 \tag{8-10a}$$

$$u\frac{\partial u}{\partial x} + v\frac{\partial u}{\partial y} = \nu\frac{\partial^2 u}{\partial y^2} + \begin{cases} \frac{1}{\rho}\left(B_x - \frac{dP}{dx}\right) = \end{cases} \begin{cases} U\dfrac{dU}{dx} & \text{forced convection} \\[2mm] \dfrac{(\rho_\infty - \rho)g}{\rho_\infty} & \text{natural convection} \end{cases}$$

$$\tag{8-10b}$$

$$u(y = 0) = 0 \tag{8-10c}$$

$$v(y = 0) = v_w \tag{8-10d}$$

$$u(y = \delta) = \begin{cases} U(x) & \text{forced convection} \\ 0 & \text{natural convection} \end{cases} \tag{8-10e}$$

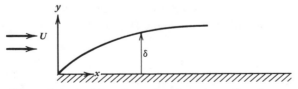

**Figure 8-4** Laminar boundary layer on a flat plate.

Before specifying natural or forced convection, the integral formulation is taken as far as possible. The x-motion equation is observed to be most difficult in the $y$ direction, as it is of second order there. Also, it is desirable to ultimately make assumptions concerning only $u$, and not $v$, because we have a better feel for the proper behavior of $u$ (and also because wall shear stress depends on $u$ and not $v$, and it is probably wall shear that is of ultimate interest).

The continuity equation [Eq. (8-10a)] permits $v$ to be found in terms of $u$. First, integrate with respect to $y$ from $y = 0$ to $y = \delta$ (note that integration beyond $y = \delta$ will add nothing to an integral since conditions are constant outside the boundary layer) to obtain

$$\int_0^\delta \frac{\partial u}{\partial y}\,dy + \int_0^\delta \frac{\partial v}{\partial y}\,dy = 0$$

which gives

$$\int_0^\delta \frac{\partial u}{\partial x}\,dy + v(y = \delta) - v(y = 0) = 0$$

or

$$v_\delta = v_w - \int_0^\delta \frac{\partial u}{\partial x}\,dy$$

Application of Leibnitz' formula [Eq. (8-1)] puts this result in the form

$$v_\delta = v_w + u(y = \delta)\frac{d\delta}{dx} - \frac{\partial}{\partial x}\left(\int_0^\delta u\,dy\right)$$

$$= v_w + U\frac{d\delta}{dx} - \frac{\partial}{\partial x}\left(\int_0^\delta u\,dy\right) \qquad (8\text{-}11)$$

Turning now to the x-motion equation [Eq. (8-10b)], it is convenient to first rearrange it into the conservation-law form

$$\frac{\partial u\,u}{\partial x} + \frac{\partial v\,u}{\partial y} - u\left(\frac{\partial u}{\partial x} + \frac{\partial v}{\partial y}\right) = \frac{B_x - dP/dx}{\rho} + \nu\frac{\partial^2 u}{\partial y^2}$$

The last term in parentheses on the left-hand side identically equals zero according to the continuity equation [Eq. (8-10a)]. Now, integration from $y = 0$ to $y = \delta$ gives

$$\int_0^\delta \frac{\partial u^2}{\partial x}\,dy + \int_0^\delta \frac{\partial vu}{\partial y}\,dy = \int_0^\delta \left(B_x - \frac{dP}{dx}\right)\frac{1}{\rho}\,dy + \nu\int_0^\delta \frac{\partial^2 u}{\partial y^2}\,dy$$

which becomes

$$\int_0^\delta \frac{\partial u^2}{\partial x}\, dy + v(y = \delta)u(y = \delta) - v(y = 0)u(y = 0)$$

$$= \frac{1}{\rho}\left[\int_0^\delta \left(B_x - \frac{dP}{dx}\right) dy + \mu \frac{\partial u(y = \delta)}{\partial y} - \mu \frac{\partial u(y = 0)}{\partial y}\right]$$

The no-slip boundary condition of Eq. (8-10c) and the relationship for $v_\delta$ given by Eq. (8-11) then permit this equation to be simplified to

$$\int_0^\delta \frac{\partial u^2}{\partial x}\, dy + Uv_w + U^2 \frac{d\delta}{dx} - U\frac{\partial}{\partial x}\left(\int_0^\delta u\, dy\right) = \frac{1}{\rho}\left[\int_0^\delta \left(B_x - \frac{dP}{dx}\right) dy - \tau_w\right]$$

Here $\tau_w = \mu\, \partial u(y = 0)/\partial y$, and it has been assumed (on the physical ground that there is no velocity gradient at the outer edge of the boundary layer) that $\tau_\delta = \mu\, \partial u(y = \delta)/\partial y = 0$. Application of the Leibnitz formula gives

$$\frac{\partial}{\partial x}\left(\int_0^\delta u^2\, dy\right) - U\frac{\partial}{\partial x}\left(\int_0^\delta u\, dy\right) = \frac{1}{\rho}\left[\int_0^\delta \left(B_x - \frac{dP}{dx}\right) dy - (\tau_w + \rho Uv_w)\right]$$

The right-hand side can be rearranged to finally give

$$\frac{\partial}{\partial x}\left[\int_0^\delta u(u - U)\, dy\right] + \left(\int_0^\delta u\, dy\right)\frac{dU}{dx} - \frac{1}{\rho}\int_0^\delta \left(B_x - \frac{dP}{dx}\right) dy = \frac{\tau_w + \rho Uv_w}{\rho}$$

$$(8\text{-}12)$$

The integral formulation [Eq. (8-12)] contains no $v$ values except for $v_w$, which can be understood as a known quantity. Note that if $v_w$ is positive, the net effect is to increase apparent wall shear; if $v_w$ is negative, the net effect is to decrease wall shear. This insight follows on physical grounds since if $v_w$ is positive, fluid with no $x$ momentum is being added to the boundary layer and must be accelerated.

Further progress requires that the type of convection considered (natural or forced) be specified. Specialization to forced convection $(B_x - dP/dx)/\rho = U\, dU/dx$ then puts Eq. (8-12) into the form

$$\frac{\partial}{\partial x}\left[U^2 \int_0^\delta \frac{u}{U}\left(1 - \frac{u}{U}\right) dy\right] + \left[\int_0^\delta \left(1 - \frac{u}{U}\right) dy\right] U\frac{dU}{dx} = \frac{\tau_w + \rho Uv_w}{\rho}$$

It is customary (and more compact) to let the displacement thickness be denoted by

$$\delta_1 = \int_0^\delta \left(1 - \frac{u}{U}\right) dy$$

and to let the momentum thickness be denoted by

$$\delta_2 = \int_0^\delta \frac{u}{U}\left(1 - \frac{u}{U}\right) dy$$

as in Section 7.2. The integral formulation for forced convection is then

$$\frac{\partial(U^2\delta_2)}{\partial x} + \delta_1 U\frac{dU}{dx} = \frac{\tau_w + \rho U v_w}{\rho} \qquad \qquad (8\text{-}13)$$

An approximate, but reasonable, velocity profile is next selected for use in the integrals of Eq. (8-13). A polynomial is assumed as

$$u = a + by + cy^2 + dy^3$$

and subjected to the conditions

from

At $y = 0$   $u = 0$ ← boundary conditions → $u = U$   at $y = \delta$

( *must* be satisfied)

physical insight since      $\dfrac{\partial u}{\partial y} = 0$

$\tau = 0$ at $y = \delta$    →

$$v_w \frac{\partial u}{\partial y} = U\frac{dU}{dx} + v\frac{\partial^2 u}{\partial y^2} \quad \xleftarrow{\text{from D. E.}} \quad \frac{\partial^2 u}{\partial y^2} = 0$$

Further specialization in the interests of simplicity and clarity is accomplished by restricting attention to the case of $U = $ const and $v_w = 0$, for which case exact solutions are given in Section 7.2. The four algebraic equations to be solved for the polynomial coefficients are

$$u(0) = 0 = a$$

$$u(\delta) = U = a + b\delta + c\delta^2 + d\delta^3$$

$$\frac{\partial u(\delta)}{\partial y} = 0 = b + 2c\delta + 3d\delta^2$$

$$\frac{\partial^2 u(\delta)}{\partial y^2} = 0 = 2c + 6d\delta$$

The resulting velocity profile is the cubic

$$\frac{u}{U} = \begin{cases} 1, & \dfrac{y}{\delta} > 1 \\[2mm] \dfrac{3}{2}\dfrac{y}{\delta} - \dfrac{1}{2}\left(\dfrac{y}{\delta}\right)^3, & \dfrac{y}{\delta} \leqslant 1 \end{cases} \qquad (8\text{-}14)$$

which is seen to have a similarity form inasmuch as the coefficients are constants. From Eq. (8-14) it follows that

$$\delta_1 = \delta \int_0^1 \left(1 - \frac{u}{U}\right) d(y/\delta) = \frac{3\delta}{8} \approx \frac{\delta}{3}$$

$$\delta_2 = \delta \int_0^1 \frac{u}{U}\left(1 - \frac{u}{U}\right) d(y/\delta) = \frac{39\delta}{280} \approx \frac{\delta}{7}$$

$$\tau_w = \mu \frac{\partial u(0)}{\partial y} = \frac{\mu}{\delta} \frac{\partial u(0)}{\partial(y/\delta)} = \frac{3\mu U}{2\delta}$$

Introduction of these results into Eq. (8-13) gives the first-order nonlinear ordinary differential equation

$$\frac{d}{dx}\left(\frac{39U^2\delta}{280}\right) = \frac{3\nu U}{2\delta}$$

Because $U = $ const here, this equation simplifies to

$$\frac{d\delta^2}{dx} = \frac{280\nu}{13U}$$

With the assumption that $\delta(x = 0) = 0$, the solution is

$$\delta = \left(\frac{280\nu x}{13U}\right)^{1/2}$$

or

$$\frac{\delta}{x} = \frac{4.64}{Re^{1/2}} \tag{8-15}$$

where $Re = Ux/\nu$. The friction coefficient is found, since the boundary-layer thickness is now known, from

$$\tau_w = \frac{C_f}{2} \frac{\rho U^2}{g_c}$$

to be

$$\frac{C_f}{2} = \frac{0.323}{Re^{1/2}} \tag{8-16}$$

Comparison with the exact answers from Eqs. (7-16) and (7-17), $\delta Re^{1/2}/x = 5$ and $(C_f/2)Re^{1/2} = 0.33206$, reveals that the integral method prediction for the boundary-layer thickness (a somewhat imaginary quantity) is 7% low

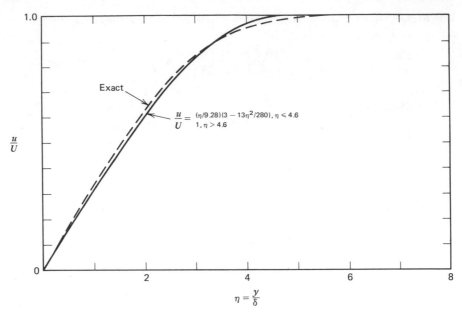

**Figure 8-5** Comparison of velocity profiles from exact and integral methods for a laminar boundary layer on a flat plate.

whereas its prediction for the friction coefficient is only 3% low. Because of the arbitrary and mildly fictitious nature of the boundary-layer thickness, it is best thought of as a parameter determined from the integral equation that allows the assumed profile to satisfy the macroscopic conservation principle to the greatest possible extent. The velocity distribution predicted by the integral method is shown in Fig. 8-5 to be in close agreement with the exact answer.

The integral form of the $x$-motion equation for a body of revolution is the subject of Problem 8-12.

**Energy Equation**

The velocity distribution in the laminar boundary layer is now known from an integral method for steady parallel flow of a constant-property fluid over an

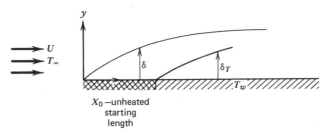

**Figure 8-6** Laminar boundary layer on a flat plate with an unheated starting length.

impermeable flat plate. Solution of the energy equation by an integral method is undertaken next with the same geometry for a case for which the exact answer is known. As shown in Fig. 8-6, the plate and free-stream temperature are constant, but there is an unheated starting length $X_0$. When the unheated starting length disappears, the thermal behavior of the laminar boundary layer is as was determined exactly in Section 7.3.

The laminar boundary-layer energy equation and boundary conditions are

$$u \frac{\partial T}{\partial x} + v \frac{\partial T}{\partial y} = \alpha \frac{\partial^2 T}{\partial y^2} \qquad (8\text{-}17a)$$

$$T(y = \infty) = T_\infty \qquad (8\text{-}17b)$$

$$T(y = 0) = \begin{cases} T_\infty, & x < X_0 \\ T_w, & x \geqslant X_0 \end{cases} \qquad (8\text{-}17c)$$

with viscous dissipation neglected.

The energy equation [Eq (8-17a)] is first rearranged into the conservation-law form of

$$\frac{\partial u\,T}{\partial x} + \frac{\partial v\,T}{\partial y} - T\left( \frac{\partial u}{\partial x} + \frac{\partial v}{\partial y} \right) = \alpha \frac{\partial^2 T}{\partial y^2}$$

The last term in parentheses on the right-hand side identically equals zero according to the continuity equation. An integration from $y = 0$ to $y = \delta_T$ then gives

$$\int_0^{\delta_T} \frac{\partial u T}{\partial x} dy + \int_0^{\delta_T} \frac{\partial v T}{\partial y} dy = \alpha \int_0^{\delta_T} \frac{\partial^2 T}{\partial y^2} dy$$

which becomes

$$\int_0^{\delta_T} \frac{\partial u T}{\partial x} dy + v(y = \delta_T)T(\delta_T) - v(y = 0)T(y = 0)$$

$$= \frac{-k\,\partial T(y = 0)/\partial y - [-k\,\partial T(y = \delta_T)/\partial y]}{\rho C_p}$$

It is presumed that there is no temperature gradient at the outer edge of the thermal boundary layer so that $\partial T(y = \delta_T)/\partial y = 0$. Adoption of the nomenclature that $v(y = 0) = v_w$, $T(y = 0) = T_w$, $T(y = \delta_T) = T_\infty$, and $-k\,\partial T(y = 0)/\partial y = q_w$ then puts the integral relation in the form

$$\int_0^{\delta_T} \frac{\partial u T}{\partial x} dy + T_\infty \left[ v_w + u_{\delta_T} \frac{d\delta_T}{dx} - \frac{\partial}{\partial x} \left( \int_0^{\delta_T} u\,dy \right) \right] = \frac{q_w + \rho C_p v_w T_w}{\rho C_p}$$

where Eq. (8-11) has been used to relate $v_\delta$ to $u_{\delta_T}$. Application of Leibnitz' formula then gives

$$\frac{\partial}{\partial x}\left(\int_0^{\delta_T} uT\, dy\right) - u_{\delta_T}T_\infty \frac{d\delta_T}{dx} + T_\infty\left[v_w + u_{\delta_T}\frac{d\delta_T}{dx} - \frac{\partial}{\partial x}\left(\int_0^{\delta_T}u\, dy\right)\right]$$
$$= (q_w + \rho C_p v_w T_w)/\rho C_p$$

$$\frac{\partial}{\partial x}\left(\int_0^{\delta_T} uT\, dy\right) - T_\infty\frac{\partial}{\partial x}\left(\int_0^{\delta_T}u\, dy\right) = \frac{q_w + \rho C_p v_w(T_w - T_\infty)}{\rho C_p}$$

One final rearrangement gives the integral form of the energy equation as

$$\frac{\partial}{\partial x}\left[\int_0^{\delta_T}u(T - T_\infty)\, dy\right] + \left(\int_0^{\delta_T}u\, dy\right)\frac{dT_\infty}{dx} = \frac{q_w + \rho C_p v_w(T_w - T_\infty)}{\rho C_p}$$

$$(8\text{-}18)$$

Note that the integral formulation of Eq. (8-18) reveals that if $v_w$ is positive, the net effect is to increase the energy flux from the wall (presuming $T_w > T_\infty$). If $v_w$ is negative, on the other hand, the net effect is to decrease the wall energy flux.

For the problem at hand, it remains only to introduce assumed temperature profiles. As before, a polynomial form is assumed

$$T = a + by + cy^2 + dy^3 + \cdots$$

and subjected to the conditions

from
boundary conditions
At $y = 0$    $T = T_w$  $\xleftarrow{\hspace{1.2cm}}$  $(must$ be satisfied$)$  $\xrightarrow{\hspace{1.2cm}}$  $T = T_\infty$    at $y = \delta_T$

$$\xrightarrow[\text{used in integral formulation}]{\text{physical insight, already}} \quad \frac{\partial T}{\partial y} = 0$$

$$v_w\frac{\partial T}{\partial y} = \alpha\frac{\partial^2 T}{\partial y^2} \xleftarrow{\text{from energy equation}} u_{\delta_T}\frac{dT_\infty}{dx} = \alpha\frac{\partial^2 T}{\partial y^2}$$

Additional, derived, conditions can also be imposed if desired. For example, the derivative of the energy equation with respect to $y$ evaluated at $y = 0$ gives the condition

$$\frac{\partial u(y = 0)}{\partial y}\frac{dT_w}{dx} = \alpha\frac{\partial^3 T(y = 0)}{\partial y^3}$$

All these conditions could be satisfied; however, it is not usually convenient to do so. If only four are satisfied, the lowest-ordered derivatives and the condition at the wall for the second derivative would be used. If it is assumed that $v_w = 0 = dT_\infty/dx$ and that

$$\frac{T - T_\infty}{T_w - T_\infty} = a + by + cy^2 + dy^3$$

these four conditions give the four algebraic equations

$$T(y = 0) = T_w = a$$

$$T(y = \delta_T) = T_\infty = a + b\delta_T + c\delta_T^2 + d\delta_T^3$$

$$\frac{\partial T(\delta_t)}{\partial y} = 0 = b + 2c\delta_T + 3d\delta_T^2$$

$$\frac{\partial^2 T(0)}{\partial y} = 0 = 2c$$

From their simultaneous solution the similarity form, as evidenced by the constant coefficients,

$$\frac{T - T_\infty}{T_w - T_\infty} = \begin{cases} 0, & \frac{y}{\delta_T} > 1 \\ 1 - \frac{3}{2}\frac{y}{\delta_T} + \frac{1}{2}\left(\frac{y}{\delta_T}\right)^3, & \frac{y}{\delta_T} \leq 1 \end{cases} \qquad (8\text{-}19)$$

results. From Eq. (8-19) it follows that

$$q_w = -k\frac{\partial T(0)}{\partial y} = -\frac{k}{\delta_T}\frac{\partial T(0)}{\partial(y/\delta_T)} = \frac{3}{2}\frac{k}{\delta_T}(T_w - T_\infty)$$

If the thermal boundary layer is thinner than· the velocity boundary layer ($\delta_T/\delta < 1$), it is found with the help of Eq. (8-14) that

$$\int_0^{\delta_T} u(T - T_\infty)\,dy = U(T_w - T_\infty)\delta_T\int_0^1 \frac{u}{U}\left(\frac{T - T_\infty}{T_w - T_\infty}\right)d\left(\frac{y}{\delta_T}\right)$$

$$= \frac{3}{20}\frac{\delta_T^2}{\delta}\left(1 - \frac{\delta_T^2}{14\delta^2}\right)U(T_w - T_\infty)$$

If $\delta_T > \delta$, $u/U = 1$ in part of the preceding integration.

Introduction of these results into the integral formulation of Eq. (8-18) gives the first-order nonlinear ordinary differential equation

$$\frac{d}{dx}\left[\frac{3}{20}\frac{\delta_T^2}{\delta}\left(1-\frac{\delta_T^2}{14\delta^2}\right)U(T_w-T_\infty)\right]=\frac{3}{2}\frac{\alpha}{\delta_T}(T_w-T_\infty)$$

which simplifies somewhat to

$$\frac{d}{dx}\left[\frac{\delta_T^2}{\delta}\left(1-\frac{\delta_T^2}{14\delta^2}\right)\right]=\frac{10\nu}{\Pr U}\frac{1}{\delta_T}$$

Of course, $\delta$ is the known function given by Eq. (8-15) so that there is only one unknown. On the basis of prior experience, it is suspected that the two boundary layers will approach a constant ratio far downstream from the leading edge. This ratio, $\zeta = \delta_T/\delta$, is formed in the differential equation to achieve

$$\frac{d}{dx}\left[\zeta^2\delta\left(1-\frac{\zeta^2}{14}\right)\right]=\frac{10\nu}{\Pr U}\frac{1}{\zeta\delta}$$

The $\zeta^2/14$ term could be discarded at this point since $\zeta < 1$. Further rearrangement gives

$$\left[\zeta^3\left(1-\frac{\zeta^2}{14}\right)\right]\frac{d(\ln\delta^{3/2})}{dx}+\frac{d}{dx}\left[\zeta^3\left(1-\frac{6}{5}\frac{\zeta^2}{14}\right)\right]=\frac{15\nu}{\Pr U\delta^2}$$

Since it has already been assumed that $\zeta < 1$, there is no appreciable additional loss in accuracy in assuming $\frac{6}{5}\approx 1$ in the last term on the left-hand side. Then a solution for $\zeta^3(1-\zeta^2/14)$ is achieved, with $\delta^{3/2}$ as an integrating factor, as

$$\zeta^3\left(1-\frac{\zeta^2}{14}\right)=\frac{15}{\Pr U}\delta^{-3/2}\int\delta^{-1/2}\,dx+c$$

On the basis that $\delta_T = 0$ when $x = X_0$, the constant of integration is then evaluated with the result that

$$\zeta\left(1-\frac{\zeta^2}{14}\right)^{1/3}=\left(\frac{13}{14}\right)^{1/3}\left[1-\left(\frac{X_0}{x}\right)^{3/4}\right]^{1/3}\Pr^{-1/3}\qquad(8\text{-}20)$$

As expected, $\zeta = \delta_T/\delta$ approaches a constant far downstream—under the assumptions of this discussion, $\zeta < 1$ always.

The local heat-transfer coefficient is attainable now that the thermal boundary-layer thickness is given by Eq. (8-20) since

$$q_w = h(T_w - T_\infty) = -k\frac{\partial T(0)}{\partial y}$$

$$= \frac{3}{2}\frac{k(T_w - T_\infty)}{\delta_T}$$

$$h = \frac{3}{2}\frac{k}{\zeta\delta}$$

Use of $\zeta$ from Eq. (8-20) and $\delta$ from Eq. (8-15) then gives

$$\frac{Nu}{Pr^{1/3}Re^{1/2}} = \frac{0.3317(1 - \zeta^2/14)^{1/3}}{\left[1 - (X_0/x)^{3/4}\right]^{1/3}} \qquad (8\text{-}20a)$$

In the limit as the unheated starting length disappears, this result becomes

$$\frac{Nu}{Pr^{1/3}Re^{1/2}} = 0.3317\left(1 - \frac{\zeta^2}{14}\right)^{1/3}$$

The $1 - \zeta^2/14$ term in this expression implies a slight Pr dependence of Nu in addition to the $Pr^{1/3}$ that is explicitly shown. For $Pr \rightarrow \infty$, it is accurate to neglect $\zeta^2/14$ and then

$$\frac{Nu}{Pr^{1/3}Re^{1/2}} = 0.3317 \qquad (8\text{-}21)$$

which is 2% below the exact answer of 0.339 from Eq. (7-29c). For $Pr \approx 1$, Eq. (8-20) reveals that $\zeta \approx 1$, so that then

$$\frac{Nu}{Pr^{1/3}Re^{1/2}} = 0.323$$

which is also 2% lower than the exact answer of 0.33206 from Eq. (7-296) [19]. For simplicity, Eq. (8-21) can be taken as a fair representation of the integral method prediction in the absence of an unheated starting length.

The integral method has been applied by Goodman [9] to determination of the effect of time-dependent wall temperature on the heat transfer to a constant-property fluid in steady boundary-layer flow over a flat plate. He shows that steady-state conditions prevail after a step change in wall temperature when

$$\frac{4t}{x} \geqslant 1.33Pr^{1/3}$$

With the response to a step change in wall temperature known, he then utilizes Duhamel's integral to obtain $q_w$ for arbitrary time variation of wall temperature; particular attention is given to sinusoidal wall-temperature variation.

The unheated starting length problem for wedge flow $U = Cx^m$ was treated by Lighthill [21] in a different approximate way. He found that the effect of a nonzero unheated starting length was provided by a term (similar to that for flat plate flow) that is

$$\left[ 1 - \left( \frac{X_0}{x} \right)^{3(m+1)/4} \right]^{-1/3}$$

The integral form of the energy equation for a body of revolution is the subject of Problem 8-13.

## 8.4   HEAT FLUX FROM PLATE OF ARBITRARY WALL TEMPERATURE

The results for a surface with an unheated starting length can be employed to evaluate the heat flux from a surface whose temperature varies in an arbitrary manner. The method is similar in derivation to that used for tube flow as sketched in Fig. 6-8. Consider a flat plate subjected to steady parallel flow of a constant-property fluid of constant free-stream temperature.

As shown in Fig. 8-7, the wall temperature can be regarded as equal to $T_\infty$ until a point $x^*$ is reached, where a step change of magnitude $\Delta(T_w - T_\infty)$ occurs that is maintained over the remainder of the plate. Let the temperature distribution following a unit step change be $f(x, x^*, y)$ as given by Eq. (8-19), which is now a known function. The temperature in the laminar boundary layer for a step change of magnitude $\Delta(T_w - T_\infty)$ is then

$$T(x, y) - T_\infty = \Delta(T_w - T_\infty)f(x, x^*, y)$$

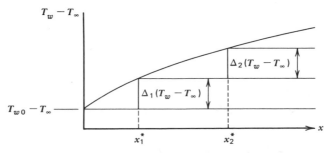

**Figure 8-7**   Representation of wall temperature variation by differential steps.

For a series of steps, it then follows that the resultant fluid temperature, relying on the linearity of the energy equation, is just the sum of the effects of the individual steps so that

$$T(x, y) - T_\infty = (T_{w_0} - T_\infty)f(x, y) + \Delta_1(T_w - T_\infty)f(x, x_1^*, y)$$

$$+\Delta_2(T_w - T_\infty)f(x, x_2^*, y) + \cdots$$

In the limit as the spacing between steps become small, the preceding sum can be represented as the integral

$$T(x, y) - T_\infty = (T_{w_0} - T_\infty)f(x, y) + \int_{x^*=0}^{x} f(x, x^*, y)\, d[T_w(x^*) - T_\infty]$$

$$(8\text{-}22)$$

Equation (8-22) allows for only one discontinuity in $T_w - T_\infty$, at the leading edge. Each point of discontinuity in $T_w - T_\infty$ requires addition of a term, $(T_{w_i} - T_\infty)f(x - x_i^*, y)$. If desired, an integration by parts can be employed to recast the integral in terms of $T_w(x^*) - T_\infty$ and $df(x - x^*, y)/dx^*$, as was done to obtain Eq. (6-42c).

The wall heat flux is obtained from Eq. (8-22) as

$$q_w(x) = -k\frac{\partial T(x, y = 0)}{\partial y}$$

$$= (T_{w_0} - T_\infty)\left[-k\frac{\partial f(x, y = 0)}{\partial y}\right]$$

$$+ \int_{x^*=0}^{x}\left[-k\frac{\partial f(x, x^*, y = 0)}{\partial y}\right] d[T_w(x^*) - T_\infty]$$

If it is recognized that $-k\,\partial f(x, x^*, y = 0)/\partial y = h(x, x^*)$, where $h(x, x^*)$ is the heat-transfer coefficient at $x$ due to a step change in surface temperature at $x^*$ as given by Eq. (8-20a), this relation can be written as

$$q_w(x) = (T_{w_0} - T_\infty)h(x) + \int_{x^*=0}^{x} h(x, x^*)\frac{d[T_w(x^*) - T_\infty]}{dx^*}dx^* \quad (8\text{-}23)$$

which is essentially an application of Duhammel's theorem.

Problem 8-7 deals with the case of a linear variation of $T_w$. Eckert et al. [10] present simplified methods to treat Eq. (8-23), pointing out in addition that if $T_w - T_\infty = Ax^\gamma$, immediate integration is possible and extending the method to turbulent boundary layers for which $h(x, x^*)/h(x,0) = [1 - (x^*/x)^{9/10}]^{-1/9}$ from Eq. (10-24). The integral method expressed by Eq. (8-23) was apparently first applied to this sort of problem by Chapman and Rubesin

[11] and was later refined by Klein and Tribus [12]. Additional discussion is provided by Schlichting [13]. The inversion of Eq. (8-23) to give the surface-temperature distribution that corresponds to a specified surface heat flux is the subject of Problem 8-9.

## PROBLEMS

**8-1**  Determine, by use of the integral method, the local heat-transfer coefficient for steady parallel flow of a constant-property fluid over a flat plate in the absence of an unheated starting length for the case in which $\delta_T > \delta$. This requires that

$$\int_0^{\delta_T} u(T - T_\infty)\, dy = \int_0^{\delta} u(T - T_\infty)\, dy + U \int_{\delta}^{\delta_T} (T - T_\infty)\, dy$$

**8-2**  Show, by use of the integral method for the conditions of Problem 8-1, that in the limit as Pr $\rightarrow$ 0

$$\frac{\mathrm{Nu}}{\mathrm{Pr}^{1/2}\,\mathrm{Re}^{1/2}} = \mathrm{const}$$

and evaluate the constant. Compare the integral method prediction with the exact answer of Eq. (7-29a).

**8-3**  An integral formulation of the two-dimensional laminar boundary-layer equations is needed for treatment of the unsteady state. For a constant-property fluid:

(a) Show that the continuity equation $\partial u/\partial x + \partial v/\partial y = 0$ gives

$$v = v_w - \frac{\partial}{\partial x}\left(\int_0^y u\, dy\right)$$

where $v_w = v(y = 0)$.

(b) Show that the x-motion equation $\partial u/\partial t + u\,\partial u/\partial x + v\,\partial u/\partial y = \partial U/\partial t + U\,\partial U/\partial x + v\,\partial^2 u/\partial y^2$ gives

$$\frac{\partial}{\partial t}\left[U\int_0^{\delta}\left(1 - \frac{u}{U}\right) dy\right] + \frac{\partial}{\partial x}\left[U^2\int_0^{\delta}\frac{u}{U}\left(1 - \frac{u}{U}\right) dy\right]$$

$$+ \left[\int_0^{\delta}\left(1 - \frac{u}{U}\right) dy\right] U\frac{\partial U}{\partial x} = \frac{\tau_w + \rho U v_w}{\rho}$$

where $\tau_w = \mu\,\partial u(y = 0)/\partial y$.

(c) Show that the energy equation $\partial T/\partial t + u\,\partial T/\partial x + v\,\partial T/\partial y = \alpha\,\partial^2 T/\partial y^2$ gives

$$\delta_T \frac{\partial T_\infty}{\partial t} + \frac{\partial}{\partial t}\left[\int_0^{\delta_T}(T - T_\infty)\,dy\right] + \frac{\partial}{\partial x}\left[\int_0^{\delta_T}u(T - T_\infty)\,dy\right]$$

$$+ \left[\int_0^{\delta_T}u\,dy\right]\frac{\partial T_\infty}{\partial x} = \frac{q_w + \rho C_p v_w(T_w - T_\infty)}{\rho C_p}$$

where $q_w = -k\,\partial T(y = 0)/\partial y$ and viscous dissipation is neglected.

8-4 Consult Goodman [9], and use his temperature and velocity profiles in the integral formulation of the energy equation as given in part c of Problem 8-3 for the case where (a) the velocity boundary-layer thickness is constant and $\delta_T$ depends only on time as a result of a step change in wall temperature that is spatially uniform and (b) the velocity boundary layer varies as $x^{1/2}$ and $\delta_T$ depends on both space and time as a result of a step change in wall temperature that is spatially uniform. Verify that the final equations to be solved are the same as Goodman's. Determine whether his integral formulation of the boundary-layer integral equation is complete.

8-5 Use the integral method to determine the skin friction and heat-transfer coefficients for steady parallel flow of a constant-property fluid over a flat plate with injection $v_w > 0$. Plot the ratio of the heat-transfer coefficient for $v_w > 0$ to that for $v_w = 0$ against $(v_w/U)(Ux/\nu)^{1/2}$ and compare with the exact results of Section 7.8.

8-6 Plot the temperature profile predicted by Eq. (8-20) for the case of no unheated starting length against $y(U/\nu x)^{1/2}$ and compare it with the exact solution for several Prandtl numbers.

8-7 (a) For the case in which the wall temperature of a flat plate varies linearly as $T_w - T_\infty = a + bx$, use Eq. (8-23) to show that the wall heat flux is given by

$$q_w = \frac{0.332}{x}\,Pr^{1/3}\,Re^{1/2}\left\{a + b\int_{x^*=0}^x\left[1 - \left(\frac{x^*}{x}\right)^{3/4}\right]^{-1/3}dx^*\right\}$$

where $Re = Ux/\nu$.

(b) Evaluate the integral of part a through use of the relationship, where $\beta(m, n)$ is the beta function [related to the gamma function by $\beta(m, n) = \Gamma(m)\Gamma(n)/\Gamma(m + n)$],

$$\beta(m, n) = \int_0^1 Z^{m-1}(1 - Z)^{n-1}\,dZ \qquad m > 0, \quad n < \infty$$

and the transformation $Z = 1 - (x^*/x)^{3/4}$ to show that

$$q_w = 0.332 \frac{k}{x} \mathrm{Pr}^{1/3} \mathrm{Re}^{1/2}(a + 1.612bx)$$

(c) Show that the local Nusselt number for $a = 0$ is 61% larger than for $b = 0$.

8-8   Use the integral method to find the surface temperature and heat-transfer coefficient for a flat plate in steady parallel flow of a constant-property fluid with constant free-stream temperature for the case in which a constant wall heat flux is imposed after an unheated starting length of $X_0$. Show that if $X_0 = 0$, the local wall temperature varies as

$$T_w - T_\infty = \frac{2.21 q_w}{k \, \mathrm{Pr}^{1/3} \, \mathrm{Re}^{1/2}} x$$

where $\mathrm{Re} = Ux/\nu$ and that the local Nusselt number is

$$\mathrm{Nu} = \frac{hx}{k} = 0.452 \mathrm{Pr}^{1/3} \, \mathrm{Re}^{1/2}$$

which is about 36% higher than for $T_w = $ const. *Note:* The insight gained from similarity solutions as expressed by Eq. (7-66) that $T_w - T_\infty = Kx^{1/2}$ if $q_w = $ const can be used in a specified $T_w$ solution by employing Eqs. (8-18) and (8-19).

8-9   It is sometimes required that the surface temperature that corresponds to a specified surface heat flux be predicted. This requires the inversion of Eq. (8-23).

(a) Show that the local heat-transfer coefficient given by Eq. (8-20a) is of the form

$$h(x, x^*) = f(x)(x^b - x^{*b})^{-a}$$

where $f(x) = 0.3317k \, \mathrm{Pr}^{1/3} \, (U/\nu)^{1/2} x^{-1/4}$, $a = \frac{1}{3}$, and $b = \frac{3}{4}$.

(b) Recognizing that $q_w(x = 0)$ will be unbounded if $T_{w_0} - T_\infty \neq 0$, write Eq. (8-23) for $T_{w_0} - T_\infty = 0$ as

$$q_w(x) = \int_{x^*=0}^{x} f(x)(x^b - x^{*b})^{-a} \frac{dy(x^*)}{dx^*} dx^*$$

where $y(x^*) = T_w(x^*) - T_\infty$. Let $\lambda^{1/b} = x^*$ and $t^{1/b} = x$, and cast the integral equation into the form

$$\frac{q_w(t)}{f(t)} = \int_{\lambda=0}^{t} (t - \lambda)^{-a} \frac{dy(\lambda)}{d\lambda} d\lambda$$

(c) Recognizing that the result of part b is a convolution integral [3], take its Laplace transform with respect to $t$ to obtain

$$\frac{\overline{q_w(t)}}{\overline{f(t)}} = \overline{t^{-a}} \, \overline{y'}$$

where the bar superscript denotes a Laplace transformed quantity. Now $\overline{t^{-a}} = \Gamma(1 - a)/s^{1-a}$ and $\overline{y'} = s\overline{y} - y(t = 0)$, where $s$ is the Laplace transform parameter. Thus

$$\frac{\overline{q_w(t)}}{\overline{f(t)}} \frac{s^{1-a}}{\Gamma(1 - a)} = s\overline{y}$$

or

$$\overline{y} = \frac{1}{\Gamma(1 - a)} \frac{\left[\overline{q_w(t)/f(t)}\right]}{s^a}$$

This result represents the convolution integral

$$y(t) = \frac{1}{\Gamma(1 - a)\Gamma(a)} \int_{\lambda=0}^{t} (t - \lambda)^{a-1} \frac{q_w(\lambda)}{f(\lambda)} \, d\lambda$$

(d) From part b recall that $t = x^b$ and $\lambda = x^{*b}$ and put the result of part c into the form

$$T_w(x) - T_\infty = \frac{b}{\Gamma(1 - a)\Gamma(a)}$$

$$\times \int_{x^*=0}^{x} (x^b - x^{*b})^{a-1} \frac{q_w(x^*)}{f(x^*)} x^{*b-1} \, dx^*$$

(e) With $\Gamma(\tfrac{1}{3}) = 2.67894$ and $\Gamma(\tfrac{2}{3}) = 1.35412$ [14], use the definitions of part a to put the result of part d into the form specific to a laminar boundary-layer on a flat plate, with $\mathrm{Re} = Ux/v$, of

$$T_w(x) - T_\infty = \frac{\tfrac{3}{4}}{0.3317\Gamma(\tfrac{2}{3})\Gamma(\tfrac{1}{3})k \, \mathrm{Pr}^{1/3} \mathrm{Re}^{1/2}}$$

$$\times \int_0^x \left[1 - \left(\frac{x^*}{x}\right)^{3/4}\right]^{-2/3} q_w(x^*) \, dx^*$$

or

$$k[T_w(x) - T_\infty]\mathrm{Pr}^{1/3}\mathrm{Re}^{1/2} = 0.623 \int_0^x \left[1 - \left(\frac{x^*}{x}\right)^{3/4}\right]^{-2/3}$$

$$\times q_w(x^*) \, dx^*$$

The integral can often be expressed in terms of beta functions by use of the transformation suggested in Problem 8-7.

**8-10**  Air flows steadily at 25 ft/sec parallel to a flat plate that is 6 in. long in the flow direction. Air temperature is 20°F, and atmospheric pressure is 14.7 lb$_f$/in$^2$. The entire surface of the plate is adiabatic, except for a strip of 1 in. width (located between 2 and 3 in. from the plate leading edge) that is electrically heated so that the heat flux from it is uniform. Determine the numerical value of the heat flux from the strip that will maintain the temperature at the plate trailing edge at 32°F. Plot the temperature distribution along the entire plate surface and discuss the relevance of this problem to aircraft wing deicing. Note that the beta functions will be needed for this analytical solution.

**8-11**  Air flows steadily at 50 ft/sec, 2000°F, and 14.7 lb$_f$/in.$^2$ parallel to a flat plate. The first 6 in. of the plate are maintained at 200°F. The next 18 in. are adiabatic. Plot the plate temperature against the position for the first 24 in. Note that the first 6 in. must be treated as a specified surface temperature problem whereas the last 18 in. must be treated as a specified heat-flux problem.

**8-12**  It is sometimes required that the velocity distribution on a body of revolution be ascertained. The integral method can be used for this purpose, and the following problem demonstrates the developments required. Figure 8P-12 illustrates the physical configuration and shows the control volume, extending above the boundary layer to a distance $Y$, to which conservation principles will be applied. This derivation from first principles is an alternative to integration of the describing partial differential equations. Steady state is assumed as are constant properties. Most importantly, the boundary layer is assumed to be thin relative to the local radius of revolution $R(x)$, an assumption that can lack accuracy in some duct flows. Wall transpiration normal to the surface is also assumed.

**(a)** Show by a detailed accounting that the conservation of mass principle applied to the control volume gives

$$\left\{ R\,\delta\phi \int_0^Y \rho u\,dy \right\}_x + \left\{ R\,\delta\phi\,\delta x\,\rho v_w \right\}_{x+\delta x/2}$$

$$- \left\{ R\,\delta\phi \int_0^Y \rho u\,dy \right\}_{x+\delta x} - \left\{ R\,\delta\phi\,\delta x\,\rho_Y v_Y \right\}_{x+\delta x/2} = 0$$

from which it follows that, for constant $\rho$,

$$v_Y = v_w - \frac{1}{R}\frac{d}{dx}\left[ R\int_0^Y u\,dy \right]$$

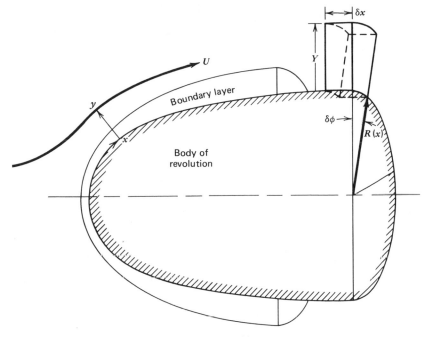

**Figure 8P-12**

**(b)** Show by a detailed accounting that the conservation of momentum principle applied to the control volume gives

$$\{R\,\delta\phi\,\delta x\,\tau_w\}_{x+\delta x/2} + \{R\,\delta\phi\,Yp\}_x$$

$$- \{R\,\delta\phi\,Y\}_{x+\delta x} + \left\{R\,\delta\phi\int_0^Y \rho u^2\,dy\right\}_x$$

$$- \left\{R\,\delta\phi\int_0^Y \rho u^2\,dy\right\}_{x+\delta x} - \{R\,\delta\phi\,\delta x\,\rho_Y v_Y u_Y\}_{x+\delta x/2} = 0$$

Then use the result of part a to eliminate $v_Y$, recognizing that $u_Y = U$, and incorporate the Bernoulli equation for inviscid flow along a stream line that gives $\rho^{-1}\,dp/dx = U\,dU/dx = dU^2/dx - U\,dU/dx$ to rearrange this equation into

$$\frac{d}{dx}\left[\int_0^Y (U^2 - u^2)\,dy\right] + \left[\int_0^Y u(U - u)\right]\frac{1}{R}\frac{dR}{dx}$$

$$- U\frac{d}{dx}\left[\int_0^Y (U - u)\,dy\right] = \frac{\tau_w + \rho v_w U}{\rho}$$

(c) Rearrange the result of part b into the form

$$\frac{d\delta_2}{dx} + \delta_2\left[\frac{1}{R}\frac{dR}{dx} + \left(2 + \frac{\delta_1}{\delta_2}\right)\frac{1}{U}\frac{dU}{dx}\right] = \frac{\tau_w + \rho v_w U}{\rho U^2}$$

where

$$\delta_2 = \int_0^\delta \frac{u}{U}\left(1 - \frac{u}{U}\right)dy$$

is the momentum thickness and

$$\delta_1 = \int_0^\delta \left(1 - \frac{u}{U}\right)dy$$

is the displacement thickness. For laminar flow, $\tau_w = \mu\, \partial u(y = 0)/\partial y$.

(d) Obtain the result of part c by the alternative method of integrating the partial differential equation form of the boundary-layer equations given by Eq. (7-69) from $y = 0$ to $y = \delta$.

(e) Show that if the Mangler transformation of Eq. (7-70),

$$\bar{x} = L^{-2}\int_0^x R^2(x)\, dx \qquad \text{and} \qquad \bar{y} = L^{-2}R(x)\, y$$

where $L$ is an arbitrary constant length, is employed the result of part c can be expressed as

$$\frac{d\delta_2'}{d\bar{x}} + \delta_2'\left(2 + \frac{\delta_1'}{\delta_2'}\right)\frac{1}{U}\frac{dU}{d\bar{x}} = \frac{\mu}{\rho U^2}\frac{\partial u(\bar{y} = 0)}{\partial y}$$

This result is the same as for flow over a wedge, transforming away the effects of the body of revolution and enabling wedge solutions to be taken over to the axisymmetric case.

8-13  It is sometimes required that the temperature distribution on a body of revolution be ascertained. The discussion in Problem 8-12 applies and Fig. 8P-12 illustrates the physical situation. It is assumed that properties are constant, steady state exists, viscous dissipation is unimportant, there is no volumetric heat source, and conduction is important only in the $y$ direction. Again, the boundary layer is taken to be thin relative to the radius of revolution $R(x)$.

(a) Show by a detailed accounting that the conservation of energy principle applied to the control volume in Fig. 8P-12 gives

$$\left\{ R\, \delta\phi \int_0^Y \rho u C_p T\, dy \right\}_x + \left\{ R\, \delta\phi\, \delta x\, \rho_w v_w C_p T_w \right\}_{x+\delta x/2}$$

$$+ \left\{ R\, \delta\phi\, \delta x\, q_w \right\}_{x+\delta x/2} - \left\{ R\, \delta\phi \int_0^Y \rho u C_p T\, dy \right\}_{x+\delta x}$$

$$- \left\{ R\, \delta\phi\, \delta x\, \rho_Y v_Y C_p T_Y \right\}_{x+\delta x/2} - \left\{ R\, \delta\phi\, \delta x\, q_Y \right\}_{x+\delta x/2} = 0$$

Here the work done by surface and body forces is neglected, in consonance with the stated assumption that viscous dissipation is negligible. For laminar flow, $q_w = -k\, \partial T(y=0)/\partial y$.

Then show in detail that this result becomes

$$\frac{1}{R}\frac{d}{dx}\left( R \int_0^Y \rho u C_p T\, dy \right) + \rho_Y v_Y C_p T_Y - \rho_w v_w C_p T_w = q_w - q_Y$$

(b) Recognize that $Y$ extends outside the boundary layer so that $T_Y = T_\infty$ and $q_Y = 0$, and incorporate the result of Problem 8-12 part a to put the result of part a into the form

$$\frac{1}{R}\frac{d}{dx}\left[ R \int_0^Y u(T - T_\infty)\, dy \right] + \left( \int_0^Y u\, dy \right)\frac{dT_\infty}{dx}$$

$$= \frac{q_w + \rho C_p v_w (T_w - T_\infty)}{\rho C_p}$$

The identity

$$\frac{d}{dx}\left\{ T_\infty R \int_0^Y u\, dy \right\} = T_\infty \frac{d}{dx}\left( R \int_0^Y u\, dy \right) + \left( R \int_0^Y u\, dy \right)\frac{dT_\infty}{dx}$$

is needed.

(c) Show in detail that the result of part b can be cast into the form

$$\frac{d}{dx}\left[ \int_0^{\delta_T} u(T - T_\infty)\, dy \right] + \left[ \int_0^{\delta_T} u(T - T_\infty)\, dy \right]\frac{1}{R}\frac{dR}{dx}$$

$$+ \left( \int_0^{\delta_T} u\, dy \right)\frac{dT_\infty}{dx} = \frac{q_w + \rho C_p v_w (T_w - T_\infty)}{\rho C_p}$$

by letting $Y$ approach the outer edge of the boundary layer. Explain how the assumption that the thermal boundary-layer thickness $\delta_T$ is less than the velocity boundary-layer thickness $\delta$ is incorporated into the preceding equation. Explain briefly the effects of Prandtl number and unheated starting length on the accuracy of this assumption, and also outline briefly the manner in which the result of part b could be used if this assumption were inaccurate (i.e., $\delta_T < \delta$).

(d) Compare the result of part c for a body of revolution with Eq. (8-18) for a flat plate.

(e) Show in detail that the result of part b can be expressed for $\delta_T < \delta$ as

$$\frac{d\Delta}{dx} + \Delta\left[\frac{1}{T_w - T_\infty}\frac{d(T_w - T_\infty)}{dx} + \frac{1}{U}\frac{dU}{dx} + \frac{1}{R}\frac{dR}{dx}\right]$$

$$+ \frac{\Delta_u}{T_w - T}\frac{dT_\infty}{dx} = \frac{q_w + \rho C_p v_w(T_w - T_\infty)}{\rho C_p U(T_w - T_\infty)}$$

with

$$\Delta = \int_0^{\delta_T} \frac{u}{U}\frac{T - T_\infty}{T_w - T_\infty}\,dy \qquad \text{and} \qquad \Delta_u = \int_0^{\delta_T}\frac{u}{U}\,dy$$

Also, show in detail that the result of part b can be alternatively expressed as

$$\frac{1}{R}\frac{d}{dx}\left[RG(T_w - T_\infty)\Delta\right] + G\Delta_u\frac{dT_\infty}{dx} = \frac{q_w + \rho C_p v_w(T_w - T_\infty)}{C_p}$$

where $G = \rho U$, a form often convenient since it suggests that the mass flux of $G = \rho U$ is important rather than $\rho$ or $U$ separately. In internal flows the total mass flow rate divided by the flow area gives $\rho U = G$.

(f) Show that the Mangler transformation of Eq. (7-70),

$$\bar{x} = L^{-2}\int_0^x R^2(x)\,dx \qquad \text{and} \qquad \bar{y} = L^{-1}R(x)y$$

where $L$ is a constant of arbitrary length, transforms the result of part e into the form

$$\frac{d\Delta'}{d\bar{x}} + \Delta'\left[\frac{1}{T_w - T_\infty}\frac{d(T_w - T_\infty)}{d\bar{x}} + \frac{1}{U}\frac{dU}{d\bar{x}}\right] + \frac{\Delta'_u}{T_w - T_\infty}\frac{dT_\infty}{d\bar{x}}$$

$$= -\frac{k}{\rho C_p U(T_w - T_\infty)}\frac{\partial T(\bar{y} = 0)}{\partial \bar{y}}$$

Here $\Delta' = R\Delta$ and $\Delta'_u = R\Delta_u$. This result is the same as for flow over a wedge, transforming away the effects of the body of revolution and enabling wedge solutions to be taken over to the axisymmetric case.

**8-14** The laminar boundary layer on a body of revolution with variable free-stream velocity is to be studied by application of the integral form of the $x$-motion equation from Problem 8-12. Constant properties, steady state, and no wall transpiration are assumed.

(a) Show that the integral form of the $x$-motion equation is

$$\frac{d\delta_2}{dx} + \delta_2\left[\frac{1}{R}\frac{dR}{dx} + \left(2 + \frac{\delta_1}{\delta_2}\right)\frac{1}{U}\frac{dU}{dx}\right] = \frac{\tau_w}{\rho U^2}$$

(b) Rearrange the result of part a into the form

$$\frac{dZ}{dx} = \frac{R^2 F}{U}$$

where $Z = \delta_2^2 R^2/\nu$ and $F = 2\tau_w\delta_2/\mu - 2(2 + \delta_1/\delta_2)K$ with $K = (\delta_2^2/\nu)dU/dx$.

(c) As discussed by Schlichting [13, pp. 192–200, 229–231], use of a fourth-order polynomial for the velocity profile leads to the realization that $F$ is a function of only $K$, $F = F(K)$. Walz [16] pointed out that the straight line $F \approx a - bK$, with $a = 0.47$ and $b = 6$, is an accurate approximation. Show that this approximation used in the result of part b gives

$$\frac{dZ}{dx} + \frac{b}{U}\frac{dU}{dx}Z = a\frac{R^2}{U}$$

(d) Show that the solution to the result of part c is

$$U^b Z = a\int_0^x R^2 U^{b-1}\,dx$$

since either $U$, $R$, or $\delta_2$ equals zero at the leading edge where $x = 0$. For the given values of $a$ and $b$, this becomes

$$U\frac{\delta_2^2}{\nu} = \frac{0.47}{R^2 U^5}\int_0^x R^2 U^5\,dx$$

(e) If $U$ and $R$ are given, the local value of $\delta_2$ can be calculated by integration. Then $K$ is known and pertinent quantities such as $\tau_w$ and $H_{12} = \delta_1/\delta_2$ can be ascertained from the literature as cited in part c. If flow separation occurs at the location for which $\tau_w = 0$, determine the accompanying value of $K$ by consulting Schlichting [12, p. 198].

8-15  The local heat-transfer coefficient due to the laminar boundary layer covering a body of revolution exposed to axisymmetric flow is sometimes needed. A simplified method that is based on the local similarity assumption that the rate of growth of the thermal boundary layer depends only on local conditions has been developed by Smith and Spalding [17]. The method is based on the suspicion, justified by either dimensional analysis or substitution of appropriate polynomial profiles for temperature and velocity into the integral energy and momentum equations for wedge flow, that

$$\frac{U}{\nu}\frac{d\Delta_4^2}{dx} = f\left(\frac{\Delta_4^2}{\nu}\frac{dU}{dx}\right)$$

where

$$\Delta_4 = \frac{k}{h} = \left[\left.\frac{\partial(T - T_w)/(T_\infty - T_w)}{\partial y}\right|_{y=0}\right]^{-1}$$

and $f$ is an as yet unknown function. From exact solutions, such as are displayed in Table 7-2 for $Pr = 0.7$, it is determined that a straight line passing through the exact answers at the stagnation and flat-plate conditions is surprisingly accurate for accelerating flows ($dU/dx \geqslant 0$), although it begins to lose accuracy for decelerating flows ($dU/dx < 0$). Then

$$\frac{U}{\nu}\frac{d\Delta_4^2}{dx} \approx 11.68 - 2.87\left(\frac{\Delta_4^2}{\nu}\frac{dU}{dx}\right)$$

(a)  Show that this differential equation has the solution

$$U^{2.87}\Delta_4^2 = 11.68\nu\int_0^x U^{1.87}\,dx$$

if either $U$ or $\Delta_4$ is zero at $x = 0$. Note: At the stagnation point where $U = x\,dU(x = 0)/dx$, $\Delta_4^2 = 4.07\nu/(dU/dx)$.

(b)  Consult the result of part f of Problem 8-13 to verify that the wedge flow result can be taken over to a body of revolution of local radius $R(x)$ if all quantities involving a $y$ dimension are multiplied by the factor $R$ and all quantities involving $\int_0^x dx$ are replaced by $\int_0^x R^2\,dx$. Then justify by a brief discussion that the wedge result of part a can be taken over for a body of revolution as

$$U^{2.87}(R\Delta_4)^2 = 11.68\nu\int_0^x U^{1.87}R^2\,dx$$

**Table 8P-15   Values of $C_{1,2,3}$ for Various Prandtl Numbers**

| Pr | $C_1$ | $C_2$ | $C_3$ |
|-----|-------|-------|-------|
| 0.7 | 0.418 | 0.435 | 1.87 |
| 0.8 | 0.384 | 0.450 | 1.90 |
| 1.0 | 0.332 | 0.475 | 1.95 |
| 5.0 | 0.117 | 0.595 | 2.19 |
| 10.0 | 0.073 | 0.685 | 2.37 |

(c) Noting that $\Delta_4 = k/h$, show that the result of part b gives (for $Pr = 0.7$)

$$\frac{Nu}{Re_x \, Pr} = \frac{0.418 \nu^{1/2} \, U^{0.435} \, R}{\left( \int_0^x U^{1.87} R^2 \, dx \right)^{1/2}}$$

For variable density, the results are [18] given in the form

$$\frac{Nu}{Re_x \, Pr} = \frac{C_1 \mu^{1/2} \, RG^{C_2}}{\left( \int_0^x G^{C_3} R^2 \, dx \right)^{1/2}}$$

with $G = \rho U$ and $C_{1,2,3}$ as shown in Table 8P-15.

**8-16** A modification of the basic integral method has been devised by Zien [20] that has been applied to both boundary-layer and heat-conduction problems. Application to one dimensional transient heat conduction in a solid initially at uniform temperature $T_0$ with its surface temperature suddenly changed to $T_s(t)$ illustrates the essential ideas. (See Fig. 8-2 for a sketch of the geometry.)

Integration of the energy equation gives

$$\rho C_p \frac{d\left[ \int_0^\infty (T - T_0) \, dx \right]}{dt} = q_w \tag{8-4}$$

with $q_w = -k \, \partial T(0, t)/\partial x$.

(a) A second integral equation is generated by integrating the moment-like equation that results from multiplying the energy equation [Eq. (8-2a)] by temperature. Show that this results in

$$\rho C_p \frac{d\left[ \int_0^\infty (T - T_0)^2 \, dx \right]}{dt} = 2(T_s - T_0) q_w - 2k \int_0^\infty \left( \frac{\partial T}{\partial x} \right)^2 dx$$

This result provides a second equation that, when used in conjunction with Eq. (8-4), allows the surface heat flux $q_w$ and temperature profile parameter $\delta$ to be determined simultaneously.

(b) For a step change of surface temperature, use the temperature profile

$$T_s - T_0 = \exp\left(\frac{-x}{\delta}\right)$$

in Eq. (8-4) and the result of part a to show that

$$\delta = \frac{4\alpha t}{3} \quad \text{and} \quad q_w = \left(\frac{k\rho C_p}{3t}\right)^{1/2}.$$

This result for $q_w$ gives $q_w/q_{w,\text{exact}} = (\pi/3)^{1/2}$ that is only 2% high. Note that $\delta$ is not a penetration depth here; rather, it is a parameter.

(c) Use the same temperature profile in the simpler integral method of Eq. (8-4) alone to show that then the prediction is $\delta = (2\alpha t)^{1/2}$ and $q_w = (k\rho C_p/2t)^{1/2}$. This result for $q_w$ gives $q_w/q_{w,\text{exact}} = (\pi/2)^{1/2}$, which is 25% high. This result illustrates the good performance of the integral method made possible by such refinements as Zien's.

## REFERENCES

1  T. von Karman, Über laminare und turbulente Reibung, *Zeitschrift für angewandte Mathematic und Mechanic* **1**, 233–252 (1921); See also NACA TM 1092, 1946.

2  K. Pohlhausen, Zur näherungsweise Integration der Differential gleichung der laminaren Reibungschicht, *Zeitschrift für angewandte Mathematic und Mechanic* **1**, 252–268 (1921).

3  T. R. Goodman, The heat balance integral and its application to problems involving a change of phase, in *Heat Transfer and Fluid Mechanics Institute*, California Institute of Technology, Pasadena, CA, June 1957, pp. 383–400.

4  T. R. Goodman, The heat balance integral—further considerations and refinements, *Transact. ASME, J. Heat Transf.* **83**, 83–86 (1961).

5  T. R. Goodman, Application of integral methods to transient nonlinear heat transfer, in *Advances In Heat Transfer*, Academic Press, New York, 1964, pp. 52–122.

6  A. A. Sfeir, The heat balance integral in steady heat conduction, *Transact. ASME, J. Heat Transf.* **98**, 466–470 (1976).

7  D. Langford, The heat balance integral method, *Internatl. J. Heat Mass Transf.* **16**, 2424–2428 (1973).

8  H. S. Carslaw and J. C. Jaeger, *Conduction Of Heat In Solids*, 2nd ed., Clarendon Press, Oxford, 1959, p. 59.

9  T. R. Goodman, Effect of arbitrary unsteady wall temperature on incompressible heat transfer, *Transact. ASME, J. Heat Transf.* **84**, 347–352 (1962).

10  E. R. G. Eckert, J. P. Hartnett, and R. Birkebak, Simplified equations for calculating local and total heat flux to nonisothermal surfaces, *J. Aeronaut. Sci.* **24**, 549–550 (1957).

11  D. R. Chapman and M. W. Rubesin, Temperature and velocity profiles in the compressible laminar boundary layer with arbitrary distribution of surface temperature, *J. Aeronaut. Sic.* **16**, 547–565 (1949).

12  J. Klein and M. Tribus, Forced convection from non-isothermal surfaces, *Heat Transfer Symposium*, Engineering Research Institute, U. of Michigan, August 1952; see also ASME Paper No. 53-5A-46, ASME Semi-Annual Meeting, 1953.

13  H. Schlichting, *Boundary-Layer Theory*, McGraw-Hill, New York, 1968, pp. 295–296.

14  R. V. Churchill, *Operational Mathematics*, McGraw-Hill, New York, 1958, pp. 57, 324.

15  M. Abramowitz and I. A. Stegun, Eds., *Handbook of Mathematical Functions*, National Bureau of Standards, Applied Mathematics Series 55, 1965.

16  A. Walz, Ein neuer Ansatz für das Geschwindigkeitsprofil der laminaren Reibungsschicht, *Lilienthal-Bericht* **141**, 8–12 (1941).

17  A. G. Smith and D. B. Spalding, Heat transfer in a laminar boundary layer with constant fluid properties and constant wall temperature, *J. Roy. Aeronaut. Soc.* **62**, 60–64 (1958).

18  W. M. Kays, *Convective Heat And Mass Transfer*, McGraw-Hill, New York, 1966, p. 226.

19  S. C. Lau and E. M. Sparrow, Average heat transfer coefficients for forced convection on a flat plate with an adiabatic starting length, *Transact. ASME, J. Heat Transf.* **102**, 364–366 (1980).

20  T. F. Zien, Approximate calculation of transient heat conduction, *AIAA J.* **14**, 404–406 (1976); Integral solutions of ablation problems with time-dependent heat flux, **16**, 1287–1295 (1978); Approximate analysis of heat transfer in transpired boundary layers with effects of Prandtl number, *Internatl. J. Heat Mass Transf.* **19**, 513–521 (1976).

21  M. J. Lighthill, Contributions to the theory of heat transfer through a laminar boundary layer, *Proc. Roy. Soc. London* **A202**, 359–377 (1950).

# 9
# TURBULENCE FUNDAMENTALS

Quick recall of the major points covered in the early treatment of the kinetic theory of gases sets the stage for a discussion of turbulent flow. As in the molecular regime, the principal theme is chaos. On a small time scale turbulent flow is characterized by irregularity and disorder with "clumps," large relative to molecular sizes, dashing madly hither and thither. There is order on a large time scale, of course, since it is observed that water does flow steadily down a pipe in response to a pressure differential. Consequently, there is hope that some sense can be made, in at least an operational way, of turbulent flow.

The very fact of completely random behavior was used to advantage in studies of the molecular motions. Although it is tempting to hope for the same sort of success for turbulent flow, that hope has not been realized to date. This does not mean that workable empirical relationships are not available; it just means that universally applicable results are not fully in hand.

Before moving into details, consider the structure of turbulent flow under typical conditions as set forth by Hinze [1]. The basic structure is one of eddies and vorticies of various sizes. All of them together constitute the turbulent flow. Within each, however, the molecular effects of molecular viscosity are important, but the interaction of all these eddies almost completely masks molecular effects, at least from the viewpoint of an observer. For air moving at about 100 m/s, these eddies have a typical size of about 1 mm, with many larger than this. This is large compared to a typical mean free path of $10^{-4}$ mm. As a matter of passing interest, a cube 1 mm on a side contains about $10^{17}$ molecules. Measurements with a hot-wire anemometer show that local velocity varies roughly 10% about the average velocity. In other words, velocity fluctuations of about 10 m/s are observed. In comparison with this, mean molecular velocities are of the order of 330 m/s. For further comparison, turbulent velocity fluctuations occur about once every $1-10^{-4}$ s, whereas molecular collisions occur about once every $10^{-10}$ s. In contrast with the short

394

distance traveled by a molecule between collisions, an eddy travels a large distance to a location where conditions are very different from those at its origin before losing its identity. It appears from these comparisons between molecular and turbulent levels of events that there is a substantial difference in their regimes.

The marked difference in orders of magnitude for the various turbulent and molecular quantities may well explain why statistical approaches, so successful for molecules, do not have complete success when applied to turbulence. Consider this a bit further. The molecular mean free path is small relative to any system dimension; thus properties characteristic of the molecules alone are found. In those cases where molecular mean free path is of the same size as a system dimension, one observes velocity slip and temperature jump at solid boundaries as discussed in Section 3.4. In such a circumstance, the system influences the "effective properties" of the molecules. Applying this line of thought to turbulent flow, one sees that the typical eddy dimension of 1 mm can very easily be of the same size as an important system dimension (e.g., a boundary-layer thickness). Accordingly, it is likely that "effective properties" of turbulent flow will be strongly influenced by the system—the fluid "effective properties" are not solely a property of the fluid. Still, the unsteady Navier–Stokes and continuity equations accurately describe turbulent flow because the eddies are much larger than a molecular mean free path and turbulent fluctuations are not excessively rapid. Since turbulent flow can be mathematically modeled, there is hope for eventual progress in separating turbulence and system effects; see Section 10.7 for a brief discussion of recent developments.


## 9.1  LAMINAR-TO-TURBULENT TRANSITION

It is experimentally observed that laminar flows often undergo transition to turbulent flow. The operational concern is to predict the position along a flat plate at which the boundary layer will become turbulent, for example, or the Reynolds number at which pipe flow can be expected to be turbulent. As intimated earlier, the requisite relations are experimentally determined—theory alone does not suffice in most cases.

Some general background is helpful. The best hope of general description is to look at things from a long-term point of view—a time-averaged viewpoint. Even in turbulent flow, all quantities have some time-averaged value about which they fluctuate. Velocity, for example, can be represented as sketched in Fig. 9-1. Thus $\mathbf{V} = \bar{\mathbf{V}} + \mathbf{V}'$. In terms of components,

$$u = \bar{u} + u'$$
$$v = \bar{v} + v'$$
$$w = \bar{w} + w'$$

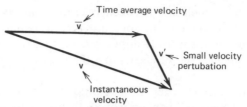

**Figure 9-1**   Representation of instantaneous velocity in turbulent flow as the vector sum of a time-averaged main component and a velocity perturbation.

where the barred quantities represent time averages and the primed quantities, instantaneous. To illustrate, $\bar{u}$ is defined as

$$\bar{u} = \frac{1}{\Delta t} \int_{t}^{t+\Delta t} u \, dt \tag{9-1}$$

where $\Delta t$ is a time interval large relative to turbulent fluctuations. Hence

$$\bar{u} = \frac{1}{\Delta t} \int_{t}^{t+\Delta t} (\bar{u} + u') \, dt = \bar{u} + \underbrace{\frac{1}{\Delta t} \int_{t}^{t+\Delta t} u' \, dt}_{\bar{u}'}$$

$$0 = \bar{u}' \tag{9-2}$$

and a time-averaged perturbation is seen to always be zero. However, $\overline{u'u'}$ is not necessarily of zero value.

The conventional way of expressing the intensity $J$ of flow turbulence is

$$J = \frac{1}{\bar{V}} \left[ \frac{1}{3} \left( \overline{u'u'} + \overline{v'v'} + \overline{w'w'} \right) \right]^{1/2} \tag{9-3}$$

Individual terms such as $\overline{u'u'}$ can be measured by a hot-wire anemometer. If $\overline{u'u'} = \overline{v'v'} = \overline{w'w'}$, the turbulence is said to be isotropic and there is no gradient in the mean velocity $\bar{V}$. Isotropy is commonly observed some distance downstream from screens in a wind tunnel, for example. When the mean velocity has a gradient, the turbulence is anisotropic.

The transition from laminar to turbulent flow is illustrated in Fig. 9-2, where a flat plate is shown in steady parallel flow of a constant-property fluid. The flow in the boundary layer adhering to the plate is laminar over the first part of the plate. Then (if there is no positive pressure gradient that might cause separation to occur) transition to turbulent flow begins at location $x_i$ and is completed by location $x_c$, the critical distance from the leading edge. The transition from laminar to turbulent flow is characterized by a rapid incease in boundary-layer thickness and a flattening of the velocity profile shown in Fig.

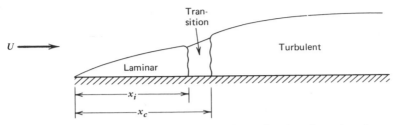

**Figure 9-2**  Transition from laminar to turbulent flow in a boundary layer.

9-3 as reported by Schubauer and Klebanoff [2] for a flat plate. The ratio of the displacement $\delta_1$ and momentum $\delta_2$ boundary-layer thicknesses also undergoes a dramatic change from $\delta_1/\delta_2 = 2.6$ for a laminar boundary layer as shown in Fig. 9-4. In the transition region there is an irregular array of laminar and turbulent regions. As sketched in Fig. 9-5, kidney-shaped turbulent patches originate at random locations and are swept downstream and, under proper conditions, amplify in size.

Transition in a boundary layer is affected by such parameters as heat transfer, free-stream pressure gradient, surface roughness, and free-stream turbulence intensity. The influence of the latter two is particularly understandable since they are intuitively seen to be sources of disturbance that can be amplified if the fluid flow has a tendency toward instability. The influence of

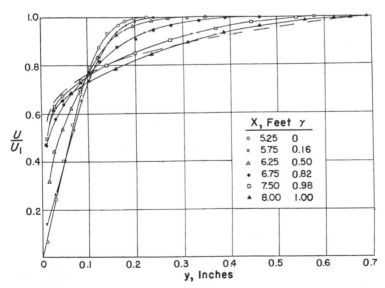

**Figure 9-3**  Measured velocity profiles in a boundary layer on a flat plate in the transition region: (1) laminar, Blasius profile; (2) turbulent, $\frac{1}{7}$th power law; $\delta = 17$ mm, $U = 27$ m/s, turbulence intensity $= 0.03\%$. (From G. B. Schubauer and P. S. Klebanoff, NACA TN 3489, 1955 and NACA Report 1289, 1956 [2].)

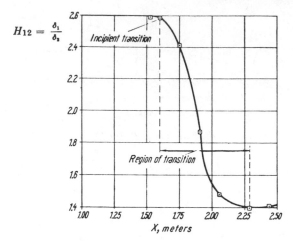

$$H_{12} = \frac{\delta_1}{\delta_2}$$

**Figure 9-4** Measured change in the shape factor $H_{12} = \delta_1/\delta_2$ for a flat plate in the transition region. (From G. B. Schubauer and P. S. Klebanoff, NACA TN 3489, 1955 and NACA Report 1289, 1956 [2].)

turbulence intensity is very pronounced, as Fig. 9-6, from Schubauer and Skramstad [3a] with early data by Hall and Hislop [3b], shows. It is apparent that at very low turbulence intensities ($J < 0.0008$), a critical Reynolds number of

$$\mathrm{Re}_{x,\,c} = \left(\frac{Ux}{\nu}\right)_{\mathrm{critical}} = 2.8 \times 10^6$$

must be exceeded for transition to begin. At the "normal" turbulence intensi-

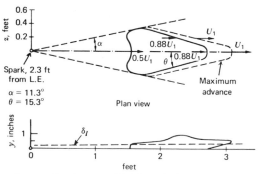

**Figure 9-5** Measured growth of an artificial turbulent spot in a laminar boundary layer on a flat plate at zero incidence: (a) plan view; (b) side view of turbulent spot artificially created at $A$ when it is about 2.4 ft from point $A$; point $A$ is 2.3 ft from the plate leading edge ($\alpha = 11.3°$, $\theta = 15.3°$, $\delta$ is boundary layer thickness, and $U = 10$ m/s). (From G. B. Schubauer and P. S. Klabanoff, NACA TN 3489, 1955 and NACA Report 1289, 1956 [2].)

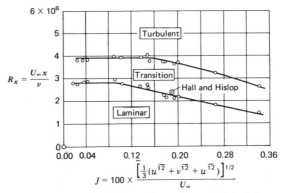

**Figure 9-6**  Measured influence of turbulence intensity on critical Reynolds number on a flat plate at zero incidence. [By permission from G. B. Schubauer and H. K. Skramstad, Laminar boundary layer oscillations and stability of laminar flow, *J. Aeronaut. Sci.* **14**, 69–78 (1947) [2].]

ties ($J \sim 0.01$) most commonly encountered, the critical Reynolds number is conventionally taken to be

$$\mathrm{Re}_{x,c} = 3.5 \times 10^5 - 5 \times 10^5 \tag{9-4}$$

If turbulence intensity is quite low, a laminar boundary layer can exist for a large distance on the plate. As the free stream becomes more turbulent, the laminar regime can exist for smaller distances. Measurement of turbulence intensities exceeding 30% and in complex recirculating flows often requires a laser velocimeter [35] rather than a hot-wire anemometer. The thickness of the transition region can be estimated by [30]

$$\mathrm{Re}_{x,c} - \mathrm{Re}_{x,i} = (60 + 4.86 M^{1.92})\,\mathrm{Re}_{x,i}^{2/3}, \qquad 0 < M < 5$$

where $M$ is the free-stream Mach number. This relationship does not require that transition be regarded as a sudden switch from laminar to turbulent flow as Eq. (9-4) intimates.

Similar results are observed for flow in a straight round duct, reported early by Osborne Reynolds [4]. In fully developed laminar flow a parabolic velocity profile exists; in fully developed turbulent flow a considerably more uniform velocity profile is found as sketched in Fig. 9-7. Under "usual" conditions the critical Reynolds number based on duct diameter, above which the flow is turbulent and below which the flow is laminar, is

$$\mathrm{Re}_{D,c} = \left(\frac{UD}{\nu}\right)_{\text{critical}} = 2300 \tag{9-5}$$

where $U$ is the velocity averaged over the duct cross-sectional area. The critical

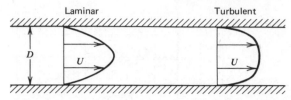

**Figure 9-7**  Comparison of laminar and turbulent velocity profiles in a circular duct.

Reynolds number is influenced by external disturbances. When particular care is taken to render the duct free of vibration and the inlet fluid free of disturbances, laminar flow has been achieved for

$$\mathrm{Re} = \frac{UD}{\nu} = 10^5$$

according to Hinze [1, p. 707]. A lower bound of

$$\mathrm{Re}_{D,c} \approx 2000$$

is reported by Schlichting [5, p. 433] below which the flow remains laminar even for strong disturbances. For noncircular ducts the critical Reynolds number can be predicted from Eq. (9-5) if the hydraulic diameter $D_H = 4$ (flow area)/(wetted perimeter) is used as the characteristic length; in triangular ducts it is possible to have laminar low in the vertices with turbulent flow in the main portion of the duct [5, p. 575–578], however.

Extension of the laminar-to-turbulent transition criterion stated in Eq. (9-5) can be extended to cover the case of a power-law fluid as Skelland [6] shows. An effective viscosity $\mu_{\mathrm{eff}}$ is first defined for use in Reynolds number as that parameter that makes Poiseuille's equation relate pressure drop to average velocity. For a power-law fluid, Eq. (6-30a) gives

$$V_{\mathrm{av}} = \frac{n}{3n+1}\left(-\frac{1}{2\mu}\frac{dP}{dx}\right)^{1/n} R^{(n+1)/n}$$

from which it follows that

$$\frac{\Delta P}{L} = \frac{2\mu\{[(3n+1)/n]V_{\mathrm{av}}\}^n}{R^{1/(n+1)}}$$

For a Newtonian fluid ($n = 1$),

$$V_{\mathrm{av}} = \frac{1}{8\mu}\frac{\Delta P}{L}R^2 = \frac{1}{32\mu}\frac{\Delta P}{L}D^2$$

or

$$\mu_{\text{eff}}\big|_{\text{Newtonian}} = \frac{1}{32}\frac{\Delta P}{L}\frac{D^2}{V_{\text{av}}}$$

For a power-law fluid then

$$\mu_{\text{eff}}\big|_{\text{power law}} = \frac{1}{32}\frac{\Delta P}{L}\frac{D^2}{V_{\text{av}}}$$

$$= \frac{\mu}{8D^{n-1}}\left(\frac{6n+2}{n}\right)^n V_{\text{av}}^{n-1}$$

Thus

$$\text{Re} = \frac{V_{\text{av}}D\rho}{\mu_{\text{eff}}} = V_{\text{av}}D\rho\frac{8D^{n-1}}{\mu V_{\text{av}}^{n-1}}\left(\frac{n}{6n+2}\right)^n$$

$$= \frac{D^n V_{\text{av}}^{2-n}}{\mu/\rho}8\left(\frac{n}{6n+2}\right)^n$$

Computation of pipe flow friction factors and transition between laminar and turbulent flow can now proceed for a power-law fluid simply by taking over all the information developed for Newtonian fluids without change—note the special case when $n = 2$.

As with the boundary layers previously discussed, laminar-to-turbulent transition in the entrance region of a duct takes place over a distance that is somewhat removed from the entrance. According to Hinze [1, p. 717], fully developed turbulent flow can be expected to require at least 40–100 diameters with the formula

$$\frac{X_{\text{entry}}}{D} = 0.693\,\text{Re}^{1/4}$$

where $\text{Re} = UD/\nu$, giving a lower bound. When the Reynolds number approaches the critical value, alternate slugs of laminar and turbulent fluid pass down the duct with the turbulent slugs decaying; when it exceeds the critical value, the turbulent slugs grow in size until they fill the duct while being swept downstream.

Pipe flow transition is similar to boundary-layer transition, but there are also dissimilarities. The major dissimilarity is that in pipe flow there is no free boundary.

Results of like nature have been found for other geometries. For Couette flow of two parallel walls, the critical Reynolds number is reported to be [1, p.

77; 5, p. 558]

$$\mathrm{Re}_{h,\,c} = \frac{U_{max}\, h}{\nu} = 1900$$

where $h$ is the wall separation; if flow is caused by a pressure gradient between parallel walls of separation $h$, then

$$\mathrm{Re}_c = \frac{U_{av}\, h}{\nu} = 1150$$

For two concentric cylinders of radial separation $d$, inner radius $R_i$, inner tangential velocity $U_i$, and the outer cylinder at rest, there are three flow regimes [5, p. 503] described by the Taylor number $\mathrm{Ta} = (U_i\, d/\nu)(d/R_i)^{1/2}$:

| | |
|---|---|
| $\mathrm{Ta} < 41.3$ | laminar Couette flow |
| $41.3 < \mathrm{Ta} < 400$ | laminar flow with Taylor vortices |
| $\mathrm{Ta} > 400$ | turbulent flow |

The destabilizing effect of centrifugal forces is illustrated by these results. Vertical stratification caused by temperature or salinity gradients also affects the critical Reynolds number. The effect of temperature on viscosity can cause early transition to turbulent boundary layers when heat flows from the surface into a gas, acting as a pressure increase in the downstream direction, whereas a cooling of a gas by the wall retards transition just as does a pressure decrease in the downstream direction—for a liquid, the effect of heat transfer would be in the opposite direction. For a fuller discussion, consult Schlichting [5, pp. 470, 483–516]; boundary-layer suction is effective in retarding transition, whereas surface roughnesses can speed transition because of the disturbances they generate.

It is understandably convenient to arrive at a single criterion for laminar-to-turbulent transition. Particular attention is focused on bringing together the criteria for round duct flow $\mathrm{Re}_{D,\,c} = U_{av} D/\nu \approx 2300$ and boundary-layer flow over a flat plate $\mathrm{Re}_{X,\,c} = U X_c/\nu \approx 3.5 \times 10^6$. To achieve the desired unification, it is necessary to base a Reynolds number for the flat-plate case on a characteristic dimension that can be the distance from the plate leading edge at which transition occurs $X_c$ (the dimension ultimately of design interest), or a local dimension less influenced by upstream events and more characteristic of the boundary layer such as the boundary-layer thickness at transition $\delta_c$ (a somewhat fictitious quantity), the displacement thickness at transition $\delta_{1c}$, or the momentum thickness at transition $\delta_{2c}$ (a real quantity related to local wall shear stress).

First, consider use of $\mathrm{Re}_{\delta,\,c} = U\delta_c/\nu$ for a boundary layer. At transition

$$\frac{U X_c}{\nu} \approx 3.5 \times 10^5$$

as experiment shows. For a laminar boundary layer, Eq. (7-16) gives $\delta/x = 5/(Ux/\nu)^{1/2}$, which, when substituted into the preceding relationship, gives the transition criterion

$$\frac{U\delta_c}{\nu} \approx 5(3.5 \times 10^5)^{1/2}$$

$$\approx 2960 \qquad\qquad (9\text{-}6)$$

In comparison, the round-duct result can be expressed in terms of the center-line velocity for a parabolic profile ($V_m = 2V_{av}$ from Eq. (6-31)] with the pipe radius taken to be the boundary-layer thickness. This procedure gives the round-duct transition criterion as

$$V_{av}\frac{D}{\nu} \approx 2300$$

$$\frac{V_m}{2}\frac{2\delta}{\nu} \approx$$

$$\frac{V_m\delta}{\nu} \approx 2300$$

which is in fair agreement with Eq. (9-6).

Second, consider use of $\mathrm{Re}_{\delta_{1c}} = U\delta_{1c}/\nu$ for a boundary layer. Equation (7-20) gives $\delta_1/\delta = 0.3442$, which, when used in Eq. (9-6), gives the transition criterion as

$$\frac{U\delta_{1c}}{\nu} \approx 2960(0.3442)$$

$$\approx 1020 \qquad\qquad (9\text{-}7)$$

In comparison, the round-duct result can be based on the center-line velocity for laminar flow ($V_m = 2V_{av}$) and the pipe displacement thickness (the radial displacement required of the wall to allow a stream of center-line velocity to flow the discrepancy between actual flow and that which would flow if all fluid had the center-line velocity). The pipe displacement thickness is calculated from

$$V_m\int_R^{R+\delta_1} 2\pi r\, dr = \int_0^R (V_m - V)2\pi r\, dr$$

with $V/V_m = 1 - (r/R)^2$ from Eq. (6-31) to be

$$\frac{\delta_1}{R} = 0.22$$

Introduction of this into the pipe transition criterion gives

$$\frac{V_{av} D}{\nu} \approx 2300$$

$$\frac{V_m}{2}(\delta_1)\frac{2R}{\nu} \approx 2300\delta_1$$

$$\frac{V_m \delta_1}{\nu} \approx 2300(0.22) = 506$$

which does not agree well with Eq. (9-7). Nevertheless, stability theory suggests that $U\delta_1/\nu$ is a proper transition criterion, and it is frequently used [5, p. 470] in a modified form. Table 9-1, based on the work of Pretsch [7], shows the influence of a pressure gradient. Furthermore, the $m = 0$ result suggests that transition begins at a value of $UX_c/\nu = 2.3 \times 10^5$ which might be the case, according to Fig. 9-6, for a highly turbulent free stream.

Third, consider the use of $\text{Re}_{\delta_{2c}} = U\delta_{2c}/\nu$ for a boundary layer. Equation (7-22) gives $\delta_1/\delta = 0.1328$, which, when used in Eq. (9-6), gives the transition criterion as

$$\text{Re}_{\delta_{2c}} = \frac{U\delta_{2c}}{\nu} = 390 \tag{9-8}$$

The momentum thickness for a pipe is the radial displacement of the pipe wall required for a stream at the center-line velocity to flow the momentum discrepancy between that actually transported by the parabolic velocity and what would have flowed had the stream all been at the center-line velocity. Thus

$$V_m^2 \int_R^{R+\delta_2} 2\pi r\, dr = \int_0^R \left(V_m^2 - V^2\right) 2\pi r\, dr$$

which gives

$$\frac{\delta_2}{R} = 0.154$$

**Table 9-1   Critical Reynolds Number for Laminar Wedge Flow [7]$^a$**

| $\beta$ | −0.1 | 0 | 0.2 | 0.4 | 0.6 | 1.0 |
|---|---|---|---|---|---|---|
| $m = (x/U)\, du/dx$ | −0.048 | 0 | 0.111 | 0.25 | 0.43 | 1.0 |
| $(U\delta_1/\nu)_{\text{critical}}$ | 126 | 660 | 3,200 | 5,000 | 8,300 | 12,600 |

$^a$See Fig. 7-7 for $m$ and $\beta$ definitions.

Introduction of this into the pipe transition criterion gives

$$\frac{V_{av} D}{\nu} \approx 2300$$

$$\frac{V_m}{2}(\delta_2)\frac{2R}{\nu} \approx 2300\delta_2$$

$$\frac{V_m \delta_2}{\nu} \approx 2300(0.154) = 350$$

which is in good agreement with Eq. (9-8). Kays [8] suggests, on the basis of studies by Moretti and Kays [9] and Schraub [10], that Eq. (9-8) be used, particularly for reversion of a turbulent boundary layer to a laminar boundary layer under a sufficiently strongly accelerating free stream. A similar criterion [36] enables prediction of the relaminarization of initially turbulent tube flows that can occur when a gas is strongly heated as in a nuclear reactor.

The general criteria to be used for transition are not known with certainty, although a body of stability theory exists [34]. Over the years, transition data have been accumulated. These tests have yielded abundant information on the effects of Mach number, surface temperature, surface roughness, bluntness, pressure gradient, transpiration, angle of attack, and so forth. These data are appraised by Reshotko [11] as having yielded neither a transition theory nor a reliable means of predicting transition. For example, wind tunnel experiments at Mach number 4.5 show no transition on a flat plate for $Re_x = Ux/\nu = 3.3 \times 10^6$ when the tunnel wall is covered with a laminar boundary layer ("quiet" operation); whereas when the tunnel wall is covered by a turbulent boundary layer ("noisy" operation), transition occurs at $Re_x = 10^6$. The difference is ascribed to acoustic noise radiated from the tunnel walls. Thus the vast body of transition data obtained in supersonic wind tunnels is suspect, and transition research continues [12–14].

The new $H - R_x$ method [37] for predicting laminar-to-turbulent transition in boundary-layer flows shows promise of being accurate, convenient, and general since it accounts for the effects of pressure gradient, heat transfer, and suction. Here $R_x$ is the Reynolds number based on distance along the surface in the direction of flow, and $H = \delta_1/\delta_2$ is the ratio of displacement $\delta_1$ to momentum $\delta_2$ boundary-layer thicknesses. The $H - R_x$ method is based on the $e^9$ method, which, in turn, is based on the observation that transition occurs at a disturbance amplification ratio of about $e^9$ where $e = 2.72$ when free-stream turbulence is low, the surface is smooth, and the flow possesses local similarity. In addition to these limitations, the $H - R_x$ method is limited to temperature differences of $T_w - T_\infty \leqslant 23°C$ since parameters other than just $H$ are involved when temperature differences are large. The $H - R_x$ method approximates the transition Reynolds number prediction of the $e^9$ method by

$$\log_{10}(R_x) = -40.4557 + 64.8066H - 26.7538H^2 + 3.3819H^3,$$

$$2.1 < H < 2.8 \quad (9-9)$$

and takes advantage of the fact that the local value of $H$ is the most direct measure of stability. To predict transition, a plot of $R_x$ versus $H$ is made from boundary-layer calculations such as are discussed in Chapter 7 (see Section 7.10 for nonsimilar calculations). This curve initially lies below the locus of Eq. (9-9); transition from laminar to turbulent flow should occur when it crosses the locus the Eq. (9-9). Such a prediction is more general than that resulting from merely taking the critical Reynolds number to be the constant value suggested by Fig. 9-6 and its associated discussion.

## 9.2  TIME-AVERAGED DESCRIBING EQUATIONS

In the foregoing, some insight into the conditions under which turbulent flow can be expected was achieved. Also, time averaging with a perturbation was introduced. The Navier–Stokes and energy equations are exact if their unsteady form is taken. But we wish to solve turbulence problems that, in their gross aspects, are steady state. Yet if the basic turbulence is as gross as experiment indicates, a steady-state approach will never be successful. To attempt to follow the random motion of turbulent eddies would be as hopeless as following the individual molecules, on the other hand.

Time averaging of the describing equations (including their unsteady terms) should give the net effect of the turbulent perturbations. In what follows it is assumed for the sake of simplicity that all molecular properties are constant, in much the spirit that guided the breakthrough to the boundary-layer equations for laminar flow. First, consider the continuity equation for the unsteady state with constant properties, which is

$$\frac{\partial u}{\partial x} + \frac{\partial v}{\partial y} + \frac{\partial w}{\partial z} = 0$$

In terms of a time-averaged and perturbation quantity, this becomes

$$\frac{\partial \bar{u} + u'}{\partial x} + \frac{\partial \bar{v} + v'}{\partial y} + \frac{\partial \bar{w} + w'}{\partial z} = 0$$

Time averaging of this equation [a bar denotes $(1/\Delta t)\int_t^{t+\Delta t} f(t)\, dt$] gives

$$\overline{\frac{\partial \bar{u} + u'}{\partial x}} + \overline{\frac{\partial \bar{v} + v'}{\partial y}} + \overline{\frac{\partial \bar{w} + w'}{\partial z}} = 0$$

Interchanging of the order of space and time differentiation gives

$$\frac{\partial \overline{\bar{u} + \bar{u}'}}{\partial x} + \frac{\partial \overline{\bar{v} + \bar{v}'}}{\partial y} + \frac{\partial \overline{\bar{w} + \bar{w}'}}{\partial z} = 0$$

But we found earlier that the time average of a perturbation quantity must be zero. So the time-averaged continuity equation is

$$\frac{\partial \bar{u}}{\partial x} + \frac{\partial \bar{v}}{\partial y} + \frac{\partial \bar{w}}{\partial z} = 0$$

or

$$\nabla \cdot \mathbf{V} = 0 \qquad (9\text{-}10)$$

Second, consider the unsteady form of the $x$-motion equation in the form of ($\tau_{ij}^m$ is a stress due to molecular level events)

$$\rho\left[\frac{\partial u}{\partial t} + u\frac{\partial u}{\partial x} + v\frac{\partial u}{\partial y} + w\frac{\partial u}{\partial z}\right] = B_x - \frac{\partial P}{\partial x} + \frac{\partial \tau_{xx}^m}{\partial x} + \frac{\partial \tau_{yx}^m}{\partial y} + \frac{\partial \tau_{zx}^m}{\partial z}$$

This is recast into the conservation form

$$\rho\left[\frac{\partial u}{\partial t} + \frac{\partial uu}{\partial x} + \frac{\partial vu}{\partial y} + \frac{\partial wu}{\partial z}\right] = B_x - \frac{\partial P}{\partial x} + \frac{\partial \tau_{xx}^m}{\partial x} + \frac{\partial \tau_{yx}^{m}}{\partial y} + \frac{\partial \tau_{zx}^{m}}{\partial z}$$

$$- \rho\left[\frac{\partial u}{\partial x} + \frac{\partial v}{\partial y} + \frac{\partial w}{\partial z}\right]u$$

Time averaging and interchanging of the order of time and space differentiation gives

$$\rho\left[\frac{\partial \bar{u}}{\partial t} + \frac{\partial \bar{u}'}{\partial t} + \frac{\partial\left(\overline{\bar{u}\bar{u}} + \overline{u\bar{u}'} + \overline{u'u'}\right)}{\partial x} + \frac{\partial\left(\overline{\bar{v}\bar{u}} + \overline{v\bar{u}'} + \overline{\bar{u}v'} + \overline{v'u'}\right)}{\partial y}\right.$$

$$\left. + \frac{\partial\left(\overline{\bar{w}\bar{u}} + \overline{w\bar{u}'} + \overline{\bar{u}w'} + \overline{w'u'}\right)}{\partial z}\right]$$

$$= \bar{B}_x + \bar{B}'_x - \frac{\partial\left(\bar{P} + \bar{P}'\right)}{\partial x} + \frac{\partial\left(\bar{\tau}_{xx}^m + \bar{\tau}'^m_{xx}\right)}{\partial x} + \cdots$$

$$\rho\left[\frac{\partial \bar{u}}{\partial t} + \frac{\partial \overline{\bar{u}\bar{u}}}{\partial x} + \frac{\partial \overline{\bar{v}\bar{u}}}{\partial y} + \frac{\partial \overline{\bar{w}\bar{u}}}{\partial z}\right] = \bar{B}_x - \frac{\partial \bar{P}}{\partial x} + \frac{\partial}{\partial x}\left[\bar{\tau}_{xx}^m - \rho\overline{u'u'}\right]$$

$$+ \frac{\partial}{\partial y}\left[\bar{\tau}_{yx}^m - \rho\overline{v'u'}\right] + \frac{\partial}{\partial z}\left[\bar{\tau}_{zz}^m - \rho\overline{u'u'}\right]$$

Realization that $\nabla \cdot \mathbf{V} = 0$ then gives

$$\rho\left(\frac{\partial \bar{u}}{\partial t} + \bar{u}\frac{\partial \bar{u}}{\partial x} + \bar{v}\frac{\partial \bar{u}}{\partial y} + \bar{w}\frac{\partial \bar{u}}{\partial z}\right) = \bar{B}_x - \frac{\partial \bar{P}}{\partial x} + \frac{\partial \bar{\tau}_{xx}}{\partial x} + \frac{\partial \bar{\tau}_{yx}}{\partial y} + \frac{\partial \bar{\tau}_{zx}}{\partial z}$$

$$(9\text{-}11)$$

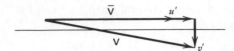

**Figure 9-8** Transport across a surface due to turbulent fluctuations of velocity.

In Eq. (9-11) it is evident that the total stress is composed of the sum of a molecular and a turbulent part $\bar{\tau} = \bar{\tau}^m + \bar{\tau}^{\text{turb}}$, both of which have the same physical origin—a particle crosses a control surface, carrying with it momentum as sketched in Fig. 9-8—with the latter called a *Reynolds stress*. In brief, time averaging has shown that effective additions to the time-averaged molecular transport are possible. We see that

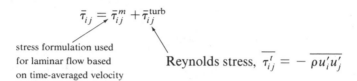

$$\bar{\tau}_{ij} = \bar{\tau}_{ij}^m + \bar{\tau}_{ij}^{\text{turb}}$$

stress formulation used
for laminar flow based           Reynolds stress, $\overline{\tau_{ij}'} = -\overline{\rho u_i' u_j'}$
on time-averaged velocity

The other two equations of motion, for $y$ and $z$, lead to the same form as given in Eq. (9-11). In each case the Reynolds stresses appear as an addition to the molecular stresses.

Third, the equations of energy and diffusion have identically the form of the $x$-motion equation, with $T$ and $\omega$ replacing $u$ (the only major complication in the energy equation is the viscous dissipation terms, which are omitted for simplicity here). It can be shown that the diffusive fluxes are given by

$$\bar{q}_i = -k\frac{\partial \bar{T}}{\partial x_i} + \rho C_p \overline{u_i' T'} = \bar{q}_i^m + \bar{q}_i^{\text{turb}}$$

$$\bar{m}_i = -\rho D_{12}\frac{\partial \bar{\omega}_1}{\partial x_i} + \rho \overline{u_i' \omega_1'} = \bar{m}_{1_i}^m + \bar{m}_{1_i}^{\text{turb}}$$

The basic task now is to calculate the turbulent transport (e.g., $\bar{\tau}_{ij}' = \overline{\rho u_i' u_j'}$ in the $x$-motion equation), from the time-averaged velocity. This cannot be done analytically from first principals in general and must rely on inspired guesswork and experiment.

## 9.3  EDDY DIFFUSIVITIES

As long as one must work without rigorous guides, it is best to try to put things into familiar and convenient forms. The laminar relations are familiar and are also convenient, particularly since they are linear relations. With boundary layer ideas in mind, plausible relations are

$$\tau_{yx} = \mu \frac{\partial \bar{u}}{\partial y} + \underbrace{\mu' \frac{\partial \bar{u}}{\partial y}}_{-\rho \overline{v'u'}} = \rho(\nu + \underset{\uparrow}{\varepsilon_m}) \frac{\partial \bar{u}}{\partial y}$$

$$\text{eddy diffusivity}$$
$$\text{for momentum}$$

$$\bar{q}_y = -k \frac{\partial \bar{T}}{\partial y} + \underbrace{\left(-k' \frac{\partial \bar{T}}{\partial y}\right)}_{\rho C_p \overline{v'T'}} = \rho C_p \left(\frac{\nu}{\text{Pr}} + \frac{\varepsilon_m}{\underset{\uparrow}{\text{Pr}'}}\right) \frac{\partial \bar{T}}{\partial y}$$

$$\text{turbulent Pr}$$

$$\bar{m}_y = -\rho D_{12} \frac{\partial \bar{\omega}_1}{\partial y} + \underbrace{\left(-\rho D'_{12} \frac{\partial \bar{\omega}_1}{\partial y}\right)}_{\rho \overline{u'\omega'}_1} = -\rho \left(\frac{\nu}{\text{Sc}} + \frac{\varepsilon_m}{\underset{\uparrow}{\text{Sc}'}}\right) \frac{\partial \bar{\omega}_1}{\partial y}$$

$$\text{turbulent Sc}$$

If the boundary-layer equations were to be considered for a constant-property fluid in turbulent flow, they would appear as

$$\frac{\partial \bar{u}}{\partial x} + \frac{\partial \bar{v}}{\partial y} = 0$$

$$\frac{D\bar{u}}{Dt} = \frac{1}{\rho}\left(\bar{B}_x - \frac{\partial \bar{P}}{\partial x}\right) + \frac{\partial}{\partial y}\left[(\nu + \varepsilon_m)\frac{\partial \bar{u}}{\partial y}\right]$$

$$\frac{D\bar{T}}{Dt} = \frac{\partial}{\partial y}\left[\left(\frac{\nu}{\text{Pr}} + \frac{\varepsilon_m}{\text{Pr}'}\right)\frac{\partial \bar{T}}{\partial y}\right]$$

$$\frac{D\bar{\omega}_1}{Dt} = \frac{\partial}{\partial y}\left[\left(\frac{\nu}{\text{Sc}} + \frac{\varepsilon_m}{\text{Sc}'}\right)\frac{\partial \bar{\omega}_1}{\partial y}\right] \qquad (9\text{-}11a)$$

Of course, the very important question remains as to what a workable expression for $\varepsilon_m$ might be. The turbulent Prandtl $\text{Pr}'$ and Schmidt $\text{Sc}'$ numbers are often assumed to be constant near unity. This assumption concerning $\text{Pr}'$ and $\text{Sc}'$ is reasonable in view of the expectation that the mechanisms for turbulent transport of momentum, heat, and mass should be the same.

One of the earliest relations for $\varepsilon_m$ was the postulate by Boussinesq [15] in 1877 that $\varepsilon_m$ is constant. This may have been guided by the constancy of molecular viscosity, but it turns out that $\varepsilon_m$ is not a constant. Particularly near solid surfaces, it is dependent on velocity and geometry. Evidently, $\varepsilon_m$ is not a property of the fluid. Nevertheless, a constant $\varepsilon_m$ is sometimes used because of its convenience.

In about 1925 Prandtl [16] devised the Prandtl mixing length relationship, which is built on ideas from kinetic theory of gases. The reasoning is that the turbulent transport is much like the molecular one; it is random and occurs because discrete lumps of matter of fixed identity (for at least short times) cross back and forth and carry their properties with them as sketched in Fig. 9-9. Near a solid surface there are "bursts" of turbulent fluid [28] similar to those shown in Fig. 9-5; these bursts are the lumps shown in Fig. 9-9. The fluid in these bursts has been visually observed [29] to break up into small eddies in the mainstream. For molecular transport, it was stated in Section 3.2 that

$$\nu = \frac{\mu}{\rho} \sim V_{av}\lambda$$

which is essentially Eq. (3-4). It is reasonable to set $\Delta L \sim \lambda$ and $|(d\bar{u}/dL)\Delta L|$ $\sim V_{av}$. Thus the eddy diffusivity for momentum is written by analogy as

$$\varepsilon_m = \frac{\mu^{turb}}{\rho} \sim (\Delta L)^2 \left|\frac{d\bar{u}}{dL}\right|$$

But what is $\Delta L$? It is analogous to the mean free molecular path and must be small near a solid surface for a turbulent eddy. Since the principal flaw of Boussinesq's hypothesis is its failure near a solid surface, attention is focused on that region. Thus it is guessed that $\Delta L = K_1 y$, where $y$ is the distance from the solid surface. Comparison with data shows that $K_1 = 0.36$. So Prandtl's mixing length relation is

$$\varepsilon_m = l^2 \left|\frac{d\bar{u}}{dy}\right| \tag{9-12}$$

where $l = K_1 y$ is the mixing length. Equation (9-12) requires that $\varepsilon_m = 0$ at the center of a flow channel where $d\bar{u}/dy = 0$, a difficulty remedied for the case of free turbulent flow (such as occurs in jets) by Prandtl's [17] use of Reichardt's [18] data to successfully devise the simpler relation

$$\varepsilon_m = K_2 b(\bar{u}_{max} - \bar{u}_{min}) \tag{9-13}$$

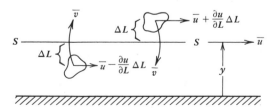

**Figure 9-9**  Turbulent transport across a surface by random movement of clumps of fixed identity.

Here $K_2$ is a constant, $b$ is the jet mixing-zone width, and $\bar{u}_{max} - \bar{u}_{min}$ is the maximum difference in velocity across the jet width. Prandtl also proposed a modification, little used because of its complexity, to Eq. (9-12) of $\varepsilon_m = l^2[(d\bar{u}/dy)^2 + L^2(d^2\bar{u}/dy^2)^2]^{1/2}$, to circumvent the difficulty. For completeness, the similarity hypothesis devised by von Karman [19], described by Schlichting [5, p. 551], is cited. He gives the mixing length to be used in Eq. (9-12) as

$$l = K_3 \left| \frac{d\bar{u}/dy}{d^2\bar{u}/dy^2} \right|$$

so that

$$\varepsilon_m = K_3^2 \left[ \frac{d\bar{u}/dy}{d^2\bar{u}/dy^2} \right]^2 \left| \frac{d\bar{u}}{dy} \right|$$

with $K_3 - 0.4$ (see Bird, et al. [32] and Problem 9-8, also).

Even with Prandtl's mixing length relation of Eq. (9-12), an unwarrantedly large discrepancy between data and prediction is observed in the region very close to the wall. Deissler [20] combined an empirical fit to data in this region with an inspired guess. He felt that the damping effect of the wall would die away exponentially as one moved away from the wall. This idea was combined with the molecular idea seen in Prandtl's relationship to the extent that $V_{av} \sim \bar{u}$ and $\lambda \sim y$ to give

$$\nu = \frac{\mu}{\rho} \sim V_{av}\lambda$$

$$\varepsilon_m = \frac{\mu^{turb}}{\rho} \sim \bar{u}y$$

The combination of these two ideas then gives

dimensionless Reynolds number

$$\varepsilon_m = n^2\bar{u}y \left[ 1 - \exp\left( -n^2\frac{\bar{u}y}{\nu} \right) \right] \tag{9-14}$$

molecular idea

exponential decay of wall damping

where $n = 0.124$. It is noteworthy that no derivatives of velocity appear, circumventing the difficulty encountered with Eq. (9-12). A similar relation was proposed at about the same time by Van Driest [25] as

$$\varepsilon_m = K^2 y^2 \left[ 1 - \exp\left( \frac{-y}{A} \right) \right]^2 \left| \frac{\partial \bar{u}}{\partial y} \right|, \qquad K = 0.4 \quad \text{and} \quad A = \frac{26\nu}{v^*}$$

$$\tag{9-14a}$$

which has been successfully applied to many boundary-layer problems. The basis for the exponential form is the subject of Problem 9-9.

Whereas molecular viscosity is a property of the fluid, the turbulent eddy viscosity is mostly a property of the flow. The specific function of the eddy viscosity is to relate turbulence stresses to gradients of the time-averaged fluid velocity. It is to be expected that any relationship that is devised will fall short of predicting all details of the flow.

The state of turbulence understanding was admirably discussed by Liepmann [21], who points out that averaging alone cannot solve turbulent flow problems. Such an attempt only introduces more unknowns and requires closure of the sequence of equations by additional measurements or by physical arguments such as have been adduced in this section. It is most likely that turbulent flow will ultimately be explainable on a fundamental basis as the long-range interaction of vortices. Once this understanding is achieved, important techniques of turbulence control may be possible. Such a state is still in the future.

The mere fact that a physical phenomenon such as turbulence has the appearance of chaos does not necessarily mean that it is random. It might still be the deterministic result of several basic physical phenomena with feedback loops. Hofstadter [38] presents a popularized survey of recent studies and discusses their application to turbulence.


## 9.4 UNIVERSAL TURBULENT VELOCITY PROFILE

This eddy diffusivity information allows turbulent velocity profiles to be computed. From these computed profiles comparisons can be made with experimental measurements and constants be evaluated. Looking ahead, the easiest experiment would be one involving steady flow in a long and smooth pipe. Measurement of wall shear stress $\tau_w$ is easy there since a momentum balance on the fluid in the pipe as suggested in Fig. 9-10 yields the relation between pressure drop and wall shear stress of

$$\tau_w = \frac{D(P_1 - P_2)}{4L}$$

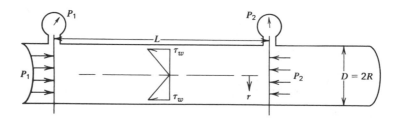

**Figure 9-10**  Shear-stress variation in a circular duct.

and also shows that shear stress varies linearly with radius as

$$\tau = \tau_w \frac{r}{R}$$

Determination of wall shear stress is one of the major reasons for finding velocity distributions, so this easy experimental method is very important.

A turbulent velocity profile can be computed by using the relationships given by Deissler and Prandtl in the region "close" to the wall and "far" from the wall, respectively, for fully developed steady turbulent flow in a smooth pipe. In the region close to the wall, molecular and turbulent contributions to shear stress are important. Thus

$$\tau = \rho(\nu + \varepsilon_m) \frac{d\bar{u}}{dy}$$

where $y$ is the distance from the wall. Use of Eq. (9-14) in this relation then yields (with $r = R - y$),

$$\tau_w \left(1 - \frac{y}{R}\right) = \rho \left\{ \nu + n^2 \bar{u} y \left[1 - \exp\left(-n^2 \frac{\bar{u}y}{\nu}\right)\right]\right\} \frac{d\bar{u}}{dy}$$

Dimensionless quantities are formed by defining $u^+ = \bar{u}/v^*$ and $y^+ = yv^*/\nu$, where $v^* = (\tau_w/\rho)^{1/2}$ is called the *friction velocity*. Then, with $\tau \approx \tau_w$ close to the wall,

$$\frac{du^+}{dy^+} = \frac{1}{1 + n^2 u^+ y^+ \left[1 - \exp\left(-n^2 u^+ y^+\right)\right]}, \qquad u^+(0) = 0 \quad (9\text{-}15)$$

Extremely close to the wall, molecular effects will predominate so that

$$\frac{du^+}{dy^+} \approx 1, \qquad u^+(0) = 0$$

giving the linear velocity profile

$$u^+ \approx y^+ \qquad (9\text{-}16)$$

which measurements show to be accurate for $0 \lesssim y^+ \lesssim 5$. Farther away from the wall both molecular and turbulent effects are important, so the nonlinear differential equation [Eq. (9-15)] must be solved numerically; it is suggested here only in integral form as

$$u^+ = 5 + \int_5^{y^+} \left\{1 + n^2 u^+ y^+ \left[1 - \exp\left(-n^2 u^+ y^+\right)\right]\right\}^{-1} dy^+$$

$$\approx 5 + 5 \ln\left(\frac{y^+}{5}\right) \qquad (9\text{-}17)$$

which is in agreement with measurements for $5 \lesssim y^+ \lesssim 26$.

Deissler's relation of Eq. (9-14) just used has given the velocity profile close to the wall where molecular and turbulent contributions must both be accounted for (a buffer zone, $5 \leqslant y^+ \leqslant 26$); the region where molecular effects are most important ($y^+ \leqslant 5$) is, accordingly, called a *laminar sublayer*—a misnomer since the flow is turbulent everywhere. Figure 9-11 illustrates this by depicting experimental measurements. Note that the local turbulence remains a substantial fraction of the local flow right up to the wall.

Farther from the wall, but still close, Prandtl's mixing length relation of Eq. (9-12) can be used. The contribution of molecular effects is ignored. Then

$$\tau = \rho \varepsilon_m \frac{d\bar{u}}{dy}$$

Introduction of Eq. (9-12) into this relation gives

$$\tau_w \left( 1 - \frac{y}{R} \right) \approx \rho K_1^2 y^2 \left( \frac{d\bar{u}}{dy} \right)^2$$

Because attention is still directed to regions near the wall, $y/R \ll 1$ in which

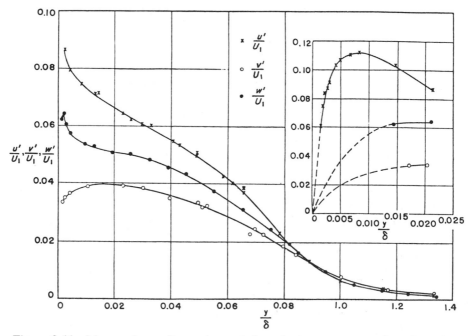

**Figure 9-11** Measured rms fluctuations of the velocity components in a boundary layer along a smooth wall with constant pressure. (From P. S. Klebanoff, NACA TN 3178, 1954 [36].)

$\tau \approx \tau_w$, this differential equation can be accurately simplified to

$$\left(\frac{\tau_w}{\rho}\right)^{1/2} = K_1 y \frac{d\bar{u}}{dy}$$

In terms of the dimensionless quantities $y^+ = yv^*/\nu$, $u^+ = \bar{u}/v^*$, and $v^* = (\tau_w/\rho)^{1/2}$, the solution is

$$u^+ = K_1^{-1} \ln y^+ + C = 2.78 \ln y^+ + 3.8 \qquad (9\text{-}18)$$

for $26 \leqslant y^+$. Measurements show that $K_1 = 0.36$ and $C = 3.8$. As the turbulent contribution is much larger than the molecular contribution, this region ($y^+ > 26$) is named the *fully turbulent core*. In the main part of the pipe, data show that

$$u^+ = 8.74(y^+)^{1/7} \qquad (9\text{-}19)$$

for $\mathrm{Re}_D = U_{av} D/\nu \approx 10^5$. As pointed out by Schlichting [5, p. 563–566] in his discussion of the data due to Nikuradse [22], the best fit is provided by $u^+ = C(y^+)^{1/n}$, where $C$ and $n$ vary with Reynolds number, as shown in Table 9-2. It is only for $\mathrm{Re} = U_{av} D/\nu \approx 10^5$ that good fit to data is found for $u^+ = 8.74(y^+)^{1/7}$; at Re substantially different from $10^5$, a different value of $n$ must be used. The power-law form is convenient for integration.

The universal turbulent velocity profile from the work of Martinelli [23] is shown in Fig. 9-12, where it is seen that the predictions given in Eqs. (9-16)–(9-19) are in good agreement with data. The arbitrary division of the flow into three basic zones (laminar sublayer, buffer, and turbulent core) is reasonable, although the velocity distribution is smooth and continuous throughout.

Efforts to devise a velocity distribution that is applicable to all regions of the pipe have been made. Reichardt [24] proposed on the basis of measurements to let the eddy diffusivity vary as

$$\frac{\varepsilon_m}{\nu} = \frac{kr_0^+}{6}\left[1 - \left(\frac{r}{r_0}\right)^2\right]\left[1 + 2\left(\frac{r}{r_0}\right)^2\right] \qquad (9\text{-}20)$$

**Table 9-2   Parameters for Power-Law Turbulent Velocity Profile in a Circular Pipe**

| $\mathrm{Re} = U_{av} D/\nu$ | $n$ | $C$ |
| --- | --- | --- |
| $4 \times 10^3$ | 6 | |
| $10^5$ | 7 | 8.74 |
| $0.8 \times 10^6$ | 8 | 9.71 |
| $\geqslant 2 \times 10^6$ | 10 | 11.5 |

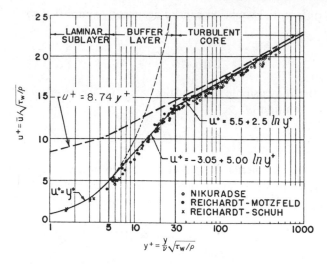

**Figure 9-12**  Generalized velocity distribution for turbulent flow in tubes. [From R. C. Martinelli, *Transact. ASME* **69**, 947–959 (1947) [23].]

with $k = 0.4$, which, when used with a linearly varying shear stress, gives the velocity distribution applicable everywhere as

$$u^+ = 5.5 + 2.5 \ln \left[ y^+ \frac{1.5(1 + r/r_0)}{1 + 2(r/r_0)^2} \right] \tag{9-20a}$$

which is not as convenient for integration as the power-law form. Relations similar to Deissler's Eq. (9-14) were proposed by Spalding [26] with

$$\frac{\varepsilon_m}{\nu} = \frac{k}{E} \left[ e^{ku^+} - 1 - ku^+ - \frac{(ku^+)^2}{2!} - \frac{(ku^+)^3}{3!} \right]$$

$$y^+ = u^+ + E^{-1} \left[ e^{ku^+} - 1 - ku^+ - \frac{(ku^+)^2}{2!} - \frac{(ku^+)^3}{3!} - \frac{(ku^+)^4}{4!} \right]$$

where $k = 0.0407$ and $E = 10$.

## 9.5  RESISTANCE FORMULAS

The flow resistance in a round duct can be deduced from the velocity distribution of Eq. (9-19). Starting with

$$u^+ = 8.74(y^+)^{1/7} \tag{9-19}$$

rearrangement gives the friction velocity as

$$v^* = u^{7/8} \left( \frac{y}{\nu} \right)^{-1/8} (8.74)^{-7/8}$$

Recalling that $v^* = (\tau_w/\rho)^{1/2}$ and solving for $\tau_w$ from this relationship, one obtains

$$\tau_w = \frac{0.0225 \rho u^2}{(uy/\nu)^{1/4}}$$

At the pipe center line, $u = U_{CL}$ and $y - R$ so that

$$\tau_w = 0.0225 \frac{\rho U_{CL}^2}{(U_{CI} R/\nu)^{1/4}}$$

or

$$\frac{C_f}{2} = \frac{\tau_w}{\rho U_{CL}^2} = \frac{0.0225}{(U_{CL} R/\nu)^{1/4}} \qquad (9\text{-}21)$$

This relationship is accurate up to a Reynolds number of only about $10^5$, as discussed in connection with Table 9-2. In terms of pressure drop, Eq. (9-21) becomes, since $U_{av} = 0.817 U_{CL}$,

$$f = \frac{2\Delta P}{(L/D)\rho U_{av}^2} = \frac{0.3164}{(U_{av} D/\nu)^{1/4}} \qquad (9\text{-}21a)$$

Although Eq. (9-21) is of a form convenient for computation and accuracy acceptable for $\text{Re} \lesssim 10^5$, Prandtl's universal law of friction for a smooth pipe

$$\frac{1}{f^{1/2}} = 2 \log_{10} \left( f^{1/2} \frac{U_{av} D}{\nu} \right) - 0.8 \qquad (9\text{-}21b)$$

is extremely accurate up to $\text{Re} = 3.4 \times 10^6$ by measurement and can be extrapolated far beyond that. See Section 11.8 for newer and more convenient relationships for $f$.

For a turbulent boundary layer on a plate in parallel flow, it is assumed that the boundary-layer thickness can be used in place of pipe radius and free-stream velocity can be used in place of the center-line velocity. Thus for smooth plates, the analogy to Eq. (9-21) is

$$\frac{C_f}{2} = \frac{\tau_w}{\rho U_\infty^2} = 0.0228 \left( \frac{U_\infty \delta}{\nu} \right)^{-1/4} \qquad (9\text{-}22)$$

for $\mathrm{Re}_\delta = U_\infty \delta/\nu \lesssim 10^5$ and where a correction for pipe curvature has been introduced in the coefficient. Less convenient friction coefficient relations that are accurate for a much wider range of Reynolds numbers are the Prandtl–Schlichting formula

$$\bar{C}_f = \frac{0.455}{\left[\log_{10}\left(U_\infty l/\nu\right)\right]^{2.58}}$$

which is accurate up to and beyond $U_\infty l/\nu = 10^9$ and the von Karman–Schoenherr formula

$$\frac{1}{\bar{C}_f^{1/2}} = 4.13 \log_{10}\left(\frac{\bar{C}_f U_\infty l}{\nu}\right) \tag{9-22a}$$

for the average friction coefficient. These two relations are based on a logarithmic velocity profile of the form of Eq. (9-18). The local friction coefficient has been correlated by Schultz-Grunow [27] as

$$C_f = \frac{0.37}{\left[\log_{10}\left(U_\infty x/\nu\right)\right]^{2.584}}$$

as discussed by Schlichting [5, p. 604], who can also be consulted for the effects of roughness. Generally speaking, surface roughness has no effect if the roughness elements are covered by the laminar sublayer. This requires that, for the wall to be hydraulically smooth,

$$\frac{v^* k}{\nu} < 5$$

where $k$ is the height of the roughness element.

The observation that turbulent drag on dolphins is lower than that expected for rigid surfaces has led to numerous investigations on the reduction of turbulent skin friction with compliant surfaces. Reductions of up to 50% are possible [33], which has importance for airborne and waterborne vehicles where turbulent boundary-layer drag is dominant.


## PROBLEMS

**9-1**  Consider Couette flow of two parallel walls moving with equal speeds in opposite directions. In such a case the shear stress is constant. Plot $\bar{u}/v^*$ against $y/h$ for both laminar and turbulent flow where $v^* = (\tau_w/\rho)^{1/2}$ and $h$ equals half the wall separation. Compare your results with the measurements of Reichardt reported by Schlichting [5, p. 557]. The methods used to achieve Eqs. (9-16)–(9-18) can be applied.

**9-2** Show that use of von Karman's similarity hypothesis for turbulent flow with constant shear stress gives the velocity profile from

$$\tau = \rho K_3^2 \frac{(d\bar{u}/dy)^4}{(d^2\bar{u}/dy^2)^2}$$

as

$$u^+ = \frac{1}{K_3} \ln(K_3 y + C_1) + C_2$$

Show that if the constant of integration $C_1$ equals zero, the result is identical to Eq. (9-18) from Prandtl's mixing-length formula.

**9-3** Consider turbulent Poiseuille flow between two parallel walls with a constant axial pressure gradient so that the shear stress varies as $\tau = \tau_w y/h$, where $y$ is the distance from the center line and $h$ is half the wall separation.

(a) Use von Karman's similarity hypothesis to obtain the velocity profile from

$$\tau = \tau_w \frac{y}{h} = \rho K_3^2 \frac{(d\bar{u}/dy)^4}{(d^2\bar{u}/dy^2)^2}$$

with the condition that $\bar{u}(y = 0) = \bar{u}_{\max}$. The result is

$$\frac{\bar{u}_{\max} - \bar{u}}{v^*} = -\frac{1}{K_3}\left\{\ln\left[1 - \left(\frac{y}{h}\right)^{1/2}\right] + \left(\frac{y}{h}\right)^{1/2}\right\}$$

where $v^* = (\tau_w/\rho)^{1/2}$. Note that $\bar{u}(y = h) \to \infty$ since molecular effects were neglected. This result is termed the *velocity-defect law*.

(b) Repeat part a with the use of Prandtl's mixing-length formula, but assume shear stress to be constant. Evaluate constants of integration with the condition that $\bar{u}(y = h) = \bar{u}_{\max}$ ($y$ is now the distance from a wall) and show that the velocity distribution is predicted to be

$$\frac{\bar{u}_{\max} - \bar{u}}{v^*} = \frac{1}{K_1} \ln\left(\frac{h}{y}\right)$$

which is of the same form as for part a.

(c) Plot the results of parts a and b and comment on the relative unimportance of the variation in shear stress, neglected in part b.

**9-4** Show that if Deissler's relation of Eq. (9-14) is modified into $\varepsilon_m = n^2 \bar{u} y$, Eq. (9-15) can be approximately solved for $y^+$. Then

$$\frac{dy^+}{du^+} - n^2 u^+ y^+ = 1$$

An integrating factor is $e^{-n^2 u^{+2}/2}$ so that the solution is

$$e^{-n^2 u^{+2}/2} y^+ = \int_0^{u^+} e^{-n^2 x^2/2} \, dx$$

$$= \frac{2^{1/2}}{n} \frac{\pi^{1/2}}{2} \left[ \frac{2}{\pi^{1/2}} \int_0^{(n^2 u^{+2}/2)^{1/2}} e^{-z^2} \, dz \right]$$

$$y^+ = n^{-1} \left( \frac{\pi}{2} \right)^{1/2} e^{n^2 u^{+2}/2} \operatorname{erf} \left( \frac{nu^+}{2^{1/2}} \right)$$

Plot this result on the curve in Fig. 9-12 and show that it fits data well for $y^+ \lesssim 26$.

**9-5** Show that if the turbulent velocity profile in a round duct of radius $R$ is given by $u/U_{CL} = (y/R)^{1/n}$, the average velocity is related to the center-line velocity $U_{CL}$ by

$$\frac{U_{av}}{U_{CL}} = 2n^2(n+1)^{-1}(2n+1)^{-1}$$

**9-6** Consider a long straight pipe through which water flows. The pipe diameter is 1 ft, and the average velocity down the pipe is 1 ft/sec. The pertinent water properties are $\rho = 62.4 \, \text{lb}_m/\text{ft}^3$ and $\nu = 10^{-5} \, \text{ft}^2/\text{sec}$. On the basis of this information:

(a) Calculate the friction factor for pressure drop $f$ from Eq. (9-21a).

(b) Calculate the shear stress at the wall $\tau_w$ from the relation that $\tau_w = D(P_1 - P_2)/4L$.

(c) Calculate the friction velocity $v^* = (\tau_w/\rho)^{1/2}$.

(d) Calculate the thicknesses of the laminar sublayer, the buffer zone, and the fully turbulent core. (*Answers:* $\delta_1 = 10^{-3}$ ft; $\delta_b = 4 \times 10^{-3}$ ft.)

(e) Determine whether the pipe can be considered to be smooth if a typical wall roughness element height $k$ is 0.045 cm.

**9-7** Estimate the extent of the transition region between laminar and turbulent flow for parallel flow of air over a flat plate with a Reynolds number based on plate total length of $10^6$.

**9-8** The similarity hypothesis of von Karman for mixing-length in turbulent flow can be obtained by considering the Taylor series for the velocity $u$ in a parallel flow in the perpendicular direction $y$ according to Cebeci and Smith [31, p. 106]. This Taylor series is

$$u(y) = y(y_0) + (y - y_0) \frac{\partial u(y_0)}{\partial y} + (y - y_0)^2 \frac{\partial^2 u(y_0)/\partial y^2}{2!} + \cdots$$

(a) Arguing that all such flows ought to be similar in shape so that scaling constants $l$ and $u_0$ must exist that make the parallel velocity independent of velocity or size of the flow field, show that the Taylor series must be expressible as

$$\frac{u}{u_0} = 1 + \left(\frac{y}{l} - \frac{y_0}{l}\right) \frac{\partial(u/u_0)}{\partial(y/l)}$$

$$+ \left(\frac{y}{l} - \frac{y_0}{l}\right)^2 \frac{\partial^2(u/u_0)/\partial^2(u/l)}{2!} + \cdots$$

(b) Considering only the first three terms of the Taylor series, and arguing that to have $u/u_0$ similar for all $y$, $\partial(u/u_0)/\partial(y/l)$ must be proportional to $\partial^2(u/u_0)/\partial^2(y/l)$, show that

$$l = \left| \frac{\partial u/\partial y}{\partial^2 u/\partial y^2} \right|$$

**9-9** The basis for the exponential form appearing in Eq. (9-14a) due to Van Driest is to be ascertained. For this purpose, consider a flat plate in a stagnant fluid with the plate oscillating in its own plane. For laminar conditions, the adjacent fluid motion is described by

$$\frac{\partial u}{\partial t} = \nu \frac{\partial^2 u}{\partial y^2}$$

with $u(0, t) = u_0 \cos(\omega t)$ and $u(y \to \infty, t) \to 0$, where $y$ is the perpendicular distance from the plate.

(a) Show that the fluid velocity varies as

$$u = u_0 \exp\left(\frac{-y}{y_s\sqrt{2}}\right) \cos\left(\omega t - \frac{yu_s}{\nu\sqrt{2}}\right)$$

with $y_s = (\nu/\omega)^{1/2}$ and $u_s = (\omega\nu)^{1/2}$.

(b) From the result of part a, show that if the fluid oscillates parallel to a stationary plate, the fluid velocity varies as

$$u' = u_0'[1 - \exp(-ny)]$$

where $n = 1/y_x\sqrt{2}$ and $u_0'$ is the fluid velocity fluctuation far from the plate.

(c) From the result of part b, show that the Reynolds stress can be plausibly expressed as

$$\bar{\tau}_{yx}^{\text{turb}} = -\rho\,\overline{v'u'} = -\rho\,\overline{v_0'u_0'}\,[1 - \exp(-ny)]^2$$

if $v_0'$ and $u_0'$ are presumed to be similarly damped by the presence of the wall. Then, since Prandtl's mixing-length idea has

$$-\overline{v_0'u_0'} = l^2\left(\frac{\partial u}{\partial y}\right)^2$$

show that it follows that

$$\varepsilon_m = l^2[1 - \exp(-ny)]^2\left|\frac{\partial \bar{u}}{\partial y}\right|$$

Van Driest took Prandtl's relation of $l = Ky$ with $K = 0.4$, and assumed $n = (\tau_\omega/\rho)^{1/2}/\nu A^+$ with $A^+ = 26$. Deissler used slightly different assumptions to achieve Eq. (9-14), but the basis for the exponential form is seen to be the damping effect of a wall on turbulent velocity fluctuations.

# REFERENCES

1  J. O. Hinze, *Turbulence*, McGraw-Hill, New York, 1975.
2  G. B. Schubauer and P. S. Klebanoff, Contributions on the mechanics of boundary layer transition, NACA TN 3489, 1955; NACA Report 1289, 1956.
3  (a) G. B. Schubauer and H. K. Skramstad, Laminar Boundary Layer Oscillations and Stability of Laminar Flow, National Bureau of Standards Research Paper 1772, reprint of confidential NACA Report, April 1943 (also published as NACA War-time Report W-8); *JAS* **14**, 69–78 (1947); also NACA Report 909; (b) A. A. Hall and G. S. Hislop, Experiments on the transition of the laminar boundary layer on a flat plate, Aeronautical Research Committee Report Memorandum 1843, 1938.
4  O. Reynolds, On the experimental investigation of the circumstances which determine whether the motion of water shall be direct or sinuous, and the law of resistance in parallel channels, *Phil. Transact. Roy. Soc.* **174**, 935–982, (1883).
5  H. Schlichting, *Boundary-Layer Theory*, McGraw-Hill, New York, 1968.

6  A. H. P. Skelland, *Nonnewtonian Flow And Heat Transfer*, Wiley, New York, 1967, pp. 74, 172.

7  J. Pretsch, Die Stabilität einer ebenen Laminarströmung bei Druckgefälle and Druckansteig, *Jb. dt. Luftfahrtforschung* **I**, 58–75, (1941).

8  W. M. Kays, *Convective Heat and Mass Transfer*, McGraw-Hill, New York, 1966, pp. 93–97.

9  P. M. Moretti and W. M. Kays, Heat transfer to a turbulent boundary layer with varying free-stream velocity and varying surface temperature—an experimental study, *Internatl. J. Heat Mass Transf.* **8**, 1187–1202 (1965).

10  F. A. Schraub, Ph.D. dissertation, Mechanical Engineering Department, Stanford University, Stanford, CA, 1965.

11  E. Reshotko, Recent developments in boundary-layer transition research, *AIAA J.* **13**, 261–265, (1975).

12  G. G. Mateer, A comparison of boundary-layer transition data from temperature-sensitive paint and thermocouple techniques, *AIAA J.* **8**, 2299–2300 (1970).

13  D. C. Reda and R. A. Leverance, Boundary-layer transition experiments on pre-ablated graphite nosetips in a hyperballistics range, *AIAA J.* **15**, 305–306, (1977).

14  R. L. Wright and E. V. Zoby, Comparison of thermal techniques for determining boundary-layer transition in flight, *AIAA J.* **15**, 1543–1544, (1977).

15  J. Boussinesq, Théorie de l'écoulement toubillant, *Mém. prés. par div. savant á l'acad. sci. Paris* **23**, 46 (1877).

16  L. Prandtl, Über die ausgebildete Turbulenz, *Zeitschrift für angewandte Mathematik und Mechanic* **5**, 136–139 (1925).

17  L. Prandtl, Bemerkungen zur Theorie der freien Turbulenz, *Zeitschrift für angewandte Mathematik und Mechanic* **22**, 241–243 (1942).

18  H. Reichardt, *Gesetzmässigkeiten der freien Trubulenz*, VDI-Forschungsheft 414, 1st ed., Berlin, 1942 (2nd ed., Berlin, 1951).

19  Th. von Karman, Mechanische Ähnlichkeit und Turbulenz, *Nach. Ges. Wiss. Göttingen, Math. Phys. Klasse* 58 (1930); *Proc. 3rd Internatl. Congr. Appl. Mech., Stockholm*, Part I, 85 (1930); NACA TM 611, 1931.

20  R. G. Deissler, Analysis of turbulent heat transfer, mass transfer, and friction in smooth tubes at high Prandtl and Schmidt numbers, NACA Report 1210, Washington, DC, 1955.

21  H. W. Liepmann, The rise and fall of ideas in turbulence, *Am. Sci.* **67**, 221–228 (1979).

22  J. Nikuradse, Gesetzmässigkeit der turbulenten Strömung in glatten Rohren, *Forsch. Arb. Ing.-Wes.*, No. 356, 1932.

23  R. C. Martinelli, Heat transfer to molten metals, *Transact. ASME* **69**, 947–959 (1947).

24  H. Reichardt, Vollstandige Darstellung der turbulenten Geschwindigkeitsverteilung in glatten Leitungen, *Zeitschrift für angewandte Mathematik und Mechanic* **31**, 208–219 (1951).

25  E. R. Van Driest, On turbulent flow near a wall, *J. Aerospace Sci.* **23**, 1007–1011 (1956).

26  D. B. Spalding, *Conference on International Developments In Heat Transfer*, ASME, Boulder, CO, Part II, 1961, pp. 439–446.

27  F. Schultz-Grunow, Neues Widerstandsgesetz für glatten Platten, *Luftfahrtforschung* **17**, 239 (1940); NASA TM 986, 1941.

28  J. T. Davies, Local eddy diffusivities related to "bursts" of fluid near solid walls, *Chem. Eng. Sci.* **30**, 996–997 (1975).

29  E. R. Corino and R. S. Brodkey, A visual investigation of the wall region in turbulent flow, *J. Fluid Mech.* **37**, 1–30 (1969).

30  K. K. Chen and N. A. Thyson, Extension of Emmons' spot theory to flows on blunt bodies, *AIAA J.* **9**, 821–825 (1971).

**31**  T. Cebeci and A. M. O. Smith, *Analysis Of Turbulent Boundary Layers*, Academic Press, New York, 1974.

**32**  R. B. Bird, W. E. Stewart, and E. N. Lightfoot, *Transport Phenomena*, Wiley, New York, 1960, p. 161.

**33**  L. M. Weinstein and M. C. Fischer, Experimental verification of turbulent skin friction reduction with compliant walls, *AIAA J.* **13**, 956–958 (1975).

**34**  M. R. Malik and S. A. Orszag, Comparison of methods for prediction of transition by stability analysis, *AIAA J.* **18**, 1485–1489 (1980).

**35**  G. D. Catalano, R. E. Walterick, and H. E. Wright, Improved measurement of turbulent intensities by use of photon correlation, *AIAA J.* **19**, 403–405 (1981).

**36**  D. M. McEligot, C. W. Coon, and H. C. Perkins, Relaminarization in tubes, *Internatl. J. Heat Mass Transf.* **13**, 431–433 (1970).

**37**  A. R. Wazzan, C. Gazley, and A. M. O. Smith, $H - R_x$ method for predicting transition, *AIAA J.* **19**, 810–812 (1981).

**38**  D. R. Hofstadter, Metamagical themas; *Scientific American* **245**, 22–43 (1981).

# 10

## TURBULENT BOUNDARY LAYERS

Most boundary layers become turbulent if their development is unimpeded. Although it is normal for the turbulent boundary layer to be preceded by a laminar one, such is not required. For example, it is possible to place a roughness element, called a *boundary-layer trip*, at the leading edge of a plate that triggers a turbulent boundary layer at the leading edge of the plate. This point is worth bearing in mind because many calculations take into account a partial coverage of the surface by a laminar boundary layer. It is possible for transition from turbulent to laminar flow to also occur under some conditions.

The $\frac{1}{7}$th power-law velocity profiles that will be frequently used do not always represent the most accurate possible description. Pipe flow results show that it is more accurate to speak of $1/n$th power profiles, where $6 \leqslant n \leqslant 10$, depending on the Reynolds number. At the same time, the $\frac{1}{7}$th power-law velocity distribution is often a quite satisfactory one, as Fig. 10-1 shows for the measurements by O'Donnell [1] on a flat plate at zero incidence.

The applicability of a logarithmic velocity distribution near the surface is shown by the measurements due to Lobb et al. [2] in Fig. 10-2 to be satisfactory, although inexact, even for compressible flow.

### 10.1  VELOCITY DISTRIBUTION IN PARALLEL FLOW

Consider parallel flow of a constant-property fluid over a flat plate. The turbulent boundary layer forming on the plate as illustrated in Fig. 10-3 is to be described as to thickness and velocity distribution. The integral form of the boundary-layer equations can be used, provided shear stress includes a turbu-

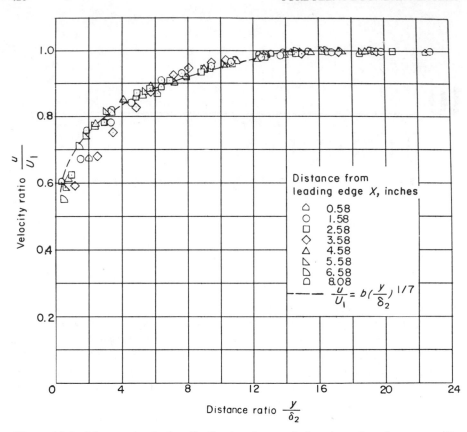

**Figure 10-1** Measured velocity distribution in a turbulent boundary layer on a flat plate at zero incidence with a supersonic free stream. (From R. M. O'Donnell, NACA TN 3122, 1954 [1].)

lent contribution. Thus

$$\frac{d(U^2\delta_2)}{dx} + \delta_1 U \frac{du}{dx} = \frac{\tau_w + \rho U v_w}{\rho} \tag{8-13}$$

where $\delta_1 = \int_0^\delta (1 - u/U)\, dy$ is the displacement thickness, $\delta_2 = \int_0^\delta (u/U)(1 - u/U)\, dy$ is the momentum thickness, and $\delta$ is the boundary-layer thickness.

The $\frac{1}{7}$th power law velocity distribution of Eq. (9-19) is used in the integral form of the momentum equation [Eq. (8-13)]. Beginning with

$$u^+ = 8.74(y^+)^{1/7} \tag{9-19}$$

from pipe flow data and using Eq. (9-21) with the free-stream velocity and

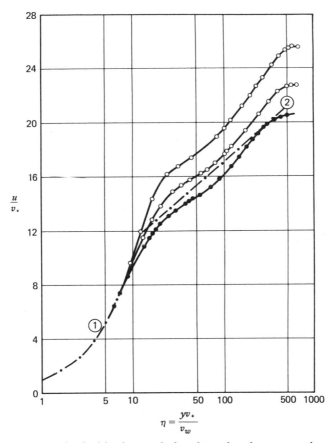

$$\eta = \frac{yv_*}{v_w}$$

**Figure 10-2** Measured velocities in a turbulent boundary layer on a channel wall with a supersonic free stream and heat transfer. Curve 1 represents the theoretical law for incompressible laminar sublayers, $u/v_* = \eta$. Curve 2 represents the theoretical law for incompressible turbulent boundary layers $u/v_* = 5.5 + 2.5 \ln(\eta)$. (From R. K. Lobb, E. M. Winkler, and J. Persh, NAVORD Report 3880, 1955 [2].)

|  | $M_\infty$ | $(T_{aw} - T_w)/T_{aw}$ | $R_2 \times 10^{-4}$ |
|---|---|---|---|
| ● | 5.75 | 0.108 | 1.16 |
| ◑ | 5.79 | 0.238 | 1.24 |
| ○ | 5.82 | 0.379 | 1.14 |

boundary-layer thickness substituted for pipe center-line velocity and radius, one obtains

$$\frac{u}{U} = \left(\frac{y}{\delta}\right)^{1/7} \tag{10-1}$$

This has a similarity form. With the velocity profile of Eq. (10-1), the

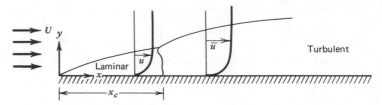

**Figure 10-3**  Transition from laminar to turbulent flow in a boundary layer on a flat plate.

displacement thickness is found to be $\delta_1/\delta = \frac{1}{8}$ (compared to $\delta_1/\delta \approx \frac{3}{8}$ for laminar flow), and the momentum thickness is found to be $\delta_2/\delta = \frac{7}{72} \approx \frac{1}{10}$ (compared to $\delta_2/\delta \approx \frac{39}{280} \sim \frac{1}{7}$ for laminar flow).

Introduction of these results into the integral form of the $x$-motion equation [Eq. (8-13)] then gives

$$\frac{d\left[U^2 7\delta/72\right]}{dx} = 0.0228\left(\frac{U\delta}{\nu}\right)^{-1/4} U^2 \tag{10-2}$$

Here the free-stream velocity is constant, and Eq. (9-22), derived from pipe flow data, has been used to relate $\tau_w$ to $U$ and $\delta$. On integration, Eq. (10-2) gives

$$\delta^{5/4} = 0.2931\left(\frac{U}{\nu}\right)^{-1/4} x + c \tag{10-3}$$

The constant of integration in Eq. (10-3) must be evaluated from knowledge of the extent of the laminar boundary layer covering the fore part of the plate. A simple answer, first essayed by Ludwig Prandtl, is obtained by saying that the turbulent boundary layer after the transition region acts as if it had begun at the leading edge with $\delta_{\text{turb}}(x = 0) = 0$. Then $c = 0$ and

$$\delta = 0.376\left(\frac{\nu}{U}\right)^{1/5} x^{4/5}$$

or

$$\frac{\delta}{x} = 0.376/\text{Re}_x^{1/5}, \qquad 5 \times 10^5 \sim \text{Re}_{x_c} \leqslant \text{Re}_x < 10^7 \tag{10-4}$$

Note that Eq. (10-4) predicts the turbulent boundary layer to vary as $x^{4/5}$, whereas a laminar boundary layer varies as $x^{1/2}$. Evaluation of the constant of integration of Eq. (10-3) is more firmly based on physical principles by noting from Eq. (8-13) that the momentum thickness must vary continuously (the subject of Problem 10-3), even though Eq. (10-4) is of quite adequate accuracy.

The turbulent wall shear stress is calculable from Eq. (9-22) now that $\delta$ is known. These two relations used together give

$$\frac{C_f}{2} = \frac{\tau_w}{\rho U_\infty^2} = 0.0228\left(\frac{U_\infty \delta}{\nu}\right)^{-1/4} = 0.0228\left[\frac{U_\infty}{\nu}0.376\left(\frac{\nu}{U_\infty}\right)^{1/5}x^{4/5}\right]^{-1/4}$$

$$\frac{C_f}{2} = \frac{\tau_w}{\rho U_\infty^2} = \frac{0.0296}{Re_x^{1/5}}, \qquad 5 \times 10^5 \sim Re_{x_c} \leqslant Re_x < 10^7 \qquad (10\text{-}5)$$

Bear in mind that this is a local turbulent friction coefficient and is valid only for $x \geqslant x_c$. The total drag on a plate must include the contribution from the laminar boundary layer, if one exists—it might not if the boundary layer is tripped at the leading edge. The total drag is the sum of that due to the laminar boundary layer on the forepart of the plate and that due to the turbulent boundary layer on the aft part. Thus

$$\text{Drag} = \int_0^{x_c}\left(\frac{C_f}{2}\right)_{\text{lam}}\rho U^2 \, dx + \int_{x_c}^x \left(\frac{C_f}{2}\right)_{\text{turb}}\rho U^2 \, dx$$

Use of Eq. (7-17) in the first integral and Eq. (10-5) in the second integral gives

$$\frac{\bar{C}_f}{2}x\rho U^2 = \int_0^{x_c}\left(0.33206\,Re_x^{-1/2}\right)\rho U^2 \, dx + \int_{x_c}^x \left(0.0296\,Re_x^{-1/5}\right)\rho U^2 \, dx$$

or

$$\frac{\bar{C}_f}{2} = \frac{0.037}{Re_x^{1/5}} - \frac{\left(-0.66412\,Re_{x_c}^{1/2} + 0.037\,Re_{x_c}^{4/5}\right)}{Re_x}$$

for $5 \times 10^5 \sim Re_{x_c} \leqslant Re_x < 10^7$. If the critical Reynolds number is taken to be the typical value, the average friction coefficient is

$$\frac{\bar{C}_f}{2} = \frac{0.037}{Re_x^{1/5}} - \frac{850}{Re_x} \qquad (10\text{-}6)$$

for $5 \times 10^5 \sim Re_{x_c} \leqslant Re_x < 10^7$.

Of interest is the thickness of the laminar sublayer $\delta_l$ whose outer edge is at

$$y^+ = \frac{v^*\delta_l}{\nu} = 5$$

which is rearranged into

$$\frac{(\tau_w/\rho)^{1/2}\delta}{\nu}\frac{\delta_l}{\delta} = 5$$

Use of Eq. (10-4) for $\delta$ and Eq. (10-5) for $\tau_w$ in the preceding relation results in

$$\frac{\delta_l}{\delta} = \frac{77.5}{\text{Re}_x^{7/10}} \tag{10-7}$$

Similarly, the outer edge of the buffer zone occurs at $y^+ = 26$. The buffer zone thickness $\delta_b$ then occupies the thickness

$$\Delta y^+ = 26 - 5 = 21$$

which is rearranged into

$$\frac{v^* \delta_b}{v} = 21$$

Proceeding as for $\delta_l$, one then finds that

$$\frac{\delta_b}{\delta} = \frac{326}{\text{Re}_x^{7/10}} \tag{10-8}$$

Equations (10-7) and (10-8) show that $\delta_l$ and $\delta_b$ increase as $x^{1/10}$, as they are nearly constant, whereas $\delta$ increases as $x^{4/5}$. Thus the turbulent core occupies increasing portions of the boundary layer at successive downstream locations.

Results for the turbulent boundary layer in the presence of a pressure gradient are attainable by the methods discussed by Schlichting [3, pp. 626–655]. In brief, the integral form of the $x$-motion boundary-layer equation is used [in a rearranged form from Eq. (8-13)] as

$$\frac{d\delta_2}{dx} + (H_{12} + 2)\frac{\delta_2}{U}\frac{dU}{dx} = \frac{\tau_w}{\rho U^2}$$

where $H_{12} = \delta_1/\delta_2$. The wall shear stress is related to $U$ and $\delta_2$ by

$$\frac{\tau_w}{\rho U^2} = \frac{\alpha}{(U\delta_2/v)^{1/n}}$$

Taking the flat-plate value with a $\frac{1}{7}$th power-law velocity profile for $H_{12} = \delta_1/\delta_2 = \frac{9}{7} \approx 1.3$ together with the preceding relation for $\tau_w$ in the integral equation, one obtains, after integration

$$\delta_2\left(\frac{U\delta_2}{v}\right)^{1/n} = U^{-b}\left(C_1 + a\int_{x=x_c}^{x} U^b \, dx\right) \tag{10-8a}$$

Where $a = \alpha(n + 1)/n$, $b = [(n + 1)(H_{12} + 2) - 1]/n$, and $C_1$ is a constant of integration to be determined from the laminar boundary layer at $x = x_c$. If it is presumed that no laminar boundary layer exists, then $C_1 = 0$ since one of $U$ or $\delta_2$ will be zero at $x = 0$. Since Eq. (10-5) can be expressed as

$$\frac{C_f}{2} = \frac{0.0128}{(U\delta_2/\nu)^{1/4}}$$

as was first done by Prandtl, it is reasonable to use $n = 4$ and $\alpha = 0.0128$; then $a = 0.016$ and $b = 4$. When the boundary layer does not separate from the surface [separation occurs when $\Gamma = (\delta_2/U)(dU/dx)(U\delta_2/\nu)^{1/4} \approx -0.06$, with $\Gamma > 0$ corresponding to accelerated and $\Gamma < 0$ corresponding to decelerated flow, respectively], knowledge of $\delta_2$ essentially completes the calculation. Local wall shear stress is available by back substitution, and drag can be obtained by its integration. Such calculations are important to nozzle design and to air foils, for example. A body of revolution is the subject of Problem 10-11.

## 10.2 TEMPERATURE DISTRIBUTION IN PARALLEL FLOW

From the velocity distribution previously determined, a heat-transfer coefficient can be determined. There are some assumptions necessary, of course. For example, it is reasonable to assume that thermal energy and momentum posses the same mechanisms of turbulent transport. This point deserves brief amplification.

The kinetic theory of gases treatment of heat, mass, and momentum transport in Chapter 3 provided similar expressions for thermal conductivity, mass diffusivity, and viscosity. A rigid-sphere model of the molecules leads to $Pr = \nu/\alpha = 1$ and $Sc = \nu/D_{12} = 1$. But the various corrections that are necessary to account for elasticity of the assumed spherical molecules, repellent forces, and persistence of velocities after collision for energy differ from those for mass or for momentum.

In turbulent motions, where large clumps of molecules are moved about together, the kinetic energy associated with the motion of these clumps of fluid contributes little to the internal energy of the fluid. Hence it is logical to assume that neither heat nor mass are affected by interaction between these clumps in turbulent motion other than by molecular-level exchanges during a time of contact between clumps. If these molecular level effects are neglected, the rate of turbulent transport of heat and mass must be equal. Although the rate of turbulent transport of momentum might be somewhat different, it would be expected to be close according to the indications of kinetic theory. Since both kinetic theory and turbulence presume random motion of the involved particles, there is a physical basis for the preceding expectations.

Measurements of the turbulent Prandtl number $Pr_t = \varepsilon_m/\alpha_t$ are surveyed by Schlichting [3] and Hinze [4]. Despite the difficulties of obtaining accurate measurements, it appears that near a solid surface $Pr_t \approx 1$, near the center of a

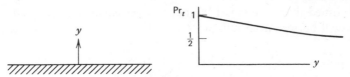

**Figure 10-4**  Variation of turbulent Prandtl number with distance from a solid surface.

duct $Pr_t \approx 0.7$, and in a free stream $Pr_t \approx \frac{1}{2}$. The $Pr_t = 1$ value corresponds to equality of mixing lengths for heat and momentum, whereas the $Pr_t = \frac{1}{2}$ value corresponds to equality of mixing lengths for heat and vorticity [4]. Efforts to develop a functional relationship between $Pr_t$ and distance from the wall have been made by Hughmark [6] and Cebeci [7]. The sketch in Fig. 10-4 illustrates the general observed behavior. In liquid metals $Pr_t \sim 1.2$ for fully developed tube flow, however [7, 22].

There are several plausible procedures that could be followed. However, since $Pr_t$ does not vary greatly, one is justified in assuming $Pr_t$ to be constant. Furthermore, when interested primarily in behavior "near" a wall, one can claim $Pr_t = 1$ without losing appreciable accuracy. Jischa and Rieke [28] have shown theoretically that $Pr_t = A + B(Pr + 1)/Pr$ with experiment giving $A = 0.825$ and $B = 0.0309$, a result fitting data for air quite well and for liquid metals as well as any other expression. Callaghan and Mason [30] measured $Pr_t \approx 1$ for reacting systems in chemical equilibrium in experiments on nitrogen dioxide gas (corrosive and poisonous but that undergoes the reaction $N_2O_4 \rightleftharpoons 2NO_2$ at 30°F).

Attention is now directed to the case where fluid flows primarily parallel to a solid wall. This corresponds to a well-developed turbulent boundary layer or to flow in a pipe well beyond the entrance region. In this way, a one-dimensional problem is posed. Figure 10-5 depicts the physical situation. The heat flux in the $y$ direction is given by

$$\bar{q} = -k\frac{\partial\bar{T}}{\partial y} - k_t\frac{\partial\bar{T}}{\partial y} = -\rho C_p(\alpha + \alpha_t)\frac{d\bar{T}}{dy}$$

For small values of $y$, it is reasonable to take $q$ as unchanged from its value at

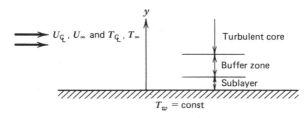

**Figure 10-5**  Division of a turbulent boundary layer into a laminar sublayer at the wall, an intermediate buffer zone, and an outer fully turbulent core.

the wall $q_w$. Together with some rearrangement, this puts the preceding equation into the form

$$-\frac{q_w}{\rho C_p} = \left(\frac{\nu}{\text{Pr}} + \frac{\varepsilon_m}{\text{Pr}_t}\right)\frac{d\bar{T}}{dy}$$

Defining dimensionless quantities as

$$v^* = \left(\frac{\tau_w}{\rho}\right)^{1/2}, \qquad T^+ = (\bar{T} - T_w)\left(\frac{\rho C_p}{q_w}v^*\right), \qquad y^+ = \frac{yv^*}{\nu}, \qquad u^+ = \frac{\bar{u}}{v^*}$$

one finally has

$$-1 = \left(\frac{1}{\text{Pr}} + \frac{\varepsilon_m/\nu}{\text{Pr}_t}\right)\frac{dT^+}{dy^+} \tag{10-9}$$

The velocity profile in Section 10.1 should give $\varepsilon_m/\nu$ so that Eq. (10-9) can be solved to relate the total imposed temperature difference to the distance required to accomplish it. The integration is divided into three portions: sublayer, $y^+ \leqslant 5$; buffer zone, $5 \leqslant y \lesssim 30$; and turbulent core, $y \gtrsim 30$. In other words,

$$T_\infty - T_w = \Delta(T - T_w)_{\text{sublayer}} + \Delta(T - T_w)_{\text{buffer zone}} + \Delta(T - T_w)_{\text{turbulent core}}$$

In dimensionless terms, this can be written as

$$\frac{T_\infty - T_w}{q_w/\rho C_p v^*} = \Delta T^+_{\text{sublayer}} + \Delta T^+_{\text{buffer zone}} + \Delta T^+_{\text{turbulent core}}$$

The region close to the wall has $\varepsilon_m/\nu$ given by Deissler's relation. As is recalled, the laminar sublayer has molecular effects predominating over turbulence so that $\varepsilon_m/\nu$ can be neglected. Thus from Eq. (10-9)

$$\frac{\partial T^+}{\partial y^+} = -\text{Pr}, \qquad y^+ < 5$$

$$\Delta T^+_{\text{sublayer}} = T^+(y^+ = 5) - T^+(y^+ = 0) = -5\,\text{Pr} \tag{10-10}$$

In the buffer zone, molecular effects and turbulence must both be considered. the form of Deissler's relation is $\varepsilon_m/\nu = n^2 u^+ y^+ [1 - \exp(-n^2 u^+ y^+)]$. Admittedly, we have already found $u^+$ as a function of $y^+$ so that Eq. (10-9) could be integrated numerically with this form of $\varepsilon_m/\nu$. However, a compact analytical result is sought. An approximation, leading to this desired direction, is made as follows. In the buffer zone

$$\tau = \rho\nu\left(1 + \frac{\varepsilon_m}{\nu}\right)\frac{d\bar{u}}{dy}$$

Assumption that $\tau \approx \tau_w$ and rearrangement then gives the dimensionless form

$$1 = \left(1 + \frac{\varepsilon_m}{\nu}\right)\frac{du^+}{dy^+} \tag{10-11}$$

The velocity distribution in the buffer zone, obtained from Deissler's relation, is approximated with reasonable accuracy by

$$u^+ \approx 5 + 5\ln\frac{y^+}{5}$$

from which it is found that

$$\frac{du^+}{dy^+} \approx \frac{5}{y^+}$$

Insertion of this information into Eq. (10-11) gives

$$\frac{\varepsilon_m}{\nu} = \frac{y^+}{5} - 1$$

Equation (10-9) then yields

$$\Delta T^+_{\text{buffer}} = -\int_5^{30} \frac{dy^+}{1/\mathrm{Pr} + (1/\mathrm{Pr}_t)(y^+/5 - 1)} = -5\,\mathrm{Pr}_t\ln\left(5\frac{\mathrm{Pr}}{\mathrm{Pr}_t} + 1\right)$$

$$\tag{10-12}$$

if $\mathrm{Pr}_t$ is constant.

In the fully turbulent core, molecular effects can be disregarded without appreciable error. Equation (10-9) then becomes

$$-1 = \frac{\varepsilon_m/\nu}{\mathrm{Pr}_t}\frac{dT^+}{dy^+}$$

which can be rewritten as

$$-1 = \frac{\varepsilon_m/\nu}{\mathrm{Pr}_t}\frac{du^+}{dy^+}\frac{dT^+}{du^+}$$

Also, the previous relation that

$$\tau_w \approx \tau = \rho\varepsilon_m\frac{d\bar{u}}{dy}$$

or

$$1 = \frac{\varepsilon_m}{\nu}\frac{du^+}{dy^+}$$

is introduced to yield

$$-\Pr_t = \frac{dT^+}{du^+}$$

From this it is found that

$$\Delta T^+_{\text{turb}} = -\int_{y^+ \approx 30}^{\text{free stream}} \Pr_t \, du^+$$

$$= -\Pr_t \left[ U^+_\infty - 5(1 + \ln 6) \right] \tag{10-13}$$

if $\Pr_t$ is constant.

Combination of Eqs. (10-10), (10-12), and (10-13) then gives the total temperature difference between the wall and the free stream as

$$(T_\infty - T_w)\frac{\rho C_p v^*}{q_w} = -5\left[ \Pr - \Pr_t + \Pr_t \ln\left( \frac{5\Pr/\Pr_t + 1}{6} \right) \right] - \Pr_t U^+_\infty$$

$$\tag{10-14}$$

A heat-transfer coefficient can now be determined from Eq. (10-14) since it is also realized that

$$h(T_w - T_\infty) = q_w$$

Rearrangement of this relationship gives

$$\left( \frac{T_\infty - T_w}{q_w} \right) \rho C_p v^* = \frac{-1}{h} \rho C_p v^*$$

$$= \frac{-k}{hL} \frac{C_p \mu}{k} \frac{\rho U_\infty L}{\mu} \frac{v^*}{U_\infty}$$

$$= -\frac{\Pr \operatorname{Re}_L}{\operatorname{Nu}_L} \frac{1}{U^+_\infty}$$

Equation (10-14), in conjunction with this result, gives

$$\frac{\operatorname{Nu}_L}{\operatorname{Re}_L \Pr^{1/3}} = \frac{1}{U^{+2}_\infty} \frac{\Pr^{2/3}}{\Pr_t + 5/U^+_\infty \left\{ \Pr - \Pr_t + \Pr_t \ln\left[ (5\Pr/\Pr_t + 1)/6 \right] \right\}}$$

By definition

$$U^{+2}_\infty = \frac{U^2_\infty}{v^{*2}} = \frac{\rho U^2_\infty}{\tau_w} = \frac{2}{C_f}$$

So

$$\frac{\mathrm{Nu}_L}{\mathrm{Re}_L \mathrm{Pr}} = \frac{C_f}{2} \frac{1}{\mathrm{Pr}_t + 5(C_f/2)^{1/2}\{\mathrm{Pr} - \mathrm{Pr}_t + \mathrm{Pr}_t \ln\left[(5\,\mathrm{Pr}/\mathrm{Pr}_t + 1)/6\right]\}}$$

$$(10\text{-}15)$$

Remember that $\mathrm{Pr}_t$ was taken to be a constant whose value must now be specified. Probably attention should center near the wall where $\mathrm{Pr}_t \approx 1$. Then

$$\frac{\mathrm{Nu}_L}{\mathrm{Re}_L \mathrm{Pr}} = \frac{C_f}{2} \frac{1}{1 + 5(C_f/2)^{1/2}\{\mathrm{Pr} - 1 + \ln\left[(5\,\mathrm{Pr} + 1)/6\right]\}} \qquad (10\text{-}16)$$

This development is attributed to von Karman [8]. The case of variable $\mathrm{Pr}_t$ was explored by Reichardt [9], Van Driest [10], and Rotta [11], who found no important improvement to the preceding form. Equation (10-16) is the turbulent form of Reynolds analogy and is not as clearcut as the laminar form. Nevertheless, it is confirmed by experiment that to close approximation the turbulent form can be considered to be $\mathrm{Nu}_L/(\mathrm{Re}_L\,\mathrm{Pr}^{1/3}) = C_f/2$ if $\mathrm{Pr} \approx 1$, as it is for many gases. A refined integral analysis utilizing more accurate approximations for velocity and temperature profiles near the wall confirms the results presented here [29]. Although a surface renewal model suggests [32] that $\mathrm{Nu}_L/\mathrm{Re}_L\,\mathrm{Pr}^{1/2} = C_f/2$ always, measurements [33] in hypersonic flow support Reynolds analogy in the form $\mathrm{Nu}_L/\mathrm{Re}_L\,\mathrm{Pr} = 1.16C_f/2$. For supersonic airflow, the coefficient of $C_f/2$ seems to vary between 1 and 1.2 [34].

If mass transfer is considered, one merely substitutes Sc for Pr and Sh for Nu in Reynolds analogy. In any case, Reynolds analogy can still be counted on to obtain mass-transfer coefficients from heat-transfer coefficients.

The analogy between heat transfer and skin friction represented by Eq. (10-16) is not restricted to parallel flow over a flat plate. It can also be used to calculate heat transfer from surfaces immersed in external streams when pressure gradients are not excessively large, even in compressible flows. It can also be applied to internal flows in circular pipes after modifications to account for curvature. Equation (10-16) is not, however, applicable to low Prandtl number fluids such as liquid metals since then molecular-level thermal conductivity effects are appreciable in the turbulent core, contrary to the assumption incorporated in Eq. (10-13).

## 10.3   FLAT-PLATE HEAT-TRANSFER COEFFICIENT

For parallel flow of a $\mathrm{Pr} \approx 1$ fluid over a constant-temperature flat plate, the turbulent heat transfer coefficient can be found from Reynolds analogy as

$$\frac{\mathrm{Nu}}{\mathrm{Re}\,\mathrm{Pr}} = \frac{C_f}{2} \qquad (10\text{-}17)$$

with $C_f/2$ given by Eq. (10-5) as $C_f/2 = 0.0296/\text{Re}_x^{1/5}$. Putting this into Reynolds analogy of Eq. (10-17), one obtains

$$\frac{\text{Nu}_x}{\text{Re}_x \text{Pr}} = \frac{0.0296}{\text{Re}_x^{1/5}}$$

$$\text{Nu}_x = 0.0296\,\text{Re}_x^{0.8}\,\text{Pr}, \qquad 5 \times 10^5 \leqslant \text{Re}_x < 10^7 \qquad (10\text{-}17a)$$

This is a local coefficient, of course. Should there be a preceeding laminar boundary layer, its contribution must be taken into account to obtain an overall film coefficient. From Eq. (10-17) an average turbulent film coefficient is obtained as

$$\bar{h} = \frac{1}{x}\int_0^x h\,dx = \frac{A}{x}\int_0^x x^{-1/5}\,dx = \frac{A}{x}\frac{5}{4}x^{4/5}$$

$$\qquad - \tfrac{5}{4}Ax^{-1/5}$$

$$= \tfrac{5}{4}h$$

$$\overline{\text{Nu}}_x = 0.037\,\text{Re}_x^{0.8}\,\text{Pr}, \qquad 5 \times 10^5 \leqslant \text{Re}_x < 10^7 \qquad (10\text{-}18)$$

Measurements suggest that in the Prandtl number range characteristic of many gases, the Prandtl number exponent is best taken to be 0.6 rather than the 1.0 shown previously.

Procedures for the calculation of heat-transfer rates in turbulent flows over nonisothermal surfaces have been devised by Spalding [12] and Kestin and co-workers [13, 14]. Measurements for such conditions were performed by Reynolds et al. [15], which also verify the turbulent Reynolds analogy of Eq. (10-16). Measurements on turbulent boundary layers on a rough surface with blowing are available [24].

The effect of free-stream turbulence intensity is surveyed by Kestin [25]. Turbulent boundary-layer heat transfer is unaffected. Laminar boundary-layer heat transfer can be substantially increased if the pressure gradient is nonzero, nearly doubling at the stagnation point of a cylinder as turbulence intensity increases from 0% to 3%. Traci and Wilcox [31] agree that the mechanism of heat-transfer enhancement at stagnation points is stretching of vortex lines.

## 10.4   FLAT PLATE WITH UNHEATED STARTING LENGTH

The effect of an unheated starting length on the turbulent heat-transfer coefficient for constant free-stream velocity can be determined by making use of an integral method. The integral form of the turbulent boundary-layer equations can be obtained by integration in the vertical direction, but by integration from $y$ to $\delta$ rather than from $y = 0$ to $y = \delta$ as was done in Chapter 8 for laminar boundary layers.

The turbulent continuity equation of Eq. (9-11a) is treated first. Integration with respect to $y$ from $y$ to $y = \delta$ gives

$$-v_\delta + v_y = \int_y^\delta \frac{\partial u}{\partial x} dy$$

$$= \frac{\partial}{\partial x}\left[\int_y^\delta u\, dy\right] - U\frac{d\delta}{dx}$$

The turbulent $x$-motion equation of Eq. (9-11a) is next integrated, again from $y$ to $y = \delta$. This operation gives

$$\int_y^\delta \frac{\partial u^2}{\partial x} dy + u_\delta v_\delta - u_y v_y = \frac{\tau_\delta - \tau_y}{\rho}$$

Now $\tau_\delta = 0$ and $U_\delta = U$, where $U$ is the free-stream velocity. So, after employing Leibnitz' formula and this integrated form of the continuity equation, the integrated $x$-motion equation becomes

$$\frac{d}{dx}\left[\int_y^\delta\left(1 - \frac{u^2}{U^2}\right) dy\right] - \left(1 - \frac{u}{U}\right)\frac{v_\delta}{U} - \frac{u}{U}\frac{d}{dx}\left[\int_y^\delta\left(1 - \frac{u}{U}\right) dy\right] = \frac{\tau_y}{\rho U^2}$$

$$(10\text{-}19)$$

The $\frac{1}{7}$th power-law velocity distribution of Eq. (10-1) is assumed, and Eq. (8-11) is used to obtain $v_\delta/U = (1/8)\,d\delta/dx$. Introduction of these items of information into Eq. (10-19) gives

$$\frac{\tau_y}{\rho U^2} = \frac{7}{72}\left[1 - \left(\frac{y}{\delta}\right)^{9/7}\right]\frac{d\delta}{dx}$$

Since the shear stress at $y = 0$ is just the wall shear stress $\tau_w$, one has $\tau_w = \frac{7}{72}\,d\delta/dx$ and so

$$\frac{\tau_y}{\tau_w} = 1 - \left(\frac{y}{\delta}\right)^{9/7} \qquad\qquad (10\text{-}20)$$

which serves as an approximate, but reasonable, estimate of the vertical variation of turbulent shear. Insertion of the $\frac{1}{7}$th power-law velocity distribution of Eq. (10-1) and the shear stress variation of Eq. (10-20) into the relation

$$\tau = \rho(\nu + \varepsilon_m)\frac{\partial u}{\partial y}$$

gives

$$\nu + \varepsilon_m = 7\delta \frac{\tau_w}{\rho U}\left(\frac{y}{\delta}\right)^{6/7}\left[1 - \left(\frac{y}{\delta}\right)^{9/7}\right] \tag{10-21}$$

Attention is next directed to the temperature distribution and vertical heat flux. First, it is assumed that Pr and $Pr_t$ approximately equal unity so that the temperature profile is similar to the velocity profile, giving

$$\frac{T - T_w}{T_\infty - T_w} = \left(\frac{y}{\delta_T}\right)^{1/7} \tag{10-22}$$

where $\delta_T$ is the thermal boundary-layer thickness. Then, with the assumption that $\nu + \varepsilon_m = \alpha + \alpha_t$ since $Pr \approx 1 \approx Pr_t$, Eqs. (10-21) and (10-22) are substituted into the relation

$$q_y = -\rho C_p(\alpha + \alpha_t)\frac{\partial T}{\partial y}$$

to obtain

$$\frac{q_y}{\rho C_p(T_w - T_\infty)} = \frac{\tau_w}{\rho U}\left(\frac{\delta}{\delta_T}\right)^{1/7}\left[1 - \left(\frac{y}{\delta}\right)^{9/7}\right]$$

Recognizing that $\tau_w/\rho U^2 = C_f/2$ and setting $y = 0$ in the preceding expression, one obtains

$$\frac{q_w}{\rho C_p U(T_w - T_\infty)} = \left(\frac{\delta}{\delta_T}\right)^{1/7}\frac{C_f}{2} \tag{10-23}$$

Next the integral form of the energy equation [Eq. (8-18)],

$$\frac{\partial}{\partial x}\left[\int_0^{\delta_T} u(T - T_\infty)\,dy\right] = \frac{q_w}{\rho C_p}$$

is utilized to solve for $\delta/\delta_T$. If Eq. (10-23) is used for $q_w$ and Eq. (10-5) for $C_f/2$, the integral form of the energy equation is

$$\frac{d}{dx}\left(\frac{7}{72}\frac{\delta_T^{8/7}}{\delta^{1/7}}\right) = \left(\frac{\delta}{\delta_T}\right)^{1/7}\frac{0.0296}{(Ux/\nu)^{1/5}}$$

It has been assumed that $\delta_T \leqslant \delta$, as is appropriate for the case of a flat plate with an unheated starting length of length $x_0$. With $\xi = \delta_T/\delta$, the above equation becomes

$$\xi^{1/7}\frac{d}{dx}(\delta\xi^{8/7}) = \frac{72}{7}\frac{0.0296}{(Ux/\nu)^{1/5}}$$

Introduction of Eq. (10-4) to relate $\delta$ to $x^{4/5}$ then allows the preceding ordinary differential equation to be recast as

$$\frac{d\xi^{9/7}}{dx} + \frac{9}{10}x^{-1}\xi^{9/7} = \frac{0.0296(9/8)(72/7)}{0.376}x^{-1}$$

from which it is found that

$$\xi^{9/7} = \frac{5}{4}\frac{(0.0296)}{0.376}\frac{72}{7} + Cx^{-9/10}$$

The constant of integration is evaluated from the condition that $\xi = \delta_T/\delta = 0$ at $x = 0$ to give

$$\xi = \left[\frac{90}{7}\left(\frac{0.0296}{0.376}\right)\right]^{7/9}\left[1 - \left(\frac{x_0}{x}\right)^{9/10}\right]^{7/9}$$

This information substituted back into Eq. (10-23) finally gives the local heat-transfer coefficient on a flat plate with an unheated starting length as

$$\frac{q_w}{\rho C_p U(T_w - T_\infty)} = \frac{\mathrm{Nu}}{\mathrm{Re\,Pr}} = \left[1 - \left(\frac{x_0}{x}\right)^{9/10}\right]^{-1/9}\frac{C_f}{2}$$

This functional form was mentioned in the discussion following Eq. (8-23). To approximately account for Pr differing from unity, the final result is taken to be

$$\frac{\mathrm{Nu}}{\mathrm{Re\,Pr}^{0.6}} = \left[1 - \left(\frac{x_0}{x}\right)^{9/10}\right]^{-1/9}\frac{C_f}{2}, \qquad 5\times10^5 \leqslant \mathrm{Re} \leqslant 10^7$$

$$= 0.0296\,\mathrm{Re}^{-1/5}\left[1 - \left(\frac{x_0}{x}\right)^{9/10}\right]^{-1/9} \tag{10-24}$$

Measurements [15] confirm Eq. (10-24). A sometimes more convenient approximation to Eq. (10-24) is

$$\frac{\mathrm{Nu}}{\mathrm{Re\,Pr}^{0.6}} = 0.0296\,\mathrm{Re}^{-1/5}\left(1 - \frac{x_0}{x}\right)^{0.12} \tag{10-25}$$

The local heat flux from a plate (to a constant velocity and temperature free stream) whose surface temperature varies can be found from these results, using the relation of Eq. (8-23) developed during a treatment of a laminar boundary-layer problem. This is the subject of Problem 10-9. The inverse problem of determining the surface temperature distribution that corresponds to a specified heat flux is the subject of Problem 10-10.

The turbulent velocity boundary layer on an axisymmetric body of revolution is treated in Problem 10-11. With knowledge of the local turbulent velocity boundary-layer thickness, Reynolds analogy in the form of Eq. (10-15) or (10-18) can then be used to obtain the local heat-transfer coefficient. Somewhat more refined procedures for determining the local heat-transfer coefficient have been proposed by Spalding [17] and by Ambrok [18], whose method is the subject of Problem 10-19. In justification of the use of Reynolds analogy, it assumes a "law of the wall" that is only weakly dependent on shear stress variation along the wall and also assumes that the ratio of shear stress to heat flux is constant in the turbulent part of the boundary layer. The first assumption is fairly accurate as is the second assumption since, for usual Prandtl values, the principal thermal resistance is in the laminar and buffer zones. Numerical methods and measurements [23] show that, in comparison to a flat surface, heat flux is greater on a concave surface and is less on a convex surface.

## 10.5   RECOVERY FACTOR FOR TURBULENT FLOW

The recovery factor $r$ for turbulent boundary-layer flow is defined, as it was for laminar flow, as

$$r = \frac{T_{aw} - T_{\infty}}{U^2/2C_p}$$

where $T_{aw}$ is the temperature an adiabatic wall would achieve. The results of measurements show that as the boundary layer changes from its laminar state near the leading edge, for which $r = \mathrm{Pr}^{1/2}$, $r$ increases to a peak ($\sim 0.89$ for air) in the laminar–turbulent transition region and then approaches a constant value predictable by

$$r = \mathrm{Pr}^{1/3} \tag{10-26}$$

Figure 10-6 and 10-7 illustrate typical experimental results. Additional support is provided by Seban and Doughty [26].

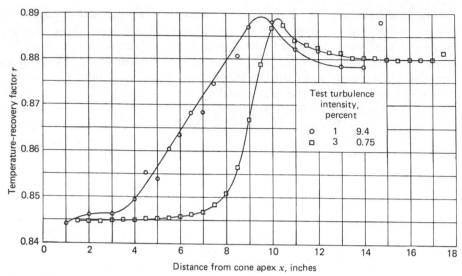

**Figure 10-6** Measured recovery factor $r$ on a cone in supersonic flow for determination of the point of laminar-to-turbulent transition. In laminar flow $r = \mathrm{Pr}^{1/2} = 0.846$; the steep slope in the recovery factor versus distance from vertex plot indicates the location of the transition region. Curve 1 is at low turbulence intensity and curve 2, at high turbulence intensity. [By permission from J. C. Evvard, M. Tucker, and W. C. Burgess, *J. Aeronaut. Sci.* **21**, 731–738 (1954) [20].]

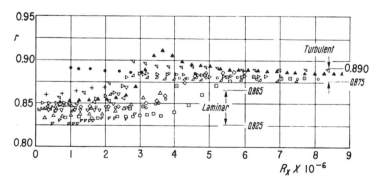

**Figure 10-7** Measured recovery factors $r$ on cones versus Reynolds number $\mathrm{Re}_x$ for Mach numbers ranging from 1.2 to 6.0. (From L. M. Mack, Jet Propulsion Laboratory Report 20–80, 1954 [21].)

The recovery factor for Pr differing appreciably from unity can be predicted by [32]

$$
r = \begin{cases}
\dfrac{4}{\pi}\left(\dfrac{\mathrm{Pr}}{2-\mathrm{Pr}}\right)^{1/2}\arctan\left(\dfrac{2-\mathrm{Pr}}{\mathrm{Pr}}\right)^{1/2}, & \mathrm{Pr}<2 \\[2em]
\dfrac{4}{\pi}, & \mathrm{Pr}=2 \\[2em]
\dfrac{2}{\pi}\left(\dfrac{\mathrm{Pr}}{2-\mathrm{Pr}}\right)^{1/2}\ln\left[\dfrac{1+(1-2/\mathrm{Pr})^{1/2}}{\left[1-(1-2/\mathrm{Pr})^{1/2}\right]}\right], & \mathrm{Pr}>2
\end{cases}
$$

## 10.6   FINITE-DIFFERENCE SOLUTIONS

As was remarked in Section 7.9 for laminar boundary layers, the conditions imposed on a problem are often so complicated that needed details of velocity and temperature distributions can be obtained only by numerical methods. For turbulent boundary layers, accurate mathematical representation of turbulent transport of energy and momentum poses difficulties additional to those discussed in connection with laminar boundary layers. A survey book by Cebeci and Smith [35] can be consulted for greater detail and depth than is presented here.

One of the difficulties that must be addressed is that the free outer boundary associated with external boundary layers makes their velocity distribution far from the wall different from the internal boundary layers of pipe flows. According to experimental observations such as those reported by Klebanoff [36] and represented in Fig. 10-8, the flow becomes intermittently turbulent to an increasing degree as the free stream is approached. The

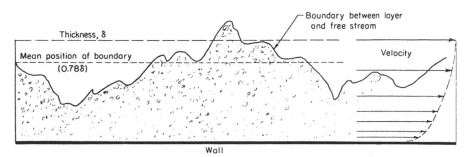

**Figure 10-8**   Sketch of a turbulent boundary layer, showing that the potential flow of the free stream can extend far into the boundary layer at times. (From P. S. Klebanoff, NACA TN 3178, 1954 [36].)

nonturbulent potential flow of the free stream penetrates the outer region of the boundary layer so that at an affected location the flow is turbulent for only a fraction $\gamma$ of the time of observation. The same phenomenon is observed for wake and jet flows that have a free boundary. Although this effect was apparently first noticed by Corrsin in 1943 with later investigation accomplished by Corrsin and Kistler [37], Klebanoff [36] gave the result useful for flat-plate boundary layers of

$$2\gamma = 1 - \text{erf}\left[5\left(\frac{y}{\delta} - 0.78\right)\right]$$

which is conveniently approximated by

$$\gamma \approx \frac{1}{1 + 5.5(y/\delta)^6} \tag{10-27}$$

with $\delta$ being the average boundary-layer thickness (where $u/U = 0.995$). Understandably, $\gamma$ is referred to as the *Klebanoff intermittency factor*. As shown in Fig. 10-8, an appreciable portion of the flow is turbulent above the average edge of the boundary layer to $y \approx 1.2\delta$, and an appreciable portion of the flow is nonturbulent as far below as $y \approx 0.4\delta$. Of course, $\gamma = 0$ entirely outside the turbulent flow and $\gamma = 1$ entirely inside.

The utility of the intermittency factor is that, in such flows as jets and boundary layers that have a free outer boundary with a nonturbulent environment, it separates turbulent transport from intermittency effects. For a turbulent boundary layer, measurements show that the eddy diffusivity is nearly constant over the outer part [36]. The effects of Mach number $0 \leqslant M \leqslant 5$ and compressibility are rendered slight in the outer part of a turbulent boundary layer [38] by the general correlation for eddy diffusivity

$$\varepsilon_{m,0} = k_2 U \delta^* \gamma \tag{10-28}$$

in which $k_2 = 0.0168$ [39], $\delta^* = \int_0^\delta (1 - u/U)\, dy$ is the displacement thickness, and $\gamma$ is given by Eq. (10-27).

Close to the solid surface the effect of intermittency is negligible. There Van Driest's modification of Prandtl's mixing-length relation described by Eq. (9-14a) can be used for this inner region to give

$$\varepsilon_{m,i} = K^2 y^2 \left[1 - \exp\left(\frac{-y}{A}\right)\right]^2 \left|\frac{\partial u}{\partial y}\right| \tag{10-29}$$

in which $K = 0.4$ and the generalized parameters [40, 44] are

$$A = \frac{26\nu}{v^*N}$$

$$N^2 = \frac{P^+\left[1 - \exp\left(11.8v_w^+\right)\right]}{v_w^+} + \exp\left(11.8v_w^+\right) \qquad (10\text{-}29a)$$

$$P^+ = \frac{\nu U}{v^{*3}}\frac{dU}{dx}$$

$$v_w^+ = \frac{v_w}{v^*}$$

where $v_w$ is the wall blowing velocity and $v^* = (\tau_w/\rho)^{1/2}$ is the friction velocity. When $v_w = 0$, $N^2 = 1 - 11.8P^+$. Equation (10-29) has been found to be accurate for compressible and incompressible turbulent flows with both mass and heat transfer (see Cebeci and colleagues [35, 44] for viscosity and density corrections not given here that improve accuracy). Further refinement is possible by multiplying Eq. (10-29) by a new intermittency factor that accounts for the intermittent turbulence downstream from a transition from laminar to turbulent flow [40, p. 246]; such a refinement avoids the need to consider transition to be a discontinuous switching from the laminar to the turbulent state.

Over the boundary-layer thickness, the eddy diffusivity is predicted by Eq. (10-29) in the inner region ($0 \leq y \leq y_c$) and by Eq. (10-28) in the outer region ($y > y_c$). The demarcation between the inner and outer regions is determined by taking $y_c$ to be the distance from the solid surface at which $\varepsilon_{m,0} = \varepsilon_{m,i}$ as illustrated in Fig. 10.9. An alternative procedure for predicting eddy diffusivity in the outer part of a turbulent boundary does not require use of the

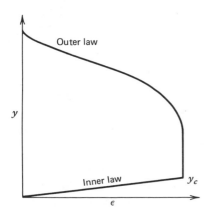

**Figure 10-9** Eddy viscosity distribution across a boundary layer in a two-region model.

intermittency factor. Instead, Prandtl's mixing length $l$ in

$$\varepsilon_m = l^2 \left| \frac{\partial u}{\partial y} \right| \qquad (9\text{-}12)$$

is used to take advantage of the fact that measurements [38] show $l/\delta$ to decrease only slowly there ($\delta$ is the boundary layer thickness at which $u/U = 0.995$) and to be unaffected by Mach number $0 \leqslant M \leqslant 5$ and compressibility. The recommended correlation [42] for the nearly constant mixing length in this outer region is

$$\frac{l_0}{\delta} = 0.089, \qquad 0 \leqslant \frac{y}{\delta} \leqslant 0.6 \qquad (10\text{-}30a)$$

A coefficient of 0.08 has also been successfully used [40]. A fit [42] to measurements [38] close to the wall is

$$\frac{l_i}{\delta} = 0.42 \left[ 1 - \exp \left( \frac{-y^+}{26} \right) \right] \frac{y}{\delta}, \qquad 0.1 \leqslant \frac{y}{\delta} \leqslant 0.6$$

$$-1.61489 \left( \frac{y}{\delta} - 0.1 \right)^2$$

$$+2.88355 \left( \frac{y}{\delta} - 0.1 \right)^3$$

$$-1.91556 \left( \frac{y}{\delta} - 0.1 \right)^4 \qquad (10\text{-}30b)$$

and

$$\frac{l_i}{\delta} = 0.42 \left[ 1 - \exp \left( \frac{-y^+}{26} \right) \right] \frac{y}{\delta}, \qquad \frac{y}{\delta} < 0.1 \qquad (10\text{-}30c)$$

Here $y^+ = y(\tau_1/\rho)^{1/2}/\nu_w$, $\tau_1 = \tau_w(1 + \nu_w y/\nu_w)$, and $\nu_w$ is the wall blowing velocity.

Regardless of whether Eqs. (10-28) and (10-29) or Eqs. (10-30)–(10-32) are used, the turbulent Prandtl number $\mathrm{Pr}_t$ is successfully assumed to be constant at 0.9 [43, 44], although descriptions of more complex variations have been essayed [35]. If $\mathrm{Pr}_t$ and $\varepsilon_m$ are known, the turbulent thermal conductivity can be obtained from

$$k^t = \frac{\rho C_p \varepsilon_m}{\mathrm{Pr}_t}$$

The case of wedge flow with constant properties is discussed to illustrate the essential features of one of the several solution procedures that have been

developed. Although velocities and enthalpies could be directly determined by
a finite difference procedure [41–43], it is computationally more efficient to
first subject the turbulent boundary-layer equations to a similaritylike coordinate transformation since the departure from similarity is seldom large. This
advantage is not as pronounced for turbulent boundary layers as it is for the
laminar case discussed in Section 7.10, however. Proceeding as for the laminar
case yields the transformed $x$-motion equation

$$\frac{\partial[(1 + \varepsilon_m^+)F'']}{\partial \eta} + P[1 - (F')^2] + \left(\frac{1+P}{2}\right)FF'' = x\left[F'\frac{\partial F'}{\partial x} - F''\frac{\partial F}{\partial x}\right]$$

$$F'(x, \eta = 0) = 0$$

$$F(x, \eta = 0) = -(\nu x U)^{-1/2}\int_0^x v_w\, dx$$

$$F'(x, \eta \to \infty) \to 1 \qquad (10\text{-}31a)$$

and energy equation

$$\frac{\partial[(1 + \varepsilon_m^+ \Pr/\Pr')\theta'/\Pr + (1 - 1/\Pr)(U^2/H_\infty)F'F'']}{\partial \eta}$$

$$= -\left(\frac{1+P}{2}\right)F\theta' + x\left[F\frac{\partial\theta}{\partial x} - \theta'\frac{\partial F}{\partial x}\right]$$

$$\theta(x, \eta = 0) = \frac{H_w - H_\infty}{H_\infty} \quad \text{or} \quad \theta'(x, \eta = 0) = -\frac{C_p q_w}{k H_\infty}$$

$$\theta(x, \eta \to \infty) \to 0 \qquad (10\text{-}31b)$$

Here quantities are as described in Section 7.10 for Eqs. (7-91a) and (7-95).
The remarks in Section 7.10 regarding finite-difference solutions also apply
here; consult Cebeci and colleagues [35, 40] for additional details of method,
such as the use of Richardson extrapolation (see also Keller and Cebeci [45]) to
improve accuracy and generalization to axisymmetric and variable property
cases. The dimensionless eddy diffusivity $\varepsilon_m^+ = \varepsilon_m/\nu$ for constant properties is
given by the dimensionless forms of Eqs. (10-28) and (10-29) as

$$\varepsilon_{m,0}^+ = k_2\gamma\left(\frac{Ux}{\nu}\right)^{1/2}\int_0^{\eta_{max}}(1 - F')\, d\eta, \qquad \eta > \eta_c$$

$$\varepsilon_{m,i}^+ = K^2\eta^2\left(\frac{Ux}{\nu}\right)^{1/2}\left\{1 - \exp\left[-\eta\left(\frac{Ux}{\nu}\right)^{1/4}(F'')^{1/2}\frac{N}{26}\right]\right\}^2 |F''|,$$

$$0 \leq \eta \leq \eta_c$$

where

$$\frac{1}{\gamma} = 1 + 5.5\left(\frac{\eta}{\eta_{\max}}\right)^6$$

$$P^+ = P\frac{(\nu/Ux)^{1/4}}{(F'')^{3/2}}$$

$$v_w^+ = (v_w/U)\frac{(Ux/\nu)^{1/4}}{(F'')^{1/2}}$$

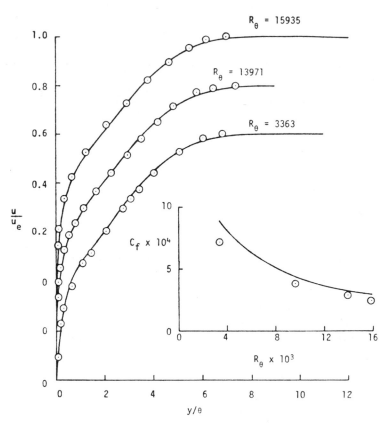

**Figure 10-10**  Comparison of calculated velocity profiles for the flow measured by Simpson et al. [46] where $v_w/u_e = 0.00784$. Symbols denote measurements and solid lines denote numerical solutions by Cebeci and Smith [35]. (By permission from T. Cebeci and P. Bradshaw, *Momentum Transfer In Boundary Layers*, Hemisphere Publishing, Washington, D.C., 1977 [40].)

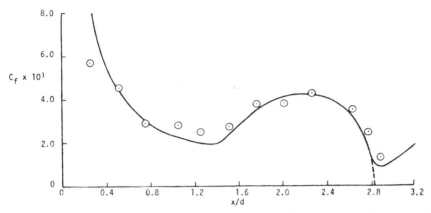

**Figure 10-11** Comparison of calculated laminar, transitional, and turbulent local skin friction coefficients $C_f$ with measurements for Schubauer's ellipse. Circles indicate measurements at $R_d = 1.18 \times 10^5$; solid and dashed lines denote numerical solutions (see Cebeci and Smith [35]) for experimental and extrapolated $u_e/u_\infty$, respectively. (By permission from T. Cebeci and P. Bradshaw, *Momentum Transfer In Boundary Layers*, Hemisphere Publishing, Washington, D.C., 1977 [40].)

where $\eta_{\max}$ is the $\eta$ value at which $u/U = 0.995$, $N$ is given in Eq. (10-29a), and $\eta_c$ is the $\eta$ value at which $\varepsilon_{m,0}^+ = \varepsilon_{m,i}^+$.

The accuracy of the finite-difference solution of the $x$-motion equation [Eq. (10-31a)], is often quite good. Figure 10-10 compares predicted [35] and measured [46] quantities for a turbulent boundary layer on a porous flat plate and shows good agreement. Good agreement for flow over an ellipse is also displayed in Fig. 10-11 for skin friction values measured [47] and predicted [35]. The minor axis of the ellipse was $d = 3.97$ in., and the laminar zone extended along the surface to $x/d = 1.25$; the transition zone occupied $1.25 \leqslant x/d \leqslant 2.27$, and separation was indicated at $x/d = 2.91$, which is in the turbulent boundary-layer region. The eddy diffusivity representations employed are really only fits to measurements for cases of local turbulent equilibrium. When departure from such equilibrium is marked, the predictions based on the equilibrium assumption lose accuracy, as Fig. 10-12 illustrates. In Fig. 10-12 measured [48] and predicted [35] velocity profiles are in fair agreement, but the skin friction coefficients are in poor agreement.

Heat-transfer coefficients predicted [39] by the numerical methods described here are often in good agreement with measurements [49] as Fig. 10-13 shows —$St = Nu/Re\,Pr$ (St represents Stanton number).

The methods discussed here (see Cebeci and Bradshaw [40] as well as Bradshaw, Cebeci, and Whitelaw [77] for implementing FORTRAN programs for both external and internal flows or Patankar and Spalding [50]) are not applicable to boundary layers that separate from the solid body. In such cases the boundary-layer assumptions often lose accuracy and the flow oscillates and

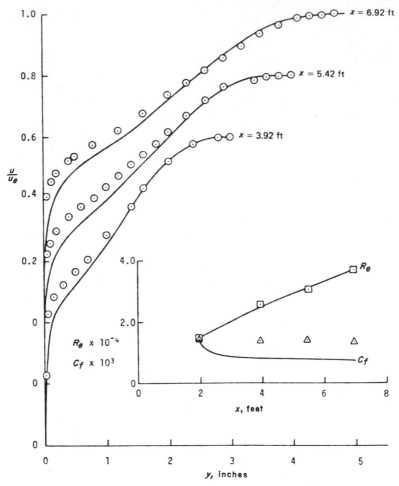

**Figure 10-12**  Comparison of calculated results for the equilibrium flow observed by Bradshaw and Ferriss [48]; symbols denote measurements and solid lines denote numerical solutions by Cebeci and Smith [35]. (By permission from T. Cebeci and P. Bradshaw, *Momentum Transfer In Boundary Layers*, Hemisphere Publishing, Washington, D.C., 1977 [40].)

recirculates. Numerical methods for separated flows are still in the developmental stage [51].

## 10.7   NEWER TURBULENCE MODELS

The Prandtl mixing-length model has been extensively and successfully applied to turbulent flows of the boundary-layer type (often referred to as *thin shear-layer flows*). The mixing-length hypothesis lacks universality, however,

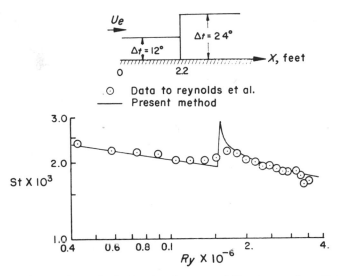

**Figure 10-13** Comparison of calculated and measured Stanton numbers St = Nu/RePr with measurements by Reynolds et al., [49]. (From T. Cebeci, A. M. O. Smith, and G. Mosinskis, *Transact. ASME, J. Heat Transf.* **92**, 133–143 (1970) [39].)

because of its assumption that turbulence is in local equilibrium, with turbulent energy locally produced and dissipated at the same rate. Hence the mixing-length model is unable to account for turbulence production at other points in the flow unless empirical changes are specifically introduced into the mixing-length distribution, an undertaking that is difficult in three-dimensional or circulating flows. As an example of the computational difficulty that can be encountered, the mixing-length hypothesis predicts [see Eq. (9-12)] that the eddy diffusivity is zero at the center of a flow channel where velocity gradients vanish, which is unfortunate when heat is transferred from one channel wall to another.

One-equation models represent an attempt to base eddy diffusivity on a quantity that is characteristic of turbulence rather than of local time-averaged velocity. The kinetic theory of gases discussed in Chapter 3 suggests that

$$\nu \sim \text{characteristic length} \times \text{characteristic velocity}$$

In the same spirit it has been proposed that eddy diffusivity would be proportional to a characteristic velocity that is $k^{1/2}$, where $k = (\overline{u'^2} + \overline{v'^2} + \overline{w'^2})/2$ is the time-averaged kinetic energy of turbulence. Thus

$$\nu^t = c'_\mu k^{1/2} L \tag{10-32}$$

which is known as the *Kolmogorov–Prandtl expression* because Kolmogorov [52] and Prandtl [53] proposed it independently at about the same time. Related suggestions have been made by others, such as Bradshaw et al. [54]. In one-equation models the characteristic length $L$ is usually determined from empirical relations and is similar to a mixing length. This similarity extends to a lack of universality that renders application to flows more general than shear-layer flows either unsuccessful or very expensive. The transport equation for the kinetic energy of turbulence $k$ needed for Eq. (10-32) is determined from the Navier–Stokes equation. The procedure followed is to multiply the unsteady equation of motion for each coordinate by its corresponding unsteady velocity and then time averaging and summing the three equations. From this is subtracted the result obtained by multiplying the time-averaged equation of motion for each coordinate by its corresponding time-averaged velocity and summing. The result is

$$\underset{\text{I}}{\frac{Dk}{Dt}} = \underset{\text{II}}{\frac{\partial\left[\nu\,\partial k/\partial x_i - \overline{u_i'(p'/\rho + k')}\right]}{\partial x_i}} + \underset{\text{III}}{\overline{(-u_i'u_j')}\,\frac{\partial u_j}{\partial x_i}} - \underset{\text{IV}}{\left[\nu\,\overline{\left(\frac{\partial u_j}{\partial x_i}\right)^2}\right]}$$

$$(10\text{-}33)$$

with primes denoting instantaneous values, overbars denoting time averaging (plain quantities are also time averaged), and employment of the Einstein convention of repeated indices (see Hinze [55] for details). For a boundary-layer flow, for example,

$$\frac{\partial k}{\partial t} + u\frac{\partial k}{\partial x} + v\frac{\partial k}{\partial y} = \frac{\partial\left[\nu\,\partial k/\partial y - \overline{v'(p'/\rho + k')}\right]}{\partial y} - \overline{(u'v')}\,\frac{\partial u}{\partial y}$$

$$- \nu\overline{\left[\left(\frac{\partial u'}{\partial x}\right)^2 + \left(\frac{\partial u'}{\partial y}\right)^2 + \left(\frac{\partial v'}{\partial x}\right)^2 + \left(\frac{\partial v'}{\partial y}\right)^2\right]}$$

Equation (10-33) can be interpreted as stating that the change (I) in $k$ due to either unsteady events or to convective transport by the mean motion equals the diffusion (II) of $k$ plus the production (III) of $k$ by turbulent stresses acting on the mean motion minus the dissipation (IV) caused by turbulent stresses acting on the turbulent motion. Detailed study [55] reveals that the dissipation interpretation of term IV is strictly correct only for homogeneous turbulence, a situation often closely approached. Now, turbulent diffusion of $k$ is presumed to be similar in mathematical form to molecular diffusion so that

$$-\overline{u'\left(\frac{p'}{\rho} + k'\right)} = \frac{\nu^t}{\sigma_k}\frac{\partial k}{\partial y}$$

In the production term it is reasonable to have

$$-\overline{(u'v')} = \nu^t\frac{\partial u}{\partial y}$$

The dissipation term is referred to by the symbol $\varepsilon$ and is given in terms of local turbulence quantities as

$$\varepsilon = \nu \sum_{i,j} \overline{\left(\frac{\partial u_j}{\partial x_i}\right)^2} = \frac{C_D k^{3/2}}{L} \tag{10-34}$$

which is dimensionally consistent, recalling that $k = \overline{(u'^2 + v'^2 + w'^2)}/2$ and that $L$ is a characteristic length. Here the dimensionless constants are commonly taken to be $\sigma_k \approx 1.0$ and $C_D \approx 0.08$. The transport equation for $k$ is then, for highly turbulent flow,

$$\frac{Dk}{Dt} = \frac{\partial\left[(\nu'/\sigma_k)\,\partial k/\partial y\right]}{\partial y} + \nu'\left(\frac{\partial u}{\partial y}\right)^2 - \frac{C_D k^{3/2}}{L} \tag{10-35}$$

which is of the same form as the boundary-layer equations of motion. Equations (10-32) and (10-35) can be solved simultaneously with the boundary-layer equations of motion. Although one-equation models such as this do account for the nonequilibrium character of turbulence through use of a transport equation for $k$, the level of universality sought is not achieved since the characteristic length $L$ is not adjusted to nonequilibrium turbulence except by empirical relations. Hence one-equation models are only marginally superior to a Prandtl mixing-length model.

Two-equation models of turbulence basically determine both the turbulent kinetic energy $k$ and the characteristic length $L$ from transport equations. The ideas underlying one-equation models are simply extended. Fortunately, the increased computational complexity is compensated by a dramatic increase in universality. The best known and most extensively tested two-equation model is the so-called $k$–$\varepsilon$ model, although others, such as the Wilcox–Rubesin model [56], have been developed. For general flows, the $k$–$\varepsilon$ model has the eddy diffusivity given by

$$\nu^t = \frac{C_\mu k^2}{\varepsilon} \tag{10-36a}$$

in which $L$ has been eliminated by combining Eqs. (10-32) and (10-34) and the turbulence stresses related to time-averaged velocities by

$$-\overline{(u_i' u_j')} = \nu^t\left(\frac{\partial u_i}{\partial x_j} + \frac{\partial u_j}{\partial x_i}\right) - \frac{2}{3} k \delta_{ij} \tag{10-36b}$$

in which $\delta_{ij}$ is the Kronecker delta ($\delta_{ij} = 0$ for $i \neq j$ and $\delta_{ij} = 1$ for $i = j$) and the transport equation for turbulent kinetic energy given by

$$\frac{Dk}{Dt} = \frac{\partial\left[(\nu'/\sigma_k)\,\partial k/\partial x_i\right]}{\partial x_i} + k\left(\frac{P}{k} - \frac{\varepsilon}{k}\right) \tag{10-36c}$$

in which $P = - \overline{(u_i'u_j')} \, \partial u_i / \partial x_j$ is the production term and the transport equation for dissipation $\varepsilon$ is given by

$$\frac{D\varepsilon}{Dt} = \frac{\partial \left[ (v'/\sigma_\varepsilon) \, \partial \varepsilon / \partial x_i \right]}{\partial x_i} + \varepsilon \left( \frac{C_1 P}{k} - \frac{C_2 \varepsilon}{k} \right) \qquad (10\text{-}36\text{d})$$

The turbulent Prandtl and Schmidt numbers are usually assumed to equal 0.9. Equation (10-36d) for $\varepsilon$ is most properly viewed as being empirically of the same form as Eq. (10-36c) for $k$; although an exact transport equation for $\varepsilon$ can be obtained by manipulating the Navier–Stokes equations, drastic assumptions are necessary to achieve a tractable form. The $k$–$\varepsilon$ model contains five empirical constants that are determined by comparing model predictions with measurements. For boundary-layer flows, common values are $C_\mu = 0.09$, $\sigma_k = 1.0$, $\sigma_\varepsilon = 1.3$, $C_1 = 1.5$, and $C_2 = 2.0$ [57]; other flows may require slightly different constants (see Rodi [58] for standard values and discussion including the effects of buoyancy). Equations (10-36c) and (10-36d) cannot be integrated up to the wall as they stand since they represent the highly turbulent case. An extended form developed by Jones and Launder [57] incorporates molecular effects so that integration to the wall can be accomplished in a continuous manner. The boundary conditions imposed on their extended form of Eqs. (10-36c) and (10-36d) are

$$k(y = 0) = 0 = \varepsilon(y = 0)$$

and

$$u_\infty \frac{dk_\infty}{dk} = -\varepsilon_\infty, \qquad u_\infty \frac{d\varepsilon_\infty}{dx} = -C_2 \exp\left( \frac{-2.5}{1 + k_\infty^2 / v\varepsilon_\infty} \right) \frac{\varepsilon_\infty^2}{k_\infty}$$

with the conditions far from the wall being just the limiting forms of Eqs. (10-36c) and (10-36d). A recommended lucid exposition of the development of turbulence models is that by Launder and Spalding [59]. Numerous examples of successful application of the $k$–$\varepsilon$ model [60–63] can be consulted for details, particularly with regard to far wakes and axisymmetric jets for which non-standard values of the constants must be used. A $k$–$\varepsilon$ model has been successfully used to calculate heat transfer from a rotating disk [76], and a modified form has been applied to turbulent wall jets [73]. Numerical solutions of two-equation models often require step sizes in the flow direction to initially be of the order of $10^{-2}$–0.5 the boundary-layer thickness; an innovation devised by Wilcox [64] allows the step size in the downstream direction to slightly exceed the boundary-layer thickness. His algorithm enables the computational efficiency of mixing-length models to be approached by a two-equation model while their accuracy and generality are surpassed. Flows with recirculation can be treated by the $k$–$\varepsilon$ model by use of otherwise standard numerical schemes [65–67].

Turbulent stress–flux equation models represent the next level of turbulence model development, qualifying for the term of second-order models of which a variety have been put forth [68–71]. The $k$–$\varepsilon$ model is based on local isotropy of turbulence in which $\overline{(u'_i u'_j)}$ is assumed to be same as $\overline{(u'_i \phi')}$, with $\phi$ as a transported scalar quantity, and this base is sometimes inaccurate. Further, it is not always the case that a single characteristic velocity $k^{1/2}$ characterizes the local turbulence. Buoyant and rotational effects also may not be accurately handled by the $k$–$\varepsilon$ model. Thus although the $k$–$\varepsilon$ model is markedly superior to a mixing-length model, it is not able to predict all turbulent flows without adjusting the constants involved. The second-order models, although more universally applicable, are substantially more complex and should not be used unless accuracy demands cannot otherwise be met. Examples of possible such cases are a symmetric channel flow with one wall rougher than the other, developed flow in a square duct, and curved mixing layers.

Algebraic stress–flux models [72–75] have been developed by simplification of differential transport equations such as those for the $k$–$\varepsilon$ model in such a manner as to reduce them to algebraic equations while retaining their basic features. Often the convective and diffusion terms are the ones simplified since they make the transport equations differential in nature. Algebraic stress–flux models are basically eddy viscosity formulas and are suitable whenever turbulent transport of turbulence quantities is not of great importance (these effects are only crudely modeled) as is in fact the case in most engineering situations. However, they are promising because they are capable of incorporating accurate representation of buoyancy, rotation, streamline curvature, nonisotropy, and wall-damping effects and so are nearly as accurate and universally applicable as second-order models while being markedly more efficient computationally.

## PROBLEMS

**10-1** Show that for the velocity profile on a flat plate $u/U = (y/\delta)^{1/n}$, the displacement thickness $\delta_1 = \int_0^\delta (1 - u/U)\, dy$, and the momentum thickness $\delta_2 = \int_0^\delta (u/U)/(1 - u/U)\, dy$ are related to the boundary-layer thickness by

$$\frac{\delta_1}{\delta} = \frac{1}{1+n} \quad \text{and} \quad \frac{\delta_2}{\delta} = n(1+n)^{-1}(2+n)^{-2}$$

**10-2** Show that, if Eq. (9-22) for turbulent shear stress on a flat plate is accurate only for $\mathrm{Re}_\delta \approx 10^5$, the range of validity of Eq. (10-4) is $5 \times 10^5 \simeq \mathrm{Re}_{x_c} < \mathrm{Re}_x < 10^7$.

**10-3** (a) Show that the integral form of the $x$-motion equation for constant free-stream velocity requires that the momentum thickness vary continuously, even though shear stress may vary discontinuously, as

$$\delta_2 = \int_0^x \frac{\tau_w}{\rho U^2} \, dx$$

**(b)** Match the laminar and turbulent momentum thickness at the transition point on a flat plate to obtain

$$\delta_{2\,\text{lam}} = \delta_{2\,\text{turb}}$$

and evaluate the constant of integration of Eq. (10-3) as

$$C = \frac{0.2931 x_c^{5/4}}{\text{Re}_{x_c}^{1/4}} \left( -1 + \frac{36.4}{\text{Re}_{x_c}^{3/8}} \right)$$

**(c)** Introduce the result for $C$ from part b into Eq. (10-13) to obtain

$$\frac{\delta}{x} = \frac{0.376}{\text{Re}_x^{1/5}} \left[ 1 - \frac{x_c}{x} \left( -1 + \frac{36.4}{\text{Re}_{x_c}^{3/8}} \right) \right]^{4/5}$$

**(d)** Show from the result of part c that at typical conditions

$$\frac{\delta}{x} = \frac{0.376}{\text{Re}_x^{1/5}} \left( 1 - \frac{0.732 x_c}{x} \right)^{4/5}$$

for $5 \times 10^5 \sim \text{Re}_{x_c} \leqslant \text{Re}_x < 10^7$. Compare this result with that of Eq. (10-3). Comment on the difficulty of matching laminar and turbulent momentum thicknesses at the transition point when the turbulent wall shear relation presumes "fully developed" turbulent flow.

**10-4** Verify that laminar-to-turbulent transition on a flat plate in parallel flow results in an increased boundary-layer thickness at the transition point.

**10-5** Compare the laminar to the turbulent wall shear stress at the transition point for parallel flow over a flat plate.

**10-6** Compare the thickness of the laminar sublayer and the buffer zone to the turbulent boundary-layer thickness for parallel flow over a flat plate at the transition point and at a position thrice as far removed from the leading edge.

**10-7** Show that Eq. (10-5) can be expressed in terms of the momentum thickness $\delta_2$ as

$$\frac{C_f}{2} = \frac{0.0128}{(U\delta_2/\nu)^{1/4}}$$

**10-8** For a turbulent boundary layer of a fluid for which $Pr \neq 1$, generalize Eq. (10-17) through use of the full turbulent Reynolds analogy expressed by Eq. (10-16). From this expression for the local heat-transfer coefficient, obtain the average coefficient, the generalization of Eq. (10-8).

**10-9** Consider a turbulent boundary layer of a $Pr \approx 1$ fluid on a flat plate whose surface temperature varies as

$$T_w - T_\infty = A + \sum_{n=1}^{N} B_n x^n \qquad n = 1, 2, \ldots, N$$

with constant free-stream velocity and temperature.

**(a)** Show that the response to a step change in plate temperature given by Eq. (10-14) together with the integral relationship given by Eq. (8-23) gives the local heat flux as

$$q_{w_x} = \left(0.0296k \frac{Pr^{0.4} Re_x^{0.8}}{x}\right)\left(A + \frac{10}{9} \sum_{n=1}^{N} n\beta_n B_n x^n\right)$$

where $\beta_n = \Gamma(8/9)\Gamma(10n/9)/\Gamma(8/9 + 10n/9)$. Here $\Gamma$ denotes the gamma function.

**(b)** A plate surface temperature varies linearly with distance from the leading edge (assume the boundary layer to be tripped at the leading edge so that there is no laminar boundary layer of consequence) with the temperature variation over the plate length being 10% of its leading edge excess over the free-stream temperature. Evaluate the accuracy of calculating the total heat loss based on the average temperature difference.

**10-10** Consider a turbulent boundary layer of a $Pr \approx 1$ fluid on a flat plate whose local heat flux is specified. The free-stream velocity and temperature are constant.

**(a)** Use the results of Problem 8-9, developed for application to laminar boundary layers, to show the plate surface temperature variation to be

$$T_w(x) - T_\infty = \frac{3.323}{k} Pr^{-0.6} Re_x^{-0.8}$$

$$\times \int_0^x \left[1 - \left(\frac{x^*}{x}\right)^{9/10}\right]^{-8/9} q_w(x^*)\, dx^*$$

(b) Show that the local heat-transfer coefficient for the case of $q_w =$ const, using the result of part a, is

$$\frac{\text{Nu}}{\text{Re}_x \, \text{Pr}^{0.6}} = \frac{0.0307}{\text{Re}_x^{1/5}}$$

and give the temperature variation that accompanies it.

(c) Show that the heat-transfer coefficient of part b for a constant plate heat flux is 4% higher than is the case for a constant plate temperature whereas for the equivalent situation with a laminar boundary layer, it is 34% higher. Check this seeming lesser sensitivity of a turbulent flow to surface temperature variation by comparing heat-transfer coefficient behavior for laminar and turbulent flow in ducts.

10-11  The turbulent boundary layer on a body of revolution for which the free-stream velocity also varies can be ascertained by application of the integral form of the x-momentum equation given in Problem 8-12 as discussed by Schlichting [3, p. 649] and first done by Millikan [16].

(a) Show that the integral form of the x-motion equation

$$\frac{d\delta_2}{dx} + \delta_2 \left[ \frac{1}{R} \frac{dR}{dx} + \left( 2 + \frac{\delta_1}{\delta_2} \right) \frac{1}{U} \frac{dU}{dx} \right] = \frac{\tau_w}{\rho U^2}$$

can be recast as, with $Z = U\delta_2/\nu$,

$$Z^{1/n} \frac{dZ}{dx} + Z^{(n+1)/n} \left[ \frac{1}{R} \frac{dR}{dx} + (1 + H_{12}) \frac{1}{U} \frac{dU}{dx} \right] = \frac{U\alpha}{\nu}$$

if $H_{12} = \text{const} \ (\approx \frac{9}{7})$ and $\tau_w/\rho U^2 = \alpha/(U\delta_2/\nu)^{1/n}$ as was the case in the development of Eq. (10-8a).

(b) Show that the result of part a can be integrated to give

$$\left( \frac{RU^{2+H}\delta_2}{\nu} \right)^{(n+1)/n} = C + \frac{\alpha}{\nu} \frac{n+1}{n}$$

$$\times \int_{x_{\text{transition}}}^{x} R^{(n+1)/n} U^{2+H+(1+H)/n} \, dx$$

(c) Specialize the result of part b for the flat-plate values of the parameters—note that $C$ can be evaluated from the laminar boundary layer at $x = x_{\text{transition}}$ and equals 0 if no laminar boundary layer exists since then one of $R$, $U$, or $\delta_2$ is zero at $x = 0$.

10-12  Compare the thermal resistance across the laminar sublayer, the buffer zone, and the turbulent core for a turbulent boundary layer on a flat plate.

**10-13** Air at a temperature of 20°F and 1 atm pressure flows along a flat surface (an idealized airfoil) at a constant velocity of 150 ft/sec. For the first 2 ft the surface is heated at a constant rate per unit of surface area, thereafter the surface is adiabatic. If the total length of the plate is 6 ft, what must the heat flux on the heated section be so that the surface temperature at the trailing edge is not below 32°F? Plot the surface temperature along the entire plate. Discuss the significance of this problem with respect to wing deicing. (A tabulation of incomplete beta functions is necessary for this problem.)

**10-14** Work the previous problem but divide the heater section into two 1-ft strips, with one at the leading edge and the other 3 ft from the leading edge. What heat flux is required such that the plate surface is nowhere less than 32°F?

**10-15** An aircraft oil cooler is to be constructed by using the skin of the wing of the cooling surface. The wing may be idealized as a flat plate over which air at 0.7 atm pressure and 25°F flows at 200 ft/sec. The leading edge of the cooler may be located 3 ft from the leading edge of the wing. The oil temperature and oil side heat-transfer resistance are such that the surface can be at approximately 130°F, uniform over the surface. How much heat can be dissipated if the cooler surface measures 2 ft × 2 ft? Would there by any substantial advantage in changing the shape to a rectangle 4 ft wide × 1 ft in the direction of flow?

**10-16** Consider a constant free-stream-velocity flow of air over a constant-surface-temperature plate. Let the boundary layer be initially a laminar one, but let a transition to a turbulent boundary layer take place in one case at $Re_x = 300,000$ and in another at $Re_x = 10^6$. Evaluate and plot (on log–log paper) Stanton number $St = Nu/Re\, Pr$ as a function of $Re_x$ out to $Re_x = 3 \times 10^6$. Assume that the transition is abrupt (which is not actually very realistic). Also, plot Stanton number for a turbulent boundary layer originating at the leading edge of the plate. Where is the "virtual origin" of the turbulent boundary layer when there is a preceding laminar boundary layer? What is the effect of changing the transition point? How high must the Reynolds number be in order for turbulent heat transfer coefficients to be calculated with 2% accuracy without considering the influence of the initial laminar portion of the boundary layer?

**10-17** A 40-ft-diameter balloon is rising vertically upward in otherwise still air at a velocity of 10 ft/sec. When it is at 5000-ft elevation, calculate the convection conductance over the entire upper hemispherical surface, making any assumptions that seem appropriate regarding the free-stream velocity distribution and the transition from a laminar to a turbulent boundary layer.

**10-18**   A round cylindrical body 4 ft in diameter has a hemispherical cap over one end. Air flows axially along the body, with a stagnation point at the center of the end cap. The air has an upstream state of 1 atm pressure, 70°F, and 200 ft/sec. Under these conditions, evaluate the local convection conductance along the cylindrical part of the surface to a point 12 ft from the beginning of the cylindrical surface, assuming a constant temperature surface. Make any assumptions that seem appropriate about an initial laminar boundary layer and about the free-stream velocity distribution around the nose. It may be assumed that the free-stream velocity along the cylindrical portion of the body is essentially constant at 200 ft/sec, although this will not be strictly correct in the region near the nose. Then calculate the conductance along the same surface by idealizing the entire system as a flat plate with constant free-stream velocity from the stagnation point. On the basis of the results, discuss the influence of the nose on the boundary layer at points along the cylindrical section, and the general applicability of the constant free-stream velocity idealization.

**10-19**   Prediction of local heat-transfer coefficients for turbulent (and laminar) boundary layers over a planar two-dimensional body of arbitrary shape with variable surface temperature can be accomplished by the relatively simple method developed by Ambrok [18] that is often in error by only about 15%. This method is to be derived.

   **(a)** Show that the integral form of the energy equation for constant properties is

$$\frac{d\left[\int_0^{\delta_t} u(T - T_\infty)\, dy\right]}{dx} = \frac{q_w}{\rho C_p}$$

   where $x$ is the distance from the leading edge.

   **(b)** Rearrange the result of part a into

$$\frac{d\left[U(T_w - T_\infty)\delta_t^*\right]}{dx} = \frac{q_w}{\rho C_p}$$

   where

$$\delta_t^* = \int_0^{\delta_t} \frac{u}{U} \frac{T - T_\infty}{T_w - T_\infty}\, dy$$

   **(c)** Since $q_w = h(T_w - T_\infty)$, rearrange the integral equation of part b into the form

$$[U(T_w - T_\infty)]^{-1} \frac{d\left[U(T_w - T_\infty)\delta_t^*\right]}{dx} = \frac{\mathrm{Nu}_x}{\mathrm{Re}_x \mathrm{Pr}} \qquad (19\mathrm{P}\text{-}1)$$

(d) Recall from previous work that for flow over a flat plate of constant wall temperature, $\mathrm{Nu}_x = A\,\mathrm{Re}_x^n$ where $A = 0.33206\,\mathrm{Pr}^{1/3}$ and $n = \frac{1}{2}$ for laminar flow while $A = 0.0296\,\mathrm{Pr}^{0.6}$ and $n = \frac{4}{5}$ for turbulent flow. Show that for such a case ($U$, $T_w$, and $T_\infty$ constant) $\mathrm{Nu}_x/\mathrm{Re}_x\,\mathrm{Pr} = A\,\mathrm{Re}_x^{n-1}/\mathrm{Pr}$ and Eq. (19P-1) gives

$$\delta_t^* = \frac{A}{\mathrm{Pr}}\,\frac{\nu}{nU}\,\mathrm{Re}_x^n$$

so that

$$\frac{\mathrm{Nu}_x}{\mathrm{Re}_x\,\mathrm{Pr}} = \left(n^{n-1}\frac{A}{\mathrm{Pr}}\right)^{1/n}\left(\frac{U\delta_t^*}{\nu}\right)^{1-1/n} \tag{19P-2}$$

(e) Ambrok reasoned that $\delta_t^*$ is a better indicator of local conditions than is $x$ and that the last result of part d should apply for all cases. Then $A$ and $n$ would have the flat-plate values for laminar and turbulent flows. Accordingly, substitute Eq. (19P-2) into the general Eq. (19P-1) to obtain

$$y^{-1+1/n}\frac{dy}{dx} = \left(n^{n-1}\frac{A}{\mathrm{Pr}}\right)^{1/n}\frac{U(T_w - T_\infty)^{1/n}}{\nu^{1-1/n}}$$

where $y = U(T_w - T_\infty)\delta_t^*$.

(f) Solve the differential equation of part e to obtain

$$\frac{U\delta_t^*}{\nu} = \left(\frac{A}{n\,\mathrm{Pr}}\right)\frac{\left[\int_0^x U(T_w - T_\infty)^{1/n}\,dx\right]^n}{\nu^n(T_w - T_\infty)}$$

(g) Substitute the result of part f into Eq. (19P-2) to obtain the predictive formula for the local heat-transfer coefficient as

$$\frac{\mathrm{Nu}_x}{\mathrm{Re}_x\,\mathrm{Pr}} = \frac{A}{\mathrm{Pr}}\frac{\left\{\int_0^x\left[U(T_w - T_\infty)^{1/n}/\nu\right]dx\right\}^{n-1}}{(T_w - T_\infty)^{1-1/n}}$$

(h) Show that for laminar boundary layers

$$\mathrm{Nu}_x = \frac{0.33206\,\mathrm{Pr}^{1/3}\,\mathrm{Re}_x(T_w - T_\infty)}{\left\{\int_0^x\left[U(T_w - T_\infty)^2/\nu\right]dx\right\}^{1/2}}$$

and for turbulent boundary layers

$$\mathrm{Nu}_x = \frac{0.0296\,\mathrm{Pr}^{0.6}\,\mathrm{Re}_x(T_w - T_\infty)^{1/4}}{\left\{\int_0^x\left[U(T_w - T_\infty)^{5/4}/\nu\right]dx\right\}^{1/5}}$$

For rocket nozzles, the review by Bartz [27] can be consulted.

**10-20**  A constant-temperature body immersed in a flowing stream has a free-stream velocity variation given by $U = U_0 + cx^{1/3}$. Use the result of Problem 10-19, part h to show that

$$\mathrm{Nu}_x = 0.332\,\mathrm{Pr}^{1/3}\,\mathrm{Re}_x^{1/2}\left(\frac{U_0 + cx^{1/3}}{U_0 + 3cx^{1/3}/4}\right)^{1/2}$$

for a laminar boundary layer (13.2% lower than the exact answer when $U_0 = 0$ [19]) while

$$\mathrm{Nu}_x = 0.0296\,\mathrm{Pr}^{0.6}\,\mathrm{Re}_x^{4/5}\left(\frac{U_0 + cx^{1/3}}{U_0 + 3cx^{1/3}/4}\right)^{1/5}$$

for a turbulent boundary layer. Comment on the effect of $c > 0$ and $c < 0$ on the local heat-transfer coefficient.

**10-21**  The velocity near the stagnation point of a constant temperature two-dimensional body immersed in a flowing stream varies as $U = cx$. Use the result of Problem 10-19, part h to show that

$$\mathrm{Nu}_x = 0.47\,\mathrm{Pr}^{1/3}\,\mathrm{Re}_x^{1/2}$$

for a laminar boundary layer (17.5% lower than the exact answer [19]) and

$$\mathrm{Nu}_x = 0.034\,\mathrm{Pr}^{0.6}\,\mathrm{Re}_x^{4/5}$$

for a turbulent boundary layer.

**10-22**  The leading edge of a gas turbine blade is a two-dimensional stagnation region with the external velocity given by $U = (8000\ \mathrm{s}^{-1})\,x$ with $x$ in meters. Gas property values are $\mathrm{Pr} = 0.69$, $\nu = 7.7 \times 10^{-5}\ \mathrm{m}^2/\mathrm{s}$, and $k = 0.059\ \mathrm{W/m\ K}$. Use the result of Problem 10-21 to calculate the numerical values of the local heat-transfer coefficient for (a) turbulent and (b) laminar boundary layers. [*Answer:* $h_{\mathrm{lam}} = 250$ $\mathrm{W/m^2\ K}$, $h_{\mathrm{turb}} = 4200(x/m)^{0.6}\ \mathrm{W/m^2\ K}$.]

## REFERENCES

1   R. M. O'Donnell, Experimental investigation at Mach-number of 2.41 of average skin friction coefficients and velocity profiles for laminar and turbulent boundary layers and assessment of probe effects, NACA TN 3122, 1954.

2   R. K. Lobb, E. M. Winkler, and J. Persh, Experimental investigation of turbulent boundary layers in hypersonic flow, NAVORD Report 3880, 1955.

3   H. Schlichting, *Boundary-Layer Theory*, McGraw-Hill, New York, 1968.

4   J. O. Hinze, *Turbulence*, 2nd ed., McGraw-Hill, New York, 1975.

5   C. J. Lawn, Turbulent heat transfer at low Reynolds numbers, *Transact. ASME, J. Heat Transf.* **91**, 532–536 (1969).

6   G. A. Hughmark, Heat and mass transfer for turbulent pipe flow, *AIChE J.* **17**, 902–909 (1971).

7   T. Cebeci, A model for eddy conductivity and turbulent Prandtl number, *Transact. ASME, J. Heat Transf.* **95**, 227–234 (1973).

8   Th. von Karman, The analogy between fluid friction and heat transfer, *Transact. ASME* **61**, 705–710 (1939).

9   H. Reichardt, Der Einfluss der wandnahen Strömung auf den turbulenten Wärmeübergang, *Reports of the Max-Planck-Institute für Strömungsforschung*, No. 3, 1950, pp. 1–63.

10   E. R. Van Driest, The turbulent boundary layer with variable Prandtl number, in *Fifty Years of Boundary Layer Research*, Braunschweig, 1955, pp. 257–271.

11   J. C. Rotta, Temperaturverteilungen in der turbulenten Grenzschicht an der ebenen Platte, *Internatl. J. Heat Mass Transf.* **7**, 215–228 (1964).

12   D. B. Spalding, Heat transfer to a turbulent stream from a surface with a step-wise discontinuity in wall temperature, in *International Developments In Heat Transfer* (proceedings of conference organized by ASME at Boulder, CO, 1961), Part II, pp. 439–446.

13   G. O. Gardner and J. Kestin, Calculation of the Spalding function over a range of Prandtl numbers, *Internatl. J. Heat Mass Transf.* **6**, 289–299 (1963).

14   J. Kestin and P. D. Richardson, Heat transfer across turbulent incompressible boundary layer, *Internatl. J. Heat Mass Transf.* **6**, 147–189 (1963).

15   O. Reynolds, W. M. Kays, and S. J. Kline, Heat Transfer In The Turbulent Incompressible Boundary Layer. I. Constant wall temperature, NASA MEMO 12-1-58 W, 1958; II. Step wall temperature distribution, NASA MEMO 12-2-58 W, 1958; III. Arbitrary wall temperature and heat flux, NASA MEMO 12-3-58 W, 1958; IV. Effect of location of transition and prediction of heat transfer in a known transition region, NASA MEMO 12-4-48 W, 1968.

16   C. B. Millikan, The boundary layer and skin friction for a figure of revolution, *Transact. AMSE, J. Appl. Mech.* **54**, 29–43 (1932), Paper APM-54-3.

17   D. B. Spalding, Heat transfer to a turbulent stream from a surface with a step-wise discontinuity in wall temperature, *International Developments in Heat Transfer* (Proceedings of the 1961–1962 Heat Transfer Conference, Boulder, CO, 1961), Part II, pp. 439–446.

18   G. S. Ambrok, Approximate solutions of the equations for the thermal boundary layer with variations in boundary layer structure, *Sov. Phys.-Tech. Phys.* **2**, 1979–1986 (1957).

19   W. M. Kays and M. E. Crawford, *Convective Heat And Mass Transfer*, McGraw-Hill, New York, 1980, p. 139.

20   J. C. Evvard, M. Tucker, and W. C. Burgess, Transition point fluctuations in supersonic flow, *JAS* **21**, 731–738 (1954); NACA TN 3100, 1954.

21   L. M. Mack, An experimental investigation of the temperature recovery-factor, Jet Propulsion Laboratory, Report 20-80, California Institute Of Technology, Pasadena, CA, 1954.

22   E. R. G. Eckert and R. M. Drake, *Analysis Of Heat And Mass Transfer*, McGraw-Hill, New York, 1972, p. 385.

23   R. E. Mayle, M. F. Blair, and F. C. Kopper, Turbulent boundary layer heat transfer on curved surfaces, *Transact. ASME, J. Heat Transf.* **101**, 521–525 (1979).

24   R. J. Moffat, J. M. Healzer, and W. M. Kays, Experimental heat transfer behavior of a turbulent boundary layer on a rough surface with blowing, *Transact. ASME, J. Heat Transf.* **100**, 134–142 (1978).

25  J. Kestin, The effect of free-stream turbulence on heat transfer rates, *Adv. Heat Transf.* **3**, 1–32 (1966).

26  R. A. Seban and D. L. Doughty, Heat transfer to turbulent boundary layers with variable free-stream velocity, *Transact. ASME* **78**, 217–223 (1956).

27  D. R. Bartz, Turbulent boundary-layer heat transfer from rapidly accelerating flow of rocket combustion gases and of heated air, *Adv. Heat Transf.* **2**, 1–108 (1965).

28  U. Müller, K. G. Rosener, and B. Schmidt, Eds., *Recent Developments In Theoretical And Experimental Fluid Mechanics*, Springer, Berlin, 1979.

29  L. Thomas, A simple integral approach to turbulent thermal boundary layer flow, *Transact. ASME, J. Heat Transf.* **100**, 744–746 (1978).

30  M. J. Callaghan and D. M. Mason, Reynolds analogy in reacting systems, heat transfer to turbulently flowing nitrogen dioxide gas, *Chem. Eng. Sci.* **19**, 763–774 (1964).

31  R. M. Traci and D. C. Wilcox, Freestream turbulence effects on stagnation point heat transfer, *AIAA J.* **13**, 890–896 (1975).

32  P. M. Gerhart, Prediction of recovery factor and Reynolds' analogy for compressible turbulent flow, *AIAA J.* **13**, 966–968 (1975).

33  E. R. Keener and T. E. Polk, Measurements of Reynolds analogy for a hypersonic turbulent boundary layer on a nonadiabatic flat plate, *AIAA J.* **10**, 845–846 (1972).

34  E. J. Hopkins and M. Inouye, An evaluation of theories for predicting turbulent skin friction and heat transfer on flat plates at supersonic and hypersonic Mach numbers, *AIAA J.* **9**, 993–1003 (1971).

35  T. Cebeci and A. M. O. Smith, Analysis of turbulent boundary layers, *Appl. Math. Mech.* **15**, 249–251 (1974).

36  P. S. Klebanoff, Characteristics of turbulence in a boundary layer with zero pressure gradient, NACA Technical Note No. 3178, 1954.

37  S. Corrsin and A. L. Kistler, The free-stream boundaries of turbulent flows, NACA Technical Note No. 3133, 1954.

38  G. Maise and H. McDonald, Mixing length and kinematic eddy viscosity in a compressible boundary layer, *AIAA J.* **6**, 73–80 (1968).

39  T. Cebeci, A. M. O. Smith, and G. Mosinskis, Solution of the incompressible turbulent boundary-layer equations with heat transfer, *Transact. ASME, J. Heat Transf.* **92**, 133–143 (1970).

40  T. Cebeci and P. Bradshaw, *Momentum Transfer In Boundary Layers*, Hemisphere, (McGraw-Hill), New York, 1977.

41  R. H. Pletcher, On a finite-difference solution for the constant-property turbulent boundary layer, *AIAA J.* **7**, 305–311 (1969).

42  R. H. Pletcher, On a solution for turbulent boundary layer flows with heat transfer, pressure gradients, and wall blowing or suction, in *Heat Transfer 1970, Proceedings Fourth International Heat Transfer Conference*, Vol. 2, Elsevier, Amsterdam, 1970, Paper FC 2.9.

43  R. H. Pletcher, Calculation method for compressible turbulent boundary flows with heat transfer, *AIAA J.* **10**, 245–246 (1972).

44  T. Cebeci, Calculation of compressible turbulent boundary layers with heat and mass transfer, *AIAA J.* **9**, 1091–1097 (1971).

45  H. B. Keller and T. Cebeci, Accurate numerical methods for boundary-layer flows. II: Two-dimensional turbulent flows, *AIAA J.* **10**, 1193–1199 (1970).

46  R. L. Simpson, W. M. Kays, and R. J. Moffat, The turbulent boundary layer on a porous plate: An experimental study of the fluid dynamics with injection and suction, *Stanford University Mechanical Engineering Department HMT-2*, 1967.

47  G. B. Schubauer, Air flow in the boundary layer of an elliptic cylinder NACA Report 652, 1939.

48  P. Bradshaw and D. H. Ferriss, The response of a retarded equilibrium boundary layer to the sudden removal of pressure gradient, NPL Aero Report 1145, 1965.

49  W. C. Reynolds, W. M. Kays, and S. J. Kline, Heat transfer in the turbulent incompressible boundary layer—III: Arbitrary wall temperature and heat flux, NASA Memo 12-3-58W, December 1958.

50  S. V. Patankar and D. B. Spalding, *Heat and Mass Transfer in Boundary Layers*, 2nd ed., Intertext, London, 1970.

51  R. Arieli and J. D. Murphy, Pseudo-direct solution to the boundary-layer equations for separated flow, *AIAA J.* **18**, 883–891 (1980).

52  A. N. Kolmogorov, Equations of turbulent motion of an incompressible fluid, *Izv. Akad. Nauk SSR Seria Fizicheska Vi.* (1–2), 1942, 56–58 (1942), English translation Imperial College, Mechanical Engineering Department Report ON/6, 1968.

53  L. Prandtl, Über ein neues Formel-system für die ausgebildete Turbulenz, Nachr. Akad. Wiss., Götingen, Math-Phys. Klasse, 1945, p. 6.

54  P. Bradshaw, D. H. Ferriss, and N. P. Atwell, Calculation of boundary layer development using the turbulent energy equation, *J. Fluid Mech.* **28**, 593–616 (1967).

55  J. O. Hinze, *Turbulence*, 2nd ed., McGraw-Hill, New York, 1975, pp. 68–74.

56  D. C. Wilcox and M. W. Rubesin, Progress in turbulence modelling for complex flowfields, NASA TP-1517, 1980.

57  W. P. Jones and B. E. Launder, The prediction of laminarization with a two-equation model of turbulence, *Internatl. J. Heat Mass Transf.* **15**, 301–314 (1972); The calculation of low-Reynolds number phenomena with a two-equation model of turbulence, *Internatl. J. Heat Mass Transf.* **16**, 1119–1130 (1973).

58  W. Rodi, *Turbulence Models and Their Application in Hydraulics*, Book Publication of the Int. Association for Hydraulic Research, Delft, The Netherlands, 1980.

59  B. E. Launder and D. B. Spalding, *Mathematical Models of Turbulence*, Academic Press, New York, 1972.

60  B. E. Launder and D. B. Spalding, The numerical computation of turbulent flow, *Comp. Meth. Appl. Mech. Eng.* **3**, 269 (1974).

61  B. E. Launder, A. P. Morse, W. Rodi, and D. B. Spalding, The prediction of free-shear flows—A comparison of the performance of six turbulence models, *Proceedings of NASA Langley Free Turbulent Shear Flows Conference*, Vol. 1, 1973, NASA SP 320.

62  E. G. Hauptman and A. Malhotra, Axial development of unusual velocity profiles due to heat transfer in variable density fluids, *Transact. ASME, J. Heat Transf.* **102**, 71–74 (1980).

63  K. Hanjalic and B. E. Launder, A Reynolds-stress model of turbulence and its application to thin shear flows, *J. Fluid Mech.* **52**, 609 (1972).

64  D. C. Wilcox, Algorithm for rapid integration of turbulence model equations on parabolic regions, *AIAA J.* **19**, 248–250 (1981).

65  A. D. Gossman, W. M. Pun, A. K. Runchal, D. B. Spalding, and M. Wolfshtein, *Heat and Mass Transfer in Recirculating Flows*, Academic Press, New York, 1969.

66  S. V. Patankar and D. B. Spalding, A calculation procedure for heat, mass and momentum transfer in three-dimensional parabolic flows, *Internatl. J. Heat Mass Transf.* **15**, 1787–1806 (1972).

67  S. V. Patankar, *Numerical Heat Transfer and Fluid Flow*, Hemisphere, Washington, D.C., 1980; A. F. Emery, P. K. Neighbors, and F. B. Gessner, Computational procedure for developing turbulent flow and heat transfer in a square duct, *Numerical Heat Transf.* **2**, 399–416 (1979).

68  B. E. Launder, G. J. Reece, and W. Rodi, Progress in the development of a Reynolds stress turbulence closure, *J. Fluid Mech.* **68**, 537–566 (1975).

69  M. M. Gibson and B. E. Launder, Ground effects on pressure fluctuation in the atmospheric boundary layer, *J. Fluid Mech.* **86**, 491 (1978).

70  J. L. Lumley and B. Khajeh-Nouri, Computation of turbulent transport, *Adv. Geophys.* **A18**, 169–192 (1974).

71  W. S. Lewellen, M. Teske, and C. du P. Donaldson, Variable density flows computed by a second-order closure description of turbulence, *AIAA J.* **14**, 382–387 (1976).

72  W. Rodi, A new algebraic relation for calculating the Reynolds stresses, *ZAMM* **56**, 219–T221 (1976).

73  M. Ljuboja and W. Rodi, Calculation of turbulent wall jets with an algebraic Reynolds stress model, *Transact. ASME, J. Fluids Eng.* **102**, 350–356 (1980).

74  W. Rodi, Examples of turbulence models for incompressible flows, *AIAA J.* **20**, 872–879 (1982).

75  A. F. Emery, P. K. Neighbors, and F. B. Gessner, The numerical prediction of developing turbulent flow and heat transfer in a square duct, *Transact. ASME, J. Heat Transf.* **102**, 51–57 (1980).

76  B. I. Sharma, Prediction of local heat transfer on a rotating disk by a two-equation model of turbulence, *Transact. ASME, J. Heat Transf.* **99**, 151–152 (1977).

77  P. Bradshaw, T. Cebeci, and J. Whitelaw, *Engineering Calculational Methods for Turbulent Flow*, Academic Press, New York, 1981.

# 11

# TURBULENT FLOW IN DUCTS

Much of the basic understanding of turbulent flow came from careful experimental study of flow in circular pipes. Chapter 9 discussed the conclusions that are drawn concerning velocity distributions. These conclusions were then applied to exterior turbulent boundary layers, leading to predictions of the local turbulent heat-transfer coefficient in Chapter 10.

Many applications require information concerning turbulent flow and heat transfer for the case in which the flow is internal to a duct that usually, but not always, has a circular cross section. No new essential difficulty is encountered in the treatment of such a flow, because the basic information came from it. The major need is to account for the effects of curvature that are largely absent in exterior flows.

The final theory developed is to be considered as a correlation of a large body of measurements. Its utility is to impart an understanding of turbulent transport phenomena and to fill in gaps and allow new areas to be explored without extensive and expensive experimentation.

## 11.1 FULLY DEVELOPED PROFILE IN CIRCULAR DUCT

As discussed in Chapter 9, measured velocity profiles for fully developed turbulent flow in smooth circular ducts are well fitted in the central region by

$$u^+ = C(y^+)^{1/n} \qquad (9\text{-}19)$$

with $c$ and $n$ being somewhat dependent on the Reynolds number $\text{Re}_D = U_{\text{av}} D / \nu$, where $D$ is the duct diameter. As before, $y$ is distance from the wall whereas $u^+ = \bar{u}/v^*$, $y^+ = yv^*/\nu$, and $v^* = (\tau_w/\rho)^{1/2}$. Table 9-2 shows that

**467**

$c = 8.74$ and $n = 7$ for a typical case of $\text{Re} \approx 10^5$. Greater accuracy is afforded by Reichardt's relationship that

$$u^+ = 5.5 + 2.5 \ln\left[y^+ \frac{1.5(1 + r/r_0)}{1 + 2(r/r_0)^2}\right]$$

$$\approx 5.5 + 2.5 \ln(y^+) \qquad (9\text{-}20a)$$

In the buffer zone, the velocity profile is approximated with good accuracy by

$$u^+ = 5 + 5 \ln\left(\frac{y^+}{5}\right), \qquad 5 \leqslant y^+ \leqslant 30 \qquad (9\text{-}17)$$

Nearer the wall in the laminar sublayer, the linear representation

$$u^+ = y^+, \qquad 0 \leqslant y^+ \leqslant 5 \qquad (9\text{-}16)$$

is quite satisfactory. In connection with flow in a circular duct, the region near the wall ($5 \leqslant y^+ \leqslant 30$) is accurately assumed to have a constant shear stress. In the central region to which the $1/n$th power-law velocity profile applies, the linear variation of shear stress

$$\tau = \tau_w\left(1\frac{-y}{r_0}\right)$$

is sufficiently important to be taken into account.

For prediction of temperature profiles, and the heat-transfer coefficient that follows from them, it will later be necessary to know the ratio of eddy diffusivity of momentum to kinematic viscosity $\varepsilon_m/\nu$. Knowledge of the velocity profile and the shear stress variation is sufficient to yield $\varepsilon_m/\nu$ since

$$\tau = (\varepsilon_m + \nu)\frac{du}{dy}$$

Accounting for the linear variation of $\tau$ and employment of dimensionless quantities allows this relation to be expressed as

$$\frac{\varepsilon_m}{\nu} = -1 + \frac{(1 - y^+/r_0^+)}{du^+/dy^+} \qquad (11\text{-}1)$$

It can be appreciated that substantial error can be incurred by employing velocity profiles that, although they give a good fit to velocity itself over a range of $y$, might give a relatively poor fit to velocity gradient. For the laminar sublayer, the linear velocity profile of Eq. (9-16) used in Eq. (11-1) results in

$$\frac{\varepsilon_m}{\nu} \approx 0, \qquad 0 \leqslant y^+ \lesssim 5 \qquad (11\text{-}2)$$

if it is recognized that $1 - y/r_0 \approx 1$. This is a reasonable result since molecular effects should overweigh turbulent effects there. For the buffer zone, the logarithmic velocity profile of Eq. (9-17) used in Eq. (11-1) yields

$$\frac{\varepsilon_m}{\nu} \approx \frac{y^+ - 5}{5}, \qquad 5 \lesssim y^+ \lesssim 30 \tag{11-3}$$

if it is again recognized that $1 - y/r_0 \approx 1$. This is also a reasonable result since at the laminar sublayer-buffer zone interface there is a match. For the fully turbulent core of the pipe, the $1/n$th power-law velocity profile of Eq. (9-19) results in

$$\frac{\varepsilon_m}{\nu} = -1 + \frac{n}{C}\left(\frac{1-y}{r_0}\right)(y^+)^{(n-1)/n}$$

which, for a $\frac{1}{7}$th power law, is

$$\frac{\varepsilon_m}{\nu} = -1 + 0.8\left(1 - \frac{y}{r_0}\right)(y^+)^{6/7} \tag{11-4}$$

In comparison, the more exact velocity profile of Eq. (9-20a) due to Reichardt for the central region of the pipe leads to

$$\frac{\varepsilon_m}{\nu} = \frac{r_0^+}{15}\left(2 - \frac{y}{r_0}\right)\left[1 + 2\left(1 - \frac{y}{r_0}\right)^2\right]\frac{y}{r_0} \tag{9-20}$$

The effect of an approximate velocity profile can be ascertained by considering the approximation to Eq. (9-20a), $u^+ \approx 5.5 + 2.5 \ln(y^+)$. Introduction of this approximation into Eq. (11-1) gives

$$\frac{\varepsilon_m}{\nu} = -1 + \left(1 - \frac{y}{r_0}\right)\frac{y^+}{2.5}$$

$$\approx \left(1 - \frac{y}{r_0}\right)\frac{y^+}{2.5} \tag{11-5}$$

since molecular effects are usually small in the turbulent central region of a pipe.

The behavior of the needed $\varepsilon_m/\nu$ is displayed in Fig. 11-1, where the mismatch at the outer edge of the buffer zone is evident. In the laminar sublayer and buffer zone it would have been possible to achieve a smooth variation of $\varepsilon_m/\nu$ at the expense of a more complex mathematical form by use of Deissler's hypothesis of Eq. (9-14); the desire for closed form solutions motivates the discontinuous representations used here. Most striking is the erroneous prediction from the simpler velocity profiles that $\varepsilon_m/\nu$ falls to low

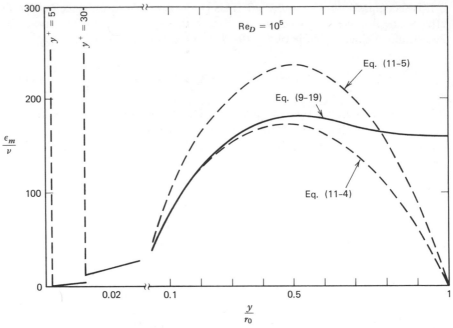

**Figure 11-1**   Variation of the ratio of eddy diffusivity to kinematic viscosity $\varepsilon_m/\nu$ with distance from the wall in a circular duct.

values near the pipe axis. The more accurate prediction of Reichardt from Eq. (9-20) is for $\varepsilon_m/\nu$ to be nearly uniform in the central region. Also, the high value of $\varepsilon_m/\nu$ in the central region testifies to the effectiveness of vigorous turbulent mixing in increasing the effective viscosity there. Since the turbulent viscosity is about two orders of magnitude greater than the molecular value, the central core very nearly behaves as a solid slug sliding down the pipe on a thin lubricating film whose thickness is about that of the laminar sublayer.

## 11.2   FULLY DEVELOPED TEMPERATURE PROFILE IN CIRCULAR DUCT FOR CONSTANT HEAT FLUX

The fully developed temperature profile is desired for a circular duct subjected to a constant wall heat flux $q_w$. From this temperature distribution the local heat-transfer coefficient is determined in much the same manner as for the laminar flow case in Chapter 6, except the viscosity and thermal conductivity are now space dependent. The situation to be studied is of importance because it yields the turbulent heat-transfer coefficient for the very long ducts that commonly occur and because, as will later be again discussed briefly, the results are very close to those for the constant wall temperature case.

The energy equation in cylindrical coordinates is

$$\rho C_p u \frac{\partial T}{\partial z} = -\frac{1}{r}\frac{\partial(rq_r)}{\partial r}$$

$$\frac{\partial T(r=r_0, z)}{\partial r} = -\frac{q_w}{k}$$

$$\frac{\partial T(r=0, z)}{\partial r} = 0 \quad \text{or} \quad q_r(r=0, z) = 0 \tag{11-6}$$

Here, the $z$ direction is down the pipe and interest is confined to regions far removed from the start of the heated region and the pipe inlet. Thus both velocity and temperature profiles are fully developed. Also, in this fully developed region the temperature must increase linearly in the $z$ direction as discussed in Chapter 6; according to Eq. (6-37c), it must then be the case that $\partial T/\partial z = dT_{\text{mixing cup}}/dz = \text{const.}$

Integration of Eq. (11-6) with respect to $y$, after making the substitution $y = r_0 - r$, gives

$$C_1 - \rho C_p \frac{dT_m}{dz}\int_0^y u(r_0 - y)\, dy = (r_0 - y)q_r$$

The constant of integration is evaluated from the condition at $y = r_0$ to give

$$-\rho C_p \frac{dT_m}{dz}\int_y^{r_0} u(r_0 - y)\, dy = (r_0 - y)q_r \tag{11-7}$$

The radial heat flux is related to temperature gradient by

$$q_r = -\rho C_p(\varepsilon_H + \alpha)\frac{\partial T}{\partial r}$$

$$= \rho C_p(\varepsilon_H + \alpha)\frac{\partial T}{\partial y}$$

$$= \rho C_p\left(\frac{\varepsilon_m}{\text{Pr}_t} + \frac{\nu}{\text{Pr}}\right)\frac{\partial T}{\partial y}$$

Insertion of this expression into Eq. (11-7) gives, with the realization that $\rho C_p dT_m/dz = -2q_w/r_0 U_{av}$, where $q_w$ is positive if leaving the fluid,

$$\frac{dT}{dy} = \frac{2q_w}{\rho C_p r_0}\frac{\int_y^{r_0}(u/U_{av})(r_0 - y)\, dy}{(\varepsilon_m/\text{Pr}_t + \nu/\text{Pr})(r_0 - y)}$$

One further integration gives, where $T_w$ is the yet unknown wall temperature,

$$T - T_w = \frac{2q_w}{\rho C_p} \int_0^y \left[ \frac{\int_y^{r_0} (u/U_{av})(1 - y'/r_0)\, dy'}{(\varepsilon_m/\text{Pr}_t + \nu/\text{Pr})(r_0 - y)} \right] dy$$

In terms of the dimensionless quantities $v^* = (\tau_w/\rho)^{1/2}$, $T^+ = (T - T_w)(\rho C_p v^*/q_w)$, $y^+ = yv^*/\nu$, and $u^+ = u/v^*$ this relationship becomes

$$T^+ = 2 \int_0^{y^+} \left[ \frac{\int_{y/r_0}^1 (u/U_{av})(1 - y'/r_0)\, d(y'/r_0)}{[(\varepsilon_m/\nu)/\text{Pr}_t + 1/\text{Pr}](1 - y/r_0)} \right] dy^+ \qquad (11\text{-}8)$$

Integration is greatly simplified by the reasonable assumption that the velocity distribution is nearly uniform, leading to $u/U_{av} \approx 1$. Then

$$T^+ = \int_0^{y^+} \frac{(1 - y^+/r_0^+)\, dy^+}{(\varepsilon_m/\nu)/\text{Pr}_t + 1/\text{Pr}} \qquad (11\text{-}9)$$

The temperature distribution will be obtained by integration of Eq. (11-9) in three parts, considering $\text{Pr}_t$ to always be a constant. Then

$$\Delta T^+ = \Delta T^+_{\text{sublayer}} + \Delta T^+_{\text{buffer zone}} + \Delta T^+_{\text{turbulent core}}$$

The laminar sublayer is considered first. Here $\varepsilon_m/\nu \approx 0$ from Eq. (11-2) and $1 - y/r_0 \approx 1$ so that

$$\Delta T^+_{\text{sublayer}} = \int_0^5 \text{Pr}\, dy^+ = 5\,\text{Pr}$$

In the buffer zone $\varepsilon_m/\nu = (y^+ - 5)/5$ from Eq. (11-3) and $1 - y/r_0 \approx 1$ so that

$$\Delta T^+_{\text{buffer}} = \int_5^{30} \frac{5\text{Pr}_t\, dy^+}{y^+ - 5(1 - \text{Pr}_t/\text{Pr})} = 5\text{Pr}_t \ln\left(1 + \frac{5\text{Pr}}{\text{Pr}_t}\right)$$

Finally, the turbulent core has $\varepsilon_m/\nu = (1 - y/r_0)y^+/2.5$ from Eq. (11-5) so that

$$\Delta T^+_{\text{core}} = 2.5\text{Pr}_t \int_{30}^{r_0^+} \frac{(1 - y^+/r_0^+)\, dy^+}{(1 - y^+/r_0^+)y^+ + 2.5\text{Pr}_t/\text{Pr}} \approx 2.5\text{Pr}_t \ln\left(\frac{r_0^+}{30}\right)$$

The neglect of $1/\text{Pr}$ is justified for $\text{Pr} \approx 1$ since $\varepsilon_m/\nu \gg 1$—this neglect makes the result inaccurate for liquid metals where $\text{Pr} \ll 1$. Realizing that

$$r_0^+ = \frac{r_0 v^*}{\nu} = \frac{r_0}{\nu}\left(\frac{\tau_w}{\rho}\right)^{1/2} = \frac{r_0}{\nu}\left(\frac{C_f}{2}U_{av}^2\right)^{1/2} = \frac{\text{Re}_D}{2}\left(\frac{C_f}{2}\right)^{1/2}$$

and adding the three temperature drops yields the temperature difference between the center line and the wall as

$$T_{CL} - T_w = \frac{q_w}{\rho C_p U_{av}(C_f/2)^{1/2}} \left\{ 5\Pr + 5\Pr_t \ln(1 + 5\Pr/\Pr_t) \right.$$

$$\left. + 2.5\Pr_t \ln\left[ \frac{\text{Re}_D(C_f/2)^{1/2}}{60} \right] \right\} \quad (11\text{-}10)$$

in which it is conventional to take $\Pr_t = 1$.

From Eq. (11-10) the heat-transfer coefficient can be determined since

$$q_w = h(T_m - T_w)$$

Rearrangement of this result gives

$$\text{Nu}_D = \frac{hD}{k} = \frac{q_w D}{k(T_{CL} - T_w)} \frac{T_{CL} - T_w}{T_m - T_w} \quad (11\text{-}11)$$

The first term on the right-hand side of Eq. (11-11) is available from Eq. (11-10), but the second term remains to be calculated. The turbulent velocity profile, as discussed in connection with Eq. (10-1), is reasonably represented by $u/U_{CL} = (y/r_0)^{1/7}$. For $\Pr \approx 1$, the temperature distribution must be of the same form so that

$$\frac{T - T_w}{T_{CL} - T_w} = \left( \frac{y}{r_0} \right)^{1/7}$$

Here $U_{CL}$ and $T_{CL}$ are center-line quantities and $y$ is the distance from the wall. The definition of the mixing-cup temperature is

$$\frac{T_m - T_w}{T_{CL} - T_w} = \frac{2\int_0^{r_0} (u/U_{CL})[(T - T_w)/(T_{CL} - T_w)]r \, dr}{2\int_0^{r_0} (u/U_{CL})r \, dr}$$

Insertion of the $\frac{1}{7}$th power-law profiles into this expression gives

$$\frac{T_m - T_w}{T_{CL} - T_w} = \frac{49/72}{49/60} = \frac{5}{6} \quad (11\text{-}12)$$

Substitution of Eqs. (11-10) and (11-12) into Eq. (11-11) gives the final result, with $\Pr_t = 1$, of

$$\text{Nu}_D = \text{Re}_D \Pr \frac{6(C_f/2)^{1/2}}{25\left\{ \Pr + \ln(1 + 5\Pr) + 0.5 \ln\left[ \text{Re}_D(C_f/2)^{1/2}/60 \right] \right\}}$$

$$(11\text{-}13)$$

Equation (11-13) follows the essential developments more accurately accomplished by Martinelli [1] and Boelter [2], who included the Reynolds number dependence of the mixing-cup temperature and allowed for application to liquid metals (Pr ≪ 1) by not ignoring molecular effects in the turbulent core. A more accurate form of Reynolds analogy for constant wall heat flux in the fully developed flow region of a smooth round duct was developed by Petukhov and Popov [3,4] and is accurate within 15% for $0.7 < \mathrm{Pr} < 50$ according to Webb [5]. Essentially, they solved Eq. (11-18) and used Reichardt's velocity and eddy diffusivity distributions and then accurately approximated the results to achieve (for rectangular ducts, too [71])

$$\frac{\mathrm{Nu}_D}{\mathrm{Re}_D \,\mathrm{Pr}} = \frac{C_f/2}{1.07 + 12.7(\mathrm{Pr}^{2/3} - 1)(C_f/2)^{1/2}} \tag{11-14}$$

where the pressure-drop friction factor $f = -(2D/\rho U_{\mathrm{av}}^2)\,dp/dx$ is calculated from the Prandtl–Karman equation

$$\frac{1}{f^{1/2}} = 2\log_{10}(\mathrm{Re}_D f^{1/2}) - 0.8 \tag{9-20b}$$

which applies only to long smooth circular ducts—$f = 4C_f$.

Not only is the recommended Reynolds analogy of Eq. (11-14) accurate, it is of the form encountered in boundary-layer applications and is satisfying for that reason. Equation (11-13) can also be put into this form, as is the subject of Problem 11-2, so the latter reason is not really unique. As outlined by Webb [5], a variety of correlations of the general form

$$\frac{\mathrm{Nu}_D}{\mathrm{Re}_D \,\mathrm{Pr}} = \frac{C_f/2}{C_{\mathrm{r}} + C_2 F(\mathrm{Pr})(C_f/2)^{1/2}}$$

have been proposed with varying $C_{1,2}$ values and functions of Pr, $F(\mathrm{Pr})$; usually $C_1 \approx 1$ and $F(\mathrm{Pr} = 1) = 0$.

Before discussing additional refinements, it is instructive to examine the results of the analysis that led to the simplified Reynolds analogy of Eq. (11-13). First, note that in turbulent flow Nu depends on Re and Pr separately in contrast to its constant value for laminar flow. Then note the temperature distributions in turbulent flow displayed in Fig. 11-2. At large Pr such as characterize water and oils, for instance, the temperature variation occurs mostly in the laminar sublayer and buffer zone, where molecular effects are important and where the approximations that led to Eq. (11-13) can be refined [e.g., by Deissler's eddy diffusivity of Eq. (9-14)]. At small Pr, which characterize liquid metals, the temperature variation occurs in the turbulent core to a major extent as a result of the large thermal conductivity that makes molecular effects important even in the central region of the pipe. It is evident

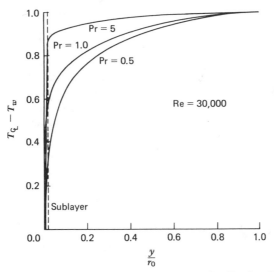

**Figure 11-2**  Effect of Prandtl number on temperature distribution in turbulent flow.

that molecular effects predominate everywhere when Pr is small; not only is the temperature profile similar to that for laminar flow, but the thermal entry length and the response to changes in wall temperature are also similar to the laminar case. At large Pr, the thermal entry length is short and there is rapid response to changes in wall temperature, both a result of rapid diffusion of heat through the fluid once it has penetrated the sublayers where the major thermal resistance occurs.

A refined and accurate analysis was accomplished by Deissler [6] with several important features, including applicability to large Prandtl numbers. Deissler assumed a turbulent Prandtl number $Pr_t$ of unity and utilized the eddy diffusivity of Eq. (9-14) near the wall to obtain

$$u^+ = \int_0^{y^+} \frac{dy^+}{1 + n^2 u^+ y^+ \left[1 - \exp\left(-n^2 u^+ y^+\right)\right]}, \qquad y^+ \leqslant 26 \qquad (9\text{-}14)$$

$$T^+ = \int_0^{y^+} \frac{dy^+}{1/Pr + n^2 u^+ y^+ \left[1 - \exp\left(-n^2 u^+ y^+\right)\right]}, \qquad y^+ \leqslant 26$$

$$(11\text{-}15)$$

Far from the wall $y^+ \geqslant 26$, he used

$$u^+ = 13 + \frac{1}{0.36} \ln\left(\frac{y^+}{26}\right) \qquad (11\text{-}16)$$

which is an excellent fit to velocity data. Because molecular effects are negligible far from the wall at large Pr, $y^+ \geq 26$, so that velocity and temperature profiles are the same, the same relation was used for $T^+$. Numerical integration was used to solve Eqs. (9-15), (11-15), and (11-16), and the ultimate Nusselt number results were in close agreement with measurements for $0.7 \leq$ Pr. Of greatest immediate interest here is the case where Pr $\to \infty$. Setting $u^+ = y^+$, since the temperature drop will occur in the laminar sublayer at large Pr, and retaining only the first two terms of the denominator series expansion gives Eq. (11-15) as

$$T^+ = \int_0^{y^+} \frac{dy^+}{1/\mathrm{Pr} + n^4(y^+)^4}$$

Evaluation of the integral at $y^+ = \infty$ gives

$$T^+ (y^+ = \infty) = \frac{\pi}{2^{2/3}n} \mathrm{Pr}^{3/4}$$

where $n = 0.124$. Now $T^+ (y^+ = \infty)$ is essentially $T^+_{\text{mixing cup}}$ because $T^+$ is very nearly constant when Pr $\to \infty$. Recognizing that

$$\mathrm{Nu}_D = \frac{2r_0^+ \, \mathrm{Pr}}{T_m^+}$$

one then finds that

$$\lim_{\mathrm{Pr}\to\infty} \mathrm{Nu}_D = \frac{32^{1/2}}{\pi} n r_0^+ \, \mathrm{Pr}^{1/4}$$

$$= \frac{2n}{\pi} \mathrm{Re}_D \, \mathrm{Pr}^{1/4} \left( \frac{C_f}{2} \right)^{1/2}$$

Here $C_f = 2\tau_w/\rho U_{\mathrm{av}}^2$. Deissler also explored entrance region effects, which are discussed later and accounted for temperature-dependent viscosity. A similar study was executed by Sparrow et al. [7], who obtained good agreement with Deissler's work.

Subsequent to the previously described investigations an analysis was performed by Kays and Leung [8] which is applicable over an extremely broad range of Prandtl numbers, from the low values of liquid metals to the high values of oils. They used Deissler's expression [Eq. (9-14)] near the wall for the eddy diffusivity of momentum near the wall and took the turbulent Prandtl number to be unity there. In the turbulent core, Reichart's expression in the form of Eq. (9-20) was used for the eddy diffusivity of momentum; the turbulent Prandtl number was not taken to be unity there. Instead, an idea suggested by Jenkins [9] and refined by Deissler [10] (both cogently discussed by Jakob [11]) was adopted in a modified form. Jenkins, as discussed by Hinze [12] in connection with an early version of the idea due to Burgers [13], postulated that an eddy can be thought of as a solid sphere that starts its flight with a uniform temperature and exchanges heat with its environment along the way. In this way the dependence of $Pr_t$ on Pr is ascertained. The results shown in Fig. 11-3 and Table 11-1 were obtained by numerical procedures, the

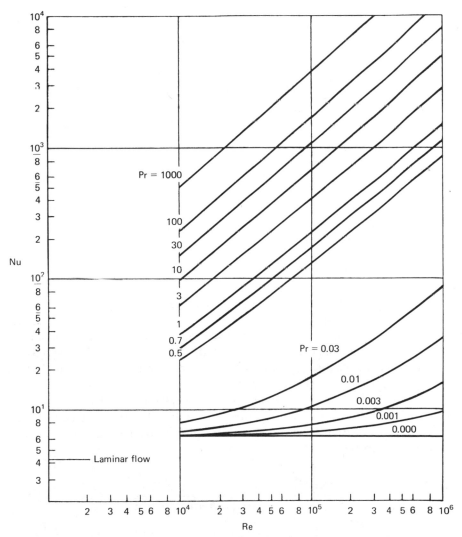

**Figure 11-3** Nusselt number for fully developed turbulent velocity and temperature profiles in a circular duct with constant wall heat flux. [From W. M. Kays and E. Y. Leung, *Internatl. J. Heat Mass Transf.* **6**, 537–557 (1963) [8].]

**Table 11-1    Nusselt Numbers for Fully Developed Turbulent Flow in a Circular Tube with Constant Heat Rate [8]**

| Pr \ Re | Laminar | $10^4$ | $3 \times 10^4$ | $10^5$ | $3 \times 10^5$ | $10^6$ |
|---|---|---|---|---|---|---|
| 0.0 | 4.364 | 6.30 | 6.64 | 6.84 | 9.95 | 7.06 |
| 0.001 | | 6.30 | 6.64 | 6.84 | 7.08 | 8.12 |
| 0.003 | | 6.30 | 6.64 | 7.10 | 8.14 | 12.8 |
| 0.01 | | 6.43 | 7.00 | 8.90 | 14.2 | 30.5 |
| 0.03 | | 6.90 | 9.10 | 15.9 | 32.4 | 80.5 |
| 0.5 | | 26.3 | 57.3 | 142 | 340 | 895 |
| 0.7 | | 31.7 | 70.7 | 178 | 430 | 1150 |
| 1.0 | | 37.8 | 86.0 | 222 | 543 | 1470 |
| 3.0 | | 61.5 | 149 | 404 | 1030 | 2900 |
| 10 | | 99.8 | 248 | 690 | 1810 | 5220 |
| 30 | | 141 | 362 | 1030 | 2750 | 8060 |
| 100 | | 205 | 522 | 1510 | 4030 | 12000 |
| 1000 | | 380 | 975 | 2830 | 7600 | 22600 |

mathematical description of the problem now being too complex to achieve a solution in closed form. Examination of Fig. 11-3 indicates that the slopes of the various curves are nearly equal on a log–log plot for $\mathrm{Pr} \gtrsim 1$, justifying the conventional correlation of turbulent data by $\mathrm{Nu}_D = F(\mathrm{Re}_D, \mathrm{Pr}) \approx F_1(\mathrm{Re}_D)F_2(\mathrm{Pr}) \approx C\,\mathrm{Re}_D^a\,\mathrm{Pr}^b$. The same result does not hold for $\mathrm{Pr} \ll 1$, where the slopes change considerably; there a correlation of the form $\mathrm{Nu}_D = C_1 + C_2(\mathrm{Re}_D\,\mathrm{Pr})^n$ is most successful.

The effort expended in achieving a quantitative understanding of eddy diffusivity for heat and momentum now can pay a handsome dividend. Because $\varepsilon_H$ and $\varepsilon_m$ are little affected by thermal boundary conditions, the $\varepsilon_H$ and $\varepsilon_m$ obtained for the constant-wall-heat-flux case can be directly applied to calculate, albeit numerically, the effects of other thermal boundary conditions. There are applications where the heat flux is nonuniform around the round pipe, as in a solar collector coolant tube. General solutions have been developed by Reynolds [14] of which only those applicable to a cosine variation $q_w(\phi) = q_{w_0}(1 + b\cos\phi)$, are presented here. For this case the local Nusselt number is given by

$$\frac{\mathrm{Nu}_D(\phi)}{\mathrm{Nu}_D} = \frac{1 + b\cos\phi}{1 + (S_1 b\,\mathrm{Nu}_D/2)\cos\phi}$$

The $S_1$ parameter is given in Table 11-2 as a function of Reynolds and Prandtl number, and $\mathrm{Nu}_D$ is the Nusselt number for a uniform peripheral temperature from Table 11-1 or Fig. 11-3.

Table 11-2   Circumferential Heat Flux Functions $S_1$ for Fully Developed Turbulent Flow In a Circular Tube with Constant Heat Rate [14]

| Pr \ Re | Laminar | $10^4$ | $3 \times 10^4$ | $10^5$ | $3 \times 10^5$ | $10^6$ |
|---|---|---|---|---|---|---|
| 0 | 1.000 | 1.000 | 1.000 | 1.000 | 1.000 | 1.000 |
| 0.001 | | 1.000 | 1.000 | 0.999 | 0.974 | 0.901 |
| 0.003 | | 0.999 | 0.994 | 0.957 | 0.831 | 0.473 |
| 0.01 | | 0.991 | 0.952 | 0.733 | 0.409 | 0.161 |
| 0.03 | | 0.923 | 0.699 | 0.348 | 0.145 | 0.0535 |
| 0.7 | | 0.121 | 0.0490 | 0.0180 | 0.00721 | 0.00275 |
| 3 | | 0.0448 | 0.0178 | 0.00629 | 0.00246 | 0.000902 |
| 10 | | 0.0239 | 0.00931 | 0.00322 | 0.00123 | 0.000438 |
| 30 | | 0.0151 | 0.00582 | 0.00199 | 0.00751 | 0.000166 |
| 100 | | 0.00994 | 0.00383 | 0.00130 | 0.000486 | 0.000166 |
| 1000 | | 0.00513 | 0.00198 | 0.000667 | 0.000248 | 0.0000841 |

## 11.3   FULLY DEVELOPED NUSSELT NUMBER FOR OTHER GEOMETRIES WITH SPECIFIED WALL HEAT FLUX

The Nusselt numbers for fully developed turbulent flow in long ducts of annular cross section and constant wall heat flux have been calculated by Kays and Leung [8]. They found excellent agreement with measurements for air,

Table 11-3   Nusselt Numbers for Fully Developed Turbulent Flow Between Parallel Planes; Constant Heat Rate; One Side Heated and the Other Side Insulated [8]

| Pr \ Re | Laminar | $10^4$ | $3 \times 10^4$ | $10^5$ | $3 \times 10^5$ | $10^6$ |
|---|---|---|---|---|---|---|
| 0.0 | 5.385 | 5.70 | 5.78 | 5.80 | 5.80 | 5.80 |
| 0.001 | | 5.70 | 5.78 | 5.80 | 5.88 | 6.23 |
| 0.003 | | 5.70 | 5.80 | 5.90 | 6.32 | 8.62 |
| 0.01 | | 5.80 | 5.92 | 6.70 | 9.80 | 21.5 |
| 0.03 | | 6.10 | 6.90 | 11.0 | 23.0 | 61.2 |
| 0.5 | | 22.5 | 47.8 | 120 | 290 | 780 |
| 0.7 | | 27.8 | 61.2 | 155 | 378 | 1030 |
| 1.0 | | 35.0 | 76.8 | 197 | 486 | 1340 |
| 3.0 | | 60.8 | 142 | 380 | 966 | 2700 |
| 10.0 | | 101 | 214 | 680 | 1760 | 5080 |
| 30.0 | | 147 | 367 | 1030 | 2720 | 8000 |
| 100.0 | | 210 | 514 | 1520 | 4030 | 12000 |
| 1000.0 | | 390 | 997 | 2880 | 7650 | 23000 |

**Table 11-4  Nusselt Numbers and Influence Coefficients for $r^* = 0.20$; Fully Developed Turbulent Flow in a Circular-Tube Annulus; Constant Heat Rate [8]**

| | Inner Wall Heated | | | | | | | | | | | |
|---|---|---|---|---|---|---|---|---|---|---|---|---|
| Re | Laminar | | $10^4$ | | $3 \times 10^4$ | | $10^5$ | | $3 \times 10^5$ | | $10^6$ | |
| Pr | $Nu_{ii}$ | $Nu_{oo}$ | $Nu_{ii}$ | $Nu_{oo}$ | $Nu_{ii}$ | $Nu_{oo}$ | $Nu_{ii}$ | $Nu_{oo}$ | $Nu_{ii}$ | $Nu_{oo}$ | $Nu_{ii}$ | $Nu_{oo}$ |
| 0.0 | 8.50 | 4.88 | 8.40 | 5.83 | 8.30 | 5.92 | 8.30 | 6.10 | 8.30 | 6.16 | 8.30 | 6.35 |
| 0.001 | | | 8.40 | 5.83 | 8.40 | 5.92 | 8.30 | 6.10 | 8.40 | 6.30 | 8.90 | 6.92 |
| 0.003 | | | 8.40 | 5.83 | 8.40 | 6.00 | 8.50 | 6.22 | 9.05 | 6.90 | 12.5 | 10.2 |
| 0.01 | | | 8.50 | 5.95 | 8.60 | 6.20 | 9.70 | 7.40 | 14.0 | 11.4 | 33.6 | 24.6 |
| 0.03 | | | 9.00 | 6.22 | 10.1 | 7.55 | 15.8 | 12.7 | 31.7 | 26.3 | 81.0 | 80.0 |
| 0.5 | | | 31.2 | 22.5 | 64.0 | 51.5 | 157 | 130 | 370 | 310 | 980 | 823 |
| 0.7 | | | 38.6 | 29.4 | 79.8 | 64.3 | 196 | 165 | 473 | 397 | 1270 | 1070 |
| 1.0 | | | 46.8 | 35.5 | 99.0 | 80.0 | 247 | 206 | 600 | 504 | 1640 | 1390 |
| 3.0 | | | 77.4 | 60.0 | 175 | 145 | 465 | 390 | 1150 | 980 | 3250 | 2760 |
| 10.0 | | | 120 | 98.0 | 290 | 243 | 800 | 680 | 2050 | 1750 | 6000 | 4980 |
| 30.0 | | | 172 | 142 | 428 | 360 | 1210 | 1030 | 3150 | 2700 | 9300 | 7850 |
| 100.0 | | | 243 | 205 | 617 | 520 | 1760 | 1500 | 4630 | 4000 | 13800 | 12000 |
| 1000.0 | | | 448 | 380 | 1140 | 980 | 3280 | 2830 | 8800 | 7500 | 26000 | 22500 |

Table 11-5　Nusselt Numbers and Influence Coefficients for $r^* = 0.50$; Fully Developed Turbulent Flow in a Circular-Tube Annulus; Constant Heat Rate [8]

| | | | | Inner Wall Heated | | | | | | | | |
|---|---|---|---|---|---|---|---|---|---|---|---|---|
| Re | Laminar | | $10^4$ | | $3 \times 10^4$ | | $10^5$ | | $3 \times 10^5$ | | $10^6$ | |
| Pr | $Nu_{ii}$ | $Nu_{00}$ | $Nu_{ii}$ | $Nu_{00}$ | $Nu_{ii}$ | $Nu_{00}$ | $Nu_{ii}$ | $\bar{Nu}_{00}$ | $Nu_{ii}$ | $Nu_{00}$ | $Nu_{ii}$ | $Nu_{00}$ |
| 0.0 | 6.18 | 5.04 | 6.28 | 5.66 | 6.30 | 5.78 | 6.30 | 5.80 | 6.30 | 5.83 | 6.30 | 5.95 |
| 0.001 | | | 6.28 | 5.66 | 6.30 | 5.78 | 6.30 | 5.80 | 6.40 | 5.92 | 6.75 | 6.40 |
| 0.003 | | | 6.28 | 5.66 | 6.30 | 5.78 | 6.40 | 5.85 | 6.85 | 6.45 | 9.40 | 9.00 |
| 0.01 | | | 6.37 | 5.73 | 6.45 | 5.88 | 7.30 | 6.80 | 10.8 | 10.3 | 23.2 | 22.6 |
| 0.03 | | | 6.75 | 6.03 | 7.53 | 7.05 | 12.0 | 11.6 | 24.9 | 24.4 | 65.5 | 64.0 |
| 0.5 | | | 24.6 | 22.6 | 52.0 | 49.8 | 130 | 125 | 310 | 298 | 835 | 795 |
| 0.7 | | | 30.9 | 28.3 | 66.0 | 62.0 | 166 | 158 | 400 | 380 | 1080 | 1040 |
| 1.0 | | | 38.2 | 34.8 | 83.5 | 78.0 | 212 | 200 | 520 | 490 | 1420 | 1340 |
| 3.0 | | | 66.8 | 60.5 | 152 | 144 | 402 | 384 | 1010 | 960 | 2870 | 2730 |
| 10.0 | | | 106 | 100 | 260 | 246 | 715 | 680 | 1850 | 1750 | 5400 | 5030 |
| 30.0 | | | 153 | 143 | 386 | 365 | 1080 | 1030 | 2850 | 2700 | 8400 | 8000 |
| 100.0 | | | 220 | 207 | 558 | 530 | 1600 | 1500 | 4250 | 4000 | 126000 | 12000 |
| 1000.0 | | | 408 | 387 | 1040 | 990 | 3000 | 2830 | 8000 | 7600 | 24000 | 23000 |

Pr $\approx$ 1, and believe the calculations to also be accurate for Pr $\gg$ 1 and Pr $\ll$ 1. The results for the limiting case of parallel planes, $r_i/r_0 = 1$, are shown in Table 11-3. For values of $r^* = r_i/r_0 = 0.2$ and 0.5, respectively, Tables 11-4 and 11-5 should be consulted; the symbols $Nu_{ii}$ and $Nu_{00}$ refer to heat applied uniformly only at the inner and outer surface, respectively. The results shown in Tables 11-3–11-5 can be used with influence coefficients (provided in the original work or in the discussion by Kays [15]) to find the effect of any ratio of wall heat fluxes. Comparison of the results in Table 11-3 and 11-5 suggests that the parallel-plane results give good estimates of Nu for annuli of $r^* > 0.5$. Sutherland and Kays [16] executed calculations to ascertain the effect of an arbitrary variation of heat flux around the periphery of either surface. The characteristic length for the Nusselt and Reynolds numbers is the hydraulic diameter in Tables 11-3–11-8.

The effect of the almost always present eccentricity of tube center lines in an annulus was studied analytically by Deissler and Taylor [17], who also give the friction factor for eccentric tubes, and experimentally by Leung et al. [18]. The analysis neglected the probable existence of eccentricity-induced secondary flow, and so only the experimental results are presented in Table 11-6. These results, in which the peripheral conduction through duct walls that has a smoothing effect on Nu was made negligible, suggest that an eccentricity decreases the average Nusselt number.

The effect of peripheral conduction through the wall was studied by Eckert and Low [19] and Baughn [76].

Generally speaking, fully developed turbulent heat-transfer coefficients in noncircular ducts are accurately predicted by use of the hydraulic diameter $D_h$ as the characteristic dimension in the circular tube predictive equations just as

**Table 11-6   Effect of Eccentricity on Turbulent-Flow Heat Transfer in Circular-Tube Annuli (Experimental Data [18])[a]**

| Radius Ratio | $e^*$ | $\dfrac{Nu_{ii,\max}}{Nu_{ii,\text{conc}}}$ | $\dfrac{Nu_{ii,\min}}{Nu_{ii,\text{conc}}}$ | $\dfrac{Nu_{00,\max}}{Nu_{00,\text{conc}}}$ | $\dfrac{Nu_{00,\min}}{Nu_{00,\text{conc}}}$ |
|---|---|---|---|---|---|
| 0.255 | 0.27 | 0.99 | 0.97 | 1.02 | 0.93 |
|  | 0.50 | 0.94 | 0.92 | 0.98 | 0.86 |
|  | 0.77 | 0.92 | 0.88 | 0.93 | 0.77 |
| 0.500 | 0.54 | 0.96 | 0.87 | 1.01 | 0.78 |
|  | 0.77 | 0.87 | 0.67 | 0.88 | 0.62 |

*Source*: By permission from W. M. Kays and M. E. Crawford, *Convective Heat and Mass Transfer*, McGraw-Hill, New York, 1980.
[a]*Notes*: $e^* = e/(r_0 - r_i)$ where $e$ is the eccentricity of the tube center lines. $Nu_{ii,\text{conc.}}$ and $Nu_{00,\text{conc.}}$ refer to the Nusselt numbers for the concentric annulus at the same Reynolds number and Prandtl number.

is the case for pressure drop calculations. Here the hydraulic diameter is defined as

$$D_h = 4 \frac{\text{cross-sectional area for flow}}{\text{wetted perimeter}}$$

For a circle of diameter $D$, $D_h = D$. This approach is of good accuracy for $\text{Pr} \gtrsim 0.5$ (it can break down in such geometries as a triangle where the corners can be substantially filled with laminar flow even though the central portion remains highly turbulent) but is inaccurate at low Prandtl numbers. As pointed out in Section 11.2, the physical basis for this Prandtl number effect is that at high Pr the major thermal resistance occurs in the thin sublayers next to the wall, which are little affected by the duct shape. At low Pr the thermal resistance of the central turbulent region is an important part of the total thermal resistance, and this region is strongly affected by the duct shape. Turbulent flow at low Pr is, therefore, similar to laminar flow and neither admits the use of a hydraulic diameter to correlate results for different geometries.

## 11.4   FULLY DEVELOPED NUSSELT NUMBER IN CIRCULAR DUCT FOR CONSTANT SURFACE TEMPERATURE

The case of fully developed turbulent flow in a long smooth circular duct of constant wall temperature occurs frequently. It is necessary to obtain the heat-transfer coefficient for this case and, additionally, to be aware of the difference caused by a constant wall temperature rather than a constant wall heat-flux boundary condition.

The iterative method utilized for the constant wall temperature case with laminar flow as discussed in connection with Eq. (6-41) can also be used for turbulent flow. In such a procedure for the turbulent flow case, temperature profiles from the constant wall heat flux case are repetitively integrated until the resultant temperature profile and Nusselt number converge. Seban and Shimazaki [20] used such a procedure. Alternatively, the solution for the thermal entry region with fully developed flow such as was developed by Sleicher and Tribus [21] can be used to determine the constant wall temperature limiting Nusselt numbers. From their results the ratio of constant wall heat flux to constant wall temperature Nusselt numbers is found to be as displayed in Fig. 11-4.

The constant wall heat flux and constant wall temperature results are essentially the same for most gases and liquids ($\text{Pr} \geqslant 0.5$). It is only for the low Prandtl number fluids, such as liquid metals, that a significant difference is found. In all cases the difference diminishes with increasing Reynolds number.

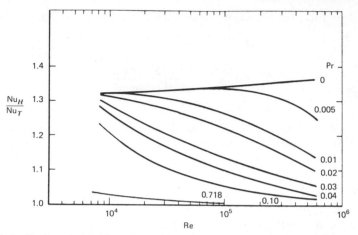

**Figure 11-4** Ratio of the Nusselt number for constant wall heat flux to that for constant wall temperature for fully developed turbulent flow in a circular duct. (By permission from C. A. Sleicher and M. Tribus, *Heat Transfer and Fluid Mechanics Institute*, Stanford University Press, (Stanford, CA,) 1956, pp. 59–78 [21].)

The physical basis for these results is as mentioned in connection with Fig. 11-1. At large Prandtl numbers, the thermal resistance mostly occurs in the sublayers near the wall, where molecular effects are important. At small Prandtl numbers, molecular effects in the central core region make the thermal resistance there a large part of the total. As Reynolds number and the turbulence level in the central core increases, molecular effects in the central core make the thermal resistance there a lesser part of the total; so, at high Reynolds numbers the ratio of the two Nusselt numbers approaches unity for all values of the Prandtl number but does so more slowly for low Prandtl number. As a consequence, the Nusselt number for a large Prandtl number fluid is rather insensitive to axial variation of either wall heat flux or wall temperature.

## 11.5  CORRELATING EQUATIONS FOR FULLY DEVELOPED FLOW IN ROUND DUCTS

Correlating equations that represent masses of experimentally or analytically obtained results are extremely convenient. The many that have been proposed have been reviewed by Sleicher and Rouse [22], who recommend, for $0.1 < \text{Pr} < 10^5$ and $10^4 < \text{Re}_D < 10^6$,

$$\text{Nu}_{D,m} = 5 + 0.015 \, \text{Re}_{D,f}^a \text{Pr}_w^b \qquad (11\text{-}17a)$$

where

$$a = 0.88 - \frac{0.24}{4 + Pr_w}$$

$$b = \frac{1}{3} + 0.5 \exp(-0.6 \, Pr_w)$$

and the subscripts $m$, $f$ and $w$ designate property evaluation at the mixing-cup, film $[T_f = (T_w + T_m)/2]$, and wall temperatures, respectively. Data is represented within 10% by Eq. (11-17a). For gases $(0.6 < Pr < 0.9)$, this relation is accurately approximated by

$$Nu_{D,m} = 5 + 0.012 \, Re_{D,f}^{0.83} (Pr_w + 0.29) \qquad (11\text{-}17b)$$

For oils and viscous fluids $(Pr > 50)$, this relation is accurately approximated by

$$Nu_{D,m} = 0.015 Re_{D,f}^{0.88} \, Pr_w^{1/3} \qquad (11\text{-}17c)$$

For liquid metals $(Pr < 0.1)$ it is tentatively recommended that

$$Nu_{D,m} = 4.8 + 0.0156 \, Re_{D,f}^{0.85} \, Pr_w^{0.93}, \qquad T_w = \text{const} \qquad (11\text{-}17d)$$

$$= 6.3 + 0.0167 \, Re_{D,f}^{0.85} \, Pr_w^{0.93}, \qquad q_w = \text{const} \qquad (11\text{-}17e)$$

which gives results within 10% of measurements.

Additional discussion is given by Kays [15], Notter and Sleicher [23], and Sleicher et al. [24]. In general, Eqs. (11-17) are based on a combination of theory (such as previously adduced) and measurement.

The Lyon–Martinelli equation [25]

$$Nu_D = 7 + 0.025(Re_D \, Pr)^{0.8}, \qquad q_w = \text{const}$$

has been used for some time but gives results that are 30–70% higher than measurements, possibly due to the nonwetting of some solids by some liquid metals and the effect of impurities. Useful information on liquid metals is given in the handbook by Lyon [26] and in the review by Stein [27]. Since flow in the first portion of the entrance region of a duct behaves much like a boundary layer on a flat plate, the result of Problem 7-13

$$\lim_{Pr \to 0} Nu_x = \left( \frac{Re_x \, Pr}{\pi} \right)^{1/2}$$

is applicable there.

A common correlating equation for long circular ducts for the usually encountered liquids and gases is the Dittus–Boelter equation [28]

$$\mathrm{Nu}_D = 0.023\,\mathrm{Re}_D^{0.8}\,\mathrm{Pr}^n \qquad (11\text{-}17\mathrm{f})$$

where $n = 0.4$ for heating and $n = 0.3$ for cooling. This equation is reasonably accurate for $10^4 < \mathrm{Re}_D < 1.2 \times 10^5$, $0.7 < \mathrm{Pr} < 120$, and when the wall temperature does not exceed the fluid mixing-cup temperature by more than $10°\mathrm{F}$ for liquids or $100°\mathrm{F}$ for gases. Fluid properties are evaluated at the average local film temperature, $T_f = (T_m + T_w)/2$. Nearly as common is the Sieder–Tate equation [29], which is applicable to fluids whose viscosity is sensitive to temperature

$$\mathrm{Nu}_{D,\,m} = 0.023\,\mathrm{Re}_{D,\,m}^{0.8}\,\mathrm{Pr}_m^{1/3}\left(\frac{\mu_m}{\mu_w}\right)^{0.14}$$

This equation is reasonably accurate for $\mathrm{Re}_D > 10^4$ and $0.7 \leqslant \mathrm{Pr} \leqslant 16{,}700$. Here fluid properties are evaluated at the local mean temperature except for the last term in which viscosity is evaluated at the mean fluid temperature $\mu_m$ and the wall temperature $\mu_w$. Equation (11-17f) gives results that can be 20% high for gases and as much as 40% low for water at high Reynolds number [22].

## 11.6  THERMAL ENTRY LENGTH

A common entrance-effect case is that in which heating begins in a duct after the velocity profile is fully developed. In such cases the general trends observed for laminar flow would be expected; Nusselt numbers would be large at the start of the heated section and would asymptotically approach the limiting value for long tubes farther down the duct.

Analyses for a circular duct have been executed for a constant wall surface temperature by Sleicher and Tribus [21] and for a constant wall heat flux by Sparrow et al. [7]. Their solutions use the same procedures as for laminar flow but now applied to Eq. (11-6) with refined turbulent contributions to the heat flux as discussed in Section 11.2. The resulting eigenvalues and eigenfunctions were numerically determined and can be used in an evaluation of entrance-region effects as is done for laminar flow. Table 11-7 presents some of the results collected by Kays [15] that can be used in Eq. (6-47a) for the constant wall temperature case. Equation (6-47a) gives the Nusselt number based on log–mean-temperature difference (LMTD) as

$$\overline{\mathrm{Nu}}_D = \frac{\lambda_0^2}{2} + \frac{\mathrm{Re}_D\,\mathrm{Pr}}{(x/D)}\ln\frac{\lambda_0^2}{8G_0} \qquad (6\text{-}47\mathrm{a})$$

**Table 11-7** **Some Function Values for the Thermal Entry Length in Turbulent Flow Through a Circular Tube [15]**

| Pr | $\mathrm{Re}_D$ | $\lambda_0^2$ | $G_0$ |
|----|------|-----|-----|
| 0.01 | $5 \times 10^4$ | 11.7 | 1.11 |
|  | $10^5$ | 13.2 | 1.3 |
|  | $2 \times 10^5$ | 16.9 | 1.7 |
| 0.7 | $5 \times 10^4$ | 235 | 28.6 |
|  | $10^5$ | 400 | 49.0 |

Recognition that $\lambda_0^2/2 = \overline{\mathrm{Nu}}_D(x = \infty)$ and rearrangement then gives

$$\frac{\overline{\mathrm{Nu}}_D(x)}{\overline{\mathrm{Nu}}_D(x = \infty)} = 1 + \frac{\mathrm{Re}_D \, \mathrm{Pr}}{(x/D) \, \lambda_0^2} \frac{2}{\lambda_0^2} \ln \frac{\lambda_0^2}{8G_0}$$

From Table 11-7 at $\mathrm{Re}_D = 5 \times 10^4$ and $\mathrm{Pr} = 0.7$ it is found that $\lambda_0^2 = 235$ and $G_0 = 28.6$. Substitution of these values into the equation written above gives

$$\frac{\overline{\mathrm{Nu}}_D(x)}{\overline{\mathrm{Nu}}_D(x = \infty)} = 1 + \frac{2}{x/D}$$

which suggests that entrance region effects be accounted for by a correction of the form

$$\frac{\overline{\mathrm{Nu}}_D(x)}{\overline{\mathrm{Nu}}_D(x = \infty)} = 1 + C_1 \left(\frac{D}{x}\right)^{C_2}$$

Such a procedure is commonly used but cannot be generally accurate because Table 11-7 indicates that the correction depends on both $\mathrm{Re}_D$ and Pr. The general behavior to be expected for local Nusselt numbers is displayed in Fig. 11-5 for constant wall heat flux. There it is seen that the entrance region effect is particularly pronounced at small Pr, but is of small importance for large Pr. Also noteworthy is the near independence of entrance-region effects on $\mathrm{Re}_D$ for $\mathrm{Pr} \approx 1$.

Eigenvalue solutions for turbulent flow in the entrance region between parallel planes was studied by Hatton and Quarmby [30]. The eigenvalue solutions they achieved were for one surface insulated and the other surface of either constant temperature or constant heat flux. Their local Nusselt number results for the one-surface-insulated, one-surface-at-constant-heat-flux case are presented in Table 11-8 as collected by Kays and Crawford [15]. In their study the velocity profile was fully developed and the entrance region was a purely thermal one. It is surprising that the Nusselt number at $\mathrm{Pr} = 1$ shows a substantial Reynolds number dependence whereas the circular duct result does not. These results are consistent with the fully developed results in Table 11-3. More recently numerical methods were employed by Emery and co-workers for the entrance regions of parallel plate ducts [74], utilizing a mixing length turbulence model and square ducts [75] utilizing an algebraic stress model. Numerical methods with a two-equation ($k$–$\varepsilon$) representation of turbulence have also been applied to variable density flow in ducts [80].

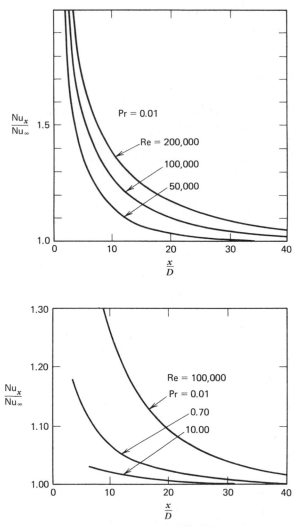

**Figure 11-5** Nusselt numbers in the thermal entry region of a circular duct for constant wall heat flux and turbulent flow with ($a$) Pr $= 0.01$ at various Reynolds numbers, ($b$) Re $= 10^5$ at various Prandtl numbers, and ($c$) Pr $= 0.7$ at various Reynolds numbers. (By permission from W. M. Kays and M. E. Crawford, *Convective Heat And Mass Transfer*, McGraw-Hill, New York, 1980 [15].)

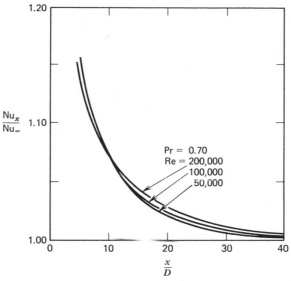

**Figure 11-5** (*Continued.*)

**Table 11-8   Nusselt Numbers for Turbulent Flow Between Parallel Planes; One Side Heated and the Other Side Insulated [15]**

| | $Nu_{11}$ | | |
|---|---|---|---|
| $x/D_h$ | $Re = 7,096$ | $Re = 73,612$ | $Re = 494,576$ |
| | | $Pr = 0.1$ | |
| 1 | 19.7 | 75.2 | 241 |
| 3 | 14.3 | 56.2 | 194 |
| 10 | 10.7 | 42.4 | 155 |
| 30 | 9.44 | 34.8 | 132 |
| 100 | 9.34 | 32.1 | 120 |
| | | $Pr = 1.0$ | |
| 1 | 47.3 | 234 | 940 |
| 3 | 37.9 | 203 | 851 |
| 10 | 31.5 | 177 | 761 |
| 30 | 28.0 | 160 | 697 |
| 100 | 27.1 | 152 | 661 |
| | | $Pr = 10$ | |
| 1 | 102 | 602 | 2925 |
| 3 | 88.6 | 575 | 2829 |
| 10 | 81.9 | 550 | 2724 |
| 30 | 78.6 | 532 | 2640 |
| 100 | 77.5 | 522 | 2590 |

*Source:* By permission from W. M. Kays and M. E. Crawford, *Convective Heat and Mass Transfer*, McGraw-Hill, New York, 1980.

In many applications the velocity and temperature profiles develop simultaneously, or nearly so. Such a case is found, for example, in heat exchangers where a coolant enters tubes through a sudden contraction from a header. Analyses have been performed, many of them using integral methods, but primary reliance is placed on measurements because of the complexities of the entrance conditions of interest (consult Hartnett [31], Latzko [32], Deissler [33], and Boelter et al. [34]). The recommended [2] modifications for the average heat-transfer coefficient, on the basis of experiments with fluids such as air for which $\mathrm{Pr} \approx 1$, are

$$\frac{\bar{h}}{h_\infty} = 1 + \frac{C}{L/D}, \qquad \frac{L_c}{D} < \frac{L}{D} < 60$$

$$= 1.11 \frac{\mathrm{Re}_D^{1/5}}{(L/D)^{4/5}}, \qquad \frac{L}{D} < \frac{L_c}{D}$$

where $h_\infty$ is the heat-transfer coefficient for an infinitely long pipe and

$$\frac{L_c}{D} = 0.625 \mathrm{Re}_D^{1/4}$$

is the number of pipe diameters required in turbulent flow for the friction factor to become constant, which is substantially less than the 40 or 50 required for the turbulent velocity profile to become fully developed. Table 11-9 gives values of $C$ for some selected inlet configurations and for $26{,}000 < \mathrm{Re}_D < 56{,}000$. The value of $C$ for a fully developed velocity profile is shown in Table 11-9 as 1.4, agreeing substantially with the analytically derived expectation of 2.0. It would be expected that entrance-region effects would be greater for low-Prandtl number fluids and lesser for high-Prandtl number fluids [70]. The scatter of measurements does not warrant distinguishing between constant

**Table 11-9   Values of $C$ for Various Inlets [2]**

| Inlet Configuration | $C$ |
| --- | --- |
| Fully developed velocity profile | 1.4 |
| Bell-mouthed with screen | 1.4 |
| Calming Section | |
| $\quad L/D = 11.2$ | 1.4 |
| $\quad L/D = 2.8$ | 3.0 |
| 45° sharp bend | 5.0 |
| Abrupt contraction | 6.0 |
| 180° round bend | 6.0 |
| 90° right angle bend | 7.0 |

wall temperature and constant wall heat flux conditions in the entrance-region correction.


## 11.7 AXIALLY VARYING SURFACE TEMPERATURE AND HEAT FLUX

When either wall surface temperature or wall heat flux vary along the axis of a duct, it is necessary in principle to account for the resultant inconstancy of the heat-transfer coefficient in the same manner as was done for laminar flow in Chapter 6. For this purpose, the eigenvalues and constants, such as are displayed in abbreviated form in Table 11-7, are needed. (For more extensive tabulations, references [7, 15, and 21] should be consulted.)

The relative importance of axially varying wall surface temperature depends on the Prandtl number. At large Prandtl numbers $1 \lesssim \mathrm{Pr}$, the effects are usually small and can be neglected. At low Prandtl numbers $\mathrm{Pr} \lesssim 0.1$, such as occur with liquid metals, the effects can be large. Figure 11-3 suggests that the effects of axially varying boundary conditions will be important for $\mathrm{Pr} \lesssim 0.03$ but can be neglected for $\mathrm{Pr} \gtrsim 0.7$. Thus cooling of a nuclear reactor with its high heat flux in the center and low heat flux at inlet and exit by use of a liquid metal coolant requires variable heat-flux calculations, whereas use of a gas or pressurized water allows a Nusselt number based on constant heat flux—the local heat flux must be used to calculate local temperature differences. These results are a consequence of the thermal resistance for large Prandtl number fluids being concentrated in the sublayers near the wall that respond quickly to changed wall conditions.


## 11.8 EFFECT OF SURFACE ROUGHNESS

Previous discussions have assumed the wall to be smooth. Manufacturing and operating conditions are often far from ideal, leading to duct walls that are rough. Usually the roughness is small relative to the duct diameter, however.

Rough walls have very little effect on laminar flow. In turbulent flow the roughness elements are found to have little effect if they are small relative to the laminar sublayer since they are then immersed in an essentially laminar flow. When the roughness elements are sufficiently large to protrude above the laminar sublayer up into the turbulent flow, on the other hand, the pressure drop is caused primarily by the pressure drag on the roughness elements rather than by viscous shear. As the pressure drag on an object is very nearly a constant fraction of the square of the velocity, the measured independence of friction coefficient for pressure drop $f = (D/L)(2/\rho U_{\mathrm{av}}^2)\Delta p$, from Reynolds number is plausible on physical grounds.

For rough tubes a practical difficulty is the influence of the shape of roughness elements on the friction factor for pressure drop, $f$ [35]. Nikuradse

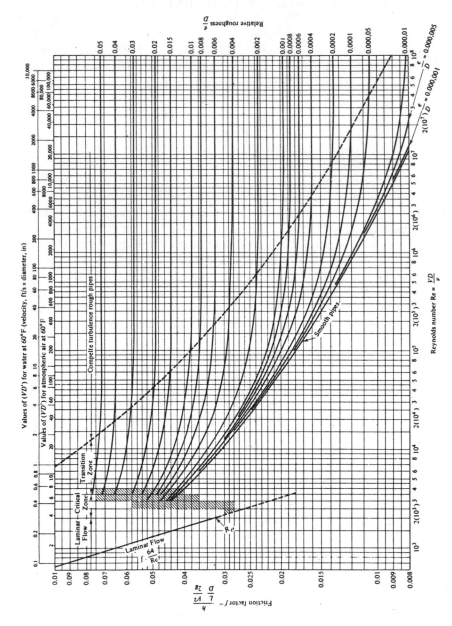

**Figure 11-6** Friction factor $f = (D/L)(2/\rho U_{av}^2)\Delta p$ for pipe flow versus Reynolds number $Re_D = U_{av}D/\nu$. [From L. F. Moody, *Transact. ASME* **66**, 671–684 (1944) [37].]

Table 11-10   Roughness of Commercial Pipes

| Pipe Type | $e$ (roughness), in. |
|---|---|
| Drawn tubing | $6 \times 10^{-5}$ |
| Brass, lead, glass, spun cement | $3 \times 10^{-4}$ |
| Commercial steel, wrought iron | $1.8 \times 10^{-3}$ |
| Cast iron (asphalt dipped) | $4.8 \times 10^{-3}$ |
| Galvanized iron | $6 \times 10^{-3}$ |
| Wood stave | $0.72 - 3.6 \times 10^{-2}$ |
| Cast iron (uncoated) | $1.02 \times 10^{-2}$ |
| Concrete | $1.2 - 12 \times 10^{-2}$ |
| Riveted steel | $3.6 - 36 \times 10^{-2}$ |

[36] glued uniformly sized sand grains as close together as possible on the inside of round pipes and made extensive measurements. Measurements on pipes with commercial roughnesses are summarized by Moody [37] in the Moody diagram in Fig. 11-6. The average height of a roughness element is denoted by $e$, some typical values of which are given in Table 11-10, but it is seen in Fig. 11-6 that the relative roughness $e/D$ is of greatest significance. Thus a small-diameter pipe can be relatively rougher than a large-diameter pipe even though both are manufactured by the same process and from the same material. For commercially rough pipes the correlation of Colebrook [38]

$$\frac{1}{f^{1/2}} = 1.74 - 2\log_{10}\left(2\frac{e}{D} + \frac{18.7}{\mathrm{Re}_D f^{1/2}}\right)$$

can be used in the turbulent range. For sand-roughened pipes the correlation given by Nedderman and Shearer [39]

$$\frac{1}{f^{1/2}} = 1.74 - 2\log_{10}\left(2\frac{e}{D} - \frac{28.3}{\mathrm{Re}_D f^{1/2}}\right)$$

with $\mathrm{Re}_D (e/D)(f/8)^{1/2} > 12$ is required since commercial roughness does not show the dip in the transition region exhibited by sand roughnesses.

An easily programmable relationship between the friction factor $f$ for pressure drop is given by Round [82] for turbulent flow as

$$\frac{1}{f^{1/2}} = 1.8\log_{10}\left(\frac{1}{0.135e/D + 6.5/\mathrm{Re}_D}\right), \qquad 4000 \lesssim \mathrm{Re}_D$$

Another such relationship valid for both laminar and turbulent flow is given by Churchill [83] as

$$\frac{f}{8} = \left[ \left( \frac{8}{\mathrm{Re}_D} \right)^{12} + \frac{1}{(A+B)^{3/2}} \right]^{1/12}$$

where

$$A = \left[ 2.457 \ln \frac{1}{(7/\mathrm{Re}_D)^{0.9} + 0.27e/D} \right]^{16}$$

$$B = \left( \frac{37{,}530}{\mathrm{Re}_D} \right)^{16}$$

Both of these relationships give the friction factor explicitly when the flow rate is specified, but still require a trial and error solution procedure when pressure drop is specified.

  For completeness, it must be mentioned that the pressure drop caused by fittings, such as valves and joints, can be significant, although they are usually referred to as *minor losses*. The loss coefficient $k_L = 2\Delta p / \rho U_{av}^2$ for some common valves and fittings is given in Table 11-11. These values are accurate

**Table 11-11   Loss Coefficients for Various Fittings**

| Fitting | $k_L$ (Loss Coefficient) |
|---|---|
| Angle valve, fully open | 3.1–5.0 |
| Ball check valve, fully open | 4.5–7.0 |
| Gate valve, fully open | 0.19 |
| Globe valve, fully open | 10.0 |
| Swing check valve, fully open | 2.3–3.5 |
| Regular-radius elbow, | |
|   Screwed | 0.9 |
|   Flanged | 0.3 |
| Long-radius elbow | |
|   Screwed | 0.6 |
|   Flanged | 0.23 |
| Close return bend, screwed | 2.2 |
| Flanged return bend, two elbows | |
|   Regular radius | 0.38 |
|   Long radius | 0.25 |
| Standard tee, screwed | |
|   Flow through run | 0.6 |
|   Flow through side | 1.8 |

only for fully developed turbulent flow upstream; if fittings are closer together than the 40–50 diameters required to achieve fully developed flow, the actual loss coefficients will be somewhat smaller.

The effect of pipe roughness on the heat-transfer coefficient is usually considerably less than on the friction factor for pressure drop. The Reynolds analogy, which suggests that $h$ is proportional to $f$ (when $\text{Pr} \approx 1$), is predicated on laminar shear providing flow resistance; this is not the case when roughness elements protrude above the laminar sublayer. A useful parameter for expressing the effect of roughness on heat transfer is $k^+ = e/\delta_L$, where $e$ is the roughness height and $\delta_L$ is the laminar sublayer thickness. As illustrated in Fig. 11-1, the laminar sublayer thickness is given by $y^+ \approx 5$, or

$$\frac{\delta_L v^*}{\nu} \approx 5$$

This relation can be rearranged in terms of the friction factor for pressure drop to give

$$k^+ = \frac{e}{\delta_L} = \text{Re}_D \frac{e}{D} \left(\frac{f}{8}\right)^{1/2}$$

where constants have been ignored. The departure from Reynolds analogy is expressed by the effectiveness parameter $\eta$, where

$$\eta = \frac{\text{Nu}/f}{\text{Nu}_s/f_s}$$

and the subscript $s$ refers to smooth-wall quantities. Burck [40] correlated the measurements of Nunner [41] and of Dipprey and Sabersky [42] as shown in Fig. 11-7. For $\text{Pr} \approx 1$ and commonly encountered turbulent Reynolds numbers, these results are approximately correlated by

$$\frac{\text{Nu}}{\text{Nu}_s} = \left(\frac{f}{f_s}\right)^{1/2}$$

Problem 11-4 illustrates the magnitude of expected roughness effects. The effectiveness $\eta$ displayed in Fig. 11-7 is correlated by

$$\eta = \log_{10}\left[\frac{\text{Pr}^{0.33}}{(k^+)^{0.243}}\right] - 0.32 \times 10^{-3} k^+ \log_{10}(\text{Pr}) + 1.25$$

Additional, but less directly useful, information is presented by Sood and Jonsson [43]. Variable property results are reported by Wassel and Mills [68].

It is expected that roughness would have greatest effects on high-Prandtl number fluids since their thermal resistance is primarily in the sublayers.

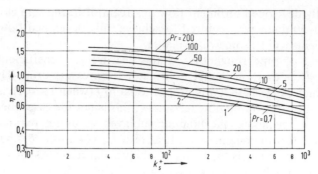

**Figure 11-7** Effectiveness parameter $\eta$ for turbulent heat transfer in tubes with sand roughness as a function of the roughness parameter $k^+$. [By permission from E. Burck, Influence of Prandtl number on heat transfer and pressure drop of artificially roughened channels, *Wärme-und Stoffübertragung* **2**, 87–98 [40] (1969).]

Low-Prandtl number fluids such as liquid metals, where molecular effects predominate throughout, would be little affected by wall roughness, on the other hand.

Recent measurements have been reported by Webb et al. [44] on the effect of discrete roughness elements. Schlichting [35, p. 667] can be consulted for roughness effects on turbulent boundary layers. Spiraled internal fins inclined at an angle with the tube axis cause increase in heat-transfer coefficients and pressure drop, according to Carnavos [72].

For equal pumping power and heat flow, 50% savings in pipe material are possible [73]. Carnavos' correlations are

$$\frac{\mathrm{Nu}}{0.023^{0.8}\,\mathrm{Pr}^{0.4}} = F$$

for heat transfer and

$$\frac{f\,\mathrm{Re}^{0.2}}{0.046} = F^*$$

for pressure drop where

$$F = \left[\frac{D_i}{D_e}\left(1 - 2\frac{e}{D_i}\right)\right]^{-0.2}\left(\frac{D_i D_h}{D_e^2}\right)^{0.5}\sec^3\alpha$$

$$F^* = \frac{D_e}{D_i}\sec^{3/4}\alpha$$

where $\alpha$ is helix angle from tube axis in degrees, $D_i$ is diameter to fin base, $D_e$ is tube diameter with fins melted down, $e$ is fin height, $D_h$ is hydraulic diameter,

$Re = D_h G / \mu$ where $G$ is mass velocity, $Nu = h D_h / k$, and sec represents the secant trigonometric function.

## 11.9  CURVED DUCTS

In curved pipes there is a secondary flow because fluid particles near the pipe axis have a higher velocity and are acted on by a larger centrifugal force than are the slower particles near the walls. The resultant secondary flow is directed outward in the center and inward (toward the center of curvature of the pipe bend) near the wall as shown in Fig. 11-8.

Curvature has a stronger effect on pressure drop in laminar than in turbulent flow. For laminar flow, the analyses by Dean [45] and Adler [46] suggest that the dimensionless Dean number De controls where

$$De = \frac{U_{av} D}{\nu} \frac{(D/D_c)^{1/2}}{2}$$

$$= \frac{Re_D}{2} \left( \frac{D}{D_c} \right)^{1/2}$$

with $D$ the pipe diameter and $D_c$ the diameter of the pipe curvature. Measurements are reasonably correlated by the simple relationship, according to Schlichting [35, p. 590], who can be consulted for other discussion and references concerning secondary flows in ducts of various cross sections,

$$\frac{f}{f_{straight}} = 0.37(De)^{0.36}, \qquad \frac{10^{1.6}}{2} < De < 500$$

originally suggested by Prandtl or by the more complex correlation due to White [47]

$$\frac{f}{f_{straight}} = \left\{ 1 - \left[ 1 - \left( \frac{5.8}{De} \right)^{0.45} \right]^{2.22} \right\}^{-1}, \qquad 5.8 < De < 1000 \quad (11\text{-}18)$$

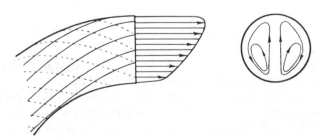

**Figure 11-8**  Flow pattern in a curved pipe.

For the augmentation of the laminar heat-transfer coefficient caused by curvature-induced secondary flow, Seban and McLaughlin [48] give

$$\frac{\text{Nu}_D}{\text{Pr}^{1/3}} = 0.13 \left( \frac{f\,\text{Re}_D^2}{8} \right)^{1/3}$$

with $f$ found from Eq. (11-18). Numerical procedures have recently been applied [69].

In turbulent flow the Dean number does not correlate measurements well. Instead, the correlation due to White [49] for the pressure-drop friction factor is

$$\frac{f}{f_{\text{straight}}} = 1 + 0.075\text{Re}_D^{1/4} \left( \frac{D}{D_c} \right)^{1/2}$$

More recently Ito [50] gave the correlation

$$\frac{f}{f_{\text{straight}}} = \left[ \text{Re}_D \left( \frac{D}{D_c} \right)^2 \right]^{1/20}, \qquad \text{Re}_D \left( \frac{D}{D_c} \right)^2 > 6 \qquad (11\text{-}19)$$

He also found that the critical value of Reynolds number above which fully turbulent flow exists in a pipe is given by

$$\text{Re}_{D,\,\text{critical}} = 2 \times 10^4 \left( \frac{D}{D_c} \right)^{0.32}$$

Seban and McLaughlin [48] found that Reynolds analogy holds for turbulent flow in curved pipes so that

$$\frac{\text{Nu}_D}{\text{Re}_D\,\text{Pr}^{0.4}} = \frac{f}{8}$$

with $f$ found from Eq. (11-19). This is only slightly different from the result found by Jeschke [51] for air

$$\frac{h}{h_{\text{straight}}} = 1 + 3.5\frac{D}{D_c}$$

and recommended by McAdams [52]. Mori and Nakayama [53] should be consulted for detailed discussion and general predictive equations regarding heat transfer and pressure drop for laminar and turbulent flow in curved pipes for constant-temperature and constant-heat-flux boundary conditions.

A related situation is that of a duct rotating about its central axis. Such rotation affects the flow and heat-transfer characteristics since Coriolis forces act on developing flow fields and centripetal forces act on density nonuniform-ities to produce secondary flows. Such effects are of interest when, for example, rotating electrical generator windings are cooled convectively. Heat-transfer coefficients are increased by as much as 25% according to Morris and Dias [79], who give experimental results for turbulent flow in tubes of square cross-section and cite additional references.

## 11.10   RENEWAL–PENETRATION HYPOTHESIS OF TURBULENCE

Analyses of turbulent heat transfer have been based mostly on the classical eddy diffusivity concept. However, Lawn [54] and Hubbard and Lightfoot [55] have suggested that turbulent transport studies should be initiated from the basis of the renewal–penetration model first suggested by Higbie in 1935 [56] to explain turbulent mass transfer at a fluid–fluid interface, particularly a liquid–gas interface with the gas being absorbed into the liquid. The renewal–penetration model provides an alternative approach and deserves description on that basis as well as on the basis that many of its predictions are correct.

As set forth by Danckwerts [57], the renewal–penetration model hypothe-sises that eddies intermittently move from the turbulent core to the wall and that the predominant transport mechanism from the wall into the eddies is by one-dimensional unsteady molecular diffusion. In a heat-transfer context, the energy equation for a fluid element in contact with the wall is

$$\alpha \frac{\partial^2 T}{\partial y^2} = \frac{\partial T}{\partial t}$$

where $t$ is the elapsed time since contact began and $y$ is the distance from the wall into the fluid. The boundary conditions imposed are

$$T(t = 0, y) = T_i$$

$$T(t, y = 0) = T_w$$

$$T(t, y = \infty) = T_i$$

where $T_i$ is the eddy temperature at the beginning of contact and $T_w$ is the wall temperature. The temperature in the eddy is then given by

$$\frac{T - T_w}{T_i - T_w} = \text{erf} \left[ \frac{y}{(4\alpha t)^{1/2}} \right] \tag{11-20}$$

The effect on the mean temperature $\bar{T}$ of eddies continually moving to and from the wall can be accounted for by use of the age distribution principle

$$\bar{T} = \int_0^\infty T\phi(t) \, dt$$

in which $\phi(t)\, dt$ represents the fraction of eddies in contact with the wall between $t$ and $t + dt$. Danckwerts suggested

$$\phi(t) = \frac{e^{-t/\tau}}{\tau}$$

where $\tau$ is the mean residence time—the form of $\phi(t)$ does not matter greatly [58].

To illustrate the essential idea, note from Eq. (11-20) that the instantaneous heat flux from the wall is

$$q_w = -k \frac{\partial T(y = 0)}{\partial y} = k \frac{T_w - T_i}{(\pi \alpha t)^{1/2}}$$

The average wall heat flux during a contact is then

$$q_{w,\,av} = \frac{1}{\tau} \int_0^\tau q_w(t)\, dt$$

$$= 2k \frac{T_w - T_i}{(\pi \alpha \tau)^{1/2}}$$

Equation of this average wall heat flux to

$$q_w = h(T_w - T_{\text{mixing cup}})$$

then yields

$$h = \left( \frac{4k\rho C_p}{\pi \tau} \right)^{1/2} \frac{T_w - T_i}{T_w - T_m} \tag{11-21}$$

The key to the successful adaptation of the renewal–penetration model is a reasonable expression for the mean frequency of renewal $1/\tau$. The renewal–penetration model is readily adapted to scraped-surface heat exchangers by merely setting $1/\tau$ equal to the rotational frequency of the blade [59, 77]. Einstein and Li [60] have developed a formulation for $1/\tau$ for turbulent tube flow. Developments such as Einstein and Li's have been pursued to also take into account the fact that some eddies do not reach the surface and have been attended by some success [61, 62]. The influence of pressure gradient on heat-transfer coefficients for fully developed turbulent flow in a pipe was studied by Gross and Thomas [63]. Other applications have been reported [64–66]; the study by Chung et al. [67] for turbulent flow in an annulus can be consulted for additional literature citations. Thomas [78] has explained the bursts of turbulence at a wall mentioned in the discussion of Fig. 9-9. The renewal–penetration mechanism predicts within a factor of 2 the heat-transfer coefficient at the rear of a cylinder in cross flow by [81]

$$Nu_D = 0.5(Pr\, Re_D)^{1/2}$$

where other simple theories are difficult to apply.

## PROBLEMS

**11-1**  Perform the calculations on which Fig. 11-1 is based.

(a) Show that wall shear stress $\tau_w$ and pressure drop $\Delta_p$ are related by

$$\tau_w = -\frac{\Delta p}{4}\frac{D}{L}$$

(b) Introduce the relationship of Eq. (9-21a) into the result of part a and rearrange to find

$$r_0^+ = \frac{r_0 v^*}{\nu} = 0.143\, Re_D^{7/8}$$

for a smooth wall.

(c) Recognizing that $y^+ = r_0^+\, y/r_0$, plot $\varepsilon_m/\nu$ against $y/r_0$ for $Re_D = 10^5$. In the central region of the pipe show the results for the $\frac{1}{7}$th power-law velocity distribution from Eq. (11-4), the approximation to Reichardt's velocity distribution from Eq. (11-5), and the very accurate velocity distribution proposed by Reichardt from Eq. (9-20).

**11-2**  (a) Make use of the relationship between the shear stress coefficient $C_f/2 = \tau_w/(\rho U_{av}^2)$ and the pressure-drop coefficient $f = (2D/L\rho U_{av}^2)\Delta p$, which is $C_f/2 = f/8$ and the Prandtl relationship for $f$ in Eq. (9-21b) to put the Reynolds analogy of Eq. (11-13) into the form

$$\frac{Nu_D}{Re_D\, Pr}$$

$$= \frac{f/2}{3.39 + 5.71(1 + 1.46\{Pr - 1 + \ln[(1 + 5Pr)/6]\})(f/2)^{1/2}}$$

Note that $f$ is more commonly given for pipe flow than is $C_f$.

**(b)** To put the result of part a in a form more familiar to Reynolds analogy, express it in terms of the shear stress coefficient $C_f$ as

$$\frac{\mathrm{Nu}}{\mathrm{Re}_D \mathrm{Pr}}$$

$$= \frac{C_f/2}{0.848 + 2.86(1 + 1.46\{\mathrm{Pr} - 1 + \ln[(1 + 5\mathrm{Pr})/6]\})(C_f/2)^{1/2}}$$

which is nearly of the form $\mathrm{Nu}_D/\mathrm{Re}_D \mathrm{Pr} = C_f/2$ that would have been expected.

**(c)** Compare the prediction of part b with Eq. (11-14) for water ($\mathrm{Pr} = 5$) for $\mathrm{Re}_D$ in the range $10^4$–$10^6$ for a long, smooth circular pipe. Comment on the apparent ability of the simplified theory of Eq. (11-13) to explain the physical processes in this situation.

**(d)** Compare the results of part c against those of the common Dittus–Boelter correlation

$$\mathrm{Nu}_D = 0.023 \, \mathrm{Re}_D^{0.8} \, \mathrm{Pr}^{1/3}$$

**11-3** Consider a long, smooth round pipe around whose periphery the heat flux varies according to $q_w(\phi) = q_{w0}(1 + b\cos\phi)$. For $\mathrm{Re}_D = 10^5$ and $b = 0.2$, determine the maximum and minimum value of $\mathrm{Nu}_D(\phi)$ and the percentage deviation of the average value of $\mathrm{Nu}_D(\phi)$ from $\mathrm{Nu}_D$ for a constant wall heat flux for:

**(a)** $\mathrm{Pr} = 0.01$, characteristic of liquid metals.

**(b)** $\mathrm{Pr} = 0.7$, characteristic of gases.

**(c)** $\mathrm{Pr} = 3.0$, characteristic of water.

Comment on the accuracy that can be expected to attend the use of $\mathrm{Nu}_D$ for constant wall heat flux as an estimate of the average $\mathrm{Nu}_D$ in a varying wall heat flux situation.

**11-4** To illustrate the effect of roughness on the friction factor for pressure drop $f$ and the heat-transfer coefficient, consider fully developed turbulent flow in a round pipe of 1-in.-diameter and made of commercial steel. Air flows such that $\mathrm{Re}_D = 10^5$.

**(a)** From Fig. 11-6 and Table 11-10, verify that $e/D = 2 \times 10^{-3}$ and $f/f_s = 0.025/0.0178 = 1.41$.

**(b)** From Fig. 11-6 verify that $k^+ = 10.1$ and $\eta = 0.9$ so that $h/h_s = \eta(f/f_s) = 1.27$. Compare this result against the simple correlation that $h/h_s = (f/f_s)^{1/2}$.

**(c)** Comment on the relative effect of roughness on pressure drop and heat transfer.

**11-5** Consider fully developed flow in a circular tube with constant heat rate per unit length. The mean flow velocity is 25 ft/sec, and the tube diameter is 1 in. Evaluate the conductance $h$ for the following cases and discuss the reasons for the differences:

(a) Air, 200°F, 1 atm pressure.

(b) Hydrogen gas, 200°F, 1 atm pressure.

(c) Liquid water, 100°F,

(d) Air, 200°F, 10 atm pressure.

**11-6** Consider a 0.5-in.-i.d. (inner diameter) tube 6 ft long wound by an electric resistance heating element. The function of the tube is to heat an organic fluid from 50 to 150°F. The mass flow rate of the fuel is 1000 $lb_m$/hr, and the following average properties may be treated as constant:

$$Pr = 10$$

$$\rho = 47 \; lb_m/ft^3$$

$$C_p = 0.5 \; Btu/lb_m \; °F$$

$$k = 0.079 \; Btu/hr \; ft \; °F$$

$$\mu = 1.6 \; lb_m/hr \; ft$$

Calculate and plot both tube surface temperature and fluid mean temperature as a function of tube length.

**11-7** Liquid potassium flows in a 1-in.-i.d. tube at a mean velocity of 8 ft/sec and a mean temperature of 1200°F. The tube is heated at a constant rate per unit of length, but the heat flux varies around the periphery of the tube in a sinusoidal manner, with the maximum heat flux twice the minimum heat flux. If the maximum surface temperature is 1500°F, evaluate the axial mean temperature gradient and prepare a plot of temperature around the periphery of the tube.

**11-8** Air at 70°F and 1 atm pressure flows at a Reynolds number of 50,000 in a 1-in.-id. circular tube. The tube wall is insulated for the first 30 in., but for the next 50 in. the tube surface temperature is constant at 100°F. Then it abruptly increases to 130°F and remains constant for another 50 in. Plot the local conductance $h$ as a function of axial distance. Does the abrupt increase in surface temperature to 130°F cause a significant change in the average conductance over the entire 100 in. of heated length? In simple heat exchanger theory, a mean conductance with respect to tube length is generally employed. In this case, how much does the mean differ from the asymptotic value of $h$?

**11-9** Starting with the constant-surface-temperature, thermal-entry-length solutions for a circular tube, calculate and compare the Nusselt numbers for constant surface temperature and for a linearly varying surface temperature, very long tubes, a Reynolds number of 50,000, and a Prandtl number of 0.01. Repeat for a Prandtl number of 0.7 and discuss the reasons for the difference noted.

**11-10** The following are the proposed specifications for the cooling tubes in a pressurized-water nuclear-power reactor: *tube configuration*—concentric circular tube annulus, with heating from the inner tube (containing the uranium fuel), and the outer tube surface having no heat flux; *tube dimensions*—inner tube diameter 1 in., outer tube diameter 2 in., tube length 15 ft; *water temperatures*—inlet 525°F, outlet 575°F; *water mean velocity*—3 ft/sec; *axial heat flux distribution*—$\dot{q}''/\dot{q}''_{max} = [1 + 2\sin(\pi x/L)]/3$. Assume that the water properties are constant at 550°F. Calculate and plot heat flux, mean water temperature, and inner and outer tube surface temperature as a function of $x$. Assume that the conductance $h$ is independent of $x$ and that the value for fully developed constant heat rate is a reasonable approximation. Justify these assumptions. Is it possible to avoid boiling with these specifications? What would be the effect of local boiling at the highest temperature parts of the system?

**11-11** Consider fully developed turbulent flow in a circular tube with heat transfer to or from the fluid at a constant rate per unit of tube length. Let there also be internal heat generation (perhaps from nuclear reaction) at a rate $S$, Btu/hr ft$^3$, which is everywhere constant. If the Reynolds number is 50,000 and the Prandtl number is 4, evaluate the Nusselt number as a function of the pertinent parameters. The conductance in the Nusselt number should be defined in the usual manner on the basis of the heat flux at the surface, the surface temperature, and the mixed mean fluid temperature. How would you handle the problem if $S$ were some specified function of tube radius?

**11-12** Water flows in a heat exchanger tube 3 ft long and 1 in. in diameter at a velocity of 15 ft/sec. The tube wall temperature is constant at 210°F as a result of condensing steam on the outside. If the inlet temperature is 60°F, what is the exit temperature? (*Answer*: $T_{out} = 85°F$.)

**11-13** Mercury at an average temperature of 200°F flows through a $\frac{1}{2}$-in.-diameter tube at the rate of $10^4$ lb$_m$/hr. Calculate the average heat transfer coefficient.

**11-14** Power generation in a nuclear reactor is principally limited by the ability to transfer heat from the reactor into a coolant. A solid-fuel reactor is to be cooled by a fluid flowing inside of 0.25-in.-diameter stainless steel tubes. If the tube wall temperature is 600°F, compare the relative merits of the use of water or liquid sodium as the coolant. In each case the velocity is 15 ft/sec and the fluid inlet temperature is 400°F.

**11-15** Reynolds analogy between the coefficients of heat transfer and wall shear can be suggested on the basis of the following simple analysis.

(a) Divide the expression for shear stress

$$\tau = \rho\left(\nu + \frac{\varepsilon_m}{\rho}\right)\frac{du}{dy}$$

by the expression for heat flux

$$q = \rho C_p(\alpha + \varepsilon_H)\frac{dT}{dy}$$

to obtain

$$\frac{\tau C_p}{q} = \frac{\nu + \varepsilon_m/\rho}{\alpha + \varepsilon_H}\frac{du}{dT}$$

(b) Assume that the molecular and turbulent Prandtl numbers are unity to set the result of part a in the form

$$\frac{\tau C_p}{q} = \frac{du}{dT}$$

(c) Recognize that if molecular and turbulent Prandtl numbers are near unity, the velocity and temperature profiles are nearly the same. Then, $du/dT \approx$ const. Replace the left-hand side by wall values and then integrate to obtain

$$\frac{\tau_w C_p}{q_w}(T_{max} - T_{wall}) = (U_{max} - U_{wall})$$

(d) Set $\tau_w = (C_f/2)\rho U_{av}^2$ and $q_w = h(T_{mixing\ cup} - T_{wall})$ in the result of part c to obtain

$$\frac{h}{\rho C_p U_{av}} = \frac{C_f}{2}\left[\frac{U_{av}}{U_{max} - U_{wall}}\frac{T_{max} - T_{wall}}{T_{mixing\ cup} - T_{wall}}\right]$$

which, since the bracketed term must be nearly unity, can be rephrased as $Nu/Re\,Pr \approx C_f/2$ and constitutes Reynolds analogy.

**11-16** Conventionally, local heat-transfer coefficients in ducts are based on the difference between the local wall and mixing-cup temperatures. If

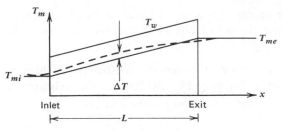

**Figure 11P-16**

axial heat conduction in the fluid is negligible, the mixing-cup temperature varies nearly linearly for constant wall heat flux as shown by solid
lines in Fig. 11P-16. The mixing-cup temperatures at inlet and exit $T_{mi}$
and $T_{me}$, respectively, can be measured directly as can the wall temperature $T_w$. Then the local temperature difference at any axial location is
readily ascertained. In liquid metals the influence of axial conduction
in the fluid requires a correction to the above-mentioned procedure.
The mixing-cup temperature is higher by the amount of $\Delta T$ than would
be expected for negligible heat conduction in the fluid.

**(a)** Make an energy balance on the fluid between a point far upstream
and a point in the midregion of the duct to find

$$\rho C_p U_{av} A(T_m - T_{mi}) = Q + kA \frac{dT_m}{dx}$$

where $Q$ is total heat added to fluid and $A$ is cross-sectional area.
Note that $dT_m/dx \approx (T_{me} - T_{mi})/L$ in the midregion.

**(b)** Show that if axial heat conduction in the fluid were negligible

$$Q = \rho C_p U_{av} A(T_m - \Delta T - T_{mi})$$

**(c)** Substitute the result of part b in that of part a to obtain the desired
temperature correction as

$$\Delta T = \frac{k}{\rho C_p U_{av}} \frac{T_{me} - T_{mi}}{L}$$

$$= (T_{me} - T_{mi}) \frac{D}{L} \frac{1}{\text{Pe}}$$

where $\text{Pe} = \text{Re}_D \text{Pr}$ is the Peclet number.

**(d)** Prepare a brief report of the discussion on this subject presented by
L. Trefethen, Measurement of mean fluid temperatures, *Transact.
ASME* **78**, 1207–1212. (1956).

**11-17**   The form of the Dittus–Boelter equation for turbulent heat transfer in a long straight duct is to be suggested by appealing to Reynolds analogy.

(a) Into Reynolds analogy in the form $\mathrm{Nu}_D/\mathrm{Re}_D\,\mathrm{Pr}^{1/3} = C_f/2$, substitute Eq. (9-21) to obtain

$$\mathrm{Nu}_D = (\mathrm{const})\,\mathrm{Re}_D^{3/4}\,\mathrm{Pr}^{1/3}$$

(b) Discuss the factors involved in the evaluation of the constant in this result.

**11-18**   Equate the heat-transfer coefficients from the renewal–penetration theory of Eq. (11-21) and the Dittus–Boelter correlation of Eq. (11-17f) for the conditions of Problem 11-8 and evaluate the mean frequency of renewal $1/\tau$. It may be assumed that $(T_w - T_i)/(T_w - T_m) = 1$. Compare this result with the range of frequency of turbulent velocity fluctuation mentioned in the introduction section of Chapter 9.

## REFERENCES

1   R. C. Martinelli, Heat transfer to molten metals, *Transact. ASME* **69**, 947–959 (1947).

2   L. M. K. Boelter, R. C. Martinelli, and F. Jonassen, Remarks on the analogy between heat transfer and momentum transfer, *Transact. ASME* **63**, 447–455 (1941).

3   B. S. Petukhov and V. N. Popov, Theoretical calculation of heat exchange and frictional resistance in turbulent flow in tubes of an incompressible fluid with variable physical properties, *Teplofiz. Vysok. Temperatur (High Temperature Heat Physics)* **1** (1), 69–83 (1963).

4   B. S. Petukhov, Heat transfer and friction in turbulent pipe flow with variable physical properties, *Adv. Heat Transf.* **6**, 503–564 (1970).

5   R. L. Webb, A critical evaluation of analytical solutions and Reynolds analogy equations for turbulent heat and mass transfer in smooth tubes, *Wärme-und Stoffübertragung* Vol. 1, Springer-Verlag, Berlin, 1971, pp. 197–204.

6   R. G. Deissler, Analysis of turbulent heat transfer, mass transfer, and friction in smooth tubes at high Prandtl and Schmidt numbers, NACA Report 1210, Washington, DC 1955.

7   E. M. Sparrow, T. M. Hallman, and R. Siegel, Turbulent heat transfer in the thermal entrance region of a pipe with uniform heat flux, *Appl. Sci. Res.* **A7**, 37–52 (1957).

8   W. M. Kays and E. Y. Leung, Heat transfer in annular passages—hydrodynamically developed turbulent flow with arbitrarily prescribed heat flux, *Internatl. J. Heat Mass Transf.* **6**, 537–557 (1963).

9   R. Jenkins, Variation of the eddy conductivity with Prandtl modulus and its use in prediction of turbulent heat transfer coefficients, *Heat Transfer and Fluid Mechanics Institute*, Stanford University Press, Stanford, CA 1951, pp. 147–158.

10   R. G. Deissler, Analysis of fully developed turbulent heat transfer at low Peclet numbers in smooth tubes with application to liquid metals, NACA Research Memo E52F05, 1955.

11   M. Jakob, *Heat Transfer*, Vol. 2, Wiley, New York, 1957, pp. 509–529.

12   J. O. Hinze, *Turbulence*, McGraw-Hill, New York, 1975, pp. 386–391.

13   J. M. Burgers, lecture notes, California Institute of Technology, Pasadena, CA, 1951.

14 W. C. Reynolds, Turbulent heat transfer in a circular tube with variable circumferential heat flux, *Internatl. J. Heat Mass Transf.* **6**, 445–454 (1963).

15 W. M. Kays and M. E. Crawford, *Convective Heat And Mass Transfer*, McGraw-Hill, New York, 1980.

16 W. A. Sutherland and W. M. Kays, Heat transfer in an annulus with variable circumferential heat flux, *Internatl. J. Heat Mass Transf.* **7**, 1187–1194 (1964).

17 R. G. Deissler and M. F. Taylor, Analysis of fully developed turbulent heat transfer and flow in an annulus with various eccentricities, NACA TN 3451, Washington, DC, 1955.

18 E. Y. Leung, W. M. Kays, and W. C. Reynolds, Report AHT-4, Department of Mechanical Engineering, Stanford University, Stanford, CA, April 15, 1962.

19 E. R. G. Eckert and G. M. Low, Temperature distribution in internally heated walls of heat exchangers composed of noncircular flow passages, NACA Report 1022, Washington, DC, 1951.

20 R. A. Seban and T. T. Shimazaki, Heat transfer to a fluid flowing turbulently in a smooth pipe with walls at constant temperature, *Transact. ASME* **73**, 803–809 (1951).

21 C. A. Sleicher and M. Tribus, Heat transfer in a pipe with turbulent flow and arbitrary wall-temperature distribution, in *Heat Transfer and Fluid Mechanics Institute*, Stanford University Press, Stanford, CA, 1956, pp. 59–78.

22 C. A. Sleicher and M. W. Rouse, A convenient correlation for heat transfer to constant and variable property fluids in turbulent pipe flow, *Internatl. J. Heat Mass Transf.* **18**, 677–683 (1975).

23 R. H. Notter and C. A. Sleicher, A solution to the turbulent Graetz problem. III. Fully developed and entry region heat transfer rates, *Chem. Eng. Sci.* **27**, 2073–2093 (1972).

24 C. A. Sleicher, A. S. Awad, and R. H. Notter, Temperature and eddy diffusivity profiles in NaK, *Internatl. J. Heat Mass Transf.* **16**, 1565–1575 (1973).

25 R. N. Lyon, Liquid metal heat-transfer coefficients, *Chem. Eng. Progr.* **47**, 75–79 (1951).

26 R. N. Lyon, Ed. *Liquid Metals Handbook*, 3rd Ed., U.S. Atomic Energy Commission and Department of the Navy, Washington, DC, 1952.

27 R. Stein, Liquid metal heat transfer, *Adv. Heat Transf.* **3**, 101–174 (1966).

28 F. W. Dittus and L. M. K. Boelter, University of California, Berkeley, California, Publications in Engineering **2**, 443 (1930).

29 E. N. Sieder and G. E. Tate, Heat transfer and pressure drop of liquids in tubes, *Indust. Eng. Chem.* **28**, 1429–1435 (1936).

30 A. P. Hatton and A. Quarmby, The effect of axially varying and unsymmetrical boundary conditions on heat transfer with turbulent flow between parallel plates, *Internatl. J. Heat Mass Transf.* **6**, 903–914 (1963).

31 J. P. Harnett, Experimental determination of the thermal entrance length for the flow of water and oil in circular pipes, *Transact. ASME* **77**, 1211–1234 (1955).

32 H. Latzko, Der Warmeübergang an einem turbulenten Flussigkeits—oder Gasstrom, *Zeitschrift fur angewandte Mathematik und Mechanik* **1**, 268–290 (1921).

33 R. G. Deissler, Turbulent heat transfer and friction in the entrance regions of smooth passages, *Transact. ASME* **77**, 1221–1234 (1955); NACA Technical Note 3016.

34 L. M. K. Boelter, G. Young, and H. W. Iverson, An investigation of aircraft heaters—XXVII. Distribution of heat transfer rate in the entrance section of a circular tube, NACA Technical Note 1451, 1948.

35 H. Schlichting, *Boundary-Layer Theory*, 6th ed., McGraw-Hill, New York, 1968, pp. 586–589.

36 J. Nikuradse, Stromungsgesetze in rauhen Rohren, *Forsch. Arb. Ing.—Wes.*, No. 361, 1933.

37 L. F. Moody, Friction factors for pipe flow, *Transact. ASME* **66**, 671–684 (1944).

38  C. F. Colebrook, Turbulent flow in pipes with particular reference to the transition region between the smooth and rough pipe laws, *J. Inst. Civ. Eng.* **11**, 133–156 (1938/1939).

39  R. M. Nedderman and G. J. Shearer, Correlations for the friction factor and velocity profile in the transition region for flow in sandroughened pipes, *Chem. Eng. Sci.* **19**, 423–428 (1964).

40  E. Burck, Influence of Prandtl number on heat transfer and pressure drop of artifically roughened channels, *Wärme-und Stoffübertragung* **2**, 87–98. (1969).

41  W. Nunner, Warmeübergang und Druckabfall in rauhen Rohren, *Verein Deutscher Ingenieure — Forschungsheft*, Nu. 4551, 1956.

42  D. F. Dipprey and R. H. Sabersky, Heat and momentum transfer in smooth and rough tubes at various Prandtl numbers, *Internatl. J. Heat Mass Transf.* **6**, 329–353 (1963).

43  N. S. Sood and V. K. Jonsson, Some correlations for resistances to heat and momentum transfer in the viscous sublayer at rough walls, *Transact. ASME, J. Heat Transf.* **91**, 488–494 (1969).

44  R. L. Webb, E. R. G. Eckert, and R. J. Goldstein, Heat transfer and friction in tubes with repeated-rib roughness, *Internatl. J. Heat Mass Transf.* **14**, 601–617 (1971).

45  W. R. Dean, The streamline motion of a fluid in a curved pipe, *Phil. Mag.* **4**, 208 (1927); **5**, 673 (1928).

46  M. Adler, Stromung in gekrummten Rohren, *Zeitschrift fur angewandte Mathematik und Mechanik* **14**, 257–275 (1934).

47  C. M. White, Streamline flow through curved pipes, *Proc. Roy. Soc. Lond. Ser. A*, **123**, 645–663 (1929).

48  R. A. Seban and E. F. McLaughlin, Heat transfer in tube coils with laminar and turbulent flow, *Internatl. J. Heat Mass Transf.* **6**, 87–95 (1963).

49  C. M. White, fluid friction and its relation to heat transfer, *Transact. Inst. Chem. Eng.* **10**, 66 (1932).

50  H. Ito, Friction factors in turbulent flow in curved pipes, *Transact. ASME, J. Basic Eng.* **81**, 123–124 (1959).

51  D. Jeschke, *Zeitschrift verein deutscher Ingenieure* **69**, 1526 (1925).

52  W. H. McAdams, *Heat Transmission*, 3rd ed., McGraw-Hill, New York, 1954, p. 228.

53  Y. Mori and W. Nakayama, Study on forced convective heat transfer in curved pipes, *Internatl. J. Heat Mass Transf.* **10**, 681–695 (1967); **10**, 37–59 (1967), **8**, 67–82 (1965).

54  C. J. Lawn, Turbulent heat transfer at low Reynolds numbers, *Transact. ASME, J. Heat Transf.* **91**, 532–536 (1969).

55  D. W. Hubbard and E. N. Lightfoot, Correlation of heat and mass transfer data for high Schmidt and Reynolds number, *Indust. Eng. Chem. Fund.* **5**, 370–379 (1966).

56  R. Higbie, The rate of absorption of a pure gas into a still liquid during short periods of exposure, *Transact. AIChE* **31**, 365–389 (1935).

57  P. V. Danckwerts, Significance of liquid-film coefficients in gas absorption, *Indust. Eng. Chem.* **43**, 1460–1467 (1951).

58  T. J. Hanratty, Turbulent exchange of mass and momentum with a boundary, *AIChE J* **2**, 359–362 (1956).

59  P. Harriott, Heat transfer in scraped-surface heat exchangers, *Chem. Eng. Progr. Symp. Ser.* **55** (29), 137–139 (1959).

60  H. A. Einstein and H. Li, The viscous sublayer along a smooth boundary, *Proc. ASCE, J. Eng. Mech. Div.* **82** (No. EM), 1–27 (1956); Shear transmission from a turbulent flow to its viscous boundary sublayer, in *Heat Transfer and Fluid Mechanics Institute Proceedings*, Vol. 13, 1955, pp. 1–16.

61  P. Harriott, A random eddy modification of the penetration theory, *Chem. Eng. Sci.* **17**, 149–154 (1962).

62  L. C. Thomas, A pseudo-surface rejuvenation model for turbulent heat transfer, Sixth Annual Southeastern Seminar On Thermal Sciences, April. 13–14, 1970; Temperature profiles for

liquid metals and moderate Prandtl number fluids, *Transact. ASME, J. Heat Transf.* **92**, 565–567 (1970); A single model for turbulent heat transfer, *Mech. Eng. News* **7** (3), 30–32 (1970).

63  R. J. Gross and L. C. Thomas, Significance of the pressure gradient on fully developed turbulent flow in a pipe, *Transact. ASME, J. Heat Transf.* **94**, 494–495 (1972).

64  H. S. Mickley and D. F. Fairbanks, Mechanism of heat transfer to fluidised beds, *AIChE J* **1**, 374–384 (1955).

65  H. L. Toor and J. M. Marchello, Film penetration model for mass and heat transfer, *AIChE J* **4**, 97–101 (1958).

66  D. W. Howard and E. N. Lightfoot, Mass transfer to falling films. Part 1. Application of the surface stretch model to uniform wave motion, *AIChE J* **14**, 458–467 (1968).

67  B. T. F. Chung, L. C. Thomas, and Y. Pang, A surface rejuvenation model for turbulent heat transfer in annular flow with high Prandtl numbers, *Transact. ASME, J. Heat Transf.* **100**, 92–97 (1978).

68  A. T. Wassel and A. G. Mills, Calculation of variable property turbulent friction and heat transfer in rough pipes, *Transact. ASME, J. Heat Transf.* **101**, 469–474 (1979).

69  G. Yee, R. Chilukuri, and J. A. C. Humphrey, Developing flow and heat transfer in strongly curved ducts of rectangular cross section, *Transact. ASME, J. Heat Transf.* **102**, 285–291 (1980).

70  S. Faggiani and F. Gori, Influence of streamwise molecular heat conduction on the heat transfer coefficient for liquid metals in turbulent flow between parallel plates, *Transact. ASME, J. Heat Transf.* **102**, 292–296 (1980).

71  F. D. Haynes and G. D. Ashton, Turbulent heat transfer in large aspect channels, *Transact. ASME, J. Heat Transf.* **102**, 384–386 (1980).

72  T. C. Carnavos, Heat transfer performance of internally finned tubes in turbulent flow, *Heat Transf. Eng.* **1**, 32–37 (1980).

73  R. L. Webb and M. J. Scott, A parametric analysis of the performance of internally finned tubes for heat exchanger applications, *Transact. ASME, J. Heat Transf.* **102**, 38–43 (1980).

74  A. F. Emery and F. B. Gessner, The numerical prediction of turbulent flow and heat transfer in the entrance region of a parallel plate duct, *Transact. ASME, J. Heat Transf.* **98**, 594–600 (1976).

75  A. F. Emery, P. K. Neighbors, and F. B. Gessner, The numerical prediction of developing turbulent flow and heat transfer in a square duct, *Transact. ASME, J. Heat Transf.* **102** 51–57 (1980).

76  J. W. Baughn, Effect of circumferential wall heat conduction on boundary conditions for heat transfer in a circular tube, *Transact. ASME, J. Heat Transf.*, **100**, 537–539 (1978).

77  J. K. Hagge and G. H. Junkhan, Mechanical augmentation of convective heat transfer in air, *Transact. ASME, J. Heat Transf.* **97**, 516–520 (1975).

78  L. C. Thomas, The surface rejuvenation model of wall turbulence: Inner laws for $u^+$ and $T^+$, *Internatl. J. Heat Mass Transf.* **23**, 1099–1104 (1980).

79  W. D. Morris and F. M. Dias, Turbulent heat transfer in a revolving square-sectioned tube, *J. Mech. Eng. Sci.* **22**, 95–101 (1980).

80  E. G. Hauptmann and A. Malhotra, Axial development of unusual velocity profiles due to heat transfer in variable density fluids, *Transact. ASME, J. Heat Transf.* **102** 71–74 (1980).

81  P. S. Virk, Heat transfer from the rear of a cylinder in transverse flow, *Transact. ASME, J. Heat Transf.* **92**, 206–207 (1970).

82  G. F. Round, An explicit approximation for the friction factor—Reynolds number relation for rough and smooth pipes, *Canadian J. Chem. Eng.* **58**, 122–123 (1980).

83  S. W. Churchill, Friction-factor equation spans all fluid-flow regimes, *Chem. Eng.* **84**, 91–92 (1977).

# 12

# NATURAL CONVECTION

In all the previous examples in this book flow has been forced. The origin of fluid motion lay outside the heat- or mass-transfer problem, and there was no need to be concerned about it. Now, however, natural convection is taken up in which there would be no motion if it were not for the heat or mass transfer. In natural convection fluid motion is caused by density variations, resulting from temperature distributions in the case of heat transfer, which are acted on by local gravitational and centrifugal forces. As might be expected, velocities are small since the forces are small; hence inertia effects are not large.

As is the case with forced convection, natural convection is conveniently divided into internal and external flows. Internal flows are those in which the fluid is enclosed by solid boundaries, whereas external flows have the fluid extending indefinitely from the solid surface from which transport occurs. In addition, a further classification as to the laminar or turbulent nature of the flow is also made.

## 12.1 LAMINAR NATURAL CONVECTION FROM A CONSTANT-TEMPERATURE VERTICAL FLAT PLATE IMMERSED IN AN INFINITE FLUID—EXACT SOLUTION

Studies are initiated by considering a vertical flat plate. It is assumed that the plate is immersed in an infinitely large body of fluid of constant properties (except insofar as density variations are important in the body force terms of the equations of motion), flow is laminar, the plate and the fluid are at constant temperature, and steady state prevails. Figure 12-1 illustrates this situation. If it is assumed that the boundary-layer description is accurate, the

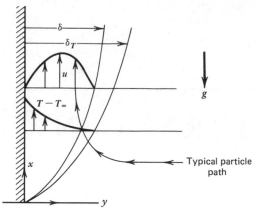

**Figure 12-1**  Laminar boundary layer in natural convection on a hot vertical flat plate.

describing equations and boundary conditions are

Continuity   $\dfrac{\partial u}{\partial x} + \dfrac{\partial v}{\partial y} = 0$

$x$ Motion   $u\dfrac{\partial u}{\partial x} + v\dfrac{\partial u}{\partial y} = \nu_\infty \dfrac{\partial^2 u}{\partial y^2} + \dfrac{B_x - dp/dx}{\rho_\infty}$

$y$ Motion   $\dfrac{dp}{\partial y} = 0$

Energy   $u\dfrac{\partial T}{\partial x} + v\dfrac{\partial T}{\partial y} = \alpha_\infty \dfrac{\partial^2 T}{\partial y^2}$

At $y = 0$   $u = 0 = v$     and     $T = T_w$

At $y = \infty$   $u = 0$     and     $T = T_\infty$

Because the expected velocities are small and temperature differences are not, viscous dissipation is ignored—its effects were studied by Gebhart [1]. The possibility of fluid cooling as it rises and expands in the lower pressure existing at higher positions is also ignored, although this is an important phenomenon in meteorology.

Before proceeding further, the pressure gradient and body force must be evaluated. The body force $B_x$ comes most easily as it is the result of gravity; thus $B_x = -\rho g$. The pressure gradient in the boundary layer requires a bit more arguing; because $dp/dy = 0$, $dp/dx$ can be evaluated at a large distance from the plate where the equations of motion show that for a stagnant fluid

$(u = 0)$, $dp/dx = -\rho_\infty g$. As a result,

$$\frac{1}{\rho_\infty}\left(B_x - \frac{dp}{dx}\right) = (\rho_\infty - \rho)\frac{g}{\rho_\infty}$$

The boundary-layer description of the problem then is

$$\frac{\partial u}{\partial x} + \frac{\partial v}{\partial y} = 0 \tag{12-1a}$$

$$u\frac{\partial u}{\partial x} + v\frac{\partial u}{\partial y} = \nu_\infty \frac{\partial^2 u}{\partial y^2} + g\frac{\rho_\infty - \rho}{\rho_\infty} \tag{12-1b}$$

$$u\frac{\partial T}{\partial x} + v\frac{\partial T}{\partial y} = \frac{\nu_\infty}{\mathrm{Pr}}\frac{\partial^2 T}{\partial y^2} \tag{12-1c}$$

and the imposed conditions are

$$u(y = 0) = 0 = v(y = 0) \qquad u(y = \infty) = 0 \qquad T(y = \infty) = T_\infty$$

$$T(y = 0) = T_w \tag{12-1d}$$

A similarity solution is sought first in order to establish the most exact solution of the laminar boundary-layer model. These exact results are then compared against measurements. Unfortunately, the basic physical idea on which a similarity solution is built for a forced convection boundary layer $[\delta \sim t^{1/2} \sim (x/U)^{1/2}]$ lacks a specified characteristic velocity $U$. As a result, a more mathematical approach is required along the lines suggested in Appendix E. The upshot of it is that the similarity variable is $\eta = yc/x^{1/4}$ and the stream function is $\psi = 4\nu cx^{3/4}F(\eta)$, where $c^4 = (g\,|\rho_\infty - \rho_w|)/(4\nu_\infty^2\rho_\infty)$, with $u = 4\nu c^2 x^{1/2}F'(\eta)$ and $v = (\nu c/x^{1/4})\,[\eta F' - 3F]$. The integral treatment presented later motivates these functional forms that were first advanced by Pohlhausen (in cooperation with Schmidt and Beckmann [2])—early partially successful analyses were proposed by Oberbeck [3] in 1879 and Lorenz [4] in 1881. Note that if a problem requiring a nonzero normal velocity at the wall is to be considered, velocity must vary as $x^{-1/4}$. The describing equations and boundary conditions of Eq. (12-1) are then given in similarity form as

$$F''' + 3FF'' - 2(F')^2 + \frac{\rho_\infty - \rho}{|\rho_\infty - \rho_w|} = 0 \tag{12-2a}$$

$$\theta'' + 3\mathrm{Pr}\,F\theta' = 0 \tag{12-2b}$$

with boundary conditions of

$$F'(0) = F'(\infty) = \theta(\infty) = 0 \tag{12-2c}$$

$$F(0) = -\frac{v_w x^{1/4}}{3vc} \tag{12-2d}$$

$$\theta(0) = 1 \tag{12-2e}$$

where $\theta = (T - T_\infty)/(T_w - T_\infty)$, $v_w$ is a possible wall "blowing" velocity (here assumed to be zero), and a prime denotes an ordinary derivative with respect to $\eta$. It can be seen that the $(\rho_\infty - \rho)/|\rho_\infty - \rho_w|$ term acts as a forcing function for velocity; if that term is zero, Eq. (12-2) would be satisfied by a constant that means that then $u = 0$ and there is no vertical motion.

There are a number of ways for the forcing function $(\rho_\infty - \rho)/|\rho_\infty - \rho_w|$ to be generated. The most common one is density dependence on temperature. Since density is only slightly dependent on temperature for most fluids and since the imposed temperature differences are expected to be moderate, the forcing function is also small and so is the resultant expected vertical velocity, as mentioned before. In a mass-transfer situation in which water, for example, evaporates from a wetted surface into stagnant air, the fluid density is dependent somewhat on the mass fraction of water vapor in the air. Since water vapor is lighter than air, the effect would be the same as if the surface were hotter than the air—there is an upward convection. In combined heat and mass transfer, it is possible for these two body forces to oppose one another; such a complication is not treated here.

Attention is here restricted to density dependence on temperature. From the definition of the coefficient of thermal expansion

$$\beta = -\frac{1}{\rho}\frac{\partial\rho}{\partial T}\bigg|_{p=\text{const}}$$

in conjunction with the series expansion

$$\rho = \rho_\infty + \frac{\partial\rho}{\partial T}(T - T_\infty) + \frac{\partial^2\rho}{\partial T^2}\frac{(T - T_\infty)^2}{2!} + \cdots$$

it is found that

$$-\left[\cdots + \frac{\partial^2\rho}{\partial T^2}\frac{(T - T_\infty)^2}{2!} + \frac{\partial\rho}{\partial T}(T - T_\infty)\right] = \rho_\infty - \rho$$

or

$$\rho_\infty\beta(T - T_\infty) \approx \rho_\infty - \rho$$

From this it is evident that $(\rho_\infty - \rho)/|\rho_\infty - \rho_w| = (T - T_\infty)/|T_w - T_\infty| = \theta$. The differential equations then are

$$F''' + 3FF'' - 2(F')^2 + \theta = 0 \qquad (12\text{-}3a)$$

$$\theta'' + 3\Pr F\theta' = 0 \qquad (12\text{-}3b)$$

with boundary conditions of

$$F'(0) = F'(\infty) = \theta(\infty) = 0 \qquad (12\text{-}3c)$$

$$F(0) = -\frac{v_w x^{1/4}}{3vc} \qquad (12\text{-}3d)$$

$$\theta(0) = 1 \qquad (12\text{-}3e)$$

where the "blowing" at the wall is assumed to be absent—$v_w = 0$.

Of interest from a mathematical point of view is the coupling of the energy and motion equations in Eq. (12-3), in spite of the constant property assumption, which requires their simultaneous solution. Refined numerical solutions to Eq. (12-3) were obtained by Ostrach [5]. The temperature and velocity profiles for various Prandtl numbers are displayed in Fig. 12-2. The velocity is seen to achieve a maximum near the plate and, like the temperature, to approach ambient conditions far from the plate. At large Pr the thermal boundary layer is considerably thinner than the velocity boundary layer. As with forced convection laminar boundary layers, the essence of the problem solution is to find the two unknown quantities—$F''(0)$ and $\theta'(0)$. Because the equations are coupled, the velocity and temperature profiles depend on two separate parameters: the Prandtl number Pr and the Grashof number Gr whose definition is given in the following paragraphs. Tabular presentation of the numerical solution requires several tables, of which only one for $\Pr = 1$ is given in Table 12-1 [5]. It should be noted there that a fluid particle far from the plate approaches the plate perpendicularly since as $\eta \to \infty$, $u = (2v/x)$ $\mathrm{Gr}_x^{1/2}F' \to 0$ and $v = (\nu/x)(\mathrm{Gr}_x/4)^{1/4}(\eta F' - 3F) \to -1.556(\nu/x)(\mathrm{Gr}_x/4)^{1/4}$. Hence a typical particle flow path is as sketched in Fig. 12-1, and the shape of the leading edge should not matter greatly.

Local heat-transfer coefficients are available from these solutions since, by definition,

$$q_w = -k\frac{\partial T(y = 0)}{\partial y} = h(T_w - T_\infty)$$

or in terms of similarity variables,

$$-k(T_w - T_\infty)\frac{\partial\theta(\eta = 0)}{\partial\eta}\frac{\partial\eta}{\partial y} = h(T_w - T_\infty)$$

$$h = -\theta'(0)\frac{kc}{x^{1/4}}$$

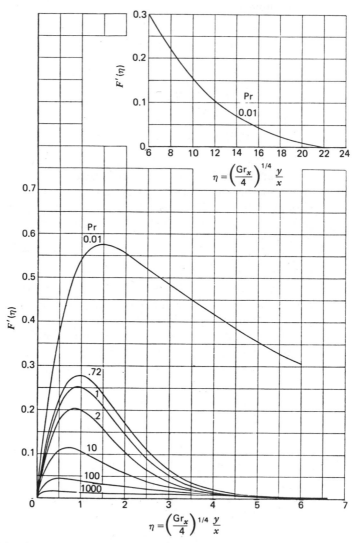

**Figure 12-2** Calculated dimensionless profiles in the laminar boundary layer on a hot vertical flat plate in natural convection of (a) velocity and (b) temperature. (From S. Ostrach, NACA Report 1111, 1953 [5].)

which shows that $h \sim x^{-1/4}$. From this the local Nusselt number is

$$\mathrm{Nu}_x = \frac{hx}{k} = -\theta'(0) \left[ \frac{g \, | \rho_\infty - \rho_w | x^3}{4 \nu^2 \rho_\infty} \right]^{1/4}$$

It is conventional to define the Grashof number Gr as

$$\mathrm{Gr}_x = g \frac{| \rho_\infty - \rho_w | x^3}{\nu^2 \rho_\infty}$$

$$= g\beta \frac{(T_w - T_\infty)x^3}{\nu^2}$$

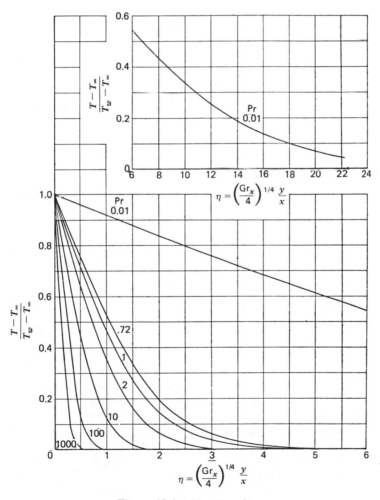

**Figure 12-2** (*Continued.*)

**Table 12-1  Functions $F$ and $\theta$ and Derivatives for Pr = 1 [5]**

| $\eta$ | $F$ | $F'$ | $F''$ | $\theta$ | $\theta'$ |
|---|---|---|---|---|---|
| 0 | 0.0000 | 0.0000 | 0.6421 | 1.0000 | −0.5671 |
| 0.1 | 0.0030 | 0.0593 | 0.5450 | 0.9433 | −0.5669 |
| 0.2 | 0.0115 | 0.1092 | 0.4540 | 0.8867 | −0.5657 |
| 0.3 | 0.0246 | 0.1503 | 0.3694 | 0.8302 | −0.5627 |
| 0.4 | 0.0413 | 0.1833 | 0.2916 | 0.7742 | −0.5572 |
| 0.5 | 0.0610 | 0.2089 | 0.2208 | 0.7189 | −0.5488 |
| 0.6 | 0.0829 | 0.2277 | 0.1572 | 0.6645 | −0.5371 |
| 0.7 | 0.1064 | 0.2406 | 0.1008 | 0.6116 | −0.5221 |
| 0.8 | 0.1308 | 0.2481 | 0.0516 | 0.5602 | −0.5038 |
| 0.9 | 0.1558 | 0.2511 | 0.0093 | 0.5109 | −0.4826 |
| 1.0 | 0.1809 | 0.2502 | −0.0263 | 0.4638 | −0.4589 |
| 1.1 | 0.2058 | 0.2461 | −0.0557 | 0.4192 | −0.4330 |
| 1.2 | 0.2300 | 0.2393 | −0.0793 | 0.3772 | −0.4056 |
| 1.3 | 0.2535 | 0.2304 | −0.0975 | 0.3381 | −0.3772 |
| 1.4 | 0.2761 | 0.2199 | −0.1110 | 0.3018 | −0.3484 |
| 1.5 | 0.2975 | 0.2083 | −0.1203 | 0.2684 | −0.3197 |
| 1.6 | 0.3177 | 0.1960 | −0.1260 | 0.2379 | −0.2915 |
| 1.7 | 0.3367 | 0.1832 | −0.1287 | 0.2101 | −0.2642 |
| 1.8 | 0.3543 | 0.1708 | −0.1288 | 0.1850 | −0.2382 |
| 1.9 | 0.3707 | 0.1575 | −0.1268 | 0.1624 | −0.2136 |
| 2.0 | 0.3859 | 0.1450 | −0.1233 | 0.1422 | −0.1907 |
| 2.1 | 0.3997 | 0.1239 | −0.1185 | 0.1242 | −0.1695 |
| 2.2 | 0.4125 | 0.1213 | −0.1127 | 0.1082 | −0.1501 |
| 2.3 | 0.4240 | 0.1104 | −0.1064 | 0.0941 | −0.1324 |
| 2.4 | 0.4346 | 0.1001 | −0.0997 | 0.0817 | −0.1164 |
| 2.5 | 0.4441 | 0.0904 | −0.0928 | 0.0708 | −0.1020 |
| 2.6 | 0.4527 | 0.0815 | −0.0859 | 0.0613 | −0.0892 |
| 2.7 | 0.4604 | 0.0733 | −0.0791 | 0.0529 | −0.0777 |
| 2.8 | 0.4673 | 0.0657 | −0.0725 | 0.0457 | −0.0676 |
| 2.9 | 0.4736 | 0.0588 | −0.0662 | 0.0392 | −0.0587 |
| 3.0 | 0.4791 | 0.0524 | −0.0602 | 0.0339 | −0.0509 |
| 3.1 | 0.4841 | 0.0467 | −0.0546 | 0.0291 | −0.0441 |
| 3.2 | 0.4885 | 0.0415 | −0.0493 | 0.0250 | −0.0381 |
| 3.3 | 0.4924 | 0.0368 | −0.0444 | 0.0215 | −0.0329 |
| 3.4 | 0.4959 | 0.0326 | −0.0399 | 0.0185 | −0.0283 |
| 3.6 | 0.5016 | 0.0254 | −0.0321 | 0.0136 | −0.0210 |
| 3.8 | 0.5061 | 0.0197 | −0.0255 | 0.0009 | −0.0155 |
| 4.0 | 0.5096 | 0.0151 | −0.0202 | 0.0072 | −0.0115 |
| 4.2 | 0.5122 | 0.0116 | −0.0158 | 0.0053 | −0.0084 |
| 4.4 | 0.5143 | 0.0087 | −0.0124 | 0.0038 | −0.0062 |
| 4.6 | 0.5158 | 0.0066 | −0.0096 | 0.0027 | −0.0045 |
| 4.8 | 0.5169 | 0.0049 | −0.0075 | 0.0020 | −0.0034 |
| 5.0 | 0.5177 | 0.0035 | −0.0057 | 0.0014 | −0.0024 |
| 5.2 | 0.5183 | 0.0025 | −0.0044 | 0.0010 | −0.0018 |
| 5.5 | 0.5189 | 0.0014 | −0.0029 | 0.0006 | −0.0011 |
| 6.0 | 0.5194 | 0.0004 | −0.0014 | 0.0002 | −0.0005 |
| 6.25 | 0.5194 | 0.0000 | −0.0010 | 0.0000 | −0.0004 |

With this definition the local Nusselt number relation is

$$\mathrm{Nu}_x = -\theta'(0)\left(\frac{\mathrm{Gr}_x}{4}\right)^{1/4}$$

(12-4)

The average heat-transfer coefficient is also obtainable from the solutions to Eq. (12-3) since

$$q_{w_{\text{total}}} = \bar{h}(T_w - T_\infty)x = \int_0^x h(T_w - T_\infty)\, dx$$

Inasmuch as $h = Bx^{-1/4}$, integration of $h$ with subsequent division by $x$ results in

$$\bar{h} = \frac{\int_0^x Bx^{-1/4}\, dx}{x} = \frac{4}{3} Bx^{-1/4}$$

$$\bar{h} = \frac{4}{3} h$$

Hence it follows from Eq. (12-4) that the average Nusselt number $\overline{\mathrm{Nu}}_x$ is given in terms of the local Grashof number as

$$\overline{\mathrm{Nu}}_x = \frac{\bar{h}x}{k} = -\theta'(0)\frac{4}{3}\left(\frac{\mathrm{Gr}_x}{4}\right)^{1/4}$$

(12-5)

In similar fashion the local shear stress on the plate is found from

$$\tau_w = \mu \frac{\partial u(y = 0)}{\partial y} = \mu \frac{\partial u}{\partial \eta}\frac{\partial \eta}{\partial y}$$

to be given in terms of dimensionless groups as

$$\tau_w = \left(4\mathrm{Gr}_x^3\right)^{1/4}\frac{\rho_\infty \nu_\infty^2}{x^2} F''(0)$$

(12-6)

Hence $\tau_w \sim x^{1/4}$ and it is seen that, in contrast to a forced convection laminar boundary layer, wall shear stress increases with distance from the leading edge. The average shear stress is found from

$$\bar{\tau}_w = x^{-1}\int_0^x \tau_w\, dx$$

to be

$$\bar{\tau}_w = \frac{4\tau_w}{5}$$

$$= \frac{2}{5}(4\mathrm{Gr}_x)^{3/4}\frac{\rho_\infty \nu_\infty^2}{x^2}F''(0) \tag{12-7}$$

Values of $\theta'(0)$ and $F''(0)$ are given in Table 12-2 for various values of Prandtl number [5, 6]. These can be used with Eq. (12-5) for the average Nusselt number and Eq. (12-7) for the average wall shear stress. Also included are values of $A$ for use in the representation for the average Nusselt number of

$$\overline{\mathrm{Nu}} = \frac{\bar{h}x}{k} = A(\mathrm{Gr}_x\,\mathrm{Pr})^{1/4}$$

$$= A(\mathrm{Ra}_x)^{1/4} \tag{12-8}$$

where the Rayleigh number Ra is defined as the product of the Grashof and

**Table 12-2**   $\theta'(0)$ **and** $F''(0)$ **versus Pr [5, 6]**

| Pr | $A$ | $\theta'(0)$ | $F''(0)$ |
|----|-----|--------------|----------|
| 0 | $0.800564\mathrm{Pr}^{1/4}$ | $0.849126\mathrm{Pr}^{1/2}$ | |
| 0.01 | 0.240279 | 0.080592 | 0.9862 |
| 0.03 | 0.308 | 0.136 | |
| 0.09 | 0.377 | 0.219 | |
| 0.5 | 0.496 | 0.442 | |
| 0.72 | 0.516492 | 0.50463 | 0.676 |
| 0.733 | 0.517508 | 0.50789 | 0.6741 |
| 1.0 | 0.534705 | 0.56714 | 0.6421 |
| 1.5 | 0.555059 | 0.651534 | |
| 2.0 | 0.568033 | 0.716483 | 0.5713 |
| 3.5 | 0.589916 | 0.855821 | |
| 5.0 | 0.601463 | 0.953956 | |
| 7.0 | 0.611035 | 1.05418 | |
| 10 | 0.619 | 1.168 | 0.4192 |
| 100 | 0.653349 | 2.1914 | 0.2517 |
| 1,000 | 0.665 | 3.97 | 0.1450 |
| 10,000 | 0.668574 | 7.0913 | |
| $\infty$ | 0.670327 | $0.710989\mathrm{Pr}^{1/4}$ | |

*Source*: By permission from A. J. Ede, Advances in free convection, *Adv. Heat Transf.* **4**, 1–64 (1967) [6]; copyright 1967, Academic Press, New York.

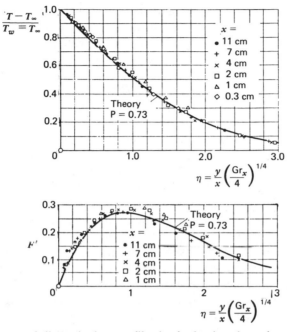

**Figure 12-3** Measured dimensionless profiles in the laminar boundary layer on a hot vertical flat plate in natural convection of (a) velocity and (b) temperature; $x$ is distance from lower edge of plate. Measurements by E. Schmidt and W. Beckman [2]. (By permission from H. Schlichting, *Boundary-Layer Theory*, 6th ed., McGraw-Hill, New York, 1968.)

Prandtl numbers $\mathrm{Ra} = \mathrm{Gr}\,\mathrm{Pr}$. The coefficient $A$ is dependent on Prandtl number, but rather weakly, as Table 12-2 shows. Use of the Rayleigh number in Eq. (12-8) takes advantage of the fact that for the creeping flow expected under normal conditions, inertial effects are negligible and the Nusselt number depends solely on the Rayleigh number rather than on Gr and Pr separately; this is the topic of Problem 12-4. For $\mathrm{Pr} > 100$, Table 12-2 confirms that $A$ varies very little and that when $\mathrm{Pr} \to \infty$

$$\overline{\mathrm{Nu}} = 0.67(\mathrm{Ra}_x)^{1/4} \tag{12-9}$$

At very small Prandtl numbers it can be argued that viscosity cannot be important since, as Fig. 12-2b suggests, the fluid velocity near the plate then tends to be imposed by events far from the plate and to resemble slug flow. This is a consequence of the high thermal conductivity that is associated with a low Pr that allows the wall temperature excess to penetrate far into the fluid. For viscosity to be unimportant, it is necessary that $\mathrm{Nu} = f(\mathrm{Gr}\,\mathrm{Pr}^2)$, and Table 12-2 confirms that this is indeed the case. So, for $\mathrm{Pr} \to 0$,

$$\overline{\mathrm{Nu}} = 0.8(\mathrm{Gr}\,\mathrm{Pr}^2)^{1/4} \tag{12-10}$$

Problem 12-5 deals in greater mathematical detail with this limiting case. Calculations for very small Pr were made by Sparrow and Gregg [7], and the limiting cases of $Pr \to 0$ and $Pr \to \infty$ were treated by LeFevre [8], who proposed the approximation for Table 12-2 that

$$A^4 = \frac{0.4Pr}{1 + 2Pr^{1/2} + 2Pr} \tag{12-11}$$

The agreement of the boundary-layer predictions with temperature and velocity measurements is quite good, as Fig. 12-3 shows. Despite the good agreement displayed in Fig. 12-3 for a vertical plate of constant temperature, measured values of the average heat-transfer coefficient seem to be somewhat higher than the boundary-layer predictions. Figure 12-4 displays a comparison for air made by Ede [6] in which the solid line represents the boundary-layer prediction [from Table 12-2 and Eq. (12-8)], and the points represent measurements. Between Rayleigh numbers of $10^5$ and $10^8$, satisfactory agreement is seen. The data deviate above the prediction, however, at both higher and lower values of Rayleigh number and, even in the region of best agreement, lie above the prediction. Since uncontrolled air currents, the commonest experimental error, lead to increased heat transfer, the deviation is generally understandable. The deviation at high Rayleigh numbers is probably due to the development of

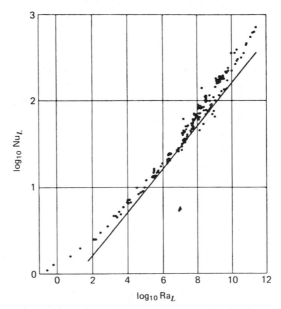

**Figure 12-4** Measured average Nusselt number versus Rayleigh number for natural convection from vertical plates to air. [By permission from A. J. Ede, Advances in Free Convection, *Adv. Heat Transf.* **4**, 1–64 (1967) [6]; copyright 1967, Academic Press, New York.]

turbulence, and that at low Rayleigh numbers may be due to increasing inaccuracy of the boundary-layer assumptions as a result of very thick boundary layers. For vertical planes immersed in a fluid whose Prandtl number is approximately unity, McAdams [9] suggests that the correlation shown in Fig. 12-5 be used. It can be seen that it agrees substantially with the data points of Fig. 12-4. For purposes of rapid estimation when Pr $\approx$ 1, the relationship

$$\overline{Nu} = \frac{\bar{h}x}{k} = 0.555(Ra_x)^{1/4}, \qquad 10 \leqslant Ra_x \leqslant 10^9 \qquad (12\text{-}12)$$

can be used [10], which was originally intended for air.

The manner in which the temperature dependence of properties should be taken into account is a question that has been addressed by a number of investigators. The conclusion drawn is that, in view of such other uncertainties as the ambient fluid being truly stagnant, all properties should be evaluated at

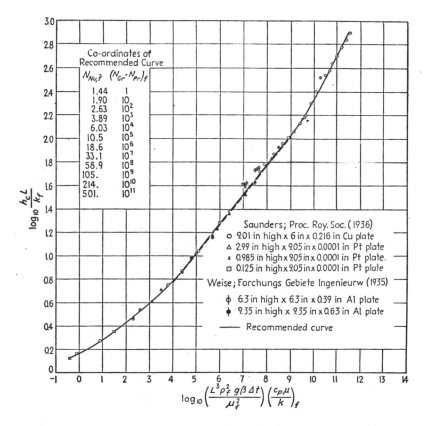

**Figure 12-5** Natural convection from short vertical plates to air. (From *Heat Transmission*, 3rd ed., by W. H. McAdams, copyright 1954, McGraw-Hill, New York; used with the permission of the McGraw-Hill Book Company.)

the film temperature $T_f$ where

$$T_f = \frac{T_w + T_\infty}{2} \qquad (12\text{-}13)$$

for use in constant-property correlation equations such as Eq. (12-8). A detailed study of the effect of variable properties on the vertical flat-plate problem was conducted by Sparrow and Gregg [11] within the framework of the boundary-layer assumptions. They found that Nusselt number correlations obtained from constant-property solutions give good accuracy when properties are all evaluated at $T_f$ from Eq. (12-13) but that slightly better accuracy results from evaluating all properties at a reference temperature $T_r$ given by

$$T_r = T_w - 0.38(T_w - T_\infty) \qquad (12\text{-}14)$$

for gases and for liquid mercury. Measurements by Eichhorn [12], who used small dust particles to determine local velocities, suggest that $T_r = T_w - 0.17(T_w - T_\infty)$, although there is no substantial superiority over $T_f$. The velocity distribution and, hence, the wall shear stress are affected a good deal more than the Nusselt number by temperature-dependent properties [10, 11].

The effect of suction or blowing on the rate of heat transfer in laminar natural convection from a vertical flat plate has been studied [13, 14]. This effect can be important in mass transfer applications, for example.

## 12.2 LAMINAR NATURAL CONVECTION FROM A CONSTANT-TEMPERATURE VERTICAL FLAT PLATE IMMERSED IN AN INFINITE FLUID—INTEGRAL APPROXIMATE SOLUTION

The problem described in Section 12.1 can be approached by an integral method. The describing boundary-layer equations in Eq. (12-1) can be put in an integral form in the manner previously explored in Chapter 8 to obtain Eqs. (8-12) and (8-18). Application of Eqs. (8-12) and (8-18) yields

$$\frac{\partial \left[ \int_0^\delta u^2 \, dy \right]}{\partial x} - g\beta \int_0^\delta (T - T_\infty) \, dy = -\frac{\tau_w}{\rho} \qquad (12\text{-}15a)$$

for the x-motion equation and

$$\frac{\partial \left[ \int_0^\delta T_u (T - T_\infty) \, dy \right]}{\partial x} = \frac{q_w}{\rho C_p} \qquad (12\text{-}15b)$$

for the energy equation. Here $\tau_w = \mu \, \partial u(y = 0)/\partial y$ and $q_w = -k \, \partial T(y = 0)/\partial y$ and the thermal $\delta_T$ and velocity $\delta$ boundary-layer thicknesses have been assumed to be nearly equal for simplicity. The integral method was apparently first applied to free convection by Squire [15]. Later Merk and Prins [16] treated free convection from vertical flat plate, vertical cone, horizontal cylinder, and sphere geometries.

In defense of the assumption that $\delta \approx \delta_T$, it can be said that it is the temperature excess that urges the fluid into a velocity different from its preferred state of rest. However, the exact solutions illustrated in Fig. 12-2 show that $\delta \approx \delta_T$ only if $Pr \approx 1$—the cases in which $\delta \neq \delta_T$ were treated by the integral method by Sugawara and Michiyoshi [17a]. The results obtained for $\delta \approx \delta_T$ will, nevertheless, serve to show the effect of Prandtl number and will also suggest the manner in which data might be correlated and will be surprisingly accurate.

A polynomial is assumed to describe the velocity distribution as

$$u = a + b\eta + c\eta^2 + d\eta^3 \qquad (12\text{-}16a)$$

where $\eta = y/\delta$ and can be subjected to the five conditions

$$u(y = 0) = 0 \qquad (12\text{-}16b)$$

$$u(y = \delta) = 0 \qquad (12\text{-}16c)$$

$$\frac{\partial u(y = \delta)}{\partial y} = 0 \qquad (12\text{-}16d)$$

$$g\beta(T_w - T_\infty) + \nu \frac{\partial^2 u(y = 0)}{\partial y^2} = 0 \qquad (12\text{-}16e)$$

$$\frac{\partial^2 u(y = \delta)}{\partial y^2} = 0 \qquad (12\text{-}16f)$$

Equations (12-16b) and (12-16c) are the same conditions imposed on the boundary-layer equations in Eq. (12-1) and reflect the no-slip condition at the plate and the stagnant fluid pool at the edge of the boundary layer, respectively. Equation (12-16d) acknowledges the absence of a shear stress $\tau$ in the fluid at the edge of the boundary layer, whereas Eqs. (12-16e) and (12-16f) are the boundary-layer x-motion equation [Eq. (12-1b)] evaluated at the wall and the edge of the boundary layer, respectively. Although all five conditions could be used, only four are employed here. The conditions with the lowest ordered derivatives are retained as is the condition involving velocity's second derivative at the wall. It is then found that

$$u = u_1\eta(1 - \eta)^2 \qquad (12\text{-}17)$$

where

$$u_1 = \frac{g\beta(T_w - T_\infty)\delta^2}{4\nu}$$

Similarly, a polynomial is selected to describe the temperature distribution as

$$\frac{T - T_\infty}{T_w - T_\infty} = A + B\eta + C\eta^2 \tag{12-18a}$$

and can be subjected to the five conditions

$$T(y = 0) = T_w \tag{12-18b}$$

$$T(y = \delta) = T_\infty \tag{12-18c}$$

$$\frac{\partial T(y = \delta)}{\partial y} = 0 \tag{12-18d}$$

$$\frac{\partial^2 T(y = 0)}{\partial y^2} = 0 \tag{12-18e}$$

$$\frac{\partial^2 T(y = \delta)}{\partial y^2} = 0 \tag{12-18f}$$

Equations (12-18b) and (12-18c) are the same conditions imposed on the boundary-layer equations in Eq. (12-1) and reflect the requirement that the fluid take on the wall and ambient values at the plate and edge of the boundary layer, respectively. Equation (12-18d) acknowledges that there is no heat flux at the edge of the boundary layer, whereas Eqs. (12-18e) and (12-18f) come from the boundary-layer energy equation [Eq. (12-1c)] evaluated at the wall and the edge of the boundary layer, respectively. Somewhat arbitrarily, the three conditions involving temperature's lowest ordered derivatives are employed to evaluate the constants of Eq. (12-18a) as

$$\frac{T - T_\infty}{T_w - T_\infty} = (1 - \eta)^2 \tag{12-19}$$

Substitution of the assumed profiles of Eqs. (12-17) and (12-19) into the integral relationships of Eq. (12-15) gives

$$\frac{d(u_1^2\delta/105)}{dx} = \left[\frac{g\beta(T_w - T_\infty)}{3}\right]\delta - \frac{u_1\nu}{\delta} \tag{12-20a}$$

$$\frac{d(u_1\delta/30)}{dx} = \frac{2\alpha}{\delta} \tag{12-20b}$$

Keep in mind that $u_1$ must be assumed to be yet unrelated to $\delta$ in order to avoid having two equations in only one unknown. This situation is the result of assuming $\delta = \delta_T$. Solution of Eq. (12-20) begins with the insight that $u_1$ and $\delta$ probably vary as some power of $x$. Thus it is guessed that $u_1 = c_1 x^m$ and $\delta = c_2 x^n$. Substitution into Eq. (12-20) then gives

$$(2m + n)c_1^2 c_2^2 x^{2m+2n-1} = 35[g\beta(T_w - T_\infty)]c_2^2 x^{2n} - 105\nu c_1 x^m$$

$$(12\text{-}21a)$$

and

$$(m + n)c_1 c_2^2 x^{m+2n-1} = 60\alpha \qquad (12\text{-}21b)$$

The only way for these two relationships to be true for all values of $x$ is to have all exponents of $x$ be equal. This requires that

$$2m + 2n - 1 = 2n = m \qquad \text{and} \qquad m + 2n - 1 = 0$$

which is satisfied by $m = \frac{1}{2}$ and $n = \frac{1}{4}$. Thus

$$u_1 = c_1 x^{1/2} \qquad (12\text{-}22a)$$

and

$$\delta = c_2 x^{1/4} \qquad (12\text{-}22b)$$

where, as a result of substitution of the $m$ and $n$ values back into Eq. (12-21),

$$c_1 = 4\left(\frac{5}{3}\right)^{1/2} \frac{[g\beta(T_w - T_\infty)]^{1/2}}{(\text{Pr} + 20/21)^{1/2}} \qquad (12\text{-}22c)$$

$$c_2 = \frac{4(15/16)^{1/4}(1 + 20/21\text{Pr})^{1/4}}{[g\beta(T_w - T_\infty)\text{Pr}/\nu^2]^{1/4}} \qquad (12\text{-}22d)$$

The integral method results give $\eta = y/\delta = y/(c_2 x^{1/4})$, according to Eq. (12-22b), which suggests that dimensionless velocity and temperature profiles depend solely on $\eta = cy/x^{1/4}$. It is seen that this $\eta$ is proportional to the similarity variable used to achieve exact solutions. In other words, the integral method can suggest the proper form of a similarity transformation to be used to obtain an exact solution. Note also that the integral method suggests that $u \sim x^{1/2}$ just as does the similarity transformation.

The local heat-transfer coefficient is obtained from

$$h(T_w - T_\infty) = -k\frac{\partial T(y = 0)}{\partial y}$$

Equation (12-19) for the approximate temperature profile used in this relationship then gives

$$h/k = \frac{2}{\delta} = \frac{2}{c_2 x^{1/4}}$$

From this the local Nusselt number follows as

$$\frac{\mathrm{Nu}_x}{(\mathrm{Ra}_x)^{1/4}} = 0.51\left(1 + \frac{20}{21\mathrm{Pr}}\right)^{-1/4}$$

where $\mathrm{Nu}_x = hx/k$ and $\mathrm{Ra}_x = [g\beta(T_w - T_\infty)x^3/\nu^2]\mathrm{Pr}$. The average Nusselt number comes from this result, since $\overline{\mathrm{Nu}} = 4\mathrm{Nu}_x/3$, as

$$\frac{\overline{\mathrm{Nu}}_x}{(\mathrm{Ra}_x)^{1/4}} = \frac{0.68\mathrm{Pr}^{1/4}}{(\mathrm{Pr} + 20/21)^{1/4}} \tag{12-23}$$

For $\mathrm{Pr} = 0.73$ as it is for air, Eq. (12-33) gives

$$\frac{\overline{\mathrm{Nu}}_x}{\mathrm{Ra}_x^{1/4}} = 0.55$$

which is only 10% below the 0.555 value of the recommended Eq. (12-12). Similar good results are observed when $\mathrm{Pr} \to \infty$; Eq. (12-33) gives

$$\frac{\overline{\mathrm{Nu}}_x}{\mathrm{Ra}_x^{1/4}} = 0.68$$

which is within about 1% of the exact Eq. (12-9). When $\mathrm{Pr} \to 0$, Eq. (12-23) predicts

$$\frac{\overline{\mathrm{Nu}}_x}{\left(G_x\mathrm{Pr}^2\right)^{1/4}} = 0.69$$

which, although agreeing in form with Eq. (12-10), is 14% low.

   The failure of the boundary-layer description of this natural convection problem at low values of the Rayleigh number is a result of the thick boundary layer that then occurs. This is demonstrated by Eqs. (12-22b) and (12-22d), which can be combined to give the local boundary-layer thickness as

$$\frac{\delta}{x} = 4\left(\frac{15}{16}\right)^{1/4}\frac{(1 + 20/21\mathrm{Pr})^{1/4}}{\mathrm{Ra}_x^{1/4}}$$

where $Ra_x = Gr_x Pr$. At low values of $Ra_x$, the boundary-layer assumption that $\delta/x \ll 1$ is inaccurate. It can also be seen that the boundary-layer thickens as Pr decreases, holding $Ra_x$ constant.

The local shear stress $\tau_w$ at the plate is given by

$$\tau_w = \mu \frac{\partial u(y=0)}{\partial y}$$

$$\approx \frac{\mu u_1}{\delta} = \mu \frac{c_1}{c_2} x^{1/4}$$

which, from Eq. (12-22), then is

$$\tau_w = \left(\frac{80}{27}\right)^{1/4} Gr_x^{3/4} \frac{\rho \nu^2}{x^2} \frac{Pr^{1/4}}{(Pr + 20/21)^{3/4}} \tag{12-24}$$

For $Pr = 1$, the integral method predicts $\tau_w = 0.79\ Gr_x^{3/4}(\rho \nu^2/x^2)$ whereas the exact Eq. (12-7) has a coefficient of 0.91, a 13% error.

The maximum velocity is found by this integral method to occur at $y_{max}/\delta = \frac{1}{3}$. Since it can be shown that

$$\frac{\eta_{similarity}}{y/\delta} = \left(\frac{60}{Pr}\right)^{1/4}\left(1 + \frac{20}{21 Pr}\right)^{1/4}$$

$$= 3.3 \quad \text{for} \quad Pr = 1$$

the exact value of $\eta_{similarity} = 0.93$ for maximum velocity given in Table 12-1 for $Pr = 1$ is in good agreement with the integral method prediction that $\eta_{similarity} = 1.1$. The maximum velocity predicted by the integral method is

$$u_{max} = \frac{16}{27}\left(\frac{5}{3}\right)^{1/2} \frac{[g\beta(T_w - T_\infty)]^{1/2} x^{1/2}}{(Pr + 20/21)^{1/2}}$$

For $Pr = 1$, the integral method predicts, therefore, that

$$u_{max}[g\beta(T_w - T_\infty)]^{-1/2} x^{-1/2} = 0.55$$

Table 12-1 for $Pr = 1$ gives the exact answer as 0.504, so the integral method is in error by about 9%.

These comparisons indicate that the integral method is quite good for heat-transfer-coefficient predictions. Its predictions of such interior details as the velocity distribution are often less accurate but are still quite useful. A refined correlation [93] is

$$\overline{Nu} = 0.68 + \frac{0.67 Ra_x^{1/4}}{\left[1 + (0.492/Pr)^{9/16}\right]^{4/9}}$$

## 12.3 LAMINAR NATURAL CONVECTION FROM A CONSTANT-HEAT-FLUX FLAT PLATE IMMERSED IN AN INFINITE FLUID

An important question is the effect of a nonuniform surface temperature on the heat-transfer coefficient for laminar natural convection from a flat plate immersed in an infinite fluid. If surface temperature variation is influential, considerable attention must be given to accurate description of surface temperature in an application. On the other hand, if surface temperature variation is not influential, no great care need be exercised to obtain an accurate heat-transfer coefficient. Fortunately, the latter case holds as will be seen.

First, a similarity analysis of the case in which the heat flux $q_w$ from the plate is a constant and all properties are constant is made. The boundary-layer description is provided by Eq. (12-1), but with the single change that at the plate the heat flux is constant so that

$$-k\frac{\partial T(y=0)}{\partial y} = q_w$$

As set forth by Sparrow and Gregg [17b], the similarity solution proceeds by selecting a stream function $\psi$ such that $u = \partial\psi/\partial y$ and $v = -\partial\psi/\partial x$ so that the continuity equation is automatically satisfied. Then it is found that a similarity solution is possible if

$$\psi = c_2 x^{4/5} F(\eta)$$

where

$$\eta = \frac{c_1 y}{x^{1/5}}$$

and

$$c_1^5 = \frac{g\beta q_w}{5k\nu^2}$$

$$c_2^5 = \frac{5^4 g\beta q_w \nu^3}{k}$$

Thus

$$u = c_1 c_2 x^{3/5} F'(\eta) \quad \text{and} \quad v = c_2 \frac{\eta F'(\eta) - 4F(\eta)}{5x^{1/5}}$$

Then, with $\theta = c_1(T_\infty - T)/(x^{1/5} q_w / k)$, the $x$-motion and energy equations become

$$F''' - 3(F')^2 + 4FF'' - \theta = 0 \qquad (12\text{-}25a)$$

$$\theta'' + \text{Pr}(4\theta' F - \theta F') = 0 \qquad (12\text{-}25b)$$

subject to the boundary conditions of

$$F(\eta = 0) = 0 = F'(\eta = 0)$$

$$\theta'(\eta = 0) = 1$$

$$\theta(\eta \to \infty) = 0 = F'(\eta \to \infty)$$

A prime is used to denote differentiation with respect to $\eta$. Equations (12-25) were solved numerically for $\text{Pr} = 0.1, 1, 10,$ and $10^2$. The temperature excess at the wall $T_w - T_\infty$ is obtained from these solutions and the definition of $\theta$ to be

$$T_w(x) - T_\infty = c_1^{-1} \frac{q_w}{k} \theta(0) x^{1/5} = - \left( \frac{5\nu^2 q_w^4 x}{k^4 g \beta} \right)^{1/5} \theta(0)$$

This relation can be rephrased in terms of the temperature excess at the top of the plate $T_w(x = L) - T_\infty$ as

$$T_w(x) - T_\infty = [T_w(L) - T_\infty] \left( \frac{x}{L} \right)^{1/5}$$

In this form it is seen that the wall temperature excess increases, gently, as the $\frac{1}{5}$th power of distance from the leading edge. Its magnitude, further, is proportional to $q_w^{4/5}$. The local heat-transfer coefficient is found in the usual way from

$$q_w = h(T_w - T_\infty)$$

to be

$$h = - \frac{[(k^4 g \beta q_w)/(5\nu^2 x)]^{1/5}}{\theta(0)}$$

varying as $q_w^{1/5}$ and $x^{-1/5}$, in contrast to the $x^{-1/4}$ variation found for constant wall temperature. In dimensionless terms, the local Nusselt number then is

$$\text{Nu}_x = \frac{hx}{k} = -\frac{(\text{Gr}_x^*/5)^{1/5}}{\theta(0)}$$

where $\text{Gr}_x^* = g\beta(q_w x/k)x^3/\nu^2$ is a modified Grashof number with $q_w x/k$ playing the role of a temperature difference. The average heat-transfer coefficient $\bar{h}$ is of interest as well. To calculate it, the temperature difference can be specified as either the mean temperature excess $L^{-1}\int_0^x (T_w - T_\infty)\, dx$ or the temperature halfway along the plate. Table 12-3 shows the results. There is little difference between the heat-transfer coefficients for constant-heat-flux and constant-temperature conditions, particularly if the driving temperature excess is evaluated at the halfway point on the plate.

Note that the modified Grashof number $\text{Gr}_x^* = g\beta(q_w x/k)x^3/\nu^2$ is related to the conventional Grashof number $\text{Gr}_x = g\beta(T_w - T_\infty)x^3/\nu^2$ by

$$\text{Gr}_x^* = \text{Gr}_x \text{Nu}_x$$

since $h(T_w - T_\infty) = q_w$. If it is presumed that the criterion for laminar flow remains as it was for the constant wall temperature case $\text{Gr}_x \text{Pr} \leqslant 10^9$, and if the integral method result of Problem 12-6, $\text{Nu}_x = 0.62[\text{Pr}^2\,\text{Gr}_x^*/(\text{Pr} + 0.8)]^{1/5}$, is used, it is found that the criterion for laminar flow in the constant heat-flux case is [18]

$$\text{Gr}_x^* \leqslant \left(\frac{0.916\text{Pr}}{\text{Pr} + 0.8}\right)^{1/4} 10^{11}$$

Application of the integral method to the case where heat flux (or surface temperature) is specified and varies over the plate surface was studied by Sparrow [19] and is the subject of Problem 12-6. Similarity solutions for plate temperature variations of the form $T_w - T_\infty = Nx^n$ and $T_w - T_\infty = Me^{mx}$

**Table 12-3  Nusselt Number Dependence on Boundary Condition for Laminar Natural Convection from a Vertical Plate [17b]**

| Pr | $\overline{\text{Nu}}_{q_w}/\overline{\text{Nu}}_{T_w}$ Mean Temperature | $\overline{\text{Nu}}_{q_w}/\overline{\text{Nu}}_{T_w}$ Halfway Temperature | $\theta(0)$ | $F''(0)$ |
|---|---|---|---|---|
| 0.1 | 1.08 | 1.02 | $-2.7507$ | 1.6434 |
| 1 | 1.07 | 1.015 | $-1.3574$ | 0.72196 |
| 10 | 1.06 | 1.01 | $-0.76746$ | 0.30639 |
| 100 | 1.05 | 1 | $-0.46566$ | 0.12620 |

were obtained by Sparrow and Gregg [20]. The integral method for specified $q_w$ was generalized for natural convection by Tribus [21] and then further extended by Bobco [22]; the results are usually within 10% of the exact solutions —for $q_w \sim x^n$, it is found that $T_w - T_\infty \sim x^{(4n+1)/5}$ as in exact solutions and

$$\text{Nu}_x = \frac{(7\text{Pr}/27)^{1/4}(n+1)^{1/2}\text{Ra}_x^{1/4}}{[(1 + 35\text{Pr}/12)(n+1) + 4/3]^{1/4}}$$

## 12.4   LAMINAR NATURAL CONVECTION FROM A CONSTANT-TEMPERATURE VERTICAL CYLINDER IMMERSED IN AN INFINITE FLUID

Laminar natural convection from a constant temperature vertical cylinder of diameter $D = 2R$ and length $L$ can be considered to be the same as from a vertical flat plate if curvature effects are negligible (i.e., $D/L$ large). When $D/L$ is sufficiently small for curvature to be significant, vertical cylinder results deviate from flat-plate results.

A similarity solution of the describing boundary-layer equations in cylindrical coordinates

$$\frac{\partial(ru)}{\partial x} + \frac{\partial(rv)}{\partial r} = 0$$

$$u\frac{\partial u}{\partial x} + v\frac{\partial u}{\partial r} = g\beta(T - T_\infty) + \frac{v}{r}\frac{\partial(r\,\partial u/\partial r)}{\partial r}$$

$$u\frac{\partial T}{\partial x} + v\frac{\partial T}{\partial r} = \frac{\alpha}{r}\frac{\partial(r\,\partial T/\partial r)}{\partial r}$$

was executed by Sparrow and Gregg [23]. It is of interest here not only because of its useful results, but also because it provides a good illustration of a perturbation technique. A stream function $\psi$ is defined such that $ru = \partial\psi/\partial r$ and $rv = -\partial\psi/\partial x$. Then a coordinate transformation is employed with $\eta = c_1(r^2 - R^2)/x^{1/4}$ and $\xi = c_2 x^{1/4}$. Then, with $\theta = (T - T_\infty)/(T_w - T_\infty)$ and $f = c_3\psi/x^{1/4}$, it is found that $u = (2c_1/c_3)\partial f/\partial\eta$ and $v = -[\xi(f + \xi\,\partial f/\partial\xi) - \eta\xi\,\partial f/\partial\eta]/(c_2 c_3 4rx)$. The boundary-layer equations of motion and energy then become

$$\xi\left(\frac{\partial f}{\partial\eta}\frac{\partial^2 f}{\partial\xi\partial\eta} - \frac{\partial f}{\partial\xi}\frac{\partial^2 f}{\partial\eta^2}\right) - f\frac{\partial^2 f}{\partial\eta^2} = \xi^2\frac{\partial}{\partial\eta}\left[(1 + \xi\eta)\frac{\partial^2 f}{\partial\eta^2}\right] + \xi^4\theta$$

$$\xi\left(\frac{\partial\theta}{\partial\xi}\frac{\partial f}{\partial\eta} - \frac{\partial\theta}{\partial\eta}\frac{\partial f}{\partial\xi}\right) - f\frac{\partial\theta}{\partial\eta} = \frac{\xi^2}{\text{Pr}}\frac{\partial}{\partial\eta}\left[(1 + \xi\eta)\frac{\partial\theta}{\partial\eta}\right]$$

which is subject to the boundary conditions of

At $\eta = 0$ $\qquad f = 0$ $\qquad$ at $\eta = \infty$ $\qquad \partial f/\partial \eta = 0$

$$\frac{\partial f}{\partial \eta} = 0 \qquad\qquad \theta = 0$$

$$\frac{\partial f}{\partial \xi} = 0$$

$$\theta = 1$$

Here

$$c_1 = \left[\frac{g\beta(T_w - T_\infty)R^3}{\nu^2}\right]^{1/4} \frac{R^{-7/4}}{2^{3/2}}$$

$$c_2 = \left[\frac{g\beta(T_w - T_\infty)R^3}{\nu^2}\right]^{-1/4} \frac{2^{3/2}}{R^{1/4}}$$

$$c_3 = \left[\frac{g\beta(T_w - T_\infty)R^3}{\nu^2}\right]^{-3/4} \frac{2^{3/2}}{\nu R^{3/4}}$$

At this point it is recognized that $\xi$ is small if $R$ is large, and interest is concentrated on that case. Therefore, $f$ and $\theta$ are expanded in series as

$$f(\xi, \nu) = \xi^2\left[f_0(\eta) + \xi f_1(\eta) + \xi^2 f_2(\eta) + \cdots\right]$$

$$\theta(\xi, \eta) = \theta_0(\eta) + \xi\theta_1(\eta) + \xi^2\theta_2(\eta) + \cdots$$

These series are then substituted into the motion and energy equations with the result that, with a prime denoting a derivative with respect to $\eta$,

$$\xi^4\left[f_0''' + 3f_0 f_0'' - 2(f_0')^2 + \theta_0\right]$$

$$+\xi^5\left[f_1''' + f_0'' + \eta f_0''' - 5f_0' f_1' + 4f_0'' f_1 + 3f_1'' f_0 + \theta_1\right] + \cdots = 0$$

$$\xi^2\left[\theta_0'' + 3\mathrm{Pr}\, f_0\theta_0'\right]$$

$$+\xi^3\left[\theta_1'' + \theta_0' + \eta\theta_0'' - \mathrm{Pr}(f_0'\theta_1 - 4f_1\theta_0' - 3f_0\theta_1')\right] + \cdots = 0$$

and the boundary conditions of

At $\eta = 0$ $\quad f_0 = f_1 = \cdots = 0$ $\qquad$ at $\eta = \infty$ $\quad f_0' = f_1' = \cdots = 0$

$\qquad\qquad f_0' = f_1' = \cdots = 0$ $\qquad\qquad\qquad \theta_0 = \theta_1 = \cdots = 0$

$\qquad\qquad \theta_0 = 1, \quad \theta_1 = \cdots = 0$

Each coefficient of $\xi^n$ must equal zero if the left-hand side of these two equations is to be satisfied for all $\xi$. Hence the solutions for $f_0$ and $\theta_0$ (the flat-plate results) can be obtained first. Then $f_1$ and $\theta_1$ can be determined numerically, by using $f_0$ and $\theta_0$ as known functions. By proceeding in this way, as many terms of the series can be evaluated as accuracy requires.

The local heat-transfer coefficient is expressed in terms of these solutions as

$$h_x = -k \frac{\partial T(R, x)/\partial r}{T_w - T_\infty} = -2Rc_1 kx^{-1/4} \frac{\partial \theta(\eta = 0, \xi)}{\partial \eta}$$

This result can be rearranged into the form

$$\mathrm{Nu}_x = \left[ -\theta_0'(0) \left( \frac{\mathrm{Gr}_x}{4} \right)^{1/4} \right] \left[ 1 + \frac{\xi \theta_1'(0)}{\theta_0'(0)} + \frac{\xi^2 \theta_2'(0)}{\theta_0'(0)} + \cdots \right]$$

in which the first bracketed term is recognized as the local Nusselt number for a vertical flat plate. Numerical results for $\mathrm{Pr} = 0.72$ and 1 were obtained and are displayed in Table 12-4. There it is seen that curvature effects make the average Nusselt number greater for the cylinder than the plate. Flat-plate results are accurate within 5% for $\mathrm{Pr} = 1$ if $\mathrm{Ra}_L^{1/4} D/L \gtrsim 33$—this corresponds to $\xi = 0.15$, so the series solution encounters no convergence difficulties. These results were extended by Cebeci [24], whose results for local coefficients are shown in Fig. 12-6 (where subscript FP denotes flat plate and cyl denotes cylinder).

Essentially the same results were obtained by application of an integral method to the describing boundary-layer equations by LeFevre and Ede [25]. The integral form of the boundary-layer equations of energy and motion is

Table 12-4  Nusselt Number Dependence on Vertical Cylinder Diameter [23]

| $\mathrm{Ra}_L^{1/4} D/L$ | $\overline{\mathrm{Nu}}_{L,\mathrm{cyl}}/\overline{\mathrm{Nu}}_{L,\mathrm{FP}}$[a] | |
|---|---|---|
| | $\mathrm{Pr} = 0.72$ | $\mathrm{Pr} = 1.0$ |
| 100 | 1.02 | 1.02 |
| 30 | 1.06 | 1.05 |
| 10 | 1.17 | 1.16 |
| 6 | 1.27 | 1.26 |

[a]FP = flat plate.

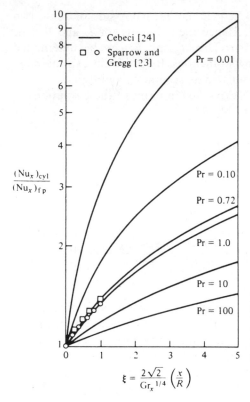

**Figure 12-6** Ratio of the local Nusselt number for a vertical cylinder to that for a vertical flat plate for laminar natural convection at constant wall temperature. (From T. Cebeci, *Fifth International Heat Transfer Conference*, Vol. 3, Section NC 1.4, 1975, pp. 15–19 [24].)

then

$$\frac{d\left[\int_R^{R+\delta} u^2 r\, dr\right]}{dx} = g\beta\int_R^{R+\delta} \theta r\, dr - \nu R\frac{\partial u(r=R)}{\partial r}$$

$$\frac{d\left[\int_R^{R+\delta} u\theta r\, dr\right]}{dx} = -\alpha R\frac{\partial \theta(r=R)}{\partial r}$$

The velocity and temperature profiles of the flat-plate situation discussed in Section 12.2 were used. The resulting two differential equations were then expanded as a power series in $1/R$ in a manner similar to that employed in the exact solution procedure to find the average Nusselt number as

$$\overline{Nu}_{x,\text{cyl}} = \frac{4}{3}\left[\frac{(7/5)Pr}{(20+21Pr)}\right]^{1/4} Ra_x^{1/4}$$

$$+ \frac{(4/35)(272+315Pr)}{64+63Pr}\frac{L}{D} + \cdots$$

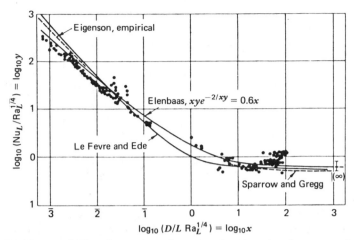

**Figure 12-7**  Measured Nusselt numbers for natural convection from a vertical cylinder in air. [By permission from A. J. Ede, Advances in Free Convection, *Adv. Heat Transf.* 4, pp. 1–64 (1967) [6]; copyright 1967, Academic Press, New York.]

It is seen again that the integral method offers the substantial advantage over an exact method of an explicit formula. The first term of this relationship is the flat plate result of Eq. (12-23). Rearrangement then gives

$$\frac{\overline{\mathrm{Nu}}_{L,\mathrm{cyl}}}{\overline{\mathrm{Nu}}_{L,\mathrm{fp}}} = 1 + K\frac{L}{D\,\mathrm{Ra}_L^{1/4}} + \cdots \tag{12-26}$$

where $K = \frac{3}{7}[(\mathrm{Pr} + 272/315)/(\mathrm{Pr} + 64/63)][15(\mathrm{Pr} + 20/21)/\mathrm{Pr}]^{1/4}$. In this form it is more clearly seen that the curvature effect tends to increase the average Nusselt number above the value it would have for a flat plate. Turbulent natural convection about a vertical cylinder was investigated with an integral method by Na and Chiou [101].

Elenbass [26, 51], on the basis of the stationary-film hypothesis of Langmuir [27], suggested that vertical cylinder average coefficients be correlated by

$$\overline{\mathrm{Nu}}_D \exp\!\left(\frac{-2}{\overline{\mathrm{Nu}}_D}\right) = 0.6\left(\frac{D}{L}\right)^{1/4}\mathrm{Ra}_D^{1/4}$$

which reduces correctly to the flat-plate result as $D/L$ becomes very large. Eigenson's [28] experiments for cylinders of widely varying sizes are included in the presentation of Fig. 12-7 due to Ede [6] where substantial agreement of investigators is found.

Additional information on this special topic is given by Kyte et al [29], whose results are shown in Fig. 12-8.

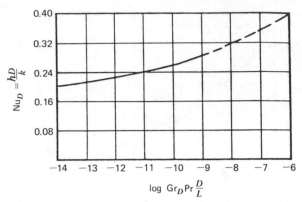

**Figure 12-8** Correlation of Nusselt number for natural convection from small vertical cylinders. [By permission from J. R. Kyte, A. J. Madden, and E. L. Piret, *Chem. Eng. Progr.*, **49**, 653–662 (1953) [29].]

During any application of these results it should be kept in mind that a slight sidewise deflection of the rising air away from the upper part of the cylinder can greatly increase the heat transfer. Accordingly, experiments must be done very carefully.

The foregoing discussion for isothermal vertical cylinders is complemented by the findings of Nagendra et al. [30] for vertical cylinders with a constant heat flux. They found

$$\overline{Nu}_D = 1.37 \left( Ra_D \frac{D}{L} \right)^{0.16}, \quad
\begin{cases}
0.6 \left( Ra_D \dfrac{D}{L} \right)^{0.25}, & Ra_D \dfrac{D}{L} \geqslant 10^4 \\[2ex]
1.37 \left( Ra_D \dfrac{D}{L} \right)^{0.16}, & 0.05 \leqslant Ra_D \dfrac{D}{L} \leqslant 10^4 \\[2ex]
0.93 \left( Ra_D \dfrac{D}{L} \right)^{0.05}, & Ra_D \dfrac{D}{L} \leqslant 0.05
\end{cases}$$

in which the average temperature difference is used in the Rayleigh number Ra. These results differ by less than 5% from the isothermal case.

## 12.5   TURBULENT NATURAL CONVECTION FROM A VERTICAL FLAT PLATE OF CONSTANT TEMPERATURE

Reference to Figs. 12-4 and 12-5 shows that the laminar relations of previous sections do not accurately predict the heat-transfer coefficient when the Rayleigh number $Ra_x = Gr_x Pr$, becomes large. Although these two figures refer to a constant temperature vertical flat plate, the same general observation

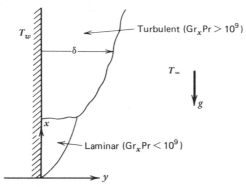

**Figure 12-9** Transition from laminar to turbulent boundary-layer flow in natural convection on a hot vertical flat plate.

applies to other geometries as well. This change in behavior is caused by the boundary layer next to the surface undergoing a transition from laminar to turbulent flow.

In the context of vertical flat plates, it appears from experimental data that the plate can be taken to be covered by a turbulent boundary layer if $Ra_L \geqslant 10^9$ [6]. However, turbulence begins gradually over a transition length whose distance from the plate leading edge is influenced by external disturbances, just as for forced convection. This point needs to be kept in mind in the discussions that follow.

To illustrate the relationship of the turbulence fundamentals studied in Chapter 9, the integral method is applied to the constant-fluid-property case as was first done by Eckert and Jackson [31]. The physical situation is as sketched in Fig. 12-9. An integral formulation of the describing boundary-layer equations is available from Section 12.2 as

$$\frac{d\left[\int_0^\delta u^2 \, dy\right]}{dx} - g\beta \int_0^\delta (T - T_\infty) \, dy = -\frac{\tau_w}{\rho} \qquad (12\text{-}15a)$$

$$\frac{d\left[\int_0^\delta u(T - T_\infty) \, dy\right]}{dx} = \frac{q_w}{\rho C_p} \qquad (12\text{-}15b)$$

where $\delta_T = \delta$ has been assumed—this assumption is justified only by the accuracy of the final results. Appropriate velocity and temperature distributions must now be selected. This endeavor is guided by the insight that vigorous turbulent mixing should yield a fairly large velocity $U_1$ near the wall. This would correspond to a turbulent boundary layer, parallel to a flat plate, of external velocity $U_1$. Thus the $\frac{1}{7}$th power-law velocity distribution of Chapter

10 is adopted here as

$$u = U_1\left(\frac{y}{\delta}\right)^{1/7}$$

Obviously this form cannot stand unchanged since, although it correctly predicts $u(y = 0) = 0$, it does not satisfy the other requirements that

$$\text{At } y = 0 \qquad g\beta(T_w - T_\infty) + \nu\frac{\partial^2 u}{\partial y^2} = 0 \qquad \text{at } y = \delta \qquad u = 0$$

$$\frac{\partial u}{\partial y} = 0$$

$$\frac{\partial^2 u}{\partial y^2} = 0$$

$$\frac{\partial^3 u}{\partial y^3} = 0$$

Hence it is finally assumed that

$$u = U_1\left[a + b\eta + c\eta^2 + d\eta^3 + e\eta^4\right]\eta^{1/7}$$

with $\eta = y/\delta$. Recognition that $a = 1$ if the $\frac{1}{7}$th power law is to hold true near the wall and imposition of the four conditions at $y = \delta$ then gives the velocity profile as

$$u = U_1(1 - \eta)^4\eta^{1/7} \qquad\qquad (12\text{-}27)$$

Equation (12-27) has $u_{max}/U_1 = 0.613$ at $\eta = \frac{1}{29}$. This suggests that $U_1$ can be interpreted as the maximum velocity in the boundary layer or proportional to it. Further, since the maximum velocity occurs near the plate, the wall shear stress and heat flux must behave nearly as for forced convection with a free-stream velocity of $U_1$. The temperature profile near the wall must be similar to the velocity profile under turbulent conditions.

Thus $T - T_w \sim u$ since both must be zero at the wall and must closely approach free-stream conditions at a distance from the wall of the order of the boundary-layer thickness. If $u \sim (y/\delta)^{1/7}$, then $T - T_w \sim (y/\delta)^{1/7}$. The conditions imposed on the temperature distribution are

At $y = 0$ $\qquad\qquad T = T_w$ $\qquad\qquad$ at $y = \delta$ $\qquad\qquad T = T_\infty$

$$g\beta(T_w - T_\infty) = -\nu \frac{\partial^2 u}{\partial y^2} \qquad\qquad\qquad \frac{\partial T}{\partial y} = 0$$

$$\frac{\partial^2 T}{\partial y^2} = 0 \qquad\qquad\qquad\qquad \frac{\partial^2 T}{\partial y^2} = 0 \qquad (12\text{-}28)$$

Of these conditions only the requirement that $T(y = 0) = T_w$ can be met simply. So

$$T - T_\infty = (T_w - T_\infty)(1 - \eta^{1/7})$$

The turbulent wall heat flux $q_w$ and shear stress $\tau_w$ cannot be obtained from the approximate profiles of Eqs. (12-27) and (12-28), unlike the laminar flow case, since these profiles do not account with sufficient accuracy for the effects of a laminar sublayer. The wall shear stress can be obtained, however, by an appeal to the forced-convection turbulence information developed in Chapter 9 in which a $\frac{1}{7}$th power-law velocity profile gave

$$\frac{\tau_w}{\rho U_\infty^2} = \frac{C_f}{2} = 0.0228\left(\frac{U_\infty \delta}{\nu}\right)^{-1/4} \qquad (9\text{-}22)$$

If it is presumed that $U_1$ of Eq. (12-27) plays the role of $U_\infty$, it follows that

$$\frac{\tau_w}{\rho} = 0.0228 U_1^2\left(\frac{\nu}{U_1 \delta}\right)^{1/4} \qquad (12\text{-}29)$$

From Reynolds analogy it is known that

$$\frac{Nu}{Re\,Pr^{1/3}} = \frac{C_f}{2}$$

Recognition that $h = q_w/(T_w - T_\infty)$, $C_f/2 = \tau_w/\rho U_1^2$, and $Re = U_1 L/\nu$ then leads to

$$\frac{q_w}{\rho C_p} = (T_w - T_\infty)Pr^{-2/3}\frac{\tau_w}{\rho U_1}$$

or

$$\frac{q_w}{\rho C_p} = 0.0228 Pr^{-2/3}(T_w - T_\infty)U_1\left(\frac{\nu}{U_1 \delta}\right)^{1/4} \qquad (12\text{-}30)$$

The three integrals required in the integral equations in Eq. (12-5) can now be evaluated as

$$\int_0^\delta u^2 \, dy = I_1 \, \delta U_1^2 \quad \text{with} \quad I_1 = 0.052315, \qquad \int_0^\delta (T - T_\infty) \, dy = I_2 (T_w - T_\infty) \delta$$

with $I_2 = \frac{1}{8}$ and $\int_0^\delta u(T - T_\infty) \, dy = I_3 U_1 (T_w - T_\infty) \delta$ with $I_3 = 0.03663$. The integral equations [Eq. (12-15)] are then

$$I_1 \frac{d(\delta U_1^2)}{dx} = g\beta(T_w - T_\infty)\delta I_2 - 0.0228 U_1^2 \left( \frac{\nu}{U_1 \delta} \right)^{1/4}$$

$$(T_w - T_\infty) I_3 \frac{d(\delta U_1)}{dx} = 0.0228 \mathrm{Pr}^{-2/3}(T_w - T_\infty) U_1 \left( \frac{\nu}{U_1 \delta} \right)^{1/4}$$

To solve these two simultaneous equations, assume that $U_1 = Ax^m$ and $\delta = Bx^n$. This assumed solution substituted into the differential equations gives

$$(2m + n)A^2 x^{2m+n-1} = g\beta(T_w - T_\infty)\frac{I_2}{I_1} x^n$$

$$-0.0228 I_1^{-1} A^{7/4} B^{-5/4} \nu^{1/4} x^{(7m-n)/4}$$

$$(m + n)x^{m+n-1} = 0.0228 I_3^{-1} \mathrm{Pr}^{-2/3} A^{-1/4} B^{-5/4} \nu^{1/4} x^{(3m-1)/4}$$

These two relationships can only be true for all values of $x$ if all exponents of $x$ are equal. Hence

$$2m + n - 1 = n = \frac{7m - n}{4} \qquad \text{and} \qquad m + n - 1 = \frac{3m - n}{4}$$

These conditions are satisfied by $m = \frac{1}{2}$ and $n = \frac{7}{10}$. Since the exponents are now known, the coefficients are readily found to be

$$A^2 = \frac{10g\beta(T_w - T_\infty)I_2/17I_1}{1 + 12I_3 \, \mathrm{Pr}^{2/3}/17I_1}$$

$$B = \left( 0.0228 \frac{0.89}{I_3} \right)^{4/5} \nu^{1/5} \mathrm{Pr}^{-8/15} \left[ \frac{1 + 12I_3 \, \mathrm{Pr}^{2/3}/17I_1}{g\beta(T_w - T_\infty)I_2/I_1} \right]^{1/10}$$

The local heat-transfer coefficient can now be computed from Eq. (12-30), by recognizing that $h = q_w/(T_w - T_\infty)$, as

$$h = 0.02979k\,\mathrm{Pr}^{1/15}\left\{\frac{\left[g\beta(T_w - T_\infty)\mathrm{Pr}/\nu^2\right]^{2/5}}{(1 + 0.494\,\mathrm{Pr}^{2/3})^{2/5}}\right\}x^{1/5} \qquad (12\text{-}31)$$

In turbulent flow $h$ increases with $x$ as $x^{1/5}$, in contrast with the laminar flow finding that $h$ decreases with $x$ as $x^{-1/4}$. Evidently, turbulent mixing becomes more intense as $x$ increases.

If it is presumed that the plate is covered by a turbulent boundary layer, the average coefficient $\bar{h}$ can be found from Eq. (12-31) by integration to be $\bar{h} = 5h/6$. In terms of dimensionless groups,

$$\overline{\mathrm{Nu}}_x = 0.0248\frac{\mathrm{Pr}^{1/15}}{(1 + 0.494\,\mathrm{Pr}^{2/3})^{2/5}}(\mathrm{Ra}_x)^{2/5} \qquad (12\text{-}32)$$

which is to be applied only if $\mathrm{Ra}_x \geqslant 10^9$. For air ($\mathrm{Pr} = 0.72$), this becomes

$$\overline{\mathrm{Nu}}_x = 0.0212(\mathrm{Ra}_x)^{2/5}$$

in which the coefficient is only 1% different from the experimental coefficient of 0.021 displayed in Fig. 12-5. Application of these results to determination of the shear stress at the wall is the subject of Problem 12-8.

The agreement of Eq. (12-32) with measurements is quite satisfactory. These measurements were made on plates of human size, however, so there is still room to question the ability of Eq. (12-32) to be extrapolated to very large Rayleigh numbers. As reviewed by Ede [6], a number of alternative suggestions have been put forth. It was early suggested that when turbulence occurs over all of the plate, $h$ should be constant, which requires $\mathrm{Nu}_x \sim \mathrm{Gr}_x^{1/3}f(\mathrm{Pr})$. Another suggestion is that in fully developed turbulence, $h$ should be independent of viscosity, which requires that $\mathrm{Nu}_x \sim f(\mathrm{Gr}_x\,\mathrm{Pr}^2)$; yet another suggestion is that thermal conductivity should also be of no influence, which, taken together with the result for negligible viscosity, requires that $\mathrm{Nu}_x \sim (\mathrm{Gr}_x\,\mathrm{Pr}^2)^{1/2}$. The $\frac{2}{5}$ exponent of $\mathrm{Gr}_x$ in Eq. (12-32) lies between the suggested values of $\frac{1}{3}$ and $\frac{1}{2}$ but is open to some criticism because it utilizes a forced convection wall shear stress relationship whose accuracy is restricted to a range of Reynolds numbers. A solution based on these lines of thought has been advanced by LeFevre [6] and is the subject of Problem 12-9.

Turbulent analyses such as these have been refined by Bayley [32]. He found that for the Prandtl numbers typical of liquid metals and for $10^{10} \leqslant \mathrm{Gr}_x \leqslant 10^{15}$,

$$\overline{\mathrm{Nu}} = 0.08\,\mathrm{Gr}_x^{1/4}$$

with the plate at constant temperature. For a fluid with $Pr \approx 1$ such as air, he found

$$\overline{Nu} = 0.1\,Ra^{1/3}, \qquad 2 \times 10^9 \leqslant Ra \leqslant 10^{12}$$

$$= 0.183\,Ra^{0.31}, \qquad 2 \times 10^9 \leqslant Ra \leqslant 10^{15} \qquad (12\text{-}32a)$$

which agrees well with measurements [33]. The exponent of the Rayleigh number is uncertain even for $Pr \approx 1$, ranging from 0.4 in Eq. (12-32) to 0.31 in Eq. (12-32b) with $\frac{1}{3}$ in Eq. (12-32a) as an intermediate value. An exponent of $\frac{1}{3}$ is appealing since it makes the choice of a characteristic length immaterial as far as $h$ is concerned and shows $h$ to be constant along the plate. If the exponent is greater than $\frac{1}{3}$, $h$ becomes very large at large Ra; if the exponent is less than $\frac{1}{3}$, $h$ becomes very small at large Ra. Available measurements are not adequate to resolve the question. A refined correlation [93] is

$$\overline{Nu}_x^{1/2} = 0.825 + \frac{0.387\,Ra_x^{1/6}}{\left[1 + (0.492/Pr)^{9/16}\right]^{8/27}} \qquad (12\text{-}32b)$$

Numerical solution of the boundary-layer equations for free convection with a mixing-length turbulence model [98] suggest that the exponent of $Ra_x$ is $\frac{1}{3}$, as is implied in Eq. (12-32b), a suggestion that is strengthened by a recent analysis [100].

Siegel [34] obtained a solution for the case of a prescribed uniform heat flux that gives the local coefficient as

$$Nu_x = \frac{0.080\,Pr^{1/3}\,Gr_x^{*2/7}}{(1 + 0.444\,Pr^{2/3})^{2/7}}$$

which is in general agreement with the measurements of Vliet and Liu [35] for water who suggested $Nu_x = 0.568(Pr\,Gr_x^*)^{0.22}$ for $2 \times 10^{13} < Gr_x^*Pr < 10^{16}$—the exponent of $Gr_x^*$ is not important to a calculation of $Nu_x$ over a restricted $Gr_x^*$ range since the coefficient is selected properly, but it is important to the question of whether $h$ is increasing, constant, or decreasing along the plate. Lemlich and Vardi [36] extended this analysis to the case where the body force is proportional to $x$, as it would be in a centrifuge, to find

$$Nu_x = \frac{0.0729\,Pr^{1/3}\,Gr_x^{*2.7}}{(1 + 0.316\,Pr^{2/3})^{2/7}}$$

In these two preceding expressions, $Gr_x^* = g\beta(q_w x/k)x^3/\nu^2$ as in the two that follow. They found in the case of laminar flow that

$$\mathrm{Nu}_x = \frac{0.546\,\mathrm{Pr}^{1/4}(\mathrm{Gr}_x^*\mathrm{Pr})^{1/4}}{(\mathrm{Pr} + 1.143)^{1/4}}$$

for uniform plate temperature and

$$\mathrm{Nu}_x = \frac{0.616\,\mathrm{Pr}^{1/4}(\mathrm{Gr}_x^*\mathrm{Pr})^{1/4}}{(\mathrm{Pr} + 1.143)^{1/4}}$$

for constant heat flux.

## 12.6 COMBINED NATURAL AND FORCED CONVECTION

Natural convection occurs whenever density gradients are present. In the absence of externally forced flow, natural convection is the only mechanism causing flow. As the externally forced flow becomes larger, it plays an appreciable role in fluid motion near the heated surface even though natural convection is still dominant. When the externally forced flow is quite large, forced-convection effects are dominant and natural convection effects are negligible. In the absence of detailed information concerning the heat-transfer coefficient in a situation of combined natural and forced convection, a rational design procedure is to calculate the coefficient for forced convection and for natural convection separately; then the larger value is adopted.

For some cases detailed studies have been made that allow a more accurate evaluation of the heat-transfer coefficient. Sometimes the demarcation between the purely forced and the combined (as well as the purely natural and combined) regimes is given in terms of the dimensionless group $\mathrm{Gr}_x/\mathrm{Re}_x^2$. To make the importance of the Grashof number/(Reynolds number)$^2$ plausible on physical grounds, an elementary consideration of the ratio of buoyancy to inertial forces $F_b/F_i$ is essayed in a mixed natural- and forced-convection case. The buoyancy force is roughly given by

$$F_b \sim g\,\Delta\rho\,L^3$$

where $\Delta\rho$ is the density differential in a cube of side $L$. The inertial force on a surface of side $L$ in a flowing stream of normal velocity $U$ is roughly given by

$$F_i \sim \rho U^2 L^2$$

Thus $F_b/F_i \sim g\,\Delta\rho\,L^3/(\rho U^2 L^2)$, which can be rearranged to give

$$\frac{F_b}{F_i} \sim \frac{g(\Delta\rho/\rho)L^3/\nu^2}{(UL/\nu)^2} = \frac{\text{Gr}}{\text{Re}^2}$$

When $\text{Gr}/\text{Re}^2$ is of the order of unity, both natural and forced convection are important. On the other hand, natural convection is dominant when $\text{Gr}/\text{Re}^2$ is large, whereas forced convection is dominant when $\text{Gr}/\text{Re}^2$ is small—Churchill [96] in a critical survey suggests that $\overline{\text{Nu}}^3_{\text{combined}} = \overline{\text{Nu}}^3_{\text{forced}} + \overline{\text{Nu}}^3_{\text{natural}}$ for aiding effects only.

The local Nusselt number for combined laminar free and forced convection from a constant temperature vertical plate was studied by Lloyd and Sparrow [37], whose results are shown in Fig. 12-10. The forced flow is in the same direction as the natural flow; the influence of natural convection is less than 5% of the forced-convection prediction if $\text{Gr}_x/\text{Re}_x^2 < 0.1$ for $\text{Pr} \approx 1$. Mori [38] analytically found that natural convection effects were less than 10% for a horizontal plate in laminar forced convection of air if $\text{Gr}_x/\text{Re}_x^{2.5} < 0.083$. The effect of forced convection on free convection is less than 10% for the local heat-transfer coefficient in laminar flow over a vertical plate if $\text{Gr}_x/\text{Re}_x^2 > 4$ for $0.73 < \text{Pr} < 10$ and $\text{Gr}_x/\text{Re}_x^2 > 60$ for $\text{Pr} \approx 100$, according to Acrivos [39]. In turbulent natural convection from a vertical plate of constant temperature,

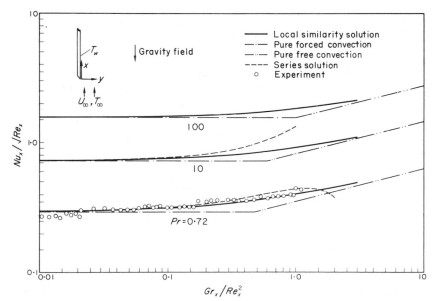

**Figure 12-10** Local Nusselt number for combined natural and forced convection from an isothermal vertical plate. [By permission from J. R. Lloyd and E. M. Sparrow, *Internatl. J. Heat Mass Transf.* **13**, 434–438 (1970) [37].]

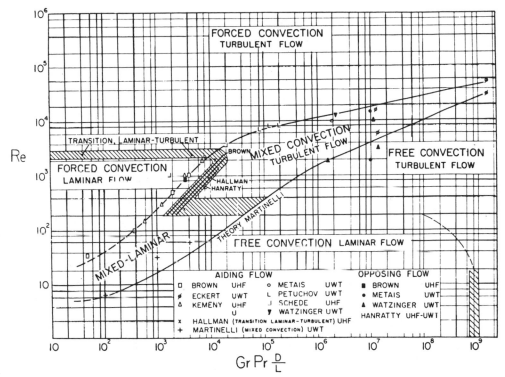

**Figure 12-11** Regimes of free, forced, and mixed convection for flow through vertical tubes (UHF, uniform heat flux; UWT, uniform wall temperature). [From B. Metais and E. R. G. Eckert, *Transact. ASME, J. Heat Transf.*, **86**, 295-296 (1964) [45].]

aiding (as opposed to cross or opposing) externally forced flow was found by Hall and Price [40] to influence the natural convection by less than 10% if $Gr_x/Re_x^2 < 10$. For laminar flow on a vertical plate, opposed natural and forced convection causes boundary-layer separation at $Gr_x/Re_x^2 = 0.2$ [82]; the effects of buoyancy forces decrease as the plate inclines toward the horizontal [83, 84].

Airflow normal to long horizontal cylinders was studied by Oosthuizen and Madan [41], who recommend, for $100 < Re_D < 3000$ and $2.5 \times 10^4 < Gr_D < 3 \times 10^5$, the correlation

$$\frac{\overline{Nu}_D}{\overline{Nu}_{D,\text{forced}}} = 1 + 0.18\frac{Gr_D}{Re_D^2} - 0.011\left(\frac{Gr_D}{Re_D^2}\right)^2$$

where $\overline{Nu}_{D,\text{forced}} = 0.464\,Re_D^{1/2} + 0.0004Re_D$. Natural-convection effects were found to be negligible if $Gr_D/Re_D^2 < 0.28$. A similar study was conducted by Collis and Williams [42]. Gebhart et al. [43] studied long horizontal wires in mixed convection and found that natural convection influence is less than 10%

of forced-convection predictions if

$$\frac{\mathrm{Gr}_D}{\mathrm{Re}_D^3} < 10.$$

Airflow about a sphere was experimentally studied by Yuge [44], who gives detailed formulations for predicting Nusselt number for combined forced and natural convection when the externally forced flow is aiding, opposing, and cross with the natural convective flow. Natural convection is of negligible influence on heat transfer if $\mathrm{Gr}_D/\mathrm{Re}_D^2 < 10^{-2}$ in aiding and cross flows. In opposing flow, a minimum Nusselt number was observed at $\mathrm{Gr}_D/\mathrm{Re}_D^2 = 2.5$.

Metais and Eckert [45] reviewed experimental information concerning forced, mixed, and natural convection in horizontal and vertical tubes. Figure 12-11 shows the various regimes for vertical tubes; Fig. 12-12 applies to horizontal tubes. In both figures the ordinate is the Reynolds number based on diameter, and the abscissa is the product of Grashof times Prandtl number times diameter divided by length—the Grashof number is based on pipe diameter

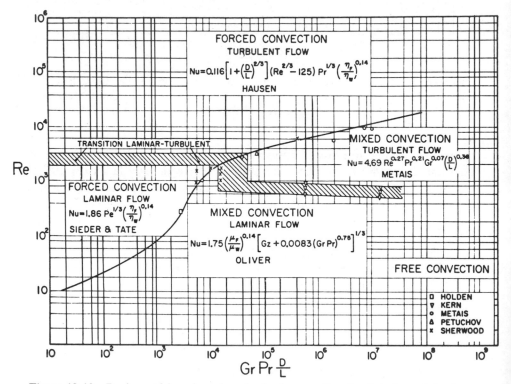

**Figure 12-12**  Regimes of free, forced, and mixed convection for flow through horizontal tubes. [From B. Metais and E. R. G. Eckert, *Transact. ASME, J. Heat Transf.* **86**, 295–296 (1964) [45].]

and difference between wall and bulk fluid temperatures. The demarcations of the forced and natural convection regimes were established at the conditions under which the actual heat flux deviated by less than 10% from the value predicted for either forced or natural convection actingly singly. In Fig. 12-11 data for forced convection both aiding and opposing natural convection was included as were data for uniform wall temperature and uniform heat flux; Fig. 12-12 includes only uniform-wall-temperature data. In Fig. 12-11 the Graetz number is defined as $Gz = Re_D \, Pr \, D/L$. These Metais–Eckert charts are for $Pr \geqslant 1$; for liquid metals consult the experimental study by Buhr et al. [46]. Recent results [92, 102] provide general confirmation.

For a heated cone with vertical axis spinning in air, Kreith [47], in a survey of convection heat transfer in rotating systems, concludes that natural convection has less than 5% influence on the flow forced by the rotation if $Gr_x/Re_x^2 < 0.052$, where $Gr_x = (g\beta \Delta T x^3/\nu^2) \cos \alpha$, where $\alpha$ is the cone half-angle and $Re_x = x^2 \omega \sin(\alpha/\nu)$ with $\omega$ cone rotational speed.

## 12.7 EMPIRICAL RELATIONS FOR NATURAL CONVECTION

A large amount of analytical and experimental information concerning natural convection is available, as is evident, for both laminar and turbulent flow. Most of this information pertains to the steady state, but a substantial amount of work has also been done [6] on both the unsteady state that occurs in the initiation of a situation that will later have a steady state and the quasi-steady state which occurs when a periodic disturbance (often an oscillatory motion of the heated surface) is imposed [99].

For the steady state, it is convenient to have a summary of results additional to those previously presented. As has already been discussed, the average Nusselt number $\overline{Nu} = \overline{h}L/k$ is dependent on the Rayleigh number $Ra = Gr \, Pr$ and is usually correlated by $\overline{Nu} = C(Ra)^n$.

### Flat Plate at Small Angle from Vertical

For a flat plate inclined at an angle $\theta$ with the horizontal, Vliet recomments [48] that the component of gravity parallel to the heated surface be used (but only for the laminar case) in vertical plate correlations for $30° < \theta < 90°$. All fluid properties are to be evaluated at the film temperature, as usual. The plate inclination affects the laminar–turbulent transition as roughly given by

$$Gr \, Pr_{trans} = 3 \times 10^5 \exp(0.1368\theta)$$

where $\theta$ is plate angle from horizontal in degrees, $30° < \theta < 90°$, with transition beginning at about $\frac{1}{10}$ this value and ending at about 10 times this value.

In turbulent heat transfer in the range of inclinations from horizontal of $30° < \theta < 90°$, use of the actual value of gravity is recommended—the component of gravity parallel to the plate surface does not seem to be important—and is to be used in vertical-plate correlations.

## Horizontal Plates

For a horizontal plate, the recommended correlations to measurements are recommended by McAdams [9] to be of the form $\overline{\text{Nu}} = C\,\text{Ra}^n$. The characteristic length $L$ to be used in this correlation for a square is the side of the square; for a rectangle it is the arithmetic average of the long and short sides, and for a circular disk, it is 0.9 times the disk diameter. The values of $C$ and $n$ are cited in Table 12-5—note that in case 2 there is no turbulent case cited.

The foregoing recommendations for $C$ and $n$ correlate measurements fairly well. However, Singh et al. [49] analyzed the problem of hot horizontal surfaces facing downward with a laminar boundary-layer formulation. An integral method of solution whose results were confirmed by careful measurements give for a square of side $2a$

$$\overline{\text{Nu}}_a = 0.816\,\text{Ra}_a^{1/5}$$

for a circle of radius $a$

$$\overline{\text{Nu}}_a = 0.818\,\text{Pr}^{0.034}\,\text{Ra}_a^{1/5}$$

and for an infinite strip of width $2a$

$$\overline{\text{Nu}}_a = 0.5\,\text{Ra}_a^{1/5}$$

The measurements leading to the correlations in Table 12-5 may have been affected by unaccounted-for heat losses from plate edges.

**Table 12-5   Parameters for Natural Convection from Horizontal Plates [9]**

| Case | Type of flow | Ra Range | $C$ | $n$ |
|------|-------------|----------|-----|-----|
| 1 | Hot surface up or cool surface down | | | |
| | Laminar | $10^5$–$2 \times 10^7$ | 0.54 | $\frac{1}{4}$ |
| | Turbulent | $2 \times 10^7$–$3 \times 10^{10}$ | 0.14 | $\frac{1}{3}$ |
| 2 | Hot surface down or cool surface up | | | |
| | Laminar | $3 \times 10^5$–$3 \times 10^{10}$ | 0.27 | $\frac{1}{4}$ |

## Horizontal Cylinders

Natural convection from a single horizontal cylinder was analyzed by Hermann [50] with the result that

$$\overline{\mathrm{Nu}}_D = 0.4\,\mathrm{Ra}_D^{1/4}$$

in the laminar case. Although the recommended fit to data requires adjustment of the coefficient, the exponent is apparently correct. McAdams' correlations [9] shown in Fig. 12-13 have been replaced by the one due to Churchill and Chu [103], which is

$$\overline{\mathrm{Nu}}_D^{1/2} = 0.6 + 0.387 B^{1/6}, \qquad 10^{-5} \leqslant B \leqslant 10^{13}$$

where $B = \mathrm{Ra}_D[1 + (0.559/\mathrm{Pr})^{9/16}]^{16/9}$. (See Morgan [104] or Ali-Arabi and Salman [105] for inclined cylinders.) Correlations for liquid metals are given by

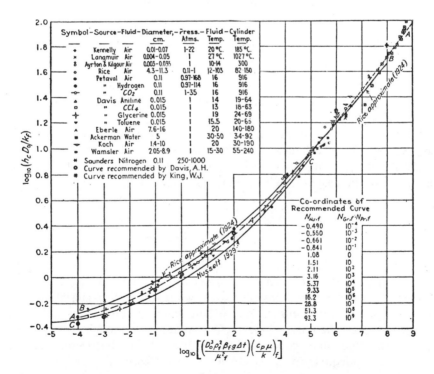

Figure 12-13 Natural convection from horizontal cylinders to gases and liquids. (From *Heat Transmission*, 3rd ed., by W. H. McAdams; copyright 1954, McGraw-Hill, New York, 1954; used with the permission of the McGraw-Hill Book Company.)

Hyman et al. [52], who recommend

$$\overline{Nu}_D = 0.53\left(\frac{Pr}{0.952 + Pr}\right)^{1/4} Ra_D^{1/4}, \qquad 10^3 \leqslant Ra_D \leqslant 10^9$$

for water, toluene, and silicones as well.

## Spheres and Miscellaneous Shapes

Natural convection from a sphere to air was measured by Yuge [44], who recommended

$$\overline{Nu}_D = 2 + 0.392\, Gr_D^{1/4}, \qquad 1 \leqslant Gr_D \leqslant 10^5$$

This is commonly generalized to, taking $Pr = 0.72$ for air,

$$\overline{Nu}_D = 2 + 0.43\, Ra_D^{1/4}, \qquad 1 \leqslant Ra_D \leqslant 10^5$$

More recently Raithby and Hollands [53] suggested the more general relation

$$\overline{Nu}_D = 2 + 0.56\left(\frac{Pr}{0.846 + Pr}\right)^{1/4} Ra_D^{1/4}$$

If sphere radius is used as the characteristic dimension, the generalized correlation due to King [54] can be used to estimate a coefficient at other values of Ra.

King's generalized correlation, mostly for air and water, shown in Fig. 12-14 can be applied to such miscellaneous shapes as blocks, spheres, horizontal cylinders, and vertical plates if a high degree of accuracy is not required. The characteristic dimension $L$ to be used is

$$\frac{1}{L} = \frac{1}{L_{\text{vertical}}} + \frac{1}{L_{\text{horizontal}}} \qquad (12\text{-}33)$$

For a sphere, $L$ is radius; for a horizontal cylinder, $L$ is diameter. As a further example, $L_v$ might be the height of a vertical cylinder and $L_h$ might be its diameter; in the turbulent range the exact value of $L$ is immaterial as a result of the $\frac{1}{3}$ slope of the curve, and in the laminar range it is still largely immaterial since the curve slope then is about $\frac{1}{4}$. Although each geometric shape might fall on a curve slightly different from the average one, this correlation is adequate for most purposes. In the case of a vertical plate, for example, its finite width might allow a three-dimensional flow pattern whose effects are roughly taken into account by use of $L$ from Eq. (12-33) in the correlations previously given for an infinitely wide vertical plate.

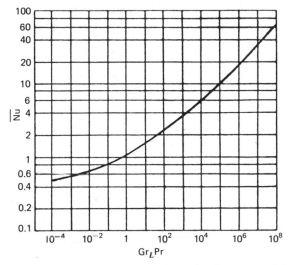

**Figure 12-14** King's generalized correlation for miscellaneous solid shapes. [From W. J. King, *Mech. Eng.*, **54**, 347–353 (1932) [54].]

Measurements of free-convection heat transfer from oblate and prolate spheriods are reported by Raithby et al. [55]. The limiting case of the thin prolate spheroid is a vertical needle, which is further discussed by Raithby and Hollands [56]. Churchill [96] gives a comprehensive correlation for laminar assisting forced and natural convection for plates, cylinders, spheres, and miscellaneous shapes.

**Enclosed Spheres and Cylinders**

Natural convection inside a spherical cavity of diameter $D$ is correlated by Kreith [57] by the empirical relation

$$\overline{Nu}_D = 0.59\, Ra_D^{1/4}, \qquad 10^4 \leqslant Ra_D \leqslant 10^9$$

$$= 0.13\, Ra_D^{1/3}, \qquad 10^9 \leqslant Ra_D \leqslant 10^{12}$$

For air enclosed between two concentric spheres, each of specified temperature, it was found by Bishop et al. [58] that the effective thermal conductivity $k_e$ is given by

$$\frac{k_e}{k} = 0.106\, Gr_b^{0.276}, \qquad 2 \times 10^4 \leqslant Gr_b \leqslant 3.6 \times 10^6 \quad 0.25 \leqslant \frac{b}{r_i} \leqslant 1.5$$

which for fluids with $Pr \approx 1$ and taking $Pr = 0.72$ for air can be generalized to $k_e/k = 0.116\, Ra_b^{0.276}$, $1.44 \times 10^4 < Ra_b < 2.6 \times 10^6$, and $0.25 \leqslant b/r_i \leqslant 1.5$.

Here $b = r_0 - r_i$, with $r_0$ being the outer sphere radius and $r_i$ the inner sphere radius. The heat flow is then found by the relation for steady heat conduction through a spherical shell

$$q = 4\pi k_e \frac{T_1 - T_2}{1/r_i - 1/r_0}$$

Natural convection in a cylindrical cavity of length $L$ and diameter $D$ was found by Evans and Stefany [59] to be correlated by

$$\overline{Nu}_D = 0.55\, Ra_L^{1/4}, \qquad 0.75 \leqslant \frac{L}{D} \leqslant 2$$

In the cylindrical annulus formed by two long and horizontal concentric cylinders, with inner radius $r_i$ and outer radius $r_0$, it is recommended [60] that the effective thermal conductivity $k_e$ be correlated as

$$\frac{k_e}{k} = 0.135\left(\frac{Ra_b\, Pr}{1.36 + Pr}\right)^{0.278}$$

for $3.2 \times 10^3 \leqslant Ra_b\, Pr/(1.36 + Pr) < 10^8$ and $0.25 \leqslant b/r_i \leqslant 3.25$ and

$$\frac{k_e}{k} = 1$$

for $Ra_b\, Pr/(1.36 + Pr) < 3.2 \times 10^3$. Here $b = r_0 - r_i$. The heat flow is then evaluated from the relation for steady heat conduction through a cylindrical shell

$$q = 2k_e L \frac{T_1 - T_2}{\ln(r_0/r_i)}$$

If $k_e/k = 1$, pure conduction takes place. These correlations were determined from measurements with air, water, and silicon oil. Inner and outer temperatures are constant. An extensive correlation [61] of data for natural convection between concentric cylinders can be consulted for additional information supported by analysis [85]. (For eccentric annuli, see Yao [86].)

## Enclosed Spaces Between Planes

Natural convection in the enclosure between two planes, each of constant temperature, inclined at an angle $\theta$ with the horizontal has been studied for many years. Recent interest in solar collector applications has led to refined correlations, most of which envision air as the fluid.

For $0° \leqslant \theta \leqslant 75°$, the correlation recommended by Hollands et al. [62] is

$$\overline{\mathrm{Nu}}_L = 1 + 1.44\left(1 - \frac{1708}{\mathrm{Ra}_L \cos \theta}\right)^{\bullet}\left[1 - \frac{1708(\sin 1.8\theta)^{1.6}}{\mathrm{Ra}_L \cos \theta}\right]$$

$$+ \left[\left(\frac{\mathrm{Ra}_L \cos \theta}{5830}\right)^{1/3} - 1\right]^{\bullet} \tag{12-33a}$$

where $\mathrm{Ra}_L = g\beta \Delta T L^3 \mathrm{Pr}/\nu^2$, $L$ is the separation of the two planes, and $|x|^{\bullet} = (|x| + x)/2$. The width and height of the planes are assumed to be quite large relative to their separation. For $75° \leqslant \theta \leqslant 90°$, the recommended [63] correlation for air is

$$\overline{\mathrm{Nu}}_L - \left[1, 0.288(A \, \mathrm{Ra}_L \sin \theta)^{1/4}, \quad 0.039(\mathrm{Ra}_L \sin \theta)^{1/3}\right]_{\max}$$

$$\tag{12-33b}$$

in which the subscript max indicates that the largest of the three quantities separated by commas should be used. Also, $A = L/H$, where $H$ is the length measured along the plane—the width of the planes does not enter into the correlation if it is large relative to $L$. Also, as long as the aspect ratio $A = L/H$ exceeds 5, it will not much affect heat transfer, as Eq. (12-33a) shows.

Ostrach's [64] survey of the literature on natural convection in fluids heated from below is recommended for detailed information and references to work prior to 1957. With regard to horizontal planes heated from below, one of the first physical descriptions of the convective motion that can result when a dense fluid overlays a lighter fluid was given by Thomson [65] in 1882. In 1900 Bénard's [66] more complete description of the flow pattern—hexagonal cells with flow ascending in the center and descending along the sides (sometimes, however, descending in the center and ascending along the sides)—has resulted in these cells being termed *Bénard cells*. The first analysis was accomplished in 1916 by Rayleigh [67], who showed that motion will first occur at $\mathrm{Ra}_L = 27\pi^4/4 = 657$, where $L$ is the plane separation, if the top and bottom surfaces of the fluid are assumed to be free. This analysis was confirmed and extended in 1926 by Jeffreys [68], who found that the critical Rayleigh number for the onset of motion is $\mathrm{Ra}_L = 1108$ if the top fluid surface is free and the bottom surface is rigid. When both surfaces are rigid, the critical Rayleigh number is 1708 and turbulence first appears at $\mathrm{Ra}_L \approx 5830$ [64]; flow becomes fully turbulent at $\mathrm{Ra}_L \approx 45,000$ for air but at lower values for water. At very small temperature differences or very small plane separations, flow is columnar—a few randomly distributed and migratory columns of fluid ascend—rather than cellular [64]; the cellular pattern emerges only as plane separation or temperature difference is greater than a certain limit. General discussion including the effects of externally imposed cross flow in a variety of configurations is given by Turner [69] as well as Ostrach [64]. More recently, the strong influence of surface tension when one surface is free has been elucidated. (See Velarde and Normand [70] for a nonmathematical discussion.) The influence of surface tension is traceable to its temperature dependence, and it is this influence that is responsible for the generally (but not always) hexagonal convection cells originally described by Bénard. In the absence of surface tension a container with one free surface is filled with rolls, each of width equal to the container depth—roll axes are parallel to the short side of a rectangular container and form concentric rings for a cylindrical container. Accurate description of a system with one free surface requires consideration of the Marangoni number

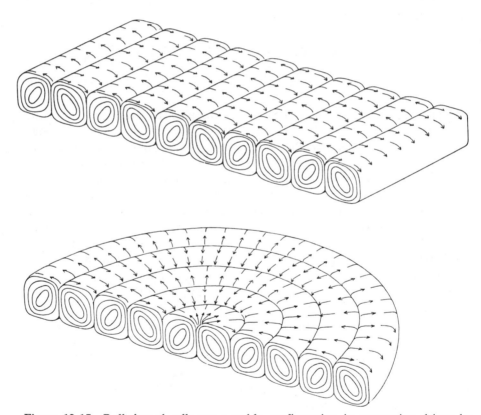

**Figure 12-15** Roll-shaped cells are a stable configuration in convection driven by buoyancy forces rather than by surface tension. The fundamental unit of the pattern consists of two rolls that rotate in opposite directions; the width of this unit is twice the depth of the fluid layer. The plan form of the pattern depends strongly on the boundaries of the layer. In a rectangular container the rolls are parallel to the shorter sides; in a circular container they form concentric rings. A stable roll pattern is usually observed only when the fluid has no free surface. [By permission from Convection, M. G. Velarde and C. Normand, *Sci. Am.* **243**, (1) 92–108 (1980) [70]; copyright 1980 by *Scientific American*, Inc., all rights reserved.]

—the same as the Rayleigh number but with the buoyant force replaced by the surface tension force. The hexagonal flow pattern occurs when the Marangoni number exceeds a critical value; other flow patterns exist prior to that. Figures 12-15 and 12-16 demonstrate the roll and hexagon flow patterns, respectively. Thus the Rayleigh number should not be solely relied on to predict the temperature difference required to initiate flow. As discussed by Velarde and Normand [70], the stability of natural convective systems is more accurately determined by the so-called Landau theory, devised in about 1937, which accounts for many terms in the series expansions considered—the older Rayleigh theory accounts for only the first term. A fuller discussion of surface tension and the Marangoni number is given in Chapter 14 in connection with condensation. Stong [71] gives instructions for the construction of a natural-convection demonstration apparatus driven by temperature differences. Natural convection driven by the difference in density between saline and fresh water is demonstrated in an interesting manner by Stong [72] and Walker [73].

With regard to the inclined enclosures to which Eq. (12-33) applies, the flow is described as follows [62, 74]. At very small $Ra_L$, the fluid motion consists of one large cell (the base flow), ascending near the hot surface and descending near the cold surface, with the streamlines parallel to the plane surfaces, except

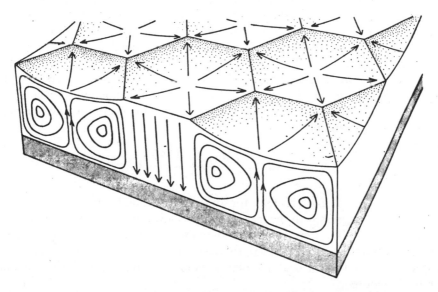

**Figure 12-16** Tessellation of the surface with hexagonal cells is a characteristic feature of convection driven by a gradient in surface tension. Where the tension is greatest the surface becomes puckered, so that its area is reduced. Over the ascending plume in the center of each cell the surface is depressed; the fluid must flow uphill before descending at the edge of the cell. [By permission from Convection, M. G. Velarde and C. Normand, *Sci. Am.* **243**, (1), 92–108 (1980) [70]; copyright 1980 by Scientific American, Inc., all rights reserved.]

at the ends, where the fluid turns as long as the inclination is not exactly zero. The heat transfer in this flow regime is consequently purely conductive, except at the extreme ends, where there is some convective heat transport associated with the fluid turning; however, as the aspect ratio $L/H$ of the slot becomes increasingly large, the contribution of this convective heat transfer to the average Nusselt number for the slot becomes vanishingly small. This conductive regime exists provided the Rayleigh number is less than a critical value given by $1708/\cos(\theta)$, in which case $\overline{\text{Nu}}_L = 1$.

For $\text{Ra} \lesssim \text{Ra}_c$, the base flow is marginally unstable, with the recipient flow theoretically consisting of steady longitudinal rolls (i.e., rolls with their axis along the upslope). For $1708 \lesssim \text{Ra}_L \cos(\theta) \lesssim 8000$, $\overline{\text{Nu}}_L = 1.44[1 - 1708/(\text{Ra}_L \cos \theta)]$ in the immediate postconductive regime. For large angles of inclination, near vertical, transverse traveling waves oriented across the slope are the first instabilities of the mean flow. The angle of inclination at which the type of instability switches from longitudinal rolls to transverse traveling waves is a function of Prandtl number and aspect ratio; for $\text{Pr} = 0.72$, if the aspect ratio $L/H$ is large ($> 25$), the switch takes place at an angle of inclination of $70°$. Decreasing Prandtl number and decreasing aspect ratio cause decrease in the angle of inclination at which transverse waves appear. The sensitivity to aspect ratio is believed to be weak: a change in the heat-transfer mechanism occurs at an angle of inclination of about $45°$ for an aspect ratio of 6 and about $5°$ when the aspect ratio is unity. Since the aspect ratios of interest to solar collectors range from 20 to 200 and angles of inclination up to $60°$, the instability will most likely manifest itself as longitudinal rolls. At very high Rayleigh numbers the flow takes up a boundary-layer structure with the resistance to heat transfer lying exclusively in two boundary layers, one on each boundary surface. The thermal resistance of each boundary layer is the same as on a single heated, facing up, inclined plate in an infinite environment, giving a Nusselt number of

$$\overline{\text{Nu}}_L = \left( \frac{\text{Ra}_L \cos \theta}{5830} \right)^{1/3}$$

The prediction of Eq. (12-33a) for heat flux through air in an enclosure inclined at $45°$ agrees with the measurements shown in Fig. 12-17 by Buchberg et al. [74]. There it is seen that a first minimum in heat flux is followed, after a local maximum, by asymptotic approach to an even lower minimum. Thus, for reduction in the flow of heat, the planes should be separated at least 5 cm and preferably more.

Natural convection between inclined planes can be suppressed by incorporating a honeycomb between the planes. For the honeycomb of square cells shown in Fig. 12-18, Buchberg et al. [75] recommend

$$\overline{\text{Nu}}_L = 1 + 0.89 \cos(\theta - 60°) \left( \frac{\text{Ra}_L}{2420 A^4} \right)^{2.88 - 1.64 \sin \theta}$$

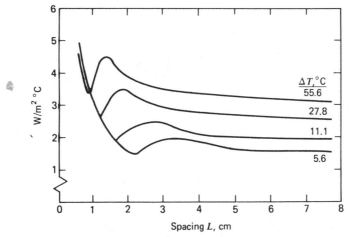

**Figure 12-17** Gap conductance $C_L$ between two isothermal planes for $T_H = 100°C$ and angle of inclination $\theta = 40°$. [From H. Buchberg, I. Catton, D. Edwards, *Transact. ASME, J. Heat Transf.* **98**, 182–188 (1976), [74].]

where $3 \leqslant A = L/D \leqslant 5$, $\theta$ is the inclination from horizontal ($30° \leqslant \theta \leqslant 90°$), and $0 \leqslant \text{Ra}_L \leqslant 6000A^4$. For noncircular honeycombs, $D$ is the hydraulic diameter. The optimal honeycomb geometry is found when $\overline{\text{Nu}}_L = 1.2$, according to Hollands [76], which occurs for air, $280 \text{ K} < T_{\text{mean}} < 370 \text{ K}$, when

$$A = C\left(1 + \frac{200}{T_m}\right)^{1/2} \frac{100}{T_m}(T_1 - T_2)^{1/4}L^{3/4}$$

with $L$ in centimeters and $T$ in kelvins. Figure 12-19 shows $C$ as a function of $\theta$. Smart et al. [90] studied rectangular honeycombs.

Equation (12-33) is rather generally applicable but can be compared against older, more specialized correlations. For parallel horizontal plates heated from below, Globe and Dropkin [77] found for water, mercury, and silicon oils

$$\overline{\text{Nu}}_L = 0.069 \, \text{Ra}_L^{1/3} \, \text{Pr}^{0.074}, \qquad 1.5 \times 10^5 < \text{Ra}_L < 10^9 \qquad (12\text{-}34)$$

For air between horizontal planes, Jakob [78] found

$$\overline{\text{Nu}}_L = \begin{matrix} 1, \\ 0.195 \, \text{Gr}_L^{1/4}, \\ 0.068 \, \text{Gr}_L^{1/3}, \end{matrix} \qquad \begin{matrix} \text{Gr}_L < 1708 \\ 10^4 \leqslant \text{Gr}_L \leqslant 4 \times 10^5 \\ 4 \times 10^5 \leqslant \text{Gr}_L \leqslant 10^7 \end{matrix} \qquad (12\text{-}35a)$$

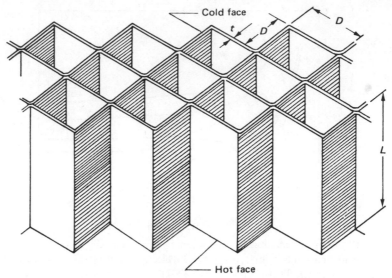

**Figure 12-18** Schematic diagram of honeycomb. (By permission from F. Kreith and J. F. Kreider, *Principles of Solar Engineering*, Hemisphere Publishing, Washington, D.C., 1978, pp. 140–144 [63].)

and for air enclosed between vertical planes

$$\overline{Nu}_L = 0.18\, Gr_L^{1/4}, \qquad \begin{cases} 1, & Gr_L < 2000 \\ 0.18\, Gr_L^{1/4}, & 2 \times 10^4 < Gr_L \leqslant 2 \times 10^5 \\ 0.065\, Gr_L^{1/3}\left(\dfrac{L}{H}\right)^{1/9}, & 2 \times 10^5 \leqslant Gr_L \leqslant 10^7 \end{cases} \qquad (12\text{-}35b)$$

for $3 \leqslant H/L$. If $H/L < 3$, the correlations for a single vertical plate in an

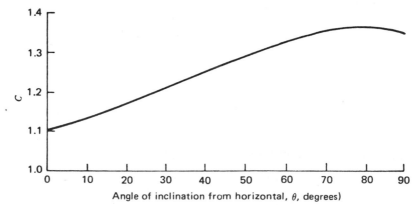

**Figure 12-19** Plot of $C$ versus angle of inclination $\theta$. (By permission from F. Kreith and J. F. Kreider, *Principles of Solar Engineering*, Hemisphere Publishing, 1978, pp. 140–144 [63].)

infinite fluid can be used to estimate the thermal resistance at each plane— where $H$ is the height of the enclosed space and $L$ is the plane separation. For liquids with $3 \leqslant \mathrm{Pr} \leqslant 3 \times 10^4$, Emery and Chu [79] found, for vertical layers,

$$\overline{\mathrm{Nu}}_L = \begin{cases} 1, & \mathrm{Ra}_L < 10^3 \\ 0.28\,\mathrm{Ra}_L^{1/4}\left(\dfrac{L}{H}\right)^{1/4}, & 10^3 \leqslant \mathrm{Ra}_L \leqslant 10^7 \end{cases} \qquad (12\text{-}36a)$$

MacGregor and Emery [80] found in vertical enclosures ($10 \leqslant H/L \leqslant 40$) with constant heat flux that

$$\overline{\mathrm{Nu}}_L = \begin{cases} 1, & \mathrm{Ra}_L < 10 \\ 0.42\,\mathrm{Ra}_L^{1/4}\,\mathrm{Pr}^{0.012}\left(\dfrac{L}{H}\right)^{0.3}, & 10^4 \leqslant \mathrm{Ra}_L \leqslant 10^7 \quad 1 \leqslant \mathrm{Pr} \leqslant 2 \times 10^4 \\ 0.046\,\mathrm{Ra}_L^{1/3}, & 10^6 \leqslant \mathrm{Ra}_L \leqslant 10^9 \quad 1 \leqslant \mathrm{Pr} \leqslant 20 \end{cases} \qquad (12\text{-}36b)$$

A survey of natural convection by Raithby and Hollands [53] gives additional details. General correlations for flat plate, horizontal cylinder, sphere, concentric cylinder, concentric and eccentric sphere, and symmetrically and unsymmetrically heated parallel vertical plate geometries are discussed for laminar flow. Turbulent flow for inclined flat plates and the effect of inclination on laminar–turbulent transition are also discussed.

The related natural convection that occurs when a heat-generating fluid is enclosed in a cavity whose walls are cooled has also been studied. Among the practical situations of importance are those of nuclear reactor safety in which molten radioactive liquid pools must be cooled. Bergholz [87] considered a rectangular cavity with cooled side walls; Emara and Kulacki [89] considered a cooled upper surface, either rigid or free.

Natural convection in freezing or melting water has some peculiar characteristics because the maximum density of water is at 4°C. Experiments for horizontal ice layers have been made [88], as have experiments on freezing paraffin [91].

Natural convection of fluid through a porous medium has been the subject of numerous studies. It is understandably important in goethermal energy utilization schemes. A representative article [94] can be consulted for background information and additional references. Measurements of the onset of convective motion have also been reported [95]. The effect of fibrous glass on natural convection in a cavity is of interest in thermal insulation of buildings, for instance, and has been studied by Seki et al. [97].

## Optimal Fin Array For Natural Convection

The ability of a finned surface to reject heat by natural convection is influenced by fin spacing and by fin dimensions. If the fins are too close together, the fluid is trapped and little heat is transferred. On the other hand, if the fins are too far apart, the surface area is so small that, again, little heat is transferred. Thus it is evident that an optimal space between fins exists. Similarly, the temperature distribution along the fin affects the heat-transfer coefficient and the fin dimensions also have optimal values.

The case of vertical rectangular fins was studied by Bar-Cohen [81]. For the geometry illustrated in Fig. 12-20, he found that the optimal fin length $b$, optimal space between fins $S$, optimal fin thickness $\delta$, and optimal heat rejection rate from the finned surface $q$ are given by

$$b = 2.89 \left( \frac{k_{\text{fin}}}{k_{\text{fluid}}} \right)^{1/2} P$$

$$S = \delta = 2.89P$$

$$\frac{q}{LW(T_w - T_\infty)} = \frac{1.3(k_{\text{fin}}k_{\text{fluid}})^{1/2}}{6P}$$

where $P^4 = L\nu^2/[g\beta(T_w - T_\infty)\text{Pr}]$.

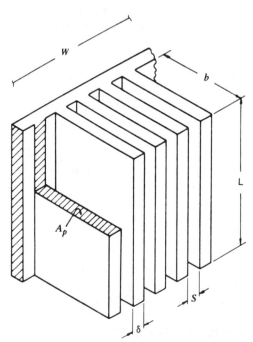

**Figure 12-20** Geometric definition of rectangular vertical fin array. [From A. Bar-Cohen, *Transact. ASME, J. Heat Transf.* **101**, 564–566 (1979) [81].]

## PROBLEMS

**12-1** Show that the similarity transformation in Section 12.1 does put the laminar boundary layer equations and boundary conditions of Eq. (12-1) into the similarity form of Eq. (12-2).

**12-2** Laminar natural convection Nusselt numbers can be of surprisingly large magnitude. To illustrate this, calculate the Nusselt number for a vertical flat plate in room air—take the Grashof number to be near its upper allowed value for laminar flow, $Gr_L \lesssim 10^9$. The result should be $Nu_L \sim 10^2$.

**12-3** Natural convection is important to cooling of gas turbine rotor blades where a fluid circulates through internal blade passages. Sometimes the maximum fluid velocity in natural convection is then surprisingly large because centrifugal force acts as the body force rather than gravity,

(a) Show for air with $g = 10^6$ ft/sec$^2$, $L = 0.25$ ft, and $T_w - T_\infty = 100°F$ that $U_{max}$ is of the order of $10^2$ ft/sec for a flat-plate geometry. The body force can be approximated as a constant over the plate length.

(b) Repeat the velocity calculation of part a for $g = 32.2$ ft/sec$^2$ and comment on the statement that velocities are usually small in natural convection.

**12-4** Velocities in natural convection from an impermeable vertical flat plate parallel to the constant local gravitational acceleration are usually small. In such a case the convective (inertia) terms in the equations of motion would be negligible, leaving body forces to be entirely countered by viscous shear forces, and "creeping motion" would exist. Since inertial forces are proportional to velocity squared whereas viscous forces are proportional to velocity itself, it can be appreciated that viscous forces are dominant at low velocities (low Reynolds numbers).

(a) Show that the boundary-layer equation of $x$-motion [Eq. (12-1b)] with convective terms deleted becomes

$$0 = \nu_\infty \frac{\partial^2 u}{\partial y^2} + g \frac{\rho_\infty - \rho}{\rho_\infty}$$

and that the other boundary-layer equations and boundary conditions are unchanged from Eq. (12-1).

(b) Use a stream function $\Psi = 4\alpha(Ra_x/4)^{1/4} F(\eta)$ such that $u = \partial\Psi/\partial y$ and $v = -\partial\Psi/\partial x$ and a similarity variable $\eta = (Ra_x/4)^{1/4}(y/x)$ to put the boundary-layer equations and boundary conditions into the similarity form

$$F''' + \theta = 0$$

$$\theta'' + 3F\theta' = 0$$

$$F'(0) = F(0) = F'(\infty) = \theta(\infty) = 0$$

$$\theta(0) = 1$$

where

$$\theta = \frac{T - T_\infty}{T_w - T_\infty}$$

and the Rayleigh number is

$$\mathrm{Ra}_x = \mathrm{Gr}_x\, \mathrm{Pr} = g\,|\rho_\infty - \rho_w|\,x^3\,\frac{\mathrm{Pr}}{\nu^2 \rho_\infty}$$

whereas

$$u = \frac{2\alpha}{x}\, \mathrm{Ra}_x^{1/2}\, F'$$

$$v = \frac{\alpha}{x}\left(\frac{\mathrm{Ra}_x}{4}\right)^{1/4}(\eta F' - 3F)$$

Here the prime superscript denotes differentiation with respect to $\eta$.

(c) Show that the similarity formulation in part b gives the local Nusselt number and shear stress as

$$\mathrm{Nu}_x = \frac{hx}{k} = -\theta'(0)\left(\frac{\mathrm{Ra}_x}{4}\right)^{1/4}$$

$$\tau_w = \left(4\,\mathrm{Ra}_x^3\right)^{1/4}\left(\frac{\rho\nu^2}{x^2\,\mathrm{Pr}}\right)F''(0)$$

Note that $\theta'(0)$ is independent of Pr since it is obtained from Pr-independent equations. Hence, in "creeping motion," $\mathrm{Nu}_x$ depends on the Gr Pr product rather than the Gr and Pr separately.

(d) Use the separation of variables technique outlined in Appendix D to obtain the similarity transformation applied in part b.

12-5  Just as inertial effects are negligible at the low velocities encountered for $\mathrm{Pr} \to \infty$ in natural convection from a vertical plate, so might viscous effects be negligible at the higher velocities that might be encountered for Pr → 0. To see this:

(a) Render the describing boundary-layer equations [Eq. (12-1)] dimensionless by letting $x' = x/L$, $y' = y/L$, $u' = u/u_r$, $v' = v/u_r$, and $T' = (T - T_w)/(T_w - T_\infty)$ to obtain

$$\frac{\partial u'}{\partial x'} + \frac{\partial v'}{\partial y'} = 0 \tag{5P-1}$$

$$u'\frac{\partial u'}{\partial x'} + v'\frac{\partial u'}{\partial y'} = \frac{\nu}{Lu_r}\frac{\partial^2 u'}{\partial y'^2}$$

$$+ \left[ g\,|\rho_\infty - \rho_w|\,\frac{L}{\rho_\infty u_r^2} \right] \frac{\rho_\infty - \rho}{|\rho_\infty - \rho_w|} \tag{5P-2}$$

$$u'\frac{\partial T'}{\partial x'} + v'\frac{\partial T'}{\partial y'} = \frac{\nu}{\text{Pr } Lu}\frac{\partial^2 T'}{\partial y'^2} \tag{5P-3}$$

Here $L$ is a constant, of arbitrary reference length, and $u_r$ is a constant reference velocity.

(b) Select $u_r = \nu/\text{Pr } L$ that puts Eqs. (5P-1)–(5P-3) into the form

$$\frac{\partial u'}{\partial x'} + \frac{\partial v'}{\partial y'} = 0 \tag{5P-4}$$

$$u'\frac{\partial u'}{\partial x'} + v'\frac{\partial u'}{\partial y'} = \text{Pr}\frac{\partial^2 u'}{\partial y'^2} + \text{Gr}_L\,\text{Pr}^2\frac{\rho_\infty - \rho}{|\rho_\infty - \rho_w|} \tag{5P-5}$$

$$u'\frac{\partial T'}{\partial x'} + v'\frac{\partial T'}{\partial y'} = \frac{\partial^2 T'}{\partial y'^2} \tag{5P-6}$$

where

$$\text{Gr}_L = g\,|\rho_\infty - \rho_w|\,\frac{L^3}{\rho\nu^2}$$

(c) Verify from the result of part b that (1) viscous effects are negligible and $\text{Gr Pr}^2$ is the controlling parameter when Pr → 0 and (2) inertial effects are negligible and $\text{Gr Pr}$ is the controlling parameter when Pr → ∞.

**12-6** An integral method is to be applied to free convection from a vertical flat plate, immersed in a constant-property fluid, whose surface heat flux is specified at the constant value $q_w$ [19].

(a) Show that use of the assumed profiles of

$$T - T_\infty = \frac{q_w \delta}{2k}(1 - \eta)^2$$

$$u = u_1 \eta (1 - \eta)^2$$

where $\eta = y/\delta$, $\delta$ is the common thermal and velocity boundary-layer thickness, and $u_1$ is a reference velocity to be determined in Eq. (12-15) gives

$$\frac{d(W^2 D/105)}{dX} = \frac{D^2}{6} - \frac{W}{D}$$

and

$$\frac{d(WD^2/30)}{dX} = \frac{2}{Pr}$$

Here $W = u_1 (g\beta q_w \nu^2/k)^{-1/4}$, $D = \delta(g\beta q_w/k\nu^2)^{1/4}$, and $X = x(g\beta q_w/k\nu^2)^{1/4}$.

(b) Solve these two differential equations to achieve

$$W = \left(\frac{6000}{Pr}\right)^{1/5} \left(Pr + \frac{4}{5}\right)^{-2/5} X^{3/5}$$

$$D = \left(\frac{360}{Pr^2}\right)^{1/5} \left(Pr + \frac{4}{5}\right)^{1/5} X^{1/5}$$

(c) From the result of part b show that

$$T_w - T_\infty = 1.622 \frac{q_w x}{k} \left(\frac{Pr + 0.8}{Pr^2 Gr_x^*}\right)^{1/5}$$

$$h_x = \frac{q_w}{T_w - T_\infty} = 0.62 \frac{k}{x} \left(\frac{Pr^2 Gr_x^*}{Pr + 0.8}\right)^{1/5}$$

where $Gr_x^*$ is a modified Grashof number defined as $Gr_x^* = (g\beta x^3/\nu^2)(q_w x/k)$. Note that $T_w$ varies as $x^{1/5}$ as predicted by the exact solution.

(d) From the results of part c show that the average Nusselt number $\overline{Nu}$ over the plate height is given by $\overline{Nu}_x = 0.775[Pr^2 Gr_x^*/(Pr + 0.8)]^{1/5}$.

(e) Recognizing that $Gr_x^* = Gr_x Nu_x$, rearrange the result of part d into the form

$$\frac{\overline{Nu}_x}{Ra_x^{1/4}} = \frac{0.683 \, Pr^{1/4}}{(Pr + 0.8)^{1/4}}$$

and compare with Eq. (12-23) for the case of constant wall temperature.

(f) Use the result of part d and the corresponding integral method result for a constant wall temperature as expressed by Eq. (12-23) to compare with the exact results given in Table 12-3.

**12-7** Show by use of Eq. (12-26) that the Nusselt number for natural convection from a vertical cylinder of constant wall temperature immersed in an infinite fluid pool first exceeds the flat-plate value by 5% when $Ra_L^{1/4} D/L \approx 33$ for $Pr = 1$.

**12-8** Employ the results found by the integral method to show that the shear stress at the wall in turbulent natural convection from a constant-temperature vertical plate immersed in a constant property fluid has:

(a) A local value given by

$$\tau_w = 0.06686 \, \rho \nu^2 \, Pr^{-23/30}(1 + 0.494 \, Pr^{2/3})^{-9/10}$$

$$\cdot \left[ g\beta(T_w - T_\infty)\frac{Pr}{\nu^2} \right]^{9/10} x^{7/10}$$

(b) An average value, assuming the plate to be wholly covered by a turbulent boundary layer, of $\bar{\tau}_w = 5\tau_w/12$. Or

$$\bar{\tau}_w = 0.02786 \, \frac{\rho \nu^2}{x^2} Pr^{-23/30}(1 + 0.494 \, Pr^{2/3})^{-9/10} Ra_x^{9/10}$$

**12-9** The Nusselt number for very large Grashof numbers for turbulent natural convection from a constant-temperature vertical flat plate can be deduced from results of investigation into turbulent forced convection [6]. It is assumed that the Reynolds number corresponds to the square root of the Grashof number. For example, in forced laminar flow $Nu \sim Re^{1/2}$ whereas in natural laminar flow $Nu \sim Gr^{1/4}$.

(a) Substitute $Gr^{1/2}$ for Re in the turbulent Reynolds analogy

$$\frac{Re \, Pr}{Nu} = \frac{2}{\bar{C}_f} + \left(\frac{2}{\bar{C}_f}\right)^{1/2} 5\left(Pr - 1 + \ln\frac{5 \, Pr + 1}{6}\right) \qquad (10\text{-}16)$$

which was derived for parallel flow over a flat plate.

(b) Solve the resulting quadratic form for $(2/\bar{C}_f)^{1/2}$, assuming Gr to be large and Pr to be approximately unity, to obtain

$$\bar{C}_f^{-1/2} \approx \left(\frac{Gr^{1/2} \, Pr}{2\bar{Nu}}\right)^{1/2} - \frac{5}{2^{3/2}}\left[Pr - 1 + \ln\frac{5Pr + 1}{6}\right]$$

(c) Recall the von Karman–Schoenherr formula for the average shear stress coefficient for parallel turbulent flow over a flat plate

$$\bar{C}_f^{-1/2} = 4.13 \log_{10}\left(\bar{C}_f \mathrm{Re}\right) \qquad (9\text{-}21a)$$

which, on replacement of Re by $\mathrm{Gr}^{1/2}$, becomes

$$\bar{C}_f^{-1/2} = 4.13 \log_{10}\left(\bar{C}_f \mathrm{Gr}^{1/2}\right)$$

Use the result of part b in this relation, with $\log_{10}(\bar{C}_f \mathrm{Gr}^{1/2}) \approx \log_{10}(2\overline{\mathrm{Nu}}/\mathrm{Pr})$, to obtain, after some rearrangement,

$$\overline{\mathrm{Nu}} = \frac{C(\mathrm{Gr}\,\mathrm{Pr}^2)^{1/2}}{\left\{\mathrm{Pr} - 1 + \ln\left[\overline{\mathrm{Nu}}(5\,\mathrm{Pr} + 1)/3\,\mathrm{Pr}\right]\right\}^2}$$

in which $C = \frac{1}{8}$ is recommended. There is no major reason to prefer this result over Eq. (12-32). Although it gives comparable results, the major benefit of such an analysis is to strengthen the idea that intense turbulence in natural convection may well result in $h$ being independent of $\mu$ and $k$, as the preceding result shows to be nearly the case.

12-10   Show by numerical example that the laminar natural convection heat-transfer coefficient from a horizontal cylinder can be computed by use of $\pi D/2$ in a vertical plate correlation. In other words, let the characteristic dimension for a horizontal cylinder be the maximum distance traveled by a fluid particle, and use an average gravitational acceleration tangent to the cylinder surface.

12-11   Demonstrate, on the basis of Eq. (12-33), that the thermal resistance to heat flow offered by vertical air layers (e.g., building construction) approaches a lower limit—nearly attained with a thickness of 2 in.—as the air gap becomes large. Comment on the increase in thermal resistance obtained by dividing a thick vertical air gap by vertical partitions. Does the thermal resistance increase nearly proportionally to the number of partitions? Also, use Eq. (12-33) to establish that the thermal resistance for horizontal air layers with a plane separation of 2–8 in. is equal to the maximum thermal resistance for a vertical air layer (at about 2 in.).

12-12   Compare the predictions in Eqs. (12-33a), (12-33b), and (12-35) for air in horizontal and vertical enclosures between two planes. Repeat for water, using Eqs. (12-36) and (12-34) in place of Eq. (12-35).

12-13   When installed, a storm window is 1 in. away from the inside window and is 2 ft high. If the inside window is at 65°F and the outside window is at 35°F, estimate the convection heat-transfer rate across the air gap between the two windows. (*Answer*: $q \approx 10$ Btu/hr ft$^2$.)

**12-14**  Estimate the average heat-transfer coefficient for natural convection between a cubical instrumentation package that is 1 ft on a side at 120°F and a large pool of water at 80°F. (*Answer*: $h \approx 230$ Btu/hr ft$^2$ °F.)

**12-15**  A short solid vertical cylinder 6 in. high and 6 in. in diameter is at 500°F and cools by natural convection in 100°F air. Estimate the average heat-transfer coefficient. (*Answer*: $h \approx 1.8$ Btu/hr ft$^2$ °F.)

**12-16**  Water in a large tank at 60°F is heated by a 1-in. diameter vertical pipe at 200°F. The water depth is 3 ft. Estimate the heat-transfer rate by natural convection from the pipe to the water. (*Answer*: $q \approx 18,000$ Btu/hr.)

**12-17**  A 6-in.-diameter sphere at 700°F cools by natural convection in 100°F air after heat treatment. Estimate the heat-transfer coefficient for natural convection and estimate the magnitude of the forced-convection velocity that would be required to appreciably affect the heat-transfer coefficient. (*Answer*: $h \approx 1.8$ Btu/hr ft$^2$ °F.)

**12-18**  A 1-ft-diameter spherical container holds electronic instruments near the bottom of a lake whose water is at 60°F. The sphere surface is maintained at constant temperature by the 1020 W dissipated by the electronics. Estimate the sphere surface temperature. (*Answer*: $T \approx$ 80°F.)

**12-19**  Water flows in forced convection past a vertical flat plate of 1 ft height so that Re $= 5 \times 10^4$. If the water temperature is 60°F, what is the highest plate temperature that can be used without natural convection effects influencing the heat-transfer coefficient by more than 5%? (*Answer*: $T \approx 80$°F.)

**12-20**  The side of a small laboratory furnace can be idealized as a vertical flat plate 2 ft high and 8 ft wide. The furnace sides are at 100°F and the surrounding air is at 70°F.

(a) Estimate the heat loss from the sides by natural convection.

(b) Estimate the lowest vertical forced velocity that would cause the heat-transfer coefficient to depart noticeably from its natural-convection value. [*Answer*: (a) $q \approx 300$ Btu/hr; (b) $V \approx 1$ ft/sec.]

**12-21**  Air at 100°F flows at Re $= 10^2$ in a horizontal tube that is 5 ft long and of 1 in. diameter. If the tube wall temperature is 500°F, determine whether the flow is purely forced convection. (*Answer*: Pure forced convection.)

**12-22**  Water at 60°F is forced to flow perpendicular to a 1-in.-diameter horizontal tube that is at 200°F. The water velocity is 10 ft/sec.

Estimate whether natural convection effects are important to a determination of the heat-transfer coefficient.

**12-23**  Air at 140°F flows through an 8-in.-diameter duct that is 15 ft long, and the duct wall temperature is 80°F. Estimate the flow velocity at which natural convection effects will becomes important with possible fluid stratification. (*Answer*: $V \approx 1.7$ ft/sec.)

**12-24**  Determine the water pressure necessary to prevent boiling of a horizontal heating element 2 ft long and 1 in. in diameter dissipates electrical energy at the rate of 4500 W in 140°F water. (*Answer*: 30 psia.)

**12-25**  A $\frac{1}{32}$-in.-diameter wire is horizontal in 60°F water. If the wire temperature is 140°F, what is the energy loss by natural convection from the wire? (*Answer*: 90 W/ft.)

**12-26**  The top of a stove approximates a square plate of 3-ft side. Estimate the heat transfer rate by natural convection from the stove top if its temperature is 110°F while the room temperature is 70°F. (*Answer*: $q = 300$ Btu/hr.)

**12-27**  A small resistance heater in the shape of a square plate of 2 in. side is to be designed for the bottom of a fish tank. The tank water must be kept at 75°F while the heater surface temperature must not exceed 85°F. Specify the wattage of the needed heating element. (*Answer*: 6 W.)

**12-28**  Square tiles of 3 in. side are heated in an oven at 1100°F. The tiles are at 900°F. Compute the natural convection heat-transfer coefficient when the tiles rest horizontally on the oven floor.

**12-29**  A block of ice is of 2 ft side and is suspended in a stagnant room with its top horizontal. The ice is at 20°F, and the room is at 80°F. Estimate the natural convective heat-transfer coefficient for the block on the (a) top, (b) bottom, and (c) average surface. (*Answers*: (a) $h \approx 0.3$ Btu/hr ft$^2$ °F; (b) $h \approx 1$ Btu/hr ft$^2$ °F.)

**12-30**  Circular sheets of plastic of 1 ft diameter and at 40°F are heated by floating on the surface of 160°F water. Estimate the heat-transfer coefficient between the sheet and the water. (*Answer*: $h = 200$ Btu/hr ft$^2$ °F.)

**12-31**  Consider a vertical isothermal surface to which rectangular aluminum fins are attached as shown in Fig. 12-20. The fins have a height $L$ of 0.13 m, and the isothermal surface is 100°C hotter than the surrounding air. Show that under these conditions (a) the optimal fin length is 60 cm, the optimal space between fins and the optimal fin thickness are both 0.6 cm, and that the optimal heat rejection rate from the vertical isothermal surface is 260 W/m$^2$ K, (b) the optimal fin dimensions are not sensitive to temperature differences, varying only with its $\frac{1}{4}$th power, and (c) the ratio of the optimal finned heat-rejection rate $q$ to the heat-rejection rate from the unfinned vertical surface $q_{uf}$ is given by

$$\frac{q}{q_{uf}} = 0.39 \left( \frac{k_{fin}}{k_{air}} \right)^{1/2}$$

which has a value of 34.

# REFERENCES

1   B. Gebhart, "Effects of viscous dissipation in natural convection," *J. Fluid Mech.* **14**, 225–232 (1962).

2   E. Schmidt and W. Beckman, Das Temperatur-und Geschwindigkeitsfeld von einer Wärme abgebenden, senkrechten Platte bei natürlicher Konvektion, *Forsch-Ing.-Wes.* **1**, 391 (1930).

3   A. Oberbeck, Ueber die Wärmeleitung der Flüssigkeiten bei Berücksichtigung der Strömungen infolge von Temperaturdifferenzen, *Annalen der Physik und Chemie* **7**, 271–292 (1879).

4   L. Lorenz, Ueber das Leitungsvermögen der Metalle für Wärme und Electricität, *Annalen der Physik und Chemie* **13**, 582–606 (1881).

5   S. Ostrach, An analysis of laminar free-convection flow and heat transfer about a plate parallel to the direction of the generating body force, NACA Report 1111, 1953.

6   A. J. Ede, Advances in free convection, *Adv. Heat Transf.* **4**, 1–64 (1967).

7   E. M. Sparrow and J. L. Gregg, Details of exact low Prandtl number boundary layer solutions for forced and for free convection, NASA Memo 2-27-59 E, 1959.

8   E. J. LeFevre, Laminar free convection from a vertical plane surface, Mechanical Engineering Research Laboratory, Heat 113 (Great Britain), 1956; *Proceedings of the 9th International Congress on Applied Mechanics, Brussels*, Vol. 4, 1956, p. 168.

9   W. H. McAdams, *Heat Transmission*, 3rd ed., McGraw-Hill, New York, 1954.

10  J. Gryzagoridis, Natural convection from a vertical flat plate in the low Grashof number range, *Internatl. J. Heat Mass Transf.* **14**, 162–164 (1971).

11  E. M. Sparrow and J. L. Gregg, The variable fluid-property problem in free convection, *Transact. ASME* **80**, 879–886 (1958).

12  R. Eichhorn, Measurement of low speed gas flows by particle trajectories: A new determination of free convection velocity profiles, *Internatl. J. Heat Mass Transf.* **5**, 915–928 (1962).

13  R. Eichhorn, The effect of mass transfer on free convection, *Transact. ASME, J. Heat Transf.* **82**, 260–263 (1960).

14  E. M. Sparrow and R. D. Cess, Free convection with blowing or suction, *Transact. ASME, J. Heat Transf.* **83**, 387–389 (1961).

15  H. B. Squire, in *Modern Developments in Fluid Dynamics*, Vol. II, S. Goldstein, Ed., Oxford University Press, 1938, pp. 801–810.

16  H. J. Merk and J. A. Prins, Thermal convection in laminar boundary layers, *Appl. Sci. Res.*, **A4**, 11–24, 195–206, 207–221 (1954).

17  (a) S. Sugawara and I. Michiyoshi, *Transact. ASME* **17**, 109 (1951); (b) E. M. Sparrow and J. L. Gregg, Laminar free convection from a vertical plate with uniform surface heat flux, *Transact. ASME* **78**, 435–440 (1956).

18  G. C. Vliet, Natural convection local heat transfer on constant-heat-flux inclined surfaces, *Transact. ASME, J. Heat Transf.* **91**, 511–516 (1969).

19  E. M. Sparrow, Laminar free convection on a vertical plate with prescribed non-uniform wall heat flux or prescribed non-uniform wall temperature, NACA TN 3508, 1955.

20 E. M. Sparrow and J. L. Gregg, Similar solutions for free convection from a non-isothermal vertical plate, *Transact. ASME* **80**, 379–386 (1958).

21 M. Tribus, Discussion on similar solutions for free convection from a non-isothermal vertical plate, *Transact. ASME* **80**, 1180–1181 (1958).

22 R. P. Bobco, "A closed-form solution for laminar free convection on a vertical plate with prescribed nonuniform, wall heat flux, *J. Aerospace Sci.* **26**, 846–847 (1959).

23 E. M. Sparrow and J. L. Gregg, Laminar free convection from the outer surface of a vertical circular cylinder, *Transact. ASME* **78**, 1823–1829 (1956).

24 T. Cebeci, Laminar-free-convective-heat transfer from the outer surface of a vertical slender cylinder, in *Fifth International Heat Transfer Conference*, V. 3, Section NC 1.4, 1975, pp. 15–19.

25 E. J. LeFevre and A. J. Ede, Laminar free convection from the outer surface of a vertical circular cylinder, *Proceedings of the 9th International Congress on Applied Mechanics Brussels*, Vol. 4, 1957, pp. 175–183.

26 W. Elenbaas (a) Heat dissipation of parallel plates by free convection, (b) The dissipation of heat by free convection of spheres and horizontal cylinders, (c) The dissipation of heat by free convection—horizontal and vertical cylinders; spheres, (d) The dissipation of heat by free convection—the inner surface of vertical tubes of different shapes of cross section, *Physica* **9**, (1942) (a) pp. 1–28, (b) pp. 285–296, (c) pp. 665–672, (d) pp. 865–874.

27 I. Langmuir, Convection and conduction of heat in gases, *Phys. Revue* **34**, 401–422 (1912).

28 L. S. Eigenson, *Zhurnal Teckhnicheskoi Fiziki* **1**, 228 (1931). *Compt. Rend. Acad. Sci. USSR* **26**, 440 (1940).

29 J. R. Kyte, A. J. Madden, and E. L. Piret, Natural-convection heat transfer at reduced pressure, *Chem. Eng. Progr.* **49**, 653–662 (1953).

30 H. R. Nagendra, M. A. Tirunarayanan, and A. Ramachandran, Laminar free convection from vertical cylinders with uniform heat flux, *Transact. ASME, J. Heat Transf.* **92**, 191–194 (1970).

31 E. R. G. Eckert and T. Jackson, Analysis of turbulent free convection boundary layer on a flat plate, NACA Report 1015, 1951.

32 F. J. Bayley, An analysis of turbulent free convection heat transfer, *Institution of Mechanical Engineers (London) Proceeding*, Vol. 169, 1955, pp. 361–370.

33 C. Y. Warner and V. S. Arpaci, An investigation of turbulent natural convection in air at low pressure along a vertical heated flat plate, *Internatl. J. Heat Mass Transf.* **11**, 397–406 (1968).

34 R. Siegel, General Electric Company Technical Information Service Report R54GL89, 1954.

35 G. C. Vliet and C. K. Liu, An experiment study of turbulent natural convection boundary layers, *Transact. ASME, J. Heat Transf.* **91**, 517–531 (1969).

36 R. Lemlich and J. Vardi, Steady free convection to a flat plate with uniform surface flux and nonuniform acceleration, *Transact. ASME, J. Heat Transf.* **86**, 562–563 (1964).

37 J. R. Lloyd and E. M. Sparrow, Combined forced and free convection flow on vertical surfaces, *Internatl. J. Heat Mass Transf.* **13**, 434–438 (1970).

38 Y. Mori, Buoyancy effects in forced laminar convection flow over a horizontal flat plate, *Transact. ASME, J. Heat Transf.* **83**, 479–482 (1961).

39 A. Acrivos, Combined laminar free- and forced-convection heat transfer in external flows, *AIChE J.* **4**, 285–289 (1958).

40 W. B. Hall and P. H. Price, Mixed forced and free convection from a vertical heated plate to air, *Fourth International Heat Transfer Conference (Paris)*, Vol. IV, August 1970, Section NC 3.3.

41 P. H. Oosthuizen and S. Madan, Combined convective heat transfer from horizontal cylinders in air, *Transact. ASME, J. Heat Transf.* **92**, 194–196 (1970).

42  D. C. Collis and M. J. Williams, Two-dimensional convection from heated wires at low Reynolds numbers, *J. Fluid Mech*. **6**, 357–384 (1959).

43  B. T. Gebhart, T. Audunson, and L. Pera, Forced, mixed and natural convection from long horizontal wires, experiments at various Prandtl numbers, *Fourth International Heat Transfer Conference (Paris)*, Vol. IV, Section NC 3.2, August 1970.

44  T. Yuge, Experiments on heat transfer from spheres including combined natural and free convection, *Transact. ASME, J. Heat Transf*. **82**, 214–220 (1960).

45  B. Metais and E. R. G. Eckert, Forced, mixed, and free convection regimes, *Transact, ASME, J. Heat Transf*. **86**, 295–296 (1964).

46  H. O. Buhr, E. A. Horsten, and A. D. Carr, The distortion of turbulent velocity and temperature profiles on heating, for mercury in a vertical pipe, *Transact. ASME, J. Heat Transf*. **96**, 152–158 (1974).

47  F. Kreith, Convection heat transfer in rotating systems, *Adv. Heat Transf*. **5**, 129–251 (1968).

48  G. C. Vliet, Natural convection local heat transfer on constant-heat-flux inclined surfaces, *Transact. ASME, J. Heat Transf*. **91**, 511–516 (1969).

49  S. N. Singh, R. C. Birkebak, and R. M. Drake, Laminar free convection from downward-facing horizontal surfaces of finite dimensions, *Prog. Heat Mass Transf. (Internatl. J. Heat Mass Transf.)* **2**, 87–98 (1969). S. N. Singh and R. C. Birkebak, Laminar free convection from a horizontal infinite strip facing downward, *Zeitschrift für Angewandte Mathematik und Physik* **20**, 454–461 (1969).

50  R. Hermann, Wärme übertragung bei freier Strömung am waagerechten Zylinder in zwei-atomigen Gasen, *VDI-Forschungsheft* 379, 1936.

51  W. Elenbaas, The dissipation of heat by free convection from vertical and horizontal cylinders, *J. Appl. Phys*. **19**, 1148–1154 (1948).

52  S. C. Hyman, C. F. Bonilla, and S. W. Ehrlich, Natural convection transfer processes: I. Heat transfer to liquid metals and nonmetals at horizontal cylinders, *Chem. Eng. Progr. Symp. Ser*. **49**, 21–31 (1953).

53  G. D. Raithby and K. G. T. Hollands, A general method of obtaining approximate solution to laminar and turbulent free convection problems, *Adv. Heat Transf*. **11**, 266–315 (1975).

54  W. J. King, The basic laws and data of heat transmission. III. Free Convection, *Mech. Eng*. **54**, 347–353 (1932).

55  G. D. Raithby, A. Pollard, K. G. T. Hollands, and M. M. Yovanovich, Free convection heat transfer from spheroids, *Transact. ASME, J. Heat Transf*. **98**, 452–458 (1976).

56  G. D. Raithby and K. G. Hollands, Free convection heat transfer from vertical needles, *Transact. ASME, J. Heat Transf*. **98**, 522–523 (1976).

57  F. Kreith, Thermal design of high altitude balloons and instrument packages, *Transact. ASME, J. Heat Transf*. **92**, 307–332 (1970).

58  E. N. Bishop, L. R. Mack, and J. A. Scanlan, Heat transfer by natural convection between concentric spheres, *Internatl. J. Heat Mass Transf*. **9**, 649–662 (1966).

59  L. B. Evans and N. E. Stefany, An experimental study of transient heat transfer to liquids in cylindrical enclosures, AIChE Paper 4, Heat Transfer Conference, Los Angeles, August 1965.

60  H. Buchberg, I. Catton, and D. K. Edwards, Natural convection in enclosed spaces: A review of application to solar energy collection, ASME Paper 74-WA/HT/12, 1974 (see also reference 74).

61  G. D. Raithby and K. G. T. Hollands, A general method of obtaining approximate solutions to laminar and turbulent free convection problems, *Adv. Heat Transf*. **11** (1975).

62  K. Hollands, T. Unny, G. Raithby, and L. Konicek, Free convective heat transfer across inclined air layers, *Transact. ASME, J. Heat Transf*. **98**, 189–193 (1976).

63  F. Kreith and J. F. Kreider, *Principles of Solar Engineering*, Hemisphere (McGraw-Hill), New York, 1978, pp. 140–144.

**64** S. Ostrach, Convective phenomena in fluids heated from below, *Transact. ASME* **79**, 299–305 (1957).

**65** James Thomson, On a changing tesselated structure in certain liquids, *Proc. Glasgow Phil. Soc.* **13**, 469 (1882).

**66** H. Bénard, Tourbillions cellulaires dan une nappe liquide, *Revue genérale des Sciences Pures et Appliĝees* **11**, 1261–1271, 1309–1328 (1900).

**67** Lord Rayleigh, On convective currents in a horizontal layer of fluid when the higher temperature is on the under side, *Philo. Mag. J. Sci.* **32**, 529–546 (1916).

**68** H. Jeffreys, The stability of a layer of fluid heated below, *Phil. Mag. J. Sci.* **2**, 833–844 (1926); Some cases of instability in fluid motion, *Proc. Roy. Soc. Lond. Ser. A* **113**, 195–208 (1928).

**69** J. S. Turner, *Buoyancy Effects in Fluids*, Cambridge University Press, 1973.

**70** M. G. Velarde and C. Normand, Convection, *Sci. Am.* **243** (1), 92–108 (1980).

**71** C. L. Stong, The study of electrostatic effects and convection currents in liquids, in "Amateur Scientist," *Sci. Am.* **216** (1), 124–128 (1967).

**72** C. L. Stong, Curious oscillators that involve salt water, flame and hot wire, in "Amateur Scientist," *Sci. Am.* **223**, (3), 221–234 (1970).

**73** J. Walker, The salt fountain and other curiosities based on the different density of fluids, in "Amateur Scientist," *Sci. Am.* **237** (4), 142–150 (1977).

**74** H. Buchberg, I. Catton, and D. Edwards, Natural convection in enclosed spaces—a review of application to solar energy collection, *Transact. ASME, J. Heat Transf.* **98**, 182–188 (1976).

**75** H. Buchberg, O. A. Lalude, and D. K. Edwards, Performance characteristics of rectangular honeycomb solar–thermal converters, *Solar Energy* **13**, 193–221 (1971).

**76** K. G. T. Hollands, Honeycomb devices in flat plate solar collectors, *Solar Energy* **9**, 159 (1965).

**77** S. Globe and D. Dropkin, Natural convection heat transfer in liquids confined by two horizontal plates and heated from below, *Transact. ASME, J. Heat Transf.* **81**, 24–29 (1959).

**78** M. Jakob, Free heat convection through enclosed plane gas layers, *Transact. ASME* **68**, 189–194 (1946).

**79** A. Emery and N. C. Chu, Heat transfer across vertical layers, *Transact. ASME, J. Heat Transf.* **87**, 110–116 (1965).

**80** R. K. MacGregor and A. F. Emery, Free convection through vertical plane layers—moderate and high Prandtl number fluids, *Transact. ASME, J. Heat Transf.* **91**, 391–403 (1969).

**81** A. Bar-Cohen, "Fin thickness for an optimized natural convection array of rectangular fins, *Transact. ASME, J. Heat Transf.* **101**, 564–566 (1979).

**82** S. Tsuruno and I. Iguchi, Mechanism of heat and momentum transfer of combined free and forced convection with opposing flow, *Transact. ASME, J. Heat Transf.* **101**, 573–575 (1979).

**83** A. Mucoglu and T. S. Chen, Mixed convection on inclined surfaces, *Transact. ASME, J. Heat Transf.* **101**, 422–426 (1979).

**84** A. Moutsoglou and T. S. Chen, Buoyancy effects in boundary layers on inclined continuous moving sheets, *Transact. ASME, J. Heat Transf.* **102**, 371–373 (1980).

**85** M. C. Jischke and M. Farshchi, Boundary layer regime for laminar free convection between horizontal circular cylinders, *Transact. ASME, J. Heat Transf.* **102**, 228–235 (1980).

**86** L. S. Yao, Analysis of heat transfer in slightly eccentric annuli, *Transact. ASME, J. Heat Transf.* **102**, 279–284 (1980).

**87** R. F. Bergholz, Natural convection of a heat generating fluid in a closed cavity, *Transact. ASME, J. Heat Transf.* **102**, 242–247 (1980).

**88** Y. Yen, Free convection heat transfer characteristics in a melt water layer, *Transact. ASME, J. Transf.* **102**, 550–556 (1980).

89  A. A. Emara and F. A. Kulacki, A numerical investigation of thermal convection in a heat-generating fluid layer, *Transact. ASME, J. Heat Transf.* **102**, 531–537 (1980).

90  D. R. Smart, K. G. T. Hollands, and G. D. Raithby, Free convection heat transfer across rectangular-celled diathermanous honeycombs, *Transact. ASME, J. Heat Transf.* **102**, 75–80 (1980).

91  E. M. Sparrow, J. W. Ramsey, and R. G. Kemink, Freezing controlled by natural convection, *Transact. ASME, J. Heat Transf.* **101**, 578–584 (1979).

92  R. R. Schmidt and E. M. Sparrow, Turbulent flow of water in a tube with circumferentially nonuniform heating, with or without buoyancy, *Transact. ASME, J. Heat Transf.* **100**, 403–409 (1978).

93  S. W. Churchill and H. H. S. Chu, Correlating equations for laminar and turbulent free convection from a vertical plate, *Internatl. J. Heat Mass Transf.* **18**, 1323–1329 (1975).

94  N. Rudraiah, B. Veerappa, S. B. Rao, Effects of nonuniform thermal gradient and adiabatic boundaries on convection in porous media, *Transact. ASME, J. Heat Transf.* **102**, 254–260 (1980).

95  S. J. Rhee, V. K. Dhir, and I. Catton, Natural convection heat transfer in beds of inductively heated particles, *Transact. ASME, J. Heat Transf.* **100**, 78–85 (1978).

96  S. W. Churchill, A comprehensive correlating equation for laminar, assisting, forced and free convection, *AIChE J.* **23**, 10–16 (1977).

97  N. Seki, S. Fukusako, and H. Inaba, Heat transfer in a cavity packed with fibrous glass, *Transact. ASME, J. Heat Transf.* **100**, 748–750 (1978).

98  T. Cebeci and A. Khattab, Prediction of turbulent-free-convective-heat transfer from a vertical flat plate, *Transact. ASME, J. Heat Transf.* **97**, 469–471 (1975).

99  Y. Kamotani, A. Prasad, and S. Ostrach, Thermal convection in an enclosure due to vibrations aboard spacecraft, *AIAA J.* **19**, 511–516 (1981).

100  E. Ruckstein and J. D. Felski, Turbulent natural convection at high Prandtl numbers, *Transact. ASME, J. Heat Transf.* **102**, 773–775 (1980).

101  T. Y. Na and J. P. Chiou, Turbulent natural convection over a slender cylinder, *Wärme-und Stoffübertragung* **14**, 157–164 (1980).

102  C. A. Hieber, Mixed convection in an isothermal horizontal tube: Some recent theories, *Internatl. J. Heat Mass Transf.* **24**, 315–322 (1981).

103  S. W. Churchill and H. H. S. Chu, Correlation equations for laminar and turbulent free convection from a horizontal cylinder, *Internatl. J. Heat Mass Transf.* **18**, 1049–1053 (1975).

104  V. T. Morgan, The overall convective heat transfer from smooth circular cylinders, *Adv. Heat Transf.* **11**, 199–264 (1975).

105  M. Ali-Arabi and Y. K. Salman, Laminar natural convection heat transfer from an inclined cylinder, *Internatl. J. Heat Mass Transf.* **23**, 45–51 (1980).

# 13

## BOILING

The change of density that accompanies a change of phase can give rise to vigorous natural convection. Heretofore, natural convection with only moderate density differences was contemplated. For example, heating or cooling air at atmospheric pressure through the temperature range 90–100°C induces a density change of $(\rho_{cold} - \rho_{hot})/\rho_{cold} = 0.027$, whereas heating of water under the same conditions induces a much larger density change of $(\rho_{cold} - \rho_{hot})/\rho_{cold} = 0.999$. The large difference in density between the liquid and the vapor state is accompanied by two other effects: (1) the vapor is much less viscous than the liquid; and (2) there is usually a well-defined interface between a vapor-filled region and a liquid-filled region.

The formation of vapor from a liquid is called *boiling*, and formation of liquid from a vapor is called *condensation*. Because of their importance to technology, a considerable body of literature has been developed concerning these two phenomena. Although boiling and condensation have individual peculiarities, they can generally be considered to be natural-convection phenomena.

### 13.1 POOL BOILING

Pool boiling, in which a surface heated above a liquid saturation temperature is immersed in a very large pool of the liquid, is the simplest of the possible boiling situations. When the liquid is at its saturation temperature, the process is termed *saturated boiling*. If the liquid is subcooled below its saturation temperature, the process is termed *subcooled boiling*. The data of Farber and Scorah [1] in Fig. 13-1a illustrate the sequence of events as the temperature of the submerged surface is increased for pool boiling from an electrically heated 0.04-in. horizontal wire in saturated water. At the small temperature differences encountered in region I there is no vapor formation since the liquid is

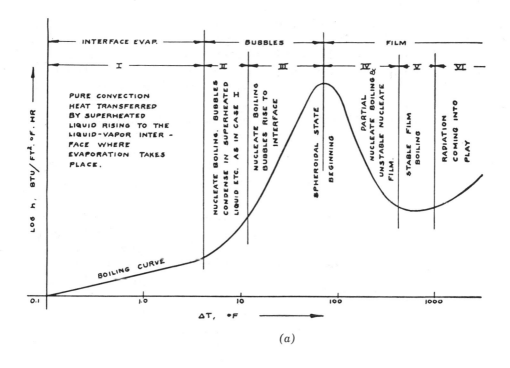

*(a)*

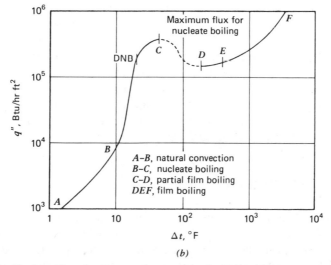

*(b)*

**Figure 13-1** Pool boiling fundamentals. (*a*) Physical interpretation of boiling curve [from E. A. Farber and R. L. Scorah, *Transact. ASME* **70**, 369–384 (1948) [1]]. (*b*) Boiling of water at 212°F on an electrically heated platinum wire measured by Nukiyama [31]; DNB signifies departure from nucleate boiling. (By permission from W. H. McAdams, *Heat Transmission*, 3rd ed., McGraw-Hill, New York, 1954.)

capable of sustaining a small amount of superheat, and natural convection without phase change occurs. At higher temperature differences encountered in region II small vapor bubbles form at a few points on the hot surface but condense when they detach from the surface and move into the liquid. At slightly higher temperature differences, such as are encountered in region III, the vapor bubbles are formed at many points and rise to the top of the pool without condensing. The boiling in regions II and III is termed *nucleate boiling*. In region IV bubble formation occurs so rapidly and at so many points on the heated surface that the surface is partially covered by an intermittent vapor film. The peak heat flux occurs at point $C$ and is called the *burnout heat flux*; hence point $C$ is often referred to as the *boiling crises*, or *burnout*. In region V the increased temperature difference results in the surface being increasingly blanketed with an insulating vapor film. The heat flux decreases, as a result, until the surface is entirely covered by a stable vapor film; a further increase in temperature difference above the value at point $D$ then again gives increased heat flux. This regime is called *stable film boiling*. At the elevated temperatures encountered in region VI, thermal radiation contributes appreciably to heat flow across the vapor film.

A striking peculiarity of boiling is that the boiling curve typified by Fig. 13-1 is multiple valued at heat fluxes between the values of point $C$ and point $D$. Above or below these values, a specified heat flux corresponds to a single temperature difference. A specified heat flux intermediate to these values can be provided by as many as three distinct temperature differences, however; such a situation tends to be unstable since the surface temperature can oscillate between the three allowed values if subjected to some small external disturbance. Most importantly, the highest of the three allowed temperatures lies in region VI, where the temperature is sufficiently high for many materials to fail. Troubles of this sort are not encountered with a surface heated by a condensing vapor since its temperature is specified—Fig. 13-1 is single valued if the abscissa is specified—but they are important to the safety of nuclear reactors and electrically heated surfaces.

## 13.2   NUCLEATE BOILING

The formation of a vapor bubble on a solid surface is observed to occur at discrete preferred sites, called *nucleation sites*, which are usually gas- or vapor-filled cavities in the heated surface as sketched in Fig. 13-2. Alternatively, a nucleation site could be a tiny sharp point, an impurity, or a boundary between crystals. The general idea for a cavity is that heat addition causes a vapor pocket to grow by evaporation from the liquid interface. Eventually the bubble grows large enough that buoyancy forces cause it to detach, leaving some vapor behind for a later repetition of these events. If the liquid is highly subcooled, the bubble may collapse without detaching, although the chain of events is otherwise unchanged. Nucleation usually occurs at a solid surface

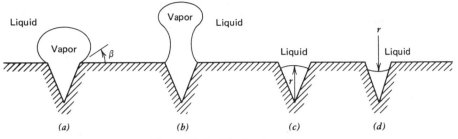

**Figure 13-2** Nucleation sites.

although it can occur in a homogeneous liquid as discussed by Leppert and Pitts [2] and can cause explosive vapor formation in a superheated liquid, called "bumping," as discussed by Westwater [3], which can be prevented by the common practice of adding "boiling stones." Cole [4] provides additional information on the rate of homogeneous nucleation and on the limits of superheat a liquid can endure; application to the vapor explosions that can occur when a molten material suddenly contacts a cooler liquid is also surveyed.

In Fig. 13-2c, the curvature of the liquid–vapor interface in a cavity is such that surface tension requires the pressure in the vapor to exceed the pressure in the liquid. As the surface cools below the liquid saturation temperature following a bubble formation and detachment, the vapor will all condense and the cavity will be completely filled with liquid, rendering it unable to form a new bubble later. Such a cavity is an inactive nucleation site. If, on the other hand, the liquid–vapor interface in the cavity is as shown in Fig. 13-2d, the pressure in the vapor is less than the pressure in the liquid. As the cavity cools, the interface recedes into the cavity, decreasing the radius of curvature and reducing the pressure in the trapped vapor. Since $p_{\text{vapor}}$ decreases, its saturation temperature also decreases, and vapor can exist in equilibrium with the liquid at the lowered temperature; this cavity always contains some vapor and can participate repetitively in cycles of bubble formation. Such a cavity is an active nucleation site. If an inert gas is trapped in a cavity, the cavity can be an active nucleation site even without any vapor initially present. Inert gases can be gradually absorbed into the liquid after a number of bubble-producing cycles, leading to a hysteresis effect [2]; after exposure to an inert atmosphere, initial conditions are restored.

The effects of radius of curvature of the liquid–vapor interface and inert-gas presence can be illustrated by imagining a vapor sphere in a liquid. The surface tension requires that the internal pressure be greater than the external pressure at equilibrium. Hence

$$(p_v + p_g) - p_L = \frac{2\sigma}{r}$$

where $p_v$ and $p_g$ are the partial pressures of vapor and inert gas, respectively, $p_L$ is the exterior pressure in the liquid, $\sigma$ is surface tension, and $r$ is the bubble radius. The Clausius–Clapeyron relationship with the perfect gas approximation relates saturation temperature $T_V$ and saturation pressure $p_v$ as

$$\frac{dp_v}{dT_v} \approx \frac{h_{fg}\rho_v}{T_v} \approx \frac{h_{fg}\,p_v}{R_v T_v^2}$$

where $R_v$ is the vapor gas constant and $h_{fg}$ is the latent heat of vaporization. Since $p_v - p_L \approx (T_v - T_{sat})dp/dT$, these two relations can be combined to obtain, with $T_{sat}$, the saturation temperature at $p_L$,

$$T_v - T_{sat} \approx \frac{R_v T_{sat}^2}{h_{fg}\, p_L}\left(-p_g + \frac{2\sigma}{r}\right) \tag{13-1}$$

This result suggests that if $T_L - T_{sat} > T_v - T_{sat}$ from this equation, a bubble will grow; if it is smaller, the bubble will collapse. Evidently, some superheat is required in the liquid for bubble growth. Also, inert gases will affect the process of bubble growth and collapse. Equation (13-1) was experimentally confirmed by Griffith and Wallis [5].

The major portion of the vapor generation responsible for bubble growth in nucleate boiling has been found to occur by evaporation from a very thin liquid layer, the microlayer, between the bubble and the hot surface. Figure 13-3 illustrates the physical situation that was first reported by Moore and Mesler [6]. Plesset and Sadhal [7] give the average microlayer thickness $\delta_0$ in terms of liquid kinematic viscosity $\nu$ and bubble lifetime $t_0$ as $\delta_0 = \frac{8}{7}(3\nu t_0)^{1/2}$. For water with $t_0 \approx 0.4 \times 10^{-1}$ s, $\delta_0 = 0.2$ mm, which is roughly the thickness of four sheets of paper.

Extensive data on nucleate pool boiling in regions II and III has been correlated within 20% by Rohsenow [8] for clean surfaces as

$$\frac{q}{A} = \mu_L h_{fg}\left[\frac{g(\rho_L - \rho_v)}{g_c \sigma}\right]^{1/2}\left[\frac{C_{p_L}(T_w - T_{sat})}{h_{fg}\,P_{r_L}^{1.7}C_{sf}}\right]^3 \tag{13-2}$$

which predicts heat flux to increase as $\Delta T^3$ and accounts for the effects of

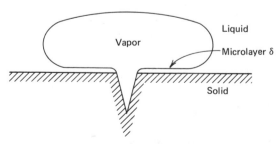

**Figure 13-3** Liquid microlayer under a vapor bubble at a nucleation site.

pressure. In Eq. (13-2):

| | |
|---|---|
| $C_{p_L}$ | saturated liquid specific heat |
| $C_{sf}$ | surface–fluid constant (see Table 13-1) |
| $g$ | gravitational acceleration |
| $g_c$ | constant, 32.17 $lb_m$ ft/$lb_f$ sec$^2$ or 1 kg m/N s$^2$ |
| $h_{fg}$ | heat of vaporization |
| $P_{r_L}$ | saturated liquid Prandtl number |
| $\dfrac{q}{A}$ | heat flux |
| $T_w - T_{sat}$ | temperature difference |
| $\mu_L$ | liquid dynamic viscosity |
| $\sigma$ | surface tension |
| $\rho_L$ | saturated liquid density |
| $\rho_v$ | saturated vapor density |

For dirty surfaces, the Prandtl number exponent varies from 0.8 to 2.0. Table 13-1 gives $C_{sf}$ for various surface–liquid combinations. There, as well as in Fig. 13-5, a substantial influence of surface condition is apparent. More detailed correlations are available both without [104] and with [109] forced convection. The surface tension of water is approximately given in units of $lb_f$/ft by $\sigma = 58 \times 10^{-4}\,(1 - 0.00142T)$ with $T$ in Fahrenheit, a linear fit to data between 212°F and the critical-point temperature of 705°F. The condition of the solid surface can substantially affect nucleate boiling heat flux as was shown by Berenson's [9, 31] measurements and the study of Vachon et al. [10]. Porous surfaces can greatly enhance nucleate boiling [102]. Nucleate pool boiling is insensitive to geometry, so Eq. (13-2) applies to horizontal and vertical surfaces as well as to wires whose diameter exceeds 0.01 cm. Subcooling of the liquid pool mostly increases the $T_w - T_{sat}$ value at which nucleation begins and thereafter has little effect [2].

When the liquid pool is so thin as to be a film, nucleate boiling heat fluxes are underpredicted by Eq. (13-2). Nishikawa [112] found that nucleate boiling heat flux from a horizontal surface to water increases when the water film thickness decreases below 3 mm; Fakeeha [113] found the increase to be a factor of 2 or greater, depending on the system pressure, at a film thickness of 0.44 mm on a vertical surface. Mesler and Mailen [114] found that this increase is due to the fact that fragments from bubbles bursting at the free surface act as secondary nucleation sites.

The burnout heat flux at point $C$ in Fig. 13-1 has been correlated from hydrodynamic considerations by Zuber [14] for a saturated liquid as

$$\frac{q_{max}}{A} = K \rho_v h_{fg} \left[ \frac{\sigma(\rho_L - \rho_v)gg_c}{\rho_v^2} \right]^{1/4} \left( \frac{\rho_L + \rho_v}{\rho_L} \right)^{1/2} \qquad (13\text{-}3)$$

where $K$ can be taken as 0.18 for water. However, a review [2] points out that

**Table 13-1    Values of $C_{sf}$ for Various Liquid–Surface Combinations** [a]

| Liquid–Surface Combination | $C_{sf}$ |
|---|---|
| Water–copper | 0.013 [11] |
| Water–scored copper | 0.0068 [10] |
| Water–emergy-polished copper | 0.0128 [10] |
| Water–emergy-polished, paraffin-treated copper | 0.0147 [10] |
| Water–chemically etched stainless steel | 0.0133 [10] |
| Water–mechanically polished stainless steel | 0.0132 [10] |
| Water–ground and polished stainless steel | 0.008 [10] |
| Water–Teflon pitted stainless steel | 0.058 [10] |
| Water–platinum | 0.013 [12] |
| Water–brass, water–nickel | 0.006 [8, 11] |
| Ethyl alcohol–chronium | 0.027 [13] |
| Benzene–chromium | 0.01 [13] |
| Carbon tetrachloride–copper | 0.013 [11] |
| Carbon tetrachloride–emery-polished copper | 0.007 [10] |
| n-Pentane–chromium | 0.015 [8] |
| n-Pentane–emery-polished copper | 0.0154 [10] |
| n-Pentane–emery-rubbed copper | 0.0074 [10] |
| n-Pentane–lapped copper | 0.0049 [10] |
| n-Pentane–emery-polished nickel | 0.0127 [10] |
| Isopropyl alcohol–copper | 0.0025 [8] |
| n-Butyl alcohol–copper | 0.003 [8] |
| 35% $K_2CO_3$–copper | 0.0054 [8] |
| 50% $K_2CO_3$–copper | 0.0027 [8] |

[a] Numbers in brackets denote references at end of chapter

Kutateladze [15] had achieved the general form of Eq. (13-3) earlier by a different approach. Borishanskii [16] included the effects of liquid viscosity in a dimensional analysis and found by comparison with data that

$$K = 0.13 + 4 \left\{ \frac{\rho_L (g_c \sigma)^{3/2} / \mu_L^2}{[g(\rho_L - \rho_v)]^{1/2}} \right\}^{-0.4}$$

Because the phenomena involved are somewhat random, burnout heat-flux data are only reproducible within about 14% [2]. Appendix F can be consulted for details of the stability analysis that leads to Eq. (13-3). A maximum value occurs at about one-third the critical pressure [2]. The burnout heat flux increases noticeably when the liquid is subcooled and the liquid flows past in forced convection as [2]

$$\frac{q_{max}}{A} - \frac{q_{max, sat}}{A} = (3.9 \text{ Btu/sec ft}^2 \, °F)(T_{sat} - T_{liq})$$

$$+ [(24 \text{ Btu/sec ft}^2)(\text{sec/ft})] V_L$$

for horizontal cylinders of 0.01–0.189 in.; burnout heat flux varies as the horizontal cylinder diameter to the $-0.15$ power [2] for diameters of 0.024–0.125 in. Information on peak boiling heat flux for horizontal plates has been given by Lienhard et al. [18]; Eq. (13-3) gives close results, but there are interesting differences. For large horizontal cylinders and spheres, the burnout heat flux is given by Eq. (13-3) with $K = 0.123$ and $K = 0.113$, respectively [17]. Burnout heat flux varies as $V_L^{1/2}$ for large $V_L$ [103].

It has been observed that boiling (nucleate boiling, in particular) can produce characteristic sounds that are sometimes called "boiling songs." Aoki and Welty [19] measured the acoustic spectra from a horizontal plate; the emitted sound pressure level increased with increasing nucleate heat flux, diminished in the peak heat-flux region as a result of attenuation by bubble coalescence above the boiling surface, and rose again when the film boiling conditions were attained. The reverse effect, the influence of acoustic vibrations on nucleate boiling, has been experimentally studied [20] and found to be noticeable under certain conditions but is seldom large. Low-frequency, mechanical, vertical vibrations of a horizontal cylinder do not increase burnout heat flux more than about 8%, according to Bergles [21], although in the nucleate boiling regime substantial heat flux increase can be obtained (a result mostly of induced motion of the liquid pool). The ability of a vertically vibrating liquid column to render bubbles motionless can influence boiling in a manner different from simple mechanical vibrations as was found by Fuls and Geiger [22], who report a heat-flux increase of about 10% in nucleate boiling in some cases.

The electric fields that are often used to heat a surface in boiling studies have been experimentally found to be able to influence the boiling curve [105]. When the surface is heated by condensing steam, electric fields can be externally applied with noticeable affect. Markels and Durfee [23] found that nucleate pool boiling from a horizontal tube was not increased much, but that burnout heat flux could be increased by a factor of 5 or more and that film boiling could be greatly delayed in its onset for applied dc (direct-current) voltages of the order of 10,000 V—similar results were found for ac (alternating-current) voltages at 60 Hz. Studies of boiling with forced convection in the presence of 60-Hz ac electric fields yielded like results [24]. The physical mechanisms involved have been discussed by Jones [25] with reference to both boiling and condensation and by Crowley [26]. A magnetic field reduces mercury nucleate boiling heat fluxes and encourages film boiling [107], a phenomenon important to heat removal from fusion reactors with magnetic confinement systems.

Forced convection of the pool liquid about the boiling surface affects the heat flux. A common approximate means for accounting for this phenomenon is the superposition suggestion due to Rohsenow [27] of

$$q = q_{conv} + q_{boil} \qquad (13\text{-}4)$$

Here $q_{conv}$ is the heat flow associated with pure forced convection, and $q_{boil}$ is the heat flow due to boiling in the absence of forced convection. Although Eq. (13-4) correlates data with some success [103], the alternative suggested by Kutateladze [28] is to relate the heat-transfer coefficients according to

$$\frac{h}{h_{conv}} = \left[ 1 + \left( \frac{h_{boil}}{h_{conv}} \right)^n \right]^{1/n} \tag{13-5}$$

He found that $n = 2$ correlates data for flow inside tubes, whereas others [29] found that 5.5 correlates nucleate boiling data for forced convection perpendicular to cylinders. (Consult Fand et al. [29] for details on calculation of $h_{conv}$ and $h_{boil}$ and Guglielmini [106] for a survey of correlations for forced-convection boiling.)

For water flowing in a pipe of diameter $D$ and length $L$ at velocity $V$, a correlation for burnout heat flux due to Lowdermilk et al. [30] is

$$\frac{q_{max}}{A} = \frac{270(\rho VD/L)^{0.85}}{D^{0.2}}, \qquad 1 < \rho V \left( \frac{D}{L} \right)^2 < 150$$

$$= \frac{1400(\rho V)^{0.5}(D/L)^{0.15}}{D^{0.2}}, \qquad 150 < \rho V \left( \frac{D}{L} \right)^2 < 10^4$$

with $q/A$ in Btu/hr ft$^2$, $\rho$ the liquid density in lb$_m$/ft$^3$, $V$ in ft/sec, and $L$ and $D$ in inches. Here 14.7 psia $< p <$ 100 psia, 0.1 ft/sec $< V <$ 98 ft/sec, $25 < L/D < 250$, 0.051 in. $< D <$ 0.188 in. This result applies to inlet subcooling from 0 to 140°F and the full range of quality. A more complete (but also more complex) correlation is available based on the work of Tippets as reviewed by Leppert and Pitts [2], which should be consulted for a more detailed account.

Although the details of the two-phase flow that occurs when a fluid boils while flowing through a tube are complex, the general trend of events is as illustrated in Fig. 13-4 for a vertical-tube evporator. There a subcooled liquid enters the bottom and contacts the hot wall where local nucleate boiling is soon experienced; the flow is bubbly when there is less than about 10% vapor. Here, the local heat-transfer coefficient is increased above its value for single-phase flow of the liquid. As the quality increases, the flow becomes annular with a thin liquid layer on the wall and a core of vapor. The vapor velocity is typically greater than the liquid's. Although some bubbles form at the wall, most evaporation occurs at the liquid–vapor interface as a result of heat conduction through the liquid film. At even higher qualities the heat transfer coefficient drops as a result of a transition to vapor (sometimes called *mist* or *fog*) flow. Burnout sometimes occurs at this point because low-conductivity vapor contacts the wall and cannot remove heat as rapidly. Vapor flow continues until the quality achieves 100%; then single-phase vapor flow exists, and the heat-transfer coefficient can be estimated by the appropriate relations for forced convection in a duct. For horizontal orientations, it is possible that vapor can congregate in the upper portion of the tube as an illustration of the

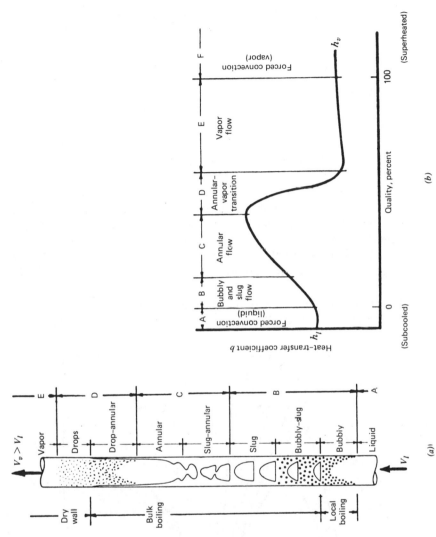

**Figure 13-4** Two-phase flow regimes of a liquid boiling during flow through a tube.

585

possible complexities that can be encountered. The survey by Kalinin et al. [63] can be consulted for additional discussion.

## 13.3 TRANSITION BOILING

At point $C$ in Fig. 13-1 the liquid intermittently touches the solid surface and is separated from the surface by a vapor film. Measurements in this region are difficult, particularly if electrically heated test sections are used; credit is usually given to Nukiyama [31] for the first studies.

Because the temperature difference at point $C$ produces vapor at a greater rate than can be removed by bubble formation during the contact phase, a vapor film forms. However, the temperature difference at point $C$ is not large enough to produce vapor at the large rate required to maintain a stable film, and so the film collapses. As the temperature difference is increased in this transition region, the liquid contacts the surface for smaller fractions of the cycle.

Typical heat flux versus temperature difference curves are shown in Fig. 13-5 from the investigations by Berenson [32]. The dependence of the boiling

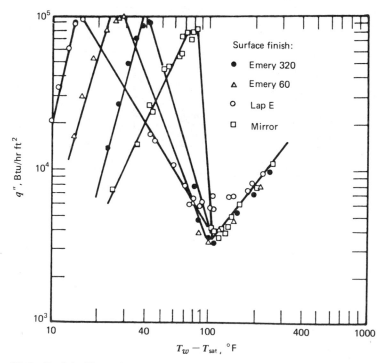

**Figure 13-5** Pool boiling of *n*-pentane from an upward-directed copper surface. [By permission from P. J. Berenson, *Internatl. J. Heat Mass Transf.* **5**, 985–999 (1962) [32].]

curve in the transition region on surface roughness suggests that the liquid contacts the surface.

Forced convection influences heat flux in the transition region [33, 34, 103]. Additives can increase transition heat flux (more than is the case for nucleate boiling) in some cases but can also cause a decrease [35, 36].

The conservative predictive procedure usually adopted is to assume that stable film boiling starts immediately after the nucleate boiling regime is passed.

## 13.4 FILM BOILING MINIMUM HEAT FLUX

At the surface temperature corresponding to point $D$ in Fig. 13-1 the liquid pool is separated from the heated surface by a continuous-vapor film. Because the vapor generation rate just equals the rate at which vapor leaves the film in detaching bubbles, the film is steadily present. The heat flux at point $D$ is termed the *film boiling minimum heat flux*.

The film boiling minimum heat flux from vertical surfaces has not been investigated either experimentally or analytically, in contrast to the great attention devoted to it for horizontal cylinders and horizontal planes. To give insight into the physical mechanisms, the horizontal plane is first considered in some detail, largely following Jordan's [37] discussion of the original work done by Zuber [14] and Chang [38].

The heat flux from a horizontal surface facing up in a pool of saturated liquid is determined by the rate at which vapor bubbles detach from the vapor film covering the surface. The unstable condition resulting from dense liquid overlaying the less dense vapor is, of course, responsible for the bubble detachment and is called *Taylor instability* after the person who first made a successful analysis (see Appendix F for details). The minimum film boiling heat flux can be computed as

$$\frac{q}{A} = \frac{\text{energy transport}}{\text{bubble}} \times \frac{\text{bubbles}}{\text{area-time}} \tag{13-6}$$

As discussed in Appendix F, the liquid–vapor interface is wavy, with the nodes and antinodes arranged in a nearly square pattern that has a characteristic wavelength $\lambda_D$. Bubbles are presumed to be spherical at detachment (photographs show this to be reasonable) of diameter proportional to $\lambda_D$, usually $\lambda_D/2$. Then, if liquid and vapor are both saturated,

$$\frac{\text{energy transport}}{\text{bubble}} = \rho_v h_{fg} \left( \frac{\lambda_D}{4} \right)^3 \frac{4\pi}{3} \tag{13-7}$$

where $\rho_v$ is vapor density and $h_{fg}$ is heat of vaporization. Each node above the interface nominal equilibrium surface grows until a bubble detaches. Then what was an antinode (below the equilibrium interface position) becomes a node and a bubble eventually detaches from there. Thus, in one cycle of interface oscillation two bubbles are detached from a nearly square area $\lambda_D^2$ of side $\lambda_D$ (the interface characteristic wavelength) and so

$$\frac{\text{bubbles}}{\text{area-oscillation}} = \frac{2}{\lambda_D^2} \tag{13-8}$$

The time required for a node to grow to a height equal to the diameter of a detaching bubble can be obtained from (the underlying Taylor stability considerations are set forth in Appendix F) the representation of the interface displacement $\eta$ from its equilibrium position

$$\eta = \eta_0 e^{b^* t} e^{\pm i(m_1 x + m_2 y)} \tag{F-21}$$

At a node the second exponential term is unity so that there

$$\eta = \eta_0 e^{b^* t}$$

The time required for the interface to achieve a specified displacement $\eta_0$ is then

$$t = (b^*)^{-1} \ln\left(\frac{\eta}{\eta_0}\right)$$

Lewis [Eq. (F-6)] in experimental confirmation of Taylor's stability analysis found that bubbles detached when $\eta = 0.4\lambda_D$. Since this is near the assumed detaching-bubble diameter $D_b$ of $\lambda_D/2$, it is plausible to speculate that at detachment the interface displacement equals $D_b$—since the initial disturbance $\eta_0$ is unknown but can be assumed to be proportional to $D_b$, now merely let $C = \ln(D_b/\eta_0)$. Then, by use of Eq. (F-27) from Appendix F to relate $b^*$ to fluid properties, the frequency of bubble emission $1/t$ is

$$\frac{[(4\pi/3)(\rho_L - \rho_v)g/(\rho_L + \rho_v)]^{1/2}}{C\lambda_D^{1/2}} \tag{13-9}$$

where the "most dangerous" wavelength $\lambda_D$ is

$$\lambda_D = 2\pi\left[\frac{3\sigma(g_c/g)}{(\rho_L - \rho_v)}\right]^{1/2} \tag{F-25}$$

Introduction of Eqs. (13-7), (13-8), (13-9) and (24-F) into Eq. (13-6) then gives the minimum film boiling heat flux as

$$\frac{q}{A} = \frac{\pi^2}{2^{3/2} 3^{5/4} C} \rho_v h_{fg} \left[\frac{\sigma(\rho_L - \rho_v)(gg_c)}{(\rho_L + \rho_v)^2}\right]^{1/4}$$

Berenson [39] empirically found that the coefficient of this equation is 0.09, from which it can be found that $C = 10$, to finally give [101]

$$\frac{q}{A} = 0.09 \rho_v h_{fg} \left[ \frac{g(\rho_L - \rho_v)}{\rho_L + \rho_v} \right]^{1/2} \left[ \frac{\sigma(g_c/g)}{\rho_L - \rho_v} \right]^{1/4} \tag{13-10}$$

for the minimum film boiling heat flux from a horizontal surface to a saturated liquid.

The temperature difference that accompanies the minimum heat flux of Eq. (13-10) was ascertained by Berenson from the physical model shown in Fig. 13-6. At any radial position in the film between bubbles the inward mass flow rate of vapor $\dot{m}$ is

$$\dot{m} = \rho_v V 2 \pi r a \tag{13-11}$$

With the assumption that heat flows from the horizontal wall to the liquid by conduction across the vapor film of thickness $a$ and that all this heat is used in vaporization (the liquid is saturated), conservation of energy requires

$$\dot{m} h_{fg} = \frac{k_v \pi (r_2^2 - r^2)(T_w - T_s)}{a} \tag{13-12}$$

provided that the flow is laminar. The surface area that generates vapor for a single bubble is taken as $\lambda_D^2/2$ (see Appendix F for details) so that $\pi r_2^2 = \lambda_D^2/2$. Then Eqs. (13-11) and (13-12) combine to yield

$$V = \frac{k_v (T_w - T_s)}{\rho_v h_{fg} a^2} \frac{\lambda^2/2 - \pi r^2}{2\pi r} \tag{13-13}$$

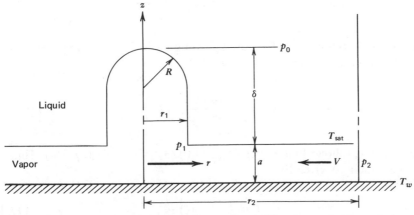

**Figure 13-6**  Physical model for minimum film boiling from a horizontal plate.

The pressure drop $p_2 - p_1$ between the interbubble position $r_2$ and the bubble position $r_1$ can be expressed in two ways. First, a Poiseuille flow assumption gives (see Appendix D)

$$0 = -\frac{\partial p}{\partial r} + \mu \frac{\partial^2 V_r}{\partial z^2} \qquad (13\text{-}14)$$

for a constant-property vapor—the inertial terms of the left-hand side can be neglected. Solution for $V_r$ gives

$$V_r = \frac{1}{2\mu} \frac{\partial p}{\partial r} z^2 + C_1 z + C_2$$

and the average velocity follows as

$$V = \frac{a}{2\mu} \frac{\partial p}{\partial r} \left( \frac{a}{3} + \frac{C_1}{z} + C_2 \right)$$

For one extreme case of a stationary liquid $V_r(z = a) = 0$,

$$V_r = \frac{1}{2\mu} \frac{\partial p}{\partial r} \left( \frac{z^2}{a} - z \right)$$

from which it follows that $\tau_w = \mu\, \partial V_r(z = 0)/\partial z = -(a/2)(\partial p/\partial r)$, which equals the shear stress at the interface $\tau_a$ by symmetry. Thus the average velocity is given by

$$V = \frac{(\tau_w + \tau_a)a}{12\mu} \qquad (13\text{-}15)$$

For another extreme case of a liquid flowing such that the shear stress at the interface is zero, $\tau_a = \mu\, \partial V_r(z = a)/\partial z = 0$,

$$V_r = \frac{a}{2\mu} \frac{\partial p}{\partial r} \left( \frac{z^2}{a} - 2z \right)$$

from which it follows that $\tau_w = \mu\, \partial V_r(z = 0)/\partial z = -(a/\mu)(\partial p/\partial r)$. Thus the average velocity is given by

$$V = (\tau_w + \tau_a) \frac{a}{3\mu} \qquad (13\text{-}16)$$

Integration of Eq. (13-14) directly between $z = 0$ and $z = a$ gives

$$\frac{\partial p}{\partial r} = \tau_w + \tau_a$$

Now Eqs. (13-15) and (13-16) suggest that

$$\tau_w + \tau_a = \frac{\beta \mu V}{a} \qquad (13\text{-}17)$$

where $3 < \beta < 12$. Incorporation of this relation into the previous relation for radial pressure gradient yields

$$\frac{\partial p}{\partial r} = \frac{\beta \mu_v V}{a^2}$$

Insertion of the average radial vapor velocity from Eq. (13-13) into this relationship gives

$$\frac{\partial p}{\partial r} = \frac{\beta \mu_v k_v (T_w - T_s)}{a^4 \rho_v h_{fg}} \frac{\lambda^2/2 - \pi r^2}{2 \pi r}$$

On integration from $r = r_1$ to $r = r_2$, the pressure difference $p_2 - p_1$ can be found. Berenson's measurements showed that

$$R = r_1 = 2.35 \left( \frac{\sigma g_c/g}{\rho_L - \rho_v} \right)^{1/2} \qquad (13\text{-}18)$$

Having assumed $\pi r_2^2 = \lambda_D^2/2$ with $\lambda_D$ available from Eq. (F-25) of Appendix F,

$$r_2 = \left( \frac{6 \pi \sigma g_c/g}{\rho_1 - \rho_v} \right)^{1/2}$$

Integration of the radial pressure gradient from $r = r_1$ to $r = r_1$ then gives

$$p_2 - p_1 = \frac{8\beta}{\pi} \frac{\mu_v k_v (T_w - T_s)}{a^4 \rho_v h_{fg}} \frac{\sigma g_c/g}{\rho_L - \rho_v} \qquad (13\text{-}19)$$

Hydrostatic considerations can be appealed to for additional information. Referring to Fig. 13-5, one can see that

$$p_2 = p_0 + \rho_L \delta \frac{g}{g_c}$$

and

$$p_1 = p_0 + \rho_v \delta \frac{g}{g_c} + \frac{2\sigma}{R}$$

Thus

$$p_2 - p_1 = (\rho_L - \rho_v)\delta \frac{g}{g_c} - \frac{2\sigma}{R} \qquad (13\text{-}20)$$

Equation (13-18) gives $R$, and the measurements of Borishansky [16] provide the information that

$$\delta = 0.68 D_{\text{bubble}} = 1.36R = 3.2\left(\sigma \frac{g_c/g}{\rho_L - \rho_v}\right)^{1/2} \qquad (13\text{-}21)$$

Combination of Eqs. (13-19)–(13-21) allows the vapor film thickness between bubbles to be obtained as

$$a = 1.68\left[\frac{\mu_v k_v (T_w - T_s)(g_c/g)}{\rho_v h_{fg}(\rho_L - \rho_v)}\left(\frac{\sigma g_c/g}{\rho_L - \rho_v}\right)^{1/2}\right]^{1/4} \qquad (13\text{-}21)$$

in which comparison with experiment shows that $(1.09\beta)^{1/4} = 1.68$—the liquid is intermediate between a rigid and a no-shear surface.

Phase-change processes in which a dense fluid overlays a lighter fluid are not confined to boiling. Dhir et al. [41] performed experiments for the case in which a horizontal slab of dry ice ($CO_2$) sublimated while under liquid water. They found that the heat transfer is predicted by Eq. (13-23) if the coefficient is reduced to 0.36, provided the water–$CO_2$ vapor interfacial temperature exceeds the freezing point. A related experimental study by Taghavi-Tafreshi et al. [42] on the melting of a horizontal slab of frozen olive oil under liquid water showed that the heat-transfer coefficient is predicted by Eq. (13-23) if the coefficient is reduced to 0.18—a good photograph of the two-dimensional interfacial wave pattern is presented. In both cases the heat-transfer coefficient is based on the temperature difference between the solid slab and the interface. Their work has application to nuclear reactor accidents [100].

The temperature difference $T_w - T_s$ can now be obtained by equating the heat conducted across the vapor film between bubbles to the total heat flow given by Eq. (13-10) since

$$\frac{q}{A} = k_v \left(\frac{A_{\text{film}}}{A_{\text{total}}}\right)\frac{T_w - T_s}{a}$$

The surface area not covered by a bubble is $A_{\text{film}} = \lambda_D^2/2 - \pi r_1^2$ and $A_{\text{total}} = \lambda_D^2/2$ so that $A_{\text{film}}/A_{\text{total}} = 0.707$. Appealing to Eq. (13-21) for $a$, one then finds

$$T_w - T_s = 0.127 \frac{\rho_v h_{fg}}{k_v} \left[ \frac{g(\rho_L - \rho_v)}{\rho_L + \rho_v} \right]^{2/3}$$

$$\times \left[ \frac{\sigma(g_c/g)}{\rho_L - \rho_v} \right]^{1/2} \left[ \frac{\mu_v(g_c/g)}{\rho_L - \rho_v} \right]^{1/3} \qquad (13\text{-}22)$$

for the minimum temperature difference for film boiling from a horizontal surface to a saturated pool with a reported [38, 39] accuracy of $\pm 10\%$. All vapor properties should be evaluated at $T_f = (T_w + T_s)/2$.

The heat-transfer coefficient is obtained by dividing the heat flux given in Eq. (13-10) by the temperature difference of Eq. (13-22) to yield

$$h = 0.709 k_v \left[ \frac{g(\rho_L - \rho_v)}{\rho_L + \rho_v} \right]^{-1/6} \left[ \frac{\sigma(g_c/g)}{\rho_L - \rho_v} \right]^{-1/4} \left[ \frac{\mu_v(g_c/g)}{\rho_L - \rho_v} \right]^{-1/3}$$

or

$$h = 0.425 \left[ \frac{\rho_v h_{fg}(\rho_L - \rho_v) k_v^3 (g/g_c)}{\mu_v(T_w - T_s)[\sigma(g_c/g)/(\rho_L - \rho_v)]^{1/2}} \right]^{1/4} \qquad (13\text{-}23)$$

with $\pm 10\%$ accuracy. To account for vapor superheat, $h_{fg} = (h_{vapor} - h_{liq}) + 0.34 C_p(T_w - T_{sat})$. Note that these results for $T_w - T_s$ and $h$ have assumed laminar vapor flow, which might not hold true at very low values of $g$.

The minimum temperature difference for film boiling is actually not necessarily determined by a Taylor instability phenomena as given by Eq. (13-22). It is possible for it to instead be limited by spontaneous nucleation on contact of the collapsing film under a detached bubble with the heated surface as explained by Yao and Henry [43]. The limiting mechanism is the one that is stable at the lowest wall temperature. Cryogenic fluids, which have a comparatively small temperature difference between their normal boiling and critical temperatures, are usually limited by rates of spontaneous nucleation on contact. Liquid metals, on the other hand, are usually limited by hydrodynamic instability. If hydrodynamic instability is the limiting mechanism, nucleation sites at the solid surface must be sufficiently numerous to generate sufficient vapor during liquid–solid contact to prevent the liquid from completely covering the wall—such a case is not possible in poorly wetting liquid metal–solid or liquid–liquid systems, such as water over mercury, since insufficient surface nucleation sites exist and in such a case the limiting mechanism might be spontaneous nucleation on contact, which can occur near saturation temperatures. Spontaneous nucleation can be on contact (bubbles form by density fluctuations at a liquid–solid interface as a result of imperfect wetting), or it can be homogeneous (bubbles form by density fluctuations entirely within the liquid bulk). Only homogeneous spontaneous nucleation is easily calculable (but serves as an upper bound and illustrates typical spontaneous nucleation behavior) by the expression

$$ J = A_1(T)\exp\left(-\frac{w}{k_1}T\right) $$

in which $J$ is homogeneous nucleations/cm$^3$ s and $T$ is the local absolute temperature. Typically, $J$ becomes appreciable only when $T$ exceeds a threshold value of approximately 195°C for ethanol and 305°C for water. Application to vapor explosions was made by Henry and Fauske [44a]. A survey of homogeneous nucleation by Springer [44b] and boiling liquid superheat in general by Afgan [45] can be consulted for greater detail.

The minimum temperature for stable film boiling is often referred to as the *Leidenfrost temperature*, after J. G. Leidenfrost who in 1756 investigated the evaporation rate of water droplets from a hot spoon [46]. He found that at sufficiently high spoon temperatures, the droplet was spherical in shape and did not touch the spoon, resting instead on a vapor film. This is variously called the *Leidenfrost phenomenon* or, because of the droplet shape, the spheroidal state—the latter term has a modern usage that includes flat disk droplets and large, bubbly liquid masses that are two forms possible in addition to a sphere [47, 48]. The maximum metastable superheat $T_{MAX}$ the liquid can sustain has been calculated by Spiegler et al. [49] as the simple relation

$$ \frac{T_{MAX}}{T_c} = \frac{27}{32} $$

for fluids with a van der Waals equation of state where $T_c$ is the absolute critical temperature. Lienhard [50] additionally imposed a Maxwell criterion and included modifications of empirical nature to find the more accurate (and more complex) relation

$$ \frac{T_{MAX}}{T_c} = 1 - 0.905\left[1 - \left(\frac{T_{sat}}{T_c}\right)^8\right] $$

where $T_{sat}$ is the absolute saturation temperature. These two relations give good results for cryogens and simple fluids but are in large error for liquid metals for whose $T_{MAX}$ Gunnerson and Cronenberg [51] should be consulted. Generally speaking, the Leidenfrost temperature lies between $T_{sat}$ and $T_{MAX}$ as

$$ T_{sat} < T_{Leid} \lesssim T_{MAX} $$

If the liquid intermittently contacts the heated surface, the surface properties are important.

The minimum film boiling heat flux from horizontal cylinders was investigated by Lienhard and Wong [52]. The analysis differs from the horizontal plate case only in the need to account for surface tension in the circumferential direction since it is in that direction that the cylinder curvature affects the equation for the motion of the liquid–vapor interface. If the cylinder radius is $R$, the effect of surface tension is to require an additional pressure difference $P_{\text{eff}}$ across the interface of

$$P_{\text{eff}} = \frac{\sigma}{\text{radius}} \approx \frac{\sigma}{R + \eta}$$

At the peak of an interfacial wave $P_{\text{eff}} = \sigma(R + \eta_0)$, whereas at the valley $P_{\text{eff}} = \sigma/R$. The average value of $P_{\text{eff}}$ is approximately $\sigma/(R + \eta_0/2)$, and the departure from this average is approximately

$$P_{\text{eff, amplitude}} \approx \frac{\sigma\eta}{2R^2}$$

Addition of this term to the right-hand side of Eq. (F-6) in Appendix F ultimately leads to the most dangerous wavelength $\lambda_D$ being given by

$$\lambda_D = 2\pi \frac{\left[3\sigma(g_c/g)/(\rho_L - \rho_v)\right]^{1/2}}{\left[1 + \sigma(g_c/g)/2R^2(\rho_L - \rho_v)\right]^{1/2}} \tag{13-24}$$

The growth rate parameter $b^*$ is then given by

$$b^* = \left(\frac{4\pi}{3\lambda_D}\right)^{1/2} \left[\frac{g(\rho_L - \rho_v)}{\rho_L + \rho_v}\right]^{1/2} \left[1 + \frac{\sigma(g_c/g)}{2R^2(\rho_L - \rho_v)}\right]^{1/2} \tag{13-25}$$

Proceeding as for the horizontal-plate case, one finds that the frequency of bubble emission is

$$\left[\frac{4\pi}{3}\frac{(\rho_L - \rho_v)g}{\rho_L + \rho_v}\right]^{1/2} \frac{\left[1 + \sigma(g_c/g)2R^2(\rho_L - \rho_v)\right]^{1/2}}{C'\lambda_D^{1/2}} \tag{13-26}$$

whereas

$$\frac{\text{bubbles}}{\text{area-oscillation}} = \frac{2}{\pi D \lambda_D} \tag{13-27}$$

where interfacial waviness is presumed to exist only along the length of the cylinder. Introduction of Eqs. (13-24)–(13-27) into Eq. (13-6) then results in the minimum heat flux being given by [101]

$$\frac{q}{A} = 0.06 \left( \frac{\rho_v h_{fg}}{D} \right) \left[ \frac{\sigma(g_c/g)}{\rho_L - \rho_v} \right]^{1/2} \left[ \frac{g g_c \sigma(\rho_L - \rho_v)}{(\rho_L + \rho_v)^2} \right]^{1/4}$$

$$\times \left[ 1 + \frac{2\sigma(g_c/g)}{D^2(\rho_L - \rho_v)} \right]^{-1/4} \tag{13-28}$$

and $C' = 20$. Equation (13-28) is, of course, intended for use with cylinders of moderate diameter for which the vapor film thickness is much smaller than the cylinder diameter and for which the cylinder diameter is less than the critical wavelength given by Eq. (F-24) of Appendix F. The effects of externally imposed electrical fields on the minimum film boiling heat flux can be substantial [25]; an applied voltage of 5000 V nearly doubles the heat flux from a 0.04-cm nichrome wire to Freon-113 for both dc and ac (50–500 Hz) conditions. The electric field, among other effects, reduces the most dangerous wavelength and increases the bubble growth rate parameter. As pointed out by Lienhard and Sun [53], the effect of the cylinder end supports alters measurements somewhat so that the minimum heat flux for a short cylinder is greater than that predicted by Eq. (13-28).

## 13.5 FILM BOILING FROM A HORIZONTAL CYLINDER

Film boiling from a horizontal cylinder immersed in a stagnant pool of saturated liquid was first successfully analyzed by Bromley [54]. The physical situation is shown in Fig. 13-7. For small to moderate cylinder diameters, heat conduction across the vapor film is the controlling physical mechanism; for large cylinder diameters, vapor removal and bubble formation are the controlling physical mechanisms. Hence the following analysis which focuses on heat conduction through the vapor film would be expected to be successful only for small to moderate cylinder diameters.

The momentum principle applied to the control volume shown in Fig. 13-6b states that

$$\overrightarrow{\text{force}} + \overrightarrow{\text{momentum}}_{\text{in}} - \overrightarrow{\text{momentum}}_{\text{out}} = \overrightarrow{\text{momentum}}_{\text{stored}} \tag{4-22}$$

Since steady state is considered, the right-hand side vanishes. Additionally, it is assumed that vapor motion is sufficiently slow that momentum fluxes are negligible compared to forces—an assumption found to be reasonable in Chapter 12 for natural convection (which film boiling incorporates as an important part). Taking the $\theta$-direction component of all forces acting on the

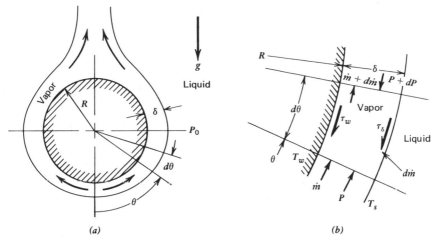

**Figure 13-7** Film boiling from a horizontal cylinder with ($a$) the coordinate system and ($b$) the control volume to which conservation principles are applied.

control volume then gives the relationship

$$[P - (P + dP)]\delta - (\tau_w + \tau_\delta)R\,d\theta - (g\sin\theta)\rho_v\frac{\delta R\,d\theta}{g_c} = 0$$

if it is assumed that $\delta/R \ll 1$ and that $\delta$ varies but slowly with $\theta$. Here $\tau_w$ and $\tau_\delta$ are shear stresses at the cylinder and interface, respectively. As in natural convection, the hydrostatic pressure variation outside the boundary layer of the vapor film gives the imposed pressure variation as

$$P = P_0 + \rho_L gR\frac{\cos\theta}{g_c}$$

Insertion of this result into Eq. (13-29) gives

$$(\rho_L - \rho_v)g\delta\frac{\sin\theta}{g_c} = (\tau_w + \tau_L) \tag{13-30}$$

Assumption that Poiseuille flow of the vapor between the stationary solid and the interface gives the relation between shear at the two surfaces and average vapor velocity $V$ as

$$\tau_w + \tau_\delta = \frac{\beta\mu_v V}{\delta} \tag{13-17}$$

in which $\beta = 3$ if the liquid exerts no shear and $\beta = 12$ if the liquid is stationary. Introduction of Eq. (13-17) into Eq. (13-30) results in

$$(\rho_L - \rho_v)g\delta\frac{\sin\theta}{g_c} = \frac{\beta\mu_v V}{\delta}$$

Recognition that the mass flow rate of the vapor $\dot{m}$ is related to the average vapor velocity by

$$\dot{m} = \rho_v \delta V$$

allows the previous relation to be restated as

$$(\rho_L - \rho_v)g\frac{\sin\theta}{g_c} = \frac{\beta\mu_v \dot{m}}{\rho_v \delta^3} \tag{13-31}$$

The conservation of energy principle applied to the control volume of Fig. 13-6b states that

$$\dot{Q}_{in} - \dot{W}_{out} = \text{internal}\ \dot{}\ \text{energy}_{stored} - \text{internal}\ \dot{}\ \text{energy}_{in} + \text{internal}\ \dot{}\ \text{energy}_{out}$$

$$\tag{4-37}$$

In terms of the particular quantities, and with the control volume outer surface on the liquid side of the liquid–vapor interface, Eq. (4-37) becomes

$$\dot{Q}_{in} - \left[ -\frac{P_1\dot{m}}{\rho_{v_1}} + \frac{P_2\dot{m}_2}{\rho_{v_2}} - \frac{P(\dot{m}_2 - \dot{m}_1)}{\rho_L} \right] = e_{v_1}\dot{m}_1 - e_{L,\text{sat}}(\dot{m}_2 - \dot{m}_1) + e_{v_2}\dot{m}_2$$

Here the subscripts 1 and 2 refer to vapor conditions at the bottom and top of the control volume, respectively, whereas $e$ is internal energy. Rearrangement results in

$$\dot{Q}_{in} = \left[ e_{v_2} + \frac{P_2}{\rho_{v_2}} - \left( e_{L,\text{sat}} + \frac{P}{\rho_L} \right) \right]\dot{m}_2 - \left[ e_{v_1} + \frac{P_1}{\rho_{v_1}} - \left( e_{L,\text{sat}} + \frac{P}{\rho_L} \right) \right]\dot{m}_1$$

$$= h_{fg_2}\dot{m}_2 - h_{fg_1}\dot{m}_1$$

$$= h_{fg}\,d\dot{m} + \dot{m}\,dh_{fg}$$

where $h_{fg}$ is latent heat of vaporization and $d\dot{m} = \dot{m}_2 - \dot{m}_1$. The latent heat of vaporization $h_{fg}$ requires accounting for a slight amount of superheating [55] as

$$h_{fg} = (h_{v,\text{sat}} - h_{L,\text{sat}}) + 0.34C_{p,v}(T_w - T_{\text{sat}}) \tag{13-32}$$

$\dot{Q}_{in}$, entering the control volume, is assumed to be controlled by conduction across the vapor film and is given by

$$\dot{Q}_{in} = k_v(T_w - T_{sat})R\frac{d\theta}{\delta} \tag{13-33}$$

It is totally expended in vaporization at the interface (no heat can be conducted away from the interface into the liquid uniformly at the saturation temperature). Furthermore, the energy required for superheating vapor is assumed to be accounted for by Eq. (13-32) so that

$$\dot{Q}_{in} = h_{fg}\,d\dot{m} \tag{13-34}$$

Equations (13-33) and (13-34) substituted into Eq. (13-31) allow $\delta$ to be eliminated and yield

$$\dot{m}^{1/3}\,d\dot{m} = \left[(\rho_L - \rho_v)\frac{g}{g_c}\frac{\rho_v}{\mu_v\beta}\right]^{1/3}\left[k_v(T_w - T_{sat})\frac{R}{h_{fg}}\right]\sin^{1/3}\theta\,d\theta$$

Integration* of this differential equation from $\theta = 0$ to $\theta = \pi$, with account for the fact that such a procedure gives only half the vapor production, results in

$$\dot{m} = 2\left[\frac{(3.45)^{3/4}}{\beta^{1/4}}\right]\left[\frac{(\rho_L - \rho_v)(g/g_c)\rho_v k_v^3(T_w - T_{sat})^3 R^3}{\left(\mu_v h_{fg}^3\right)}\right]^{1/4}$$

An average heat-transfer coefficient is defined by

$$q = \dot{m}h_{fg} = \bar{h}_c 2R(T_w - T_{sat})$$

which gives

$$\bar{h}_c = C\left[\frac{(g/g_c)(\rho_L - \rho_v)\rho_v k_v^3 h_{fg}}{\mu_v D(T_w - T_{sat})}\right]^{1/4}$$

or

$$\overline{Nu}_c = \bar{h}_c\frac{D}{k_v} = C\left[Ra\frac{h_{fg}}{C_{p_v}(T_w - T_{sat})}\right]^{1/4} \tag{13-35}$$

*Note that, with $\Gamma(m)$ being the gamma function,

$$\int_0^{\pi/2}\sin^n x\,dx = \frac{\pi^{1/2}}{2}\frac{\Gamma(n/2 + 1/2)}{\Gamma(1 + n/2)} \quad \text{for} \quad n > -1$$

For $n = 1/3$, this definite integral equals 1.2946.

where the Rayleigh number Ra is

$$Ra = \left[ g \frac{\rho_L - \rho_v}{\rho_v} \frac{D^3}{\nu_v^2} \right] Pr_v$$

and $C = 0.958/\beta^{1/4}$. Comparison with experiment shows that $C = 0.62$ within 15%, in good agreement with the expected variation (for $3 < \beta < 12$) of $0.515 \leqslant C \leqslant 0.728$. As expected, a natural-convection-like correlation results. Problem 13-18 contains additional discussion. Pitschmann and Grigull [56] extended the Bromley model to obtain

$$\overline{Nu}_c = \overline{h}_c \frac{D}{k} = 0.9(Ra^*)^{0.08} + 0.8(Ra^*)^{0.2} + 0.02(Ra^*)^{0.4}$$

where $Ra^* = Ra\, Pr^*$ and $Pr^* = Pr[1 + 3.33 h_{fg}/C_p(T_w - T_s)]$; experimental verification was accomplished by Hesse et al. [57].

The effect of thermal radiation must be included at the elevated temperatures that often accompany film boiling. Order of magnitude arguments led Bromley to

$$\frac{\overline{h}}{\overline{h}_c} = \left( \frac{\overline{h}}{\overline{h}_c} \right)^{-1/3} + \frac{h_r}{\overline{h}_c} \qquad (13\text{-}36)$$

which is approximately represented by

$$\overline{h} \approx \overline{h}_c + \frac{3h_r}{4}$$

Here $\overline{h}_c$ is the heat-transfer coefficient for conduction across the vapor acting alone as given by Eq. (13-35), $h_r$ is the radiative coefficient for the assumed parallel plate geometry, and $\overline{h}$ is the net heat-transfer coefficient. With the liquid taken to be a perfect absorber ($\epsilon_L = 1$),

$$h_r = \frac{\sigma'}{1/\epsilon_w + 1/\epsilon_L - 1} \frac{T_w^4 - T_{sat}^4}{T_w - T_{sat}}$$

where $\sigma' = 0.1714 \times 10^{-8}$ Btu/hr ft$^2$ R$^4$ is the Stefan–Boltzmann constant, $\epsilon_w$ is the wall thermal emittance, and $T$ is absolute temperature. The theoretical basis for Eq. (13-36) was set forth by Lubin [58].

The assumptions underlying Eq. (13-35) restrict its applicability to cylinder diameters above 0.1 cm. At large diameters waves at the liquid–vapor interface make the prediction of Eq. (13-35) too low. To account for this latter effect Breen and Westwater [59] correlated measurements over a wide range of cylinder diameters, $2.2 \times 10^{-4} - 1.895$ in., by

$$\frac{\overline{h}_c \lambda_c^{1/4}}{F} = 0.59 + 0.069 \frac{\lambda_c}{D}, \qquad 0.15 \leqslant \frac{\lambda_c}{D} \leqslant 300 \qquad (13\text{-}37)$$

in which the critical wavelength is $\lambda_c = 2\pi[\sigma(g_c/g)/(\rho_L - \rho_v)]^{1/2}$ and $F^4 = (g/g_c)(\rho_L - \rho_v)\rho_v k_v^3 h_{fg}/[\mu_v(T_w - T_{sat})]$. Bromley's Eq. (13-35) is accurate for $0.8 \leqslant \lambda_c/D \leqslant 8$. A slightly better fit to data is provided by

$$\frac{\bar{h}_c \lambda_c^{1/4}}{F} = \begin{cases} 0.6, & \lambda_c/D < 0.8 \\ 0.62\left(\dfrac{\lambda_c}{D}\right)^{1/4}, & 0.8 \leqslant \dfrac{\lambda_c}{D} \leqslant 8 \\ 0.16\left(\dfrac{\lambda_c}{D}\right)^{0.83}, & 8 < \dfrac{\lambda_c}{D} \end{cases} \qquad (13\text{-}37a)$$

Pomerantz [60] verified Breen and Westwater's correlation under increased gravitational fields. Thermal radiation is still accounted for by Eq. (13-36).

The effect of upward forced convection at velocity $V$ on film boiling from a horizontal cylinder to saturated pool was studied by Bromley et al. [61]. They found that Eq. (13-35) applies when $V/(gD)^{1/2} < 1$ while for $V/(gD)^{1/2} > 2$ the heat-transfer coefficient is given by

$$\overline{Nu}_c = \bar{h}_c\frac{D}{k} = 2.7\left[\text{Re}_{D_v} \text{Pr}_v \frac{h_{fg}}{C_p(T_w - T_{sat})}\right]^{1/2} \qquad (13\text{-}38)$$

in which $\text{Re}_{D_v} = VD/\nu_v$ and $\text{Pr}_v = \mu_v C_{pv}/k_v$. The effect of thermal radiation is accounted for in an approximate way by

$$\bar{h} = \bar{h}_c + \frac{7h_r}{8}$$

The effect of subcooling for upward directed flow was examined by Motte and Bromley [62]. A comprehensive survey of the pertinent literature was made by Kalinin et al. [63]. Witte [64] recommends that the lead coefficient of Eq. (13-38) be changed to 2.98 for flow around a sphere. Recent measurements [103] confirm Eq. (13-38) but suggest a better correlation.

Film boiling from a horizontal plate has been correlated within 25% by Klimenko [108] over a wide range of conditions and for many fluids. The correlation for laminar conditions $\text{Ga}(\rho_L/\rho_v - 1) < 10^8$ is

$$Nu = 0.19\left[\text{Ga}\left(\frac{\rho_L}{\rho_v} - 1\right)\right]^{1/3} \text{Pr}_v^{1/3} f_1$$

with

$$f_1 = \begin{cases} 1, & \dfrac{h_{fg}}{C_{p,v}(T_w - T_{sat})} \leqslant 1.4 \\ 0.89\left[\dfrac{h_{fg}}{C_{p,v}(T_w - T_{sat})}\right]^{1/3} & \text{otherwise} \end{cases}$$

whereas for turbulent conditions $\mathrm{Ga}(\rho_L/\rho_v - 1) > 10^8$, it is

$$\mathrm{Nu} = 0.0086 \left[ \mathrm{Ga} \left( \frac{\rho_L}{\rho_v} - 1 \right) \right]^{1/2} \mathrm{Pr}_v^{1/3} f_2$$

with

$$f_2 = \begin{cases} 1, & \dfrac{h_{\mathrm{fg}}}{C_{p,v}(T_w - T_{\mathrm{sat}})} < 2 \\[4mm] 0.71 \left[ \dfrac{h_{\mathrm{fg}}}{C_{p,v}} (T_w - T_{\mathrm{sat}}) \right]^{1/2} & \text{otherwise} \end{cases}$$

where $\mathrm{Ga} = g\lambda_c^3/2\nu_v^2$ is the Galileo number, $\mathrm{Nu} = h\lambda_c/k$ is the Nusselt number, $\lambda_c = 2\pi[\sigma(g_c/g)/(\rho_L - \rho_v)]^{1/2}$ is the critical wavelength, and all vapor properties are evaluated at $(T_w + T_{\mathrm{sat}})/2$. This correlation is tested for the parameter ranges of $7 \times 10^4 < \mathrm{Ga}(\rho_L/\rho_v - 1) < 3 \times 10^8$, $0.69 < \mathrm{Pr} < 3.45$, $0.031 < h_{\mathrm{fg}}/C_{p,v}(T_w - T_{\mathrm{sat}}) < 7.3$, $0.0045 < P/P_{\mathrm{critical}} < 0.98$, and $1 < g/g_{\mathrm{earth\ normal}} < 21.7$. The effect of plate size is accounted for by the correlation

$$\frac{h}{h_\infty} = \begin{cases} 1, & \dfrac{L}{\lambda_c} > 5 \\[4mm] 2.9 \left( \dfrac{\lambda_c}{L} \right)^{0.67}, & \dfrac{L}{\lambda_c} < 5 \end{cases}$$

where $L$ is the minimum dimension of the plate.

## 13.6   FILM BOILING FROM A VERTICAL SURFACE

Film boiling from a vertical surface immersed in a stagnant pool is successfully approached as a boundary-layer problem if vapor flows up in a film between the solid and the liquid rather than detaching as bubbles at the liquid–vapor interface. As shown in Fig. 13-8, heat flows primarily by conduction across the vapor film to the interface where it enters the liquid. If the liquid is subcooled, only part of the heat is used to form vapor; otherwise, all the heat entering the liquid forms vapor. The vapor formed enters the vapor film, "blowing" toward the solid surface, and forms a boundary layer whose thickness $\delta$ is a real physical quantity. Near the bottom edge of the solid, flow is laminar, with eventual transition to turbulent flow if the solid is sufficiently high.

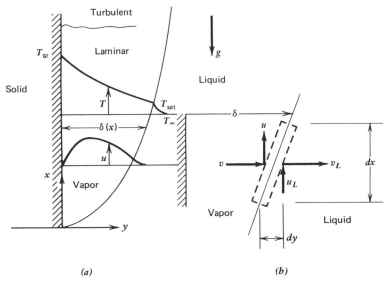

(a)                                                              (b)

**Figure 13-8**  Film boiling from a vertical plate with (a) the coordinate system and (b) a control volume fixed in the liquid–vapor interface used to obtain boundary conditions.

Assuming steady state and constant vapor and liquid properties, the laminar boundary-layer equations are

$$\frac{\partial u}{\partial x} + \frac{\partial v}{\partial y} = 0 \qquad\qquad (13\text{-}39a)$$

$$u\frac{\partial u}{\partial x} + v\frac{\partial u}{\partial y} = (\rho_L - \rho)\frac{g}{\rho} + \nu\frac{\partial^2 u}{\partial y^2} \qquad\qquad (13\text{-}39b)$$

$$u\frac{\partial T}{\partial x} + v\frac{\partial T}{\partial y} = \alpha\frac{\partial^2 T}{\partial y^2} \qquad\qquad (13\text{-}39c)$$

The liquid's stagnant condition far from the plate (near the plate it is slightly dragged along by the vapor) gives $\partial P/\partial x = -\rho_L g/g_c$, where the subscript $L$ denotes a liquid quantity. The body force $B_x$ is $-\rho g/g_c$. The assumption that there is no pressure change across the interface is accurate but inexact, as was discussed in Problem 4-13—the fluid is accelerated because of its density change, and this requires a pressure difference.

Boundary conditions to be imposed at the interface are different from those heretofore treated and are considered first. The conservation of mass principle applied to a control volume straddling the interface as shown in Fig. 13-8b requires that

$$\rho v \, dx - \rho u \, dy = \rho_L v_L \, dx - \rho_L u_L \, dy$$

With the recognition that at the interface $y = \delta$, this relation is rearranged as

$$\rho\left(v - u\frac{d\delta}{dx}\right) = \rho_L\left(v_L - u_L\frac{d\delta}{dx}\right) \tag{13-40a}$$

at $y = \delta$.

A similar application of the conservation of mass principle requires that, with $e$ representing internal energy per unit mass and K.E., kinetic energy per unit mass,

$$\left[-k\frac{\partial T}{\partial y} + (e + \text{K.E.})\rho v + Pv\right]dx$$

$$-\left[-k\frac{\partial T}{\partial x} + (e + \text{K.E.})\rho u + Pu\right]dy + q_r\,dx$$

$$=\left[-k_L\frac{\partial T_L}{\partial y} + (e_L + \text{K.E.}_L)\rho_L v_L + P_L v_L\right]dx$$

$$-\left[-k_L\frac{\partial T_L}{\partial x} + (e_L + \text{K.E.}_L)\rho_L u_L + P_L u_L\right]dy$$

where $q_r$ is the net radiative heat flux between the plate and the interface. Incorporating Eq. (13-40a), neglecting changes in kinetic energy, defining the latent heat of vaporization as $h_{\text{fg}} = (e + P/\rho) - (e_L + P_L/\rho_L)$, and recognizing again that $y = \delta$ at the interface, one obtains

$$q_r - k\left(\frac{\partial T}{\partial y} - \frac{\partial T}{\partial x}\frac{d\delta}{dx}\right) = -k_L\left(\frac{\partial T_L}{\partial y} - \frac{\partial T_L}{\partial x}\frac{d\delta}{dx}\right)$$

$$-h_{\text{fg}}\rho\left(v - u\frac{d\delta}{dx}\right)$$

Since $\partial T/\partial x \ll \partial T/\partial y$ in the boundary-layer assumptions, this relation is accurately approximated (except at the origin) by

$$q_r - k\frac{\partial T}{\partial y} = -k_L\frac{\partial T_L}{\partial y} - h_{\text{fg}}\rho\left(v - u\frac{d\delta}{dx}\right) \tag{13-40b}$$

at $y = \delta$. In the same way the momentum principle applied to the interfacial control volume leads to two equations, one for the perpendicular direction, and one for the parallel direction. The one in the perpendicular direction is of little utility since it has already been assumed that there is no pressure difference across the interface. Convective momentum transport is negligible at the interface as a result of the low velocities, and the interface curvature is also negligible (except near the origin where the boundary-layer equations are inaccurate anyway); therefore, the parallel direction yields only the requirement that the tangential shears on the liquid and the vapor sides are equal, giving

$$\mu \frac{\partial u}{\partial y} = \mu_L \frac{\partial u_L}{\partial y} \qquad (13\text{-}40c)$$

at $y = \delta$. Additionally, if there is to be no interfacial velocity slip or temperature jump, it is necessary that at $y = \delta$

$$u - u_L \qquad (13\text{-}40d)$$

$$T = T_{\text{sat}} \qquad (13\text{-}40e)$$

The conditions at the plate where $y = 0$ require that

$$u = 0 \qquad (13\text{-}40f)$$

$$v - 0 \qquad (13\text{-}40g)$$

if there is no velocity slip and the plate is impermeable and

$$T = T_w \qquad (13\text{-}40h)$$

if there is no temperature jump.

An integral method of solution of Eqs. (13-39) and (13-40) will first be essayed. Not only does it give very reasonable results, it also indicates the functional forms that must appear in the more exact solution, which is discussed later. First the continuity equation [Eq. (13-39a)] is integrated across the vapor film to obtain

$$v_\delta = \int_0^\delta \frac{\partial u}{\partial x} dy$$

$$= -\frac{\partial \left( \int_0^\delta u \, dy \right)}{\partial x} + u_\delta \frac{d\delta}{dx} \qquad (13\text{-}41)$$

in which Eq. (13-40g) has been used. Second, the x-motion equation [Eq.

(13-39b)] is rearranged into the conservation form

$$\frac{\partial(uu)}{\partial x} + \frac{\partial(vu)}{\partial y} - u\left(\overset{0}{\cancel{\frac{\partial u}{\partial x}}} + \frac{\partial v}{\partial y}\right) = (\rho_L - \rho)\frac{g}{\rho} + v\frac{\partial^2 u}{\partial y^2}$$

which is integrated across the vapor film to obtain

$$\int_0^\delta \frac{\partial(u^2)}{\partial x}\,dy + v_\delta u_\delta = (\rho_L - \rho)g\frac{\delta}{\rho} + v\left[\frac{\partial u(\delta)}{\partial y} - \frac{\partial u(0)}{\partial y}\right]$$

$$\frac{\partial\left(\int_0^\delta u^2\,dy\right)}{\partial x} - u_\delta^2\frac{d\delta}{dx} + v_\delta u_\delta = \frac{(\rho_L - \rho)g\delta}{\rho} + v\left[\frac{\partial u(\delta)}{\partial y} - \frac{\partial u(0)}{\partial y}\right]$$

in which Eq. (13-40f) has been used. Insertion of Eq. (13-41) into this relationship yields

$$\frac{d\left(\delta\int_0^1 u^2\,d\eta\right)}{dx} - u_\delta\frac{d\left(\delta\int_0^1 u\,d\eta\right)}{dx} = (\rho_L - \rho)\frac{g\delta}{\rho} + \frac{v[du(1)/d\eta - du(0)/d\eta]}{\delta}$$

$$(13-42)$$

with $\eta = y/\delta$. Third, the energy equation [Eq. (13-39c)] is rearranged into the conservation form

$$\frac{\partial(uT)}{\partial x} + \frac{\partial(vT)}{\partial y} - T\left(\overset{\text{0 by continuity}}{\cancel{\frac{\partial u}{\partial x}} + \frac{\partial v}{\partial y}}\right) = \alpha\frac{\partial^2 T}{\partial y^2}$$

which is integrated across the vapor film to obtain

$$\int_0^\delta \frac{\partial(uT)}{\partial x}\,dy + v_\delta T_\delta = \alpha\left[\frac{\partial T(\delta)}{\partial y} - \frac{\partial T(0)}{\partial y}\right]$$

$$\frac{d\left(\int_0^\delta uT\,dy\right)}{dx} - u_\delta T_\delta\frac{d\delta}{dx} + v_\delta T_\delta =$$

in which Eq. (13-40g) has been used. Insertion of Eq. (13-41) into this relationship yields

$$\frac{d\left(\delta\int_0^1 uT\,d\eta\right)}{dx} - T_0\frac{d\left(\delta\int_0^1 u\,d\eta\right)}{dx} = \frac{\alpha[dT(1)/d\eta - dT(0)/d\eta]}{\delta}$$

with $\eta = y/\delta$. This can be rearranged further, with the help of Eq. (13-40e), into

$$\frac{d\left[\delta\int_0^1 u(T - T_s)\,d\eta\right]}{dx} = \frac{\alpha[dT(1)/d\eta - dT(0)/d\eta]}{\delta} \qquad (13\text{-}43)$$

A velocity distribution is assumed of the form

$$u = a_0 + a_1\eta + a_2\eta^2 + a_3\eta^3 + \cdots$$

which can be subjected to the conditions of

At $\eta = 0$
$$0 = (\rho_L - \rho)\frac{g}{\rho} + \nu\frac{\partial^2 u}{\partial y^2} \qquad (13\text{-}39a)$$

$$u = 0 \qquad (13\text{-}40f)$$

At $\eta = 1$
$$\left[u_\delta\frac{d\delta}{dx} - \frac{d\left(\delta\int_0^1 u\,d\eta\right)}{dx}\right]\frac{\partial u}{\partial y} = (\rho_L - \rho)\frac{g}{\rho} + \nu\frac{\partial^2 u}{\partial y^2} \qquad (13\text{-}39a)$$

$$\mu\frac{\partial u}{\partial y} = \mu_L\frac{\partial u_L}{\partial y} \qquad (13\text{-}40c)$$

$$u = u_L \qquad (13\text{-}40d)$$

At this point it is desirable to introduce the simplifications that the liquid is saturated ($\partial T_L/\partial y = 0$), thermal radiation is unimportant ($q_r = 0$), and the liquid is so viscous compared to the vapor that the liquid is immovable ($u_\delta = 0$). Meeting the first, second, and fifth of the preceding possible conditions gives

$$u = \frac{(\rho_L - \rho)g\delta^2}{2\rho\nu}[\eta - \eta^2 - a(\eta - \eta^3)] \qquad (13\text{-}44)$$

The remaining parameter $a$ is evaluated by substituting Eq. (13-44) into the integral x-motion equation [Eq. (13-42)] to obtain

$$\left(1 - 3a + \frac{16a^2}{7}\right)\frac{d\delta^4}{dx} = 144\frac{\rho\nu^2}{(\rho_L - \rho)g}a$$

Now $d\delta^4/dx$ is expected to be constant. For convenience, let

$$C = \frac{(\rho_L - \rho)gh_{fg}}{16\nu k(T_w - T_s)}\frac{d\delta^4}{dx} \qquad (13\text{-}44a)$$

Note that if $C$ is constant, it is necessary that $\delta \sim x^{1/4}$. The algebraic relation for $a$ is

$$\frac{16a^2}{7} - \left[3 + 9\mathrm{Pr}\frac{h_{fg}/C_p(T_w - T_s)}{C}\right]a + 1 = 0$$

The coupling between the velocity and temperature distributions is evident here since $C$ is unknown and must be determined by simultaneous solution of the energy equation. Usually, the coupling is weak since it is commonly the case that $C_p(T_w - T_s)/h_{fg} \ll 1$ and $C \approx 1$. Then it is seen that

$$a \approx \frac{C_p(T_w - T_s)/h_{fg}}{9\mathrm{Pr}} \qquad (13\text{-}44b)$$

The "blowing" of stagnant vapor toward the plate from the interface is seen in Eq. (13-44) to cause the velocity profile to be displaced downward and toward the plate from the parabolic distribution that is its essential characteristic. A temperature distribution is next assumed of the form

$$T - T_s = b_0 + b_1\eta + b_2\eta^2 + b_3\eta^3 + b_4\eta^4 + \cdots$$

which can be subjected to the conditions

At $\eta = 0 \qquad \dfrac{d^2T}{d\eta^2} = 0 \qquad (13\text{-}39c)$

$$T = T_w \qquad (13\text{-}40h)$$

At $\eta = 1 \qquad \dfrac{1}{\delta}\left[u_\delta\dfrac{d\delta}{dx} - \dfrac{d\left(\delta\int_0^1 u\,d\eta\right)}{dx}\right]\dfrac{dT}{d\eta} = \dfrac{\alpha}{\delta^2}\dfrac{d^2T}{d\eta^2} \qquad (13\text{-}39c)$

$$T = T_s \qquad (13\text{-}40e)$$

$$q_r - \frac{k}{\delta}\frac{dT}{d\eta} = -k_L\frac{\partial T_L}{\partial y} + \rho h_{fg}\frac{d\left(\delta\int_0^1 u\,d\eta\right)}{dx} \qquad (13\text{-}40b)$$

Compliance with the first, second, fourth, and fifth of these possible five conditions gives, with $q_r = 0$,

$$(T - T_s)(T_w - T_s) = 1 - \left(1 + \frac{1 - C}{2}\right)\eta + \frac{1 - C}{2}\eta^3 \qquad (13\text{-}45)$$

where $C$ is the constant defined by Eq. (13-44a). Substitution of Eqs. (13-44) and (13-45) into the integrated energy Eq. (13-43) gives

$$\left[1 - \frac{7a}{5} - (1 - C)\left(\frac{3}{10} - \frac{16a}{35}\right)\right]\frac{d\delta^4}{dx} = 48\left[\frac{\rho\nu\alpha}{(\rho_L - \rho)g}\right](1 - C)$$

Elimination of $d\delta^4/dx$ from this expression by use of the definition of $C$ yields

$$\frac{1}{3}\frac{C_p(T_w - T_s)}{h_{fg}}\left[1 - \frac{7a}{5} - (1 - C)\left(\frac{3}{10} - \frac{16a}{35}\right)\right]C = 1 - C$$

It is convenient to note that for water, $C_p(T_w - T_s)/h_{fg} \sim \frac{1}{10}$, as is the subject of Problem 13-20. Then $C \approx 1$ and the corrected value is readily found to be

$$1 - C \approx \frac{1}{3}\frac{C_p(T_w - T_s)}{h_{fg}} \qquad (13\text{-}46)$$

The "blowing" toward the plate from the interface is seen in Eq. (13-45) to bow the temperature profile slightly below the linear variation that is its essential characteristic. From the definition of $C$ one can next obtain the vapor film thickness $\delta$ as

$$\delta = 2\left[1 - \frac{1}{3}\frac{C_p(T_w - T_s)}{h_{fg}}\right]^{1/4}\left[\frac{x(T_w - T_s)\nu k}{gh_{fg}(\rho_L - \rho)}\right]^{1/4} \qquad (13\text{-}47)$$

which shows, as expected, that $\delta \sim x^{1/4}$. The heat flux at the wall $q_w$ is now available from

$$q_w = -k\frac{\partial T(0)}{\partial y}$$

$$= -\frac{k}{\delta}\frac{dT(0)}{d\eta}$$

$$= \frac{k}{\delta}(T_w - T_s)\left(1 + \frac{1 - C}{2}\right)$$

From this the heat-transfer coefficient follows, with the use of Eqs. (13-46) and

(13-47), as

$$h = \frac{q_w}{T_w - T_s} = \frac{1}{2}k\left[\frac{g(\rho_L - \rho)\text{Pr}}{\rho \nu^2 x}\frac{h_{fg}}{C_p(T_w - T_s)}\right]^{1/4} K$$

where $K \approx \left[1 + \dfrac{C_p(T_w - T_s)}{h_{fg}}\right]^{1/4}$.

Interfacial "blowing" is responsible for $K$ differing from unity, a variation that is only approximately determined here. Problem 13-27 illustrates that if "blowing" were originally neglected in the analysis, a rougher correction could easily be made, neglecting details of convective momentum and energy transport.

The average heat-transfer coefficient $\bar{h}$ is related to the local coefficient by $\bar{h} = 4h/3$. Hence

$$\bar{N}u = \frac{\bar{h}x}{k} = \frac{2}{3}\left[\text{Ra}\frac{h_{fg}}{C_p(T_w - T_s)}\right]^{1/4} K \qquad (13\text{-}48)$$

where the Rayleigh number is $\text{Ra} = g(\rho_L - \rho)\text{Pr}\,x^3/\rho\nu^2$. Bromley [54] suggested that the lead coefficient should range between $\frac{2}{3}$ for an immovable liquid and 0.943 for a liquid that exerts no shear; Ellion [65] correlated measurements with the lead coefficient equal to 0.714.

A detailed integral analysis of laminar film boiling from a vertical surface to a saturated liquid has been accomplished by Frederking [68].

It must be noted that if $x$ is sufficiently large, the vapor will undergo transition to turbulent flow and Eq. (13-48) will not give accurate predictions. Hsu and Westwater [37, 66] found that transition begins at a Reynolds number of

$$\text{Re} = \frac{u_{max}\delta}{\nu} \approx 100$$

Their results, although accurate predictors, are cumbersome, and those with specific interest are referred to the given references for details. For turbulent film boiling from a vertical surface, it is recommended [63] that the correlation by Labuntsov [67]

$$\bar{N}u = \frac{\bar{h}x}{k} = 0.25\,\text{Ra}_x^{1/3}$$

be used for $\text{Ra} \geqslant 2 \times 10^7$.

Exact solutions to the describing boundary-layer Eqs. (13-39) and associated boundary conditions of Eq. (13-40) are in good agreement with those of the foregoing integral method. Koh [69] treated the saturated-liquid case and accounted for the liquid boundary-layer motion that is described by the parameter of $(\rho\mu/\rho_L\mu_L)^{1/2}$, accounting for vapor Prandtl numbers of 0.5 and 1.0. The subcooled liquid case was solved by Sparrow and Cess [70] that required that the boundary-layer energy and $x$-motion equations be solved simultaneously with those for the vapor film. The calculations are summarized by the example shown in Fig. 13-9 to show the change in local Nusselt number for laminar film boiling of water with $T_s = 467°F$ at a vertical surface with $T_w = 767°F$. Figure 13-9 shows that subcooling can have a substantial effect— the pure free convection limit at large liquid subcooling represents the fact that the liquid seems to be in laminar free convection at an impermeable vertical plate of temperature $T_s$. Experimental support for these trends of liquid subcooling is exhibited in Fig. 13-10. An integral analysis of the same problem was performed by Frederking and Hopenfeld [71], whose work is discussed in the survey by Kalinin et al. [63]. The effect of intermediate subcooling can be taken in account by employing the superposition idea that

$$q_w^4 = q_{w_{sat}}^4 + q_{w_{large\ subcool}}^4$$

which leads to

$$\overline{Nu} = \frac{hx}{k} = \frac{2}{3}\left\{ Ra_{x,v}\left[\frac{h_{fg}}{C_p(T_w - T_s)}\right] + \frac{k_L(T_s - T_\infty)}{k_v(T_w - T_s)}\frac{Ra_{x,l}\ Pr_L}{(1 + Pr_L)}\right\}^{1/4}$$

Parallel forced-convection laminar film boiling from a constant temperature plate to a saturated liquid was treated by Cess and Sparrow [72]. They solved

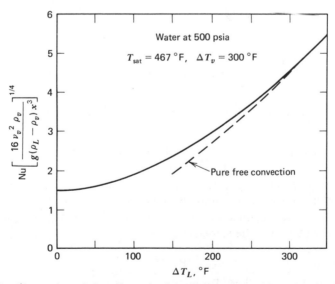

**Figure 13-9** Illustration of the effect of subcooling on Nusselt number for film boiling from a vertical plate. [From E. M. Sparrow and R. D. Cess, *Transact. ASME, J. Heat Transf.* **84**, 149–156 (1962) [70].]

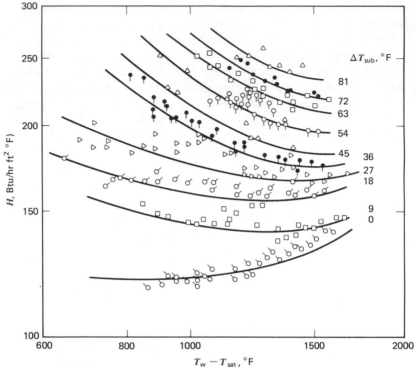

**Figure 13-10** Effect of liquid subcooling on the heat-transfer coefficient for film boiling of distilled water from a 0.0197-in.-diameter wire. (From F. Tachibana and S. Fukui, *International Developments In Heat Transfer*, ASME, 1961, p. 219 [110].)

the describing laminar boundary-layer equations in the vapor film and the adjacent affected liquid. Their results are well represented by

$$\text{Nu}\,\text{Re}^{-1/2}\frac{\mu}{\mu_L}\left(1 + \pi^{1/2}\,\text{Nu}\,\text{Re}^{-1/2}\frac{\mu}{\mu_L}\right)^{1/2} = \tfrac{1}{2}K^{1/2} \qquad (13\text{-}49a)$$

for $10^{-3} \leqslant K = (\rho\mu/\rho_L\mu_L)\text{Pr}\,h_{\text{fg}}/C_p(T_w - T_s) \leqslant 1$, where the local Nusselt number is $\text{Nu} = hx/k$ and $\text{Re} = U_\infty x/\nu_L$ whereas $x$ is the distance from the plate leading edge and $U_\infty$ is the parallel approach velocity of the liquid far from the plate. This result cannot be taken to the limit of $U_\infty \to 0$ since no body-force term was retained in the $x$-motion equations. The local wall shear $\tau$ is well represented over the same range by

$$\frac{(\tau/\rho_L U_\infty^2)\text{Re}^{1/2}}{1 - \pi^{1/2}\,\text{Re}^{1/2}(\tau/\rho_L U_\infty^2)} = 0.5K^{1/2} \qquad (13\text{-}49b)$$

This last result is useful in assessing the benefit of film boiling in drag reduction of waterborne vehicles [73] (see Problem 13-24). Cess [74] used an integral method to show that a constant plate heat flux increases the Nusselt number and plate shear by 41% above the constant plate temperature results, agreeing roughly with the 38% value found for laminar single-phase results at a flat plate (at Pr ≈ 1), as is the subject of Problem 13-25. Subcooled forced convection film boiling from a constant temperature plate was studied by Cess and Sparrow [75], who found liquid subcooling to increase both heat transfer and plate shear.

## 13.7 OTHER GEOMETRIES AND CONDITIONS

Other geometries and conditions have been studied. Pilling and Lynch [76] investigated the quenching of hot objects of various shapes. For film boiling of cryogenic fluids from spheres, the data due to Merte and Clark [77] give the same heat flux versus $T_w - T_s$ as predicted by Bromley's Eq. (13-35) for normal Earth-gravitation and near-zero accelerations. Frederking et al. [78] give similar data for liquid nitrogen and liquid helium at normal Earth-gravitation acceleration. Frederking and Clark [79] utilized an integral analysis for the lower part of the sphere where no bubbles are released; Frederking and Daniels [80] studied the frequency of bubble release from the upper part of the sphere and bubble diameter. A successful analysis of film boiling from fully submerged spheres was executed by Hendricks and Baumeister [81]. Their approach was adopted by Marschall and Farrar [82] in an analysis of half-submerged spheres; the results were applied [83] to film boiling in a scaling liquid. The presence of scale on the sphere suggests that liquid–solid contact occurs, perhaps by liquid droplets ejected into the vapor film from the wavy interface.

The effects of externally imposed pressure fluctuations on film boiling from a wire were measured by DiCicco and Schoenhals [84]. They found that pressure excursions of 14.7–90 psig above atmospheric pressure at frequencies of 11–26 Hz resulted in a film boiling heat flux that was nearly double that expected from steady-state boiling at the peak pressure. An analytical study in the framework of the unsteady boundary-layer equations by Burmeister and Schoenhals [85] for laminar film boiling from a vertical surface suggests that 10% departure of the time-averaged local heat flux from steady-state results can be expected if the imposed pressure sinusoidal amplitude and frequency combine as shown in Fig. 13-11.

Oscillation of the heated surface during film boiling has also been studied. Measurements for film boiling from a sphere of diameter $D$ oscillating sinusoidally at amplitude $X$ and frequency $f$ in hertz in saturated Freon-11 and liquid nitrogen are correlated by Schmidt and Witte [86] by

$$\overline{Nu} = \frac{\bar{h}D}{k} = 0.14 \left[ \text{Ra} \frac{h_{fg}^*}{C_p(T_w - T_s)} (1 + \text{Fr}) \right]^{1/3}$$

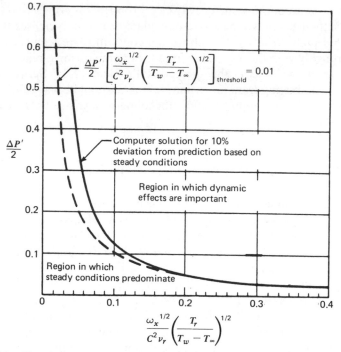

**Figure 13-11** Threshold conditions for local heat flux in film boiling of water from a vertical plate with sinusoidal unsteady system pressure ($T_w = 612°F$, $T_\infty = 212°F$, $p_0 = 1_{atm}$, $\Delta p' = (p_{max} - p_{min})/p_{min}$, $Tr = (T_w + T_\infty)/2$, $C^4 = g(\rho_L - \rho_u)/4\nu_r^2\rho_v$, $\nu_r$ is vapor kinematic viscosity, and $\omega$ is angular frequency of pressure oscillation). [From L. C. Burmeister and R. J. Schoenhals, *Progr. Heat Mass Transf.* **2**, pp. 371–394 (1969) [85].]

for the turbulent case of

$$Ra\frac{h_{fg}^*}{C_p(T_w - T_s)} \geqslant 6 \times 10^7$$

Here $Ra = g(\rho_L - \rho)Pr\, D^3/\rho\nu^2$, $h_{fg}^* = h_{fg} + C_p(T_w - T_s)/2$, and $Fr = X^2f^2/1.5gD$ is a Froude number and $Xf/(1.5)^{1/2}$ is the rms (root-mean-square) average velocity. The effect of diameter on film boiling from a sphere is the same as for cylinders [63]. For laminar film boiling from a sphere [79],

$$\overline{Nu} = \frac{\bar{h}D}{k} = 0.586\left[Ra\frac{h_{fg}}{C_p(T_w - T_s)}\right]^{1/4}$$

A detailed study of film boiling from large and small spheres by Gunnerson and Cronenberg [87] can be consulted for a deeper treatment.

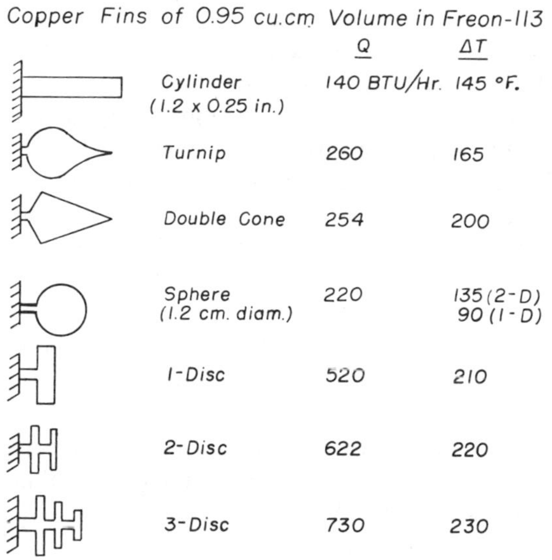

Copper Fins of 0.95 cu.cm Volume in Freon-113

| | | Q | ΔT |
|---|---|---|---|
| | Cylinder (1.2 x 0.25 in.) | 140 BTU/Hr. | 145 °F. |
| | Turnip | 260 | 165 |
| | Double Cone | 254 | 200 |
| | Sphere (1.2 cm. diam.) | 220 | 135 (2-D) / 90 (1-D) |
| | 1-Disc | 520 | 210 |
| | 2-Disc | 622 | 220 |
| | 3-Disc | 730 | 230 |

**Figure 13-12** Fins for boiling liquids. (By permission from Professor J. W. Westwater, University of Illinois at Champaign–Urbana.)

## 13.8 FINS FOR BOILING LIQUIDS

Fins (sometimes called *extended* or *sculptured surfaces*) are often used to increase heat-transfer rates to fluids that have low heat-transfer coefficients. Less common is the use of fins in boiling situations since the heat-transfer coefficient is usually considered to be so large that the addition of fins would improve heat-transfer rates little, if any—their internal resistance would overweigh the increased surface area they provide. However, fins can increase heat-transfer rates in a boiling situation if the wall temperature is sufficiently high to cause film boiling and all other resistances are small, as Problem 13-28 suggests.

The last condition implies that fouling must be small. Probable applications would be in cooling of nuclear fuel rods, cooling of electrical equipment such as radio-power tubes, and some evaporators (e.g., as a spiral tape insert for tubes).

The variable heat-transfer coefficient that occurs on a fin whose temperature varies from values at the root that cause film boiling (with moderate coefficients) to midvalues that cause nucleate boiling (with high coefficients) to tip values that cause natural convection without boiling (with low coefficients) lead to optimal fin shapes that are somewhat unusual. As shown in Fig. 13-12 for pool boiling to Freon-113 for the case of copper fins at 0.95 cm³ volume, the optimal shape is substantially different from a cylinder. The cylinder, although transferring about 100 times more heat than could the bare wall (which would be in film boiling if the fin were absent for the conditions of the cited results), is less effective than the turnip shape. The three-disk fin is the best shape of those shown, transferring about five times more heat than the cylinder. The optimal shape provides small surface area where the heat-transfer coefficient is low, the regions of film boiling at the root and natural convection at the tip, and large surface area where the coefficient is large—the midregion where nucleate boiling occurs. The stem in the root region serves to reduce the temperature to a value corresponding to the maximum coefficient.

Detailed information of analytical and experimental considerations is available [88–93], as are mathematical treatments [94–97]. Additional test data are reported by Klein and Westwater [98]. The use of coatings of low thermal conductivity (e.g., bismuth on copper) to improve fin performance still further is reported by Shih and Westwater [99]; they also give results for spheres, hemispheres, and disks [83a].

To avoid trapping vapor between adjacent fins, fin spacing should be about equal to the diameter of a bubble formed in nucleate boiling. This is about $\frac{1}{16}$ in. for organic liquids at atmospheric pressure.

The evolution of enhanced surface geometries for nucleate boiling is surveyed by Webb [111]. The effects of surface roughness formed by mechanical working or chemical etching of the base surface, metallic and nonmetallic coatings applied to the base surface, and promoters such as pierced cover sheets and wires or screens are reported. In some cases the nucleate boiling heat flux is nearly twice that from a smooth surface. Nucleate boiling heat-transfer coefficients of enhanced surfaces in a tube bundle approximately equal those measured in single-tube tests. Liquid superheats of as much as 8°C are sometimes required for initiation of nucleate boiling on enhanced surfaces.

## PROBLEMS

**13-1**  Use Eq. (13-2) to predict the heat flux and the coefficient of heat transfer for nucleate pool boiling of saturated water at 212°F on a 242°F copper heater surface. (*Answer:* $q/A = 6 \times 10^4$ Btu/hr ft$^2$, $h = 2 \times 10^3$ Btu/hr ft$^2$ °F.)

**13-2**  Compare the results of Problem 13-1 with those of Fig. 13-1.

**13-3**  Use Eq. (13-3) to predict the maximum heat flux for nucleate pool boiling of saturated water at 212°F. (*Answer:* $q_{max}/A = 4 \times 10^5$ Btu/hr ft$^2$.)

**13-4**  Equate Eqs. (13-2) and (13-3) to derive an expression for the maximum temperature difference $T_w - T_{sat}$ in nucleate pool boiling of a saturated liquid. For saturated water at 212°F and a copper heater surface, show that the heater temperature at which the peak heat flux occurs is $T_w = 270$°F. Use the results of Problems 13-3 and 13-4 to show that $h = 1.5 \times 10^3$ Btu/hr ft$^2$ °F.

**13-5** Predict the maximum heat flux for nucleate pool boiling of saturated water at 380 psia and compare the result with that of Problem 13-3 at 14.7 psia. (*Answer:* $q_{max}/A = 10^6$ Btu/hr ft$^2$.)

**13-6** Estimate the maximum heat flux for nucleate pool boiling of water at 14.7 psia that is subcooled 20°C and compare with the result for saturated water.

**13-7** A frying pan is electrically heated, dissipating 100 W across a surface of 520 cm$^2$. The frying pan is filled with saturated water. Estimate the temperature of the frying pan. (*Answer:* 230°F.)

**13-8** A stainless steel kettle whose 0.3-m-diamcter bottom is maintained at 250°F contains saturated water at 212°F. Estimate the time required to boil away a 20-cm water layer. (*Answer:* 23 min.)

**13-9** A long brass rod of 1 cm diameter is quenched in saturated water at 212°F. At a certain instant the rod surface temperature is 225°F.

(a) Estimate the temperature gradient of the rod surface.

(b) Estimate the amount of water subcooling required to increase the temperature gradient by 10%.

**13-10** Repeat Problem 13-3 for a gravitational acceleration 1/6 that at the Earth's surface.

**13-11** An immersion heater for a coffee mug is to be made of stainless steel. It must heat a water-filled mug to its saturation temperature in 5 min, and, to avoid possible carly material failure, the heater must operate in the nucleate boiling regime. Estimate the least heater area required and stipulate the wattage of the electrical heating element.

**13-12** A horizontal plate is exposed to a large stagnant pool of saturated water at 212°F. Determine the numerical values of:

(a) The minimum film boiling heat flux.

(b) The temperature difference between plate liquid $T_w - T_s$ associated with the minimum film boiling heat flux.

(c) The heat-transfer coefficient for this situation.

(d) Compare the results for parts a–c with those at the point of burnout boiling heat flux.

[*Answers:* (a) $q/A = 11,000$ Btu/hr ft$^2$, (b) 280°F, (c) $h = 40$ Btu/hr ft$^2$ °F.]

(e) Compare the results for parts a and c for natural convection to liquid water.

**13-13** For the conditions stated in Problem 13-12, determine the numerical values of:

(a) The vapor film thickness between bubbles.

(b) The "most dangerous" wavelength $\lambda_D$.

(c) The distance between detaching bubbles.

(d) The average vapor velocity in the film between bubbles at an average distance from a bubble center of $\lambda_D/3$.

(e) The Reynolds number $Va/\nu_v$ in the vapor film (is flow laminar or turbulent?).

(f) The diameter of a detaching bubble.

(g) The pressure difference $p_2 - p_1$ in the vapor film between the bubble position and a position between bubbles.

**13-14** A horizontal cylinder of 0.1 in. diameter is placed horizontally in a large stagnant pool of saturated water. The temperature difference is such that the minimum film boiling heat flux occurs.

(a) Determine whether the cylinder diameter is less than the critical wavelength.

(b) Determine the minimum film boiling heat flux.

**13-15** Estimate the Leidenfrost temperature for water at atmospheric pressure. Comment on the ease of obtaining the Leidenfrost temperature on a kitchen stove or a skillet.

**13-16** Show that the minimum film boiling heat flux varies as $q_{min}/A = \lambda_D b^*$ for a horizontal plate, according to Eq. (13-10), and as $q_{min}A = \lambda_D^2 b^*$ for a horizontal cylinder, according to Eq. (13-28). Here $\lambda_D$ is the most dangerous wavelength and $b^*$ is the maximum growth rate. Review the formulation for bubbles/area–oscillation given by Eqs. (13-8) and (13-27) to trace the origin of the difference in $q_{min}/A$ formulas.

**13-17** A horizontal stainless steel cylinder of $\frac{1}{4}$ inch outside diameter is maintained at 600°F while immersed in a pool of stagnant water saturated at 212°F.

(a) Determine the numerical value of the film boiling heat-transfer coefficient $\bar{h}$, neglecting radiation.

(b) Determine the numerical value of the coefficient, taking thermal radiation into account. What is the percentage influence of thermal radiation?

(c) Determine the heat flux and the vapor production rate.

(d) Estimate the thickness of the vapor film and whether the assumption that $\delta/R \ll 1$ is violated.

**13-18** (a) Compare the correlation of Eq. (13-35) for film boiling from a horizontal cylinder with the correlation for laminar natural convection from a horizontal cylinder. Note that if an equivalent specific heat were defined as $C_{p_{eff}} = h_{fg}/(T_w - T_s)$, Eq. (13-35) could be expressed as $\overline{Nu}_c = 0.62 \, Ra*^{1/4}$. Are the coefficients and exponents of Rayleigh number nearly equal?

(b) Show that the heat-transfer coefficient for minimum film boiling for a horizontal plate given by Eq. (13-23) can be expressed as

$$Nu = \frac{hL}{k_v} = 0.425 \left[ Ra \frac{h_{fg}}{C_{p_v}(T_w - T_{sat})} \right]^{1/4}$$

where the Rayleigh number is defined as $Ra = g[(\rho_L - \rho_v)/\rho_v]L^3 Pr_v/v_v^2$ and $L = [\sigma(g_c/g)/(\rho_L - \rho_v)]^{1/2}$, which is related to the most dangerous wavelength $\lambda_D$ by $L = \lambda_D/[2\pi(3^{1/2})]$. Show that an alternative expression is

$$\frac{h\lambda_c^{1/4}}{F} = 0.673 \pm 10\%$$

where $\lambda_c = 2\pi \left[ \frac{\sigma(g_c/g)}{\rho_L - \rho_v} \right]^{1/2}$ and

$$F^4 = \rho_v h_{fg}(\rho_L - \rho_v)k_v^3 \frac{g}{\mu_v(T_w - T_{sat})}.$$

Compare this result with the prediction of Eq. (13-37a) for large-diameter horizontal cylinders.

**13-19** Determine the film boiling heat-transfer coefficient for saturated water at atmospheric pressure on an electrically heated 0.08-in.-diameter horizontal tube at a wall temperature of 1200°F. Also determine the energy supply rate in watts required for a 6-in. length of this tube.

**13-20** Show that for film boiling of water, the typical value of $C_p(T_w - T_s)/h_{fg} \sim \frac{1}{10}$. The results of Problem 13-12 can be used to estimate typical $T_w - T_s$ values.

**13-21** In film boiling of a saturated liquid at a vertical surface show that:

(a) "Blowing" of vapor into the film toward the plate slightly increases the temperature gradient at the wall—plot $(T - T_s)/(T_w - T_s)$ against $y/\delta$ for $C_p(T_w - T_s)/h_{fg} = \frac{1}{10}$.

(b) The maximum velocity in the vapor film is

$$u_{max} = 2^{-1} \left[ 1 - \frac{C_p(T_w - T_s)}{3h_{fg}} \right]^{1/2} \left[ \frac{x(T_w - T_s)(\rho_L - \rho)gk}{\rho^2 v h_{fg}} \right]^{1/2}$$

which occurs at $y/\delta = \frac{1}{2}$—plot $u/u_{max}$ against $y/\delta$ for $C_p(T_w - T_s)/h_{fg} = \frac{1}{10}$. Calculate a typical numerical value of $u_{max}$ for saturated water. Calculate the maximum value of $x$ for which laminar vapor flow can be expected for saturated water.

**13-22** To explore the effect of liquid motion on film boiling of a saturated liquid at a vertical plate in a simple way:

**(a)** Neglect convective terms in the boundary-layer $x$-motion equation [Eq. (13-39b)] to obtain $d^2u/dy^2 = -(\rho_L - \rho)g/\rho\nu$ and, assuming $u(y = 0) = 0$ and no interfacial shear [$du(y = \delta)/dy = 0$], show that the velocity profile is (with $\eta = y/\delta$)

$$u = \frac{(\rho_L - \rho)g\delta^2}{\rho\nu}\left(\eta - \frac{\eta^2}{2}\right)$$

**(b)** Neglect convective terms in the boundary-layer energy equation [Eq. (13-39c)] to obtain $d^2T/dy^2 = 0$ and, assuming $T(y = 0) = T_w$ and $T(y = \delta) = T_s$, show that the temperature profile is

$$\frac{T - T_s}{T_w - T_s} = 1 - \eta$$

**(c)** Substitute the results of part b into the interfacial energy boundary condition of Eq. (13-40b) to find

$$\frac{k(T_w - T_s)}{\delta} = \rho h_{fg}\frac{d\left[(\rho_L - \rho)g\delta^3/3\rho\nu\right]}{dx}$$

or

$$\delta = 2^{1/2}\left[\frac{x(T_w - T_s)\nu k}{gh_{fg}(\rho_L - \rho)}\right]^{1/4}$$

**(d)** For the result of part c, evaluate the heat-transfer coefficient $h = q_w/(T_w - T_s) = k/\delta$ to finally obtain the average coefficient $\bar{h}$ as

$$\overline{Nu} = \frac{\bar{h}x}{k} = \frac{2^{3/2}}{3}\left[Ra\frac{h_{fg}}{C_p(T_w - T_{sat})}\right]^{1/4}$$

where $Ra = g(\rho_L - \rho)Pr\,x^3/\rho\nu^2$. Compare the coefficient found with the one that Ellion found to best correlate data and with that of Eq. (13-48).

**13-23** Compare the expression for $\overline{Nu}$ for film boiling of a saturated liquid at a vertical surface with that for laminar free convection from a vertical surface without phase change.

**13-24**  Compare the laminar drag of a vehicle submerged in water with and without a vapor film enveloping the surface that would otherwise contact liquid water.

**13-25**  Consider parallel forced-convection laminar film boiling from a constant-heat-flux plate to a saturated liquid.

(a) Assuming the velocity at the liquid–vapor interface to be the same as the free-stream velocity $U_\infty$, show that the vapor velocity profile is $u = U_\infty y/\delta$, where $\delta$ is the vapor film thickness.

(b) Substitute the result of part a into the integral form of the interface energy boundary condition

$$-k\frac{\partial T(y=\delta)}{\partial y} = \rho h_{fg}\frac{d\left(\delta\int_0^1 u\, d\eta\right)}{dx} \qquad (13\text{-}40b)$$

where $\eta = y/\delta$, and show that this leads to

$$2q_w/\rho U_\infty h_{fg} = \frac{d\delta}{dx}$$

if the vapor temperature profile is assumed to be linear, or

$$\delta = \frac{2q_w x}{\rho U_\infty h_{fg}}$$

(c) Since $q_w \approx k(T_w - T_s)/\delta$, show that

$$T_w - T_s = \frac{2q_w^2 x}{k\rho U_\infty h_{fg}}$$

(d) Since $h \approx k(T_w - T_s)/\delta$, show that the local Nusselt number is given by

$$\mathrm{Nu} = \frac{hx}{k} = \frac{\rho U_\infty h_{fg}}{2q_w}$$

$$= 0.707\left[\frac{x\rho U_\infty h_{fg}}{k(T_w - T_s)}\right]^{1/2}$$

(e) Compare the result of part d with Eq. (13-49a) for constant plate temperature and show that $h$ is 41% higher for constant heat flux than for constant temperature.

**13-26** Consider laminar film boiling from a vertical plate to a saturated liquid in which the liquid is assumed to be immovable (as a result of its large viscosity relative to the vapor). A similarity solution to the describing boundary-layer equations [Eqs. (13-39)] and boundary conditions of Eqs. (13-40) for the vapor film is to be attempted.

**(a)** Show that if a stream function $\psi$ such that $u = \partial\psi/\partial y$ and $v = -\partial\psi/\partial x$ is chosen to be

$$\psi = 4\nu Cx^{4/3}f$$

where $f = f(\eta)$ and $\eta = Cy/x^{1/4}$ with $C^4 = g(\rho_L - \rho)/4\nu^2\rho$, then $u = 4C^2\nu x^{1/2}f'$ and $v = C\nu x^{-1/4}(\eta f' - 3f)$. Here a prime denotes a derivative with respect to $\eta$.

**(b)** On the basis of the results of part a, show that the vapor $x$-motion and energy boundary-layer equations become

$$f''' + 2ff'' - 2(f')^2 + 1 = 0, \qquad 0 \leqslant \eta \leqslant \eta_\delta$$

$$\theta'' + 3\mathrm{Pr}\, f\theta' = 0, \qquad 0 \leqslant \eta \leqslant \eta_\delta$$

where the dimensionless temperature is $\theta = (T - T_s)/(T_w - T_s)$.

**(c)** Also show that the boundary conditions to be imposed on the results of part b are

$$f(\eta = 0) = 0 = f'(\eta = 0) \qquad \text{and} \qquad \theta(\eta = 0) = 1$$

$$f'(\eta = \eta_\delta) = 0 = \theta(\eta = \xi)$$

and

$$\left[ \mathrm{Pr}\frac{C_p(T_w - T_s)}{h_{fg}} \right]\theta'(\eta = \eta_\delta) = 3f(\eta = \eta_\delta)$$

where $\eta_\delta = C\delta/x^{1/4}$. The solution to the ordinary nonlinear differential equations of part b thus requires that the unknown values of $f''(0)$, $\theta'(0)$, and $\eta_\delta$ be determined simultaneously in such a manner as to meet the stated conditions at the interface (where $\eta = \eta_\delta$).

**(d)** Show that, if the unknown location of the interface is set at unity by employing the transformation $Z = \eta/\eta_\delta$, the results of parts b and c are obtained in the more convenient form of

$$\frac{d^3f}{dZ^3} + 3\eta_\delta f \frac{d^2f}{dZ^2} - 2\eta_\delta \left(\frac{df}{dZ}\right)^2 + \eta_\delta^3 = 0, \qquad 0 \leqslant Z \leqslant 1$$

$$\frac{d^2\theta}{dZ^2} + 3\mathrm{Pr}\,\eta_\delta f \frac{d\theta}{dZ} = 0, \qquad 0 \leqslant Z \leqslant 1$$

$$f(0) = 0 = \frac{df(0)}{dZ} \qquad \text{and} \qquad \theta(0) = 1$$

$$\frac{df(1)}{dZ} = 0 = \theta(1)$$

and

$$\left[ \mathrm{Pr} \frac{C_p(T_w - T_s)}{h_{fg}} \right] \frac{d\theta(1)}{dZ} = -3\eta_\delta f(1)$$

in which $d^2f(0)/dZ^2$, $d\theta(0)/dZ$, and $\eta_\delta$ are to be determined. Note that despite the complexity of this description, liquid motion is still neglected. Nevertheless, the influence of the two vapor parameters of Pr and $C_p(T_w - T_s)/h_{fg}$ are elucidated and the exact solution of a limiting case can be obtained numerically that is useful to ascertain the integral method accuracy. (Consult Koh [69] for details of the treatment of drag-induced liquid motion.)

(e) Show that the local heat-transfer coefficient is given by

$$\mathrm{Nu} = \frac{hx}{k} = -\left(\frac{Cx^{3/4}}{\eta_\delta}\right) \frac{d\theta(0)}{dZ}$$

13-27 The latent heat of vaporization $h_{fg}$ used in Problem 13-22 needs correction to account for the fact that the flowing vapor film is superheated. The upward enthalpy flow in the vapor film at a vertical plate can be investigated to ascertain the required correction since

$$\int_0^\delta u\left[h_{fg} + C_p(T - T_s)\right] dy = \delta \bar{u}\left[h_{fg} + aC_p(T_w - T_s)\right]$$

in which $\bar{u}$ is the average vapor velocity and $a$ is the fraction of the wall superheat that constitutes the correction. Show that the corrected latent heat of vaporization is $h_{fg} + C_p(T_w - T_s)/2$.

13-28 Consider a very hot surface of constant temperature that is immersed in a liquid. Discuss the manner in which the heat flux from the hot surface varies as the thickness of a poorly conducting material bonded onto the surface is increased. Note that if film boiling originally ensues, the intermediate insulating material can have a surface temperature sufficiently low to allow nucleate boiling to occur with a resultant increase in heat flux.

13-29   Show that relationships for film boiling to a stagnant saturated liquid pool, [e.g., Eq. (13-35)] are of the same form as for natural convection without phase change $\overline{\mathrm{Nu}} = C(\mathrm{Ra}^*)^n$ if the vapor specific heat $C_{p_v}$ is replaced by $h_{\mathrm{fg}}/(T_w - T_s)$, which gives $\mathrm{Ra}[h_{\mathrm{fg}}/C_{p_v}(T_w - T_s)] = \mathrm{Ra}^*$.

# REFERENCES

1   E. A. Farber and R. L. Scorah, Heat transfer to water boiling under pressure, *Transact. ASME* **70**, 369–384 (1948).

2   G. Leppert and C. C. Pitts, Boiling, *Adv. Heat Transf.* **1**, 185–266 (1964).

3   J. W. Westwater, Boiling of liquids, *Adv. Chem. Eng.* **1**, 1–31 (1958).

4   R. Cole, Boiling nucleation, *Adv. Heat Transf.* **10**, 85–166 (1974).

5   P. Griffith and J. D. Wallis, The role of surface conditions in nucleate boiling, *Chem. Engr. Prog. Symp.*, *Ser. 30* **56**, 49–63 (1960).

6   F. D. Moore and R. B. Mesler, The measurement of rapid surface temperature fluctuations during nucleate boiling of water, *AIChE J.* **7**, 620–624 (1961); An experimental study of surface cooling by bubbles during nucleate boiling of water, *AIChE J.* **10**, 656–660 (1964); A. Kovacs and R. B. Mesler, Making and testing small surface thermocouples for fast response, *Rev. Sci. Instr.* **35**, 485–488 (1964); R. R. Olander and R. G. Watts, An analytical expression of microlayer thickness in nucleate boiling, *Transact. ASME, J. Heat Transf.* **91**, 178–180 (1969).

7   M. S. Plesset and S. S. Sadhal, An analytical estimate of the microlayer thickness in nucleate boiling, *Transact. ASME, J. Heat Transf.* **101**, 180–182 (1979).

8   W. M. Rohsenow, A method of correlating heat transfer data for surface boiling of liquids, *Transact. ASME* **74**, 969–976 (1952).

9   P. Berenson, Transition boiling heat transfer from a horizontal surface, *Transact. ASME, J. Heat Transf.* **83**, 351–358 (1961).

10  R. I. Vachon, G. H. Nix, and G. E. Tanger, Evaluation of constants for the Rohsenow pool-boiling correlation, *Transact. ASME, J. Heat Transf.* **90**, 239–247 (1968).

11  E.L. Piret and H. S. Isbin, Natural circulation evaporation two-phase heat transfer, *Chem. Eng. Progr.* **50**, 305–311 (1954).

12  J. N. Addoms, Heat transfer at high rates to water boiling outside cylinders, D.Sc. thesis, Massachusetts Institute of Technology, Department of Chemical Engineering, Cambridge, MA, 1948.

13  M. T. Chichelli and C. F. Bonilla, Heat transfer to liquids boiling under pressure, *Transact. AIChE* **41**, 755–787 (1945).

14  N. Zuber, On the stability of boiling heat transfer, *Transact. ASME* **80**, 711–720 (1958).

15  S. S. Kutateladze, *Zh. Tekh. Fiz.* **20**, 1389 (1950).

16  V. M. Borishanskii, Heat transfer to a liquid freely flowing over a surface heated to a temperature above the boiling point, *Zh. Tekh. Fiz.* **26**, 452 (1956) (see also AEC-tr-3405 translation from Problems of heat transfer during a change of state, by S. S. Kutateladze, 1959).

17  J. H. Lienhard and M. M. Hasan, On predicting boiling burnout with the mechanical energy stability criterion, *Trans. ASME, J. Heat Transf.* **101**, 276–279 (1979).

18  J. H. Lienhard, V. K. Dhir, and D. M. Riherd, Peak pool boiling heat-flux measurements on finite horizontal flat plates, *Transact. ASME, J. Heat Transf.* **95**, 477–482 (1973).

19  T. Aoki and J. R. Welty, Frequency distribution of boiling-generated sound, *Transact. ASME, J. Heat Transf.* **92**, 542–544 (1970).

20  F. W. Schmidt, D. F. Torok, and G. E. Robinson, Experimental study of the effects of an ultrasonic field in a nucleate boiling system, *Transact. ASME, J. Heat Transf.* **89**, 289–294 (1967).

21  A. E. Bergles, The influence of heated-surface vibration on pool boiling, *Transact. ASME, J. Heat Transf.* **91**, 152–154 (1969).

22  G. M. Fuls and G. E. Geiger, Effect of bubble stabilization on pool boiling heat transfer, *Transact. ASME, J. Heat Transf.* **92**, 635–640 (1970).

23  M. Markels and R. L. Durfee, The effect of applied voltage on boiling heat transfer, *AIChE J.* **10**, 106–110 (1964).

24  M. Markels and R. L. Durfee, Studies of boiling heat transfer with electrical fields, *AIChE J.* **11**, 716–723 (1965).

25  T. B. Jones, Electrodynamically coupled film boiling, ASME Paper 76-WA/HT-48 (1976); Electrodynamically enhanced heat transfer in liquids—a review, *Adv. Heat Transf.* **14**, 107–148 (1978); T. B. Jones and R. C. Schaeffer, Electrodynamically coupled minimum film boiling in dielectric liquids, *AIAA J.* **14**, 1759–1765 (1976).

26  J. M. Crowley, Onset of boiling in electrohydrodynamic spraying, *AIAA J.* **15**, 734–736 (1977).

27  W. M. Rohsenow, Heat transfer with evaporation, in *Heat Transfer*, University of Michigan Press, Ann Arbor, MI, 1953.

28  S. S. Kutateladze, Boiling heat transfer, *Internatl. J. Heat Mass Transf.* **4**, 31–45 (1961).

29  R. M. Fand, K. K. Keswani, M. M. Jotwani, and R. C. C. Ho, Simultaneous boiling and forced convection heat transfer from a horizontal cylinder to water, *Transact. ASME, J. Heat Transf.* **98**, 395–400 (1976).

30  W. H. Lowdermilk, C. D. Lanzo, and B. L. Siegel, Investigation of boiling burnout and flow stability for water flowing in tubes, NACA Technical Note 4382, 1958.

31  S. Nukiyama, The maximum and minimum values of the heat, $Q$, transmitted from metal to boiling water under atmospheric pressure, *J. Soc. Mech. Engrs. (Jap.)* **37**, 367 (1934).

32  P. J. Berenson, Experiments on pool-boiling heat transfer, *Internatl. J. Heat Mass Transf.* **5**, 985–999 (1962).

33  J. B. McDonough, W. Milich, and E. C. King, An experimental study of partial film boiling region with water at elevated pressures in a round vertical tube, *Chem. Eng. Progr. Symp.*, Ser. 32 **57**, 197–208 (1961).

34  F. S. Pramuk and J. W. Westwater, Effect of agitation on the critical temperature difference for a boiling liquid, *Chem. Eng. Progr. Symp.*, Ser. 18 **52**, 79–83 (1956).

35  A. J. Lowery and J. W. Westwater, Heat transfer to boiling methanol—effect of added agents, *Ind. Eng. Chem.* **49**, 1445–1448 (1957).

36  T. Dunskus and J. W. Westwater, The effect of trace additives on the heat transfer to boiling isopropanol, *Chem. Engr. Progr. Symp.*, Ser. 32 **57**, 173–181 (1961).

37  D. P. Jordan, Film and transition boiling, *Adv. Heat Transf.* 55–128 (1968).

38  Y. P. Chang, Wave theory of heat transfer in film boiling, *Transact. ASME, J. Heat Transf.* **81**, 1–12 (1959).

39  P. J. Berenson, Film boiling heat transfer from a horizontal surface, *Transact. ASME, J. Heat Transf.* **83**, 351–358 (1961).

40  C. T. Sciance and C. P. Colver, Minimum film-boiling point for several light hydrocarbons, *Transact. ASME, J. Heat Transf.* **92**, 659–660 (1970).

41  V. K. Dhir, J. N. Castle, and I. Catton, Role of Taylor instability on sublimation of a horizontal slab of dry ice, *Transact. ASME, J. Heat Transf.* **99**, 411–418 (1977).

42  K. Taghavi-Tafreshi, V. K. Dhir, and I. Catton, Thermal and hydrodynamic phenomena associated with melting of a horizontal substrate plate beneath a heavier immiscible liquid, *Transact. ASME, J. Heat Transf.* **101** 318–325 (1979).

43  S. Yao and R. E. Henry, An investigation of the minimum film boiling temperature on horizontal surfaces, *Transact. ASME, J. Heat Transf.* **100** 260–267 (1978).

44  (a) R. E. Henry and H. K. Fauske, Nucleation processes in large scale vapor explosions, *Transact. ASME, J. Heat Transf.* **101**, 280–287 (1979); (b) G. S. Springer, Homogeneous nucleation, *Adv. Heat Transf.* **14**, 281–346 (1978).

45  N. H. Afgan, Boiling liquid superheat, *Adv. Heat Transf.* **11**, 1–49 (1975).

46  J. G. Leidenfrost, De aquae comminis nonnullis qualitatibus tractatus, Duisburg, 1756; J. G. Leidenfrost, On the fixation of water in diverse fire, (translator C. Wares), *Internatl. J. Heat Mass Transf.* **9**, 1153–1166 (1966).

47  K. J. Baumeister, T. D. Hamill, F. L. Schwartz, and G. J. Schoessow, Film boiling heat transfer to water drops on a flat plate, *Chem. Eng. Progr. Symp., Ser. 64* **62**, 52–61 (1966).

48  B. M. Patell and K. J. Bell, The Leidenfrost phenomenon for extended liquid masses, *Chem. Eng. Progr. Symp., Ser. 64* **62**, 62–71 (1966).

49  P. Spiegler, J. Hopenfeld, M. Silberberg, C. F. Bumpus, and A. Norman, Onset of stable film boiling and the foam limit, *Internatl. J. Heat Mass Transf.* **6**, 987–994. (1963).

50  J. H. Lienhard, Correlation for the limiting liquid superheat, *Chem. Eng. Sci.* **31**, 847–849 (1976).

51  F. S. Gunnerson and A. W. Cronenberg, On the thermodynamic superheat limit for liquid metals and its relation to the Leidenfrost temperature, *Transact. ASME, J. Heat Transf.* **100**, 734–737 (1978).

52  J. H. Lienhard and P. T. Y. Wong, The dominant unstable wavelength and minimum heat flux during film boiling on a horizontal cylinder, *Transact. ASME, J. Heat Transf.* **86**, 220–226 (1964).

53  J. H. Lienhard and K. Sun, Effects of gravity and size upon film boiling from horizontal cylinders, *Transact. ASME, J. Heat Transf.* **92**, 292–298 (1970).

54  L. A. Bromley, Heat transfer in stable film boiling, *Chem. Eng. Progr.* **46**, 221–227 (1950).

55  W. H. Rohsenow, Heat transfer and temperature distribution in laminar film condensation, *Transact. ASME* **78**, 1654–1648 (1956).

56  P. Pitschmann and U. Grigull, Film boiling on horizontal cylinders, *Wärme-und Stoffübertragung* **3**, 75–84 (1970).

57  G. Hesse, E. M. Sparrow, and R. J. Goldstein, Influence of pressure on film boiling heat transfer, *Transact. ASME, J. Heat Transf.* **98**, 166–172 (1976).

58  B. T. Lubin, Analytical derivation for total heat transfer coefficient in stable film boiling from vertical plate, *Transact. ASME, J. Heat Transf.* **91**, 452–253 (1969).

59  B. P. Breen and J. W. Westwater, Effect of diameter of horizontal tubes on film boiling heat transfer, *Chem. Eng. Progr.* **58** (1), 67–72 (1962).

60  M. L. Pomerantz, Film boiling on a horizontal tube immersed in increased gravity fields, *Transact. ASME, J. Heat Transf.* **86**, 213–219 (1964).

61  L. A. Bromley, N. R. LeRoy, and J. A. Robbers, Heat transfer in forced convection film boiling, *Ind. Eng. Chem.* **45**, 2639–2646 (1953).

62  E. I. Motte and L. A. Bromley, Film boiling of flowing subcooled liquids, *Ind. Eng. Chem.* **49** 1921–1928 (1957).

63 E. K. Kalinin, I. I. Berlin, and V. V. Kostyuk, Film-boiling heat transfer, *Adv. Heat Transf.* **11**, 51–197 (1975).

64 L. C. Witte, Film boiling from a sphere, *Ind. Eng. Chem., Fundamentals* **7**, 517–518 (1968).

65 M. E. Ellion, A study of the mechanism of boiling heat transfer, Memo 20-88, Jet Propulsion Laboratory, Pasadena, CA, 1954.

66 Y. Y. Hsu and J. W. Westwater, Approximate theory for film boiling on vertical surfaces, *Chem. Eng. Progr. Symp., Ser. 30* **56**, 15–24 (1960).

67 D. A. Labuntsov, Calculation of heat transfer for film boiling of liquid on vertical heating surfaces, *Teploenergetika* **10**, 60 (1963).

68 T. H. K. Frederking, Laminar two-phase boundary layers in natural convection film boiling, *Zeitschrift für angewandte Mathematik und Physik* **14**, 207–218 (1963).

69 J. C. Y. Koh, Analysis of film boiling on vertical surfaces, *Transact. ASME, J. Heat Transf.* **84**, 55–62 (1962).

70 E. M. Sparrow and R. D. Cess, The effect of subcooled liquid on laminar film boiling, *Transact. ASME, J. Heat Transf.* **84**, 149–156 (1962).

71 T. H. K. Frederking and J. Hopenfeld, Laminar two-phase boundary layers in natural convection film boiling of subcooled liquid, *Zeitschrift für angewandte Mathematik und Physik* **15**, 388–399 (1964).

72 R. D. Cess and E. M. Sparrow, Film boiling in a forced-convection boundary layer flow, *Transact. ASME, J. Heat Transf.* **83**, 370–376 (1961).

73 W. S. Bradfield, R. O. Barkdoll, and J. T. Byrne, Some effects of film boiling on hydrodynamic drag, *Internatl. J. Heat Mass Transf.* **5**, 615–622 (1962).

74 R. D. Cess, Forced-convection film boiling on a flat plate with uniform surface heat flux, *Transact. ASME, J. Heat Transf.* **84**, 395 (1962).

75 R. D. Cess and E. M. Sparrow, Subcooled forced-convection film boiling on a flat plate, *Transact. ASME, J. Heat Transf.* **87**, 377–379 (1961).

76 N. B. Pilling and T. D. Lynch, Cooling properties of technical quenching liquids, *Trans. Am. Inst. Mining Met. Eng.* **62**, 665–688 (1920).

77 H. Merte and J. A. Clark, Boiling heat-transfer data for liquid nitrogen at standard and near-zero gravity, *Adv. Cryog. Eng.* **7**, 546–550 (1961).

78 T. H. K. Frederking, R. C. Chapman, and S. Wang, Heat transport and fluid motion during cooldown of single bodies to low temperatures, in *International Advances In Cryogenic Engineering*, K. D. Timmerhaus, Ed., Plenum Press, New York, 1965, Paper T-3.

79 T. H. K. Frederking and J. A. Clark, Natural convection film boiling on a sphere, *Adv. Cryog. Eng.* **8**, 501–506 (1963).

80 T. H. K. Frederking and D. J. Daniels, The relation between bubble diameter and frequency of removal from a sphere during film boiling, *Transact. ASME, J. Heat Transf.* **88**, 87–93 (1966).

81 R. C. Hendricks and K. J. Baumeister, Film boiling from submerged spheres, NASA TN D-5124, 1969.

82 E. Marschall and L. C. Farrar, Film boiling from a partly submerged sphere, *Internatl. J. Heat Mass Transf.* **18**, 875–878 (1975).

83 L. C. Farrar and E. Marschall, Film boiling in a scaling liquid, *Transact. ASME, J. Heat Transf.* **98**, 173–177 (1976).

83a. C. C. Shih and J. W. Westwater, Spheres, hemispheres and discs as high-performance fins for boiling heat transfer, *Internatl. J. Heat Mass Transf.* **17**, 125–133 (1974).

84 D. A. DiCicco and R. J. Schoenhals, Heat transfer in film boiling with pulsating pressures, *Transact. ASME, J. Heat Transf.* **86**, 457–461 (1964).

85    L. C. Burmeister and R. J. Schoenhals, Effects of pressure fluctuations on laminar film boiling, *Progr. Heat Mass Transf.* **2**, 371–394 (1969).

86    W. E. Schmidt and L. C. Witte, Oscillation effects upon film boiling from a sphere, *Transact. ASME, J. Heat Transf.* **94**, 491–493 (1972).

87    F. S. Gunnerson and A. W. Cronenberg, On the minimum film boiling conditions for spherical geometries, *Transact. ASME, J. Heat Transf.* **102**, 335–341 (1980).

88    K. W. Haley and J. W. Westwater, Heat transfer from a fin to a boiling liquid, *Chem. Eng. Sci.* **20**, 711 (1965).

89    K. W. Haley and J. W. Westwater, Boiling heat transfer from single fins, *Proceedings of the Third International Heat Transfer Conference*, Vol. 3, 1966, pp. 245–253.

90    D. R. Cash, G. J. Klein, and J. W. Westwater, Approximate optimum fin design for boiling heat transfer, *Trans. ASME, J. Heat Transf.* **93**, 19–24 (1971).

91    D. L. Bondurant and J. W. Westwater, Performance of transverse fins for boiling heat transfer, *Chem. Eng. Progr. Symp., Ser. 113* **67**, 30–37 (1971).

92    S. A. Kovalev, V. M. Zhukov, and G. M. Kazakov, *Teplofizika Vysokikh Temperatur* **8**, 217–219 (1970).

93    G. R. Rubin, L. I. Royzen, and I. N. Dul'kin, Heat transfer in liquid boiling on insulation-coated fins, *Heat Transf. Sov. Res.*, **3**, 130–134 (1971); from *Inzehenerno-Fizicheskii Zhurnal* **20**, 26–30 (1971).

94    M. Cumo, S. Lopez, and G. C. Pinchera, Numerical calculation of extended surface efficiency, *Chem. Eng. Progr. Symp., Ser. 59* **61**, 225–233 (1965).

95    F. S. Lai and Y. Y. Hsu, Temperature distribution in a fin partially cooled by nucleate boiling, *AIChE J.* **13**, 817–821 (1967).

96    Y. Y. Hsu, Analysis of boiling on a fin, NASA TN D-4797, 1968.

97    M. Siman-Tov, Analysis and design of extended surfaces in boiling liquids, *Chem. Eng. Progr. Symp., Ser. 102* **66**, 174–184 (1970).

98    G. J. Klein and J. W. Westwater, Heat transfer from multiple spines to boiling liquids, *AIChE J.* **17**, 1050–1056 (1971).

99    C. C. Shih and J. W. Westwater, Use of coatings of low thermal conductivity to improve fins used in boiling liquids, *Internatl. J. Heat Mass Transf.* **15**, 1965–1968 (1972).

100   V. K. Dhir, Sublimation of a horizontal slab of dry ice: An analog of pool boiling on a flat plate, *Transact. ASME, J. Heat Transf.* **102**, 380–382 (1980).

101   J. H. Lienhard and V. K. Dhir, On the prediction of the minimum pool boiling heat flux, *Transact. ASME, J. Heat Transf.* **102**, 457–460 (1980).

102   W. Nakayama, T. Daikoku, H. Kuwahara, and T. Nakajima, Dynamic model of enhanced boiling heat transfer on porous surfaces, *Transact. ASME, J. Heat Transf.* **102**, 445–456 (1980).

103   S. Yilmaz and J. W. Westwater, Effect of velocity on heat transfer to boiling Freon-113, *Transact. ASME, J. Heat Transf.* **102**, 26–31 (1980).

104   K. Stephan and M. Abdelsalam, Heat-transfer correlations for natural convection boiling, *Internatl. J. Heat Mass Transf.* **23**, 73–87 (1980).

105   J. Berghmans, Electrostatic fields and the maximum heat flux, *Internatl. J. Heat Mass Transf.* **19**, 791–797 (1976).

106   G. Guglielmini, E. Nannei, and C. Pisoni, Survey of heat transfer correlations in forced convection boiling, *Wärme-und Stoffübertragung* **13**, 177–185 (1980).

107   L. Y. Wagner and P. S. Lykoudis, Mercury pool boiling under the influence of a magnetic field, *Internatl. J. Heat Mass Transf.* **24**, 635–643 (1981).

108   V. V. Klimenko, Film boiling on a horizontal plate—new correlation, *Internatl. J. Heat Mass Transf.* **24**, 69–79 (1981).

109  K. Stephan and H. Auracher, Correlations for nucleate boiling heat transfer in forced convection, *Internatl. J. Heat Mass Transf.* **24**, 99–107 (1981).

110  F. Tachibana and S. Fukui, Heat transfer in film boiling to subcooled liquids, in *International Developments in Heat Transfer*, American Society of Mechanical Engineers, New York, 1961, p. 219.

111  R. L. Webb, The evolution of enhanced surface geometries for nucleate boiling, *Heat Transf. Eng.* **2**, 46–69 (1981).

112  N. Nishikawa, H. Kusuda, K. Yamasaki, and K. Tanaka, Nucleate boiling at low liquid level, *Bull. JSME* **10**, 328–338 (1967).

113  A. H. Fakeeha, Heat transfer on a vertical surface, M.S. Thesis, Chemical and Petroleum Engineering Department, University of Kansas, Lawrence.

114  R. B. Mesler and G. Mailen, Nucleate boiling in thin liquid films, *AICHE J.* **23**, 954–957 (1977).

# 14

## CONDENSATION

In many respects the heat flow that accompanies condensation on a surface closely resembles natural convection in a single-phase fluid, just as is the case for boiling. As was remarked in Chapter 13 for boiling, the change in density that accompanies a change from the vapor to the liquid phase gives larger body forces per unit volume for the liquid than for the vapor. Hence fluid motion can occur by natural means.

Because natural convection generally is characterized by low velocities, a number of forces other than gravity can play an important role. Centrifugal forces and shear stresses due to externally forced vapor motion can induce motion of a condensed liquid film, for example, as can electromagnetic fields. Surface tension also can play a role in the generation of liquid motion when the liquid–vapor interface has finite curvature or when surface tension varies over the interface due to gradients in temperature or chemical composition.

When condensation occurs on a cooled solid surface, as is often the case in engineering applications, the condensed liquid is observed to either form a continuous film on the surface or else to coalesce into droplets. The former behavior, called *film condensation*, is the more common; the latter, called *dropwise condensation*, is accompanied by higher condensation rates and is difficult to achieve continuously. Examples of equipment in which condensation occurs on a cooled solid surface are condensers for Rankine power generation cycles and vapor compression refrigeration cycles, dehumidifiers for air conditioning, and heat pipes. It should be recognized that condensation sometimes changes a vapor into a solid as in vacuum deposition of metals and in cryopumping.

It should be noted that the high heat-transfer coefficients that typically accompany boiling and condensation seldom lead to heat power cycles that nearly have Carnot efficiency $\eta = 1 - T_{sink}/T_{source}$. As Problem 14-1 suggests, operation of a Carnot cycle at maximum *power* with finite heat-transfer coefficients in the isothermal expansion and compression processes gives an efficiency of $\eta' = 1 - (T_{sink}/T_{source})^{1/2}$, according to Curzon and Ahlborn [1].

Condensation can occur away from solid boundaries by homogeneous bulk nucleation in a pure fluid or by heterogeneous nucleation on small foreign particles. Expansion in a stream turbine nozzle accompanied by spontaneous formation of liquid nuclei that grow rapidly (leading to a condensation shock), the events in a Wilson cloud chamber, and expansion of a vapor in space are examples of the former. Water droplets that form around dust, salt particles, or AgI particles in fogs, clouds, and rocket exhaust plumes [2] are examples of the latter.

An excellent survey by Merte [3] can be consulted for details and references beyond those presented in the following treatment.

## 14.1 SURFACE TENSION

Liquids differ from solids by the freedom of motion enjoyed by the molecules of liquids. Liquids differ from gases by the much larger cohesive forces between molecules that tend to restrict separation of molecules. One of the common consequences of this difference between liquids and gases is that a body of liquid tends to contract to the configuration of smallest surface area as sketched in Fig. 14.1. At the surface of the liquid mass a molecule is subjected to tangential forces $F_s$ that cancel and a much larger inward attractive force $F_i$ to the many nearby liquid molecules than the small outward attractive force $F_0$ to the few distant gas molecules. This net inward force causes inward movement of surface molecules until the maximum possible number are in the interior, leading to a surface of minimum area. In water, these forces have their origin in the fact that hydrogen atoms in one molecule attract oxygen atoms of neighboring molecules.

As might be expected, increase of the surface area above this minimum amount requires exertion of external force. As this external force moves through the required displacement, it does the work needed to bring additional

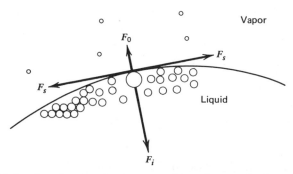

**Figure 14-1** Origin of surface tension in strong inward attractive force between many nearby interior liquid molecules and a surface liquid molecule greatly exceeding weak outward attractive force between few distant exterior vapor molecules.

molecules to the surface from the interior. The liquid surface can, as a result, be said to possess a surface energy that could do work in the process of moving back to the minimum-surface-area configuration. The surface free energy per unit area is commonly called the *surface tension* which has dimensions of force/length, which is equivalent to energy/area—force–length/length$^2$.

It can be appreciated from the foregoing discussion that a flat liquid surface can be due only to the predominant effect of some force, such as gravitational attraction, other than surface tension forces. The Bond number *Bo* characterizes the relative magnitude of gravitational and body forces [4] according to

$$Bo = \frac{F_{gravity}}{F_{surface\ tension}} = \frac{(\rho_L - \rho_v)L^3 g/g_c}{\sigma L}$$

$$= \frac{(\rho_L - \rho_v)L^2 g}{\sigma g_c}$$

where $L$ is a characteristic length of the system considered. If a liquid surface is curved, a pressure difference exists across it. The magnitude of this pressure difference can be ascertained by consideration of the situation sketched in Fig. 14-2 where a small interfacial area is shown. At equilibrium the forces due to pressure difference and to surface tension must be balanced. The normal force exerted by the pressure difference acting on the projected area of the interface is $(P_1 - P_2)R_1 \sin(2\,\Delta\theta)R_2 \sin(2\,\Delta\phi)$. The normal force due to surface tension pulling tangentially is $\sigma R_1\,\Delta\theta \sin\Delta\phi + \sigma R_2\,\Delta\phi \sin\Delta\theta$. Equation of these two normal forces and realization that small interfacial areas require that $\Delta\phi,\,\Delta\theta \to 0$ leads to

$$P_1 - P_2 = \sigma\left(\frac{1}{R_1} + \frac{1}{R_2}\right) \tag{14-1}$$

A more precise derivation of Eq. (14-1) [5] would have allowed the interface to move a small distance and then equated the work done by the pressure forces to the change in surface free energy. In such a derivation the question of the variation of surface tension $\sigma$ with interface radius occurs. Tolman, as well as Kirkwood and Buff, proposed [6], on the basis of statistical mechanics considerations, that

$$\sigma = \sigma_\infty\left(1 + \frac{\delta}{r}\right)$$

where $\sigma_\infty$ is the surface tension at a flat surface, $r$ is the interface curvature, and $\delta$ is a length between 0.25 and 0.6 of the liquid molecular radius.

For most substances, the surface tension decreases as temperature decreases. The derivation of Eq. (14-1) suggests that a temperature variation along an

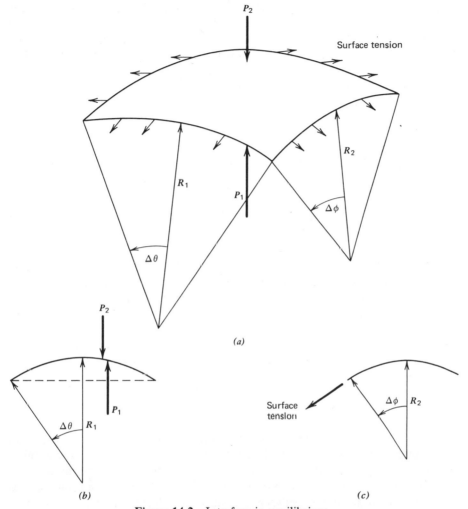

Figure 14-2   Interface in equilibrium.

interface would result in surface tension variation and a consequent force imbalance that could induce liquid motion. The Marangoni number Ma is the ratio of the imbalance in surface tension forces to liquid tangential viscous forces according to

$$\text{Ma} = \frac{\Delta F_{\text{surface tension}}}{F_{\text{tang viscous}}} = \frac{(L\, d\sigma/dT)(L\, \nabla T)}{\mu_L V L^2/L}$$

where $L$ is a characteristic system dimension (which could be a boundary-layer thickness) and $\nabla T$ is a temperature gradient on the interface. Two excellent reviews on surface-tension effects are available [7, 8]. Inasmuch as the velocity $V$ is not specified, it is reasonable to relate it to the stabilizing effect of the liquid thermal conductivity, which tends to equalize surface temperatures. The distance $\delta$ a step change in temperature propogates by diffusion through a medium was found in Chapter 5 to be given by

$$\delta \sim (\alpha_L t)^{1/2} \qquad (5\text{-}53)$$

If the derivative is taken with respect to time and allowed to be the characteristic velocity, then $\dot{\delta} \sim \alpha_L/\delta$, which suggests that the characteristic velocity is represented by $V \sim \alpha_L/L$. Hence the Marangoni number is [4]

$$\text{Ma} = \frac{L^2 \nabla T (d\sigma/dT)}{\mu_L \alpha_L}$$

Fluid flows driven by surface tension effects at low gravity are particularly observable at low-gravity conditions such as would be encountered in the processing of materials in outer space. There bubble migration through a liquid could be influenced by temperature gradients [9] that could cause liquid motion whenever a free surface exists [10, 127, 128].

Surface tension $\sigma$ is closely related to latent heat of vaporization $h_{fg}$. To demonstrate this, surface tension is interpreted as a surface energy per unit area. Then

$$\sigma = u(z - z')n'$$

where $u$ is the mutual potential energy between two molecules of mean spacing $r$, $z$ is the number of nearest neighbors within the bulk liquid, $z'$ is the number of nearest neighbors at the interface $z' \sim z/2$, and $n'$ is the number of molecules per unit area at the interface ($n' \sim 1/r^2$). Also, evaporation requires that the potential energy between the surface molecules and those beneath must be overcome. This energy is the latent heat of vaporization and can be expressed as

$$h_{fg} = \frac{\rho n z u}{2}$$

where $\rho$ is the mass density of the liquid, $n$ is the number of molecules per unit volume ($n \sim 1/r^3$), and the factor of 2 accounts for the attractive forces only acting from below. Division of this expression for $h_{fg}$ into that for $\sigma$ gives

$$\frac{\sigma}{\rho h_{fg}} \approx r \sim 10^{-10} m \qquad (14\text{-}2)$$

since the intermolecular spacing $r$ of most liquids is of the order of $10^{-10}$ m.

Energy considerations can be used to estimate the tensile stress required to rupture a liquid. For this purpose, it is convenient to consider surface tension $\sigma$ to be surface energy per unit area. Formation of a surface area $A$ consequently requires an energy of $2\sigma A$. If the rupture is considered to occur when liquid molecules are separated by a distance that is of the order of the molecular spacing, the work done by the rupturing force $-P_{rupt}Ar$ must equal that stored on the newly created two surfaces. Hence

$$-P_{rupt} = \frac{2\sigma}{r} \qquad\qquad (14\text{-}3)$$

The fact that cavitation occurs at considerably lower values than the upper bound of Eq. (14-3) is due to the effects of dissolved gases, oil films, and differing intermolecular forces between liquid and solid when cavitation occurs on a solid surface.

## 14.2  BULK CONDENSATION

Surface tension plays an important role in bulk condensation. In particular, it is involved in the determination of the critical liquid droplet size, which is just in equilibrium with the surrounding vapor. To see this, consider a liquid droplet of radius $r$ at pressure $P_L$ and temperature $T$ in equilibrium with a vapor at pressure $P_v$ and temperature $T$, which is also in equilibrium with a pool of flat-surface liquid as sketched in Fig. 14-3. From Eq. (14-1) the pressure in the droplet exceeds that in the vapor, according to

$$P_L = P_v + \frac{2\sigma}{r}$$

in which $2\sigma/r$ can be considered to be a pressure differential $\Delta P$.

The total free energy $G$ of the system is [11]

$$G = n_v g_v(P_v, T) = n_L g_L(P_v, T) + 4\pi r^2 \sigma \qquad\qquad (14\text{-}4)$$

in which $n_v$ and $n_L$ are the number of moles of vapor and liquid whereas $g_v$

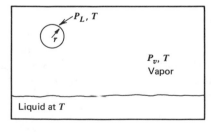

Liquid at $T$

**Figure 14-3**  Vapor in equilibrium with flat liquid pool and a liquid droplet.

and $g_L$ are Gibbs free energy per mole of vapor and liquid, respectively. The last term is the surface free energy, representing the work required to form a surface area $4\pi r^2$ with constant surface tension. At thermodynamic equilibrium the system free energy is at a minimum. Hence

$$dG = n_v \, dg_v + g_v \, dn_v + n_L \, dg_L + g_L \, dn_L + 4\pi\sigma \, dr^2 = 0$$

Since the number of moles is constant, $dn_v = -dn_L$. Also, $n_L = 4\pi r^3/3v_L$. At constant temperature and pressure neither $g_v$ nor $g_L$ varies. As a result of these three observations, the expression for $dG$ shows that

$$g_v(P_v, T) = g_L(P_v, T) + \frac{2v_L\sigma}{r} \qquad (14\text{-}5)$$

at equilibrium. It can be appreciated from Eq. (14-5) that the value of $P_v$ required to maintain the drop in equilibrium at the specified temperature $T$ depends on the drop radius $r$. To ascertain this dependence, Eq. (14-5) is differentiated with $T$ constant to obtain

$$dg_v - dg_L = 2v_L\sigma \, d\left(\frac{1}{r}\right)$$

The thermodynamic relation that $dg = v \, dP$ is then introduced to obtain

$$(v_v - v_L)dP_v = 2v_L\sigma \, d\left(\frac{1}{r}\right)$$

Recognizing that $v_v \gg v_L$ and taking the vapor to be a perfect gas $v_v = RT/P_v$, where $R$ is the universal gas constant, one can restate this expressions as

$$\frac{1}{P_v} dP_v = \frac{2v_L\sigma}{RT} d\left(\frac{1}{r}\right)$$

Integration between the limits of $r = \infty$, where $P_v = P_{sat}$, the saturation pressure normally associated with $T$, and the equilibrium radius of $r = r*$ gives

$$\frac{P_v}{P_{sat}} = \exp\left(\frac{2v_L\sigma}{r*RT}\right) \qquad (14\text{-}6)$$

Equation (14-6) shows that a drop is in equilibrium at a specified temperature only if the pressure of the surrounding vapor exceeds what is normally considered to be its saturation pressure $P_{sat}$. Restated, the drop is in equilibrium only if the vapor is supersaturated. If the surrounding vapor pressure is lower than that required by Eq. (14-6), the drop will shrink by evaporation. If the surrounding vapor pressure is higher than that required by Eq. (14-6), the drop will grow by condensation.

Equation (14-6) gives the supersaturation pressure ratio for a drop of size $r*$ to be in equilibrium. If the supersaturation pressure ratio is specified, the equilibrium drop size is obtained from a rearrangement of Eq. (14-6) as

$$r* = \frac{2v_L \sigma / RT}{\ln(P_v / P_{sat})} \tag{14-7}$$

It is apparent that when a liquid–vapor interface is curved, equilibrium requires that the vapor be supersaturated. The amount of supersaturation is illustrated for water at 68°F in Table 14-1. There it is seen that if condensation is to occur with negligible supersaturation of the vapor, foreign nuclei of some type must be made available since it is unlikely that the number of molecules required (about $2.7 \times 10^{11}$ for a $10^{-4}$-in. drop) would come together sponta- neously. On the other hand, when the vapor is extremely supersaturated only a few (approximately eight) liquid molecules comprise a drop of the correspond- ing equilibrium size and could come together spontaneously. The minimum drop size cannot be smaller than the average distance between liquid mole- cules; such a limitation allows an estimate of the maximum supersaturation ratio to be $P_v / P_{sat} \approx 10^2$ as is the subject of Problem 14-4. An equilibrium of a drop with its vapor is unstable since the drop will grow if its diameter suddenly experiences either an infinitesimal size increase as a result of condensation or a sudden infinitesimal size decrease due to evaporation.

The formation of vapor cavities in a bulk liquid (a boiling phenomenon) is similar to the formation of liquid drops in a bulk vapor (a condensation phenomenon). This similarity is explored in Problem 14-7.

An application in which bulk condensation is important is the flow of a vapor through a turbine nozzle since droplets can seriously erode turbine blades. The fluids involved are often water but are occasionally liquid metals. Hill et al. [15] reviewed the classical liquid drop nucleation theory and applied it to the calculation of the rate per unit volume $J$ at which molecules are added to a drop of equilibrium size. They found, the derivation of which is the subject of Problem 14-8,

$$J = \left(\frac{P_v}{KT}\right)^2 \frac{M}{N_A \rho_L} \left(\frac{2\sigma}{\pi m}\right)^{1/2} \exp\left(\frac{-4\pi\sigma r*^2}{3KT}\right) \tag{14-8}$$

**Table 14-1   Water Drop Diameter and Supersaturation Pressure Ratio at 68°F**

| Diameter, in. | Number of Molecules | Supersaturation Ratio |
|---|---|---|
| $10^{-2}$ | $2.7 \times 10^{17}$ | 1.000009 |
| $10^{-4}$ | $2.6 \times 10^{11}$ | 1.00086 |
| $10^{-6}$ | $2.7 \times 10^{5}$ | 1.0905 |
| $10^{-7}$ | $2.7 \times 10^{2}$ | 2.38 |

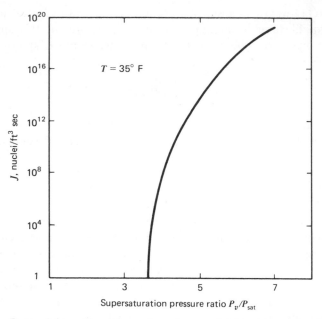

**Figure 14-4** Rate of formation of condensation nuclei in water vapor. [From P. G. Hill, H. Witting, and E. P. Demetri, *Transact. ASME, J. Heat Transf.* **85**, 303–317 (1963) [15].]

where $K = 1.38054 \times 10^{-23}$ J/molecule K is the Boltzmann constant, $M$ is molecular weight, $N_A = 6.03 \times 10^{26}$ molecules/kgmol is Avogadro's number, and $r^*$ is given by Eq. (14-7). Figure 14-4 shows the rate per unit volume at which molecules are added to a drop for water at 35°F versus the supersaturation pressure ratio $P_v/P_{sat}$. There it is seen that the curve is so steep at low values of $J$ that a critical supersaturation pressure ratio can be defined above which condensation occurs spontaneously, a ratio that equals 4.2 for water at

**Table 14-2   Supersaturation Pressure Ratio [16]** $P_v/P_{sat}$

| Vapor | Temperature, K | Measured | Calculated |
|---|---|---|---|
| Water | 275.2 | $4.2 \pm 0.1$ | 4.2 |
| Water | 261.0 | 5.0 | 5.0 |
| Methanol | 270.0 | 3.0 | 1.8 |
| Ethanol | 273.0 | 2.3 | 2.3 |
| I-Propanol | 270.0 | 3.0 | 3.2 |
| Isopropyl alcohol | 265.0 | 2.8 | 2.9 |
| $n$-Butyl alcohol | 270.0 | 4.6 | 4.5 |
| Nitromethane | 252.0 | 6.0 | 6.2 |
| Ethyl acetate | 242.0 | 8.6–12.3 | 10.4 |

$35°F$. This corresponds to a critical drop radius of $6.5 \times 10^{-10}$ m that is substantially less than a mean free path in the vapor $10^{-8}$ m, having about 40 molecules of liquid. The data due to Volmer and Flood [16], who expanded vapors to visible condensation in a cylinder, shown in Table 14-2 verify the prediction of Eq. (14-8) surprisingly well. More recent data on condensation in nozzles [17] shows that although water and many other fluids are accurately treated by Eq. (14-8), the Lothe–Pound theory [18] (which accounts for surface tension variation with drop size and includes quantum-statistical effects) must be used for $NH_3$. The condensation of water from still moist air is not accurately predicted by these methods [15]. A comprehensive review of nucleation theory [19] can be consulted for additional details.

## 14.3  DROPWISE CONDENSATION ON SOLID SURFACES

A vapor condensing on a solid surface can either form widely separated liquid drops, a mode termed *dropwise condensation*, or a uniform liquid film, termed *film condensation*. In this respect there are many similarities between condensation and boiling at solid surfaces. In both nucleate boiling and dropwise condensation, the nucleation occurs at preferred nucleation sites, the number of which increases with increasing temperature difference—as does the heat flux, as Fig. 14-5 (due to Welch and Westwater [20]) illustrates. In both film

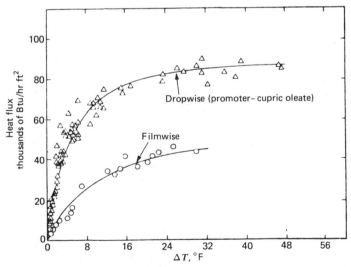

**Figure 14-5**  Comparison of drop and film condensation of steam at atmospheric pressure on a vertical copper surface. (By permission of the American Society of Mechanical Engineers and J. F. Welch and J. W. Westwater, *Proceedings of the International Heat Transfer Conference, University of Colorado*, Part II, 1961, pp. 302–309 [20].)

boiling and film condensation, a layer of usually laminarly flowing fluid acts as an insulator between the solid surface and the pool of fluid of unchanged phase. One difference is that the existence of a peak condensation heat flux, similar to the burnout heat flux in boiling, as temperature difference increases has yet to be observed. A second difference is that vapor bubbles in nucleate boiling are readily detached from their surface of formation since buoyant forces are larger than surface forces holding the bubble to the surface whereas liquid drops in dropwise condensation find gravity forces for possible removal smaller than the surface forces holding them to the surface. It has been speculated [3] that with a suitable liquid drop removal mechanism such as centrifugal force, a peak condensation heat flux could be achieved, beyond which individual drops will merge to eventually form a continuous liquid film just as nucleate boiling gradually merges into film boiling as the imposed temperature difference exceeds burnout conditions. A third difference is that film condensation rather than dropwise condensation might occur at low imposed temperature differences, in contrast to the case for boiling in which nucleate boiling always occurs at low-temperature differences.

Generally, condensation on solid surfaces of industrial interest is film condensation. For this reason it is recommended that design of heat exchange equipment conservatively assume film condensation. As Fig. 14-5 illustrates, film condensation heat fluxes are lower than those for dropwise condensation. Efforts to consistently have dropwise condensation in industrial equipment for lengthy periods have not been successful, although the potential improvement in heat fluxes continues to motivate effort in this area [21, 22, 122].

Prediction of the possible occurrence of dropwise condensation requires a determination of the liquid's ability to wet the solid surface. Dropwise condensation is likely to occur when the liquid cannot wet the surface, whereas film condensation is likely to occur when the liquid can wet the surface. The judgment of wetting ability proceeds as follows. For the drop resting on a solid and surrounded by vapor illustrated in Fig. 14-6, there are three forces, due to three surface tensions, which must be in equilibrium: (1) $\sigma_{Lv}$, which acts in the liquid–vapor interface that has a contact angle $\theta$ with the solid; (2) $\sigma_{Ls}$, which is determined by the properties of the solid surface and the liquid and acts

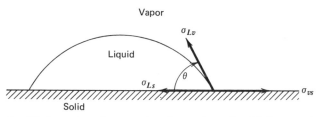

**Figure 14-6** Equilibrium of surface tension forces for a liquid drop resting on a solid surface in a vapor.

along the solid surface on the liquid side; and (3) $\sigma_{vs}$, is determined by the properties of the solid surface and the vapor and acts along the solid surface on the vapor side. Horizontal equilibrium of these three forces requires that

$$\sigma_{vs} = \sigma_{Ls} + \sigma_{Lv}\cos\theta$$

which suggests that the surface properties of the solid are important to the determination of $\theta$. A liquid is considered not to wet a surface if $\theta > 90°$ and to wet a surface if $\theta < 90°$. Measurement of the contact angle for a series of homologous liquids on different surfaces results in an approximately linear relation between $\cos(\theta)$ and $\sigma_{Lv}$ as Fig. 14-7, due to Shafrin and Zisman [23], illustrates. Extrapolation of these data to $\theta = 0$, where the liquid can be considered to completely wet the surface, yields a critical surface tension $\sigma_{cr}$ that is a characteristic of the surface alone. Dropwise condensation occurs when $\sigma_{Lv} > \sigma_{cr}$. Table 14-3 [23] gives critical surface tensions for several surfaces. The fact that film condensation occurs with water on most metals suggests that most metals have $\sigma_{cr} > \sigma_{Lv}$, although so-called promoters have been found to be effective on metals in promoting dropwise condensation. Probably the promoters form a layer that reduces $\sigma_{cr}$ below $\sigma_{Lv}$.

As Table 14-3 suggests, it is possible to coat a metal surface on which film condensation normally occurs with a substance on which dropwise condensa-

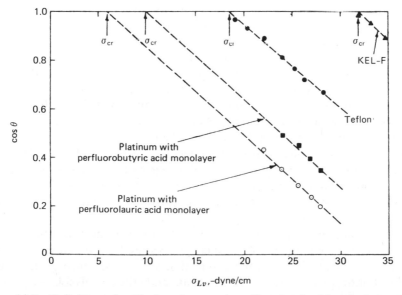

**Figure 14-7** Definition of critical surface tension. (Reprinted with permission from E. G. Shafrin and W. A. Zisman, *J. Phys. Chem.* **64**, 519–524 (1960) [23]; copyright 1960, American Chemical Society.)

**Table 14-3 Critical Surface Tensions of Some Solids [23]**

| Solid | $\sigma_{cr}$, dyne/cm |
|---|---|
| TFE (Teflon) | 18 |
| Polyethylene | 31 |
| Kel-F | 31 |
| Polystyrene | 33 |
| Polyvinyl chloride | 39 |
| Nylon | 46 |
| Platinum with perfluorobutyric acid monolayer | 10 |
| Platinum with perfluorolauric acid monolayer | 6 |

tion naturally occurs. Tests by Brown and Thomas [24] showed that the change from film to dropwise condensation allowed by a 0.0001-in. film of polytetra-fluorethylene outside a horizontal 0.75-in. brass tube increased the overall heat-transfer coefficient for steam condensation by a factor of nearly 3. This procedure is similar to that of coating a very hot surface on which film boiling occurs with a poorly conducting material in order to achieve nucleate boiling to achieve a net increase in heat flux as is the subject of Problem 13-28. Noble metal coatings have also been tried [25] but are usually rejected on economic grounds. The fluoroplastic Emralon 300™ shows promise [125] as a coating, potentially having a 20-year lifetime.

Instead of the mechanical coating described previously, it is also possible to alter the properties of the solid surface by a promoter. For this to be successful, the entire system must be saturated with the promoter (not just the condensing surface) since the vapor becomes contaminated and the promoter concentration must be maintained. Promoters tried include oleic, stearic, or linoleic acids, benzyl mercaptan, sulfur or selenium compounds, and long-chain hydrocarbons [26]. None of these promoters have proved to be satisfactory for operation intervals of several years, failing by surface fouling or oxidation, although operation for up to 1 year has been achieved [20]. It is important that noncondensable gases be removed for dropwise condensation heat fluxes to achieve their possible large values [27].

The measurements of dropwise condensation heat-transfer coefficients have been correlated by a number of workers [3]. Naturally, accurate use of any of the correlations requires that the application and measurement conditions be similar. For dropwise condensation of steam and ethylene glycol, Peterson and Westwater [28] recommend

$$Nu = 1.46 \times 10^{-6} \, Re^{-1.63} \, Pr^{0.5} \, \pi_k^{1.16} \tag{14-9}$$

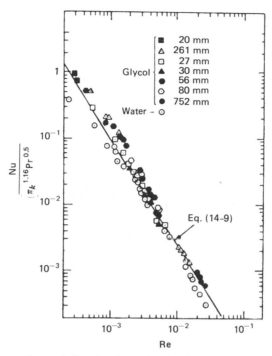

**Figure 14-8**  Improved correlation for dropwise condensation on vertical surfaces. [By permission from A. C. Peterson and J. W. Westwater, *Chem. Eng. Progr. Symp. Ser.* **62**, 135–142 (1966) [28].]

for $1.65 \leqslant \mathrm{Pr} \leqslant 23.6$ and $7.8 \times 10^{-4} \leqslant \pi_k \leqslant 2.65 \times 10^{-2}$ where

$$\mathrm{Nu} = \frac{2\sigma T_v h}{h_{fg}\rho_L k_L(T_w - T_v)} \qquad \mathrm{Pr} = \mu_L \frac{C_{pL}}{k_L}$$

$$\mathrm{Re} = \frac{k_L(T_w - T_v)}{\mu_L h_{fg}} \qquad \pi_k = \frac{2\sigma(d\sigma/dT)T_v}{\mu_L^2 h_{fg}}$$

where $h_{fg}$ is the latent heat of vaporization, the subscript $L$ refers to liquid properties, and all properties are evaluated at the saturation temperature. Figure 14-8 shows that good correlation is achieved for the conditions of these measurements. For additional discussion of dropwise condensation, the survey of Merte [3] can be consulted.

## 14.4  LAMINAR FILM CONDENSATION ON VERTICAL PLATES

As mentioned in Section 14.3, it is difficult to cause dropwise condensation to occur continuously over long periods of time. Most commonly, then, a solid condensing surface is covered by a liquid film that separates the solid surface from the vapor.

Initial discussion of film condensation begins with a vertical flat plate as illustrated in Fig. 14-9. Close similarity to the film boiling discussion in Section 13.6, and illustration in Fig. 13-8 is apparent—the vapor and the liquid have merely exchanged positions. It is not surprising that the successful lines of analysis are the same for both cases since film condensation is also a boundary-layer phenomenon in its essential aspects. Indeed, the analysis of laminar film condensation accomplished by Nusselt [30] in 1916 was later successfully applied to film boiling.

The constant-property boundary-layer equations for the film in differential form are as given by Eq. (13-39) for film boiling, but with the liquid and vapor interchanged. They are repeated here for convenience as

$$\frac{\partial u_L}{\partial x} + \frac{\partial v_L}{\partial y} = 0 \qquad (14\text{-}10a)$$

$$u_L \frac{\partial u_L}{\partial x} + v_L \frac{\partial u_L}{\partial y} = \frac{(\rho_L - \rho)g}{\rho_L} + \nu_L \frac{\partial^2 u_L}{\partial y^2} \qquad (14\text{-}10b)$$

$$u_L \frac{\partial T_L}{\partial x} + v_L \frac{\partial T_L}{\partial y} = \alpha_L \frac{\partial^2 T_L}{\partial y^2} \qquad (14\text{-}10c)$$

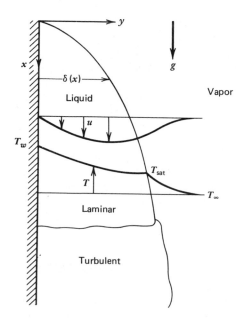

**Figure 14-9** Boundary layer of condensate on a vertical flat plate.

Unsubscripted variables refer to vapor quantities. The boundary conditions imposed are that at the liquid–vapor interface ($y = \delta$)

$$\rho_L\left(v_L - u_L\frac{d\delta}{dx}\right) = \rho\left(v - u\frac{d\delta}{dx}\right) \qquad (14\text{-}11a)$$

$$-k_L\frac{\partial T_L}{\partial y} = -k\frac{\partial T}{\partial y} - h_{fg}\rho_L\left(v_L - u_L\frac{d\delta}{dx}\right) \qquad (14\text{-}11b)$$

$$\mu_L\frac{\partial u_L}{\partial y} = \mu\frac{\partial u}{\partial y} \qquad (14\text{-}11c)$$

$$u_L = u \qquad (14\text{-}11d)$$

$$T_L = T_s \qquad (14\text{-}11e)$$

whereas at the solid surface ($y = 0$)

$$u_L = 0 \qquad (14\text{-}11f)$$

$$v_L = 0 \qquad (14\text{-}11g)$$

$$T_L = T_w \qquad (14\text{-}11h)$$

If vapor temperature and velocity distributions are to be determined, the boundary-layer equations for the vapor must also be written and solved. Here it is assumed for simplicity that the vapor is saturated and exerts no shear on the liquid so that there is no need to determine the temperature and velocity distributions in the vapor.

Solutions of high accuracy can be obtained from Eqs. (14-10) and (14-11) by employing similarity transformations to reduce them to ordinary differential equations that are then solved by numerical methods. Here, however, insight is desired first and adequate accuracy is attainable from an integral method, a simple version of which is the subject of Problem 14-26. In general, as for the film boiling case, the integral form of the boundary-layer equations is given in the same form as Eqs. (13-42) and (13-43) as, with $\eta = y/\delta$,

$$\frac{d\left(\delta\int_0^1 u_L^2\,d\eta\right)}{dx} - u_{L,\delta}\frac{d\left(\delta\int_0^1 u_L\,d\eta\right)}{dx}$$

$$= \frac{(\rho_L - \rho)g\delta}{\rho} + \frac{\nu_L}{\delta}\left[\frac{du_L(1)}{d\eta} - \frac{du_L(0)}{d\eta}\right]$$

which is the $x$-motion equation

$$\frac{d\left[\delta\int_0^1 u_L(T - T_s)d\eta\right]}{dx} = \frac{\alpha_L}{\delta}\left[\frac{dT(1)}{d\eta} - \frac{dT(0)}{d\eta}\right] \qquad (14\text{-}13)$$

which is the energy equation. The integral form of the continuity equation is, in analogy to Eq. (13-41) for film boiling,

$$v_{L,\delta} = -\frac{d\left(\delta\int_0^1 u_L\,d\eta\right)}{dx} + u_{L,\delta}\frac{d\delta}{dx} \qquad (14\text{-}14)$$

An approximate liquid velocity distribution is obtained by requiring the assumed polynomial

$$u_L = a_0 + a_1\eta + a_2\eta^2 + a_3\eta^3 + \cdots$$

to meet the conditions of no-slip at the wall, Eq. (14-11f), no shear at the liquid–vapor interface [Eq. (14-11c)] and to satisfy the $x$-motion equation [Eq. (14-10b)] at the wall. The result is

$$u_L = \frac{(\rho_L - \rho)g\delta^2}{\mu_L}\left[\left(\eta - \frac{\eta^2}{2}\right) - a(3\eta - \eta^3)\right] \qquad (14\text{-}15a)$$

The remaining parameter $a$ in Eq. (14-15a) is evaluated by substitution of the velocity profile into the integral $x$-motion equation [Eq. (14-12)] to obtain

$$\left(1 - \frac{29a}{4} + \frac{93a^2}{7}\right)\frac{d\delta^4}{dx} = 72\frac{\rho_L v_L^2}{(\rho_L - \rho)g}a \qquad (14\text{-}15b)$$

Now $d\delta^4/dx$ is expected to be a constant. For convenience, let

$$C = (\rho_L - \rho)g\frac{h_{fg}}{T_s - T_w}(4v_L k_L)^{-1}\frac{d\delta^4}{dx}$$

and introduce this relationship into Eq. (14-15b) to find

$$\frac{93a^2}{7} - \left[\frac{29}{4} + 18\,\mathrm{Pr}_L\frac{h_{fg}/C_p(T_s - T_w)}{C}\right]a + 1 = 0$$

Since $C$ is yet unknown, it is evident that the energy equation must be solved simultaneously to provide a relation for $C$—the energy and $x$-motion equations are coupled. A simplification is adopted here, however, since it is recognized that $C_p(T_s - T_w)/h_{fg} \ll 1$ and $C \approx 1$ in most cases. Then

$$a \approx \frac{C_{p_L}(T_s - T_s)/h_{fg}}{18\,\mathrm{Pr}_L}$$

An approximate polynomial temperature distribution is assumed as

$$T_L - T_s = b_0 + b_1\eta + b_2\eta^2 + b_3\eta^3 + b_4\eta^4 + \cdots$$

which is required to meet the conditions of no thermal jump at the wall [Eq. (14-11h)], equilibrium at the liquid–vapor interface [Eq. (14-11e)], conservation of energy at the liquid–vapor interface [Eq. (14-11b)], and to satisfy the energy equation [Eq. (14-10c)] at the wall. The result is

$$\frac{T_L - T_s}{T_w - T_s} = 1 - \left(1 + \frac{1-C}{2}\right)\eta + \left(\frac{1-C}{2}\right)\eta^3 \qquad (14\text{-}15c)$$

Introduction of the temperature and velocity profiles of Eq. (14-15) into the integral energy equation [Eq. (14-13)] gives

$$\frac{(\rho_L - \rho)g}{8\rho_L\nu_L}\left[1 - \frac{18a}{5} + (1-C)\left(-\frac{11}{30} + \frac{48a}{35}\right)\right]\delta\frac{d\delta^3}{dx} = \frac{3\alpha_L(1-C)}{2}$$

This relation is rearranged, using the previously found approximation for $a$, into

$$\frac{1}{4}\frac{C_p(T_s - T_w)}{h_{fg}}\left[1 - \frac{18a}{5} + (1-C)\left(-\frac{11}{30} + \frac{48a}{35}\right)\right]C = 1 - C$$

Again, the relationship $C \approx 1$ for small $C_{p_L}(T_s - T_w)/h_{fg}$, the most commonly encountered case, leads to the close approximation that $1 - C \approx C_p(T_s - T_w)/4h_{fg}$. The essentially linear temperature profile and parabolic velocity profile are shown by Eqs. (14-15c) and (14-15a), respectively. In both cases the effect of condensation at the outer edge of the laminar boundary layer is seen to be small, causing the temperature profile to be "blown" toward the plate and causing the maximum velocity to be diminished as stagnant fluid is added.

The condensate film thickness follows from the definition of $C$ as

$$\delta = 2^{1/2}\left[1 - \frac{1}{4}\frac{C_{p_L}(T_s - T_w)}{h_{fg}}\right]^{1/4}\left[\frac{x(T_s - T_w)\nu_L k_L}{gh_{fg}(\rho_L - \rho)}\right]^{1/4} \qquad (14\text{-}16)$$

The heat flux at the wall is now available as

$$q_w = -k\frac{\partial T(0)}{\partial y}$$

$$= -\frac{k}{\delta}\frac{\partial T(0)}{\partial \eta}$$

$$= \frac{k}{\delta}(T_w - T_s)\left[1 + \frac{(1 - C)}{2}\right]$$

From this and Eq. (14-16) the local heat-transfer coefficient for laminar condensation is found to be

$$h = \frac{q_w}{T_w - T_s} = \frac{1}{2^{1/2}}k\left[\frac{g(\rho_L - \rho)\mathrm{Pr}}{x\rho_L \nu_L^2}\frac{h_{\mathrm{fg}}}{C_{p_L}(T_s - T_w)}\right]^{1/4} K \quad (14\text{-}17a)$$

Here $K \approx [1 + \frac{3}{4}C_{p_L}(T_s - T_w)/h_{\mathrm{fg}}]^{1/4}$ is in close agreement with Koh [31] and Rohsenow [32], who found that it is more accurate to use a coefficient of 0.68 instead of $\frac{3}{4}$. The average heat-transfer coefficient $\bar{h}$ is related to the local value by $\bar{h} = 4h/3$. Hence the average Nusselt number $\overline{\mathrm{Nu}} = \bar{h}x/k$ is given by

$$\overline{\mathrm{Nu}} = 0.943\left[\mathrm{Ra}\frac{h_{\mathrm{fg}}}{C_{p_L}(T_s - T_w)}\right]^{1/4} \quad (14\text{-}47)$$

whose coefficient best fits measurements if set to 1.13 as discussed later and where the Rayleigh number is $\mathrm{Ra} = g(\rho_L - \rho)\mathrm{Pr}\,x^3/\rho_L \nu_L^2$ and $K \approx 1$ has been assumed. Equation (14-17) is the result found by Nusselt. Liquid property variations with temperature are taken into account by evaluating liquid properties at $T_{\mathrm{ref}} = T_w + (T_s - T_w)/4$, according to Drew [36] and confirmed by Poots and Miles [44] and Lott and Parker [45].

The temperature difference on which the transfer coefficient depends in Eq. (14-16) might not be specified. In such cases a form that employs the condensate flow rate, which often is specified when temperature difference is not, is convenient. This form, the derivation of which is the subject of Problem 14-13, is

$$\frac{\bar{h}}{k}\left(\frac{\nu_L^2}{g}\right)^{1/3} = 1.47\left(\frac{\rho_L - \rho}{\rho_L}\right)^{1/3}\mathrm{Re}^{-1/3} \quad (14\text{-}18)$$

where $\mathrm{Re} = 4\dot{m}/\mu_L$ is a Reynolds number and $\dot{m}$ is the condensate mass flow rate per unit width of the vertical plate. Best agreement with data is found with a coefficient of 1.88, as discussed later.

Equations (14-17) and (14-18) can be applied to condensation from the upper surface of plates inclined at an angle $\theta$ from the horizontal if the gravitational component parallel to the plate $g \sin \theta$ is used. Gerstmann and Griffith [33] treat laminar film condensation on the lower surface of inclined plates, whereas a related study by Leppert and Nimmo [34] gives results for the extreme case of condensation atop a horizontal strip, cases in which surface tension can be expected to have an effect.

The accuracy of Eqs. (14-18) and (14-17) can be judged from the measurements for vertical plates shown in Fig. 14-10, due to Spencer and Ibele [35], for steam of 0–100°F superheat. These measurements show that actual laminar heat-transfer coefficients can be as much as 50% above or below the predicted value. It is believed that rough surfaces often retain the condensate, thus thickening the boundary layer and reducing the coefficient of heat transfer below the predicted value. Heat-transfer coefficients above the predicted value may be due to surface waves, an instability that not only increases surface area and intermittently thins the liquid film but also promotes mixing. These surface waves are formed when $\mathrm{Re} = 4\dot{m}/\mu_L \geqslant 33$ and are influenced by surface tension, as Kapitsa [46] found. Much more plentiful measurements exist for vertical tubes, for which Eqs. (14-17) and (14-18) should apply for either exterior or interior condensation as long as the condensate film is thin relative to the tube diameter. For these vertical tubes, McAdams [36] recom-

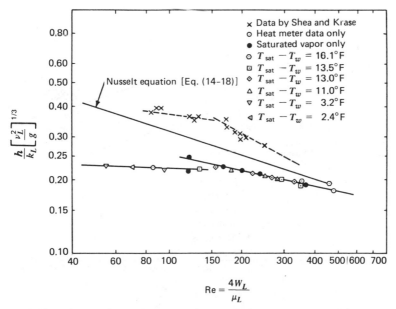

**Figure 14-10** Measured condensation heat-transfer coefficients of saturated and superheated steam on a vertical surface. (By permission from D. L. Spencer and W. E. Ibele, *Proceedings of the Third International Heat Transfer Conference*, Vol. 2, 1966, pp. 337–347 [35].)

mends that the coefficients of Eq. (14-17) and (14-18) be changed to 1.13 and
1.88, respectively. As illustrated in Fig. 14-11, due to Selin [37], such an
increase in coefficient substantially improves the agreement between prediction
and measurement, suggesting that the effects of surface waves are usually
predominant.

The integral analysis that led to Eq. (14-17) reveals the essential features of
the condensation phenomenon. Its assumption of negligible vapor drag can
cause serious overestimation of condensation heat-transfer coefficients in some
cases, however. Koh et al. [39] accounted for interfacial vapor drag during
laminar film condensation of a stagnant saturated vapor on a vertical plate in
their execution of a similarity solution of the boundary-layer equations.
Problem 14-14 deals with this topic. Vapor drag has little effect for large liquid
Prandtl numbers, as Fig. 14-12 shows, reducing local transfer coefficients by
less than 10% percent for $Pr = 1$ and having a diminishing effect for increasing
$Pr$. For small liquid Prandtl numbers, such as are characteristic of liquid
metals (see Fig. 14-13), vapor drag is quite influential. The effect of vapor drag
on temperature distribution in the condensate film is slight, but its effect on
the velocity distribution can be substantial, particularly for small values of $Pr$.
Figure 14-14, for $Pr = 1$, shows that vapor drag reduces the interfacial
velocity, causing the maximum velocity to occur within the film. A similar

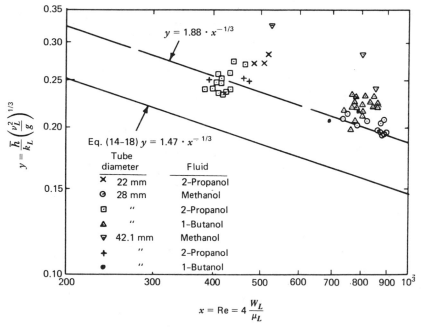

**Figure 14-11** Variation of condensing heat transfer coefficient with Reynolds number
for vertical tubes. (By permission from the American Society of Mechanical Engineers
and G. Selin, *Proceedings of the International Heat Transfer Conference, University of
Colorado*, Part II, 1961, pp. 279–289 [37].)

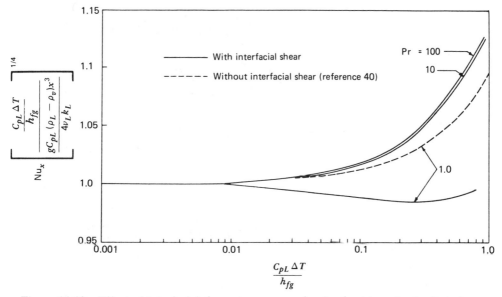

**Figure 14-12**  Effect of interfacial shear stress on condensing heat transfer for Pr > 1. [By permission from J. C. Y. Koh, E. M. Sparrow, and J. P. Hartnett, *Internatl. J. Heat Mass Transf.* **2**, 69–82 (1961) [39].]

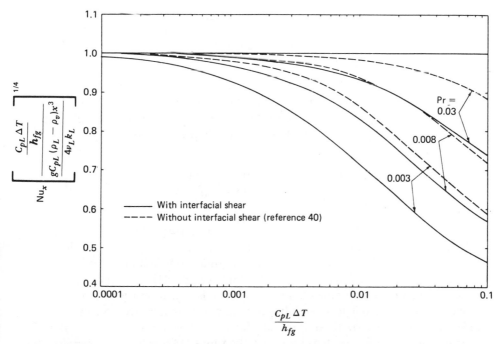

**Figure 14-13**  Effect of interfacial shear stress on condensing heat transfer, liquid metal range of Pr. [By permission from J. C. Y. Koh, E. M. Sparrow, and J. P. Hartnett, *Internatl. J. Heat Mass Transf.* **2**, 69–89 (1961) [39].]

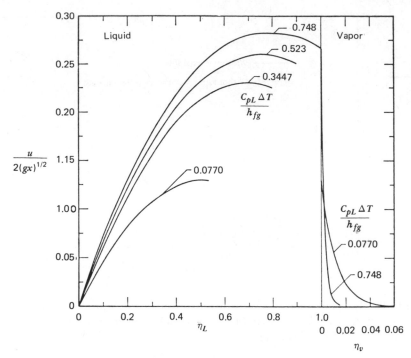

**Figure 14-14** Velocity profiles in the laminar boundary layer for condensation on a vertical flat plate with vapor drag at Pr = 1 and $[(\rho\mu)_L/(\rho\mu)_v]^{1/2} = 600$. [By permission from J. C. Y. Koh, E. M. Sparrow, and J. P. Hartnett, *Internatl. J. Heat Mass Transf.* **2**, 69–82 (1961) [39].]

study by Chen [41], who employed a perturbed integral technique to solve the describing equation for the same problem, found that the effect of all factors on the average heat transfer coefficient is closely predicted by

$$\left(\frac{\bar{h}}{\bar{h}_{\text{Nusselt}}}\right)^4 = \frac{1 + (0.68 + 0.02/\text{Pr})A}{1 + (0.85 - 0.15A)A/\text{Pr}} \qquad (14\text{-}19)$$

subject to the restrictions that $A \leqslant 2$ for $\text{Pr} \geqslant 1$ and $A \leqslant 20\,\text{Pr}$ for $\text{Pr} \leqslant 0.05$—the cases of $0.05 < \text{Pr} < 1$ are excluded—in which $A = C_p(T_s - T_w)/h_{\text{fg}}$; it is noteworthy that the $(\rho\mu)_L/(\rho\mu)_v$ parameter that arises in an exact analysis has only a second-order affect on the heat transfer [39, 41]. Chen [41] compared predictions with measurements for liquid metals and found that the measurements were from 0.3 to 0.9 of his predictions, a discrepancy seemingly not explained by the hypothesis that the fraction of vapor molecules striking a surface and actually condensing is less than unity [42]. The data compiled by Sukhatme and Rohsenow [29] shown in Fig. 14-15 are augmented elsewhere [43].

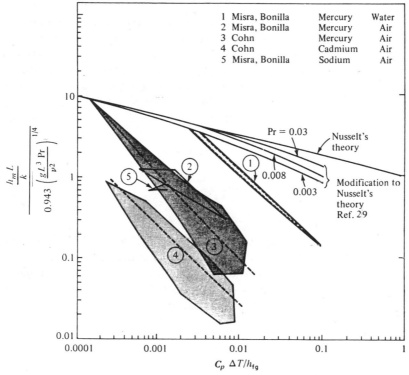

**Figure 14-15** Comparison of measurements of liquid metal condensation rates with Nusselt's theory and modifications of it. [From S. P. Sukhatme and W. M. Rohsenow, *Transact. ASME, J. Heat Transf.* **88**, 19–28 (1966) [29].]

The time $t_s$ required to achieve steady condensation after a step change in wall temperature from $T_s$ to $T_w$ was found by Sparrow and Siegel [129] for vertical plates to be

$$t_s = \left[ \frac{h_{fg}\rho\mu x}{kg(T_s - T_w)(\rho_L - \rho)} \right]^{1/2} \left[ 1 + \frac{C_p(T_s - T_w)}{8h_{fg}} \right] \qquad (14\text{-}20)$$

During this transition period the condensate film thickness $\delta$ and the local heat-transfer coefficient $h$ vary as

$$\delta = \begin{cases} \delta_0 \left( \dfrac{t}{t_s} \right)^{1/2} \\ \delta_0 \end{cases} \qquad h = \begin{cases} h_0 \left( \dfrac{t}{t_s} \right)^{1/2}, & t \leqslant t_s \\ h_0, & t \geqslant t_s \end{cases}$$

where $\delta_0$ and $h_0$ are given by Eqs. (14-16) and (14-17), respectively. An integral method and the method of characteristics, the subject of Problem 14-16, were employed to achieve these approximate results. Later Wilson [47] employed the method of characteristics to achieve solutions for transient condensation on horizontal cylinders, steady condensation and transient evaporation from inclined cylinders, and steady condensation and transient evaporation from inclined cylinders, and steady condensation on upward-pointing cones. For a horizontal cylinder, it is found that the instantaneous film thickness atop the cylinder approaches its steady-state value $\delta_0$ (see Problem 14-27) asymptotically as

$$\frac{\delta^2}{\delta_0^2} = \tanh\frac{t}{2t_s}$$

where $t_s = 3\nu_L D/3g\delta_0^2$.

Vapor superheat generally increases condensation by less than 10%, an effect that is appreciable only if the wall and saturation temperatures differ by less than about 3°C. This has been confirmed by analytical studies for a vertical flat plate [48, 49] and by experiment [36]. To account for the minor effect of vapor superheat, an effective heat of vaporization can be employed as

$$h_{fg_{eff}} = h_{fg} + C_{p_v}(T_v - T_s)$$

and used in relations for saturated vapor.

The effect of forced convection of the condensing vapor is of interest and has received attention. Admission of vapor into a condenser is often accompanied by fairly high vapor velocities because of the great disparity between vapor and liquid densities. Hence vapor shear at the interface can be appreciable and can either aid or oppose condensate motion. The effect of vapor shear can be seen in the following simple way following Rohsenow et al. [50]. The conditions are as shown in Fig. 14-9 with the addition of a vapor shear $\tau_v$ at the outer edge of the condensate film. For laminar conditions the liquid $x$-motion equation is

$$\frac{d^2u}{dy^2} = -\frac{(\rho_L - \rho)g}{\rho_L\nu_L}$$

subject to $u(0) = 0$ and $du(\delta)/dy = \tau_v/\mu_L$, neglecting convective effects. The energy equation, also neglecting convective effects, is

$$\frac{d^2T_L}{dy^2} = 0$$

subject to $T_L(0) = T_w$ and $T_L(\delta) = T_s$. The approximate velocity and temperature distributions are then

$$u = \frac{(\rho_L - \rho)g\delta^2}{\rho_L \nu_L}\left(\eta - \frac{\eta^2}{2}\right) + \frac{\tau_v}{\mu_L \delta}\eta$$

and

$$\frac{T - T_s}{T_w - T_s} = 1 - \eta$$

An energy balance at the liquid–vapor interface requires

$$-k_I \frac{\partial T_L(\delta)}{\partial y} = -h_{fg}\rho_L \frac{d\left(\delta \int_0^1 u\, d\eta\right)}{dx}$$

Introduction of the velocity and temperature profiles into this relation yields

$$\delta_0^4 = \left[1 + \frac{4\tau_v}{3g(\rho_L - \rho)\delta}\right]\delta^4 \tag{14-20a}$$

where $\delta_0$ is the film thickness with $\tau_v = 0$ given by Eq. (14-16). If it is assumed for the moment that $\tau_v/g(\rho_L - \rho)\delta \ll 1$, the simplification that

$$\frac{\delta}{\delta_0} \approx \left[1 + \frac{4\tau_v}{3g(\rho_L - \rho)\delta_0}\right]^{-1/4}$$

follows. On the basis of this result, the local heat flux at the wall is

$$q_w = -k\frac{\partial T(0)}{\partial y} = \frac{k(T_s - T_s)}{\delta}$$

$$= q_{w0}\left[1 + \frac{4\tau_v}{3g(\rho_L - \rho)\delta_0}\right]^{1/4}$$

where $q_{w0}$ is the wall heat flux with $\tau_v = 0$ given by Eq. (14-17). Although rough, this estimate suggests that vapor drag in the direction of the body force can substantially increase condensation rates, reducing them if its direction is opposite that of the body force. For laminar flow in both the condensate and vapor boundary layers, Fig. 14-16, due to Jacobs [51], shows the results for cocurrent vapor drag and body force where $\mathrm{Re}_{Lx} = U_\infty x/\nu_L$, $\mathrm{Fr}_x = U_\infty^2/gx$, $\mathrm{Nu}_x = hx/k_L$, $H = C_{pL}(T_s - T_w)/\mathrm{Pr}_L h_{fg}$, $Z = 1/\mathrm{Fr}_x$, and the ordinate is $\mathrm{Nu}_x \mathrm{Fr}_x^{1/2} H^{1/4} \mathrm{Re}_x^{-1/2}$. A detailed similarity solution, including the effect of

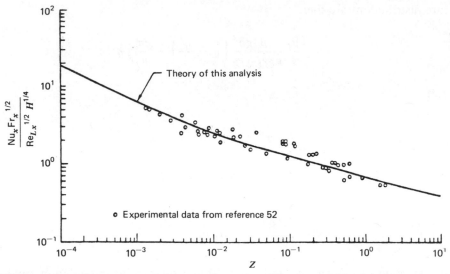

**Figure 14-16** Comparison of theory and measurement for combined body force and forced convection for Freon-113. [By permission from H. R. Jacobs, *Internatl. J. Heat Mass Transf.* **9**, 637–648 (1966) [52].]

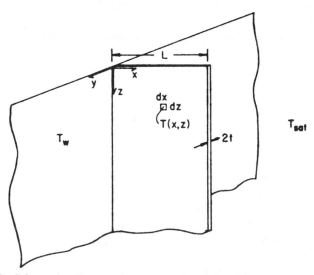

**Figure 14-17** Schematic diagram for fin condensation. [From S. V. Patankar and E. M. Sparrow, *Transact. ASME, J. Heat Transf.* **102**, 186–187 (1980) [56].]

noncondensable gases, was executed by Sparrow et al. [53]. The extreme case of vapor drag with negligible body force (e.g., applicable to a horizontal plate) was solved by Cess [54] and Koh [55].

Laminar film condensation on a vertical fin, whose physical configuration is depicted in Fig. 14-17, was studied analytically by Patankar and Sparrow [56]. They found that condensation rates on such a fin are a fraction of that which would occur on a vertical plate of uniform temperature as

$$\frac{Q_{\text{fin}}}{Q_{\text{isothermal}}} = 0.9257 Z^{1/8}$$

where $Z = [4k_L \mu_L (T_s - T_w)/h_{\text{fg}} g \rho_L^2][k_{\text{fin}} t/k_L L^2]^4 Z$ and $Q_{\text{isothermal}}$ is computed from Eq. (14-17). Problem 14-37 treats a variation of this situation.

## 14.5  TURBULENT FILM CONDENSATION ON VERTICAL PLATES

The information in Section 14-4 pertains to a condensate film whose flow is laminar, which is the usual case. Occasionally, however, the condensate film flows in a turbulent manner. This circumstance can occur on tall vertical surfaces or on banks of horizontal tubes (the latter are discussed in more detail in a later section).

It is reasonable to base the criterion for the transition from laminar to turbulent flow on the Reynolds number $\text{Re}_x = 4\dot{m}/\mu_L$, with terms as defined following Eq. (14-18). Since the condensing heat flow into the wall must all have come from the released heat of vaporization,

$$\dot{m} h_{\text{fg}} = \bar{h} x (T_s - T_w) \tag{14-21}$$

which allows the Reynolds number to be alternatively given as $\text{Re} = 4\bar{h}x(T_s - T_w)/\mu_L h_{\text{fg}}$. For a laminar condensate film, Eq. (14-17) substituted into the latter Re expression gives

$$\frac{\text{Re}}{4} = 0.9426 \left[ \frac{g^{1/3}\rho_L^{2/3}k(T_s - T_w)x}{h_{\text{fg}}\mu_L^{5/3}} \right]^{3/4}$$

Figure 14-18 due to Grober et al. [57] shows that this laminar relation fails at $\text{Re} \gtrsim 1400$, which is the sought-after transition criterion in the absence of vapor drag due to forced convection. On the basis of turbulent flow principles, the form of the turbulent correlation was deduced; their suggested fit to the data is

$$\frac{\text{Re}}{4} = 0.003 \left[ \frac{g^{1/3}\rho_L^{2/3}k(T_s - T_w)x}{h_{\text{fg}}\mu_L^{5/3}} \right]^{3/2} \tag{14-22}$$

Solution of Eq. (14-22) for the temperature difference and insertion of that

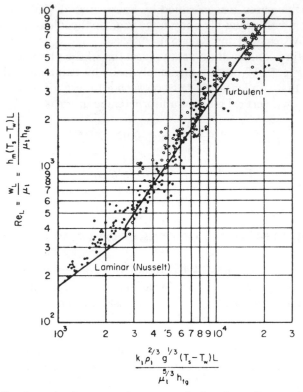

**Figure 14-18**  Condensation heat transfer on a vertical surface. (By permission from H. Grober, S. Erk, and U. Grigull, *Fundamentals of Heat Transfer*, McGraw-Hill, New York, 1961 [57].)

result into Eq. (14-21) gives an expression for the average turbulent heat-transfer coefficient as

$$\frac{\bar{h}\left(\nu_L^2/g\right)^{1/3}}{k} = 0.0131\,\mathrm{Re}^{1/3}, \qquad \mathrm{Re} \geqslant 1400 \qquad (14\text{-}23)$$

or

$$\frac{\bar{h}x}{k} = 0.003\left[\frac{gk(T_s - T_w)x^3\rho_L^2}{h_{fg}\mu_L^3}\right]^{1/2}, \qquad \mathrm{Re} \geqslant 1400$$

Equation (14-23) is fitted to data whose Prandtl number ranges from 1 to 5.

The condensing heat-transfer coefficient for turbulent flow of a Prandtl number condensate such as a liquid metal is less satisfactorily correlated. Dukler [58] applied the eddy diffusivity described by Deissler and the von Karman mixing length (both discussed in Section 9.2) to such a calculation and

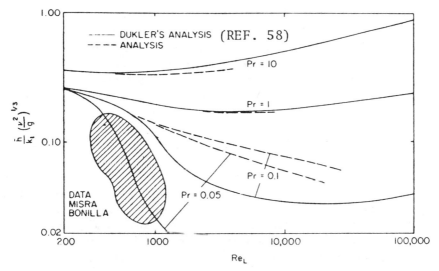

**Figure 14-19**  Turbulent condensation on vertical surfaces for various Prandtl numbers. [By permission from J. Lee, *AIChE J.* **10**, 540–544 (1964) [59].]

obtained good agreement with measurements, despite neglecting molecular conductivity relative to eddy conductivity and having some inconsistencies in the velocity profile. Lee [59] removed these assumptions and repeated the calculations to find substantially less satisfactory agreement with measurements, as Fig. 14-19 shows. As a consequence, no fully satisfactory correlation of turbulent condensation heat-transfer coefficients can be presented, a difficulty that may ultimately be traced to either imprecise measurements or such factors as the unknown presence of noncondensable gases.

It is noteworthy that condensing heat-transfer coefficients are often roughly the same for laminar and turbulent cases. This is illustrated by Fig. 14-20 from McAdams [36].

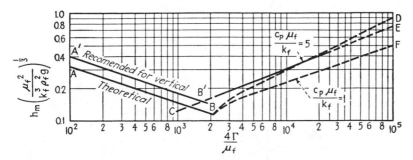

**Figure 14-20**  McAdams' recommended curves $A'B'$ and $CE$ for film condensation of single vapors on vertical tubes of plates. (By permission from W. H. McAdams, *Heat Transmission*, 3rd ed., McGraw-Hill, New York, 1954 [36].)

## 14.6  LAMINAR FILM CONDENSATION ON TUBES

The relationships presented for condensation on vertical plates are usually applicable to condensation of vertical tubes. One condition for such applicability is that the condensate film be thin compared to the tube radius, a condition that is usually met, as Problem 14-18 illustrates. For condensation on the exterior of a vertical tube the application of vertical plate results is straightforward. Condensation on the interior surface of a vertical tube is occasionally complicated somewhat by the need to account for the vapor draft down the tube.

For a horizontal tube, modification of vertical plate results is necessary. This modification, which is relatively minor, was first done by Nusselt [30] and proceeds as outlined in Section 13.5 for film boiling from a horizontal cylinder —historically, Nusselt's condensation analysis preceded the film boiling analysis. Figure 14-21 provides the geometric basis for the developments that follow. An energy balance on the condensate film between $\theta$ and $\theta + d\theta$ gives

$$h_{fg}\frac{d\dot{m}}{dx} = \frac{k_L(T_w - T_s)}{\delta} \tag{14-24}$$

Note that $\theta = x/R$. The component of gravity acting tangentially to the tube surface is $g\sin\theta$; thus, with the parabolic velocity distribution used for a vertical flat plate, the condensate mass flow rate is given by

$$\dot{m} = \rho_L\int_0^\delta u\,dy$$

$$= \left[\rho_L(\rho_L - \rho)\delta^2 g\frac{\sin\theta}{\mu_L}\right]\left[\delta\int_0^1\left(\eta - \frac{\eta^2}{2}\right)d\eta\right]$$

$$= \rho_L(\rho_L - \rho)\delta^3 g\frac{\sin\theta}{3\mu_L}$$

Insertion of this relationship into Eq. (14-24) gives

$$\dot{m}^{1/3}\,d\dot{m} = \frac{Rk(T_w - T_s)}{h_{fg}}\left[\frac{g(\rho_L - \rho)}{3\nu_L}\right]^{1/3}\sin^{1/3}\theta\,d\theta \tag{14-24a}$$

Integration of this equation from $\theta = 0$ to $\theta = \pi$ gives the condensate production from one side as

$$\dot{m} = 1.924\left[\frac{R^3 k^3(T_w - T_s)^3 g(\rho_L - \rho)}{h_{fg}^3\nu_L}\right]^{1/4} \tag{14-24b}$$

Here the footnote following Eq. (13-35) has been used. Accounting for the fact

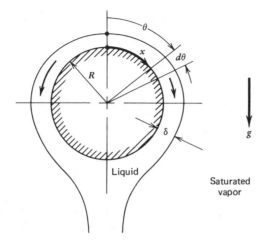

**Figure 14-21** Film condensation on a horizontal tube.

that this relation gives only half of the condensate flow from the tube, one finds that an energy balance on the entire tube yields

$$2h_{fg}\dot{m} = \bar{h}2\pi R(T_s - T_w)$$

Solution for $\bar{h}$ gives the overall Nusselt number as

$$\overline{\mathrm{Nu}} = 0.728\left[\mathrm{Ra}\,\frac{h_{fg}}{C_{p_L}(T_s - T_w)}\right]^{1/4} \tag{14-25}$$

where $\overline{\mathrm{Nu}} = \bar{h}D/k$ and $\mathrm{Ra} = g(\rho_L - \rho)\mathrm{Pr}\,D^3/\rho_L\nu_L^2$. A fuller discussion of the analysis leading to Eq. (14-25) is provided by Jakob [60] and by Chen [61], who points out that this problem, when treated in detail, does not possess similarity. Nusselt numbers for arbitrary axisymmetric bodies in variable gravity were obtained by Dhir and Lienhard [62, 63] by similar methods and by Yang [64] for spheres.

For those cases in which it is more convenient to have a correlation in terms of condensate flow rate than temperature difference, a rearrangement of Eq. (14-25) is possible. First, a Reynolds number is defined as for a vertical plate as $\mathrm{Re} = 4\dot{M}/\mu_L$, where $\dot{M}$ is now the rate of condensate flow per unit length of tube. Expressing the temperature difference in terms of other variables by use Eq. (14-24b), substituting that result into Eq. (14-25), and appealing to the definition of the Reynolds number written previously, one obtains

$$\frac{\bar{h}}{k}\left[\frac{\nu_L^2\rho_L}{g(\rho_L - \rho)}\right]^{1/3} = \frac{1.52}{\mathrm{Re}^{1/3}}, \qquad \mathrm{Re} \leqslant 2800 \tag{14-26}$$

for laminar flow of condensate. Viewing flow around each half of the tube as the equivalent of flow down a single vertical plate for which laminar flow occurs when $Re = 4\dot{m}/\mu_L \le 1400$, it is estimated that laminar flow would exist on the tube when $Re = 4\dot{M}/\mu_L \le 2800$. Because of its small vertical dimension a horizontal tube rarely has turbulent condensate flow; use Eq. (14-23) according to McAdams [36] if turbulence occurs.

The similarity between the Reynolds number relations for heat-transfer coefficients for vertical plates and horizontal cylinders is striking, as revealed by comparison of Eqs. (14-26) and (14-18). The study by Selin [37] (see Fig. 14-22) shows that better predictions can be made if the coefficients in Eqs. (14-25) and (14-26) are changed to 0.61 and 1.27, respectively. This modification is in the direction opposite from that made for vertical plates, weakening the initially perceived similarity. The similarity is yet strong enough, however, to enable use of Fig. 14-13 to ascertain the effect of low Prandtl numbers such as are characteristic of liquid metals. As with vertical plates, measurements for liquid metals are 0.3–0.9 of predicted values.

The effect of surface tension on the manner in which condensate leaves the bottom of the tube has been studied by Henderson and Marchello [38].

For condensation on the outside of tubes inclined at an angle with the horizontal, Selin [37] correlated measurements within 15% for $0° \le \theta \le 60°$ by taking the component of gravity perpendicular to the tube, $g \cos \theta$, for use in Eq. (14-25) in place of $g$. The success of this simple strategem may be due to the fact that most of the condensation occurs atop the tube, where the condensate film is thin and surface tension effects are small; the cross flow that occurs down the top of the inclined tube has negligible effect compared to the

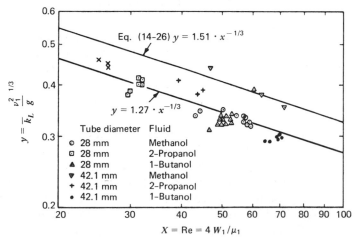

**Figure 14-22** Condensation heat-transfer measurements for horizontal tubes. (By permission from the American Society of Mechanical Engineers and G. Selin, *Proceedings of the International Heat Transfer Conference, University of Colorado*, Part II, 1961, pp. 279–289 [37].)

**Figure 14-23** Film condensation on a bank of horizontal tubes.

flow around the circumference of the tube. The bottom of the tube, where the condensate film is thick and surface tension effects are large, experiences a minor part of the total condensation. Hence condensate drainage down the bottom of the inclined tube is unlikely to have a large effect on the overall condensation rate. Sheynkman and Linetskiy [65] present a more refined treatment of condensation on inclined tubes, treating the top as covered by circumferential flow without surface tension effects and treating separately the thick film along the bottom where surface tension effects are considered.

In many technical applications, such as some condensers, banks of horizontal tubes are arranged in vertical rows as illustrated in Fig. 14-23. In this situation, first treated by Nusselt [30], the condensate from the topmost tube drains onto the second tube, and so on. The thicker condensate film on lower tubes renders them less effective condensing surfaces than the upper tubes whose condensate films are relatively thin. For each tube, the analysis leading to Eq. (14-24a) applies. Hence for each tube it follows that

$$\dot{m}^{4/3}_{bottom,\,n} = \dot{m}^{4/3}_{top,\,n} + B$$

in which $B^{3/4} = 1.924[R^3 k^3 (T_w - T_s)^3 g(\rho_L - \rho)/h^3_{fg} \nu_L]^{1/4}$. For tube number 1,

$$\dot{m}^{4/3}_{b,\,1} = B$$

For tube number 2,

$$\dot{m}_{b,2}^{4/3} = \dot{m}_{b,1}^{4/3} + B$$

$$= 2B$$

Generalization to the $n$th tube results in

$$\dot{m}_{b,n}^{4/3} = nB$$

As before, the overall heat-transfer coefficient is found from the requirement that

$$2h_{\mathrm{fg}}\dot{m}_{b,n} = 2\pi R n \bar{h}\,(T_s - T_w)$$

to be

$$\frac{\bar{h}\,(nD)}{k} = \frac{2h_{\mathrm{fg}}}{\pi(T_s - T_w)}(nB)^{3/4}$$

This result shows that the relationship for a single horizontal tube [Eq. (14-25)] can be taken over directly if $nD$ is used in place of $D$. Specifically, for $n$ tubes in a vertical row,

$$\frac{\bar{h}\,(nD)}{k} = 0.728\left[\frac{g(\rho_L - \rho)(nD)^3 h_{\mathrm{fg}}}{k_L \nu_L(T_s - T_w)}\right]^{1/4} \qquad (14\text{-}27)$$

a result that suggests that the condensation rate will vary as $n^{3/4}$. The various effects of splashing, uneven bowing of tubes, external vibrations, surface ripples, and so forth make it unnecessary to modify (in a major way) the coefficient of Eq. (14-27) to correlate measurements. A minor modification to Eq. (14-27) developed by Chen [61], accounting for additional condensation on the subcooled condensate stream between tubes, consists of multiplying it by the term

$$\frac{\left[1 + 0.2A(n-1)\right]\left[1 + (0.68 + 0.02A)A/\mathrm{Pr}\right]}{1 + (0.95 - 0.15A)A/\mathrm{Pr}}$$

Here $A = C_{p_L}(T_s - T_w)/h_{\mathrm{fg}}$, and restrictions are that $A(n-1) \leqslant 2$, $A \leqslant 2$, $\mathrm{Pr} \leqslant 0.05$ or $\mathrm{Pr} \geqslant 1$, and $A/\mathrm{Pr} \leqslant 20$ for $n = 1$ whereas $A/\mathrm{Pr} \leqslant 0.1$ for $n > 1$.

    Laminar film condensation of a vapor flowing perpendicular to a horizontal cylinder has been treated by Denny and Mills [66]. Their lengthy resulting equation for the local heat-transfer coefficient involves the Reynolds number $\mathrm{Re} = U_{v\infty}D/\nu_L$, developed from an integral method of solving the boundary-layer equations. For water with $T_{sat} = 110°F$, $T_{sat} - T_{wall} = 40°F$, and $D_{tube} = 0.25$ in., a vapor approach velocity of 25 ft/sec increases the average condensing heat-transfer coefficient by about 40%; a velocity of 4 ft/sec increases the average heat-transfer coefficient by about 8%.

## 14.7 CONDENSATION INSIDE TUBES

From time to time a vapor is condensed on the inside of tubes, rather than on the outside. In such a case the flow of vapor down the tube leads to an important shear stress on the condensate film that separates the tube wall from the vapor. The general trend, suggested in Section 14.4 by a simple treatment for flat-plate geometry, is that vapor-induced shear increases condensation rate if the vapor motion aids the local body force in causing condensate motion. Depending on the magnitude and orientation of friction and momentum exchange with the flowing vapor and gravity forces, the condensate film may form an annulus of uniform thickness, or it may stratify with a much thicker liquid film along the bottom of the tube if gravity acts perpendicular to the tube axis. Because the high density ratios commonly encountered produce a low liquid fraction and the vapor velocity is usually high, an annular condensate film of uniform thickness is most common.

Nusselt [30] extended his vertical-plate analysis to laminar film condensation inside a vertical tube with forced vapor flow. He assumed that the shear at the liquid–vapor interface (the condensate formed an annular film on the tube wall) could be predicted from pipe flow relations as $\tau_v = f\rho_v U_v^2 / 8g_c$, where $f$ is the usual friction factor. Equation (14-20a) resulted and was solved for vapor flow both aiding and opposing the body force. His results, clearly developed and discussed by Jakob [60], yield low predictions for heat-transfer coefficients at either high vapor velocities or when the condensate flow is turbulent as a result of the inaccurate manner of shear calculation and the assumption of laminar condensate flow. This sort of analysis was extended by Rohsenow et al. [50] to include turbulent condensate films.

Over the years a substantial amount of data has been collected; many of these works are cited and discussed in the survey by Merte [3]. For example, Rosson and Myers [67] and Akers and Rosson [68] measured local heat-transfer coefficients inside horizontal tubes. However, the correlations of Carpenter and Colburn [69] appear to have been most generally applicable in the refined form developed by Soliman et al. [70].

Soliman et al. [70] applied the correlation by Lockhart and Martinelli [71] for friction in two-phase adiabatic flow to the basic ideas set forth by Carpenter and Colburn [69]. Their resulting correlation, predicting measurements within 40%, for the local heat-transfer coefficient is

$$\frac{h\mu_L}{k_L \rho_L^{1/2}} = 0.036 \, Pr_L^{0.65} F_0^{1/2} \tag{14-28}$$

where $F_0$ is an effective stress given by the sum of three parts as

$$F_0 = F_f + F_m \pm F_a$$

and the positive sign is used for downward vapor flow (assisting gravity) whereas the negative sign is used for upward vapor flow (opposing gravity). The friction term $F_f$ dominates at high to intermediate quality. The momentum term $F_m$ arises from the momentum given to the slow liquid by the fast vapor on condensation. The gravity term $F_a$ dominates at low quality. Here

$$F_f \frac{\pi^2 \rho_v D^4}{8W^2} = 0.045 \, \mathrm{Re}^{-0.2}\left[ X^{1.8} + 5.7\left(\frac{\mu_L}{\mu_v}\right)^{0.0523} (1 - X)^{0.47} X^{1.33} \left(\frac{\rho_v}{\rho_L}\right)^{0.261} \right.$$

$$\left. +8.11\left(\frac{\mu_L}{\mu_v}\right)^{0.105} (1 - X)^{0.94} X^{0.86} \left(\frac{\rho_v}{\rho_L}\right)^{0.522} \right]$$

$$F_m \frac{\pi^2 \rho_v D^4}{8W^2} = 0.5\left( D\frac{dX}{dz}\right)\left[ 2(1 - X)\left(\frac{\rho_v}{\rho_L}\right)^{2/3} \right.$$

$$+ \left(2X - 3 + \frac{1}{X}\right)\left(\frac{\rho_v}{\rho_L}\right)^{4/3} + (2X - 1 - \beta X)\left(\frac{\rho_v}{\rho_L}\right)^{2/3}$$

$$\left. + \left(2\beta - \beta X - \frac{\beta}{X}\right)\left(\frac{\rho_v}{\rho_L}\right)^{5/3} + 2(1 - X - \beta + \beta X)\frac{\rho_v}{\rho_L} \right]$$

$$F_a \frac{\pi^2 \rho_v D^4}{8W^2} = \frac{0.5}{\mathrm{Fr}}\left\{ 1 - \left[ 1 + \frac{(1 - X)(\rho_v/\rho_L)^{2/3}}{X} \right]^{-1} \right\}$$

where $\mathrm{Fr} = 16W^2/[\pi^2 D^5 a(\rho_L - \rho_v)\rho_v]$ is a Froude number, $a$ is the gravity component along the tube axis, $W$ is the total mass flow rate of liquid and vapor, $D$ is tube diameter, $\mathrm{Re} = 4W/\pi D\mu_v$ is a Reynolds number, $X$ is local quality, $\beta$ is 1.25 for a turbulent film and 2.0 for a laminar film, and (assuming uniform heat removal along the tube with unity entering quality) $dX/dz = -1/L_0$ if $L_0$ is the total length of the tube.

These relations for the components of $F_0$ allow the conditions for "runback" to be predicted. For vapor flow upward against gravity, slugging and plugging can occur as a result of folding and runback of the condensate film when vapor-induced friction and momentum exchanges are insufficient to drag the condensate against the pull of gravity. The criterion for the onset of such a condition is that $F_0 = F_f + F_m - F_a = 0$; from the foregoing relationships, the quality at which this condition occurs can be ascertained.

Fortunately for ease of prediction, Eq. (14-28) reduces at high quality ($X \approx 1$) to

$$\mathrm{Nu} = 0.0054\,\mathrm{Pr}_L^{0.65}\,\mathrm{Re}_v^{0.9}\frac{\mu_v}{\mu_L}\left(\frac{\rho_L}{\rho_v}\right)^{1/2} \tag{14-29}$$

with $\mathrm{Nu} = hD/k_L$, $\mathrm{Re}_v = 4W_v/\pi D\mu_v$, where $W$ is the mass flow rate of vapor. Equation (14-29) resembles the correlation for Nusselt number in single-phase turbulent flow, as comparison with Eq. (11-17e) shows. Shah [121] gives a refined correlation.

## 14.8   NONGRAVITATIONAL CONDENSATE REMOVAL

Actual condensation rates are only about 1% of the maximum possible condensate rate that would result if all vapor molecules incident on a cooled surface were condensed. This realization, the subject of Problem 14-33, has led many to seek practical means of thinning the insulating condensate film by a nongravitational means of causing condensate flow. When gravity is absent, nongravitational condensate removal is very important, of course.

Condensation on a horizontal rotating disc was studied by Sparrow and Gregg [72], who achieved a similarity solution of the laminar boundary-layer equations. They found that the condensate film is of a constant thickness $\delta$ predictable by

$$\delta\left(\frac{\omega}{\nu}\right)^{1/2} = 1.107\left[\frac{C_{p_L}(T_s - T_w)}{\mathrm{Pr}_L\,h_{\mathrm{fg}}}\right]^{1/4}$$

where $\omega$ is the angular velocity of the disk. The heat-transfer coefficient, neglecting vapor drag that was studied later [73], is also a constant predictable by

$$\frac{h(\nu/\omega)^{1/2}}{k} = 0.904\left[\frac{\mathrm{Pr}_L\,h_{\mathrm{fg}}}{C_{p_L}(T_s - T_w)}\right]^{1/4}$$

These results are accurate for $\mathrm{Pr} \approx 1$ and $C_{p_L}(T_s - T_w)/h_{\mathrm{fg}} < 0.1$; see the original work for other conditions. Laminar flow of condensate is expected for $\mathrm{Re} = r^2\omega/\nu \lesssim 3 \times 10^5$. It can be shown that $h$ for a rotating disk $h_{\mathrm{rot}}$ is related to $h$ for a vertical plate under the sole influence of gravity $h_{\mathrm{vert}}$ as

$$\frac{h_{\mathrm{rot}}}{h_{\mathrm{vert}}} = \left(\frac{8x\omega^2}{3g}\right)^{1/4}$$

The measurements by Nadapurkar and Beatty [74] are about 30% less than those predicted analytically. Condensation on a rotating cone [75] with its apex uppermost was found by the same methods to have nearly the same coefficient as for a rotating disk since

$$\frac{h_{\text{cone}}}{h_{\text{disk}}} = (\sin \phi)^{1/2}$$

where $\phi$ is the half-angle of the cone. A rotating cone has been proposed as part of a system for rendering saline water potable and was studied by Bromley [76]. (For condensation on the inside of rotating cones, consult the study by Marto [77].)

A vertical tube spinning about its own axis was studied experimentally by Nicol and Gacesa [78]. They found that at low angular velocity $\omega$ the overall heat-transfer coefficient measurements were correlated by

$$\frac{\overline{\text{Nu}}}{\left[gL^3 h_{\text{fg}}\rho_L/\nu_L k_L(T_s - T_w)\right]^{1/4}} = \begin{array}{ll} 0.0943, & \text{We} \leqslant 250 \\ 0.00923\,\text{We}^{0.39}, & \text{We} > 250 \end{array}$$

where $\text{We} = \rho_L \omega^2 D^3/4\sigma$ is a dimensionless Weber number and $\overline{\text{Nu}} = \bar{h}D/k$. This correlation holds until the Nusselt number is nearly trebled. At higher angular velocities, the effect of gravity is unimportant and the correlation of measurements is

$$\overline{\text{Nu}} = 12.26\,\text{We}^{0.496}$$

These measurements for a vertical cylinder are compared with those of others for a rotating horizontal tube and additional references are cited.

Condensation for the case of a vertical tube rotating about a vertical axis (displaced from the tube axis) was studied by Mochizuki and Shiratori [79] for internal condensation and Suryanarayana [80] for external condensation. Their results are given in Fig. 14-24, where $\text{Nu}_M = \bar{h}_M D/k_L$ is the measured overall Nusselt number, $\text{Nu}_g = \bar{h}D/k$ is given by Eq. (14-17), and $G = a/g$, where $a$ is the centrifugal acceleration at the tube axis. Increases by a factor of as much as 4 in heat-transfer rate are evident.

The frequently marked effect of the body force exerted on a fluid by a nonuniform electric field was noted in Chapter 13 in connection with boiling. A number of investigations have shown a similar large effect in the case of condensation—the survey by Jones [82] discusses the physical mechanisms involved for both boiling and condensation as well as for natural convection, fluidized beds, solidification, and heat pipes. Velkoff and Miller [83] found experimentally that the condensation rate of Freon-113 on a plate could be increased by a factor of as much as 3 by placing electrodes with dc voltage parallel to the plate. Choi [84] condensed Freon-113 in an electrically grounded vertical tube with a central electrode to which dc voltage was applied. At voltages of about 30,000 V, the condensation rate was doubled. The efficacy of an ac electrical field was established by Holmes and Chapman [85], who

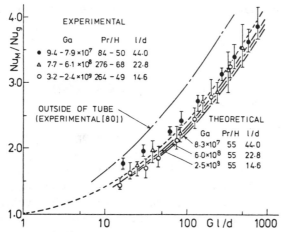

**Figure 14-24** Comparison of measurements and theory for condensation on a vertical tube rotating about a vertical axis displaced from the tube axis. [From S. Mochizuki and T. Shiratori, *Transact. ASME, J. Heat Transf.* **102**, 158–162 (1980) [79].]

applied up to 50,000 V to electrodes parallel to a plate to increase condensation rates by a factor of 10. Seth and Lee [86] found that a more modest increase in condensation rate of Freon-113 on a horizontal cylinder resulted as dc voltage was applied when air was present as a noncondensable gas. In all cases a voltage threshold must be exceeded before the effect of electrical fields is manifested.

Suction of condensate through the porous surface on which it condenses has been considered. Jain and Bankoff [87] applied a perturbation technique to solve an integral form of the laminar boundary-layer equations for condensation on a vertical plate with a uniform suction velocity. As much as a 50% increase in local heat-transfer coefficient could be achieved, but only after a large fraction of the condensate layer is sucked away. These results were confirmed by Yang [88] in a more rigorous manner and by Lienhard and Dhir [89] with a straightforward integral method. The same general conclusions apply to horizontal tubes [90].

The effect of mechanical vibrations on a vertical tube was explored by Dent [91], who reported experiments for steam condensation on a vertical tube [1 in. o.d. (outer diameter), 41 in. length] with pinned ends that was vibrated transversely at 22–98 Hz at an amplitude of 0.5 in. by a Scotch-yoke mechanism attached to the middle. Heat-transfer coefficients were increased up to 60%, not by throwing off condensate, but rather by causing condensate to slosh from side to side. His correlation is

$$\frac{h_x}{h_{\text{no vibr}}} = \left[1 + \left(\frac{a_0\omega}{\pi}\right)^2 \frac{\mu_L h_{\text{fg}}}{gk_L x(T_s - T_w)}\right]^{1/4}$$

where $a_0$ is vibration amplitude and $\omega$ is vibration frequency.

Capillary forces have also been extensively used to move condensate in heat pipes [92, 93].

## 14.9   EFFECT OF NONCONDENSABLE GAS AND MULTICOMPONENT VAPORS

The presence of a noncondensable gas in a vapor can significantly lower the condensation rate below that for a pure vapor. An early experimental study by Othmer [94], more recently augmented by the measurements of Henderson and Marchello [95], showed that only a few parts by volume of air could reduce condensation heat transfer coefficients by up to 50%. Such effects are important in power plant condensers where buildup of such noncondensable gases as air is unavoidable and where provisions to remove noncondensable gases are not perfectly effective. Similarly, separation of ammonia from air by condensation requires that the presence of air be tolerated. In such cases design procedures must allow for the presence of noncondensables.

Approximate procedures to account for the effect of a noncondensable gas on condensation rates were early set forth by Colburn and Drew [96] and by Kern [97] in more modern times. More recently, boundary-layer approximations have been successfully applied.

Before presenting details, it is helpful to acquire a physical understanding of the physical mechanisms at work. This is readily accomplished by consideration of the results for Stefan's diffusion problem presented in Section 5.3 and illustrated in Figs. 5-6 and 5-7. As the vapor flows to the condensate–vapor interface, the noncondensable gas is swept along with it. Since the interface is impermeable to the noncondensable gas, its concentration there greatly increases above its value at more distant locations. Correspondingly, the partial pressure of the vapor at the interface is reduced below its value at more distant locations. Because of this reduced partial vapor pressure, the equilibrium temperature at the condensate–vapor interface is reduced and the driving temperature difference for heat transfer $(T_s - T_w)$ is reduced, causing reduction in the condensation rate. Figure 14-25$a$ illustrates this situation. The equilibrium level of the noncondensable gas at the interface requires that noncondensable gas be removed by (1) diffusion, (2) drag from the falling condensate, or (3) free convection caused by density variations as a result of varying composition either aiding (if the noncondensable gas is heavier than the vapor) or opposing (if the noncondensable gas is lighter than the vapor) item 2, or forced convection. The more effective the mechanism of noncondensable removal from the interface, the less will be the reduction of condensation rate.

The boundary-layer description of this problem considers the condensate film characteristics to be as given in Section 14.4 from a so-called Nusselt analysis of condensation on a vertical plate. Specifically, the condensate film

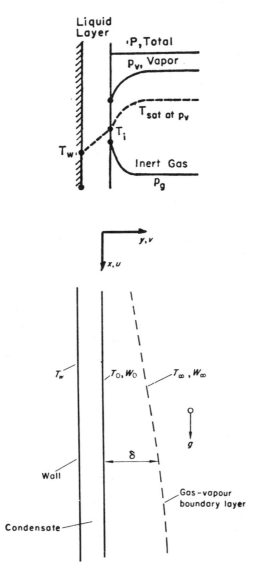

**Figure 14-25** (*a*) Film condensation in the presence of noncondensable gases. (*b*) Coordinate system for integral analysis of film condensation in the presence of noncondensable gases [by permission from J. W. Rose, *Internatl. J. Heat Mass Transf.* **12**, 233–237 (1969) [99].]

thickness $\delta$ is

$$\delta = \left[\frac{C_{p_L}(T_s - T_w)}{h_{fg}\,\mathrm{Pr}_L}\right]^{1/4}\left(\frac{4\nu_L^2 x}{g}\right)^{1/4} \tag{14-16}$$

the condensate interface velocity $u_0$ is

$$u_0 = 2\left[\frac{C_{p_L}(T_s - T_w)}{h_{fg}\,\mathrm{Pr}_L}\right]^{1/2}\left(\frac{gx}{4}\right)^{1/2} \tag{14-15a}$$

and the local heat flux $q$ is

$$q = \left[\frac{C_{p_L}(T_s - T_w)}{h_{fg}\,\mathrm{Pr}_L}\right]^{3/4}\left(\frac{gh_{fg}^4\mu_L^4}{4\nu_L^2 x}\right)^{1/4} \tag{14-17a}$$

The challenge now is to calculate the saturation temperature $T_s$ since then all would follow as before inasmuch as the heat flux is available from Eq. (14-17a) or the heat-transfer coefficient is available from Eq. (14-17).

The situation in the vapor–noncondensable mixture adjacent to the condensate film is illustrated in Fig. 14-25b, where the existence of a boundary layer in the gaseous region is shown. Its behavior is described by the boundary-layer equations for natural convection of continuity

$$\frac{\partial u}{\partial x} + \frac{\partial v}{\partial y} = 0$$

x-motion

$$u\frac{\partial u}{\partial x} + v\frac{\partial u}{\partial y} = g\left(1 - \frac{\rho_\infty}{\rho}\right) + \nu\frac{\partial^2 u}{\partial y^2}$$

and diffusion

$$u\frac{\partial \omega}{\partial x} + v\frac{\partial \omega}{\partial y} = D\frac{\partial^2 \omega}{\partial y^2}$$

The effect of temperature variations is ignored, with deletion of the energy equation from the group to be solved, and density is taken to be constant, except for the body force term in the x-motion equation. Exact solutions to these equations were achieved by Sparrow and Lin [98] for some common cases; the validity of the underlying assumptions was established by the later detailed study of Minkowycz and Sparrow [49]. A more general result was obtained by Rose [99] in approximate form by application of the integral method. The boundary-layer equations in integral form are

$$\frac{d\left(\delta_v \int_0^1 u^2 \, d\eta\right)}{dx} - v_0 u_0 + \frac{\nu}{\delta_v} \frac{\partial u(0)}{\partial \eta} - gX\delta_v \int_0^1 W \, d\eta = 0$$

and

$$\frac{d\left(\delta_v \int_0^1 uW \, d\eta\right)}{dx} + \frac{D}{\delta_v} \frac{\omega_\infty}{\omega_0} \frac{\partial W(0)}{\partial \eta} = 0$$

where $\eta = y/\delta_v$, $W = \omega - \omega_\infty$, $X = (M_g - M_v)/[M_g - \omega_\infty(M_g - M_v)]$, and $W = \omega - \omega_\infty$. Here $\omega = \rho_{gas}/(\rho_{vapor} + \rho_{gas})$ is the mass fraction of the noncondensable gas. The solution of these two equations, with use of the assumed profiles of

$$u = u_0(1 - \eta)^2 + \bar{u}\eta(1 - \eta)^2$$

$$\frac{\omega - \omega_\infty}{\omega_0 - \omega_\infty} = (1 - \eta)^2$$

gives

$$10F \, \mathrm{Sc} \frac{\mu_L \rho_L}{\mu \rho} \left(\frac{\omega_\infty}{W_0}\right)^2 \left(\frac{20}{21} + \mathrm{Sc} \frac{\omega_0}{\omega_\infty}\right) + \frac{8}{F^2 \, \mathrm{Sc}} \frac{\mu \rho}{\mu_L \rho_L} \left(\frac{W_0}{\omega_0}\right)^2 \left(\frac{5F}{28} - X \frac{W_0}{3}\right)$$

$$= \frac{100 \omega_\infty}{21 \omega_0} - \frac{2W_0}{\omega_0} + 8\mathrm{Sc} \qquad (14\text{-}30)$$

The accuracy of Eq. (14-30), for which $F = C_{p_L}(T_s - T_w)/h_{fg} \, \mathrm{Pr}_L$ and $W_0 = \omega_0 - \omega_\infty$ and $\mathrm{Sc} = \nu/D$ is the Schmidt number for the gas mixture, was confirmed by the measurements of Al-Diwany and Rose [100]. Best agreement resulted when the noncondensable gas was heavier than the vapor. Possible fogging in the boundary layer due to cooling, mentioned early by Jakob [60] and McAdams [36], is not taken into account by Eq. (14-30) but will not always occur since some supersaturation is tolerable. Such fogging is, however, taken into account in the integral analysis by Mori and Hijikata [101].

Suppose now that the wall temperature $T_w$, total pressure $p$, and mass fraction of noncondensable gas in the bulk mixture $\omega_\infty$ are specified and the heat transfer rate is to be computed. One begins by guessing an equilibrium interface temperature $T_s$ from which a value of $F$ is computed for use in Eq.

(14-30). Equation (14-30) is solved for the one unknown $\omega_0$, the noncondensable gas mass fraction at the interface. From this result, the mass fraction of the vapor at the interface $1 - \omega_0$ is known and is substituted into Eq. (5-30) to obtain the partial pressure of the vapor at the interface. If the partial pressure of the vapor is known, the corresponding saturation temperature $T_s$ can be ascertained from steam tables. If this value of $T_s$ differs appreciably from that initially assumed, the procedure is repeated until no appreciable difference results. Once $T_s$ has been found, the heat flux is obtained by using $T_s$ in Eq. (14-17a).

Clearly, removal of noncondensable gases from the interface by forced convection will be beneficial. Such a conclusion was reached by Denny et al. [102], Denny and Jusionis [103], Citakoglu and Rose [104], and Rose [120], who gives the corresponding form of Eq. (14-30). Liquid metals were studied by Turner et al. [105].

Condensation of two condensable vapors whose condensates are miscible was studied by Sparrow and Marschall [106]. Marschall and Hickman [107] treated condensation of two condensable vapors whose condensates are immiscible.

Sage and Estrin [108] executed an analytical study of vapor condensation on a vertical plate in the presence of two noncondensable gases. Such a model is applicable to fission product removal from contaminated vapors as well as a number of industrial processes. Additional effort on the same topic is reported by Taitel and Tamir [109]. Exact treatments of multicomponent condensation are complicated by the fact that the mass diffusivities involved are dependent on concentration (see Bird et al. [110] for an introductory discussion and Sparrow and Niethammer [111] for a recent application). A common linearization of these relations involves use of an effective binary diffusivity. An expanded development is given by Toor [112, 113].

## 14.10  SCULPTURED SURFACES

It has been seen that the rate of heat transfer in film condensation is limited by heat conduction through the condensate film. Consequently, augmentation of film-condensation heat-transfer coefficients depends on the ability to find some way to thin the condensate film. The sculptured surfaces that result from such an attempt are analogous (in a weak way) to the fins discussed in Chapter 13 to improve boiling heat flow rates.

In 1953 Gregorig [114] studied condensation on wavy surfaces and noted that surface tension could cause large pressure gradients in a condensate film whose curvature varied with position. The typical wavy surface shown in Fig. 14-26 in a top view is vertical. At the crest, the convex curvature requires the pressure in the condensate to exceed the vapor pressure; at the trough, the concave curvature requires the pressure in the condensate to be less than the vapor pressure. Hence the condensate is sucked from the crest down to the

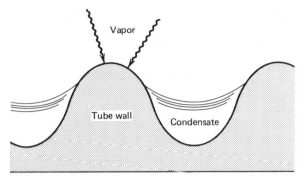

**Figure 14-26** Section through a vertical tube with axial flutes, showing the effects of surface tension on the configuration of the condensate film. [From E. Lusterader, R. Richter, and R. N. Neugebauer, *Transact. ASME, J. Heat Transf.* **81**, 297–307 (1959) [115].]

trough by this pressure gradient. The thin condensate film at the crest is accompanied by high heat-transfer coefficients that more than compensate for the diminished heat-transfer coefficients in the trough resulting from condensate accumulation that drains down the trough. Tests performed by Lusterader et al. [115] revealed a heat-transfer coefficient for film condensation about four times larger than could be achieved on a smooth surface. This general finding has been applied to horizontal condenser tubes [116] and to vertical-tube preheaters and evaporators [117, 118]. Webb [124] developed a generalized design procedure for so-called Gregorig surfaces.

On downward-facing surfaces, a two-dimensional wavy surface must be replaced by a three-dimensional (or doubly rippled) surface. Such a case was studied by Markowitz et al. [119], who discuss its application to the design of pool boilers for electronic equipment cooling. Their analysis suggested a possible fivefold improvement of heat-transfer coefficient over that achievable with a smooth surface. The measured twofold improvement was considered to be adversely affected by the presence of noncondensable gases that would tend to accumulate near the high-heat-flux crests.

Twisted-tape inserts and internal fins can increase average condensing heat-transfer coefficients by as much as 30 and 150%, respectively, over values for condensation in smooth horizontal tubes [123]. Sculptured exterior surfaces (externally augmented) can increase condensing heat-transfer coefficients by about a factor of 5 over smooth-tube values for horizontal tubes, according to the measurements by Carnavos [126].

## PROBLEMS

**14-1**  Consider a Carnot cycle operating between a heat source at $T_1$ and a heat sink at $T_2$. The *power* produced is to be maximized, subject to the realization that heat must flow at a rate proportional to temperature difference during the isothermal expansion and compression (see Curzon and Ahlborn [1] for additional discussion). The input energy $w_1$ during expansion of the working substance at constant temperature $T_{1w}$ is $w_1 = \alpha t_1(T_1 - T_{1w})$, in which $\alpha$ is a heat-transfer coefficient and $t_1$ is the duration of the expansion. Similarly, the rejected energy $w_2$ during compression of the working substance at constant temperature $T_{2w}$ is $w_2 = \beta t_2(T_{2w} - T_2)$, in which $\beta$ is a heat-transfer coefficient and $t_2$ is the duration of the compression. The two adiabatic processes connecting the isothermal expansion and compression are reversible, which requires that $w_1/T_{1w} = w_2/T_{2w}$.

(a) Show, from this information, that the ratio of expansion and compression durations is

$$\frac{t_1}{t_2} = \frac{\beta T_{1w}(T_{2w} - T_2)}{\alpha T_{2w}(T_1 - T_{1w})}$$

(b) Next, assume that the time required to complete a cycle is proportional to the sum of expansion and compression durations $t_1 + t_2$ to show that the cycle power $P$ is

$$P = \frac{w_1 - w_2}{\gamma(t_1 + t_2)}$$

in which $\gamma > 1$ is an unspecified constant. Introduce the result of part (a) into this equation to obtain

$$P = \frac{\alpha\beta xy(T_1 - T_2 - x - y)}{\gamma[\beta T_1 y + \alpha T_2 x + xy(\alpha - \beta)]}$$

where $x = T_1 - T_{1w}$ and $y = T_{2w} - T_2$.

(c) Maximize $P$ by setting $\partial P/\partial x = 0 = \partial P/\partial y$ to obtain the two relationships

$$\beta T_1 y(T_1 - T_2 - x - y) = x[\beta T_1 y + \alpha T_2 x + xy(\alpha - \beta)]$$

and

$$\alpha T_2 x(T_1 - T_2 - x - y) = y[\beta T_1 y + \alpha T_2 x + xy(\alpha - \beta)]$$

From the latter equation show that $y = x(\alpha T_2/\beta T_1)^{1/2}$ and substitute it into the former equation to find

$$\left(1 - \frac{\alpha}{\beta}\right)z^2 - 2\left[1 + \left(\frac{\alpha T_2}{\beta T_1}\right)^{1/2}\right]z + \left(1 - \frac{T_2}{T_1}\right) = 0$$

with $z = x/T_1$.

**(d)** Solve the last equation of part c to find

$$\frac{x}{T_1} = \frac{1 - (T_2/T_1)^{1/2}}{1 + (\alpha/\beta)^{1/2}}$$

**(e)** Show that the Carnot cycle efficiency at maximum power $\eta'$ is given by

$$\eta' = \frac{w_1 - w_2}{w_1}$$

$$= 1 - \frac{T_{2w}}{T_{1w}}$$

$$= 1 - \frac{T_2 + y}{T_1 - x}$$

$$= 1 - \left(\frac{T_2}{T_1}\right)^{1/2}$$

which well represents actual steam-point plant efficiencies [1].

**(f)** Show that the Carnot cycle efficiency at maximum power $P$ is given by

$$P = \frac{\alpha\beta}{\gamma}\left(\frac{T_1^{1/2} - T_2^{1/2}}{\alpha^{1/2} + \beta^{1/2}}\right)^2$$

Note that the power produced depends on the heat-transfer coefficients even though the efficiency does not.

**14-2** **(a)** Confirm Eq. (14-2) for water by substituting property values into the equation and estimating $r$ from $r^3 = M/\rho N_A$, where $M$ is molecular weight and $N_A$ is Avogadro's number.

**(b)** Use the result of part a to show that the tensile stress that would rupture water is of the order of $10^4$ times atmospheric pressure.

**14-3** Show that the pressure inside a liquid drop of radius $r^*$ in equilibrium with a supersaturated vapor is

$$P_L = \frac{2\sigma}{r^*} + P_{sat}\exp\frac{2v_L\sigma}{r^*RT}$$

Evaluate $P_L/P_{sat}$ for water at $68°F$ when $2r^* = 10^{-6}$ in. and compare it with the corresponding $P_v/P_{sat}$ from Table 14-1.

**14-4** To estimate the limiting vapor supersaturation (or subcooling) that can exist without condensation, introduce the drop radius, essentially the distance between liquid molecules, that corresponds to about eight liquid molecules into Eq. (14-7) for water at 68°F. (*Answer:* $P_v/P_{sat} \approx 10^2$, which was experimentally obtained in a nozzle [12], although only $P_v/P_{sat} \approx 8$ was experimentally obtained in a nonflow device [13].)

**14-5** The change in free energy $\Delta G$ of a liquid–vapor system such as shown in Fig. 14-3 is to be ascertained.

(a) Note that when the system is all vapor, the total free energy is given by Eq. (14-4) as $G_0 = (n_v + n_L)g_v$.

(b) Subtract the result of part a from the total free energy of the system when part of the system is a liquid drop to obtain the change in free energy from the all-vapor case as

$$\Delta G = G - G_0 = n_L(g_L - g_v) + 4\pi\sigma r^2$$

(c) Note from Eq. (14-5) that $g_L - g_v = -2v_L\sigma r^*$ and that $n_L v_L = 4\pi r^3/3$, and substitute these results into the result of part b to obtain

$$\Delta G = 4\pi\sigma \left( r^2 - \frac{2r^3}{3r^*} \right)$$

which is the free energy of formation of a drop of radius $r$.

(d) Show that the maximum value of $G$, the free energy of formation of a drop of radius $r^*$, is

$$\Delta G_{max} = \frac{4\pi\sigma r^{*2}}{3}$$

**14-6** A supersaturated vapor at temperature $T$ and pressure $P_v$ that is in equilibrium with a liquid drop of radius $r^*$ can be considered to be subcooled below the saturation temperature $T_{sat}$ [similar to the manner in which $P_v$ exceeds $P_{sat}$, shown by Eq. (14-6)]. The subcooling $T_v - T_{sat}$ is to be estimated. To do this:

(a) Utilize the Clausius-Clapeyron equation, relating saturation pressure and temperature,

$$T\frac{dP}{dT} = \frac{h_{fg}}{v_v - v_L}$$

to obtain, assuming the vapor to be a perfect gas and $v_v \gg v_L$,

$$\ln\left(\frac{P_v}{P_{sat}}\right) = h_{fg}\frac{1/T_v - 1/T_{sat}}{R}$$

since $P = P_v$ when $T = T_{\text{sat}}$ and $P = P_{\text{sat}}$ when $T = T_v$. Here $R$ is the universal gas constant and $h_{\text{fg}}$ is the latent heat of vaporization per mole.

**(b)** Combine the result of part a with Eq. (14-6) to find

$$r^* = \frac{2v_L\sigma}{h_{\text{fg}}} \frac{T_{\text{sat}}}{T_{\text{sat}} - T_v}$$

**(c)** The result of part b is the smallest equilibrium drop and also the smallest drop that could be expected in dropwise condensation. Show that for water at 100°C and atmospheric pressure,

$$r^* = \frac{2 \times 10^{-6} \text{ cm °C}}{T_{\text{sat}} - T_0}$$

which gives $r^* = 10^{-6}$ cm for 2°C subcooled vapor.

**14-7** The formation of vapor cavities in a bulk liquid is similar to the formation of liquid drops in a bulk vapor. To see this:

**(a)** Show by application of Eq. (14-3), where $r$ is taken as the distance between liquid molecules ($r \approx 10^{-10}m$), that the formation of a vapor cavity in a liquid requires

$$P_\infty - P_{\text{rupture}} \approx 10^4 \text{ atmospheric pressure}$$

which represents not only a negative pressure, but also a substantial superheating (just as considerable subcooling is required for condensation in the limiting case).

**(b)** Show by application of the Clausius–Clapeyron equation, relating saturation pressure and temperature,

$$T\frac{dP}{dT} = \frac{h_{\text{fg}}}{v_v - v_L}$$

into which are substituted the approximations $dP \approx 2\sigma/r$ and $dT \approx T_v - T_{\text{sat}}$ that the liquid superheat $T_v - T_{\text{sat}}$ required for a vapor cavity of radius $r^*$ to be in equilibrium with the liquid is

$$T_v - T_{\text{sat}} = \frac{2\sigma v_v}{r^* h_{\text{fg}}}$$

Superheat limits for liquid mixtures are treated by Pinnes and Mueller [14].

**(c)** Noting that the result of part b determines the minimum size of vapor bubble whose preexistence is required for bulk nucleation to start, show that for water at 100°C and atmospheric pressure

$$r^* = \frac{4 \times 10^{-3} \text{ cm } °C}{T_v - T_{sat}}$$

which gives $r^* = 2 \times 10^{-3}$ cm for 2°C superheated liquid. Compare these results with the corresponding ones for condensation from Problem 14-6.

14-8  The rate per unit volume $J$ at which molecules are added to a liquid drop from a supersaturated vapor is given by Eq. (14-8), which is to be derived [15].

(a) Show by consideration of Boltzmann's law [Eq. (2-30a)] and the free energy for formation of a liquid drop of radius $r$ from vapor, shown in Problem 14-5(c) to be

$$\Delta G = 4\pi\sigma\left(r^2 - \frac{2r^3}{3r^*}\right)$$

with $r^*$ given by Eq. (14-6), that the equilibrium distribution of drop sizes is

$$f(r) = Fe^{-\Delta G/KT}$$

where $K$ is the Boltzmann constant, $f$ is the number of drops of radius $r$ or number of molecules $n$, and $F$ is the total number of drops or total number of molecules, depending on whether $\Delta G$ is expressed in terms of drop radius or number of molecules.

(b) Recognize that the net rate $J$ at which drops with $n$ molecules increase in number of molecules to $n + 1$ is

$$J = \left(\begin{array}{c}\text{rate of molecule} \\ \text{addition to drops} \\ \text{with } n \text{ molecules}\end{array}\right) - \left(\begin{array}{c}\text{rate of molecule} \\ \text{loss from drops} \\ \text{with } n + 1 \text{ molecules}\end{array}\right)$$

$$= \beta A_n \eta_n - \gamma A_{n+1}\eta_{n+1}$$

Here $\beta = N(KT/2\pi m)^{1/2}$ from Eqs. (2-6) and (2-32a) is the flux of added molecules across a surface, $A_n$ is the surface area of a drop with $n$ molecules, $\eta_n$ is the actual number (different, in general, from the equilibrium number) of drops having $n$ molecules, $\gamma$ is the flux of lost molecules across a surface, and $m$ is the mass of a molecule. Show that at equilibrium ($J = 0$ and $\eta_n = f_n$)

$$\gamma A_{n+1} = \frac{\beta A_n f_n}{f_{n+1}}$$

and use this result to obtain, for near-equilibrium conditions,

$$J = \beta A_n f_n \left( \frac{\eta_n}{f_n} - \frac{\eta_{n+1}}{f_{n+1}} \right)$$

$$\approx -\beta A_n f_n \frac{\Delta(\eta_n/f_n)}{\Delta n}$$

(c) Show that since it is reasonable to expect $\eta/f = 0$ for $n \to \infty$ (there are no really large drops in a nonequilibrium case even though there might be some in an equilibrium case) and $\eta/f = 1$ for $n \to 1$ (equilibrium predictions are accurate except for very large drops), it follows that

$$J = \frac{\beta}{\int_{n=1}^{\infty} \frac{dn}{f_n A_n}}$$

(d) Show that $1/f_n = e^{\Delta G/KT}/F$ has such a sharp maximum in the vicinity of $r^*$ that it is accurate to use the approximation

$$J = \frac{\beta A_n N}{\int_{n=1}^{\infty} e^{\Delta G/KT} dn}$$

where $F = N$ is the total number of fluid molecules considered.

(e) Show that $\Delta G$ from part a is expressible, since $n = 4\pi r^3 \rho_L/3m$, as

$$\Delta G = an^{2/3} - bn$$

where $a = 4\pi\sigma(3m/4\pi\rho_L)^{2/3}$ and $b = 2\sigma m/\rho_L r^*$ and that the series expansion of $\Delta G$ about the equilibrium point (where $r = r^*$) is

$$\Delta G = \frac{bn^*}{2} - a(n^*)^{-4/3} \frac{(n - n^*)^2}{9} + \cdots$$

Use this result to find

$$\int_{n=1}^{\infty} e^{\Delta G/KT} dn = e^{bn^*/2KT} \left( \frac{9\pi KTn^{*4/3}}{a} \right)^{1/2}$$

(f) Substitute the result of part e into the result of part d to obtain Eq. (14-8).

(g) Show that the dimensions of Eq. (14-7) are molecules$/m^3$ s.

(h) Comment on the accuracy of applying kinetic theory to extremely small drops.

**14-9**    Compare the heat-flux–temperature-difference variation for dropwise condensation shown in Fig. 14-5 with that predicted by Eq. (14-9) for steam condensing at atmospheric pressure.

**14-10**  Compare the heat-flux variation with imposed temperature difference for dropwise condensation against that for nucleate boiling.

**14-11**  Compare the magnitude of the heat-transfer coefficient for dropwise condensation with that for nucleate boiling for water at 1 atm of pressure.

**14-12**  The maximum possible condensation heat flux is to be estimated on the basis of the assumption that all vapor molecules arriving at a surface condense and yield their latent heat of vaporization $h_{fg}$ whereas none are reemitted by evaporation. From the kinetic theory of gases developments embodied in Eqs. (2-7) and (2-32a), show that the estimate is

$$\frac{q_{\text{cond, max}}}{A} = h_{fg} P \frac{M}{2\pi RT}$$

where $P$ is the vapor pressure, $T$ is the vapor temperature, $M$ is the vapor molecular weight, and $R$ is the universal gas constant. (Related discussion is given by Sukhatme and Rohsenow [29].)

**14-13**  The local heat-transfer coefficient for laminar condensation on a vertical plate given in terms of temperature difference by Eq. (14-17) is to be cast into the form of Eq. (14-18), involving condensate flow rate rather than temperature difference.

    **(a)** Show that the Reynolds number

$$\text{Re} = \frac{\text{average velocity} \times \text{hydraulic diameter}}{\text{dynamic viscosity}}$$

    is $\text{Re} = 4u_{av}\delta/\mu_L = 4\dot{m}/\rho_L\mu_L$, where $\dot{m}$ is the condensate mass flow rate per unit width.

    **(b)** Introduce into the result of part a the fact that $\dot{m}h_{fg} = \bar{h}x(T_s - T_w)$, since the heat of condensation was absorbed by the plate, and eliminate $T_s - T_w$ by making use of Eq. (14-17) to obtain Eq. (14-18).

**14-14**  A similarity transformation for the problem of laminar film condensation of a stagnant saturated vapor on a vertical plate is to be obtained. The boundary-layer equations for the liquid are given in Eq. (14-10), whereas for the vapor only the continuity and x-motion equations are needed as $\partial u/\partial x + \partial v/\partial y = 0$ and $u\,\partial u/\partial x + v\,\partial u/\partial y = \nu\,\partial^2 u/\partial y^2$

subject to the boundary conditions in Eq. (14-11) and $u_{\text{vapor}}(y \to \infty) \to 0$.

(a) Show that the similarity variable $\eta_L = C_L y / x^{1/4}$ with $C_L^4 = g(\rho_L - \rho_v)/4\nu_L^2\rho_L$ and the stream function $\psi_L = 4\nu_L C_L x^{3/4} F(\eta_L)$ give the liquid velocity components as $u_L = \partial\psi_L/\partial y = 4\nu_L C_L^2 x^{1/2} F'$ and $v_L = -\partial\psi_L/\partial x = \nu_L C_L x^{-1/4}(\eta_L F' - 3F)$, where $F' = dF/d\eta_L$.

(b) Show that the similarity variable $\eta_v = C_v y / x^{1/4}$ with $C_v^4 = g/4\nu_v^2$ and the stream function $\psi_v = 4\nu_v C_v x^{3/4} f(\eta_v)$ give the vapor velocity components as $u_v = \partial\psi_v/\partial y = 4\nu_v C_v^2 x^{1/2} f'$ and $v_v = -\partial\psi_v/\partial x = \nu_v C_v x^{-1/4}(\eta_v f' - 3f)$.

(c) Use the results of parts a and b to show that the ordinary differential equations (which can be solved numerically) constituting the boundary-layer equations are

$$F''' + 3FF'' - 2(F')^2 + 1 = 0 \quad \text{liquid } x \text{ motion}$$
$$\theta'' + 3\text{Pr } F\theta' = 0 \quad \text{liquid energy}$$
$$f''' + 3ff'' - 2(f')^2 = 0 \quad \text{vapor } x \text{ motion}$$

with $\theta = (T - T_s)/(T_w - T_s)$ and subject to the boundary conditions

$$F(0) = 0 = F'(0), \qquad \theta(0) = 1$$

$$f(0) = \left[\frac{(\rho\mu)_L}{(\rho\mu)_v}\right]^{1/2}\left(\frac{\rho_L - \rho_v}{\rho_L}\right)^{1/2} F(\eta_{L\delta})$$

$$f''(0) = \left[\frac{(\rho\mu)_L}{(\rho\mu)_v}\right]^{1/2}\left(\frac{\rho_L - \rho_v}{\rho_L}\right)^{1/2} F''(\eta_{L\delta})$$

$$\theta(\eta_{L\delta}) = 0$$

$$f'(\eta_v \to \infty) \to 0$$

where the unknown quantities to be determined are $F''(0)$, $\theta'(0)$, $f'(0)$, and $\eta_{L\delta} = C_L\delta/x^{1/4}$.

(d) Show that the local Nusselt number is obtainable from the solution to the equations of part c as

$$\text{Nu}_x\left[\frac{C_{p_L}(T_s - T_w)/h_{\text{fg}}}{gC_p(\rho_L - \rho_v)x^3/4\nu_L k_L}\right]^{1/4} = \frac{C_{p_L}(T_s - T_w)}{\text{Pr}_L\, h_{\text{fg}}}[-\theta'(0)]$$

Note that the three physical parameters involved are $\text{Pr}_L$, $(\rho\mu)_L/(\rho\mu)_v$, and $C_{p_L}(T_s - T_w)/h_{\text{fg}}$.

(e) Comment on the agreement between the functional form for $\eta_{L\delta}$ of part a and the result of a simpler integral analysis [Eq. (14-16)].

14-15   A vertical flat plate is exposed to stagnant saturated water vapor at 100°C. The plate is initially at 100°C. Suddenly the plate temperature is permanently changed to 38°C, and laminar film condensation begins. Show that steady condensate rates are achieved after 0.7 s at 15 cm from the top of the plate and after 1 s at 30 cm from the top of the plate. Show that the average heat-transfer coefficient over the entire plate $\bar{h}$ varies with time as

$$\frac{\bar{h}}{\bar{h}_0} = 1 + \frac{3t_s}{5t}, \qquad t \geqslant t_s$$

where $\bar{h}_0$ is the steady state average coefficient for the entire plate and $t_s$ is the transient period for the bottom of the plate.

14-16   The subject problem of Problem 14-15 is to be solved by an integral technique.

(a) Show that the energy equation in integral form is

$$\frac{\partial\left(\delta\int_0^1 \theta\,d\eta\right)}{\partial t} + \frac{\partial\left(\delta\int_0^1 u\theta\,d\eta\right)}{\partial x} = \frac{h_{fg}}{C_p(T_s - T_w)}\frac{\partial(M/\rho)}{\partial z} - \frac{\alpha}{\delta}\frac{\partial\theta(0)}{\partial\eta}$$

where $M$ is the rate of condensation per unit width of plate and $\theta = (T - T_s)/(T_w - T_s)$, whereas the conservation of mass gives

$$\frac{\partial\left[\delta\int_0^1 u\,d\eta\right]}{\partial x} + \frac{\partial\delta}{\partial t} = \frac{\partial(M/\rho)}{\partial z}$$

(b) Assume the velocity and temperature profiles to be the steady-state relations

$$u = \frac{(\rho_L - \rho)g\delta^2}{\mu}\left(\eta - \frac{\eta^2}{2}\right)$$

$$\theta = 1 - \eta$$

(c) Use the profiles of part b to find, from the energy equation of part a, that

$$\frac{\partial(M/\rho)}{\partial x} = \frac{C_p(T_s - T_w)}{h_{fg}}\left[\frac{\alpha}{\delta} - \frac{3(\rho_L - \rho_v)g\delta^2}{8\mu}\delta^2\frac{\partial\delta}{\partial x} - \frac{1}{2}\frac{\partial\delta}{\partial t}\right]$$

(d) Use the velocity profile of part b in the continuity equation of part a to obtain an expression for $\partial(M/\rho)/\partial x$. Combine this result with that of part c to find

$$P\frac{\partial \delta}{\partial x} + Q\frac{\partial \delta}{\partial t} = R$$

where $P = [\rho_L(\rho_L - \rho_v)gh'_{fg}/k\mu(T_s - T_w)]\delta^3$, $Q = [\rho h'_{fg}(1 + C_p$ $(T_s - T_w)/8h'_{fg})/k(T_s - T_w)]\delta$, $R = 1$, and $h'_{fg} = h_{fg} + 3C_p$ $(T_s - T_w)/8$.

(e) The method of characteristics* requires

$$\underbrace{\frac{dx}{P}}_{1} = \underbrace{\frac{dt}{Q}}_{2} = \underbrace{\frac{d\delta}{R}}_{3}$$

subject to $\delta(x = 0, t) = 0 = \delta(x, t = 0)$. This result follows from a geometric interpretation of the partial differential equation of part d. Since $\partial \delta/\partial x$, $\partial \delta/\partial t$, and $-1$ are direction numbers of the normal $N$ to the surface $\delta(x, t)$, the equation of part d states that $N$ is perpendicular to a line $L$ through the same surface and with direction numbers $P$, $Q$, and $R$. Now let the plane containing $N$ and $L$ cut the surface in the curve $C$ that has direction numbers $dx$, $dt$, and $d\delta$. Since $C$ and $L$ have the same direction, the two sets of direction numbers are proportional as stated previously. If groups 1 and 2 are equated, the characteristic line separating the steady-state and transient regions in the $x - t$ plane is

$$\frac{dt}{dx} = \frac{Q}{P} = \frac{\mu(1 + C_p(T_s - T_w)/8h'_{fg}}{g(\rho_L - \rho_v)\delta^2}$$

(f) On each characteristic, if groups 1 and 3 are equated, it follows that

$$\delta^3 \, d\delta = \frac{k\mu(T_s - T_w)}{\rho_L(\rho_L - \rho_v)gh'_{fg}} dx$$

and if groups 2 and 3 are equated, then

$$\delta \, d\delta = \frac{k(T_s - T_w)}{\rho h'_{fg}(1 + C_p(T_s - T_w)/8h'_{fg})} dt$$

Combine these relationships with the result of part e to obtain Eq. (14-20) and relationships for $\delta(x, t)$ and $h(x, t)$ (see Fig. 14P-16).

---

*F. B. Hildebrand, *Advanced Calculus For Engineers*, Prentice-Hall, Englewood Cliffs, NJ, 1949; F. H. Miller, *Partial Differential Equations*, Wiley, New York, 1958.

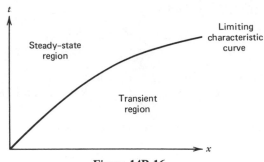

**Figure 14P-16**

**14-17** Determine the time required to achieve steady condensation on a horizontal tube of 4 cm diameter and 50°C below the temperature of the saturated steam to which it is suddenly exposed. (*Answer: t ≈* 0.5 s.)

**14-18** Saturated steam at 54°C condenses on the outside surface of a 3.7-m-long vertical tube of 2.5 cm o.d. whose outerwall temperature is held constant at 43°C. For film condensation, the total condensation rate on the tube is to be determined.

(a) Assuming a laminar condensate film (to be checked later), show that the heat-transfer coefficient for the entire tube is $\bar{h} = 4600$ W/m² K from Eq. (14-17) with a coefficient of 1.13.

(b) Show that the condensate flow at the bottom of the tube is $\dot{m} = \pi DL\bar{h}(T_s - T_w)/h_{fg} = 0.0062$ kg/s.

(c) Show that Re $= 4\dot{m}/\mu_L = 560 < 1400$ to confirm the original laminar flow assumption.

(d) Show that the condensate film thickness at the bottom of the tube is $\delta = 0.019$ cm (appeal to Eq. (14-16) if necessary, or realize $h = k/\delta$). Comment on the applicability of flat-plate relations to vertical tubes at this value of $\delta/D$.

(e) On the basis of the fact that $\rho \bar{u} \delta = \dot{m}$, show that the average velocity is $\bar{u} = 0.42$ m/s.

**14-19** Saturated steam at 54°C condenses on the outside surface of a vertical tube of 2.5 cm o.d. whose outer wall temperature is held at 43°C. For film condensation the tube length required to have a condensate flow of 0.0032 kg/s from the bottom of the tube is to be determined.

(a) First, show that laminar conditions exist since Re $= 4\dot{m}/\mu_L = 280 < 1400$.

(b) Show that, with the use of Eq. (14-18) with a coefficient of 1.88, the average heat-transfer coefficient is $\bar{h} = 5400$ W/m² K.

(c) Realizing that $\dot{m} = \pi DL\bar{h}(T_s - T_w)/h_{fg}$, show that the required tube length is $L = 1.6$ m.

(d) If the Reynolds number had been less than 33, would it have been necessary to modify the coefficient of Eq. (14-18) to 1.88 from 1.47?

14-20 A square vertical plate 1 ft on a side and at 208°F contacts saturated steam at 212°F.

(a) Show from Eq. (14-17) with a coefficient of 1.13 that the average heat-transfer coefficient is $\bar{h} = 2700$ Btu/hr ft$^2$ °F if condensate flow is laminar.

(b) Noting that the condensate flow rate $\dot{m}$ from the bottom of the plate is related to the heat-transfer coefficient by $\dot{m} = \bar{h}A(T_s - T_w)/h_{fg}$, show that $\dot{m} = 11$ lb$_m$/hr.

(c) Show that the flow in the condensate film is laminar since Re $= 4\dot{m}/\mu_L = 64 < 1400$.

14-21 Consider laminar film condensation of a saturated and stagnant vapor on the inside of a vertical circular tube of uniform wall temperature, neglecting vapor drag at the interface. Estimate the distance from the top of the tube at which liquid fills the tube.

14-22 A shallow pan is exposed to stagnant saturated steam at 100°C. The sides of the pan are insulated and the bottom is kept at 90°C. Estimate the rate of condensate accumulation in the pan.

14-23 A saturated stagnant vapor condenses on the outside of a constant-temperature cone, with its apex pointing vertically up. Derive the differential equation relating local condensate thickness to temperature difference, fluid properties, and distance from cone apex. Solve the differential equation if possible (reference 47 may be helpful).

14-24 Stagnant saturated steam at 100°C condenses on a vertical tube whose temperature is 65°C. At what tube length would turbulent condensate flow occur?

14-25 Saturated steam at 54°C condenses on the outside of a 3.7-m-long vertical tube of 2.5 cm o.d. whose outer wall temperature is held at 43°C. For film condensation, the total condensation rate on the tube is to be determined.

(a) Show from Eq. (14-22) that the Reynolds number for the condensate is Re $= 4\dot{m}/\mu_L = 5824 > 1400$ and thus turbulent conditions exist. Show that turbulence begins about 14 m from the top of the tube.

(b) From the result of part a show that $\dot{m} = 159$ kg/s.

(c) On the basis of Eq. (14-23), show that the average turbulent heat-transfer coefficient is $\bar{h} = 4743$ W/m$^2$ K.

(d) Compare the result of part c with the result of part a of Problem 14-18 for laminar condensation—are the two $\bar{h}$ values approximately equal?

**14-26** A simple determination of the heat-transfer coefficient for laminar film condensation of a stagnant vapor on a vertical plate is to be made. Referring to Fig. 14-9, an energy balance over the film gives

$$\frac{d\left\{\int_0^\delta \rho_L u\left[C_{p_L}(T - T_s) + h_{fg}\right] dy\right\}}{dx} = \frac{k_L(T_s - T_w)}{\delta}$$

(a) Use linear temperature and parabolic velocity distributions for the liquid [Eqs. (14-15a) and (15-15c)] in this differential equation to find

$$\delta^4 = \frac{4\nu_L k_L(T_w - T_s)x/gh_{fg}(\rho_L - \rho)}{1 + 3C_{p_L}(T_w - T_s)/8h_{fg}}$$

which is nearly identical to Eq. (14-16).

(b) Proceed with the assumption of a linear temperature distribution in the liquid film to find the local heat-transfer coefficient as

$$h = \frac{k}{\delta} = 2^{-1/2}k\left[\frac{g(\rho_L - \rho)\,\text{Pr}}{x\rho_L \nu_L^2}\frac{h_{fg}}{C_{p_L}(T_w - T_s)}\right]^{1/4} K$$

where $K = [1 + 3C_{p_L}(T_w - T_s)/8h_{fg}]^{1/4}$, in close agreement with Eq. (14-17a).

**14-27** The condensate film thickness is to be estimated at the top of a horizontal tube on which stagnant saturated vapor condenses.

(a) Show that near the top of the tube ($\theta \approx 0$), $\dot{m} = C^{3/4}\theta$, where

$$C = \frac{Rk(T_w - T_s)}{h_{fg}}\left[\frac{g(\rho_L - \rho)}{3\nu_L}\right]^{1/3}.$$

(b) Utilize the result of part a and Eq. (14-24) to finally obtain

$$\frac{\delta(\theta = 0)}{D} = \frac{1.11}{\left\{\text{Ra}\left[h_{fg}/C_p(T_s - T_w)\right]\right\}^{1/4}}$$

**14-28** Explain briefly how to estimate the condensation rate on the underside of a cool horizontal plate exposed to a stagnant saturated vapor. *Hint:* Utilize relations for film boiling from the topside of a hot horizontal plate (references 33 and 34 may also be helpful).

**14-29** For horizontal condenser tubes arranged vertically as sketched in Fig. 14-23, the lower tubes have a condensation rate that is substantially

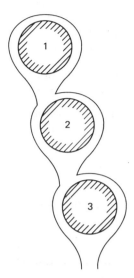

**Figure 14P-29**

diminished below that of the upper tubes. A staggered scheme designed to circumvent this diminution has been proposed as sketched in Fig. 14P-29. All the condensate from an overlying tube passes onto the left side of the tube below, leaving the right side bare of condensate for rapid condensation.

**(a)** Show that in the limit as the number of tubes becomes large and condensation of the left side makes a negligible contribution, the average heat-transfer coefficient is half that of a single horizontal tube.

**(b)** Show that the total condensate flow rate from each tube divided by the condensate flow rate from the topmost tube varies as 3.28, 4.77, and 5.21 for the second, third, and fourth tubes, respectively.

**14-30** A horizontal 2-in.-o.-d. tube is surrounded by saturated steam at 2 psia. The tube is maintained at 90°F. Show that the average heat-transfer coefficient for this condensation process is $\bar{h} = 1246$ Btu/hr ft² °F and that the condensate film flows in a laminar fashion.

**14-31** What length : diameter ratio will provide the same condensation rate in both horizontal and vertical orientations, assuming that condensate flow is laminar, for a tube?

**14-32** Compare the average heat-transfer coefficient for condensation of a saturated stagnant vapor on a 0.61-m-long vertical surface with that for condensation on twenty-four 2.5-cm-o.d. horizontal tubes arranged in a vertical tier.

**14-33** Compare the laminar film condensation rate on a horizontal tube with the maximum possible condensation rate that would result if all vapor molecules incident on a surface were condensed. Take the conditions to be those of Problem 14-30. (*Answer:* $\dot{m}_{max}/\dot{m}_{actual} \approx 150$.)

**14-34** Estimate the angular velocity at which the horizontal tube described in Problem 14-30 must spin about its own axis in order to double the condensation rate. (Information from reference 78 may be needed.)

**14-35** Estimate the angular velocity at which the vertical tube specified in Problem 14-25 must spin about its own axis in order to double the condensation rate.

**14-36** Estimate the condensation rate for saturated steam at 54°C condensing on the outside of a 3.7-m-long vertical tube of 2.5 cm o.d. whose outer wall temperature is 43°C and that spins about an axis perpendicular to the tube length and located at one end of the tube. The angular velocity is 120 rpm.

**14-37** Laminar film condensation occurs on a vertical flat plate fin. The physical situation is sketched in Fig. 14P-37. Fin temperature $T$ varies only along the fin length, with the value $T_w$ at the root. The vapor is saturated at $T_s$ and is stagnant. The fin tip is insulated.

**(a)** Assume that the local condensate film velocity and temperature distributions are those given in Section 14.4 for a vertical plate of constant temperature, and show that the condensate film thickness $\delta$ variation is described by

$$\frac{d\Delta^4}{dz} = 4B\theta, \qquad 0 \leqslant z \leqslant 1 \tag{37P-1}$$

$$\Delta(0) = 0 \tag{37P-2}$$

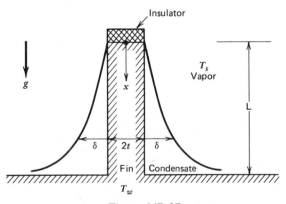

**Figure 14P-37**

Prepare an energy balance on a differential-sized control volume of the fin and show that the fin temperature variation is described by

$$\frac{d^2\theta}{dz^2} = \frac{A\theta}{\Delta}, \qquad 0 \leqslant z \leqslant 1 \qquad \text{(37P-3)}$$

$$\frac{d\theta(0)}{dz} = 0 \qquad \text{(37P-4)}$$

$$\theta(1) = 1 \qquad \text{(37P-5)}$$

where $A = k_L L / k_f t$, $B = \mu_L k_L (T_s - T_w)/h_{fg} \rho_L^2 g L^3$, $z = x/L$, $\Delta = \delta/L$, and $\theta = (T - T_s)/(T_w - T_s)$.

(b) Obtain a single equation for fin temperature by solving Eq. (37P-3) for $\delta$ and substituting the result into Eq. (37P-1) to obtain

$$\frac{d(\theta/\theta'')^4}{dz} = \frac{4B}{A^4}\theta$$

The primes denote differentiation with respect to $z$.

(c) Show that the result of part b can be rewritten as

$$\left(\frac{\theta}{\theta''}\right)^2 \frac{d(\theta/\theta'')}{dz} = \frac{B}{A^4}\frac{d\theta'}{dz}$$

and integrated to obtain

$$\left(\frac{\theta}{\theta''}\right)^3 = \frac{3B}{A^4}(\theta' + C_1)$$

(d) Note that $\theta/\theta'' = \Delta/A$ from Eq. (37P-3), and consider Eqs. (37P-2) and (37P-4) to show that $C_1 = 0$.

(e) Rearrange the result of part c into

$$(\theta')^{1/3} \frac{d\theta'}{dz} = \left(\frac{A^4}{3B}\right)^{1/3} \theta$$

Substitute $\theta' \, d\theta'/d\theta$ for $d\theta'/dz$ and integrate to obtain

$$\theta' = \left(\frac{7}{6}\right)^{3/7}\left(\frac{A^4}{3B}\right)^{1/7}(\theta^2 - \theta_0^2)^{3/7}$$

where $\theta_0 = \theta(z = 0)$ is to be determined later.

(f) Employ the transformation $\theta = \theta_0 \cosh(y)$ in the result of part e followed by integration from $z = 0$ to $z = 1$ to obtain

$$\theta_0^{1/7} \int_0^{\text{arc cosh} \frac{1}{\theta_0}} \sinh^{1/7}(y)\, dy = \left(\frac{7}{6}\right)^{3/7} \left(\frac{A^4}{3B}\right)^{1/7}.$$

(g) Since accurate determination of $\theta_0$ is required only for $\theta_0 \approx 1$, when $\sinh(y) \approx y$, show that the result of part f yields

$$\theta_0 = \frac{1}{\cosh ML}$$

where $ML = (16\sqrt{2}/21)^{1/2}(A^4/B)^{1/8}$.

(h) Show that the heat flow rate per unit length at the fin root is given for one side of the fin by

$$Q = \left(\frac{7}{6}\right)^{3/7} k_L(T_s - T_w) A^{-1} \left(\frac{A^4}{B}\right)^{1/7} \tanh^{6/7} ML$$

whereas if the fin were everywhere at the root temperature, the maximum heat flow rate would be

$$Q_{\max} = \frac{2^{3/2}}{3} \frac{k_L(T_s - T_w)}{B^{1/4}}$$

(i) From the results of part h show that the fin efficiency is given by

$$\eta = \frac{Q}{Q_{\max}} = \left[\frac{\tanh(ML)}{ML}\right]^{6/7}$$

which is within 2% of the exact answer determined by a numerical method (W. K. Nader, Extended surface heat transfer with condensation, in *Proceedings of the Sixth International Heat Transfer Conference, Toronto*, August 1978, pp. 407–412, Paper CS-5). Note that, except for the $\frac{6}{7}$ exponent caused by variable condensate thickness, this expression for fin efficiency is of the same form as that for a constant heat-transfer coefficient. Lienhard and Dhir [130] can be consulted for a variety of cases involving condensation on fins. Burmeister [131] gives more discussion.

(j) Compare the condensation rate of stagnant saturated steam at 20°C on a steel fin ($t = 0.32$ cm, $L = 2.5$ cm, and $T_s - T_w = 6$ K) with the condensation rate if the fin were uniformly maintained at the root temperature. Estimate the accuracy of neglecting temperature variation across the fin width by evaluating the Biot number $tk_L/\delta k_f$ at the fin tip and root. (*Answer:* $\eta = 0.3$.)

**14-38** Consult reference 72 and execute a solution by the integral method of the problem of laminar film condensation on a rotating disk.

**14-39** Consult reference 81 and execute a solution by the integral method of the problem of laminar film condensation on a flat plate with a linearly varying body force.

**14-40** Determine the condensation rate for the conditions stated in Problem 14-20 if air is present in the bulk mixture as a mass fraction of $\omega_\infty = 0.02$. Express it as a fraction of the condensation rate in the absence of a noncondensable gas.

# REFERENCES

1   F. L. Curzon and B. Ahlborn, Efficiency of a Carnot engine at maximum power output, *Am. J. Phys.* **43**, 22–24 (1975).

2   J. W. Meyer, Kinetic model for aerosol formation in rocket contrails, *AIAA J.* **17**, 135–144 (1979).

3   H. Merte, Condensation heat transfer, *Adv. Heat Transf.* **9**, 181–272 (1973).

4   H. J. Palmer, The hydrodynamic stability of rapidly evaporating liquids at reduced pressure, *J. Fluid Mech.* **75**, 487–511 (1976).

5   E. A. Guggenheim, The thermodynamics of interfaces in systems of several components, *Transact. Faraday Soc.* **36**, 397–412 (1940).

6   R. C. Tolman, The effect of droplet size on surface tension, *J. Chem. Phys.* **17**, 333–337 (1949); J. G. Kirkwood and F. P. Buff, The statistical mechanical theory of surface tension, *J. Chem. Phys.* **17**, 338–343 (1949).

7   L. E. Scriven and C. V. Sternling, The Marangoni effects, *Nature* **187**, 186–188 (1960).

8   D. B. R. Kenning, Two-phase flow with nonuniform surface tension, *Appl. Mech. Rev.* **21**, 1101–1111 (1968).

9   J. M. Papazian, Bubble behavior during solidification in low gravity, *AIAA J.* **17**, 1111–1117 (1979); J. M. Papazian and R. L. Kosson, Effect of internal heat transport on the thermal migration of bubbles, *AIAA J.* **17**, 1279–1280 (1979).

10  S. Ostrach and A. Pradhan, Surface-tension induced convection at reduced gravity, *AIAA J.* **16**, 419–424 (1978).

11  J. Frenkel, *Kinetic Theory Of Liquids*, Clarendon Press, Oxford, 1946, pp. 366–374.

12  G. L. Goglia and G. J. Van Wylen, Experimental determination of supersaturation of nitrogen vapor expanding in a nozzle, *Transact. ASME, J. Heat Transf.* **83**, 27–32 (1961).

13  C. T. R. Wilson, *Proc. Roy. Soc. Lond.* **61**, 240–243 (1897).

14  E. L. Pinnes and W. K. Mueller, Homogeneous vapor nucleation and superheat limits of liquid mixtures, *Transact. ASME, J. Heat Transf.* **101**, 617–621 (1979).

15  P. G. Hill, H. Witting, and E. P. Demetri, Condensation of metal vapors during rapid expansion, *Transact. ASME, J. Heat Transf.* **85**, 303–317 (1963).

16  M. Volmer and H. Flood, Tröpfchenbildung in Dämpfen, *Zeitschrift für Physikalische Chemie, Abteilung A* **170**, 273–285 (1934).

17  H. L. Jaeger, E. D. Wilson, P. G. Hill, and K. C. Russell, Nucleation of supersaturated vapors in nozzles. I. $H_2O$ and $NH_3$, *J. Chem. Phys.* **51**, 5380–5388 (1969).

18  J. Lothe and G. M. Pound, On the statistical mechanics of nucleation theory, *J. Chem. Phys.* **45**, 630–634 (1966).

19  J. Feder, K. C. Russell, J. Lothe, and G. M. Pound, Homogeneous nucleation and growth of droplets in vapours, *Adv. Phys.* **15**, 111–178 (1966).

20  J. F. Welch and J. W. Westwater, Microscopic study of dropwise condensation, *International Developments in Heat Transfer, Proceedings of the International Heat Transfer Conference*, University of Colorado, 1961, Part II, 1961, pp. 302–309.

21  H. Tanaka, Further developments of dropwise condensation theory, *Transact. ASME, J. Heat Transf.* **101**, 603–611 (1979).

22  C. Bonacina, S. Del Giudice, and G. Comini, Dropwise evaporation, *Transact. ASME, J. Heat Transf.* **101**, 441–452 (1979).

23  E. G. Shafrin and W. A. Zisman, Constitutive relations in the wetting of low energy surfaces and the theory of the retraction method of preparing monolayers, *J. Phys. Chem.* **64**, 519–524 (1960).

24  A. R. Brown and M. A. Thomas, Filmwise and dropwise condensation of steam at low pressures, in *Proceedings of the Third International Heat Transfer Conference, Chicago, Vol.* 2, 1966, *pp.* 300–305.

25  R. A. Erb and E. Thelen, Promoting permanent dropwise condensation, *Indust. Eng. Chem.* **57**, 49–52 (1965).

26  H. Hampson and N. Ozisik, An investigation into the condensation of steam, *Proc. Inst. Mech. Eng., Lond.* **1B**, 282–294 (1952); L. C. F. Blackman and M. J. S. Dewar, Promoters for dropwise condensation of steam, Pts. I–IV, *J. Chem. Soc.*, 162–176 (January–March 1957); L. C. F. Blackman, M. J. S. Dewar, and H. Hampson, An investigation of compounds promoting dropwise condensation of steam, *J. Appl. Chem.* **7**, 160–171 (1957); B. D. J. Osment, D. Tudor, R. M. M. Speirs, and W. Rugman, Promoters for the dropwise condensation of steam, *Transact. Inst. Chem. Eng.* **40**, 152–160 (1962); E. G. Lefevre and J. Rose, An experimental study of heat transfer by dropwise condensation, *Internatl. J. Heat Mass Transf.* **8**, 1117–1133 (1965); J. L. McCormick and J. W. Westwater, Nucleation sites for dropwise condensation, *Chem. Eng. Sci.* **20**, 1021–1036 (1965).

27  E. Citakoglu and J. W. Rose, Dropwise condensation—some factors influencing the validity of heat transfer measurements, *Internatl. J. Heat Mass Transf.* **11**, 523–537 (1968).

28  A. C. Peterson and J. W. Westwater, Dropwise condensation of ethylene glycol, *Chem. Eng. Progr., Symp. Ser.* **62**, 135–142 (1966).

29  S. P. Sukhatme and W. M. Rohsenow, Heat transfer during film condensation of a liquid metal vapor, *Transact. ASME, J. Heat Transf.* **88**, 19–28 (1966).

30  W. Nusselt, Die Oberflachenkondensation des Wasser dampfes, *Zeitschrift des Vereins deutscher Inginuere* **60**, 541–575 (1916).

31  J. C. Y. Koh, An integral treatment of two-phase boundary layer in film condensation, *Transact. ASME, J. Heat Transf.* **83**, 359–362 (1961).

32  W. M. Rohsenow, Heat transfer and temperature distribution in laminar-film condensation, *Transact. ASME* **78**, 1645–1648 (1956).

33  J. Gerstmann and P. Griffith, Laminar film condensation on the underside of horizontal and inclined surfaces, *Internatl. J. Heat Mass Transf.* **10**, 567–580 (1967).

34  G. Leppert and B. Nimmo, Laminar film condensation on surfaces normal to body or inertial forces, *Transact. ASME, J. Heat Transf.* **90**, 178–179 (1968).

35  D. L. Spencer and W. E. Ibele, Laminar film condensation of a saturated and superheated vapor on a surface with a controlled temperature distribution, in *Proceedings of the Third International Heat Transfer Conference, Chicago*, Vol. 2, 1966, pp. 337–347.

36  W. H. McAdams, *Heat Transmission*, 3rd ed., McGraw-Hill, New York, 1954.

37  G. Selin, Heat transfer by condensing pure vapours outside inclined tubes, *International Developments in Heat Transfer, Proceedings of the International Heat Transfer Conference, University of Colorado, 1961*, Part II, 1961, pp. 279–289.

38  C. L. Henderson and J. M. Marchello, Role of surface tension and tube diameter in film condensation on horizontal tubes, *AIChE J.* **13**, 613–614 (1967).

39  J. C. Y. Koh, E. M. Sparrow, and J. P. Hartnett, The two phase boundary layer in laminar film condensation, *Internatl. J. Heat Mass Transf.* **2**, 69–82 (1961).

40  E. M. Sparrow and J. L. Gregg, A boundary-layer treatment of laminar film condensation, *Transact. ASME, J. Heat Transf.* **81**, 13–23 (1959).

41  M. M. Chen, An analytical study of laminar film condensation: Part I—Flat plates, *Transact. ASME, J. Heat Transf.* **83**, 48–54 (1961).

42  W. M. Rohsenow, Film condensation of liquid metals, *Progress In Heat and Mass Transfer*, Vol. 7, Pergamon Press, New York, 1973.

43  D. G. Kroger and W. M. Rohsenow, Film condensation of saturated potassium vapor, *Internatl. J. Heat Mass Transf.* **10**, 1891–1894 (1967).

44  G. Poots and R. G. Miles, Effect of variable physical properties on laminar film condensation of saturated vapor on a vertical flat plate, *Internatl. J. Heat Mass Transf.* **10**, 1677–1692 (1967).

45  R. L. Lott and J. D. Parker, The effect of temperature-dependent viscosity on laminar condensation heat transfer, *Transact. ASME, J. Heat Transf.* **95**, 267–268 (1973).

46  P. L. Kapitsa, Wave flow of thin layers of a viscous fluid, *Zh. Eksp. Teoret. Fiz.* **18**, 1 (1948).

47  S. D. R. Wilson, Unsteady and two-dimensional flow of a condensate film, *Transact. ASME, J. Heat Transf.* **98**, 313–315 (1976).

48  W. J. Minkowycz and E. M. Sparrow, The effect of superheating on condensation heat transfer in a forced convection boundary layer flow, *Internatl. J. Heat Mass Transf.* **12**, 147–154 (1969).

49  W. J. Minkowycz and E. M. Sparrow, Condensation heat transfer in presence of noncondensables, interfacial resistance, superheating, variable properties, and diffusion, *Internatl. J. Heat Mass Transf.* **9**, 1125–1144 (1966).

50  W. M. Rohsenow, J. H. Webber, and A. T. Ling, Effect of vapor velocity on laminar and turbulent-film condensation, *Transact. ASME* **78**, 1637–1644 (1956).

51  H. R. Jacobs, An integral treatment of combined body force and forced convection in laminar film condensation, *Internatl. J. Heat Mass Transf.* **9**, 637–648 (1966).

52  H. R. Jacobs, Combined body force with forced convection in laminar film condensation heat transfer of Freon 113, Ph.D. dissertation, Department of Mechanical Engineering, Ohio State University, 1965.

53  E. M. Sparrow, W. J. Minkowycz, and M. Saddy, Forced convection condensation in the presence of noncondensables and interfacial resistance, *Internatl. J. Heat Mass Transf.* **10**, 1829–1845 (1967).

54  R. D. Cess, Laminar-film condensation on a flat plate in the absence of a body force, *Zeitschrift für angewandte Mathematik und Physik* **11**, 426–433 (1960).

55  J. C. Y. Koh, Film condensation in a forced-convection boundary layer flow, *Internatl. J. Heat Mass Transf.* **5**, 941–954 (1962).

56  S. V. Patankar and E. M. Sparrow, Condensation on an extended surface, *Transact. ASME, J. Heat Transf.* **101**, 434–440 (1979); J. E. Wilkins, Condensation on an extended surface, *Transact. ASME, J. Heat Transf.* **102**, 186–187 (1980).

57  H. Grober, S. Erk, and U. Grigull, *Fundamentals Of Heat Transfer*, McGraw-Hill, New York, 1961.

58  A. E. Dukler, Fluid mechanics and heat transfer in vertical falling-film systems, *Chem. Eng. Progr. Symp. Ser.* **56** (30), 1–10 (1960).

59  J. Lee, Turbulent condensation, *AIChE J.* **10**, 540–544 (1964).

60  M. Jakob, *Heat Transfer*, Vol. 1, Wiley, New York, 1949.

61  M. M. Chen, An analytical study of laminar film condensation: Part 2—Single and multiple horizontal tubes, *Transact. ASME, J. Heat Transf.* **83**, 55–60 (1961).

62  V. Dhir and J. Lienhard, Laminar film condensation on plane and axisymmetric bodies in nonuniform gravity, *Transact. ASME, J. Heat Transf.* **93**, 97–100 (1971).

63  V. K. Dhir and J. H. Lienhard, Similar solutions for film condensation with variable gravity or body shape, *Transact. ASME, J. Heat Transf.* **95**, 483–486 (1973).

64  J. W. Yang, Laminar film condensation on a sphere, *Transact. ASME, J. Heat Transf.* **95**, 174–178 (1973).

65  A. G. Sheynkman and V. N. Linetskiy, Hydrodynamics and heat transfer by film condensation of stationary steam on an inclined tube, *Heat Transf.—Sov. Res.* **1**, 90–97 (1969).

66  V. E. Denny and A. F. Mills, Laminar film condensation of a flowing vapor on a horizontal cylinder at normal gravity, *Transact. ASME, J. Heat Transf.* **91**, 495–510 (1969).

67  H. F. Rosson and J. A. Myers, Point values of condensing film coefficients inside a horizontal pipe, *Chem. Eng. Progr. Symp. Ser.* **61** (59), 190–199 (1965).

68  W. W. Akers and H. F. Rosson, Condensation inside a horizontal tube, *Chem. Eng. Progr. Symp. Ser.* **56** (30), 145–154 (1960).

69  F. G. Carpenter and A. P. Colburn, The effect of vapor velocity on condensation inside tubes, in *Proceedings of the General Discussion of Heat Transfer, The Institute of Mechanical Engineers and the ASME*, July 1951, pp. 20–26.

70  M. Soliman, J. R. Schuster, and P. J. Berenson, A general heat transfer correlation for annular flow condensation, *Transact. ASME, J. Heat Transf.* **90**, 267–276 (1968).

71  R. W. Lockhart and R. C. Martinelli, Proposed correlation of data for isothermal two-phase two-component flow in pipes, *Chem. Eng. Progr.* **45**, 39–48 (1949).

72  E. M. Sparrow and J. L. Gregg, A theory of rotating condensation, *Transact. ASME* **81**, 113–120 (1959).

73  E. M. Sparrow and J. L. Gregg, The effect of vapor drag on rotating condensation, *Transact. ASME* **82**, 71–72 (1960).

74  S. S. Nadapurkar and K. O. Beatty, Condensation on a horizontal rotating disc, *Chem. Eng. Progr. Symp. Ser.* **56**, 129–137 (1960).

75  E. M. Sparrow and J. P. Hartnett, Condensation on a rotating cone, *Transact. ASME, J. Heat Transf.* **83**, 101–102 (1961).

76  L. A. Bromley, Prediction of performance characteristics of Hickman–Badger centrifugal boiler compression still, *Indus. Eng. Chem.* **50**, 233–236 (1958).

77  P. J. Marto, Laminar film condensation on the inside of slender, rotating truncated cones, *Transact. ASME, J. Heat Transf.* **95**, 270–272 (1973).

78  A. A. Nicol and M. Gacesa, Condensation of steam on a rotating vertical cylinder, *Transact. ASME, J. Heat Transf.* **92**, 144–152 (1970).

79  S. Mochizuki and T. Shiratori, Condensation heat transfer within a circular tube under centrifugal acceleration field, *Transact. ASME, J. Heat Transf.* **102**, 158–162 (1980).

80  N. V. Suryanarayana, Condensation heat transfer under high gravity condition, *Proceedings of the 5th International Heat Transfer Conference*, Vol. III, 1974, pp. 279–285.

81  J. C. Chato, Condensation in a variable acceleration field and the condensing thermosyphon, *Transact. ASME, J. Eng. Power* **87**, 355–360 (1965).

82  T. B. Jones, Electrodynamically enhanced heat transfer in liquids—a review, *Adv. Heat Transf.* **14**, 107–148 (1978).

83   H. R. Velkoff and J. H. Miller, Condensation of vapor on a vertical plate with a transverse electrostatic field, *Transact. ASME, J. Heat Transf.* **87**, 197–201 (1965).

84   H. Y. Choi, Electrohydrodynamic condensation heat transfer, *Transact. ASME, J. Heat Transf.* **90**, 98–102 (1968).

85   R. E. Holmes and A. J. Chapman, Condensation of Freon-114 in the presence of a strong nonuniform alternating electric field, *Transact. ASME, J. Heat Transf.* **92**, 616–620 (1970).

86   A. K. Seth and L. Lee, The effect of an electric field in the presence of noncondensible gas on film condensation heat transfer, *Transact. ASME, J. Heat Transf.* **94**, 237–260 (1974).

87   K. C. Jain and S. G. Bankoff, Laminar film condensation on a porous vertical wall with uniform suction velocity, *Transact. ASME, J. Heat Transf.* **86**, 481–489 (1964).

88   J. W. Yang, Effect of uniform suction on laminar film condensation on a porous vertical wall, *Transact. ASME, J. Heat Transf.* **92**, 252–256 (1970).

89   J. Lienhard and V. Dhir, A simple analysis of laminar film condensation with suction, *Transact. ASME, J. Heat Transf.* **94**, 334–336 (1972).

90   N. A. Frankel and S. G. Bankoff, Laminar film condensation on a porous horizontal tube with uniform suction velocity, *Transact. ASME, J. Heat Transf.* **87**, 95–102 (1965).

91   J. C. Dent, The calculation of heat transfer coefficient for condensation of steam on a vibrating vertical tube, *Internatl. J. Heat Mass Transf.* **12**, 991–996 (1969).

92   P. Dunn and D. A. Reay, *Heat Pipes*, 2nd ed., Pergamon Press, New York, 1978.

93   S. W. Chi, *Heat Pipe Theory And Practice*, McGraw-Hill, New York, 1976.

94   D. F. Othmer, The condensation of steam, *Indust. Eng. Chem.* **21**, 576–583 (1929).

95   C. L. Henderson and J. M. Marchello, Film condensation in the presence of a noncondensable gas, *Transact. ASME, J. Heat Transf.* **91**, 447–450 (1969).

96   A. P. Colburn and T. B. Drew, The condensation of mixed vapors, *Transact. AIChE* **33**, 197–208 (1937).

97   D. Q. Kern, *Process Heat Transfer*, McGraw-Hill, New York, 1950.

98   E. M. Sparrow and S. H. Lin, Condensation heat transfer in the presence of a noncondensable gas, *Transact. ASME, J. Heat Transf.* **86**, 430–436 (1964).

99   J. W. Rose, Condensation of a vapour in the presence of a non-condensing gas, *Internatl. J. Heat Mass Transf.* **12**, 233–237 (1969).

100  H. K. Al-Diwany and J. Rose, Free convection film condensation of steam in the presence of non condensable gases, *Internatl. J. Heat Mass Transf.* **16**, 1359–1369 (1973).

101  Y. Mori and K. Hijikata, Free convective condensation heat transfer with non condensable gas on a vertical surface, *Internatl. J. Heat Mass Transf.* **16**, 2229–2240 (1973).

102  V. E. Denny, A. F. Mills, and V. J. Jusionis, Laminar film condensation from a steam–air mixture undergoing forced flow down a vertical surface, *Transact. ASME, J. Heat Transf.* **93**, 297–304 (1971).

103  V. E. Denny and V. J. Jusionis, Effect of noncondensable gas and forced flow on laminar film condensation, *Internatl. J. Heat Mass Transf.* **15**, 315–326 (1972).

104  E. Citakoglu and J. W. Rose, Dropwise condensation—some factors influencing the validity of heat-transfer measurements, *Internatl. J. Heat Mass Transf.* **11**, 523–537 (1968).

105  R. H. Turner, A. F. Mills, and V. E. Denny, The effect of noncondensable gas on laminar film condensation of liquid metals, *Transact. ASME, J. Heat Transf.* **95**, 6–11 (1973).

106  E. M. Sparrow and E. Marschall, Binary, gravity-flow film condensation, *Transact. ASME, J. Heat Transf.* **91**, 205–211 (1969).

107  E. Marschall and R. S. Hickman, Laminar gravity-flow film condensation of binary vapor mixtures of immiscible liquids, *Transact. ASME, J. Heat Transf.* **95**, 1–5 (1973).

108  F. E. Sage and J. Estrin, Film condensation from a ternary mixture of vapors upon a vertical surface, *Internatl. J. Heat Mass Transf.* **19**, 323–333 (1976).

109  Y. Taitel and A. Tamir, Film condensation of multicomponent mixtures, *Internatl. J. Multiphase Flow* **1**, 697 (1974).

110  R. B. Bird, W. E. Stewart, and E. N. Lightfoot, *Transport Phenomena*, Wiley, New York, 1960.

111  E. M. Sparrow and J. E. Niethammer, Natural convection in a ternary gas mixture—application to the naphthalene sublimation technique, *Transact. ASME, J. Heat Transf.* **101**, 404–410 (1979).

112  H. L. Toor, Diffusion in three-component gas mixtures, *AIChE J.* **3**, 198–207 (1957).

113  H. L. Toor, Solution of the linearized equations of multicomponent mass transfer, *AIChE J.* **10**, 448–455, 460–465 (1964).

114  R. Gregorig, Hautkondensation an Feingewellten Oberflächen bei Berücksichtigung der Oberflächen spannungen, *Zeitschrift für angewandte Mathematik und Physik* **5**, 36–49 (1954).

115  E. L. Lusterader, R. Richter, and F. N. Neugebauer, The use of thin films for increasing evaporation and condensation rates in process equipment, *Transact. ASME, J. Heat Transf.* **81**, 297–307 (1959).

116  K. Nabavian and L. A. Bromley, Condensation coefficient of water, *Chem. Eng. Sci.* **18**, 651–660 (1963).

117  D. G. Thomas, Enhancement of film condensation heat transfer rates on vertical wires, *Indust. Eng. Chem. Funda.* **6**, 97–103 (1967).

118  D. G. Thomas, Enhancement of film condensation rate on vertical tubes by longitudinal fins, *AIChE J.* **14**, 644–649 (1968).

119  A. Markowitz, B. B. Mikic, and A. E. Bergles, Condensation on a downward-facing horizontal rippled surface, *Transact. ASME, J. Heat Transf.* **94**, 315–320 (1972).

120  J. W. Rose, Approximate equations for forced-convection condensation in the presence of a non-condensing gas on a flat plate and horizontal tube, *Internatl. J. Heat Mass Transf.* **23**, 539–546 (1980).

121  M. M. Shah, A general correlation for heat transfer during film condensation inside pipes, *Internatl. J. Heat Mass Transf.* **22**, 547–556 (1979).

122  S. A. Stylianou and J. W. Rose, Dropwise condensation on surfaces having different conductivities, *Transact. ASME, J. Heat Transf.* **102**, 477–482 (1980).

123  J. H. Royal and A. E. Bergles, Augmentation of horizontal in-tube condensation by means of twisted-tape inserts and internally finned tubes, *Transact. ASME, J. Heat Transf.* **100**, 17–24 (1978).

124  R. L. Webb, A generalized procedure for the design and optimization of fluted Gregorig condensing surfaces, *Transact. ASME, J. Heat Transf.* **101**, 335–339 (1979).

125  R. M. Desmond and B. V. Karlekar, Experimental observations of a modified condenser tube design to enhance heat transfer in a steam condenser, ASME Paper 80-HT-53.

126  T. C. Carnavos, An experimental study: Condensing R-11 on augmented tubes, ASME Paper 80-HT-54 (1980).

127  N. O. Young, J. S. Goldstein, an M. J. Block, The motion of bubbles in a vertical temperature gradient, *J. Fluid Mech.* **6**, 350–356 (1959).

128  J. R. A. Pearson, On convection cells induced by surface tension, *J. Fluid Mech.* **4**, 489–500 (1958).

129  E. M. Sparrow and R. Siegel, Transient film condensation, *Transact. ASME, J. Appl. Mech.* **26**, 120–121 (1959).

130  J. H. Liehard and V. K. Dhir, Laminar film condensation on nonisothermal and arbitrary-heat-flux-surfaces, *Transact. ASME, J. Heat Transf.* **96**, 197–203 (1974).

131  L. C. Burmeister, Vertical fin efficiency with film condensation, *Transact. ASME, J. Heat Transf.* **104**, 391–393 (1982).

# APPENDIX

# A

# VECTOR ANALYSIS

A vector differs from a scalar (a quantity completely characterized by its magnitude) in that a vector is characterized by both its magnitude *and* its direction. Examples of a scalar are mass, temperature, time, and volume; examples of vectors are velocity, force, acceleration, and displacement from a fixed origin.

A vector can be represented in terms of its components along coordinate axes. In rectangular coordinates a vector $\mathbf{V}$ is represented by its $V_x$, $V_y$, and $V_z$ components along the $x$, $y$, and $z$ axes, respectively. This is shown in Fig. A-1, where it is evident that $V_x = |\mathbf{V}| \cos \phi_x$, $V_y = |\mathbf{V}| \cos \phi_y$, and $V_z = |\mathbf{V}| \cos \phi_z$, where $|\mathbf{V}| = (V_x^2 + V_y^2 + V_z^2)^{1/2}$ is the magnitude of the vector. These three components can be used together with unit vectors along the coordinate axes to more fully describe the vector. In rectangular coordinates

$$\mathbf{V} = V_x \hat{\mathbf{i}} + V_y \hat{\mathbf{j}} + V_z \hat{\mathbf{k}}$$

Each unit vector points in the positive coordinate direction and is of unit magnitude, or length; for example, $\hat{\mathbf{i}}$ points in the positive $x$ direction and $|\hat{\mathbf{i}}| = 1$. It can be shown that

$$\cos \phi_x = V_x/|\mathbf{V}|, \qquad \cos \phi_y = \frac{V_y}{|\mathbf{V}|}, \qquad \cos \phi_z = \frac{V_z}{|\mathbf{V}|}$$

The sum of two vectors, $\mathbf{C} = \mathbf{A} + \mathbf{B}$, is defined as the vector whose components are the sums of the corresponding components of the original vectors, $\mathbf{A}$ and $\mathbf{B}$. In rectangular coordinates,

$$C_x = A_x + B_x, \qquad C_y = A_y + B_y, \qquad C_z = A_z + B_z$$

**699**

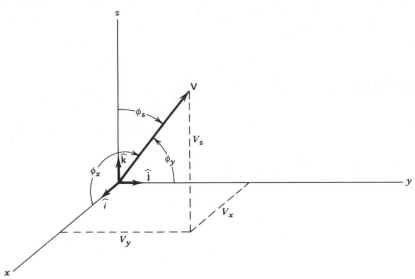

**Figure A-1**   A vector in rectangular coordinates.

This definition of the sum of two vectors is equivalent to the parallelogram rule illustrated in Fig. A-2.

Vector subtraction is defined in terms of the negative of a vector. By definition, the product of a scalar $c$ with a vector $\mathbf{A}$ is $\mathbf{B} = c\mathbf{A}$, where $B_x = cA_x$, $B_y = cA_y$, and $B_z = cA_z$ in rectangular coordinates. Thus, with $c = -1$, the negative of a vector $\mathbf{C} = -\mathbf{V}$ is defined as the vector whose components are the negative corresponding components of the original vector. In rectangular coordinates,

$$C_x = -V_x, \qquad C_y = -V_y, \qquad C_z = -V_z$$

The addition and subtraction of vectors is associative since those operations are associative for real numbers. Thus

$$\mathbf{A} + (\mathbf{B} + \mathbf{C}) = (\mathbf{A} + \mathbf{B}) + \mathbf{C} = \mathbf{A} + \mathbf{B} + \mathbf{C}$$

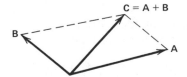

**Figure A-2**   Vector addition.

and no parentheses are needed to clarify the meaning. Similarly, the addition and subtraction of vectors is commutative, so that

$$\mathbf{A} + \mathbf{B} = \mathbf{B} + \mathbf{A}$$

The scalar product (also called the *dot product*, or *inner product*) of two vectors is a scalar quantity that is defined in general as

$$c = \mathbf{A} \cdot \mathbf{B} = |\mathbf{A}||\mathbf{B}| \cos \theta$$

where $\theta$ is the angle between the two vectors. Alternatively, in rectangular coordinates

$$\mathbf{A} \cdot \mathbf{B} = A_x B_x + A_y B_y + A_z B_z$$

which gives

$$\cos \theta = \frac{A_x B_x + A_y B_y + A_z B_z}{|\mathbf{A}||\mathbf{B}|}$$

For example, $\hat{\mathbf{i}} \cdot \hat{\mathbf{i}} = 1$ whereas $\hat{\mathbf{i}} \cdot \hat{\mathbf{j}} = 0$ in rectangular coordinates. The scalar product is commutative

$$\mathbf{A} \cdot \mathbf{B} = \mathbf{B} \cdot \mathbf{A}$$

and distributive

$$\mathbf{A} \cdot (\mathbf{B} + \mathbf{C}) = \mathbf{A} \cdot \mathbf{B} + \mathbf{A} \cdot \mathbf{C}$$

The physical significance of the scalar product might be, for example, the work $w$ done by a constant force $\mathbf{F}$ moving through a distance $\mathbf{r}$ and acting at an angle $\theta$ that is

$$w = |\mathbf{F}||\mathbf{r}| \cos \theta = \mathbf{F} \cdot \mathbf{r}$$

The vector product (also called the *cross product*, or *outer product*) of two vectors is a vector quantity that is defined in general as

$$\mathbf{C} = \mathbf{A} \times \mathbf{B} = |\mathbf{A}||\mathbf{B}| \sin \theta \, \hat{\mathbf{e}}$$

where $\theta$ is the angle swept as $\mathbf{A}$ is rotated toward $\mathbf{B}$ through the smallest possible angle and the direction $\hat{\mathbf{e}}$ of the resultant vector is perpendicular to the plane containing $\mathbf{A}$ and $\mathbf{B}$ and pointing in the direction that a *right-hand* screw would advance as $\mathbf{A}$ is rotated toward $\mathbf{B}$. In rectangular coordinates it is often convenient to evaluate the vector product in terms of a determinant as

$$
\mathbf{A} \times \mathbf{B} = \begin{vmatrix} \hat{\mathbf{i}} & \hat{\mathbf{j}} & \hat{\mathbf{k}} \\ A_x & A_y & A_z \\ B_x & B_y & B_z \end{vmatrix}
$$

$$
= (A_y B_z - A_z B_y)\hat{\mathbf{i}} + (A_z B_x - A_x B_z)\hat{\mathbf{j}} + (A_x B_y - A_y B_x)\hat{\mathbf{k}} \quad \text{(A-1)}
$$

Also, for example, $\hat{\mathbf{i}} \times \hat{\mathbf{i}} = 0$, $\hat{\mathbf{i}} \times \hat{\mathbf{j}} = \hat{\mathbf{k}}$, $\hat{\mathbf{i}} \times \hat{\mathbf{k}} = -\hat{\mathbf{j}}$ in rectangular coordinates. The vector product is not commutative because of the $\sin\theta$ term in its definition, which makes

$$
\mathbf{A} \times \mathbf{B} = -\mathbf{B} \times \mathbf{A}
$$

It is distributive, however, so that

$$
\mathbf{A} \times (\mathbf{B} + \mathbf{C}) = \mathbf{A} \times \mathbf{B} + \mathbf{A} \times \mathbf{C}
$$

It is conventional to adopt the convenience of orthogonal (mutually perpendicular) coordinates, of which rectangular coordinates are only one example, even though this is not strictly necessary. What is necessary is that the vector operations give the same results (the same magnitude and direction of vector and scalar results) in all coordinate systems; as the conventional coordinate systems are defined, this requirement is met. Because a vector operation such as $\mathbf{A} \times \mathbf{B}$ involves the idea of *rotating* $\mathbf{A}$ toward $\mathbf{B}$, a positive direction of rotation must be arbitrarily designated. There are only two choices, a right-handed (dextral) system or a left-handed (sinistral) system. Because most humans are right-handed, the right-handed convention is universal; it signifies that a positive direction is the direction that the pointed end of a right-handed screw will advance when turned clockwise as viewed from its head end. A right-handed coordinate system is one such that, when looking at the series of axes

$$
x\,y\,z\,x\,y\,z\,x\,y\,z\,x\,y\,z\,x\,y \;\cdots
$$

any axis rotated into its successor advances a right-handed screw in the direction of the next-following axis. In rectangular coordinates, for example, rotation of $x$ into $y$ gives $z$, $y$ into $z$ gives $x$, and so forth, as Fig. A-1 shows. For the cylindrical and spherical coordinate systems illustrated in Fig. A-3, the unit vectors point in directions that are dependent upon the coordinates of the point in question, which is not the case for rectangular coordinates. It can be seen that the $r$, $\theta$, $z$ grouping forms a dextral set in cylindrical coordinates that is valid also for the $r$, $\theta$, $\phi$ set in spherical coordinates.

Two triple products deserve particular mention. The triple scalar product $D = \mathbf{A} \cdot \mathbf{B} \times \mathbf{C}$ physically represents the volume of a parallelepiped with the

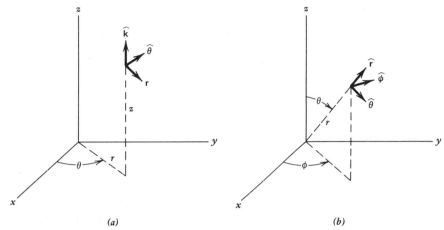

*(a)*                                                    *(b)*

**Figure A-3**   A vector in (*a*) cylindrical coordinates and (*b*) spherical coordinates.

**A, B, C** vectors as sides. It can be shown that

$$D = \mathbf{A} \cdot \mathbf{B} \times \mathbf{C} = -\mathbf{B} \cdot \mathbf{A} \times \mathbf{C} \tag{A-2}$$

The triple vector product $\mathbf{D} = \mathbf{A} \times \mathbf{B} \times \mathbf{C}$ can be shown to also be given as

$$\mathbf{D} = \mathbf{A} \times \mathbf{B} \times \mathbf{C} = \mathbf{B}(\mathbf{A} \cdot \mathbf{C}) - \mathbf{C}(\mathbf{A} \cdot \mathbf{B})$$

which is known as the "back cab rule." Note that the parentheses are needed to make the meaning of this rule clear. Both of these triple product identities can be demonstrated to be true in rectangular coordinates and, therefore, in all coordinates.

## GRADIENT

The gradient of a scalar function is most simply understood by considering a rectangular coordinate system in which the change of a scalar $f$ caused by a displacement, $d\mathbf{s} = dx\,\hat{\mathbf{i}} + dy\,\hat{\mathbf{j}} + dz\,\hat{\mathbf{k}}$, is to be evaluated. By the chain rule, the change in the scalar is

$$df = \frac{\partial f}{\partial x}\,dx + \frac{\partial f}{\partial y}\,dy + \frac{\partial f}{\partial z}\,dz$$

But this is seen to be equivalent to the scalar product

$$df = \left( \hat{\mathbf{i}}\,\frac{\partial f}{\partial x} + \hat{\mathbf{j}}\,\frac{\partial f}{\partial y} + \hat{\mathbf{k}}\,\frac{\partial f}{\partial z} \right) \cdot \left( \hat{\mathbf{i}}\,dx + \hat{\mathbf{j}}\,dy + \hat{\mathbf{k}}\,dz \right)$$

The first right-hand term in parentheses is a vector called the *gradient of the scalar function* and written as either $\nabla f$ or grad($f$). The $\nabla$ operator is called "del" and is given in rectangular coordinates as $\nabla = \hat{\mathbf{i}}\,\partial/\partial x + \hat{\mathbf{j}}\,\partial/\partial y + \hat{\mathbf{k}}\,\partial/\partial z$. With these definitions

$$df = \nabla f \cdot d\mathbf{s}$$

The gradient is perpendicular to lines of constant scalar value (contour lines or surfaces). This can be seen by considering an infinitesimal displacement, $d\mathbf{s}$ that moves tangentially to a nearby point on a contour line so that $df = 0$. Then

$$df = \nabla f \cdot d\mathbf{s} = 0$$

Because a zero scalar product implies perpendicularity of the two vectors and because $d\mathbf{s}$ is tangent to the contour line, this result shows that $\nabla f$ is perpendicular to the contour line. Also, from its definition it can be seen that the gradient points in the direction of increasing functional values. The derivative of a scalar function in the direction of the displacement $d\mathbf{s}$ (whose magnitude is $d\mathbf{s}$) is seen to be

$$\frac{df}{ds} = \nabla f \cdot \frac{d\mathbf{s}}{ds}$$

$$= |\nabla f| \cos\theta$$

where $\theta$ is the angle between $\nabla f$ and $d\mathbf{s}$. Thus $df/ds$ is just the projection of $\nabla f$ in the desired direction.

In cylindrical coordinates $\nabla f$ is defined in a manner that parallels that used for rectangular coordinates. For cylindrical coordinates

$$d\mathbf{s} = \hat{\mathbf{r}}\,dr + \hat{\boldsymbol{\theta}}r\,d\theta + \hat{\mathbf{k}}\,dz$$

and

$$df = \frac{\partial f}{\partial r}dr + \frac{\partial f}{\partial \theta}d\theta + \frac{\partial f}{\partial z}dz = \nabla f \cdot d\mathbf{s}$$

Comparison of the components of $\nabla f \cdot d\mathbf{s}$ with the chain rule expansion of $df$ gives

$$(\nabla f)_r\,dr = \frac{\partial f}{\partial r}dr, \qquad (\nabla f)_\theta r\,d\theta = \frac{\partial f}{\partial \theta}d\theta, \qquad (\nabla f)_z\,dz = \frac{\partial f}{\partial z}dz$$

from which it follows that

$$\nabla f = \frac{\partial f}{\partial r}\hat{\mathbf{r}} + \left(r^{-1}\frac{\partial f}{\partial \theta}\right)\hat{\boldsymbol{\theta}} + \frac{\partial f}{\partial z}\hat{\mathbf{k}}$$

In spherical coordinates one has

$$ds = \hat{r}\, dr + \hat{\theta}\, r\, d\theta + \hat{\phi}\, r \sin\theta\, d\phi$$

Proceeding similarly results, finally, in

$$\nabla f = \frac{\partial f}{\partial r}\,\hat{r} + \left(r^{-1}\frac{\partial f}{\partial \theta}\right)\hat{\theta} + (r\sin\theta)^{-1}\frac{\partial f}{\partial \phi}\,\hat{\phi}$$

in spherical coordinates.

## DIVERGENCE

The scalar product of the gradient operator $\nabla$ with a vector $\mathbf{V}$ is called the *divergence* of the vector and is written as

$$\operatorname{div}(\mathbf{V}) = \nabla \cdot \mathbf{V}$$

In rectangular coordinates one has

$$\operatorname{div}(\mathbf{V}) = \left(\frac{\partial}{\partial x}\,\hat{i} + \frac{\partial}{\partial y}\,\hat{j} + \frac{\partial}{\partial z}\,\hat{k}\right) \cdot \left(V_x\hat{i} + V_y\hat{j} + V_z\hat{k}\right)$$

which leads to

$$\operatorname{div}(\mathbf{V}) = \frac{\partial V_x}{\partial x} + \frac{\partial V_y}{\partial y} + \frac{\partial V_z}{\partial z}$$

since $\hat{i} \cdot \hat{i} = 1$, $\hat{i} \cdot \hat{j} = 0$, and so on.

If the vector is the gradient of a scalar $f$, one has the scalar product as $\nabla \cdot (\nabla f) = \nabla^2 f$. In rectangular coordinates one has

$$\nabla^2 f = \left(\frac{\partial}{\partial x}\,\hat{i} + \frac{\partial}{\partial y}\,\hat{j} + \frac{\partial}{\partial z}\,\hat{k}\right) \cdot \left(\frac{\partial f}{\partial x}\,\hat{i} + \frac{\partial f}{\partial y}\,\hat{j} + \frac{\partial f}{\partial z}\,\hat{k}\right)$$

which leads to the scalar quantity

$$\nabla^2 f = \frac{\partial^2 f}{\partial x^2} + \frac{\partial^2 f}{\partial y^2} + \frac{\partial^2 f}{\partial x^2}$$

The important divergence theorem, also referred to as *Gauss' theorem*, relates surface and volume integrals of a vector as

$$\int_A \mathbf{V} \cdot d\mathbf{A} = \int_V \operatorname{div}(\mathbf{V})\, dV$$

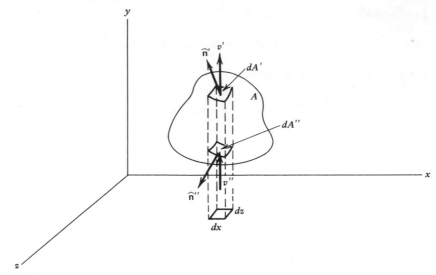

**Figure A-4**  Derivation of the divergence (or Gauss') theorem.

where $A$ is a closed surface that encloses the volume $V$. To demonstrate the truth of this relation, consider a rectangular coordinate system for which $\mathbf{V} = u\hat{\mathbf{i}} + v\hat{\mathbf{j}} + w\hat{\mathbf{k}}$ and $d\mathbf{A} = \hat{\mathbf{n}}\,dA$ in which $\hat{\mathbf{n}}$ is an outward-drawn unit vector on $A$. Then

$$\int_A \mathbf{V} \cdot d\mathbf{A} = \int_A (\hat{\mathbf{n}} \cdot \hat{\mathbf{i}}u + \hat{\mathbf{n}} \cdot \hat{\mathbf{j}}v + \hat{\mathbf{n}} \cdot \hat{\mathbf{k}}w)\,dA$$

The second term on the right-hand side of this equation is selected for detailed consideration. Referring to Fig. A-4, one sees that

$$\int_A \hat{\mathbf{n}} \cdot \hat{\mathbf{j}}v\,dA = \int_A v\hat{\mathbf{j}} \cdot d\mathbf{A} = \int_{A_{x-z}} (v' - v'')\,dA_{x-z} = \int_{A_{x-z}} (v' - v'')\,dx\,dz$$

or

$$\int_A \hat{\mathbf{n}} \cdot \hat{\mathbf{j}}v\,dA = \int_{A_{x-z}} \int_{y''}^{y'} \frac{\partial v}{\partial y}\,dy\,dx\,dz$$

$$= \int_V \frac{\partial v}{\partial y}\,dV$$

By parallel treatment of the remaining two terms it is seen that

$$\int_A \mathbf{V} \cdot d\mathbf{A} = \int_V \text{div}(\mathbf{V})\,dV$$

which can be used as the general definition for the divergence

$$\text{div}(\mathbf{V}) = \lim_{V \to 0} V^{-1} \int_A \mathbf{V} \cdot d\mathbf{A}$$

The divergence of a vector can be shown to physically represent the net rate of flow out of a small control volume.

## CURL

The vector product of the gradient operator $\nabla$ with a vector $\mathbf{V}$ is called the *curl* of the vector and is written as

$$\text{curl}(\mathbf{V}) = \nabla \times \mathbf{V}$$

In rectangular coordinates one has

$$\text{curl}(\mathbf{V}) = \left( \frac{\partial}{\partial x} \hat{\mathbf{i}} + \frac{\partial}{\partial y} \hat{\mathbf{j}} + \frac{\partial}{\partial z} \hat{\mathbf{k}} \right) \times \left( V_x \hat{\mathbf{i}} + V_y \hat{\mathbf{j}} + V_z \hat{\mathbf{k}} \right)$$

which leads to

$$\text{curl}(\mathbf{V}) = \left( \frac{\partial V_z}{\partial y} - \frac{\partial V_y}{\partial z} \right) \hat{\mathbf{i}} + \left( \frac{\partial V_x}{\partial z} - \frac{\partial V_z}{\partial x} \right) \hat{\mathbf{j}} + \left( \frac{\partial V_y}{\partial x} - \frac{\partial V_x}{\partial y} \right) \hat{\mathbf{k}}$$

Since $\hat{\mathbf{i}} \times \hat{\mathbf{i}} = 0$, $\hat{\mathbf{i}} \times \hat{\mathbf{k}} = -\hat{\mathbf{j}}$, and so forth.

The important Stokes' theorem relates line and surface integrals of a vector as

$$\int_L \mathbf{V} \cdot d\mathbf{L} = \int_A \text{curl}(\mathbf{V}) \cdot d\mathbf{A}$$

where $L$ is a closed circuit that is covered by the simply connected surface $A$. To demonstrate the truth of this relation, consider the situation illustrated in Fig. A-5. The surface $A$ bounded by the closed curve $L$ is divided into small rectangular meshes. For the small rectangular mesh denoted $a$–$b$–$c$–$d$, a Tayor series expansion of the vector whose value is $\mathbf{V}$ at the center of the mesh gives

$$\mathbf{V}_{ab} = \mathbf{V} - \tfrac{1}{2}(\mathbf{L}_2 \cdot \nabla \mathbf{V})$$

$$\mathbf{V}_{cd} = \mathbf{V} + \tfrac{1}{2}(\mathbf{L}_2 \cdot \nabla \mathbf{V})$$

$$\mathbf{V}_{bc} = \mathbf{V} + \tfrac{1}{2}(\mathbf{L}_1 \cdot \nabla \mathbf{V})$$

$$\mathbf{V}_{da} = \mathbf{V} - \tfrac{1}{2}(\mathbf{L}_1 \cdot \nabla \mathbf{V})$$

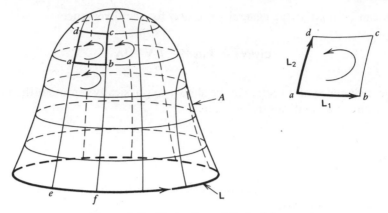

**Figure A-5**   Derivation of Stokes' theorem.

Then

$$\int_{L_{abcd}} \mathbf{V} \cdot d\mathbf{L} = \mathbf{V}_{ab} \cdot \mathbf{L}_1 + \mathbf{V}_{bc} \cdot \mathbf{L}_2 - \mathbf{V}_{cd} \cdot \mathbf{L}_1 - \mathbf{V}_{da} \cdot \mathbf{L}_2$$

$$= \tfrac{1}{2}(2\mathbf{L}_1 \cdot \nabla \mathbf{V}) \cdot \mathbf{L}_2 - \tfrac{1}{2}(2\mathbf{L}_2 \cdot \nabla \mathbf{V}) \cdot \mathbf{L}_1$$

$$= \mathbf{L}_2 \cdot (\mathbf{L}_1 \cdot \nabla \mathbf{V}) - \mathbf{L}_1 \cdot (\mathbf{L}_2 \cdot \nabla \mathbf{V})$$

$$= \left[ \mathbf{L}_2(\mathbf{L}_1 \cdot \nabla) - \mathbf{L}_1(\mathbf{L}_2 \cdot \nabla) \right] \cdot \mathbf{V}$$

Utilizing the "back cab rule" with $\mathbf{A} = \nabla$, $\mathbf{B} = \mathbf{L}_2$, and $\mathbf{C} = \mathbf{L}_1$ allows this result to be rephrased as

$$\int_{L_{abcd}} \mathbf{V} \cdot d\mathbf{L} = \left[ (\mathbf{L}_1 \times \mathbf{L}_2) \times \nabla \right] \cdot \mathbf{V}$$

Next, the preceding expression is recast into

$$\int_{L_{abcd}} \mathbf{V} \cdot d\mathbf{L} = (\mathbf{L}_1 \times \mathbf{L}_2) \cdot (\nabla \times \mathbf{V})$$

by appealing to the triple vector product relationship of Eq. A-2. Since the vector $\mathbf{L}_1 \times \mathbf{L}_2$ is just the area of the mesh $d\mathbf{A}$, it follows that

$$\int_{L_{abcd}} \mathbf{V} \cdot d\mathbf{L} = (\nabla \times \mathbf{V}) \cdot d\mathbf{A}$$

Each small rectangular mesh is treated in similar fashion, cancellation of the line integral occurring at each side of a mesh since it is traversed once in each direction, except for the *e–f* edge along the original circuit **L**. Each mesh, however, contributes the amount $(\nabla \times \mathbf{V}) \cdot d\mathbf{A}$ to the integral. Therefore, in summing the results for all the meshes, one has

$$\int_{\mathbf{L}} \mathbf{V} \cdot d\mathbf{L} = \int_A (\nabla \times \mathbf{V}) \cdot d\mathbf{A}$$

$$= \int_A \text{curl}(\mathbf{V}) \cdot d\mathbf{A}$$

which was to be proved.

Stokes' theorem can be employed to give a general definition of the component of the curl of a vector in a direction $\hat{\mathbf{n}}$ as

$$\text{curl}(\mathbf{V}) \cdot \hat{\mathbf{n}} = \lim_{A \to 0} A^{-1} \int_{\mathbf{L}} \mathbf{V} \cdot d\mathbf{L}$$

where the curve **L** that bounds the surface $A$ is in a plane perpendicular to the direction $\hat{\mathbf{n}}$. An alternative general definition of the curl can be obtained by applying the divergence theorem to the vector $\mathbf{V} \times \mathbf{c}$, where $\mathbf{c}$ is an arbitrary constant vector. The divergence theorem provides that

$$\int_A (\mathbf{V} \times \mathbf{c}) \cdot \mathbf{n} \, dA = \int_V \nabla \cdot (\mathbf{V} \times \mathbf{c}) \, dV$$

which in view of the relation for a triple scalar product of Eq. (A-2) can be rearranged into

$$\int_A \mathbf{c} \cdot (\hat{\mathbf{n}} \times \mathbf{V}) \, dA = \int_V \mathbf{c} \cdot (\nabla \times \mathbf{V}) \, dV$$

Since $\mathbf{c}$ is a constant vector, it can be seen that it must be the case that

$$\int_A (\hat{\mathbf{n}} \times \mathbf{V}) \, dA = \int_V (\nabla \times \mathbf{V}) \, dV$$

$$= \int_V \text{curl}(\mathbf{V}) \, dV$$

Therefore, in general

$$\text{curl}(\mathbf{V}) = \lim_{V \to 0} V^{-1} \int_A (\hat{\mathbf{n}} \times \mathbf{V}) \, dA$$

in which $A$ is the surface enclosing the volume $V$ and $\hat{\mathbf{n}}$ is the outward drawn unit vector normal to $A$. Application of this result to a small rectangular volume in rectangular coordinates gives

$$\text{curl}(\mathbf{V}) = \lim_{\substack{\Delta x \\ \Delta y \\ \Delta z}\to 0} \frac{\begin{aligned}\hat{\mathbf{i}} \times (u\hat{\mathbf{i}} + v\hat{\mathbf{j}} + w\hat{\mathbf{k}})\,|_{x+\Delta x/2,\,y,\,z}\Delta y\Delta z - \hat{\mathbf{i}} \\ \times (u\hat{\mathbf{i}} + v\hat{\mathbf{j}} + w\hat{\mathbf{k}})\,|_{x-\Delta x/2,\,y,\,z}\Delta y\Delta z + \cdots\end{aligned}}{\Delta x\,\Delta y\,\Delta z}$$

$$= \frac{\partial u}{\partial x}(\hat{\mathbf{i}} \times \hat{\mathbf{i}}) + \frac{\partial v}{\partial x}(\hat{\mathbf{i}} \times \hat{\mathbf{j}}) + \frac{\partial w}{\partial x}(\hat{\mathbf{i}} \times \hat{\mathbf{k}}) + \cdots$$

$$= \frac{\partial v}{\partial x}\hat{\mathbf{k}} - \frac{\partial w}{\partial x}\hat{\mathbf{j}} + \cdots$$

Evaluation of the remaining terms gives the result (for rectangular coordinates only) that is expressible in determinant form as

$$\nabla \times \mathbf{V} = \text{curl}(\mathbf{V}) = \begin{vmatrix} \hat{\mathbf{i}} & \hat{\mathbf{j}} & \hat{\mathbf{k}} \\ \dfrac{\partial}{\partial x} & \dfrac{\partial}{\partial y} & \dfrac{\partial}{\partial z} \\ u & v & w \end{vmatrix}$$

in parallel with Eq. (A-1).

A physical interpretation of curl($\mathbf{V}$) can be obtained by considering a rotating flywheel as illustrated in Fig. A-6. The angular velocity of the flywheel is given in magnitude as $\omega$ and may be considered to be a vector if it is directed in the direction that a right-hand screw would advance. The peripheral velocity of a point on the wheel rim at $\mathbf{r}$ from the origin is given in magnitude by

$$|\mathbf{V}| = |\omega||\mathbf{r}|\sin\theta$$

$$= |\omega \times \mathbf{r}|$$

It is evident from Fig. A-6 that the correct direction for $V$ is given by

$$\mathbf{V} = \omega \times \mathbf{r}$$

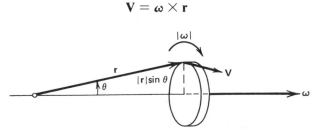

**Figure A-6**  Physical interpretation of the curl of a vector.

The curl of $V$ is next found to be

$$\text{curl}(\mathbf{V}) = \nabla \times (\boldsymbol{\omega} \times \mathbf{r})$$

This result is most simply evaluated by realizing that in rectangular coordinates centered at 0 in Fig. A-6, $\mathbf{r} = x\hat{\mathbf{i}} + y\hat{\mathbf{j}} + z\hat{\mathbf{k}}$, $\boldsymbol{\omega} = w_x\hat{\mathbf{i}} + w_y\hat{\mathbf{j}} + w_k\hat{\mathbf{k}}$, and $\nabla = \partial/\partial x\,\hat{\mathbf{i}} + \partial/\partial y\,\hat{\mathbf{j}} + \partial/\partial z\,\hat{\mathbf{k}}$. Performance of the indicated vector operations then finally yields

$$\text{curl}(\mathbf{V}) = 2\boldsymbol{\omega}$$

The aptness of the "curl" appellation lies in the fact that the $\nabla \times$ operation applied to the local velocity yields twice the angular velocity there. In fluid mechanics, it is customary to speak of the vorticity vector, $\boldsymbol{\zeta} = 2\boldsymbol{\omega}$, to avoid carrying the factor of 2.

## ADDITIONAL USEFUL RESULTS

The curl of the gradient of a scalar field is identically zero. In other words,

$$\text{curl}(\nabla f) = \nabla \times \nabla f = 0$$

Thus in a fluid mechanics problem, if the velocity in a fluid is derivable from the gradient of a scalar, the fluid flow is irrotational. This result is demonstrated to be true by considering the rectangular coordinate case for which

$$\text{curl}(\nabla f) = \begin{vmatrix} \hat{\mathbf{i}} & \hat{\mathbf{j}} & \hat{\mathbf{k}} \\ \dfrac{\partial}{\partial x} & \dfrac{\partial}{\partial y} & \dfrac{\partial}{\partial z} \\ \dfrac{\partial f}{\partial x} & \dfrac{\partial f}{\partial y} & \dfrac{\partial f}{\partial z} \end{vmatrix} = \left( \frac{\partial^2 f}{\partial y\,\partial z} - \frac{\partial^2 f}{\partial z\,\partial y} \right)\hat{\mathbf{i}} + \cdots = 0$$

The divergence of the curl of a vector is identically zero. In other words,

$$\text{div}(\text{curl}\,\mathbf{V}) = \nabla \cdot (\nabla \times \mathbf{V}) = 0$$

This result can be demonstrated to be true in rectangular coordinates for which

$$\text{div}(\text{curl}\,\mathbf{V}) = \frac{\partial}{\partial x}\left( \frac{\partial w}{\partial y} - \frac{\partial v}{\partial z} \right) + \frac{\partial}{\partial y}\left( \frac{\partial u}{\partial z} - \frac{\partial w}{\partial x} \right) + \cdots = 0$$

The Laplacian of a vector $\nabla^2 \mathbf{V}$ is obtainable from application of the "back cab rule" to

$$\nabla \times \nabla \times \mathbf{V}$$

which yields

$$\nabla \times \nabla \times \mathbf{V} = \nabla(\nabla \cdot \mathbf{V}) - (\nabla \cdot \nabla)\mathbf{V}$$

This result can be rephrased as

$$\text{curl}(\text{curl}\,\mathbf{V}) = \text{grad}(\text{div}\,\mathbf{V}) - \nabla^2\mathbf{V}$$

From this result it generally follows that

$$\nabla^2\mathbf{V} = \text{grad}(\text{div}\,\mathbf{V}) - \text{curl}(\text{curl}\,\mathbf{V})$$

The important Green theorems can be derived from the divergence theorem. For example, let $f_1$ and $f_2$ be two scalars and let $\mathbf{F} = f_1 \nabla f_2$. Then, since it is generally true that

$$\nabla \cdot (f\mathbf{V}) = f\nabla \cdot \mathbf{V} + \mathbf{V} \cdot \nabla f$$

as can readily be demonstrated in rectangular coordinates, it follows that

$$\nabla \cdot \mathbf{F} = \nabla \cdot (f_1 \nabla f_2)$$

$$= f_1 \nabla^2 f_2 + \nabla f_1 \cdot \nabla f_2$$

Now, application of the divergence theorem to $\mathbf{F} = f_1 \nabla f_2$ gives

$$\int_A \mathbf{F} \cdot d\mathbf{A} = \int_V \text{div}(\mathbf{F})\,dV$$

$$\int_A f_1 \nabla f_2 \cdot d\mathbf{A} = \int_V \left( f_1 \nabla^2 f_2 + \nabla f_1 \cdot \nabla f_2 \right) dV$$

which is the *first form of Green's theorem.*

Interchanging the roles of $f_1$ and $f_2$, so that $\mathbf{F} = f_2 \nabla f_1$, and proceeding in a fashion parallel to that used previously gives

$$\int_A f_2 \nabla f_1 \cdot d\mathbf{A} = \int_V \left( f_2 \nabla^2 f_1 + \nabla f_1 \cdot \nabla f_2 \right) dV$$

Subtraction of this result from the first form of Green's theorem gives

$$\int_A (f_1 \nabla f_2 - f_2 \nabla f_1) \cdot d\mathbf{A} = \int_V \left( f_1 \nabla^2 f_2 - f_2 \nabla^2 f_1 \right) dV$$

which is the *second or symmetric form of Green's theorem.*

**Table A-1    Some Useful Vector Relationships**

$\nabla(f_1 + f_2) = \nabla f_1 + \nabla f_1$

$\nabla(f_1 f_2) = f_1 \nabla f_2 + f_2 \nabla f_1$

$\text{div}(\mathbf{F} + \mathbf{G}) = \text{div}\,\mathbf{F} + \text{div}\,\mathbf{G}$

$\text{curl}(\mathbf{F} + \mathbf{G}) = \text{curl}\,\mathbf{F} + \text{curl}\,\mathbf{G}$

$\nabla(\mathbf{F} \cdot \mathbf{G}) = (\mathbf{F} \cdot \nabla)\mathbf{G} + (\mathbf{G} \cdot \nabla)\mathbf{F} + \mathbf{F} \times \text{curl}\,\mathbf{G} + \mathbf{G} \times \text{curl}\,\mathbf{F}$

$\text{div}(f\,\mathbf{F}) = f\,\text{div}\,\mathbf{F} + \mathbf{F} \cdot \nabla f$

$\text{div}(\mathbf{F} \times \mathbf{G}) = \mathbf{G} \cdot \text{curl}\,\mathbf{F} - \mathbf{F} \cdot \text{curl}\,\mathbf{G}$

$\text{div}\,\text{curl}\,\mathbf{F} = 0$

$\text{curl}(f\,\mathbf{F}) = f\,\text{curl}\,\mathbf{F} + \nabla f \times \mathbf{F}$

$\text{curl}(\mathbf{F} \times \mathbf{G}) = \mathbf{F}\,\text{div}\,\mathbf{G} - \mathbf{F}\,\text{div}\,\mathbf{G} + (\mathbf{G} \cdot \nabla)\mathbf{F} - (\mathbf{F} \cdot \nabla)\mathbf{G}$

$\text{curl}\,\text{curl}\,\mathbf{F} = \text{grad}\,\text{div}\,\mathbf{F} - \nabla^2 \mathbf{F}$

$\text{curl}\,\nabla f = 0$

$$\int_A \mathbf{F} \cdot d\mathbf{A} = \int_V \text{div}\,\mathbf{F}\,dV$$

$$\int_L \mathbf{F} \cdot d\mathbf{L} = \int_A \text{curl}\,\mathbf{F} \cdot d\mathbf{A}$$

$$\int_A f\,d\mathbf{A} = \int_V \nabla f\,dV$$

$$\int_A \mathbf{F}(\mathbf{G} \cdot d\mathbf{A}) = \int_V [\mathbf{F}\,\text{div}\,\mathbf{G} + (\mathbf{G} \cdot \nabla)\mathbf{F}]\,dV$$

$$\int_A \hat{\mathbf{n}} \times \mathbf{F}\,dA = \int_V \text{curl}\,\mathbf{F}\,dV$$

$$\int_L f\,d\mathbf{L} = \int_A \hat{\mathbf{n}} \times \nabla f\,dA$$

## Table A-2  Vector Operations in Several Coordinate Systems

*Rectangular Coordinates*

$$\nabla f = \hat{\mathbf{i}}\,\frac{\partial f}{\partial x} + \hat{\mathbf{j}}\,\frac{\partial f}{\partial y} + \hat{\mathbf{k}}\,\frac{\partial f}{\partial z}$$

$$\nabla \cdot \mathbf{V} = \frac{\partial u}{\partial x} + \frac{\partial v}{\partial y} + \frac{\partial w}{\partial z}$$

$$\nabla \times \mathbf{V} = \left( \frac{\partial w}{\partial y} - \frac{\partial v}{\partial z} \right)\hat{\mathbf{i}} + \left( \frac{\partial u}{\partial z} - \frac{\partial w}{\partial x} \right)\hat{\mathbf{j}} + \left( \frac{\partial v}{\partial x} - \frac{\partial u}{\partial y} \right)\hat{\mathbf{k}}$$

$$\nabla^2 f = \frac{\partial^2 f}{\partial x^2} + \frac{\partial^2 f}{\partial y^2} + \frac{\partial^2 f}{\partial z^2}$$

*Cylindrical Coordinates*

$$\nabla f = \left( \frac{\partial f}{\partial r} \right)\hat{\mathbf{r}} + \left( r^{-1}\frac{\partial f}{\partial \theta} \right)\hat{\boldsymbol{\theta}} + \left( \frac{\partial f}{\partial z} \right)\hat{\mathbf{k}}$$

$$\nabla \cdot \mathbf{V} = r^{-1}\frac{\partial (r v_r)}{\partial r} + r^{-1}\frac{\partial v_\theta}{\partial \theta} + \frac{\partial v_z}{\partial z}$$

$$\nabla \times \mathbf{V} = \left( r^{-1}\frac{\partial v_z}{\partial \theta} - \frac{\partial v_\theta}{\partial z} \right)\hat{\mathbf{r}} + \left( \frac{\partial v_r}{\partial z} - \frac{\partial v_z}{\partial r} \right)\hat{\boldsymbol{\theta}} + r^{-1}\left[ \frac{\partial (r v_\theta)}{\partial r} - \frac{\partial v_r}{\partial \theta} \right]\hat{\mathbf{k}}$$

$$\nabla^2 f = r^{-1}\frac{\partial (r\,\partial f/\partial r)}{\partial r} + r^{-2}\frac{\partial^2 f}{\partial \theta^2} + \frac{\partial^2 f}{\partial z^2}$$

*Spherical Coordinates*

$$\nabla f = \frac{\partial f}{\partial r}\hat{\mathbf{r}} + \left( r^{-1}\frac{\partial f}{\partial \theta} \right)\hat{\boldsymbol{\theta}} + \frac{\partial f/\partial \theta}{r \sin \theta}\hat{\boldsymbol{\phi}}$$

$$\nabla \cdot \mathbf{V} = r^{-2}\frac{\partial (r v_r)}{\partial r} + \frac{\partial (v_\theta \sin \theta)/\partial \theta}{r \sin \theta} + \frac{\partial v_\phi/\partial \phi}{r \sin \theta}$$

$$\nabla \times \mathbf{V} = \left[ \frac{\partial (v_\phi \sin \theta)/\partial \theta - \partial v_\theta/\partial \phi}{r \sin \theta} \right]\hat{\mathbf{r}} + \frac{\partial v_r/\partial \phi}{r \sin \theta}$$

$$- r^{-1}\frac{\partial (r v_\phi)}{\partial r}\hat{\boldsymbol{\theta}} + r^{-1}\left[ \frac{\partial (r v_\theta)}{\partial r} - \frac{\partial v_r}{\partial \theta} \right]\hat{\boldsymbol{\phi}}$$

$$\nabla^2 f = r^{-2}\frac{\partial (r^2\,\partial f/\partial r)}{\partial r} + \frac{\partial (\sin \theta\,\partial f/\partial \theta)/\partial \theta}{r^2 \sin \theta} + \frac{\partial^2 T/\partial \phi^2}{r^2 \sin^2 \theta}$$

# B

# HEAT AND MASS DIFFUSION

The steady diffusive fluxes of heat and mass through a homogeneous material are related to gradients of temperature and mass fraction by

$$\mathbf{q} = -k \, \nabla T \quad \text{and} \quad \dot{\mathbf{m}}_1 = -\rho D_{12} \, \nabla \omega_1$$

where the thermal conductivity $k$ and the density–mass diffusivity product $\rho D_{12}$ are considered to be properties of the material that are independent of direction.

There occasionally arise circumstances in which there is such a large departure from either the steady state or the homogeneous material case, that the diffusive flux of heat and mass is not accurately predicted by the preceding two relations. It is also found under certain extreme conditions that the diffusive heat and mass fluxes influence each other to a significant extent. To account for these effects, somewhat more complicated predictive relations are needed than the two preceding ones.

## STEADY HEAT AND MASS DIFFUSION IN ANISOTROPIC MATERIALS

Materials whose properties do not depend on direction are said to be *isotropic*, and those whose properties do depend on direction are said to be *anisotropic*. For example, the thermal conductivity of an oak log [1] is 0.12 Btu/hr ft °F perpendicular to the grain and 0.23 Btu/hr ft °F parallel to the grain, a twofold variation. Other materials, notably pyrolytic graphite with a 200-fold thermal conductivity ratio in its two principal direction and individual crystals, also exhibit substantial anisotropy. Such anisotropic behavior is experimentally observed for mass diffusion as well as for heat conduction. Almost all fluids can be considered to be isotropic, although the preferential orientation of fluids such as liquid crystals, blood, lubricants, and suspensions can exhibit anisotropy in both viscosity and thermal conductivity [2, 3]—a thermal conductivity ratio in the two principal directions of about 30% has been measured for some lubricants, for example.

Because both heat and mass diffusion are affected in a similar manner by material anisotropy, only heat conduction is discussed in further detail. In general, the component of the heat flux vector $\mathbf{q}$ in a specified direction will depend not only on the temperature gradient in that direction, but also on the temperature gradients in the other two coordinate directions. Thus in rectangular coordinates

$$-q_x = k_{xx} \frac{\partial T}{\partial x} + k_{xy} \frac{\partial T}{\partial y} + k_{xz} \frac{\partial T}{\partial z}$$

$$-q_y = k_{yx} \frac{\partial T}{\partial x} + k_{yy} \frac{\partial T}{\partial y} + k_{yz} \frac{\partial T}{\partial z}$$

$$-q_z = k_{zx} \frac{\partial T}{\partial x} + k_{zy} \frac{\partial T}{\partial y} + k_{zz} \frac{\partial T}{\partial z} \qquad \text{(B-1)}$$

in which $k_{xy} = k_{yx}$, $k_{xz} = k_{zx}$, and $k_{yz} = k_{zy}$ by symmetry considerations. In an isotropic material only $k_{xx}$, $k_{yy}$, and $k_{zz}$ are nonzero. Although these three values of thermal conductivity need not necessarily be equal in an isotropic material, they usually are.

By a coordinate transformation it is always possible to find a new set of coordinate axes, called the *principal axes* and denoted by $\zeta_1$, $\zeta_2$, and $\zeta_3$ for which the heat flux components are more simply given by

$$q_{\zeta_1} = -k_{\zeta_1} \frac{\partial T}{\partial \zeta_1}$$

$$q_{\zeta_2} = -k_{\zeta_2} \frac{\partial T}{\partial \zeta_2}$$

$$q_{\zeta_3} = -k_{\zeta_3} \frac{\partial T}{\partial \zeta_3} \qquad \text{(B-2)}$$

where $k_{\zeta_1}$, $k_{\zeta_2}$, and $k_{\zeta_3}$ are called the *principal conductivities*. The principal axes are dictated by the material structure; for instance, they are specified for crystals. Alternatively, two-dimensional laminated materials such as plywood and transformer cores would have one principal axis parallel to the laminae and one principal axis perpendicular to them. When the principal axes coincide with the shape of the object, Eqs. (B-2) can be employed; when this does not conveniently occur, the more complicated Eqs. (B-1) must be used. Equations (B-1) and (B-2) are based on steady-state measurements. Additional information on anisotropic diffusion is available [4–6].

Fortunately, the random orientation of anisotropic crystals causes most metal objects of appreciable size to behave in an isotropic manner. Similarly, whatever crystalline nature a liquid might possess that would tend toward anisotropy is masked by random orientations for liquid bodies of appreciable size with consequent effective isotropic behavior.

## STEADY HEAT AND MASS DIFFUSION IN BINARY FLUIDS

For a pure fluid in the steady state, the separate rates at which heat and mass diffuse relative to the mass-averaged velocity are accurately found to be proportional to the temperature and mass fraction gradients, respectively. If gradients become extremely large, the linear relationships begin to lose accuracy. Likewise, if the fluid simultaneously experiences both heat and mass diffusion, the two fluxes influence each other in a manner that can be accurately predicted by a linear combination of the driving gradients of temperature and mass fraction. On physical grounds, this interdependence is plausible since the motion of the particles that transfers mass also transfers energy and vice versa. The Onsager relations of thermodynamics give information regarding the linearized coupling between effects that can be expected.

The general expressions for a multicomponent fluid are complex [7–9]. For a two component fluid, they are of considerably simpler form and display the essential interplay of effects more comprehendibly. In the binary fluid case [10] the diffusive heat and mass fluxes are given by

$$\dot{m}_1 = \underbrace{\left[-\rho D_{12}\nabla\omega_1\right]}_{\text{Fick's diffusion}} + \underbrace{\left[\frac{M_1 M_2 D_{12}\omega_1}{p}\nabla p\right]}_{\text{pressure diffusion}} + \underbrace{\left[\frac{M_1 M_2 \omega_1 \omega_2 D_{12}}{RT}(\mathbf{B}_2 - \mathbf{B}_1)\right]}_{\text{body-force diffusion}}$$

$$+ \underbrace{\left[\frac{-\rho D_{12}\omega_1\omega_2\alpha}{T}\nabla T\right]}_{\substack{\text{thermal diffusion}\\ \text{(Soret effect)}}} \qquad\qquad (\text{B-3})$$

$$\mathbf{q} = \underbrace{\left[-k\nabla T\right]}_{\substack{\text{Fourier}\\ \text{conduction}}} + \underbrace{\left[(H_1 - H_2)\dot{m}_1\right]}_{\substack{\text{interdiffusional}\\ \text{convection}}} + \underbrace{\left[\frac{-RTM^2\alpha}{M_1 M_2}\dot{m}_1\right]}_{\substack{\text{diffusion thermo}\\ \text{(Dufour effect)}}} \qquad (\text{B-4})$$

where $\alpha$ is the thermal diffusion factor. Applications of such refined expressions for fluxes as Eq. (B-4) to boundary-layer problems have been made by

Baron [18] and Hoshizaki [19]. The viscous stresses in a multicomponent fluid are affected by diffusive mass fluxes because the velocities used in the usual relations for a Newtonian fluid employ mass-averaged velocities whereas viscous stresses are actually connected with momentum fluxes. In rectangular coordinates, the surface stresses are given by

$$\tau_{a_{ij}} = \tau_{ij} + \underbrace{\left[ -\frac{\rho}{g_c}\left( \omega_1 \frac{\dot{m}_{1i}}{\rho_1}\frac{\dot{m}_{1j}}{\rho_1} + \omega_2 \frac{\dot{m}_{2i}}{\rho_2}\frac{\dot{m}_{2j}}{\rho_2} \right) \right]}_{\text{diffusional momentum transport}} \qquad \text{(B-5)}$$

where $\tau_{a_{ij}}$ is the apparent surface stress, $\tau_{ij}$ is the surface stress on the $i$th face and acting in the $j$th direction computed for a Newtonian fluid with the use of mass-averaged velocities, and $m_{1i}$ is the diffusive flux of species in the $i$th direction relative to the mass-averaged velocity in the $i$th direction.

The pressure diffusion term in Eq. (B-3) indicates that a net movement of species 1 can occur if a pressure gradient is imposed. Although this is normally a negligible effect, it can be important in swirling flows where tremendous pressure gradients occasionally exist, such as is the case in some centrifuges. The body-force diffusion term is nonzero only when different body forces act on the two components. This case occurs in plasma technology where the fluid interacts with electric and magnetic forces and in ionic systems, generally. If gravity is the only body force, then $\mathbf{B}_1 = \mathbf{B}_2$ and the body-force diffusion term vanishes. The thermal diffusion term describes the tendency of a species of mass to diffuse in an imposed temperature gradient and is negligible unless very large gradients are encountered. Thermal diffusion is often called the *Soret effect* and was independently predicted by Chapman and Enskog from a refined kinetic theory analysis. The thermal diffusion effect has been utilized for isotope separation in the Clusius-Dickel column [9, 11], which combines convection to achieve continuous separation. A gas mixture is placed in a vertical column whose walls are cool and with a heated wire in the center. Thermal diffusion enriches the fraction of one species in the hot central region and enriches the fraction of the other species in the cool wall region. Natural convection moves the cool species to the bottom and the hot species to the top, thereby achieving continuous species separation.

The diffusion thermo term in Eq. (B-4) is often called the *Dufour effect* in recognition of his discovery of this influence in 1873. It indicates that a diffusive mass flux also gives rise to an energy flux. The thermal diffusion factor $\alpha$ is also important to this effect. Although normally negligible, the diffusion thermo effect can be appreciable when, for example, helium is blown through a porous surface into a hot gas stream in an effort to protect the surface from the hot gas [12, 13]. The interdiffusional convection term of Eq. (B-4), also usually negligible, indicates that diffusional mass transfer leads to a net energy flux, even when the net mass diffusion is zero if the different species of mass particles carry different amounts of energy at the same temperature.

As seen in Eq. (B-4), this effect, like the other coupling effects, can be regarded as second order inasmuch as it involves the product of two small terms (the difference in species enthalpy times the mass diffusion rate relative to the mass-averaged velocity).

The diffusional momentum transport term in Eq. (B-5) is usually negligible. As discussed previously in a related case, it is a second-order term and is usually negligible since it involves the square of diffusional mass fluxes.

As pointed out by Eckert [12], among others, the definition of properties in a multicomponent fluid requires a careful statement of conventions. To illustrate this, consider Eqs. (B-3) and (B-4), which when taken together give the heat flux, neglecting pressure diffusion and body-force diffusion, as

$$\mathbf{q} = -\left[k + \frac{\rho D_{12}\omega_1\omega_2\alpha}{T}\left(H_1 - H_2 - \frac{RTM^2\alpha}{M_1 M_2}\right)\right]\nabla T$$

$$- \rho D_{12}\left[H_1 - H_2 - \frac{RTM^2\alpha}{M_1 M_2}\right]\nabla\omega_1 \qquad (B-6)$$

For a binary mixture, therefore, the thermal conductivity which describes diffusional heat flux under the condition of zero net diffusional mass flux is given in Eq. (B-4) as $k$, the conventionally understood property for a pure fluid. The thermal conductivity that describes diffusional heat flux under the condition of zero concentration gradient is somewhat different, as it is the coefficient of $\nabla T$ in Eq. (B-6). Only if the thermal diffusion and diffusion thermo effects are negligible are the two thermal conductivities really equivalent.

## TRANSIENT HEAT, MASS, AND MOMENTUM DIFFUSION

The partial differential equations normally used to describe heat, mass, and momentum transfer provide predictions that are in excellent agreement with experimental observations in most physical cases encountered. It is usually true that gradients in space and time are not large, and so there is no sacrifice in accuracy in letting the time interval and control volume size become infinitesimally small in the derivation of the equations of continuity, motion, energy, and mass diffusion. Rate equations (e.g., Fourier's law) that are really accurate only for small gradients in the steady state are then substituted into the equations resulting from this step. Most often, departure from local equilibrium is so slight that excellent accuracy is still achieved, but the limiting behavior of the resulting equations for short times in the face of sudden disturbances (e.g., as might occur in shock waves, detonations, and pulsed laser heating of a solid) is incorrect.

As pointed out by Weymann [14], whose discussion guides the following remarks, the essential difficulty can be understood by consideration of the parabolic equation that describes unsteady heat, mass, and viscous shear diffusion. This equation has the form, for a one-dimensional case, of

$$\frac{\partial \psi}{\partial t} = D \frac{\partial^2 \psi}{\partial x^2} \qquad (B\text{-}7)$$

and the fundamental solution of

$$\psi(x, t) = (4\pi Dt)^{-1/2} e^{-x^2/4Dt}$$

At $t \equiv 0$, this solution gives $\psi \equiv 0$ everywhere except at the origin, where it is large. For $t > 0$, $\psi$ differs from zero at all $x$ regardless of how nearly $t$ approaches zero. Hence a disturbance is apparently propagated with an infinitely large speed, and this is a physically unreasonable result. Of course, this unreasonable result is usually of no interest since the predictions at "large" times are the ones of real interest and are made accurately.

On physical grounds, it is apparent that the "grainy" structure of matter gives a lower limit on space resolutions. This lower limit $\lambda$ is appropriately selected to be the mean free path in gases and the intermolecular spacing in liquids and solids. Similarly, the time needed for a particle of mass to adjust to changed conditions provides a physical lower limit on time resolutions. This lower limit $\tau$ is appropriately given by the inverse of the collision frequency for a simple billiard ball gas, the relaxation time needed to distribute energy in an equilibrium fashion among the various internal storage modes of a complex molecule, or the time required for a disturbance to move the appropriate least space interval at the appropriate characteristic velocity (the mean thermal velocity in a gas and the speed of sound in a liquid or a solid.) Although the precise value of these lower limits cannot be found, these thoughts are usefully employed in achieving a criterion by which the applicability of the usually used describing equations [of the form of Eq. (B-7)] can be judged. In their derivation, either explicitly or implicitly, the approximation is made that, for example,

$$\psi(x_0 \pm \lambda, t) = \psi(x_0, t) \pm \left( \frac{\partial \psi}{\partial x} \bigg|_{x_0, t} \right) \lambda$$

An approximation of the same form is made for time. For these approximations to be accurate and for the solutions to the usual describing equations to be accurate, it is necessary that

$$\left| \frac{1}{\psi} \frac{\partial \psi}{\partial x} \right| \ll \frac{1}{\lambda} \quad \text{and} \quad \left| \frac{1}{\psi} \frac{\partial \psi}{\partial t} \right| \ll \frac{1}{\tau} \qquad (B\text{-}8)$$

Solutions from Eq. (B-7) can be judged for satisfaction of physical assumptions by insertion into the criteria of Eq. (B-8).

The form of the equation more suited than Eq. (B-7) to a description of physical events at short times is derivable from a one-dimensional random walk. A walker moves along the $x$ axis in discrete steps of size $\lambda$, each step taking a time interval $\tau$. At the end of each step, a new one is started immediately. The probability of a step in the $+x$ direction is $p$, and the probability of a step in the $-x$ direction is $q$. The probabilities must sum to unity, $p + q = 1$. The probability that a walker is at the location $x = k\lambda$ after $n$ steps is denoted by $u_{n,k}$. To reach $x = k\lambda$ after the $(n + 1)$th step, the walker must have been either at $x = (k - 1)\lambda$ or $x = (k + 1)\lambda$ after the $n$th step. In symbolic form, this is expressed as

$$u_{n+1,k} = pu_{n,k-1} + qu_{n,k+1} \tag{B-9}$$

For sufficiently large $n$, the discrete function $u_{n,k}$ can be approximated well by a continuous function as

$$u_{n,k} \approx u(t = n\tau, x = k\lambda) \tag{B-10}$$

Insertion of Eq. (B-10) in Eq. (B-9) gives

$$u(t + \tau, x) = pu(t, x - \lambda) + qu(t, x + \lambda)$$

Introduction of a Taylor series expansion about $t$, $x$ into this relationship gives

$$\cdots + \frac{\partial^3 u}{\partial t^3}\frac{\tau^3}{3!} + \frac{\partial^2 u}{\partial t^2}\frac{\tau^2}{2} + \frac{\partial u}{\partial t}\tau + u(t, x)$$

$$= p\left[u(t, x) - \frac{\partial u}{\partial x}\lambda + \frac{\partial^2 u}{\partial x^2}\frac{\lambda^2}{2!} - \frac{\partial^3 u}{\partial x^3}\frac{\lambda^3}{3!} + \frac{\partial^4 u}{\partial x^4}\frac{\lambda^4}{4!} + \cdots\right]$$

$$+ q\left[u(t, x) + \frac{\partial u}{\partial x}\lambda + \frac{\partial^2 u}{\partial x^2}\frac{\lambda^2}{2!} + \frac{\partial^3 u}{\partial x^3}\frac{\lambda^3}{3!} + \frac{\partial^4 u}{\partial x^4}\frac{\lambda^4}{4!} + \cdots\right]$$

which can be slightly simplified into

$$\cdots + \frac{\partial^3 u}{\partial t^3}\frac{\tau^3}{3!} + \frac{\partial^2 u}{\partial t^2}\frac{\tau^2}{2!} + \frac{\partial u}{\partial t}\tau = (q - p)\left[\lambda\frac{\partial u}{\partial x} + \frac{\partial^3 u}{\partial x^3}\frac{\lambda^3}{3!} + \cdots\right]$$

$$+ \frac{\partial^2 u}{\partial x^2}\lambda^2 + \frac{\partial^4 u}{\partial x^4}\frac{2\lambda^4}{4!} + \cdots$$

It is seen that any desired accuracy can be achieved by retaining more terms in the series expansion. Restriction of attention to the accuracy improvement over Eq. (B-7) offered by a single additional term in the time series of this expression results in, for the additional reasonable condition that $q = p$,

$$\frac{\partial^2 u}{\partial t^2} + \frac{2\partial u}{\tau \partial t} = 2\frac{\lambda^2}{\tau^2}\frac{\partial^2 u}{\partial x^2} \tag{B-11}$$

Equation (B-11) is a modified wave equation that has the property that the speed of disturbance propagation is less than the average velocity of the walker $c = \lambda/\tau$. It can be shown from kinetic theory relations that $\lambda^2/\tau \approx 2D$; thus Eq. (B-11) can also be written as

$$\frac{D}{c^2}\frac{\partial^2 u}{\partial t^2} + \frac{\partial u}{\partial t} = D\frac{\partial^2 u}{\partial x^2}$$

in which form it is seen that the diffusion equation form of Eq. (B-7) is accurate if the characteristic speed $c$ is large.

It should be pointed out that a plausible alternative derivation can be made in which, taking heat conduction in a solid as an example, it is first assumed that conductive heat flux is given by

$$\tau\frac{\partial \mathbf{q}}{\partial t} + \mathbf{q} = -k\,\nabla T \tag{B-12}$$

Now, the result of applying the energy conservation principle to an infinitesimal control volume is

$$\rho C_p \frac{\partial T}{\partial t} = -\nabla \cdot \mathbf{q} \tag{B-13}$$

Taking the time derivative of Eq. (B-13) in combination with the result of a $\nabla \cdot$ operation applied to Eq. (B-12), one obtains

$$\tau\frac{\partial^2 T}{\partial t^2} + \frac{\partial T}{\partial t} = \alpha\,\nabla^2 T$$

The discussion presented by Eckert and Drake [6, pp. 23–27] gives historical details, pointing out that unsteady (non-Fourier) effects might be important at low temperatures in liquid helium. Among the numerous findings is that the imposition of a step change in surface heat flux $q_s$ on the surface of a stationary material gives an instantaneous surface temperature jump of

$$\lim_{t \to 0} \Delta T_s = q_s\left(\frac{\tau}{\rho C_p k}\right)^{1/2}$$

for a relaxation time of $\tau$—only 1% departure from Fourier model results is achieved after $t \sim 50\tau$ [15]. Baumeister and Hamill [20] point out that the maximum conductive heat flux that can be achieved by a step change of a surface temperature is

$$q_{max} = q(t = \tau) = \frac{\rho C_p}{3} \Delta T_s c$$

where $c$ is the velocity of propagation. From this result it is found that the maximum convective heat-transfer coefficient is

$$h_{max} = \frac{q_{max}}{\Delta T_s} = \frac{\rho C_p}{3} c$$

which for water at room temperature ($c \approx c_{sound} = 1600$ m/s and $\tau = 2D/c^2 \approx 1.5 \times 10^{-13}$ s) amounts to

$$h_{max} = 3.4 \times 10^8 \text{ Btu/hr ft}^2 \text{ °F} = 19.3 \times 10^8 \text{ W/m}^2 \text{ °C}$$

whereas for air at room temperature ($c \approx c_{sound} = 1100$ ft/sec and $\tau \approx 0.5 \times 10^{-9}$ sec), one finds

$$h_{max} = 7.5 \times 10^4 \text{ Btu/hr ft}^2 \text{ °F} = 42.5 \times 10^4 \text{ W/m}^2 \text{ °C}$$

The possibility of resonant amplification that is presented by a describing equation of the wave-equation form was analytically explored by Lumsdaine [16]. It was found that thermal energy could be amplified to a higher temperature in a solid by imposing on one face a sinusoidally oscillating surface temperature whose frequency nearly equals the resonant frequency $c/2L$, where $L$ is the characteristic size of the solid. The nonlinear behavior of temperature waves in a solid was analytically determined by Lindsay and Straughan [17].

# REFERENCES

1   W. M. Rohsennow and H. Y. Choi, *Heat, Mass, And Momentum Transfer*, Prentice-Hall, Englewood Cliffs, 1961, p. 517.

2   L. N. Novichenok, G. V. Gnilitsky, and L. N. Khokhlenkov, Study of anisotropy of thermal conductivity of flowing systems with shear, *Progress In Heat And Mass Transfer*, Vol. 4, Pergamon Press, New York, 1971, pp. 159–164.

3   M. S. Khader and R. I. Vachon, Heat Transfer in micropolar laminar radial channel flow, *Transact. ASME, J. Heat Transf.* **99**, 684–687 (1977).

4   H. S. Carslaw and J. C. Jaeger, *Conduction Of Heat In Solids*, Clarendon Press, Oxford, 1959, pp. 38–49.

5   M. N. Ozisik, *Boundary Value Problems Of Heat Conduction*, International Textbook Company, Scranton, PA, 1968, pp. 455–479.

6   E. R. G. Eckert and R. M. Drake, *Analysis Of Heat And Mass Transfer*, McGraw-Hill, New York, 1972, pp. 12–17.

7   J. O. Hirschfelder, C. F. Curtis, and R. B. Bird, *Molecular Theory of Gases And Liquids*, Wiley, New York, 1954, pp. 514–523, 584–585.

8   S. Chapman and T. G. Cowling, *The Mathematical Theory of Non-Uniform Gases*, Cambridge University Press, 1958, pp. 244–258, 399–409.

9   R. B. Bird, W. E. Stewart, and E. N. Lightfoot, *Transport Phenomena*, Wiley, New York, 1960, pp. 563–580.

10  E. R. G. Eckert and R. M. Drake, *Analysis Of Heat And Mass Transfer*, McGraw-Hill, New York, 1972, p. 721.

11  R. D. Present, *Kinetic Theory Of Gases*, McGraw-Hill, New York, 1958, pp. 117–127.

12  E. M. Sparrow, W. J. Minkowycz, E. R. G. Eckert, and W. E. Ibele, The effect of diffusion thermo and thermal diffusion for helium injection into plane and axisymmetric stagnation flow of air, *Transact. ASME, J. Heat Transf.* **86**, 311–319 (1964).

13  E. R. Eckert, Diffusion thermo effects in mass transfer cooling, in *Proceedings of the Fifth U.S. National Congress on Applied Mechanics*, 1966, p. 639.

14  H. D. Weymann, Finite speed of propagation in heat conduction, diffusion, and viscous shear motion, *Am. J. Phys.* **35**, 448–496 (1967).

15  M. J. Maurer and H. A. Thompson, NonFourier effects at high heat flux, *Transact. ASME, J. Heat Transf.* **95**, 284–286 (1973).

16  E. Lumsdaine, Thermal resonance, *Mech. Eng. News*, **9**, 34–37 (1972).

17  K. A. Lindsay and B. Straughan, Temperature waves in a rigid heat conductor, *Zeitschrift für angewandte Mathematik und Physik* **27**, 653–662 (1976).

18  J. R. Baron, Thermodynamic coupling in boundary layers, *ARS J.* **32**, 1053–1059 (1962).

19  H. Hoshizaki, Heat transfer in planetary atmospheres at supersatellite speeds, *ARS J.* **32**, 1544–1552 (1962).

20  K. J. Baumeister and T. D. Hamill, Hyperbolic heat-conduction equation—a solution for the semi-infinite body problem, *Transact. ASME, J. Heat Transfer* **93**, 126–127 (1977).

# C

# NEWTONIAN AND NON-NEWTONIAN FLUIDS

The linear relations between velocities and the stresses acting on a fluid element which characterize a Newtonian fluid can either be determined on fundamental grounds from a detailed consideration of molecular behavior or by analogy with the relationships that describe the behavior of solids. Fortunately, both lines of inquiry yield the same result. The latter procedure is more readily comprehended and thus is pursued here.

First, consider an initially rectangular element of material that is subjected to shear as illustrated in Fig. C-1. For a solid, stresses are proportional to strains, and it is experimentally observed that

$$\tau_{yx} = G\gamma_{yx}$$

for small strains, where $G$ is the shear modulus of the solid material and $\gamma_{yx}$ is the angle in the $y$–$x$ plane through which the vertical side of the material element has displaced as a consequence of the imposed shear stress $\tau_{yx}$. For a fluid, stresses are proportional to the rate of strain, and it is experimentally observed that

$$\tau_{yx} = \mu \frac{du}{dy} \tag{C-2}$$

That these two relations are analogous can be demonstrated by realizing that, for the fluid-flow case, the change in angular displacement that occurs in a

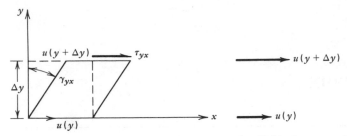

**Figure C-1** Deformation rate of an initially rectangular fluid element subjected to shear.

time interval $\Delta t$ is

$$\Delta\gamma_{yx} = \arctan\left\{[u(y+\Delta y) - u(y)]\frac{\Delta t}{\Delta y}\right\}$$

In the limit as all differential quantities become small, this result yields

$$\dot{\gamma}_{yx} = \frac{du}{dy} \tag{C-3}$$

Insertion of this result into Eq. (C-2) shows that for a fluid

$$\tau_{yx} = \mu\dot{\gamma}_{yx} \tag{C-4}$$

which is directly analogous to the relationship for a solid [Eq. (C-1)], with viscosity replacing shear modulus and angular strain *rate* replacing angular strain.

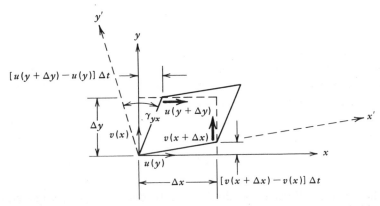

**Figure C-2** Rotation of adjoining sides of an initially rectangular fluid element subjected to shear.

In general a fluid element experiences rotation of both the initially vertical and the initially horizontal sides. As shown in Fig. C-2, the included angle at the lower left corner of the fluid element undergoes an angular displacement given by

$$\frac{\pi}{2} - \arctan\left\{ [u(y + \Delta y) - u(y)]\frac{\Delta t}{\Delta y} \right\}$$

$$- \arctan\left\{ [v(x + \Delta x) - v(x)]\frac{\Delta t}{\Delta x} \right\} - \frac{\pi}{2}$$

Referring this angular displacement to the rotated $x'-y'$ axes and taking the limit as all differential quantities become small, one finds that

$$\dot{\gamma}_{yx} = \frac{\partial u}{\partial y} + \frac{\partial v}{\partial x}$$

In similar fashion it is found that the rates of angular strain viewed in the $y-z$ and $x-z$ planes are given by

$$\dot{\gamma}_{yz} = \frac{\partial v}{\partial z} + \frac{\partial w}{\partial y}$$

$$\dot{\gamma}_{xz} = \frac{\partial w}{\partial x} + \frac{\partial u}{\partial z}$$

As a consequence of the analogy previously developed, it then follows that

$$\tau_{yx} = \mu\left( \frac{\partial u}{\partial y} + \frac{\partial v}{\partial x} \right) = \tau_{xy} \tag{C-5a}$$

$$\tau_{yz} = \mu\left( \frac{\partial v}{\partial z} + \frac{\partial w}{\partial y} \right) = \tau_{zy} \tag{C-5b}$$

$$\tau_{xz} = \mu\left( \frac{\partial w}{\partial x} + \frac{\partial u}{\partial z} \right) = \tau_{zx} \tag{C-5c}$$

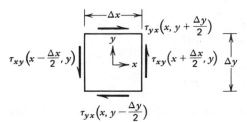

**Figure C-3** Free-body diagram to demonstrate that $\tau_{ij} = \tau_{ji}$.

The need for equality of $\tau_{ij}$ and $\tau_{ji}$ can be demonstrated by summing the moments shown in the free-body diagram of Fig. C-3 about the $z$ axis. Such a procedure gives

$$\Sigma \text{ torque} = \text{moment of inertia about } z \text{ axis} \times \text{angular acceleration}$$

$$\left[ \tau_{yx}\left( x, y + \frac{\Delta y}{2} \right) + \tau_{yx}\left( x, y - \frac{\Delta y}{2} \right) \right] \frac{\Delta x \, \Delta y}{2}$$

$$- \left[ \tau_{xy}\left( x + \frac{\Delta x}{2}, y \right) + \tau_{xy}\left( x - \frac{\Delta x}{2}, y \right) \right] \frac{\Delta x \, \Delta y}{2}$$

$$= \left( \rho \, \Delta x \, \Delta y \frac{\Delta x^2 + \Delta y^2}{12} \right) \frac{\text{angular acceleration}}{g_c}$$

In the limit as all differential quantities become small, this relation becomes

$$\tau_{yx} - \tau_{xy} = \rho(\Delta x^2 + \Delta y^2) \frac{\text{angular acceleration}}{12 g_c}$$

Since it is not experimentally observed that a small fluid element spins rapidly, it must be the case that

$$\tau_{yx} = \tau_{xy}$$

or, in general, $\tau_{ij} = \tau_{ji}$. The existence of couple stresses has been postulated, however, that can invalidate this equality [1].

The relationship between the normal stresses and strain is slightly more complicated in form than for the shear cases of Eqs. (C-1)–(C-5), although an analogy is still developed here. For a solid, imposition of a balanced normal stress in only the $x$ direction $\tau_{xx}$ will cause a strain in the $x$ direction $\varepsilon_x$ whose magnitude is

$$\varepsilon_x = \frac{\tau_{xx}}{E}$$

where $E$ is the modulus of elasticity. In addition, the imposed normal stress in the $x$ direction will cause strains in the other perpendicular directions, $\varepsilon_y$ and $\varepsilon_z$, which are proportional to $\varepsilon_x$ so that

$$\varepsilon_y = -\frac{\varepsilon_x}{m} = -\frac{\tau_{xx}}{mE}$$

$$\varepsilon_z = -\frac{\varepsilon_x}{m} = -\frac{\tau_{xx}}{mE}$$

where $m$ denotes Poisson's ratio, which is about $3/10$ for many solids. The net strain in the $x$ direction for normal stresses acting in all three coordinated directions is then

$$\varepsilon_x = \frac{\tau_{xx}}{E} - \frac{\tau_{yy}}{mE} - \frac{\tau_{zz}}{mE} \tag{C-6a}$$

Similarly,

$$\varepsilon_y = \frac{\tau_{yy}}{E} - \frac{\tau_{xx}}{mE} - \frac{\tau_{zz}}{mE} \tag{C-6b}$$

$$\varepsilon_z = \frac{\tau_{zz}}{E} - \frac{\tau_{xx}}{mE} - \frac{\tau_{yy}}{mE} \tag{C-6c}$$

On addition of Eqs. (C-6), the volume dilation $dV/V = e = \varepsilon_x + \varepsilon_y + \varepsilon_z$ is found to be

$$e = \frac{m-2}{mE}(\tau_{xx} + \tau_{yy} + \tau_{zz}) \tag{C-7}$$

Rearrangement of Eq. (C-6a) into the form

$$E\varepsilon_x = \tau_{xx}\left(1 + \frac{1}{m}\right) - \frac{\tau_{xx} + \tau_{yy} + \tau_{zz}}{m}$$

with a following substitution from Eq. (C-7) gives the more convenient relationship

$$E\varepsilon_x = \tau_{xx}\frac{m+1}{m} - e\frac{E}{m-2}$$

In this way the normal stresses are expressed in terms of the strains as

$$\tau_{xx} = 2G\left(\varepsilon_x + \frac{e}{m-2}\right) \tag{C-8a}$$

$$\tau_{yy} = 2G\left(\varepsilon_y + \frac{e}{m-2}\right) \tag{C-8b}$$

$$\tau_{zz} = 2G\left(\varepsilon_z + \frac{e}{m-2}\right) \tag{C-8c}$$

where $G = (mE/2)/(m+1)$ relates the shear modulus $G$ to the modulus of elasticity $E$ and Poisson's ratio $m$.

To develop the basis for the analogy a bit further, recall that the volume dilatation is defined as

$$\varepsilon = \varepsilon_x + \varepsilon_y + \varepsilon_z$$

and that the strains in the $x$, $y$, and $z$ directions can be considered to be given by

$$\varepsilon_x = \frac{\partial \xi}{\partial x}, \qquad \varepsilon_y = \frac{\partial \eta}{\partial y}, \qquad \varepsilon_z = \frac{\partial \zeta}{\partial z}$$

in which $\xi$, $\eta$, and $\zeta$ are displacements of a point located at $x$, $y$, $z$ before deformation. With this understanding, it is seen then

$$\varepsilon = \mathrm{div}(\mathbf{s}) \tag{C-9}$$

where $\mathbf{s} = \xi \hat{\mathbf{i}} + \eta \hat{\mathbf{j}} + \zeta \hat{\mathbf{k}}$ is the vector displacement described previously.

It was demonstrated earlier that the viscous shear stress relations for a fluid can be obtained from those for a solid by replacing $G$ with $\mu$ and the angular strain with the rate of angular strain. In the same vein, the analogy is extended to obtain the normal stress relations for a fluid from those for a solid by replacing $G$ with $\mu$ (as before) and the strain by the rate of strain. Substitution of this idea into Eqs. (C-8) and (C-9) gives

$$\tau_{xx} = -p + 2\mu \frac{\partial u}{\partial x} - \mu_2 \mathrm{div}(\mathbf{V}) \tag{C-10a}$$

$$\tau_{yy} = -p + 2\mu \frac{\partial v}{\partial y} - \mu_2 \mathrm{div}(\mathbf{V}) \tag{C-10b}$$

$$\tau_{zz} = -p + 2\mu \frac{\partial w}{\partial x} - \mu_2 \mathrm{div}(\mathbf{V}) \tag{C-10c}$$

Here $\mu_2$ is the second coefficient of viscosity that cannot be evaluated by analogy with a solid's behavior. The pressure $p$ has been added* to Eqs. (C-10) because the normal stresses must be equal to the pressure in a static situation, with the analogy to a solid's behavior giving only the viscous stresses.

Evaluation of the second coefficient of viscosity $\mu_2$ can proceed by summing Eqs. (C-10) to obtain

$$p = -\frac{\tau_{xx} + \tau_{yy} + \tau_{zz}}{3} + \left( \frac{2\mu}{3} - \mu_2 \right) \mathrm{div}(\mathbf{V})$$

*It should be noted that addition of Eqs. (C-8) yields

$$\bar{\tau} = \frac{\tau_{xx} + \tau_{yy} + \tau_{xx}}{3} = \frac{2G(m+1)e}{3(m-2)}$$

Addition and substraction of $\bar{\tau}$ to and from Eq. (C-8a), for example, then yields

$$\tau_{xx} = \bar{\tau} + 2G\left( \varepsilon_x - \frac{e}{3} \right)$$

in which the coefficient $2\mu/3 - \mu_2$ is termed the *bulk viscosity*. The average of the normal stresses in a static situation ($\mathbf{V} = 0$) is defined as the pressure. Pursuit of this reasonable convention into the unsteady or compressible situation then requires that there be a zero-valued coefficient of the nonzero div($\mathbf{V}$) term. This requires that the bulk viscosity, $2\mu/3 - \mu_2$, be zero; or

$$\mu_2 = \frac{2\mu}{3} \qquad\qquad (C\text{-}11)$$

which is referred to as *Stoke's hypothesis*. The accuracy of Eq. (C-11) is confirmed by statistical analysis for gases of low density; in dense gases these methods show that bulk viscosity is quite small. However, it must be remembered that the developments of this appendix all presume steady-state conditions; if markedly unsteady conditions are encountered, the bulk viscosity can differ importantly from zero. Processes with a relatively slow transfer of energy between translational, vibrational, and rotational degrees of freedom (e.g., as occur in rapid chemical reactions or ultrasonic vibrations) would be characterized by a nonzero bulk viscosity, in which case $p$ cannot be the average normal stress when density variations are very large. Experiment shows that it is possible for $\mu_2$ to be negative and of a magnitude that is 200 times greater than $\mu$.

An excellent discussion of the second coefficient of viscosity is given by Rosenhead [2], and the effect of bubbles in a liquid on the second coefficient of viscosity is discussed by Taylor [3] and Davies [4]. A parallel derivation of Stokes' viscosity relation is given by Schlichting [5]. The tensor calculus is used (more rigorously than the method of analogy of this Appendix, but also less readably) by Schlichting [6] as well as by Deissler [7], who points out that the validity of assuming a linear relation between stresses and velocities is as important a question as the value of bulk viscosity to be used in a linear model. By analogy for a fluid then

$$\tau_{xx} = \bar{\tau} + 2\mu\frac{\partial\mu}{\partial x} - (2/3)\mu\,\text{div}\,\mathbf{V}$$

with similar results for $\tau_{yy}$ and $\tau_{zz}$. Thus, if there is no wish to deal with thermodynamic pressure (which is really well defined only for equilibrium conditions), use of the average normal stress $\bar{\tau}$ makes consideration of the second coefficient of viscosity unnecessary. The departure of $\mu_2$ from $2\mu/3$ is, accordingly, a measure of the lack of equilibrium in the considered process—$\mu_2$ is then seen to not be entirely a property of the fluid.

Externally imposed electrical fields can influence the apparent viscosity of a fluid [8], although such effects are usually small.

Not all fluids possess a linear relation between stresses and velocities as mentioned earlier in this appendix and in Chapter 4. Those possessing the linear relationship are called *Newtonian fluids* and those without, *non-*

*Newtonian fluids.* Non-Newtonian fluids are commonly encountered in industry and everyday life, and their flow properties are of great interest and importance. Figure 4-5 illustrates the general state of affairs, showing only Newtonian fluids to have a constant viscosity. Because few have experience with non-Newtonian fluids, this discussion emphasizes qualitative description rather than mathematical description. A brief discussion of mathematical descriptions and of further references is given by Bird et al. [9]. In general, mathematical models can be characterized by the number of adjustable parameters in the relationship between shear and velocity gradient. Thus, a Newtonian fluid is a one-parameter model. Worthwhile qualitative discussions are given by Walker [10] and Collyer [11].

An ideal Bingham plastic fluid is one for which the shear rate is proportional to the shear stress only after a yield stress $\tau_0$ has been surpassed. A simple two-parameter representation for Couette flow is

$$\tau_{yx} \pm \tau_0 = \mu \frac{du}{dy}, \qquad |\tau_{yx}| > \tau_0$$

$$0 = \frac{du}{dy}, \qquad |\tau_{yx}| < \tau_0$$

and the proper sign is to be affixed to $\tau_0$ in keeping with the direction of flow. Examples of fluids accurately characterized as a Bingham plastic are cements (typically $\mu \sim 2400$ cP and $\tau_0 \sim 48$ N/m$^2$), toothpaste, drilling mud, sewage sludge, and aqueous nuclear fuel slurries. An actual plastic fluid may be better characterized as having a somewhat curved early portion to its stress–velocity gradient curve. An explanation of plastic fluid behavior that has been advanced is that such fluids have a three-dimensional structure that can withstand shear stress below the yield stress but that break down above this value. Many plastic fluids are not ideal Bingham plastic, having a nonstraight stress–velocity gradient curve after surpassing the yield stress.

Shear-thinning (pseudoplastic) fluids have a shear stress–velocity gradient curve whose slope decreases with increasing velocity gradient. The apparent viscosity decreases, and hence the fluid is "thinned" by the shear. A logarithmic plot of the curve for such a fluid is generally linear, giving rise to the two-parameter power-law mathematical model for Couette flow of

$$\tau_{yx} = \mu \left| \frac{du}{dy} \right|^{n-1} \frac{du}{dy}$$

For shear-thinning fluids, $n < 1$. When $n = 1$, the fluid is Newtonian. Examples of shear-thinning fluids are dilute solutions of high polymers (e.g., polyethylene oxide or polyacrylamide), most printing inks, paper pulp, and napalm. Several explanations for this sort of behavior have been advanced. In one, asymmetric particles or molecules are progressively aligned with streamlines, an alignment that responds nearly instantaneously to changes in imposed shear; after complete alignment at high shear, the apparent viscosity becomes constant and the curve is straight. In another explanation, a shear-thinning fluid has a structure when undisturbed that is broken down as shear stresses are increased. A characteristic of shear-thinning fluids is that they seem to be "lumpy" when poured.

Shear-thickening (dilatant) fluids are less common than shear-thinning fluids and behave conversely. Their apparent viscosity increases with increasing velocity gradient. In terms of the power-law model for Couette flow

$$\tau_{yx} = \mu \left| \frac{du}{dy} \right|^{n-1} \frac{dy}{dy}$$

with $n > 1$. Examples of shear-thickening fluids are close packed suspensions such as some paints, quicksand, wet sand, and starch in water. A mixture of 66.7 g of wheat starch in 100 ml of water provides a good demonstration; slow stirring with a rod meets little resistance, but no vigorous or rapid motion occurs without great resistance. Shear-thickening fluids were apparently first investigated by Osborne Reynolds in 1885. He called them dilatant fluids because his explanation of their behavior required dilation (volume increase) on shearing. In this explanation it is helpful to visualize wet sand. At low shear rates, water lubricates the movement of sand particles past one another and forces are small. At high shear rates, the close packing of sand grains is disrupted because they must lift over one another to move past and the water lubrication is insufficient, and thus, forces are large. Such a physical model requires that all suspensions of solids in liquids should be dilatant at high solid content. Although this expectation is confirmed by observation, it has not been shown that *all* shear-thickening fluids undergo a volume increase when sheared. Hence the term "dilatant" is less descriptive than "shear-thickening." One alternative physical explanation is that shearing forces fluid particles together so that bonds and structure can be formed—rather than disrupted, as would be the case for a shear-thinning fluid. Hence increase of shear leads to a thicker fluid. Alternatively, it has been suggested that fluid particles rubbing against one another can acquire an electrical charge; the resultant electrical attraction leads to an increase in apparent viscosity. Walker [18] discusses the physical phenomena involved and points out that the effects of electrical charges explain the viscous properties of sand, muds, and powders. Yet another physical explanation is advanced for shear-thickening fluids that are suspensions or solutions of long-chain molecules. When the long-chain molecules are at rest, they are coiled. When sheared, the molecules are stretched and the apparent viscosity increases. When completely uncoiled, further shear merely orients the molecules along a streamline and shear thinning begins.

Thixotropic fluids are shear-thinning fluids whose apparent viscosity depends on both the shearing rate and the length of time the shearing has been applied. When shear is removed, a substantial amount of time is required to regain initial conditions. Margarine, some paints, shaving cream, and catsup are examples of a thixotropic fluid. One physical explanation for this effect in the case of asymmetric particles is that, as discussed before for shear-thinning, the structure possessed by an undisturbed fluid is broken down as shear stresses are increased, only now the bonds would take a longer time to be ruptured. For suspensions of clay, however, an alternative explanation seems more likely. The initial structure might be a gel; under shearing action, the gel is transformed into a less solid colloidal fluid. Third, a long-chain polymer may be aligned along streamlines with accompanying uncoiling, disentangling, and stretching to decrease viscosity in a noninstantaneous manner. Negative thixotropic fluids (rheopectic) increase their apparent viscosity and the direction of the preceding physical explanations is reversed.

Viscoelastic fluids have the peculiar property of being elastic to an important extent as well as being viscous. In general, the elasticity is due to the coiling of long-chain polymers in the fluid; shearing compresses or extends these coils, which then behave somewhat like springs. In the case of solutions of polyethylene oxide in water, hydrogen bonding between water molecules and the oxygen atoms of the coiled polyethylene chain gives a structural order that is the origin of the elasticity of the solution. An emulsion of one Newtonian fluid in another is also viscoelastic; distortion of the shape of the dispersed droplets by shearing causes their surface energy to increase, an increase that is released when shear stress is removed. Some soap solutions are observed to be viscoelastic, as well. Silicone putty, STP Oil Treatment, some condensed soups, the thick part of egg white, Slime (a toy manufactured by the Mattel Corporation), some shampoo, some condensed milk, and gelatin in water also exhibit viscoelastic effects. A viscoelastic fluid, when strained a small amount, will spring back to its original condition when it is unstressed. A related swelling after issuing from the orifice of a die is called the *die-swell*, or *Barus*, *effect*. Such fluids also have the capability of being drawn out into long fine threads, a phenomenon called *spinnability*. Self-siphoning occurs when, for example, a stream is poured from a container—after the container is again placed level, the flow continues until the container is emptied. In the Weissenberg effect a viscoelastic fluid climbs up a spinning rod in the center of a container, a behavior that is contrary to that of a Newtonian fluid (which would be depressed in the vicinity of the central rod into a parabola). The Weissenberg effect is a consequence of the shear stress created by the spinning rod leading in turn to an inward-directed radial force that pushes the fluid inward and then up the rod. The Kaye effect refers to the fact that when a viscoelastic fluid is poured onto a surface, small streamers of fluid occasionally emerge from the heap of fluid formed at the point of impact. The Kaye effect is believed to be due to the fact that the poured stream initially falls slowly and, because of this low shearing rate, has a high apparent viscosity and thus a heap forms at the impact point; after the heap is formed, later portions of the impacting stream can experience higher shearing rates and low apparent viscosity, enabling it to bounce elastically from the relatively rigid surface of the heap.

An application of non-Newtonian fluids that deserves separate mention is that of drag reduction and increased stability of liquid jets. It has been found that addition of small amounts of polymers, such as polyethylene oxide at 30 ppm (parts per miilion), can reduce turbulent drag in pipes and on ship hulls by as much as 60%. This effect is confined to turbulent flow, the viscosity for laminar flow being slightly increased by the polymer that would increase drag. This effect is believed by Peterline [12] and Gordon [13] to be caused by suppression of part of the spectrum of turbulence by stretching and partial uncoiling of macromolecules of polymer. Also, the breakup of a water jet can be greatly delayed by addition of 0.01% polyethylene oxide. The reduction of drag is generally accompanied by a reduction in heat transfer, an effect discussed in the survey by Dimant and Poreh [14] (see also Ghajar and Tiederman [15]). Polymeric and fibrous additives sometimes reduce drag more when used in combination than when either is used alone [16].

A general critical review of the heat transfer and flow characteristics by Metzner [17] is available to augment and amplify the preceding comments.

# REFERENCES

1   V. K. Stokes, On some effects of couple stresses in fluids on heat transfer, *Transact. ASME, J. Heat Transf.* **91**, 182–184.

2   L. Rosenhead, The second coefficient of viscosity: A brief review of fundamentals, *Proc. Roy. Soc. Lond.* **A226**, 1–6, (1954).

3   G. I. Taylor, The two coefficients of viscosity for an incompressible fluid containing air bubbles, *Proc. Roy. Soc. Lond.* **A226**, 34–39 (1954).

4   R. O. Davies, A note on Sir Geoffrey Taylor's paper, *Proc. Roy. Soc. Lond.* **A226**, 39 (1954).

5   H. Schlichting, *Boundary Layer Theory*, 4th ed., McGraw-Hill, New York, 1960, pp. 46–51.

6   H. Schlichting, *Boundary-Layer Theory*, 6th ed., McGraw-Hill, New York, 1968, pp. 49–60.

7   R. G. Deissler, Derivation of the Navier–Stokes equations, *Am. J. Phys.*, **44** (11), 1129 (1976).

8   A. A. Kim, Application of electrorheological effect for mechanical treatment of mechanolabile materials, *Prog. Heat Mass Transf.* (*Internatl. J. Heat Mass Transf.*) **4**, 127–132 (1971).

9   R. B. Bird, W. E. Stewart, and E. N. Lightfoot, *Transport Phenomena*, Wiley, New York, 1965, pp. 10–15, 101–106.

10   J. Walker, Serious fun with polyox, Silly Putty, Slime, and other non-Newtonian fluids, in "Amateur Scientist," *Sci. Am.* **239**, 186–196 (1978).

11   A. A. Collyer, Time independent fluids, *Physics Ed.* **8**, 333–338 (1973); Demonstrations with viscoelastic liquids, ibid., **8**, 111–116 (1973); Viscoelastic fluids, ibid., **9**, 313–321 (1974).

12   A. Peterline, Molecular model of drag reduction by polymer solutes, *Nature*, **227**, 598–599 (1970).

13   R. J. Gordon, Mechanism for turbulent drag reduction in dilute polymer solutions, *Nature* **227**, 599–600 (1970).

14   Y. Dimant and M. Poreh, Heat transfer in flows with drag reduction, *Adv. Heat Transf.* **12**, 77–113 (1976).

15   A. J. Ghajar and W. G. Tiederman, Prediction of heat transfer coefficients in drag reducing turbulent pipe flows, *AIChE J.* **23**, 128–131 (1977).

16   D. D. Kale and A. B. Metzner, Turbulent drag reduction in dilute fiber suspensions: Mechanistic considerations, *AIChE J.* **22**, 669–674 (1976).

17   A. B. Metzner, Heat transfer in non-Newtonian fluids, *Adv. Heat Transf.* **2**, 357–397 (1965).

18   J. Walker, Why do particles of sand and mud stick together when they are wet?, in "Amateur Scientist," *Sci. Am.* **246** (N.I), 174–179 (1982).

# CONTINUITY, MOTION, ENERGY, AND DIFFUSION EQUATIONS

Equations of continuity, motion, energy, and diffusion in several coordinate systems are presented in this appendix.

## CONTINUITY

Rectangular coordinates $(x, y, z)$ are

$$\frac{\partial \rho}{\partial t} + \frac{\partial}{\partial x}(\rho v_x) + \frac{\partial}{\partial y}(\rho v_y) + \frac{\partial}{\partial z}(\rho v_z) = 0$$

Cylindrical coordinates $(r, \theta, z)$ are

$$\frac{\partial \rho}{\partial t} + \frac{1}{r}\frac{\partial}{\partial r}(\rho r v_r) + \frac{1}{r}\frac{\partial}{\partial \theta}(\rho v_\theta) + \frac{\partial}{\partial z}(\rho v_z) = 0$$

Spherical coordinates $(r, \theta, \phi)$ are

$$\frac{\partial \rho}{\partial t} + \frac{1}{r^2}\frac{\partial}{\partial r}(\rho r^2 v_r) + \frac{1}{r\sin\theta}\frac{\partial}{\partial \theta}(\rho v_\theta \sin\theta) + \frac{1}{r\sin\theta}\frac{\partial}{\partial \phi}(\rho v_\phi) = 0$$

## MOTION

### Motion In Rectangular Coordinates

In terms of viscous stresses:

$x$ Component $\quad \rho\left(\dfrac{\partial v_x}{\partial t} + v_x\dfrac{\partial v_x}{\partial x} + v_y\dfrac{\partial v_x}{\partial y} + v_z\dfrac{\partial v_x}{\partial z}\right)$

$$= -\frac{\partial p}{\partial x} + \left(\frac{\partial \tau_{xx}}{\partial x} + \frac{\partial \tau_{yx}}{\partial y} + \frac{\partial \tau_{zx}}{\partial z}\right) + \rho g_x$$

$y$ Component $\quad \rho\left(\dfrac{\partial v_y}{\partial t} + v_x\dfrac{\partial v_y}{\partial x} + v_y\dfrac{\partial v_y}{\partial y} + v_z\dfrac{\partial v_y}{\partial z}\right)$

$$= -\frac{\partial p}{\partial y} + \left(\frac{\partial \tau_{xy}}{\partial x} + \frac{\partial \tau_{yy}}{\partial_y} + \frac{\partial \tau_{zy}}{\partial z}\right) + \rho g_y$$

$z$ Component $\quad \rho\left(\dfrac{\partial v_z}{\partial t} + v_x\dfrac{\partial v_z}{\partial x} + v_y\dfrac{\partial v_z}{\partial y} + v_z\dfrac{\partial v_z}{\partial z}\right)$

$$= -\frac{\partial p}{\partial z} + \left(\frac{\partial \tau_{xz}}{\partial x} + \frac{\partial \tau_{yz}}{\partial y} + \frac{\partial \tau_{zz}}{\partial z}\right) + \rho g_z$$

In terms of velocity gradients for a Newtonian fluid with constant $\rho$ and $\mu$:

$x$ Component $\quad \rho\left(\dfrac{\partial v_x}{\partial t} + v_x\dfrac{\partial v_x}{\partial x} + v_y\dfrac{\partial v_x}{\partial y} + v_z\dfrac{\partial v_x}{\partial z}\right)$

$$= -\frac{\partial p}{\partial x} + \mu\left(\frac{\partial^2 v_x}{\partial x^2} + \frac{\partial^2 v_x}{\partial y^2} + \frac{\partial^2 v_x}{\partial z^2}\right) + \rho g_x$$

$y$ Component $\quad \rho\left(\dfrac{\partial v_y}{\partial t} + v_x\dfrac{\partial v_y}{\partial x} + v_y\dfrac{\partial v_y}{\partial y} + v_z\dfrac{\partial v_y}{\partial z}\right)$

$$= -\frac{\partial p}{\partial y} + \mu\left(\frac{\partial^2 v_y}{\partial x^2} + \frac{\partial^2 v_y}{\partial y^2} + \frac{\partial^2 v_y}{\partial z^2}\right) + \rho g_y$$

$z$ Component $\quad \rho\left(\dfrac{\partial v_z}{\partial t} + v_x\dfrac{\partial v_z}{\partial x} + v_y\dfrac{\partial v_z}{\partial y} + v_z\dfrac{\partial v_z}{\partial z}\right)$

$$= -\frac{\partial p}{\partial z} + \mu\left(\frac{\partial^2 v_z}{\partial x^2} + \frac{\partial^2 v_z}{\partial y^2} + \frac{\partial^2 v_z}{\partial z^2}\right) + \rho g_z$$

## Motion In Cylindrical Coordinates

In terms of viscous stresses:

$r$ Component $\quad \rho\left(\dfrac{\partial v_r}{\partial t} + v_r\dfrac{\partial v_r}{\partial r} + \dfrac{v_\theta}{r}\dfrac{\partial v_r}{\partial \theta} - \dfrac{v_\theta^2}{r} + v_z\dfrac{\partial v_r}{\partial z}\right)$

$$= -\dfrac{\partial p}{\partial r} + \left(\dfrac{1}{r}\dfrac{\partial}{\partial r}(r\tau_{rr}) + \dfrac{1}{r}\dfrac{\partial \tau_{r\theta}}{\partial \theta} - \dfrac{\tau_{\theta\theta}}{r} + \dfrac{\partial \tau_{rz}}{\partial z}\right) + \rho g_r$$

$\theta$ Component $\quad \rho\left(\dfrac{\partial v_\theta}{\partial t} + v_r\dfrac{\partial v_\theta}{\partial r} + \dfrac{v_\theta}{r}\dfrac{\partial v_\theta}{\partial \theta} + \dfrac{v_r v_\theta}{r} + v_z\dfrac{\partial v_\theta}{\partial z}\right)$

$$= -\dfrac{1}{r}\dfrac{\partial p}{\partial \theta} + \left(\dfrac{1}{r^2}\dfrac{\partial}{\partial r}(r^2\tau_{r\theta}) + \dfrac{1}{r}\dfrac{\partial \tau_{\theta\theta}}{\partial \theta} + \dfrac{\partial \tau_{\theta z}}{\partial z}\right) + \rho g_\theta$$

$z$ Component $\quad \rho\left(\dfrac{\partial v_z}{\partial t} + v_r\dfrac{\partial v_z}{\partial r} + \dfrac{v_\theta}{r}\dfrac{\partial v_z}{\partial \theta} + v_z\dfrac{\partial v_z}{\partial z}\right)$

$$= -\dfrac{\partial p}{\partial z} + \left(\dfrac{1}{r}\dfrac{\partial}{\partial r}(r\tau_{rz}) + \dfrac{1}{r}\dfrac{\partial \tau_{\theta z}}{\partial \theta} + \dfrac{\partial \tau_{zz}}{\partial z}\right) + \rho g_z$$

In terms of velocity gradients for a Newtonian fluid with constant $\rho$ and $\mu$:

$r$ Component $\quad \rho\left(\dfrac{\partial v_r}{\partial t} + v_r\dfrac{\partial v_r}{\partial r} + \dfrac{v_\theta}{r}\dfrac{\partial v_r}{\partial \theta} - \dfrac{v_\theta^2}{r} + v_z\dfrac{\partial v_r}{\partial z}\right)$

$$= -\dfrac{\partial p}{\partial r} + \mu\left[\dfrac{\partial}{\partial r}\left(\dfrac{1}{r}\dfrac{\partial}{\partial r}(rv_r)\right)\right.$$

$$\left. + \dfrac{1}{r^2}\dfrac{\partial^2 v_r}{\partial \theta^2} - \dfrac{2}{r^2}\dfrac{\partial v_\theta}{\partial \theta} + \dfrac{\partial^2 v_r}{\partial z^2}\right] + \rho g_r$$

$\theta$ Component $\quad \rho\left(\dfrac{\partial v_\theta}{\partial t} + v_r\dfrac{\partial v_\theta}{\partial r} + \dfrac{v_\theta}{r}\dfrac{\partial v_\theta}{\partial \theta} + \dfrac{v_r v_\theta}{r} + v_z\dfrac{\partial v_\theta}{\partial z}\right)$

$$= -\dfrac{1}{r}\dfrac{\partial p}{\partial \theta} + \mu\left[\dfrac{\partial}{\partial r}\left(\dfrac{1}{r}\dfrac{\partial}{\partial r}(rv_\theta)\right)\right.$$

$$\left. + \dfrac{1}{r^2}\dfrac{\partial^2 v_\theta}{\partial \theta^2} + \dfrac{2}{r^2}\dfrac{\partial v_r}{\partial \theta} + \dfrac{\partial^2 v_\theta}{\partial z^2}\right] + \rho g_\theta$$

$z$ Component $\quad \rho\left(\dfrac{\partial v_z}{\partial t} + v_r\dfrac{\partial v_z}{\partial r} + \dfrac{v_\theta}{r}\dfrac{\partial v_z}{\partial \theta} + v_z\dfrac{\partial v_z}{\partial z}\right)$

$$= -\dfrac{\partial p}{\partial z} + \mu\left[\dfrac{1}{r}\dfrac{\partial}{\partial r}\left(r\dfrac{\partial v_z}{\partial r}\right) + \dfrac{1}{r^2}\dfrac{\partial^2 v_z}{\partial \theta^2} + \dfrac{\partial^2 v_z}{\partial z^2}\right] + \rho g_z$$

## Motion In Spherical Coordinates

In terms of viscous stresses:

$r$ component

$$\rho\left(\frac{\partial v_r}{\partial t} + v_r\frac{\partial v_r}{\partial r} + \frac{v_\theta}{r}\frac{\partial v_r}{\partial \theta} + \frac{v_\phi}{r\sin\theta}\frac{\partial v_r}{\partial \phi} - \frac{v_\theta^2 + v_\phi^2}{r}\right)$$

$$= -\frac{\partial p}{\partial r} + \left(\frac{1}{r^2}\frac{\partial}{\partial r}\left(r^2\tau_{rr}\right) + \frac{1}{r\sin\theta}\frac{\partial}{\partial \theta}\left(\tau_{r\theta}\sin\theta\right)\right.$$

$$\left. + \frac{1}{r\sin\theta}\frac{\partial \tau_{r\phi}}{\partial \phi} - \frac{\tau_{\theta\theta} + \tau_{\phi\phi}}{r}\right) + \rho g_r$$

$\theta$ Component

$$\rho\left(\frac{\partial v_\theta}{\partial t} + v_r\frac{\partial v_\theta}{\partial r} + \frac{v_\theta}{r}\frac{\partial v_\theta}{\partial \theta} + \frac{v_\phi}{r\sin\theta}\frac{\partial v_\theta}{\partial \phi} + \frac{v_r v_\theta}{r} - \frac{v_\phi^2\cot\theta}{r}\right)$$

$$= -\frac{1}{r}\frac{\partial p}{\partial \theta} + \left(\frac{1}{r^2}\frac{\partial}{\partial r}\left(r^2\tau_{r\theta}\right) + \frac{1}{r\sin\theta}\frac{\partial}{\partial \theta}\left(\tau_{\theta\theta}\sin\theta\right)\right.$$

$$\left. + \frac{1}{r\sin\theta}\frac{\partial \tau_{\theta\phi}}{\partial \phi} + \frac{\tau_{r\theta}}{r} - \frac{\cot\theta}{r}\tau_{\phi\phi}\right) + \rho g_\theta$$

$\phi$ Component

$$\rho\left(\frac{\partial v_\phi}{\partial t} + v_r\frac{\partial v_\phi}{\partial r} + \frac{v_\theta}{r}\frac{\partial v_\phi}{\partial \theta} + \frac{v_\phi}{r\sin\theta}\frac{\partial v_\phi}{\partial \phi} + \frac{v_\phi v_r}{r} + \frac{v_\theta v_\phi}{r}\cot\theta\right)$$

$$= -\frac{1}{r\sin\theta}\frac{\partial p}{\partial \phi} + \left(\frac{1}{r^2}\frac{\partial}{\partial r}\left(r^2\tau_{r\phi}\right) + \frac{1}{r}\frac{\partial \tau_{\theta\phi}}{\partial \theta} + \frac{1}{r\sin\theta}\frac{\partial \tau_{\phi\phi}}{\partial \phi}\right.$$

$$\left. + \frac{\tau_{r\phi}}{r} + \frac{2\cot\theta}{r}\tau_{\theta\phi}\right) + \rho g_\phi$$

In terms of velocity gradients for a Newtonian fluid with constant $\rho$ and $\mu$:

$r$ Component

$$\rho\left(\frac{\partial v_r}{\partial t} + v_r\frac{\partial v_r}{\partial r} + \frac{v_\theta}{r}\frac{\partial v_r}{\partial \theta} + \frac{v_\phi}{r\sin\theta}\frac{\partial v_r}{\partial \phi} - \frac{v_\theta^2 + v_\phi^2}{r}\right)$$

$$= -\frac{\partial p}{\partial r} + \mu\left(\nabla^2 v_r - \frac{2}{r^2}v_r - \frac{2}{r^2}\frac{\partial v_\theta}{\partial \theta} - \frac{2}{r^2}v_\theta\cot\theta\right.$$

$$\left. - \frac{2}{r^2\sin\theta}\frac{\partial v_\phi}{\partial \phi}\right) + \rho g_r$$

$\theta$ Component

$$\rho\left(\frac{\partial v_\theta}{\partial t} + v_r\frac{\partial v_\theta}{\partial r} + \frac{v_\theta}{r}\frac{\partial v_\theta}{\partial \theta} + \frac{v_\phi}{r\sin\theta}\frac{\partial v_\theta}{\partial \phi} + \frac{v_r v_\theta}{r} - \frac{v_\phi^2\cot\theta}{r}\right)$$

$$= -\frac{1}{r}\frac{\partial p}{\partial \theta} + \mu\left(\nabla^2 v_\theta + \frac{2}{r^2}\frac{\partial v_r}{\partial \theta} - \frac{v_\theta}{r^2\sin^2\theta}\right.$$

$$\left. - \frac{2\cos\theta}{r^2\sin^2\theta}\frac{\partial v_\phi}{\partial \phi}\right) + \rho g_\theta$$

$\phi$ Component $\quad \rho \left( \dfrac{\partial v_\phi}{\partial t} + v_r \dfrac{\partial v_\phi}{\partial r} + \dfrac{v_\theta}{r} \dfrac{\partial v_\phi}{\partial \theta} + \dfrac{v_\phi}{r \sin \theta} \dfrac{\partial v_\phi}{\partial \phi} + \dfrac{v_\phi v_r}{r} + \dfrac{v_\theta v_\phi}{r} \cot \theta \right)$

$$= -\dfrac{1}{r \sin \theta} \dfrac{\partial p}{\partial \phi} + \mu \left( \nabla^2 v_\phi - \dfrac{v_\phi}{r^2 \sin^2 \theta} + \dfrac{2}{r^2 \sin \theta} \dfrac{\partial v_r}{\partial \phi} \right.$$

$$\left. + \dfrac{2 \cos \theta}{r^2 \sin^2 \theta} \dfrac{\partial v_\theta}{\partial \phi} \right) + \rho g_\phi$$

where

$$\nabla^2 = \dfrac{1}{r^2} \dfrac{\partial}{\partial r} \left( r^2 \dfrac{\partial}{\partial r} \right) + \dfrac{1}{r^2 \sin \theta} \dfrac{\partial}{\partial \theta} \left( \sin \theta \dfrac{\partial}{\partial \theta} \right) + \dfrac{1}{r^2 \sin^2 \theta} \left( \dfrac{\partial^2}{\partial \phi^2} \right)$$

## VISCOUS STRESS COMPONENTS FOR NEWTONIAN FLUIDS

### Rectangular Coordinates $(x, y, z)$

$$\tau_{xx} = \mu \left[ 2 \dfrac{\partial v_x}{\partial x} - \dfrac{2}{3} (\nabla \cdot \mathbf{v}) \right]$$

$$\tau_{yy} = \mu \left[ 2 \dfrac{\partial v_y}{\partial y} - \dfrac{2}{3} (\nabla \cdot \mathbf{v}) \right]$$

$$\tau_{zz} = \mu \left[ 2 \dfrac{\partial v_z}{\partial z} - \dfrac{2}{3} (\nabla \cdot \mathbf{v}) \right]$$

$$\tau_{xy} = \tau_{yx} = \mu \left[ \dfrac{\partial v_x}{\partial y} + \dfrac{\partial v_y}{\partial x} \right]$$

$$\tau_{yz} = \tau_{zy} = \mu \left[ \dfrac{\partial v_y}{\partial z} + \dfrac{\partial v_z}{\partial y} \right]$$

$$\tau_{zx} = \tau_{xz} = \mu \left[ \dfrac{\partial v_z}{\partial x} + \dfrac{\partial v_x}{\partial z} \right]$$

### Cylindrical Coordinates $(r, \theta, z)$

$$\tau_{rr} = \mu \left[ 2 \dfrac{\partial v_r}{\partial r} - \dfrac{2}{3} (\nabla \cdot \mathbf{v}) \right]$$

$$\tau_{\theta\theta} = \mu \left[ 2 \left( \dfrac{1}{r} \dfrac{\partial v_\theta}{\partial \theta} + \dfrac{v_r}{r} \right) - \dfrac{2}{3} (\nabla \cdot \mathbf{v}) \right]$$

$$\tau_{zz} = \mu \left[ 2 \frac{\partial v_z}{\partial z} - \frac{2}{3} (\nabla \cdot \mathbf{v}) \right]$$

$$\tau_{r\theta} = \tau_{\theta r} = \mu \left[ r \frac{\partial}{\partial r} \left( \frac{v_\theta}{r} \right) + \frac{1}{r} \frac{\partial v_r}{\partial \theta} \right]$$

$$\tau_{\theta z} = \tau_{z\theta} = \mu \left[ \frac{\partial v_\theta}{\partial z} + \frac{1}{r} \frac{\partial v_z}{\partial \theta} \right]$$

$$\tau_{zr} = \tau_{rz} = \mu \left[ \frac{\partial v_z}{\partial r} + \frac{\partial v_r}{\partial z} \right]$$

## Spherical Coordinates $(r, \theta, \phi)$:

$$\tau_{rr} = \mu \left[ 2 \frac{\partial v_r}{\partial r} - \frac{2}{3} (\nabla \cdot \mathbf{v}) \right]$$

$$\tau_{\theta\theta} = \mu \left[ 2 \left( \frac{1}{r} \frac{\partial v_\theta}{\partial \theta} + \frac{v_r}{r} \right) - \frac{2}{3} (\nabla \cdot \mathbf{v}) \right]$$

$$\tau_{\phi\phi} = \mu \left[ 2 \left( \frac{1}{r \sin \theta} \frac{\partial v_\phi}{\partial \phi} + \frac{v_r}{r} + \frac{v_\theta \cot \theta}{r} \right) - \frac{2}{3} (\nabla \cdot \mathbf{v}) \right]$$

$$\tau_{r\theta} = \tau_{\theta r} = \mu \left[ r \frac{\partial}{\partial r} \left( \frac{v_\theta}{r} \right) + \frac{1}{r} \frac{\partial v_r}{\partial \theta} \right]$$

$$\tau_{\theta\phi} = \tau_{\phi\theta} = \mu \left[ \frac{\sin \theta}{r} \frac{\partial}{\partial \theta} \left( \frac{v_\phi}{\sin \theta} \right) + \frac{1}{r \sin \theta} \frac{\partial v_\theta}{\partial \phi} \right]$$

$$\tau_{\phi r} = \tau_{r\phi} = \mu \left[ \frac{1}{r \sin \theta} \frac{\partial v_r}{\partial \phi} + r \frac{\partial}{\partial r} \left( \frac{v_\phi}{r} \right) \right]$$

## ENERGY

## Energy In Rectangular Coordinates

In terms of viscous stresses and heat fluxes:

$$\rho C_p \left( \frac{\partial T}{\partial t} + u \frac{\partial T}{\partial x} + v \frac{\partial T}{\partial y} + w \frac{\partial T}{\partial x} \right)$$

$$= - \left( \frac{\partial q_x}{\partial x} + \frac{\partial q_y}{\partial y} + \frac{\partial q_z}{\partial z} \right) + q'''$$

$$+\beta T\left(\frac{\partial p}{\partial t} + u\frac{\partial p}{\partial x} + v\frac{\partial p}{\partial y} + w\frac{\partial p}{\partial z}\right) - p\left(\frac{\partial u}{\partial x} + \frac{\partial v}{\partial y} + \frac{\partial w}{\partial z}\right)$$

$$+\left[\left(\tau_{xx}\frac{\partial u}{\partial x} + \tau_{yx}\frac{\partial u}{\partial y} + \tau_{zx}\frac{\partial u}{\partial z}\right) + \left(\tau_{xy}\frac{\partial v}{\partial x} + \tau_{yy}\frac{\partial v}{\partial y} + \tau_{zy}\frac{\partial v}{\partial z}\right)\right.$$

$$\left.+\left(\tau_{xz}\frac{\partial w}{\partial x} + \tau_{yz}\frac{\partial w}{\partial y} + \tau_{zz}\frac{\partial w}{\partial z}\right)\right]$$

In terms of velocity and temperature gradients for $\rho$, $k = \text{const}$:

$$\rho C_p\left(\frac{\partial T}{\partial t} + u\frac{\partial T}{\partial x} + v\frac{\partial T}{\partial y} + w\frac{\partial T}{\partial z}\right)$$

$$= k\left(\frac{\partial^2 T}{\partial x^2} + \frac{\partial^2 T}{\partial y^2} + \frac{\partial^2 T}{\partial z^2}\right) + q'''$$

$$+\beta T\left(\frac{\partial p}{\partial t} + u\frac{\partial p}{\partial x} + v\frac{\partial p}{\partial y} + w\frac{\partial p}{\partial z}\right) + \mu\Phi$$

## Energy In Cylindrical Coordinates

In terms of viscous stresses and heat fluxes:

$$\rho C_p\left(\frac{\partial T}{\partial t} + v_r\frac{\partial T}{\partial r} + \frac{v_\theta}{r}\frac{\partial T}{\partial \theta} + v_z\frac{\partial T}{\partial z}\right)$$

$$= -\left[\frac{1}{r}\frac{\partial(rq_r)}{\partial r} + \frac{1}{r}\frac{\partial q_\theta}{\partial \theta} + \frac{\partial q_z}{\partial z}\right]$$

$$+q''' + \beta T\left(\frac{\partial p}{\partial t} + v_r\frac{\partial p}{\partial r} + \frac{v_\theta}{r}\frac{\partial p}{\partial \theta} + v_z\frac{\partial p}{\partial z}\right)$$

$$-p\left[\frac{1}{r}\frac{\partial(rv_r)}{\partial r} + \frac{1}{r}\frac{\partial(v_\theta)}{\partial \theta} + \frac{\partial(v_z)}{\partial z}\right]$$

$$+\left\{\tau_{rr}\frac{\partial v_r}{\partial r} + \tau_{\theta\theta}\frac{1}{r}\left(\frac{\partial v_\theta}{\partial \theta} + v_r\right) + \tau_{zz}\frac{\partial v_z}{\partial z}\right.$$

$$\left.+\tau_{r\theta}\left[r\frac{\partial(v_\theta/r)}{\partial r} + \frac{1}{r}\frac{\partial v_r}{\partial \theta}\right] + \tau_{rz}\left(\frac{\partial v_z}{\partial r} + \frac{\partial v_r}{\partial z}\right)\right\}$$

In terms of velocity and temperature gradients for $\rho$, $k = $ const:

$$\rho C_p \left( \frac{\partial T}{\partial t} + v_r \frac{\partial T}{\partial r} + \frac{v_\theta}{r} \frac{\partial T}{\partial \theta} + v_z \frac{\partial T}{\partial z} \right)$$

$$= k \left[ \frac{1}{r} \frac{\partial (r \partial T / \partial r)}{\partial r} + \frac{1}{r^2} \frac{\partial^2 T}{\partial \theta^2} + \frac{\partial^2 T}{\partial z^2} \right]$$

$$+ q''' + \beta T \left( \frac{\partial p}{\partial t} + v_r \frac{\partial p}{\partial r} + \frac{v_\theta}{r} \frac{\partial p}{\partial \theta} + v_z \frac{\partial p}{\partial z} \right) + \mu \Phi$$

## Energy In Spherical Coordinates

In terms of viscous stresses and heat fluxes:

$$\rho C_p \left( \frac{\partial T}{\partial t} + v_r \frac{\partial T}{\partial r} + \frac{v_\theta}{r} \frac{\partial T}{\partial \theta} + \frac{v_\phi}{r \sin \theta} \frac{\partial T}{\partial \phi} \right)$$

$$= - \left[ \frac{1}{r^2} \frac{\partial (r^2 q_r)}{\partial r} + \frac{1}{r \sin \theta} \frac{\partial (q_\theta \sin \theta)}{\partial \theta} + \frac{1}{r \sin \theta} \frac{\partial q_\phi}{\partial \phi} \right]$$

$$+ q''' + \beta T \left( \frac{\partial p}{\partial t} + v_r \frac{\partial T}{\partial r} + \frac{v_\theta}{r} \frac{\partial T}{\partial \theta} + \frac{v_\phi}{r \sin \theta} \frac{\partial T}{\partial \phi} \right)$$

$$- p \left[ \frac{1}{r^2} \frac{\partial (r^2 v_r)}{\partial r} + \frac{1}{r \sin \theta} \frac{\partial (v_\theta \sin \theta)}{\partial \theta} + \frac{1}{r \sin \theta} \frac{\partial v_\phi}{\partial \phi} \right]$$

$$+ \left\{ \tau_{rr} \frac{\partial v_r}{\partial r} + \tau_{\theta\theta} \left( \frac{1}{r} \frac{\partial v_\theta}{\partial \theta} + \frac{v_r}{r} \right) + \tau_{\phi\phi} \left( \frac{1}{r \sin \theta} \frac{\partial v_\phi}{\partial \phi} + \frac{v_r}{r} + \frac{v_\theta \cot \theta}{r} \right) \right.$$

$$+ \tau_{r\theta} \left( \frac{\partial v_\theta}{\partial r} + \frac{1}{r} \frac{\partial v_r}{\partial \theta} - \frac{v_\theta}{r} \right) + \tau_{r\phi} \left( \frac{\partial v_\phi}{\partial r} + \frac{1}{r \sin \theta} \frac{\partial v_r}{\partial \phi} - \frac{v_\phi}{r} \right)$$

$$+ \tau_{\theta\phi} \left( \frac{1}{r} \frac{\partial v_\phi}{\partial \theta} + \frac{1}{r \sin \theta} \frac{\partial v_\theta}{\partial \phi} - \frac{\cot \theta}{r} v_\phi \right) \right\}$$

In terms of velocity and temperature gradients for $\rho$, $k = $ const:

$$\rho C_p \left( \frac{\partial T}{\partial t} + v_r \frac{\partial T}{\partial r} + \frac{v_\theta}{r} \frac{\partial T}{\partial \theta} + \frac{v_\phi}{r \sin \theta} \frac{\partial T}{\partial \phi} \right)$$

$$= k \left[ \frac{1}{r^2} \frac{\partial (r^2 \partial T / \partial r)}{\partial r} + \frac{1}{r^2 \sin \theta} \frac{\partial (\sin \theta \, \partial T / \partial \theta)}{\partial \theta} + \frac{1}{r^2 \sin \theta} \frac{\partial^2 T}{\partial \phi^2} \right] + q'''$$

$$+ \beta T \left( \frac{\partial p}{\partial t} + v_r \frac{\partial T}{\partial r} + \frac{v_\theta}{r} \frac{\partial T}{\partial \theta} + \frac{v_\phi}{r \sin \theta} \frac{\partial T}{\partial \phi} \right) + \mu \Phi$$

## VISCOUS DISSIPATION FUNCTION

### Rectangular

$$\Phi = 2\left[\left(\frac{\partial u}{\partial x}\right)^2 + \left(\frac{\partial v}{\partial y}\right)^2 + \left(\frac{\partial w}{\partial z}\right)^2\right]$$

$$+ \left[\frac{\partial v}{\partial x} + \frac{\partial u}{\partial y}\right]^2 + \left[\frac{\partial w}{\partial y} + \frac{\partial v}{\partial z}\right]^2 + \left[\frac{\partial u}{\partial z} + \frac{\partial w}{\partial x}\right]^2$$

$$- \frac{2}{3}\left[\frac{\partial u}{\partial x} + \frac{\partial v}{\partial y} + \frac{\partial w}{\partial z}\right]^2$$

### Cylindrical

$$\Phi = 2\left[\left(\frac{\partial v_r}{\partial r}\right)^2 + \left(\frac{1}{r}\frac{\partial v_\theta}{\partial \theta} + \frac{v_r}{r}\right)^2 + \left(\frac{\partial v_z}{\partial z}\right)^2\right]$$

$$+ \left[r\frac{\partial}{\partial r}\left(\frac{v_\theta}{r}\right) + \frac{1}{r}\frac{\partial v_r}{\partial \theta}\right]^2 + \left[\frac{1}{r}\frac{\partial v_z}{\partial \theta} + \frac{\partial v_\theta}{\partial z}\right]^2$$

$$+ \left[\frac{\partial v_r}{\partial z} + \frac{\partial v_z}{\partial r}\right]^2 - \frac{2}{3}\left[\frac{1}{r}\frac{\partial}{\partial r}(rv_r) + \frac{1}{r}\frac{\partial v_\theta}{\partial \theta} + \frac{\partial v_z}{\partial z}\right]^2$$

### Spherical

$$\Phi = 2\left[\left(\frac{\partial v_r}{\partial r}\right)^2 + \left(\frac{1}{r}\frac{\partial v_\theta}{\partial \theta} + \frac{v_r}{r}\right)^2 + \left(\frac{1}{r\sin\theta}\frac{\partial v_\phi}{\partial \phi} + \frac{v_r}{r} + \frac{v_\theta\cot\theta}{r}\right)^2\right]$$

$$+ \left[r\frac{\partial}{\partial r}\left(\frac{v_\theta}{r}\right) + \frac{1}{r}\frac{\partial v_r}{\partial \theta}\right]^2 + \left[\frac{\sin\theta}{r}\frac{\partial}{\partial \theta}\left(\frac{v_\phi}{\sin\theta}\right) + \frac{1}{r\sin\theta}\frac{\partial v_\theta}{\partial \phi}\right]^2$$

$$+ \left[\frac{1}{r\sin\theta}\frac{\partial v_r}{\partial \phi} + r\frac{\partial}{\partial r}\left(\frac{v_\phi}{r}\right)\right]^2$$

$$- \frac{2}{3}\left[\frac{1}{r^2}\frac{\partial}{\partial r}(r^2 v_r) + \frac{1}{r\sin\theta}\frac{\partial}{\partial \theta}(v_\theta\sin\theta) + \frac{1}{r\sin\theta}\frac{\partial v_\phi}{\partial \phi}\right]^2$$

## HEAT-FLUX COMPONENTS

### Rectangular

$$q_x = -k\frac{\partial T}{\partial x}$$

$$q_y = -k\frac{\partial T}{\partial y}$$

$$q_z = -k\frac{\partial T}{\partial z}$$

### Cylindrical

$$q_r = -k\frac{\partial T}{\partial r}$$

$$q_\theta = -k\frac{1}{r}\frac{\partial T}{\partial \theta}$$

$$q_z = -k\frac{\partial T}{\partial z}$$

### Spherical

$$q_r = -k\frac{\partial T}{\partial r}$$

$$q_\theta = -k\frac{1}{r}\frac{\partial T}{\partial \theta}$$

$$q_\phi = -k\frac{1}{r\sin\theta}\frac{\partial T}{\partial \phi}$$

## BINARY DIFFUSION

In terms of mass fluxes:

$$\text{Rectangular} \quad \rho\left(\frac{\partial \omega_1}{\partial t} + u\frac{\partial \omega_1}{\partial x} + v\frac{\partial \omega_1}{\partial y} + w\frac{\partial \omega_1}{\partial z}\right)$$

$$= r_1''' - \left(\frac{\partial \dot{m}_{1x}}{\partial x} + \frac{\partial \dot{m}_{1y}}{\partial y} + \frac{\partial \dot{m}_{1z}}{\partial z}\right)$$

Cylindrical $\quad \rho\left(\dfrac{\partial \omega_1}{\partial t} + v_r \dfrac{\partial \omega_1}{\partial r} + \dfrac{v_\theta}{r} \dfrac{\partial \omega_1}{\partial \theta} + v_z \dfrac{\partial \omega_1}{\partial z}\right)$

$$= r_1''' - \left[\dfrac{1}{r} \dfrac{\partial(r\dot{m}_{1r})}{\partial r} + \dfrac{1}{r} \dfrac{\partial \dot{m}_{1\theta}}{\partial \theta} + \dfrac{\partial \dot{m}_{1z}}{\partial z}\right]$$

Spherical $\quad \rho\left(\dfrac{\partial \omega_1}{\partial t} + v_r \dfrac{\partial \omega_1}{\partial r} + \dfrac{v_\theta}{r} \dfrac{\partial \omega_1}{\partial \theta} + \dfrac{v_\phi}{r \sin \theta} \dfrac{\partial \omega_1}{\partial \phi}\right)$

$$= r_1''' - \left[\dfrac{1}{r^2} \dfrac{\partial(r^2 \dot{m}_{1r})}{\partial r} + \dfrac{1}{r \sin \theta} \dfrac{\partial(\dot{m}_{1\theta} \sin \theta)}{\partial \theta} + \dfrac{1}{r \sin \theta} \dfrac{\partial(\dot{m}_{1\phi})}{\partial \phi}\right]$$

In terms of mass fraction gradients for $\rho$, $D_{12} =$ const:

Rectangular $\quad \dfrac{\partial \omega_1}{\partial t} + u \dfrac{\partial \omega_1}{\partial x} + v \dfrac{\partial \omega_1}{\partial y} + w \dfrac{\partial \omega_1}{\partial z}$

$$= \dfrac{r_1'''}{\rho} + D_{12}\left[\dfrac{\partial^2 \omega_1}{\partial x^2} + \dfrac{\partial^2 \omega_1}{\partial y^2} + \dfrac{\partial^2 \omega_1}{\partial z^2}\right]$$

Cylindrical $\quad \dfrac{\partial \omega_1}{\partial t} + v_r \dfrac{\partial \omega_1}{\partial r} + \dfrac{v_\theta}{r} \dfrac{\partial \omega_1}{\partial \theta} + v_z \dfrac{\partial \omega_1}{\partial z}$

$$= \dfrac{r_1'''}{\rho} + D_{12}\left[\dfrac{1}{r} \dfrac{\partial(r\partial \omega_1/\partial r)}{\partial r} + \dfrac{1}{r^2} \dfrac{\partial^2 \omega_1}{\partial \theta^2} + \dfrac{\partial^2 \omega_1}{\partial z^2}\right]$$

Spherical $\quad \dfrac{\partial \omega_1}{\partial t} + v_r \dfrac{\partial \omega_1}{\partial r} + \dfrac{v_\theta}{r} \dfrac{\partial \omega_1}{\partial \theta} + \dfrac{v_\phi}{r \sin \theta} \dfrac{\partial \omega_1}{\partial \phi}$

$$= \dfrac{r_1'''}{\rho} + D_{12}\left[\dfrac{1}{r^2} \dfrac{\partial(r^2 \partial \omega_1/\partial r)}{\partial r} + \dfrac{1}{r^2 \sin \theta} \dfrac{\partial(\sin \theta\, \partial \omega_1/\partial \theta)}{\partial \theta}\right.$$

$$\left. + \dfrac{1}{r^2 \sin \theta} \dfrac{\partial^2 \omega_1}{\partial \phi^2}\right]$$

# E

# SIMILARITY TRANSFORMATION BY A SEPARATION OF VARIABLES

The various methods by which similarity transformations can be discovered for reduction of the number of variables in engineering problems are treated by Hansen [1]. Those he discusses are the free parameter, separation of variables (largely refined by Abbott and Kline [2]), group theory, and dimensional analysis methods. To this list should be added the method due to Moore [3] that is particularly suited to unsteady problems. Similarity transformation of a boundary-layer motion equation by a separation of variables method is discussed in this appendix.

Generally speaking, two types of problem are encountered. One type is the *well-posed* problem, which has a completely prescribed set of boundary and initial conditions. For a general form of the similarity transformation, such a problem has either one similarity variable or none. A second type is the problem in which some, but not all, boundary and initial conditions are given. None, one, or many similarity variables may be found to exist for a general form of the similarity transformation. The laminar boundary-layer equations are usually of the second type inasmuch as the free-stream conditions are not specified precisely and a velocity distribution is not specified at one particular position.

The result of the similarity transformation usually, but not always, is a reduction from a partial differential to an ordinary differential equation that, because of the nonlinearity of the physical phenomena described, is nonlinear. The ordinary differential equation need only be solved once, albeit by numerical methods, to enable application to all situations encompassed by the original partial differential equation.

Because the method of separation of variables is familiar to many in the context of heat-conduction problem solution, it is used to show in detail the manner in which a similarity form can be found for the $x$-motion laminar boundary-layer equation for steady flow of a constant-property fluid over a wedge.

The equation considered is

$$u \frac{\partial u}{\partial x} + v \frac{\partial u}{\partial y} = U \frac{dU}{dx} + v \frac{\partial^2 u}{\partial y^2} \tag{E-1}$$

$$u(x, y = 0) = 0 = v(x, y - 0)$$

$$u(x, y \rightarrow \infty) \rightarrow U(x) \tag{E-2}$$

The stream function, satisfying the continuity equation since $u = \partial \psi / \partial y$ and $v = -\partial \psi / \partial x$, is assumed to have the separation of variables form

$$\psi = H(\zeta) F(\eta) \tag{E-3}$$

where the coordinate transformation

$$x, y \rightarrow \zeta(x, y), \qquad \eta(x, y)$$

has been employed. For specific treatment let

$$\zeta = x \qquad \text{and} \qquad \eta = yg(x) \tag{E-4}$$

where $g(x)$ is a yet-unknown function of $x$.

The coordinate transformation specified by Eq. (E-4) gives

$$\frac{\partial \zeta}{\partial x} = 1, \qquad \frac{\partial \zeta}{\partial y} = 0, \qquad \frac{\partial \eta}{\partial x} = y \frac{dg}{dx} = \eta d \frac{\ln(g)}{dx}, \qquad \text{and} \qquad \frac{\partial \eta}{\partial y} = g$$

From this the velocities and their derivatives are found by the chain rule to be given in the transformed coordinates in terms of the stream function as

$$u = \frac{\partial \psi}{\partial y} = \frac{\partial \psi}{\partial \zeta} \frac{\partial \zeta}{\partial y} + \frac{\partial \psi}{\partial \eta} \frac{\partial \eta}{\partial y} = g \frac{\partial \psi}{\partial \eta} \tag{E-5}$$

$$-v = \frac{\partial \psi}{\partial x} = \frac{\partial \psi}{\partial \zeta} \frac{\partial \zeta}{\partial x} + \frac{\partial \psi}{\partial \eta} \frac{\partial \eta}{\partial x}$$

$$= \frac{\partial \psi}{\partial \zeta} + \eta \frac{d \ln(g)}{dx} \frac{\partial \psi}{\partial \eta} \tag{E-6}$$

$$\frac{\partial u}{\partial x} = \frac{\partial}{\partial x}\frac{\partial \psi}{\partial y} = \frac{\partial}{\partial x}\left(g\frac{\partial \psi}{\partial \eta}\right) = \frac{dg}{dx}\frac{\partial \psi}{\partial \eta} + g\frac{\partial}{\partial x}\frac{\partial \psi}{\partial \eta}$$

$$= \frac{dg}{dx}\frac{\partial \psi}{\partial \eta} + g\left(\frac{\partial^2 \psi}{\partial \zeta \partial \eta}\frac{\partial \zeta}{\partial x} + \frac{\partial^2 \psi}{\partial \eta^2}\frac{\partial \eta}{\partial x}\right)$$

$$= \frac{dg}{dx}\frac{\partial \psi}{\partial \eta} + g\left[\frac{\partial^2 \psi}{\partial \zeta \partial \eta} + \frac{\partial^2 \psi}{\partial \eta^2}\eta\frac{d\ln(g)}{dx}\right] \tag{E-7}$$

$$\frac{\partial u}{\partial y} = \frac{\partial}{\partial y}\frac{\partial \psi}{\partial y} = \frac{\partial}{\partial y}\left(g\frac{\partial \psi}{\partial \eta}\right)$$

$$= g\left[\frac{\partial^2 \psi}{\partial \zeta \partial \eta}\frac{\partial \zeta}{\partial y} + \frac{\partial^2 \psi}{\partial \eta^2}\frac{\partial \eta}{\partial y}\right] = g^2\frac{\partial^2 \psi}{\partial \eta^2} \tag{E-8}$$

$$\frac{\partial^2 u}{\partial y^2} = \frac{\partial}{\partial y}\frac{\partial^2 \psi}{\partial y^2} = \frac{\partial}{\partial y}\left(g^2\frac{\partial^2 \psi}{\partial \eta^2}\right)$$

$$= g^2\left[\frac{\partial^3 \psi}{\partial \zeta \partial \eta^2}\frac{\partial \zeta}{\partial y} + \frac{\partial^3 \psi}{\partial \eta^3}\frac{\partial \eta}{\partial y}\right] = g^3\frac{\partial^3 \psi}{\partial \eta^3} \tag{E-9}$$

Substitution of Eqs. (E-5)–(E-9) into Eq. (E-1) with $\psi = H(\zeta) F(\eta)$ according to Eq. (E-3) gives, after rearrangement and elimination of the distinction between $\zeta$ and $x$,

$$\frac{d^3 F}{d\eta^3} + \left[\frac{1}{vg}\frac{dH}{dx}\right]F\frac{d^2 F}{d\eta^2} - \frac{1}{vg}\left[\frac{dH}{dx} + \frac{d\ln(g)}{dx}H\right]\left(\frac{dF}{d\eta}\right)^2 = -\frac{U\,dU/dx}{vg^3 H}$$

$$\tag{E-10}$$

A separation of variables will be possible in Eq. (E-10) if

$$\frac{1}{vg}\frac{dH}{dx} = c_1 \tag{E-11}$$

and

$$\frac{1}{vg}\left[\frac{dH}{dx} + \frac{d\ln(g)}{dx}H\right] = c_2 \tag{E-12}$$

Then Eq. (E-10) has the form

$$F''' + c_1 FF'' - c_2(F')^2 = \lambda = -\frac{U\,dU/dx}{vg^3 H} \tag{E-13}$$

where $\lambda$ is a separation constant of yet unknown value as required by the fact that the left-hand side is solely a function of $\eta$ whereas the right-hand side depends solely on $x$.

To determine allowable terms for $H(x)$ and $g(x)$, first consider Eq. (E-11) that, when differentiated once, shows that

$$\frac{d^2H}{dx^2} = c_1\nu\,\frac{dg}{dx}$$

Insertion of this relation, together with Eq. (E-11), into Eq. (E-12) to eliminate $g(x)$ gives

$$H\frac{d^2H}{dx^2} = \left(\frac{c_2}{c_1} - 1\right)\left(\frac{dH}{dx}\right)^2$$

On letting $n = c_2/c_1 - 1$, one finds the solution to this differential equation to be

$$H = \left[(1-n)(c_3x + c_4)\right]^{1/(1-n)}$$

$$g = \frac{c_3}{c_1\nu}\left[(1-n)(c_3x + c_4)\right]^{n/(1-n)} \qquad\qquad \text{(E-14)}$$

for $n \neq 1$ and

$$H = c_4 e^{c_3x}$$

$$g = \frac{c_3 c_4}{c_1\nu}e^{c_3x} \qquad\qquad \text{(E-15)}$$

for $n = 1$. From Eq. (E-13) the corresponding free-stream velocity is then found to be, for $n \neq 1$,

$$U^2 = -\lambda\nu\int g^3 H\,dx + c_6$$

$$= -\frac{\lambda}{\nu^2}\frac{c_3^2}{c_1^3}\frac{(1-n)}{1+n}^{2(n+1)/(1-n)}(c_3x + c_4)^{2(n+1)/(1-n)} + c_6$$

If $c_6 = 0$ arbitrarily, then

$$U = \left[-\frac{\lambda}{\nu^2}\frac{c_3^2}{c_1^3}\frac{1}{n+1}\right]^{1/2}\left[(1-n)(c_3x + c_4)\right]^{(n+1)/(1-n)} \qquad\qquad \text{(E-16)}$$

with only a slight loss of generality. For $n = 1$ this procedure gives

$$U^2 = -\frac{\lambda}{2\nu^2} \frac{c_3^2 c_4^4}{c_1^3} e^{4c_3 x} + c_6$$

which becomes, with $c_6 = 0$ arbitrarily,

$$U = \left[ -\frac{\lambda}{2\nu^2} \frac{c_3^2 c_4^4}{c_1^3} \right]^{1/2} e^{2c_3 x} \qquad (E-17)$$

Major interest is focused on the case in which $n \neq 1$.

Additional information can be deduced by considering the boundary condition of Eq. (E-2) that

$$u(x, y \rightarrow \infty) \rightarrow U(x)$$

In transformed variables this condition is expressed as

$$gH \frac{dF(\infty)}{d\eta} = U$$

Now $dF(\infty)/d\eta$ is a number that is here arbitrarily set equal to unity. Then

$$gH = U$$

Elimination of $g$ by use of Eq. (E-11) puts this relation into the form

$$\frac{dH^2}{dx} = 2c_1 \nu U$$

Now if $U = c(x + K)^m$ as suggested by Eq. (E-16), this relation requires that

$$H = \left[ \frac{2c_1 \nu}{m + 1} c(x + K)^{m+1} \right]^{1/2} \qquad (E-18)$$

and the corresponding form for $g$ is then available from Eq. (E-11) as

$$g = \left[ \frac{m + 1}{2} \frac{c}{c_1 \nu} (x + K)^{m-1} \right]^{1/2} \qquad (E-19)$$

The constant $c_2$ can now be evaluated in terms of the $m$ of the assumed free-stream velocity variation $U = c(x + K)^m$. From Eq. (E-12) it is found with the help of Eqs. (E-18) and (E-19) that

$$c_2 = \frac{2c_1 m}{1 + m} \qquad (E-20)$$

Similarly, the separation constant $\lambda$ is found from Eq. (E-13)

$$\lambda = -\frac{dU^2/dx}{2\nu g^3 H}$$

with the aid of Eqs. (E-18) and (E-19) to be

$$\lambda = -\frac{2c_1 m}{1 + m} \tag{E-21}$$

At this point the similarity equation Eq. (E-13), has been found to have the form

$$F''' + c_1 FF'' + \frac{2c_1 m}{1 + m}\left[1 - (F')^2\right] = 0$$

where

$$\eta = y\left[\frac{m + 1}{2}\frac{c}{c_1\nu}(x + K)^{m-1}\right]^{1/2}$$

$$= \frac{y}{x + K}\left(\frac{m + 1}{2c_1}\right)\mathrm{Re}^{1/2}$$

with

$$\mathrm{Re} = \frac{U(x + K)}{\nu}$$

There are no other conditions by which to evaluate constants, so $c_1$ could have any value. Although a variety of choices have been adopted in the literature, $c_1$ is here taken to equal $(m + 1)/2$. Then

$$F''' + \frac{m + 1}{2}FF'' + m\left[1 - (F')^2\right] = 0 \tag{E-22}$$

where

$$\eta = y\left[\frac{c}{\nu}(x + K)^{m-1}\right]^{1/2}$$

The constant $K$ cannot be evaluated without additional information. Experimental measurements suggest that it is usually small. However, because it cannot be shown to identically equal zero, the similarity equation expressed by Eq. (E-20) is only accurate away from the leading edge if $K$ is assumed to equal zero. In free convection it has been suggested [4] that use of $x + K$ rather than

$x$ (with $K$ an experimentally determined constant) provides a better correlation of Nusselt number versus Rayleigh number. The effect of a nonzero $K$ on the similarity solution is to displace downstream a distance $K$ all solutions obtained on the assumption of $K = 0$.

## REFERENCES

1   A. G. Hansen, *Similarity Analyses Of Boundary Value Problems in Engineering*, Prentice-Hall, Englewood Cliffs, 1964.

2   D. E. Abbot and S. J. Kline, Simple methods for classification and construction of similarity solutions of partial differential equations, Report MD-6, Department of Mechanical Engineering, Stanford University, Stanford, California. See also AFOSR-TN-60-1163.

3   F. K. Moore, Unsteady laminary boundary layer flow, NACA TN 2471, Washington, DC, 1951.

4   A. J. Ede, Advances in free convection, *Adv. Heat Transf.* **4**, 46 (1967).

# APPENDIX

# F

# BOILING AND LIQUID–VAPOR INTERFACE STABILITY

In this appendix the presentations of the stability of a liquid-vapor interface set forth by Jordan [1] and Leppert and Pitts [2] are largely followed. The classical works of Milne-Thompson [3] and Lamb [4] can be consulted for additional reading on the subject.

## TAYLOR INSTABILITY

Two fluids, one over the other, flow horizontally, as illustrated in Fig. F-1. The upper fluid lies in region $R$ that is of indefinite $x$ and $y$ extent and that extends a distance $a$ above the nominal interface. The lower fluid occupies region $R'$ that is likewise of indefinite $x$ and $y$ extent and that extends a distance $a'$ below the nominal interface. The fluids are incompressible, inviscid, and immiscible, and the flow is irrotational. The constant major fluid velocity is along the $x$ axis of magnitude $U$ and $U'$ for the upper and lower fluid, respectively. Superimposed on this major velocity is a small perturbation due to the wavy character of the interface. Hence the velocity of the upper fluid is

$$\mathbf{V} = (U + u)\hat{\mathbf{i}} + v\hat{\mathbf{j}} + w\hat{\mathbf{k}} \qquad (\text{F-1a})$$

and the velocity of the lower liquid is

$$\mathbf{V}' = (U' + u')\hat{\mathbf{i}} + v'\hat{\mathbf{j}} + w'\hat{\mathbf{k}} \qquad (\text{F-1b})$$

755

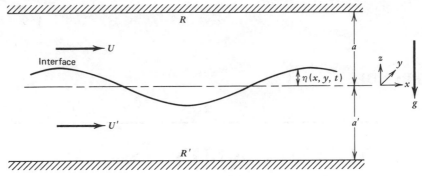

**Figure F-1**  Taylor instability in which one fluid overlays another less dense fluid.

where $u$, $v$, and $w$ are velocity perturbations along the $x$-, $y$-, and $z$ axes. The local instantaneous deviation of the interface above the nominal interface is $\eta(x, y, t)$.

Since the flow is irrotational, the fluid velocity is derivable from the gradient of a scalar (see Appendix A for further elaboration). Let the negative of this scalar be termed the velocity potential $\Phi$. Then $\mathbf{V} = -\nabla\Phi$ in $R$ and $\mathbf{V}' = -\nabla\Phi'$ in $R'$. Or

$$U + u = -\frac{\partial\Phi}{\partial x}, \qquad v = -\frac{\partial\Phi}{\partial y}, \qquad w = -\frac{\partial\Phi}{\partial z} \quad \text{in} \quad R \qquad \text{(F-2a)}$$

$$U' + u' = -\frac{\partial\Phi'}{\partial x}, \qquad v' = -\frac{\partial\Phi'}{\partial y}, \qquad w' = -\frac{\partial\Phi'}{\partial z} \quad \text{in} \quad R' \text{ (F-2b)}$$

The continuity equations, $\nabla \cdot \mathbf{V} = 0$ as Eq. (4-7) shows, for the upper and lower fluids then are

$$\nabla^2\Phi = 0 \quad \text{in } R \qquad\qquad\qquad \text{(F-3a)}$$

$$\nabla^2\Phi' = 0 \quad \text{in } R' \qquad\qquad\qquad \text{(F-3b)}$$

where $\nabla^2 = \partial^2/\partial x^2 + \partial^2/\partial y^2 + \partial^2/\partial z^2$ in rectangular coordinates. The inviscid equation of motion is $(\rho/g_c)[\partial\mathbf{V}/\partial t + (\mathbf{V}\cdot\nabla)\mathbf{V}] = -\nabla p - \rho g\hat{\mathbf{k}}/g_c$ as Eqs. (4-25) and (4-35) show. Reference to Table A-1 in Appendix A shows additionally that

$$\nabla(\mathbf{V}\cdot\mathbf{V}) = 2(\mathbf{V}\cdot\nabla)\mathbf{V} + 2\mathbf{V}\times\text{curl}\,\mathbf{V}$$

For irrotational flow (curl $\mathbf{V} = 0$) and in view of the fact that $\mathbf{V}\cdot\mathbf{V} = V^2_{\text{magnitude}}$, it then follows that the inviscid equation of motion can be rewritten as

$$\frac{\rho}{g_c}\left[\frac{\partial\mathbf{V}}{\partial t} + \nabla\left(\frac{V^2_{\text{mag}}}{2} + \frac{g_c p}{\rho} + gz\right)\right] = 0$$

Introducing Eqs. (F-3) into this result leads to

$$\frac{g_c p}{\rho} = \frac{\partial \Phi}{\partial t} - \frac{V_{mag}^2}{2} - gz + C \quad \text{in} \quad R \tag{F-4a}$$

$$\frac{g_c p'}{\rho'} = \frac{\partial \Phi'}{\partial t} - \frac{V_{mag}'^2}{2} - gz + C' \quad \text{in} \quad R' \tag{F-4b}$$

where $C$ and $C'$ are arbitrary constants.

The inviscid assumption prevents the no-slip condition from being enforced at either the interface or the upper and lower surfaces. However, the impermeability of the upper and lower surfaces can be satisfied by imposing the two boundary conditions that

$$w(x, y, z = a, t) = 0 = w'(x, y, z = -a', t) \tag{F-5}$$

A third boundary condition is obtained by requiring the wavy interface between the two fluids to be in equilibrium with pressure forces balanced by surface tension forces. For small interface deviations $\eta$, from the nominal interface position the difference between the two fluid pressures is

$$p - p' = \sigma \left( \frac{\partial^2 \eta}{\partial x^2} + \frac{\partial^2 \eta}{\partial y^2} \right) \quad \text{at} \quad z = \eta(x, y, t) \tag{F-6}$$

where $\sigma$ is surface tension.

A fourth boundary condition is achievable by recognizing that at the wavy interface

$$z = \eta(x, y, t)$$

or, defining a function as the difference between $z$ and $\eta$,

$$F(x, y, z, t) = z - \eta(x, y, t) = 0 \tag{F-7}$$

Expansion of $F$ in series about the point $x_0$, $y_0$, $t_0$ gives

$$F(x, y, z, t) = F(x_0, y_0, t_0) + dx \frac{\partial F(x_0, y_0, t_0)}{\partial x} + dy \frac{\partial F(x_0, y_0, t_0)}{\partial y}$$

$$+ dz \frac{\partial F(x_0, y_0, t_0)}{\partial z} + dt \frac{\partial F(x_0, y_0, t_0)}{\partial t} + \cdots \tag{F-8}$$

Realize that $F = 0$ always and that $dx/dt$, $dy/dt$, and $dz/dt$ can be interpreted as the three velocity components of a point on the wavy interface (subject to the understanding of Section 4.2) that must be the velocity of the fluids on either side of the interface. Then $dx/dt = -\partial\Phi/\partial x$, for example, and Eq. (F-8) can be written as

$$0 = -\frac{\partial\Phi}{\partial x}\frac{\partial F}{\partial x} - \frac{\partial\Phi}{\partial y}\frac{\partial F}{\partial y} - \frac{\partial\Phi}{\partial z}\frac{\partial F}{\partial z} + \frac{\partial F}{\partial t}$$

$$0 = -\nabla\Phi \cdot \nabla F$$

Emplacing Eq. (F-7) into this result then yields the fourth boundary condition as

$$-\frac{\partial\Phi}{\partial z} = \frac{\partial\eta}{\partial t} - \frac{\partial\Phi}{\partial x}\frac{\partial\eta}{\partial x} - \frac{\partial\Phi}{\partial y}\frac{\partial\eta}{\partial y} \quad \text{at} \quad z = \eta \qquad (F\text{-}9a)$$

Since $dx/dt = -\partial\Phi'/\partial x$ as well, it follows in parallel fashion that

$$-\frac{\partial\Phi'}{\partial z} = \frac{\partial\eta}{\partial t} - \frac{\partial\Phi'}{\partial x}\frac{\partial\eta}{\partial x} - \frac{\partial\Phi'}{\partial y}\frac{\partial\eta}{\partial y} \quad \text{at} \quad z = \eta \qquad (F\text{-}9b)$$

A fifth boundary condition is obtained by combining Eqs. (F-4) and (F-6) to achieve

$$\frac{\rho}{g_c}\left(\frac{\partial\Phi}{\partial t} - \frac{V_{\text{mag}}^2}{2} - g\eta + C\right) - \frac{\rho'}{g_c}\left[\frac{\partial\Phi'}{\partial t} - \frac{V_{\text{mag}}'^2}{2} - g\eta + C'\right]$$

$$= \sigma\left(\frac{\partial^2\eta}{\partial x^2} + \frac{\partial^2\eta}{\partial y^2}\right) \qquad (F\text{-}10)$$

The velocity potentials are taken to be

$$\Phi = \phi - Ux \qquad (F\text{-}11a)$$

$$\Phi' = \phi' - U'x \qquad (F\text{-}11b)$$

in which $\phi$ and $\phi'$ are the small perturbations in velocity potential that give the small perturbations in velocity cited in Eq. (F-1). Introduction of Eq. (F-11) into Eqs. (F-3), (F-5), (F-9), and (F-10) gives the final formulation of the mathematical problem to be solved as

$$\nabla^2\phi = 0 \quad \text{in} \quad R, \qquad \nabla^2\phi' = 0 \quad \text{in} \quad R' \qquad (F\text{-}12a)$$

$$\frac{\partial\phi(z = a)}{\partial z} = 0, \qquad \frac{\partial\phi'(z = -a)}{\partial z} = 0 \qquad (F\text{-}12b)$$

$$-\frac{\partial\phi(z = 0)}{\partial z} = \frac{\partial\eta}{\partial t} + U\frac{\partial\eta}{\partial x}, \qquad -\frac{\partial\phi'(z = 0)}{\partial z} = \frac{\partial\eta}{\partial t} + U'\frac{\partial\eta}{\partial x}$$

$$(F\text{-}12c)$$

—such higher-order terms as $(\partial\phi/\partial x)(\partial\eta/\partial x) + (\partial\phi/\partial y)(\partial\eta/\partial y) +$

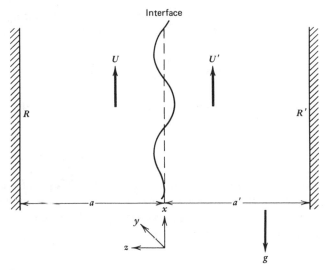

**Figure F-2** Helmholtz instability in which two fluids flow parallel to a common wavy interface.

$U_{av}$ with speeds given by $\pm C$, where

$$C^2 = C_0^2 - \frac{\rho\rho'(U - U')^2}{(\rho + \rho')^2}$$

As before, a stable interface requires that the frequency of oscillation $\omega$ have no imaginary part. Hence the stability criterion is that

$$\frac{\sigma g_c L^3}{m_1^2} \geq \frac{\rho\rho'(U - U')^2}{\rho + \rho'} \tag{F-29}$$

If this criterion is not satisfied the interface oscillates with increasing amplitude until it disrupts entirely, a condition termed *Helmholtz instability* [8].

Application of these results to the peak heat-flux boiling condition leads to the identification of the vapor as being in region $R'$ and flowing upward at velocity $U' = V_v$ and the liquid being in region $R$ and flowing downward at velocity $U = -V_L$. The downflowing liquid must compete with the upflowing vapor; at a certain upward velocity of the vapor, the interface between the two counterflowing streams will be disrupted by Helmholtz instability. At incipient instability the equality of Eq. (F-29) holds, giving

$$g_c m_1 = \frac{\rho\rho'(V_v + V_L)^2}{\rho + \rho'} \tag{F-30}$$

where it has been assumed for simplicity that $L = m_1$—a one-dimensional assumption. A further condition is that, for the square array of vapor columns described at the end of the Taylor Instability section, equality of upward vapor mass flow and downward liquid mass flow requires that

$$\rho V_L\left(\lambda^2 - \frac{\pi\lambda^2}{16}\right) = \rho'V_v\frac{\pi\lambda^2}{16} \tag{F-31}$$

Equations (F-30) and (F-31) then yield the critical vapor velocity as

$$V_{v,\text{critical}} = \left(\frac{\sigma g_c m_1}{\rho'}\right)^{1/2} \frac{[(\rho + \rho')/\rho]^{1/2}}{1 + \pi(\rho'/\rho)/(16 - \pi)}$$

the last term of which is nearly unity. Rayleigh's analysis [9] of a circular gas jet in a liquid shows that axially symmetric disturbances with wavelengths larger than the jet circumference are unstable for all jet velocities. Recalling that wavelength $\lambda$ is related to wave number $m$ as

$$\lambda = \frac{2\pi}{m}$$

and that the vapor jet has been taken to have a diameter of $\lambda/2$, one obtains $m = 4/\lambda$ and this expression for $V_{v,\text{critical}}$ as

$$V_{v,\text{critical}} = \left(4\sigma\frac{g_c}{\lambda\rho'}\right)^{1/2}\left(\frac{\rho + \rho'}{\rho}\right)^{1/2} \tag{F-32}$$

The heat flux carried away by the vapor (in saturated boiling) is given by

$$\frac{q}{A} = \rho'V_v\frac{A_v}{A}h_{\text{fg}}$$

which is further reduced at the critical conditions of peak heat flux, with the use of Eqs. (F-26) and (F-32), to

$$0.12 \leqslant \frac{(q/A)}{\rho_v h_{\text{fg}}\left[\sigma(\rho_L - \rho_v)gg_c/\rho_v^2\right]^{1/4}\left[(\rho_L + \rho_v)/\rho_L\right]^{1/2}} \leqslant 0.16 \tag{F-33}$$

which is identical in form with the recommended Eq. (12-39) due to Zuber.

## REFERENCES

1   D. P. Jordan, Film and transition boiling, *Adv. Heat Transf.* **5**, 117–122, (1968).
2   G. Leppert and C. C. Pitts, Boiling, *Adv. Heat Transf.* **1**, 234–238 (1964).
3   L. M. Milne-Thompson, *Theoretical Hydrodynamics*, 3rd ed., Macmillian, New York, 1955, pp. 374–431.
4   H. Lamb, *Hydrodynamics*, 6th ed., Dover, New York, 1945, pp. 370–375.

5   G. Taylor, The instability of liquid surfaces when accelerated in a direction perpendicular to their plane, I, *Proc. Roy. Soc. Lond.* **A201**, 192 (1950).

6   D. J. Lewis, The instability of liquid surfaces when accelerated in a direction perpendicular to their plane, II, *Proc. Roy. Soc. Lond.* **A201**, 81 (1950).

7   R. Bellman and R. H. Pennington, Effects of surface tension and viscosity on Taylor instability, *Quart. Appl. Math.* **12**, 151 (1954).

8   H. von Helmholtz, Ueber discontinuirliche Flüssigkeitsbewegungen, *Berl. Monatsber.* (April 1868); *Phil. Mag.* (November 1868); *Wissenschaftliche Abhandlungen*, Leipzig, **14b** (1882–1883).

9   Lord Rayleigh, *Theory Of Sound*, Dover, New York, 1945.

10  K. Taghavi-Tafreshi and V. K. Dhir, Taylor instability in boiling, melting and condensation or evaporation, *Internatl. J. Heat Mass Transf.* **23**, 1433–1445 (1980).

# G

# UNITS, CONVERSION FACTORS, AND FUNDAMENTAL CONSTANTS

The International System of Units (SI) is based on the seven arbitrarily chosen dimensional quantities expressed in terms of the seven base units listed in Table G-1. The unit of length is the meter, which is 1,650,763.73 wavelengths in vacuum of the radiation corresponding to the transition between levels $2p_{10}$ and $5d_5$ of the krypton-86 atom. The unit of mass is the kilogram, which is the mass of the international prototype of the kilogram located at the headquarters of the International Bureau of Weights and Measures. The unit of time is the second, which is the duration of 9,192,631,770 periods of the radiation corresponding to the transition between two hyperfine levels of the ground state of the cesium-133 atom. The unit of electric current is the ampere, which is the constant current which would exert a force of $2 \times 10^{-7}$ newtons per meter of length between two straight parallel conductors of infinite length and negligible cross section placed 1 meter apart in a vacuum. The unit of thermodynamic temperature is the kelvin, which is $1/273.16$ of the thermodynamic temperature of the triple point of water. The unit of the amount of matter is the mole, which is the amount of matter of a system that contains as many elementary entities as there are atoms in 0.012 kilogram of carbon-12. The unit of luminous intensity is the candela, which is the luminous intensity in the direction perpendicular to a surface of $1/600,000$ square meter of a blackbody at the freezing temperature of platinum under a pressure of 101,325 newtons

Table G-1    Base Units for SI

| Quantity | Name | Symbol |
|---|---|---|
| Length | meter | m |
| Mass | kilogram | kg |
| Time | second | s |
| Electric current | ampere | A |
| Thermodynamic temperature | kelvin | K |
| Amount of matter | mole | mol |
| Luminous intensity | candela | cd |

per square meter. Two supplementary units are the radian for plane angles (symbol rad) and the steradian for solid angles (symbol sr). The newton is a derived unit.

The units of other dimensional quantities are derived from these base units. The units of the dimensional quantity of force are derived from the base units by considering Newton's second law of motion in the form

$$F = ma \qquad \text{(G-1)}$$

In the SI system the unit of force is taken to be the newton (symbol N) whose magnitude is such that it produces an acceleration of one meter per second per second when acting on a mass of one kilogram. Substituting this definition into Eq. (G-1) expresses the unit of force in terms of the base units as

$$N = kg\ m/s^2$$

An older view (which is not the official SI view) is that force is a dimensional quantity separate and distinct from the dimensional quantities of mass, length, and time. In this older view it is not possible to derive the unit of force from the units of mass, length, and time so Newton's second law must be written with a dimensional constant of proportionality $g_c$ as

$$F = \frac{ma}{g_c} \qquad \text{(G-2)}$$

Substitution of the definition of the newton into Eq. (G-2) gives $g_c = 1$ kg m/N s². In the English system the unit of force is the pound force (symbol $lb_f$) whose magnitude is such that it produces an acceleration of 32.174 feet per second per second when acting on a mass of one pound mass (symbol $lb_m$). Substitution of the definition of the pound force into Eq. (G-2) gives $g_c = 32.174\ lb_m\ ft/lb_f\ s^2$. In the English system the numerical value of $g_c$ differs from unity and considerable numerical error results from its omission in calculations. The advantage of the SI system is that $g_c$ has a numerical value of

**Table G-2   Some Derived Units in SI**

| Quantity | Name | Symbol | Units | SI Base Units |
|---|---|---|---|---|
| Frequency | hertz | Hz | $1/s$ | $1/s$ |
| Force | newton | N | $m\,kg/s^2$ | $m\,kg/s^2$ |
| Energy | joule | J | $N\,m$ | $m^2\,kg/s^2$ |
| Power | watt | W | $J/s$ | $m^2\,kg/s^3$ |
| Electric charge | coulomb | C | $A\,s$ | $A\,s$ |
| Electric potential | volt | V | $W/A$ | $m^2\,kg/s^3\,A$ |
| Electric resistance | ohm | $\Omega$ | $V/A$ | $m^2\,kg/s^3\,A^2$ |
| Electric capacitance | farad | F | $C/V$ | $s^4\,A^2/m^2\,kg$ |
| Magnetic flux | weber | Wb | $V\,s$ | $m^2\,kg/s^2\,A$ |
| Pressure | pascal | Pa | $N/m^2$ | $kg/m\,s^2$ |
| Inductance | henry | H | $Wb/A$ | $m^2\,kg/s^2\,A^2$ |
| Luminous flux | lumen | lm | $cd\,sr$ | $cd\,sr$ |
| Illuminance | lux | lx | $1m/m^2$ | $cd\,sr/m^2$ |

unity and no numerical error results from its omission in calculations. Hence the SI system adopts the simplicity of entirely dispensing with the older view and its relatively cumbersome $g_c$ to have

$$N = kg\,m/s^2$$

Some derived units are listed in Table G-2.

**Table G-3   SI Prefixes**

| Multiplier | Symbol | Prefix |
|---|---|---|
| $10^{18}$ | E | exa |
| $10^{15}$ | P | peta |
| $10^{12}$ | T | tera |
| $10^9$ | G | giga |
| $10^6$ | M | mega |
| $10^3$ | k | kilo |
| $10^2$ | h | hecto |
| $10^1$ | da | deka |
| $10^{-1}$ | d | deci |
| $10^{-2}$ | c | centi |
| $10^{-3}$ | m | milli |
| $10^{-6}$ | $\mu$ | micro |
| $10^{-9}$ | n | nano |
| $10^{-12}$ | p | pico |
| $10^{-15}$ | f | femto |
| $10^{-18}$ | a | atto |

**Table G-4   Conversion Factors**

| Physical quantity | Symbol | Conversion Factor |
|---|---|---|
| Area | | $1 \text{ ft}^2 = 0.0929 \text{ m}^2$ |
| | | $1 \text{ in.}^2 = 6.452 \times 10^{-4} \text{ m}^2$ |
| | | $1 \text{ mi}^2 = 2.59 \text{ km}^2$ |
| | ha | $1 \text{ hectare} = 10^4 \text{ m}^2$ |
| | a | $\text{are} = 10^2 \text{ m}^2$ |
| | b | $\text{barn} = 10^{-28} \text{ m}^2$ |
| Density | | $1 \text{ lb}_m/\text{ft}^3 = 16.018 \text{ kg}/\text{m}^3$ |
| | | $1 \text{ slug}/\text{ft}^3 = 515.379 \text{ kg}/\text{m}^3$ |
| Diffusivity | | $1 \text{ ft}^2/\text{hr} = 2.581 \times 10^{-5} \text{ m}^2/\text{s}$ |
| | | $1 \text{ ft}^2/\text{sec} = 9.029 \times 10^{-2} \text{ m}^2/\text{s}$ |
| Force | | $1 \text{ lb}_f = 4.448 \text{ N}$ |
| Heat, energy, or work | | $1 \text{ Btu} = 1055.1 \text{ J}$ |
| | | $1 \text{ cal} = 4.186 \text{ J}$ |
| | | $1 \text{ ft lb}_f = 1.3558 \text{ J}$ |
| | | $1 \text{ hp hr} = 2.685 \times 10^6 \text{ J}$ |
| | | $1 \text{ Therm} = 105.5 \text{ MJ}$ |
| | eV | $1 \text{ electron volt} = 1.60219 \times 10^{-19} \text{ J}$ |
| Heat flow rate | | $1 \text{ Btu}/\text{hr} = 0.2931 \text{ W}$ |
| | | $1 \text{ Btu}/\text{sec} = 1055.1 \text{ W}$ |
| Heat flux | | $1 \text{ Btu}/\text{hr ft}^2 = 3.1525 \text{ W}/\text{m}^2$ |
| Heat-transfer coefficient | | $1 \text{ Btu}/\text{hr ft}^2 \, ^\circ\text{F} = 5.678 \text{ W}/\text{m}^2 \text{ K}$ |
| Internal energy or enthalpy | | $1 \text{ Btu}/\text{lb}_m = 2326.0 \text{ J}/\text{kg}$ |
| | | $1 \text{ cal}/\text{g} = 4184 \text{ J}/\text{kg}$ |
| Length | | $1 \text{ ft} = 0.3048 \text{ m}$ |
| | | $1 \text{ in.} = 2.54 \text{ cm}$ |
| | | $1 \text{ mi} = 1.6093 \text{ km}$ |
| | | $1 \text{ nautical mi} = 1.852 \text{ km}$ |
| | Å | $1 \text{ ångstrom} = 10^{-10} \text{ m}$ |
| | AU | $1 \text{ astronomical unit} = 1.496 \times 10^8 \text{ km}$ |
| | pc | $1 \text{ parsec} = 30.857 \times 10^{12} \text{ km}$ |
| Mass | | $1 \text{ lb}_m = 0.4536 \text{ kg}$ |
| | | $1 \text{ slug} = 14.594 \text{ kg}$ |
| | | $1 \text{ metric ton} = 10^3 \text{ kg}$ |
| | | $1 \text{ ton} = 907.2 \text{ kg}$ |
| Mass flow rate | | $1 \text{ lb}_m/\text{hr} = 0.000126 \text{ kg}/\text{s}$ |
| | | $1 \text{ lb}_m/\text{sec} = 0.4536 \text{ kg}/\text{s}$ |
| Power | | $1 \text{ hp} = 745.7 \text{ W}$ |
| | | $1 \text{ ft lb}_f/\text{sec} = 1.3558 \text{ W}$ |
| | | $1 \text{ Btu}/\text{sec} = 1055.1 \text{ W}$ |
| | | $1 \text{ Btu}/\text{hr} = 0.293 \text{ W}$ |
| Pressure | | $1 \text{ lb}_f/\text{in}^2 = 6894.8 \text{ Pa}$ |
| | | $1 \text{ lb}_f/\text{ft}^2 = 47.88 \text{ Pa}$ |
| | | $1 \text{ atm} = 101,325 \text{ Pa}$ |
| | | $1 \text{mm } H_2O = 9.80665 \text{ Pa}$ |
| | | $1 \text{ mm Hg} = 133.3 \text{ Pa}$ |
| | | $1 \text{ in. } H_2O = 249.1 \text{ Pa}$ |
| | | $1 \text{ bar} = 10^5 \text{ Pa}$ |

**Table G-4**  (*Continued*)

| Physical quantity | Symbol | Conversion Factor |
|---|---|---|
| Specific heat capacity | | 1 Btu/lb$_m$ °F = 4187 J/kg K |
| Temperature | | $T(R) = \frac{9}{5}T(K)$ |
| | | $T(°F) = \frac{9}{5}[T(°C)] + 32$ |
| | | $T(°F) = \frac{9}{5}[T(K) - 273.15] + 32$ |
| Thermal conductivity | | 1 Btu/hr ft °F = 1.731 W/m K |
| Velocity | | 1 ft/sec = 0.3048 m/s |
| | | 1 mi/hr = 0.44703 m/s |
| | | 1 knot = 0.51444 m/s |
| Viscosity, dynamic | | 1 lb$_m$/ft sec = 1.488 N s/m$^2$ |
| | | 1 cP = 0.00100 N s/m$^2$ |
| Viscosity, kinematic | | 1 centistoke = $10^{-6}$ m$^2$/s |
| Volume | | 1 ft$^3$ = 0.02832 m$^3$ |
| | | 1 in$^3$ = 1.6387 × $10^{-5}$ m$^3$ |
| | | 1 gal (U.S.) = 0.003785 m$^3$ |
| | | 1 gal (U.K.) = 0.004546 m$^3$ |
| | | 1 liter = $10^{-3}$ m$^3$ |
| Volumetric flow rate | | 1 ft$^3$/min = 0.000472 m$^3$/s |
| | | 1 U.S. gpm = 0.06301 liter/s |
| | | 1 U.K. gpm = 0.07577 liter/s |

Multiples of SI units are designated by the standard prefixes listed in Table G-3. Some conversion factors are listed in Table G-4 and some physical constants, in Table G-5.

To illustrate the arbitrariness involved in selecting a system of units, consider the additional use of Newton's gravitational law

$$F = G\frac{m_1 m_2}{r^2}$$

to express the unit of mass in terms of length and time. Then the constant

**Table G-5  Some Physical Constants in SI Units**

| Quantity | Symbol | Value |
|---|---|---|
| Avogadro constant | $N_A$ | 6.022169 × $10^{26}$ kmol$^{-1}$ |
| Boltzmann constant | $K$ | 1.380622 × $10^{-23}$ J/K |
| Gas constant | $R$ | 8314.34 J/kmol K |
| Planck constant | $h$ | 6.626196 × $10^{-34}$ J s |
| Speed of light in a vacuum | $c$ | 2.997925 × $10^8$ m/s |
| Stefan–Boltzmann constant | $\sigma$ | 5.66961 × $10^{-8}$ W/m$^2$ K$^4$ |

$G = 6.685 \times 10^{-5}$ m$^3$/kg s$^2$ is neglected to yield

$$F = \frac{m_1 m_2}{r^2}$$

Combining this result with Newton's second law

$$F = m_1 a$$

shows that

$$m_2 = ar^2$$

Hence, mass and force could be expressed in terms of length and time as

$$\text{kg} = \text{m}^3/\text{s}^2$$

and

$$\text{N} = \text{m}^4/\text{s}^4$$

# Author Index

# Subject Index

$(\partial\phi/\partial z)(\partial\eta/\partial z)$ have been neglected—

$$\frac{\rho}{g_c}\left[\frac{\partial\phi(z=0)}{\partial t} + U\frac{\partial\phi(z=0)}{\partial x} - g\eta\right]$$

$$-\frac{\rho'}{g_c}\left[\frac{\partial\phi'(z=0)}{\partial t} + U'\frac{\partial\phi'(z=0)}{\partial x} - g\eta\right] = \sigma\left(\frac{\partial^2\eta}{\partial x^2} + \frac{\partial^2\eta}{\partial y^2}\right) \qquad \text{(F-12d)}$$

In Eq. (F-12d) the unperturbed condition, $\eta = 0$ and $\phi = 0$ and $\phi' = 0$, is satisfied by letting $C = U^2/2$ and $C' = U'^2/2$. It has also been assumed that $\eta \approx 0$ and that

$$\left(U - \frac{\partial\phi}{\partial x}\right)^2 + \left(\frac{\partial\phi}{\partial y}\right)^2 + \left(\frac{\partial\phi}{\partial z}\right)^2 \approx 2U\frac{\partial\phi}{\partial x}$$

which is accurate if $u, v, w \ll U$.

The boundary condition in Eq. (F-12) is satisfied by

$$\phi = Af(x, y, t)\frac{\cosh[L(z-a)]}{\sinh La} \qquad \text{(F-13a)}$$

$$\phi' = A'f(x, y, t)\frac{\cosh[L(z+a')]}{\sinh La'} \qquad \text{(F-13b)}$$

Substitution of the functional form of Eq. (F-13) into the continuity equation for each fluid, Eq. (F-12), reveals that it is necessary that $f(x, y, t)$ be a solution to

$$\frac{\partial^2 f}{\partial x^2} + \frac{\partial^2 f}{\partial y^2} + L^2 f = 0 \qquad \text{(F-14)}$$

Inasmuch as it is expected that the interface will be wavy in both space and time, it is reasonable to assume the interface deviation $\eta$ to have the functional form

$$\eta = \eta_0 \exp[\pm i(\omega t + m_1 x + m_2 y)] \qquad \text{(F-15)}$$

where $\eta_0$ is an amplitude of oscillation, $\omega$ is a frequency of oscillation, and $m_{1,2}$ is a wave number such that the wavelength $\lambda$ is given by $\lambda = 2\pi/m$. Since the fluid velocities are expected to have a similar wavy characteristic in space and time, a similar functional form is assumed for $f$ as

$$f(x, y, t) = \exp[\pm i(\omega t + m_1 x + m_2 y)] \qquad \text{(F-16)}$$

Substitution of Eq. (F-16) in Eq. (F-14) gives

$$L^2 = m_1^2 + m_2^2 \qquad \text{(F-17a)}$$

whereas substitution of Eq. (F-16) into Eq. (F-12c) gives

$$A = \pm \frac{i\eta_0(\omega + m_1 U)}{L}, \qquad A' = \mp \frac{i\eta_0(\omega + m_1 U')}{L} \qquad \text{(F-17b)}$$

and substitution into Eq. (F-12d) gives

$$\pm i\rho A(\omega + Um_1)\frac{\coth aL}{g_c} \mp i\rho' A'(\omega + U'm_1)\frac{\coth(a'L)}{g_c}$$

$$= -\sigma\eta_0(m_1^2 + m_2^2) + \eta_0(\rho - \rho')\frac{g}{g_c} \qquad \text{(F-17c)}$$

Combination of Eqs. (F-17) yields the single relation between frequency $\omega$ and the wave numbers $m_1$ and $m_2$ of

$$\rho(\omega + Um_1)^2\frac{\coth aL}{g_c} + \rho'(\omega + U'm_1)^2\frac{\coth a'L}{g_c}$$

$$= \sigma L^3 - (\rho - \rho')\frac{Lg}{g_c} \qquad \text{(F-18)}$$

A general condition for the stability of the interface can be obtained from Eq. (F-18). For present purposes, the specialization of negligible gross motion, $U \approx 0 \approx U'$, and deep fluid layers, $a \to \infty$ and $a' \to \infty$, is admissible and puts Eq. (F-18) into the simpler form of

$$\omega^2 = g_c L \frac{\sigma L^2 - (\rho - \rho')g/g_c}{\rho + \rho'} \qquad \text{(F-19)}$$

Note from Eq. (F-15) that an imaginary part to $\omega$ results in an interface that experiences increasingly large undulations. In other words, the interface is stable only if

$$\sigma L^2 \geqslant (\rho - \rho')\frac{g}{g_c} \qquad \text{(F-20)}$$

Violation of the stability criterion of Eq. (F-20) is called *Taylor instability* and is illustrated by the air–water interface below an inverted tumbler of water. Although such an interface can be stable (e.g., through the influence of a gauze), it is usually metastable, and any small disturbance usually grows and results in a two-dimensional pattern of vapor jets flowing upward and liquid flowing downward. The unstable arrangement of dense liquid above a lighter vapor is characteristic of the peak heat flux condition in boiling.

Taylor's [5] original analysis, experimentally confirmed by Lewis [6], neglected surface tension and the density of one liquid. Effects of surface tension and viscosity were later incorporated by Bellman and Pennington [7]—viscosity usually has little effect.

No unique specification of the boundaries of the interface has been made, so there is no unique solution for the wave pattern and spacing of nodes. An admissible assumption, however, is that there is a square pattern ($x$ and $y$ directions are interchangeable) and a coordinate axis passes through a node, an antinode, and so on. For this to occur, $m_1 = m_2$ and $L^2 = 2m^2$. Furthermore, in a boiling situation the stability criterion of Eq. (F-20) is violated so that Eq. (F-19) is more conveniently rearranged into

$$b^2 = (i\omega)^2 = \frac{g_c L[(\rho - \rho')g/g_c - \sigma l^2]}{\rho + \rho'} \tag{F-21}$$

and it is seen then that Eq. (F-15) becomes

$$\eta = \eta_0 \exp(bt) \exp[\pm i(m_1 x + m_2 y)]$$

In such a case $b$ is appropriately termed the *growth-rate parameter*.

Equation (F-21) reveals that the largest value of $b$—which is $b^*$—occurs when $L$ achieves the "most dangerous" value $L_D$ of

$$L_D = \left[(\rho - \rho')\frac{g}{g_c}3\sigma\right]^{1/2} \tag{F-22}$$

which gives

$$b^* = \left[\frac{(2/3)(\rho - \rho')g}{\rho + \rho'}\right]^{1/2}\left[\frac{(\rho - \rho')(g/g_c)}{3\sigma}\right]^{1/4} \tag{F-23}$$

Next, note that the barely stable case of $b = 0$ is found from Eq. (F-21) to occur at $L_0^2 = (\rho - \rho')(g/g_c)/\sigma$—which, since $L^2 = 2m^2$ and wavelength $\lambda$ is related to wave number $m$ by $\lambda = 2\pi/m$, gives the associated wavelength as

$$\lambda_0 = 2\pi\left[\frac{\sigma(g_c/g)}{\rho - \rho'}\right]^{1/2} \tag{F-24}$$

where the $2^{1/2}$ factor is deleted to account for measurement of distance between nodes along the side of the presumed square array rather than along a diagonal. Similarly, the "most dangerous" wavelength corresponding to $L_D$ is found to be

$$\lambda_D = 2\pi\left[\frac{3\sigma(g_c/g)}{\rho - \rho'}\right]^{1/2} \tag{F-25}$$

Although the distance $\lambda$ between nodes along the sides of a square is not known, it is clear that

$$\lambda_0 \leqslant \lambda \leqslant \lambda_D \tag{F-26}$$

Also, Eq. (F-23) can be recast into

$$b^* = \left[ \frac{(4\pi/3\lambda_D)(\rho - \rho')g}{\rho + \rho'} \right]^{1/2} \tag{F-27}$$

As suggested by Eq. (F-26), there is likely to be a spectrum of unstable wavelengths and, therefore, a spectrum of growth rate parameters rather than the single one given by Eq. (F-27). (see reference [10] for further details and photographs.)

Because of Taylor instability a horizontal liquid-over-vapor interface (as occurs at peak heat flux boiling conditions) tends to break up in a square pattern of side length $\lambda$. From each unit cell of area $\lambda^2$ rises a circular vapor column of diameter $\lambda/2$ and area $\pi\lambda^2/16$. This diameter is reasonable, although arbitrary, since it is the distance between a node and an antinode. Consequently, the rate of vapor release from the interface depends only on the maximum frequency of vapor bubble emission from the interface that is limited by Helmholtz instability.

## HELMHOLTZ INSTABILITY

The preceding analysis for the condition illustrated in Fig. F-1 that led to Eq. (F-18) can be directly applied to the variant condition illustrated in Fig. F-2. Here, the primary specialization adopted is that gravity acts parallel to the interface and plays no first-order role in its deformation—$g = 0$ in Eq. (F-18). Further simplifications are that the fluid layers are very wide, $a \to \infty$ and $a' \to \infty$, so that Eq. (F-18) becomes

$$\rho\left(\omega^2 + Um_1\right)^2 + \rho'\left(\omega + U'm_1\right)^2 - \sigma L^3 g_c = 0$$

Solution of this relation for $\omega$ yields

$$\omega = -m_1 U_{av} \pm m_1 \left[ \frac{C_0^2 - \rho\rho'(U - U')^2}{(\rho + \rho')^2} \right]^{1/2} \tag{F-28}$$

where $U_{av} = (\rho U + \rho' U')/(\rho + \rho')$ is the average velocity of the two flows and $C_0 = [(\sigma g_c L^3/m_1^2)/(\rho + \rho')]^{1/2}$ is the velocity of a surface wave in the absence of bulk currents ($U_1 = 0 = U'$). It is seen that surface waves travel relative to